Quantum Physics

Quantum physics allows us to understand the nature of the physical phenomena which govern the behavior of solids, semiconductors, lasers, atoms, nuclei, subnuclear particles, and light. In *Quantum Physics*, Le Bellac provides a thoroughly modern approach to this fundamental theory.

Throughout the book, Le Bellac teaches the fundamentals of quantum physics using an original approach which relies primarily on an algebraic treatment and on the systematic use of symmetry principles. In addition to the standard topics such as one-dimensional potentials, angular momentum and scattering theory, the reader is introduced to more recent developments at an early stage. These include a detailed account of entangled states and their applications, the optical Bloch equations, the theory of laser cooling and of magneto-optical traps, vacuum Rabi oscillations, and an introduction to open quantum systems. This is a textbook for a modern course on quantum physics, written for advanced undergraduate and graduate students.

MICHEL LE BELLAC is Emeritus Professor at the University of Nice, and a well-known elementary particle theorist. He graduated from Ecole Normale Supérieure in 1962, before conducting research with CNRS. In 1967 he returned to the University of Nice, and was appointed Full Professor of Physics in 1971, a position he held for over 30 years. His main fields of research have been the theory of elementary particles and field theory at finite temperatures. He has published four other books in French and three other books in English, including *Thermal Field Theory* (Cambridge 1996) and *Equilibrium and Non-equilibrium Statistical Thermodynamics* with Fabrice Mortessagne and G. George Batrouni (Cambridge 2004).

Quantum Physics

Michel Le Bellac

University of Nice

Translated by

Patricia de Forcrand-Millard

CAMBRIDGE
UNIVERSITY PRESS

CAMBRIDGE
UNIVERSITY PRESS

University Printing House, Cambridge CB2 8BS, United Kingdom

One Liberty Plaza, 20th Floor, New York, NY 10006, USA

477 Williamstown Road, Port Melbourne, VIC 3207, Australia

314-321, 3rd Floor, Plot 3, Splendor Forum, Jasola District Centre, New Delhi - 110025, India

79 Anson Road, #06-04/06, Singapore 079906

Cambridge University Press is part of the University of Cambridge.

It furthers the University's mission by disseminating knowledge in the pursuit of
education, learning and research at the highest international levels of excellence.

www.cambridge.org
Information on this title: www.cambridge.org/9781107602762

© Cambridge University Press 2006

Originally published in French as *Physique quantique* by EDP Sciences 2003, © EDP Sciences/CNRS
Editions 2003. First published in English by Cambridge University Press 2006.
Reprinted 2007

A catalogue record for this publication is available from the British Library

ISBN 978-0-521-85277-7 Hardback
ISBN 978-1-107-60276-2 Paperback

Contents

Foreword by Claude Cohen-Tannoudji   *page* xiii
Preface   xv
Table of units and physical constants   xix

1 Introduction   1
  1.1   The structure of matter   1
    1.1.1   Length scales from cosmology to elementary particles 1
    1.1.2   States of matter   2
    1.1.3   Elementary constituents   5
    1.1.4   The fundamental interactions   7
  1.2   Classical and quantum physics   9
  1.3   A bit of history   13
    1.3.1   Black-body radiation   13
    1.3.2   The photoelectric effect   16
  1.4   Waves and particles: interference   17
    1.4.1   The de Broglie hypothesis   17
    1.4.2   Diffraction and interference of cold neutrons   18
    1.4.3   Interpretation of the experiments   21
    1.4.4   Heisenberg inequalities I   24
  1.5   Energy levels   27
    1.5.1   Energy levels in classical mechanics and classical models
        of the atom   27
    1.5.2   The Bohr atom   29
    1.5.3   Orders of magnitude in atomic physics   31
  1.6   Exercises   33
  1.7   Further reading   40

2 The mathematics of quantum mechanics I: finite dimension   42
  2.1   Hilbert spaces of finite dimension   42
  2.2   Linear operators on $\mathcal{H}$ 44
    2.2.1   Linear, Hermitian, unitary operators   44
    2.2.2   Projection operators and Dirac notation   46

2.3 Spectral decomposition of Hermitian operators 48

    2.3.1 Diagonalization of a Hermitian operator 48

    2.3.2 Diagonalization of a 2×2 Hermitian matrix 50

    2.3.3 Complete sets of compatible operators 51

    2.3.4 Unitary operators and Hermitian operators 52

    2.3.5 Operator-valued functions 53

2.4 Exercises 54

2.5 Further reading 60

3 Polarization: photons and spin-1/2 particles **61**

3.1 The polarization of light and photon polarization 61

    3.1.1 The polarization of an electromagnetic wave 61

    3.1.2 The photon polarization 68

    3.1.3 Quantum cryptography 73

3.2 Spin 1/2 75

    3.2.1 Angular momentum and magnetic moment in classical physics 75

    3.2.2 The Stern–Gerlach experiment and Stern–Gerlach filters 77

    3.2.3 Spin states of arbitrary orientation 80

    3.2.4 Rotation of spin 1/2 82

    3.2.5 Dynamics and time evolution 87

3.3 Exercises 89

3.4 Further reading 95

4 Postulates of quantum physics **96**

4.1 State vectors and physical properties 96

    4.1.1 The superposition principle 96

    4.1.2 Physical properties and measurement 98

    4.1.3 Heisenberg inequalities II 104

4.2 Time evolution 105

    4.2.1 The evolution equation 105

    4.2.2 The evolution operator 108

    4.2.3 Stationary states 109

    4.2.4 The temporal Heisenberg inequality 111

    4.2.5 The Schrödinger and Heisenberg pictures 114

4.3 Approximations and modeling 115

4.4 Exercises 116

4.5 Further reading 124

5 Systems with a finite number of levels **125**

5.1 Elementary quantum chemistry 125

    5.1.1 The ethylene molecule 125

    5.1.2 The benzene molecule 128

5.2	Nuclear magnetic resonance (NMR)		132
	5.2.1	A spin 1/2 in a periodic magnetic field	132
	5.2.2	Rabi oscillations	133
	5.2.3	Principles of NMR and MRI	137
5.3	The ammonia molecule		139
	5.3.1	The ammonia molecule as a two-level system	139
	5.3.2	The molecule in an electric field: the ammonia maser	141
	5.3.3	Off-resonance transitions	146
5.4	The two-level atom		149
5.5	Exercises		152
5.6	Further reading		157

6 Entangled states — **158**
6.1	The tensor product of two vector spaces		158
	6.1.1	Definition and properties of the tensor product	158
	6.1.2	A system of two spins 1/2	160
6.2	The state operator (or density operator)		162
	6.2.1	Definition and properties	162
	6.2.2	The state operator for a two-level system	164
	6.2.3	The reduced state operator	167
	6.2.4	Time dependence of the state operator	169
	6.2.5	General form of the postulates	171
6.3	Examples		171
	6.3.1	The EPR argument	171
	6.3.2	Bell inequalities	174
	6.3.3	Interference and entangled states	179
	6.3.4	Three-particle entangled states (GHZ states)	182
6.4	Applications		185
	6.4.1	Measurement and decoherence	185
	6.4.2	Quantum information	191
6.5	Exercises		198
6.6	Further reading		207

7 Mathematics of quantum mechanics II: infinite dimension — **209**
7.1	Hilbert spaces		209
	7.1.1	Definitions	209
	7.1.2	Realizations of separable spaces of infinite dimension	211
7.2	Linear operators on $\mathcal{H}$		213
	7.2.1	The domain and norm of an operator	213
	7.2.2	Hermitian conjugation	215
7.3	Spectral decomposition		216
	7.3.1	Hermitian operators	216
	7.3.2	Unitary operators	219

7.4   Exercises   220
7.5   Further reading   221

8   Symmetries in quantum physics   **222**
8.1   Transformation of a state in a symmetry operation   223
      8.1.1   Invariance of probabilities in a symmetry operation   223
      8.1.2   The Wigner theorem   225
8.2   Infinitesimal generators   227
      8.2.1   Definitions   227
      8.2.2   Conservation laws   228
      8.2.3   Commutation relations of infinitesimal generators   230
8.3   Canonical commutation relations   234
      8.3.1   Dimension $d = 1$ 234
      8.3.2   Explicit realization and von Neumann's theorem   236
      8.3.3   The parity operator   237
8.4   Galilean invariance   240
      8.4.1   The Hamiltonian in dimension $d = 1$ 240
      8.4.2   The Hamiltonian in dimension $d = 3$ 243
8.5   Exercises   245
8.6   Further reading   249

9   Wave mechanics   **250**
9.1   Diagonalization of X and P and wave functions   250
      9.1.1   Diagonalization of X 250
      9.1.2   Realization in $L_x^{(2)}(\mathbb{R})$ 252
      9.1.3   Realization in $L_p^{(2)}(\mathbb{R})$ 254
      9.1.4   Evolution of a free wave packet   256
9.2   The Schrödinger equation   260
      9.2.1   The Hamiltonian of the Schrödinger equation   260
      9.2.2   The probability density and the probability
              current density   261
9.3   Solution of the time-independent Schrödinger equation   264
      9.3.1   Generalities   264
      9.3.2   Reflection and transmission by a potential step   265
      9.3.3   The bound states of the square well   270
9.4   Potential scattering   273
      9.4.1   The transmission matrix   273
      9.4.2   The tunnel effect   277
      9.4.3   The S matrix   280
9.5   The periodic potential   283
      9.5.1   The Bloch theorem   283
      9.5.2   Energy bands   285

9.6   Wave mechanics in dimension $d = 3$ 289
        9.6.1   Generalities   289
        9.6.2   The phase space and level density   291
        9.6.3   The Fermi Golden Rule   293
9.7   Exercises   297
9.8   Further reading   306

10 **Angular momentum**   **307**
10.1   Diagonalization of $\vec{J}^2$ and J_z 307
10.2   Rotation matrices   311
10.3   Orbital angular momentum   316
        10.3.1   The orbital angular momentum operator   316
        10.3.2   Properties of the spherical harmonics   319
10.4   Particle in a central potential   323
        10.4.1   The radial wave equation   323
        10.4.2   The hydrogen atom   327
10.5   Angular distributions in decays   331
        10.5.1   Rotations by π, parity, and reflection with respect
                 to a plane   331
        10.5.2   Dipole transitions   332
        10.5.3   Two-body decays: the general case   337
10.6   Addition of two angular momenta   339
        10.6.1   Addition of two spins 1/2   339
        10.6.2   The general case: addition of two angular momenta $\vec{J}_1$ and $\vec{J}_2$ 341
        10.6.3   Composition of rotation matrices   344
        10.6.4   The Wigner–Eckart theorem (scalar and vector operators) 345
10.7   Exercises   347
10.8   Further reading   357

11 **The harmonic oscillator**   **358**
11.1   The simple harmonic oscillator   359
        11.1.1   Creation and annihilation operators   359
        11.1.2   Diagonalization of the Hamiltonian   360
        11.1.3   Wave functions of the harmonic oscillator   362
11.2   Coherent states   364
11.3   Introduction to quantized fields   367
        11.3.1   Sound waves and phonons   367
        11.3.2   Quantization of a scalar field in one dimension   371
        11.3.3   Quantization of the electromagnetic field   375
        11.3.4   Quantum fluctuations of the electromagnetic field   380
11.4   Motion in a magnetic field   384
        11.4.1   Local gauge invariance   384
        11.4.2   A uniform magnetic field: Landau levels   387

11.5   Exercises   390
11.6   Further reading   402

12   Elementary scattering theory   **404**
12.1   The cross section and scattering amplitude   404
       12.1.1   The differential and total cross sections   404
       12.1.2   The scattering amplitude   406
12.2   Partial waves and phase shifts   409
       12.2.1   The partial-wave expansion   409
       12.2.2   Low-energy scattering   413
       12.2.3   The effective potential   417
       12.2.4   Low-energy neutron–proton scattering   419
12.3   Inelastic scattering   420
       12.3.1   The optical theorem   420
       12.3.2   The optical potential   423
12.4   Formal aspects   425
       12.4.1   The integral equation of scattering   425
       12.4.2   Scattering of a wave packet   427
12.5   Exercises   429
12.6   Further reading   437

13   Identical particles   **438**
13.1   Bosons and fermions   438
       13.1.1   Symmetry or antisymmetry of the state vector   438
       13.1.2   Spin and statistics   441
13.2   The scattering of identical particles   446
13.3   Collective states   448
13.4   Exercises   450
13.5   Further reading   454

14   Atomic physics   **455**
14.1   Approximation methods   455
       14.1.1   Generalities   455
       14.1.2   Nondegenerate perturbation theory   457
       14.1.3   Degenerate perturbation theory   458
       14.1.4   The variational method   459
14.2   One-electron atoms   460
       14.2.1   Energy levels in the absence of spin   460
       14.2.2   The fine structure   461
       14.2.3   The Zeeman effect   463
       14.2.4   The hyperfine structure   465
14.3   Atomic interactions with an electromagnetic field   467
       14.3.1   The semiclassical theory   467
       14.3.2   The dipole approximation   469

14.3.3   The photoelectric effect   471

14.3.4   The quantized electromagnetic field: spontaneous emission   473

14.4   Laser cooling and trapping of atoms   478

    14.4.1   The optical Bloch equations   478

    14.4.2   Dissipative forces and reactive forces   482

    14.4.3   Doppler cooling   484

    14.4.4   A magneto-optical trap   489

14.5   The two-electron atom   491

    14.5.1   The ground state of the helium atom   491

    14.5.2   The excited states of the helium atom   493

14.6   Exercises   495

14.7   Further reading   506

15   Open quantum systems   **507**

15.1   Generalized measurements   509

    15.1.1   Schmidt's decomposition   509

    15.1.2   Positive operator-valued measures   511

    15.1.3   Example: a POVM with spins 1/2   513

15.2   Superoperators   517

    15.2.1   Kraus decomposition   517

    15.2.2   The depolarizing channel   522

    15.2.3   The phase-damping channel   523

    15.2.4   The amplitude-damping channel   524

15.3   Master equations: the Lindblad form   526

    15.3.1   The Markovian approximation   526

    15.3.2   The Lindblad equation   527

    15.3.3   Example: the damped harmonic oscillator   529

15.4   Coupling to a thermal bath of oscillators   530

    15.4.1   Exact evolution equations   530

    15.4.2   The Markovian approximation   533

    15.4.3   Relaxation of a two-level system   535

    15.4.4   Quantum Brownian motion   538

    15.4.5   Decoherence and Schrödinger's cats   542

15.5   Exercises   544

15.6   Further reading   550

Appendix A   The Wigner theorem and time reversal   **552**

    A.1   Proof of the theorem   553

    A.2   Time reversal   555

Appendix B   Measurement and decoherence   **561**

    B.1   An elementary model of measurement   561

    B.2   Ramsey fringes   564

B.3   Interaction with a field inside the cavity   567
B.4   Decoherence   569

Appendix C   The Wigner–Weisskopf method   **573**

References   578
Index   579

Foreword

Quantum physics is now one hundred years old, and this description of physical phenomena, which has transformed our vision of the world, has never been found at fault, which is exceptional for a scientific theory. Its predictions have always been verified by experiment with impressive accuracy. The basic concepts of quantum physics such as probability amplitudes and linear superpositions of states, which seem so strange to our intuition when encountered for the first time, remain fundamental. However, during the last few decades an important evolution has occurred. The spectacular progress made in observational techniques and methods of manipulating atoms now makes it possible to perform experiments so delicate that they were once considered as only "thought experiments" by the founders of quantum mechanics. The existence of "nonseparable" quantum correlations, which forms the basis of the Einstein–Podolsky–Rosen "paradox" and which violates the famous Bell inequalities, has been confirmed experimentally with high precision. "Entangled" states of two systems which manifest such quantum correlations are now better understood and even used in practical applications such as quantum cryptography. The entanglement of a measuring device with its environment reveals an interesting new pathway to better understanding of the measurement process.

In parallel with these conceptual advances, our everyday world is being invaded by devices which function on the basis of quantum phenomena. The laser sources used to read compact disks, in ophthalmology, and in optical telecommunications are based on light amplification by atomic systems with population inversion. Nuclear magnetic resonance is widely used in hospitals to obtain ever more detailed images of the organs of the human body. Millions of transistors are incorporated in the chips which allow our computers to perform operations at phenomenal speeds.

It is therefore clear that any modern course in quantum physics must cover these recent developments in order to give the student or researcher a more accurate idea of the progress that has been made and to motivate the better understanding of physical phenomena whose conceptual and practical importance is increasingly obvious. This is the goal that Michel Le Bellac has successfully accomplished in the present work.

Each of the fifteen chapters of this book contains not only a clear and concise description of the basic ideas, but also numerous discussions of the most recent conceptual and experimental developments which give the reader an accurate idea of the advances in

xiii

the field and the general trends in its evolution. Chapter 6 on entangled states is typical of this method of presentation. Instead of stressing the mathematical properties of the tensor product of two spaces of states, which is rather austere and forbidding, this chapter is oriented on discussion of the idea of entanglement, and introduces several examples of theoretical and experimental developments (some of them very new) such as the Bell inequalities, tests of these inequalities and in particular the most recent ones based on parametric conversion, GHZ (Greenberger, Horne, Zeilinger) states, the idea of decoherence illustrated by modern experiments in cavity quantum electrodynamics (discussed in more detail in an appendix), and teleportation. It is difficult to imagine a more complete immersion in one of the most active current areas of quantum physics. Numerous examples of this modern presentation can be found in other chapters, too: interference of de Broglie waves realized using slow neutrons or laser-cooled atoms; tunnel-effect microscopy; quantum field fluctuations and the Casimir effect; non-Abelian gauge transformations; the optical Bloch equations; radiative forces exerted by laser beams on atoms; magneto-optical traps; Rabi oscillations in a cavity vacuum, and so on.

I greatly admire the effort made by the author to give the reader such a modern and compelling view of quantum physics. Of course, not all subjects can be treated in great detail, and the reader must make some effort to obtain a deeper comprehension of the subject. This is aided by the detailed bibliography given in the form of both footnotes to the text and a list of suggested reading at the end of each chapter. I am sure that this text will lead to better comprehension of quantum physics and will stimulate greater interest in this absolutely central discipline. I would like to thank Michel Le Bellac for this important contribution which will certainly give physics a more exciting image.

Claude Cohen-Tannoudji

Preface

This book has grown out of a course given at the University of Nice over many years for advanced undergraduates and graduate students in physics. The first ten chapters correspond to a basic course in quantum mechanics for advanced undergraduates, and the last four could serve to complement a graduate course in, for example, atomic physics. The book contains about 130 exercises of varying length and difficulty, most of which have actually been used in homework or exams.

This book should be interesting not only to students in physics and engineering, but also to a wider group of physicists: graduate students, researchers, and secondary-school teachers who wish to update their knowledge of quantum physics. It discusses recent developments not covered in the classic texts such as entangled states, quantum cryptography and quantum computing, decoherence, interactions of a laser with a two-level atom, quantum fluctuations of the electromagnetic field, laser manipulation of atoms, and so on, and it also includes a concise discussion of the current ideas about measurement in quantum mechanics as an appendix.

The organization of this book differs greatly from that of the classic texts, which typically begin with the Schrödinger equation and then proceed to study its solution in various situations. That approach makes it necessary to introduce the basic principles of quantum mechanics in a relatively complicated situation, and they end up being obscured by calculations which are often rather complex. Instead, I have striven to present the fundamentals of quantum mechanics using the simplest examples, and the Schrödinger equation appears only in Chapter 9. I follow the approach of pushing the logic adopted by Feynman (Feynman *et al.* [1965]) to its limit: developing the algebraic approach as far as possible and exploiting the symmetries, so as to present quantum mechanics within an autonomous framework without reference to classical physics. There are several advantages to this logic.

- The algebraic approach allows the solution of simple problems in finite-dimensional (for example, two-dimensional) spaces, such as photon polarization, spin 1/2, two-level atoms, and so on.
- This approach leads to the clearest statement of the postulates of quantum mechanics, as the fundamental issues are separated from the less fundamental ones (for example, the correspondence principle is not a fundamental postulate).

• The use of the symmetry properties leads to the most general introduction to fundamental physical properties such as momentum, angular momentum, and so on as the infinitesimal generators of these symmetries, without resorting to the correspondence principle or classical analogies.

Another advantage of this approach is that the reader wishing to learn about the recent developments in quantum information theory need consult only the first six chapters. These are sufficient for comprehension of the basics of quantum information, without passing through the stages of expansion of the wave function in spherical harmonics and solving the Schrödinger equation in a central potential!

I have given special attention to the pedagogical aspects. The order of chapters was carefully chosen: the early ones use only finite-dimensional spaces, and only after the basic principles have been covered do I go on to the general case in Chapter 7. Chapters 11 to 14 and the appendices involve more advanced techniques which may be of interest to professional physicists. An effort has been made regarding the vocabulary, in order to avoid certain historically dated expressions which can obstruct the understanding of quantum mechanics. Following the modernization proposed by J.-M. Lévy-Leblond (Quantum words for a quantum world, in *Epistemological and Experimental Perspectives on Quantum Physics*, D. Greenberger, W. L. Reiter and A. Zeilinger (eds.) Dordrecht: Kluwer (1999)), I use "physical property" instead of "observable" and "Heisenberg inequality" instead of "uncertainty principle," and I avoid expressions such as "complementarity" and "wave–particle duality."

The key chapters of this book, that is, those which diverge most obviously from the traditional treatment, are Chapters 3, 4, 5, 6, and 8. Chapter 3 introduces the space of states for the example of photon polarization and shows how to go from a wave amplitude to a probability amplitude. Spin 1/2 takes the reader directly to a problem without a classical analog. The essential properties of spin 1/2, namely the algebra of the Pauli matrices, the rotation matrices, and so on, are obtained using only two hypotheses: (1) two-dimensionality of the space of states and (2) rotational invariance. The Larmor precession of the quantum spin allows us to introduce the evolution equation. This chapter prepares the reader for the statement of the postulates of quantum mechanics in the following chapter, and it is possible to illustrate each postulate in a concrete fashion by returning to the examples of Chapter 3. The distinction between the general conceptual framework of quantum mechanics and the modeling of a particular problem is carefully explained. In Chapter 5 quantum mechanics is applied to some simple and physically important systems with a finite number of levels, a particular case being the diagonalization of the Hamiltonian in the presence of a periodic symmetry. This chapter also uses the example of the ammonia molecule to introduce the interaction of a two-level atomic or molecular system with an electromagnetic field, and the fundamental concepts of emission and absorption.

Chapter 6 is devoted to entangled states. The practical importance of these states dates from the early 1980s, but they are often ignored by textbooks. This chapter also deals with fundamental applications such as the Bell inequalities, two-photon interference, and measurement theory, as well as potential applications such as quantum computing.

Chapter 8 is devoted to the study of symmetries using the Wigner theorem, which is generally ignored in textbooks despite its crucial importance. Rotational symmetry allows the angular momentum to be defined as an infinitesimal generator, and the commutation relations of $\vec{J}$ can be demonstrated immediately with emphasis on their geometrical origin. The canonical commutation relations of X and P are derived from the identification of the momentum as the infinitesimal generator of translations. Finally, I obtain the most general form of the Hamiltonian compatible with Galilean invariance using a hypothesis about the velocity transformation law. This Hamiltonian will be reinterpreted later on within the framework of local gauge invariance.

The other chapters can be summarized as follows. Chapter 1 has the triple goal of (1) introducing the basic notions of microscopic physics which will be used later on in the text; (2) introducing the behavior of quantum particles, conventionally called "wave–particle duality"; and (3) presenting a simple explanation, with the aid of the Bohr atom, of the notion of energy level and of level spectrum. Chapter 2 presents the essential ideas about Hilbert space in the case of finite dimension. Chapter 7 gives some information about Hilbert spaces of infinite dimension; the goal here is of course not to present a mathematically rigorous treatment, but rather to warn the reader of certain pitfalls in infinite dimension.

The final chapters are devoted to more classic applications. Chapter 9 presents wave mechanics and its usual applications (the tunnel effect, bound states in the square well, periodic potentials, and so on). The angular momentum commutation relations already presented in Chapter 8 reappear in Chapter 10 in the construction of eigenstates of $\vec{J}^2$ and J_z, and lead to the Wigner–Eckart theorem for vector operators. Chapter 11 develops the theory of the harmonic oscillator and motion in a constant magnetic field, which provides the occasion for explaining local gauge invariance. An important section in this chapter deals with quantized fields: the vibrational field and phonons, and the electromagnetic field and its quantum fluctuations. Chapters 12 and 13 are devoted to scattering and identical particles. In Chapter 14 I present a brief introduction to the physics of one-electron atoms, the main objective being to calculate the forces on a two-level atom placed in the field of a laser and to discuss applications such as Doppler cooling and magneto-optical traps.

The appendices deal with subjects which are a bit more technically demanding. The proof of the Wigner theorem and the time-reversal operation are explained in detail in Appendix A. Some complementary information about the theory and experiments on decoherence can be found in Appendix B along with a discussion of some current ideas about measurement. Finally, Appendix C contains a discussion of the method of Wigner and Weisskopf for unstable states.

Acknowledgments

I have benefited from the criticism and suggestions of Pascal Baldi, Jean-Pierre Farges, Yves Gabellini, Thierry Grandou, Jacques Joffrin, Christian Miniatura, and especially Michel Brune (to whom I am also indebted for Figs. 6.9, B.1, and B.2), Jean Dalibard,

Fabrice Mortessagne, Jean-Pierre Romagnan, and François Rocca, who have read large parts or in some cases all of the manuscript. I also wish to thank David Wilkowski, who provided the inspiration for the text in some of the exercises of Chapter 14. Of course, I bear sole responsibility for the final text. The assistance of Karim Bernardet and Fabrice Mortessagne, who initiated me into XFIG and installed the software, was crucial for realizing the figures, and I also thank Christian Taggiasco for competently installing and maintaining all the necessary software. Finally, this book would never have seen the light of day were it not for the encouragement and unfailing support of Michèle Leduc, and I am very grateful to Claude Cohen-Tannoudji for writing the Preface.

Addendum for the English edition

In addition to minor corrections, I have included a few new exercises, partly rewritten Chapters 5 and 6, and added a new chapter on open quantum systems. I am grateful to Jean Dalibard and Christian Miniatura for their careful reading of this new chapter and for their useful comments. I would like to thank Simon Capelin and Vincent Higgs for their help in the publication and, above all, Patricia de Forcrand-Millard for her excellent translation and for her patience in our many email exchanges in order to find the right word.

Units and physical constants

The physical constants below are given with a relative precision of 10^{-3} which is sufficient for the numerical applications in this book.

Speed of light in vacuum	$c = 3.00 \times 10^8 \text{ m s}^{-1}$
Planck constant	$h = 6.63 \times 10^{-34} \text{ J s}$
Planck constant divided by 2π	$\hbar = 1.055 \times 10^{-34} \text{ J s}$
Electronic charge (absolute value)	$q_e = 1.602 \times 10^{-19} \text{ C}$
Fine structure constant	$\alpha = q_e^2/(4\pi\varepsilon_0 \hbar c) = e^2/(\hbar c) = 1/137$
Electron mass	$m_e = 9.11 \times 10^{-31} \text{ kg} = 0.511 \text{ MeV} c^{-2}$
Proton mass	$m_p = 1.67 \times 10^{-27} \text{ kg} = 938 \text{ MeV} c^{-2}$
Bohr magneton	$\mu_B = q_e \hbar/(2m_e) = 5.79 \times 10^{-5} \text{ eV T}^{-1}$
Nuclear magneton	$\mu_N = q_e \hbar/(2m_p) = 3.15 \times 10^{-8} \text{ eV T}^{-1}$
Bohr radius	$a_0 = \hbar^2/(m_e e^2) = 0.529 \times 10^{-8} \text{ m}$
Rydberg constant	$R_\infty = m_e e^4/(2\hbar^2) = 13.61 \text{ eV}$
Boltzmann constant	$k_B = 1.38 \times 10^{-23} \text{ J K}^{-1}$
Electron volt and temperature	$1 \text{ eV} = 1.602 \times 10^{-19} \text{ J} = k_B \times 11\,600 \text{ K}$
Gravitational constant	$G = 6.67 \times 10^{-11} \text{ N m}^2 \text{ kg}^{-2}$

1

Introduction

The first objective of this chapter is to briefly review some of the basic ideas about the structure of matter, in particular the concepts of microscopic physics, in order to recall the knowledge gained in previous physics (and chemistry) courses and make it more precise. Our review will be very concise, and most statements will be made without any proof or detailed discussion. A second objective is to give a brief description of some of the crucial stages in the early development of quantum physics. We shall not follow the strict historical order of this development or present the arguments used at the beginning of the last century by the founding fathers of quantum mechanics; rather, we shall stress the concepts which we shall find useful later on. Our last objective is to give an elementary introduction to some of the basic ideas, like those of a quantum particle or energy level, that will reappear throughout this text. We shall base our review on the Bohr theory, which provides a simple, though far from convincing, explanation of how energy levels are quantized and how the spectrum of the hydrogen atom arises. This chapter should be reread later on, once the basic ideas of quantum mechanics have been made explicit and illustrated by examples. From the practical point of view, it is possible to skip the general considerations of Sections 1.1 and 1.2 at the first reading and begin with Section 1.3, returning to those two sections later on as needed.

1.1 The structure of matter

1.1.1 Length scales from cosmology to elementary particles

Table 1.1 gives the length scales in meters of some typical objects, ranging from the size of the known Universe to the subatomic scale. A unit of length convenient for measuring astrophysical distances is the light-year (l.y.): 1 l.y. $= 0.95 \times 10^{16}$ m. The submeter scales commonly used in physics are the micrometer $1\,\mu$m $= 10^{-6}$ m, the nanometer 1 nm $=$ 10^{-9} m, and the femtometer (or fermi, F) 1 fm $= 10^{-15}$ m. Objects at the microscopic scale are often studied using electromagnetic radiation of wavelength of the order of the characteristic size of the object under study (by means of a microscope, X-rays, etc.).[1] It is well known that

---

[1] Other techniques are neutron scattering (Exercise 1.6.4), electron microscopy, tunneling microscopy (Section 9.4.2), and so on.

Table 1.1 *Some typical distance scales*

	Size (m)
Known Universe	1.3×10^{26}
Radius of the Milky Way	$\sim 5 \times 10^{20}$
Sun–Earth separation	1.5×10^{11}
Radius of the Earth	6.4×10^{6}
Man	~ 1.7
Insect	0.01 to 0.001
E. coli (bacterium)	$\sim 2 \times 10^{-6}$
HIV (virus)	1.1×10^{-7}
Fullerene C_{60}	0.7×10^{-9}
Atom	$\sim 10^{-10}$
Lead nucleus	7×10^{-15}
Proton	0.8×10^{-15}

the limiting resolution is determined by the wavelength used: it is fractions of a micrometer for a microscope using visible light, or fractions of a nanometer when X-rays are used. The wavelength spectrum of electromagnetic radiation (infrared, visible, etc.) is summarized in Fig. 1.1.

1.1.2 States of matter

We shall be particularly interested in phenomena occurring at the microscopic scale, and so it is useful to recall some of the elementary ideas about the microscopic description of matter. Matter can exist in two different forms: an ordered form, namely a crystalline solid, and a disordered form, namely a liquid, a gas, or an amorphous solid.

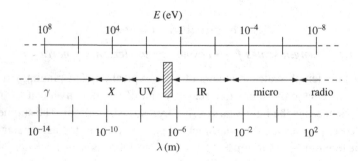

Fig. 1.1. Wavelengths of electromagnetic radiation and the corresponding photon energies. The boundaries between different types of radiation (for example, between γ-rays and X-rays) are not strictly defined. A photon of energy $E = 1$ eV has wavelength $\lambda = 1.24 \times 10^{-6}$ m, frequency $\nu = 2.42 \times 10^{14}$ Hz, and angular frequency $\omega = 1.52 \times 10^{15}$ rad s^{-1}.

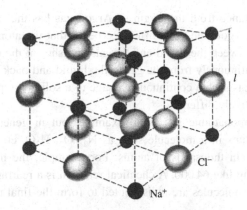

Fig. 1.2. Arrangement of atoms in a crystal of sodium chloride. The chlorine ions Cl⁻ are larger than the sodium ions Na⁺.

A crystalline solid possesses long-range order. As an example, in Fig. 1.2 we show the microscopic structure of sodium chloride. The basic crystal pattern is repeated with periodicity $l = 0.56$ nm, forming the *crystal lattice*. Starting from a chlorine ion or a sodium ion and moving along one of the links of the cubic structure, we again reach a chlorine ion or a sodium ion after a distance $n \times 0.56$ nm, where n is an integer. This is what we mean by long-range order.

Liquids, gases, and amorphous solids do not possess long-range order. Let us take as an example a monatomic liquid, namely liquid argon. To a first approximation the argon atoms can be represented as impenetrable spheres of diameter $\sigma \simeq 0.36$ nm. In Fig. 1.3 we schematically show an atomic configuration for a liquid in which the spheres practically touch each other, but are arranged in a disordered fashion. Taking the center of one atom as the origin, the probability $\mathsf{p}(r)$ of finding the center of another atom at a distance r from the former is practically zero for $r \lesssim \sigma$. However, this probability reaches a maximum at $r = \sigma, 2\sigma, \ldots$ and then oscillates before becoming stable at a constant value, whereas in the case of a crystalline solid the function $\mathsf{p}(r)$ possesses peaks

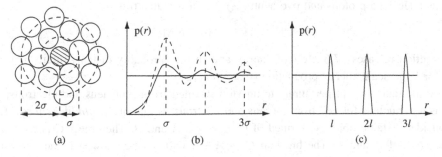

Fig. 1.3. (a) Arrangement of atoms in liquid argon. (b) Probability $\mathsf{p}(r)$ for a liquid (dashed line) and for a gas (solid line). (c) Probability $\mathsf{p}(r)$ for a simple crystal.

no matter what the distance from the origin is. Argon gas has the same type of atomic configuration as liquid argon, the only difference being that the atoms are much farther apart. The difference between the liquid and the gas vanishes at the critical point, and it is possible to move continuously from the gas to the liquid and back while going around the critical point, whereas such a continuous passage to a solid is impossible because the type of order is qualitatively different.

We have chosen a monatomic gas as an example, but in general the basic object is a combination of atoms in a molecule such as N_2, O_2, H_2O, etc. Certain molecules like proteins may contain thousands of atoms. For example, the molecular weight of hemoglobin is something like 64 000. A chemical reaction is a rearrangement of atoms – the atoms of the initial molecules are redistributed to form the final molecules:

$$H_2 + Cl_2 \rightarrow 2HCl.$$

An atom is composed of a positively charged atomic nucleus (or simply nucleus) and negatively charged electrons. More than 99.9% of the mass of the atom is in the nucleus, because the ratio of the electron mass m_e to the proton mass m_p is $m_e/m_p \simeq 1/1836$. The atom is ten thousand to a hundred thousand times larger than the nucleus: the typical size of an atom is 1 Å (where 1 Å $= 10^{-10}$ m $= 0.1$ nm), while that of a nucleus is several fermis (or femtometers).[2]

An atomic nucleus is composed of protons and neutrons. The former are electrically charged and the latter are neutral. The proton and neutron masses are identical to within 0.1%, and this mass difference can often be neglected in practice. The *atomic number* Z is the number of protons in the nucleus, and also the number of electrons in the corresponding atom, so that the atom is electrically neutral. The *mass number* A is the number of protons plus the number of neutrons N: $A = Z + N$. The protons and neutrons are referred to collectively as *nucleons*. Nuclear reactions involving protons and neutrons are analogous to chemical reactions involving atoms: a nuclear reaction is a redistribution of protons and neutrons to form nuclei different from the initial ones, while a chemical reaction is a redistribution of atoms to form molecules different from the initial ones. An example of a nuclear reaction is the fusion of a deuterium nucleus (^{2}H, a proton and a neutron) and a tritium nucleus (^{3}H, a proton and two neutrons) to form a helium-4 nucleus (^{4}He, two protons and two neutrons) plus a free neutron:

$$^2H + {}^3H \rightarrow {}^4He + n + 17.6 \text{ MeV}.$$

The reaction releases 17.6 MeV of energy and in the (probably distant) future may be used for large-scale energy production (fusion energy).

An important concept pertaining to an atom formed from a nucleus and electrons, as well as to a nucleus formed from protons and neutrons, is that of the *binding energy*. Let us consider a stable object C formed of two objects A and B. The object C is termed a *bound state* of A and B. The breakup $C \rightarrow A + B$ will not be allowed if the mass m_C

---

[2] We shall often use the Ångström (Å), which is the characteristic atomic scale, rather than nm.

of C is less than the sum of the masses m_A and m_B of A and B, that is, if the binding energy E_b

$$E_b = (m_A + m_B - m_C)c^2 \qquad (1.1)$$

is positive.[3] Here c is the speed of light and E_b is the energy needed to dissociate C into $A + B$. In atomic physics this energy is called the ionization energy, and it is the energy necessary to break up an atom into a positive ion and an electron, or, stated differently, to remove an electron from the atom. In the case of molecules E_b is the dissociation energy, or the energy needed to break up the molecule into atoms. A particle or a nucleus that is unstable in a particular configuration may be perfectly stable in a different configuration. For example, a free neutron (n) is unstable: in about fifteen minutes on average it disintegrates into a proton (p), an electron (e), and an electron antineutrino ($\overline{\nu}_e$); this is the basic decay of β-radioactivity:

$$n^0 \rightarrow p^+ + e^- + \overline{\nu}_e^0, \qquad (1.2)$$

where we have explicitly indicated the charge of each particle. This decay is possible because the masses[4] of the particles in (1.2) satisfy

$$m_n c^2 > (m_p + m_e + m_{\overline{\nu}})c^2,$$

where

$$m_n \simeq 939.5\,\mathrm{MeV}\,c^{-2}, \quad m_p \simeq 938.3\,\mathrm{MeV}\,c^{-2}, \quad m_e \simeq 0.51\,\mathrm{MeV}\,c^{-2}, \quad m_{\overline{\nu}_e} \simeq 0.$$

On the other hand, a neutron in a stable atomic nucleus does not decay; taking as an example the deuterium nucleus (the deuteron, ^{2}H), we have

$$m_{^2\mathrm{H}}c^2 \simeq 1875.6\ \mathrm{MeV} < (2m_p + m_e + m_{\overline{\nu}_e})c^2 \simeq 1878.3\ \mathrm{MeV},$$

and so the decay

$$^2\mathrm{H} \rightarrow 2p + e + \overline{\nu}_e$$

is impossible: the deuteron is a proton–neutron bound state.

1.1.3 Elementary constituents

So far, we have broken up molecules into atoms, atoms into electrons and nuclei, and nuclei into protons and neutrons. Can we go even farther? For example, can we break

---

[3] According to the celebrated Einstein relation $E = mc^2$; by simple dimensional analysis we can relate mass and energy to each other, so that, for example, masses can be expressed in $J\,c^{-2}$ or in $\mathrm{eV}\,c^{-2}$.

[4] Three recent experiments, those of S. Fukuda *et al.* (SuperKamiokande Collaboration), Solar B8 and hep neutrino measurements from 1258 days of SuperKamiokande data, *Phys. Rev. Lett.* **86**, 5651 (2001), Q. Ahmad *et al.* (SNO Collaboration), Interactions produced by B8 solar neutrinos at the Sudbury Neutrino Observatory, *Phys. Rev. Lett.*, **87**, 071301 (2001), and K. Eguchi *et al.* (Kamland Collaboration), First results from Kamland: evidence from reactor antineutrino disappearance, *Phys. Rev. Lett.* **90**, 021802 (2003), demonstrate convincingly that the neutrino mass is not zero, but is probably of order $10^{-2}\ \mathrm{eV}\,c^{-2}$; cf. Exercise 4.4.6 on neutrino oscillations. For a review, see D. Wark, Neutrinos: ghosts of matter, *Physics World* **18**(6), 29 (June 2005).

up a proton or an electron into more elementary constituents? Is it possible, for example, that a neutron is composed of a proton, an electron, and an antineutrino, as Eq. (1.2) suggests? A simple argument based on the Heisenberg inequalities shows that the electron cannot pre-exist inside the neutron (Exercise 9.7.4), but instead is *created* at the moment the decay occurs. Therefore, we cannot say that a neutron is composed of a proton, an electron, and a neutrino. One could also imagine "breaking" a proton or a neutron into more elementary constituents by bombarding it with energetic particles, just as, for example, happens when a deuteron is bombarded by electrons of several MeV in energy:

$$e + {}^2\mathrm{H} \to e + p + n.$$

The deuteron ${}^2\mathrm{H}$ is broken up into its constituents, a proton and a neutron. However, the situation is not repeated when a proton is bombarded by electrons. When low-energy electrons are used, the collisions are elastic:

$$e + p \to e + p,$$

and when the electron energy is high enough (several hundred MeV), the proton does not break up; instead, other particles are created, for example in reactions like

$$e + p \to e + p + \pi^0,$$

$$e + p \to e + n + \pi^+ + \pi^0,$$

$$e + p \to e + K^+ + \Lambda^0,$$

where the π and K mesons and the Λ^0 hyperon are new particles whose nature is not important for the present discussion. The crucial point is that these particles do not exist *ab initio* inside the proton, but are created at the instant the reaction occurs.

It therefore appears that at some point it is not possible to decompose matter into constituents which are more and more elementary. We can then ask the following question: what is the criterion for a particle to be elementary? The current idea is that a particle is elementary if it behaves as a point particle in its interactions with other particles. According to this idea, the electron, neutrino, and photon are elementary, while the proton and neutron are not: they are "composed" of quarks. These quotation marks are important, because quarks do not exist as free states,[5] and the quark "composition" of the proton is very different from the proton and neutron composition of the deuteron. Only indirect (but convincing) evidence of this quark composition exists.

As far as is known at present,[6] there exist three families of elementary particles or "particles of matter" of spin 1/2.[7] They are listed in Table 1.2, where the electric charge q is expressed in units of the proton charge. Each family is composed of leptons and quarks,

---

[5] What exactly is meant by the quark "mass" is quite complicated, at least for the so-called "light" quarks – the up, down, and strange quarks. Something close to the mass defined in the usual way is obtained for the heavy b and t quarks.

[6] There is a very strong argument for limiting the number of families to three. In 1992 experiments at CERN showed that the number of families is limited to three on the condition that the neutrino masses are less than 45 GeV c^{-2}. The actual experimental value of the number of families is 2.984 ± 0.008.

[7] Spin 1/2 is defined in Chapter 3 and spin in general in Chapter 10.

Table 1.2 *Matter particles. The electric charges are measured in units of the proton charge.*

	Lepton $q = -1$	Neutrino $q = 0$	Quark $q = 2/3$	Quark $q = -1/3$
Family 1	electron	neutrino$_e$	up quark	down quark
Family 2	muon	neutrino$_\mu$	charmed quark	strange quark
Family 3	tau	neutrino$_\tau$	top quark	bottom quark

and each particle has a corresponding antiparticle of the opposite charge. The leptons of the first family are the electron and its antiparticle the positron e$^+$, as well as the electron neutrino ν_e and its antiparticle the electron antineutrino $\bar{\nu}_e$. The quarks of this family are the up quark u of charge 2/3 and the down quark d of charge $-1/3$ plus, of course, the corresponding antiquarks $\bar{u}$ and $\bar{d}$, with charges $-2/3$ and 1/3, respectively. The proton is the combination uud and the neutron is the combination udd. This first family is sufficient for our everyday life, as all ordinary matter is composed of these particles. The neutrino is essential for the cycle of nuclear reactions occurring in the normally functioning Sun. While the existence of this first family is justified by an anthropocentric argument (if the family did not exist, we would not be here to talk about it), the reason for the existence of the other two families remains obscure.[8]

To these particles we need to add those that "carry" the interactions: the photon for electromagnetic interactions, the W and Z bosons for weak interactions, the gluons for strong interactions, and the graviton for gravitational interactions.[9] Now let us discuss these interactions.

1.1.4 The fundamental interactions

There are four types of fundamental interaction (forces): strong, electromagnetic, weak, and gravitational.[10] The *electromagnetic interaction* will play a leading role in this book, as it governs the behavior of atoms, molecules, solids, etc. The electrical forces obeying Coulomb's law dominate. We recall that a charge q fixed at the coordinate origin exerts a force on a charge q' at rest located at a point $\vec{r}$

$$\vec{F} = \frac{qq'}{4\pi\varepsilon_0} \frac{\hat{r}}{r^2}, \tag{1.3}$$

---

[8] As I. I. Rabi reputedly said of the muon: "Who ordered that?" Nevertheless, we know that each family must be complete: this is how the existence of the top quark and the value of its mass were predicted several years before its experimental discovery in 1994. Owing to its high mass, about 175 times that of the proton, the top quark was not discovered until the proton–antiproton collider known as the Tevatron was in operation in the USA.

[9] More rigorously, the electromagnetic and weak interactions have by now been unified as the electroweak interaction. The gluon, just like the quark, does not exist as a free state. Finally, the existence of the graviton is still hypothetical.

[10] Every once in a while a "fifth force" is "discovered," but it soon disappears again!

where $\hat{r}$ is a unit vector $\vec{r}/r$, $r = |\vec{r}|$, and ε_0 is the vacuum permittivity.[11] If the charges move with speed v, we must also take into account the magnetic forces. However, they are weaker than the Coulomb force by a factor $\sim (v/c)^2$ (we are using $\sim$ in the sense "of the order of"). For the electrons of the outer shells of an atom $(v/c)^2 \approx (1/137)^2 \ll 1$, but, owing to the extremely high precision of atomic physics experiments, the effects of magnetic forces are easily seen in phenomena such as the fine structure or the Zeeman effect (Section 14.2.3). The Coulomb force (1.3) is characterized by

- the $1/r^2$ force law. This is called a *long-range* force law;
- the strength of the force as measured by the *coupling constant $qq'/4\pi\varepsilon_0$*.

The modern, field-theoretic, point of view is that electromagnetic forces are generated by the exchange of "virtual" photons between charged particles.[12] Quantum field theory is the result of the (conflicting![13]) marriage between quantum mechanics and special relativity. The interactions between atoms or between molecules are represented as effective forces, for example van der Waals forces (Exercise 14.6.1). These forces are not fundamental because they are derived from the Coulomb force – they are actually the Coulomb force in disguise in the case of complex, electrically neutral systems.

The *strong interaction* is responsible for the cohesion of the atomic nucleus. In contrast to the Coulomb force, it falls off exponentially with distance according to the law $\simeq (1/r^2)\exp(r/r_0)$ with $r_0 \simeq 1\,\mathrm{F}$, and therefore is termed a *short-range* force. For $r \lesssim r_0$ this force is very strong, such that the typical energies inside the nucleus are of the order of MeV, while for the outer-shell electrons of an atom they are of the order of eV. In reality, the forces between nucleons are not fundamental, because, as we have seen, nucleons are composite particles. The forces between nucleons are analogous to the van der Waals forces between atoms, and the fundamental forces are actually those between the quarks. However, the quantitative relation between the nucleon–nucleon force and the quark–quark force is far from understood. The gluon, a particle of zero mass and spin 1 like the photon, plays the same role in the strong interaction as the photon plays in the electromagnetic one. The charge is replaced by a property conventionally referred to as color, and the theory of strong interactions is therefore called *(quantum) chromodynamics*. The *weak interaction* is responsible for radioactive β-decay:

$$(Z, N) \rightarrow (Z+1, N-1) + \mathrm{e}^- + \overline{\nu}_{\mathrm{e}}. \tag{1.4}$$

A special case is that of (1.2), which is written in the notation of (1.4) as

$$(0, 1) \rightarrow (1, 0) + \mathrm{e}^- + \overline{\nu}_{\mathrm{e}}.$$

Like the strong interaction, the weak interaction is short-range; however, as suggested by its name, it is much weaker than the former. The carriers of the weak interaction are

---

[11] We shall systematically use the notation $\hat{r}$, $\hat{n}$, $\hat{p}$ etc. for unit vectors in ordinary space.

[12] The term "virtual photons" will be explained in Section 4.2.4.

[13] The combination of quantum mechanics and special relativity leads to infinities, which must be controlled by a procedure called renormalization. The latter was not fully understood and justified until the 1970s.

spin-1 bosons: the charged $W^{\pm}$ and the neutral Z^0 with masses 82 MeV c^{-2} and 91 MeV c^{-2}, respectively (about 100 times the proton mass). The leptons, quarks, spin-1 bosons (also referred to as gauge bosons: the photon, gluons, $W^{\pm}$, and Z^0; see Exercise 11.5.11 for some elementary explanations), as well as a hypothetical spin-0 particle called the Higgs boson which gives masses to all the particles, are the particles of the *Standard Model* of particle physics. This model has been tested experimentally with a precision of better than 0.1% over the past ten years.

Last of all, we have the *gravitational interaction* between two masses m and m', which, in contrast to the Coulomb interaction, is always attractive:

$$\vec{F} = -Gmm'\frac{\hat{r}}{r^2}. \tag{1.5}$$

Here the notation is the same as in (1.2) and G is the gravitational constant. The force law (1.5) is, like the Coulomb law, a long-range law, and since the two forces have the same form we can form the ratio of these forces between an electron and a proton:

$$\frac{F_C}{F_{gr}} = \left(\frac{q_e^2}{4\pi\varepsilon_0}\right)\left(\frac{1}{Gm_e m_p}\right) \sim 10^{39}.$$

In the hydrogen atom the gravitational force is negligible; in general, this force is completely negligible for all the phenomena of atomic, molecular, and solid-state physics. General relativity, the relativistic theory of gravity, predicts the existence of gravitational waves.[14] These are the gravitational analog of electromagnetic waves, and the spin-2, massless graviton is the analog of the photon. Nevertheless, at present there is no quantum theory of gravity. The unification of quantum mechanics and general relativity and the explanation of the origin of mass and the three particle families are major challenges of theoretical physics in the twenty-first century.

Let us summarize our presentation of the elementary constituents and the fundamental forces. There exist three families of matter particles, the leptons and quarks, plus the carriers of the fundamental forces: the photon for the electromagnetic interaction, the gluon for the strong interaction, the W and Z bosons for the weak interaction, and, finally, the hypothetical graviton for the gravitational interaction.

1.2 Classical and quantum physics

Before introducing quantum physics, let us briefly review the fundamentals of classical physics. There are three main branches of classical physics, and each has different ramifications.

---

[14] At present, there is only indirect, but convincing, evidence for gravitational waves from observations of binary pulsars (neutron stars). Such waves may some day be detected on Earth in the VIRGO, LIGO, and LISA experiments. The graviton will probably be observed only in the very distant future.

1. The first branch is *mechanics*, where the fundamental law is Newton's law. Newton's law is the fundamental law of dynamics; it states that in an inertial frame the force $\vec{F}$ on a point particle of mass m is equal to the derivative of its momentum $\vec{p}$ with respect to time:

$$\vec{F} = \frac{d\vec{p}}{dt}. \tag{1.6}$$

This form of the fundamental equation of dynamics remains unchanged when the modifications due to special relativity, introduced by Einstein in 1905, are taken into account. In the general form of (1.6) we must use the relativistic expression for the momentum as a function of the particle velocity $\vec{v}$ and mass m:

$$\vec{p} = \frac{m\vec{v}}{\sqrt{1 - v^2/c^2}}. \tag{1.7}$$

2. The second branch is *electromagnetism*, summarized in the four Maxwell equations which give the electric field $\vec{E}$ and magnetic field $\vec{B}$ as functions of the charge density ρ_{em} and the current density $\vec{j}_{em}$, which are referred to as the *sources* of the electromagnetic field:

$$\vec{\nabla} \cdot \vec{B} = 0, \qquad \vec{\nabla} \times \vec{E} = -\frac{\partial \vec{B}}{\partial t}, \tag{1.8}$$

$$\vec{\nabla} \cdot \vec{E} = \frac{\rho_{em}}{\varepsilon_0}, \qquad c^2 \vec{\nabla} \times \vec{B} = \frac{\partial \vec{E}}{\partial t} + \frac{1}{\varepsilon_0} \vec{j}_{em}. \tag{1.9}$$

These equations lead to a description of the propagation of electromagnetic waves in a vacuum at the speed of light:

$$\left(\frac{1}{c^2} \frac{\partial^2}{\partial t^2} - \nabla^2 \right) \left\{ \begin{matrix} \vec{E} \\ \vec{B} \end{matrix} \right. = 0. \tag{1.10}$$

Maxwell's equations allow us to make the connection to optics, which becomes a special case of electromagnetism. The connection between mechanics and electromagnetism is supplied by the *Lorentz law* giving the force on a particle of charge q and velocity $\vec{v}$:

$$\vec{F} = q(\vec{E} + \vec{v} \times \vec{B}). \tag{1.11}$$

3. The third branch is *thermodynamics*, in which the main consequences are derived from the second law:[15] there exists no transformation whose sole effect is to extract a quantity of heat from a reservoir and convert it entirely to work. This second law leads to the concept of entropy which lies at the base of all of classical thermodynamics. The microscopic origin of the second law was understood at the end of the nineteenth century by Boltzmann and Gibbs, who were able to relate this law to the fact that a macroscopic sample of matter is made up of an enormous ($\sim 10^{23}$) number of atoms; this allows us to use probability arguments, on which statistical mechanics is founded. The principal result of statistical mechanics is the *Boltzmann law*: the

---

[15] The first law is just energy conservation, while the third is fundamentally of quantum origin.

probability $p(E)$ for a physical system in equilibrium at absolute temperature T to have energy E includes a factor called the *Boltzmann weight* $p_B(E)$:[16]

$$p_B(E) = \exp\left(-\frac{E}{k_B T}\right) = \exp(-\beta E), \qquad (1.12)$$

where k_B is the Boltzmann constant (the gas constant R divided by Avogadro's number), and we have introduced the usual notation $\beta = 1/k_B T$. However, classical statistical mechanics is not in fact a consistent theory, and it is sometimes necessary to resort to questionable arguments to obtain a sensible result, for example in computing the entropy of a perfect gas. Quantum physics removes all these difficulties.

4. To be completely rigorous, we should mention a fourth branch of classical physics: the relativistic theory of gravity, which in effect is not included in the three branches listed above. This theory is called *general relativity*, and is a geometrical description in which gravitational forces arise from the curvature of spacetime.

Equations (1.6)–(1.11) represent the fundamental laws of classical physics, which can be summarized in only seven equations! The reader may wonder what happened to all the other familiar laws of physics such as Ohm's law, Hooke's law, the laws of fluid dynamics, etc. Some of these laws are derived directly from the fundamental ones; for example, Coulomb's law is a consequence of the Maxwell equations and the Lorentz force (1.11) for static charges, and the Euler equation for a perfect fluid is a consequence of the fundamental law of dynamics. Many other laws are *phenomenological.*[17] They are not universally valid, in contrast to the fundamental laws. For example, some media do not obey Ohm's law; the relation between the induction $\vec{D}$ and the electric field $\vec{D} = \varepsilon \vec{E}$ (for an isotropic medium) does not hold when the electric field becomes strong, giving rise to the phenomena of nonlinear optics. Hooke's law does not apply if the tension becomes too large, and so on. The mechanics of solids, elasticity and fluid mechanics follow from (1.6) and various phenomenological laws like the law that relates the force, velocity gradient, and viscosity in fluid mechanics. It is important to clearly distinguish between the small number of fundamental laws and the large number of phenomenological laws which, for lack of anything better, are used in classical physics to describe matter.

Although there is no doubt that classical physics is useful, it does possess a serious shortcoming: although physics claims to be a theory of matter, classical physics is completely incapable of *explaining* the behavior of matter given its constituents and the forces between them.[18] It cannot predict the existence of atoms, because it is not possible to construct a length scale using the constants of classical physics: the masses and charges

---

[16] The probability $p(E)$ is the product of $p_B(E)$ (1.12) and the factor $\mathcal{D}(E)$, the "energy-level density," which in classical physics is obtained by integrating over phase space; see Footnote 21. The quantum calculation of the level density is described in Section 9.6.2.

[17] Quite often a phenomenological law is nothing but the first term of a Taylor series.

[18] This statement should be qualified slightly. There do exist good microscopic models in classical physics: for example, the kinetic theory of gases permits reliable calculation of the transport coefficients (viscosity, thermal conductivity) of a gas. However, neither the existence of the molecules making up the gas nor the value of the effective cross section needed in the calculation can be explained by classical physics.

of the nucleus and electrons.[19] It cannot explain why the Sun shines or why sodium vapor emits yellow light, and it has nothing to say about the chemical properties of the alkalines, about the fact that copper conducts electricity while sulfur is an insulator, and so on. When the classical physicist needs a property of a material such as an electrical resistance or a specific heat, he or she has no choice but to measure it experimentally. In contrast, quantum mechanics attempts to explain the behavior of matter starting from the constituents and forces. Naturally, it is not possible to make precise predictions based on first principles except for the simplest systems, like the hydrogen or helium atoms. The complexity of the calculations does not allow, for example, prediction of the crystal structure of silver based on the data for this atom, but given the crystal structure it can explain why silver is a conductor, which classical physics is incapable of doing.

It should not be concluded from this discussion that classical physics can no longer be interesting and innovative. On the contrary, during the past twenty years classical physics has taken on new life with the development of new ideas about chaotic dynamical systems, instabilities, nonequilibrium phenomena, and so on. Moreover, such familiar problems as turbulence and friction remain poorly understood and extremely interesting. There simply exist problems that by their nature are not *suitable* for study using classical physics.

Quantum physics aspires to explain the behavior of matter on the basis of its constituents and forces, but there is a price to pay: quantum objects display radically new behavior which defies our intuition developed from the behavior of classical objects. That said, quantum mechanics proves to be a remarkable tool which so far has always given correct results and is capable of coping with problems ranging from quark physics to cosmology and all scales in between. Without quantum mechanics, most of modern technology would never have seen the light of day. All of information technology is based on our quantum understanding of solids and, in particular, semiconductors. The miniaturization of electronic devices will make quantum mechanics more and more omnipresent in modern technology.

The vast majority of physicists do not worry about the puzzling aspects of quantum mechanics, but simply use it as a tool without asking questions of principle. Nevertheless, the theoretical and, especially, experimental progress made over the past twenty years have led to a better grasp of certain aspects of the behavior of quantum objects. Although things are still far from clear, we shall see in Chapter 6 and Appendix B that we are certainly on the path to a more satisfactory understanding of quantum mechanics. Perhaps in a few years Feynman's statement, "I think it can be stated today that no one understands quantum mechanics," will become obsolete. Before discussing the recent developments, let us go back a few years to the beginning of quantum physics.

---

[19] If we include the speed of light, we can construct a length scale, the *classical electron radius*

$$r_e = \frac{q_e^2}{4\pi\varepsilon_0} \frac{1}{m_e c^2} \simeq 2.8 \times 10^{-15}\,\text{m},$$

but it is four orders of magnitude too small to be related to atomic dimensions. Another way of saying all this is to invoke the scale invariance of the classical equations; cf. Wichman [1967], Chapter 1.

1.3 A bit of history

1.3.1 Black-body radiation

A hot object such as a red-hot iron or the Sun emits electromagnetic radiation with a frequency spectrum that depends on temperature. The power emitted $u(\omega, T)$ per unit frequency ω and unit area depends on the absolute temperature T of the object. Purely thermodynamical arguments can be made to show that if the object is perfectly absorbing, that is, if it is a *black body*, then $u(\omega, T)$ is a universal function independent of the object at a given temperature. An excellent realization of a black body for visible light is a small opening in a cavity whose interior is painted black. A light ray which enters the cavity has practically no chance of getting out, because at each reflection there is a high probability of being absorbed by the inner wall of the cavity (Fig. 1.4).

Let us suppose that the cavity is heated to a temperature T. The atoms of the inner wall emit and absorb electromagnetic radiation, and a system of standing waves in thermodynamical equilibrium is established in the cavity. If the cavity is a parallelepiped of sides L_x, L_y, and L_z and we use periodic boundary conditions, the electric field will have the form $\vec{E}_0 \exp[i(\vec{k} \cdot \vec{r} - \omega t)]$, with the wave vector $\vec{k}$ perpendicular to $\vec{E}_0$ and of the form

$$\vec{k} = \left(\frac{2\pi}{L_x} n_x, \frac{2\pi}{L_y} n_y, \frac{2\pi}{L_z} n_z \right), \tag{1.13}$$

where (n_x, n_y, n_z) are positive or negative integers and $\omega = c|\vec{k}| = ck$. It can be shown that each standing wave behaves like a harmonic oscillator[20] of frequency ω with energy proportional to the squared amplitude $\vec{E}_0^2$. According to the Boltzmann law (1.12), the probability that this oscillator has energy E involves the factor

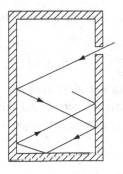

Fig. 1.4. Cavity for black-body radiation.

---

[20] This will be explained in Section 11.3.3.

$\exp(-E/k_B T) = \exp(-\beta E)$. In fact, in this case the level density $\mathcal{D}(E)$ (cf. Footnote 16) is a constant,[21] and the average energy of this oscillator is simply

$$\langle E \rangle = \frac{\int dE\, E \exp(-\beta E)}{\int dE \exp(-\beta E)} = -\frac{\partial}{\partial \beta} \ln \left(\int dE \exp(-\beta E) \right)$$

$$= -\frac{\partial}{\partial \beta} \ln \frac{1}{\beta} = \frac{1}{\beta} = k_B T. \tag{1.14}$$

The average energy of each standing wave is $k_B T$. Since there are an infinite number of possible standing waves, the energy inside the cavity is infinite!

The emitted power $u(\omega, T)$ has a simple relation to the energy density $\epsilon(\omega, T)$ per unit frequency in the cavity (Exercise 1.6.2):

$$u(\omega, T) = \frac{c}{4} \epsilon(\omega, T), \tag{1.15}$$

so that we need to compute $\epsilon(\omega, T)$, from which we obtain the energy density:

$$\epsilon(T) = \int_0^\infty d\omega\, \epsilon(\omega, T). \tag{1.16}$$

Thermodynamics gives the scaling law

$$\epsilon(\omega, T) = \omega^3 \varphi\left(\frac{\omega}{T}\right), \tag{1.17}$$

but tells us nothing about the explicit form of the function φ except that it is independent of the shape of the cavity. Let us try to find it up to a multiplicative factor by means of dimensional analysis. A priori, $\epsilon(\omega, T)$ can only depend on ω, c, the energy $k_B T$, and a dimensionless constant A which cannot be fixed by dimensional analysis. The only possible solution is (Exercise 1.6.2)

$$\epsilon(\omega, T) = A c^{-3} (k_B T) \omega^2 = \omega^3 \left[A c^{-3} \left(\frac{k_B T}{\omega} \right) \right], \tag{1.18}$$

which has the form (1.17). We rediscover the fact that the energy density in the cavity is infinite:

$$\epsilon(T) = \int_0^\infty d\omega\, \epsilon(\omega, T) = A c^{-3} (k_B T) \int_0^\infty \omega^2 d\omega = +\infty.$$

The constant A can be calculated in statistical mechanics (Exercise 1.6.2), but this does not resolve the problem of the infinite energy, and the dimensional analysis strongly suggests that black-body radiation cannot be explained unless a new physical constant is introduced.

---

[21] The integration over phase space for a one-dimensional harmonic oscillator gives, for an arbitrary function $f(E)$ (Exercise 1.6.2),

$$\int dx\, dp\, \delta\left(E - \frac{p^2}{2m} - \frac{1}{2} m\omega^2 x^2 \right) f(E) = \frac{2\pi}{\omega} f(E),$$

where x and p are the position and momentum, and δ is a Dirac delta function.

Out of all the hypotheses that could lead to the unacceptable result of infinite energy, Planck chose the one on which the calculation (1.14) of the average oscillator energy is based.[22] Instead of allowing E to take all possible values between zero and infinity, he assumed that it can take only discrete values E_n which are integer multiples of the oscillator frequency ω with proportionality coefficient $\hbar$:

$$E_n = n(\hbar\omega), \quad n = 0, 1, 2, \ldots \tag{1.19}$$

The constant $\hbar$ is called *Planck's constant*; more precisely, it is Planck's constant h divided by 2π: $\hbar = h/2\pi$.[23] Planck's constant is measured in joule seconds (J s), and it has dimensions $\mathcal{M}\mathcal{L}^2\mathcal{T}^{-1}$ and numerical value

$$\hbar \approx 1.054 \times 10^{-34}\,\text{J s} \quad \text{or} \quad h \approx 6.63 \times 10^{-34}\,\text{J s}.$$

According to the Boltzmann law, the normalized probability of observing an energy E_n is

$$\mathsf{p}(E_n) = e^{-\beta n \hbar\omega}\left(\sum_{n=0}^{\infty} e^{-\beta n \hbar\omega}\right)^{-1} = \exp(-\beta n \hbar\omega)[1 - \exp(-\beta\hbar\omega)]. \tag{1.20}$$

In obtaining (1.20) we have used the fact that the summation over n is that of a geometrical series. Setting $x = \exp(-\beta\hbar\omega)$, we easily find the average oscillator energy $\langle E \rangle$:

$$\langle E \rangle = (1-x)\sum_{n=0}^{\infty}(n\hbar\omega)x^n = (1-x)\hbar\omega x\frac{\mathrm{d}}{\mathrm{d}x}\sum_{n=0}^{\infty}x^n$$

$$= (1-x)\hbar\omega x\frac{\mathrm{d}}{\mathrm{d}x}\frac{1}{1-x} = \frac{\hbar\omega x}{1-x} = \frac{\hbar\omega}{\exp(\beta\hbar\omega)-1}. \tag{1.21}$$

This expression can be used to calculate the energy density (Exercise 1.6.2)

$$\epsilon(\omega, T) = \frac{\hbar}{\pi^2 c^3}\frac{\omega^3}{\exp(\beta\hbar\omega)-1} \tag{1.22}$$

and then $u(\omega, T)$, in perfect agreement with experiment for a suitably chosen value of $\hbar$ and with the result (1.17) of thermodynamics. We note that the classical approximation (1.18) is valid if $k_B T \gg \hbar\omega$, that is, for low frequencies.

The best-known example of black-body radiation is the relic 3 K background radiation filling the Universe, also called the cosmic microwave background (CMB).[24] The frequency distribution of this radiation is in remarkable agreement with the Planck

---

[22] In reality, Planck applied his arguments to a "resonator," the nature of which remains obscure, and the present argument follows that of Einstein (1905). Dealing with electromagnetic field oscillations is simpler and more direct, but it does distort the historical truth. Our "historical" presentation, like that of many textbooks, is more reminiscent of a fairy tale (H. Kragh, Max Planck: the reluctant revolutionary, *Physics World* **13** (12), 31 (December 2000)) than actual history. Likewise, it does not appear that the physicists of the late nineteenth century were troubled by the infinite energy or the absence of a fundamental constant.

[23] We shall systematically use $\hbar$ rather than h, and somewhat carelessly refer to $\hbar$ as Planck's constant; the relation $E = \hbar\omega$ is of course the same as $E = h\nu$, where ν is the ordinary frequency measured in hertz and ω is the angular or rotational frequency measured in rad s^{-1}: $\omega = 2\pi\nu$. Since we nearly always use ω rather than ν, we shall just refer to ω as the frequency.

[24] A particularly good account of the Big Bang is given by S. Weinberg in, *The First Three Minutes: A Modern View of the Origin of the Universe*, New York: Basic Books (1977).

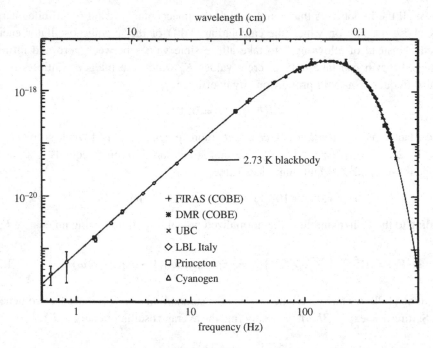

Fig. 1.5. The 3 K black-body radiation. On the vertical axis is the radiation intensity in W m^{-2} sr^{-1} Hz^{-1}. The remarkable agreement with Planck's law for $T = 2.73$ K is clearly seen. Taken from J. Rich, *Fundamentals of Cosmology*, New York: Springer (2001).

law (1.22) for the temperature 2.73 K $\approx$ 3 K (Fig. 1.5), but this radiation is no longer in thermodynamical equilibrium. It was decoupled from matter about 380 000 years after the Big Bang, that is, after the birth of the Universe. At the instant of decoupling the temperature was about 10^4 K. The subsequent expansion of the Universe has reduced this value to the present one of 3 K. Deviations from a fully isotropic black-body radiation, of the order of 10^{-3}, arise from the motion of the Solar System with respect to the cosmic microwave background, owing to the Doppler effect. There are also angular dependent temperature fluctuations, $\sim 10^{-5}$, which are much more interesting as they give us important information on the early history of the Universe.

1.3.2 The photoelectric effect

The integer n in (1.19) has a particularly important physical interpretation: the reason that the energy of a standing wave of frequency ω is an integer multiple $n\hbar\omega$ of $\hbar\omega$ is that it corresponds to precisely n *photons* (or "particles of light") of energy $\hbar\omega$. It is this interpretation that led Einstein to introduce the concept of photon in order to explain the photoelectric effect. When a metal is illuminated by electromagnetic radiation, some electrons escape from it and there is a threshold effect that depends on the frequency

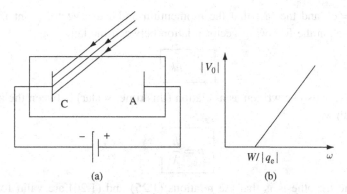

Fig. 1.6. The Millikan experiment. (a) Schematic view of the experiment. (b) $|V_0|$ as a function of ω.

and not the intensity of the radiation. The Millikan experiment (Fig. 1.6) confirms the Einstein interpretation: the electrons emitted from the metal have kinetic energy E_k

$$E_k = \hbar\omega - W, \tag{1.23}$$

where W is the work function. An electron of charge q_e does not reach the cathode if $|q_e V| > E_k$. If V_0 is the potential at which the current vanishes, then

$$|V_0| = \frac{\hbar}{|q_e|}\omega - \frac{W}{|q_e|}. \tag{1.24}$$

The potential $|V_0|$ as a function of ω has a constant slope $\hbar/|q_e|$, and the value of $\hbar$ coincides with that for black-body radiation, thus confirming the Einstein hypothesis[25] that electromagnetic radiation is composed of photons.[26] The fact that the value of $\hbar$ is the same as in the case of black-body radiation strongly suggests that one must introduce a new fundamental constant.

1.4 Waves and particles: interference

1.4.1 The de Broglie hypothesis

From Eq. (1.19) for $n = 1$ we find $E = \hbar\omega$, the *Planck–Einstein relation* between the energy and frequency of a photon. The photon possesses momentum

$$p = \frac{E}{c} = \frac{\hbar\omega}{c},$$

---

[25] Another rewriting of history! Some qualitative results on the photoelectric effect were obtained by Lenard in the early 1900s, but the precise measurements of Millikan were made 10 years after the Einstein hypothesis. Einstein seems to have been motivated not by the photoelectric effect, but by thermodynamic considerations. See G. Margaritondo, *Physics World* **14**(4), 17 (April 2001).

[26] The argument is not completely convincing, because the photoelectric effect can be explained within the framework of a semiclassical theory, where the electromagnetic field is not quantized and where there is no concept of photon; cf. Section 14.3.3. However, it is not possible to explain the photoelectric effect without introducing $\hbar$. The fact that a photomultiplier whose operation is based on the photoelectric effect registers isolated counts can be attributed to the quantum nature of the device rather than the arrival of isolated photons.

but using $\omega = ck$ and the fact that the momentum and wave vector point in the same direction we obtain the following vector relation between the latter:

$$\boxed{\vec{p} = \hbar\vec{k}} \; . \tag{1.25}$$

This equation can also be written as a relation (this time, scalar) between the momentum and wavelength λ:

$$\boxed{p = \frac{h}{\lambda}} \; . \tag{1.26}$$

The de Broglie hypothesis is that the relations (1.25) and (1.26) are valid for all particles. According to this hypothesis, a particle of momentum $\vec{p}$ possesses wave properties characterized by the de Broglie wavelength $\lambda = h/p$. If $v \ll c$ we can use $\vec{p} = m\vec{v}$, while otherwise we use the general expression (1.7), except for $m = 0$, when $p = E/c$. If this hypothesis is correct, particles must have observable wave properties; in particular, they must undergo interference and diffraction.

1.4.2 Diffraction and interference of cold neutrons

Since the 1980s, modern experimental techniques have allowed interference and diffraction of particles to be verified in experiments based on simple principles and admitting direct interpretation. Such experiments have been performed using photons, electrons, atoms, molecules, and neutrons. Here we have chosen, a bit arbitrarily, to discuss neutron experiments, as they are particularly elegant and clear. Neutron diffraction by crystals has been around for fifty years now and is a classic experiment (Exercise 1.6.4), but modern experiments are carried out using macroscopic devices with slits that can be viewed by the naked eye, rather than a crystal lattice with a spacing of a few angstroms.

The experiments were performed in the 1980s by a group in Innsbruck using the research nuclear reactor of the Laue-Langevin Institute in Grenoble. Neutrons of mass m_n are produced in the fission of uranium-235 in the reactor core, and then channeled to the experiments. The order of magnitude of their kinetic energy is $k_B T$, where $T \approx 300 \, \text{K}$ is the ambient temperature. Such neutrons are termed thermal and have kinetic energy $\sim k_B T \approx 1/40 \, \text{eV}$ for $T = 300 \, \text{K}$. The momentum $p = \sqrt{2m_n k_B T}$ corresponds to a speed $v = p/m_n$ of about 1000 m s^{-1}, and according to (1.26) the associated wavelength λ_{th} is $h/\sqrt{2m_n k_B T} \approx 1.8 \, \text{Å}$. The wavelength is increased when the neutrons are made to pass through a low-temperature material. For example, if the temperature of the material is 1 K, the wavelength will increase to $\lambda = \lambda_{th}\sqrt{300} \approx 31 \, \text{Å}$. Such neutrons are termed "cold." In the experiments of the Innsbruck group, the neutrons were cooled to 25 K using liquid deuterium.[27] This produced neutrons with an average wavelength of about 20 Å.

---

[27] Deuterium was chosen over hydrogen, as the latter inconveniently absorbs neutrons in the reaction $n + p \to {}^2\text{H} + \gamma$ (see Exercise 14.6.8). This is why in a nuclear reactor heavy water is a better moderator than ordinary water.

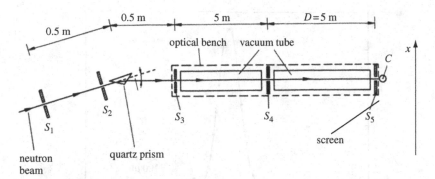

Fig. 1.7. Experimental setup for neutron diffraction and interference: S_1 and S_2 are collimating slits, S_3 is the entrance slit, S_4 is the object slit, and S_5 is the slit at the location of the counter C. From A. Zeilinger *et al.*, *Rev. Mod. Phys.* **60**, 1067 (1988).

The experimental setup is shown schematically in Fig. 1.7. The neutrons are detected by means of BF_3 counters, in which the boron absorbs neutrons in the reaction

$$^{10}B + n \rightarrow {}^{7}Li + {}^{4}He$$

with an efficiency of nearly 100%. The counter is placed behind the screen at S_5, and counts the number of neutrons arriving in the neighborhood of S_5.

In the diffraction experiment the slit S_4 has a width of $a = 93\,\mu m$, which leads to a diffraction maximum of angular size

$$\theta = \frac{\lambda}{a} \approx 2 \times 10^{-5} \text{ rad}.$$

On the screen located $D = 5\,m$ from the slit the linear size of the diffraction peak is of order $100\,\mu m$. It is possible to calculate the diffraction pattern precisely, taking into account, for example, the spread of wavelengths about the average value of 20 Å. The theoretical result is in excellent agreement with experiment (Fig. 1.8).

In the interference experiment, two 21-μm slits have their centers separated by a distance $d = 125\,\mu m$. The separation between fringes on the screen is

$$i = \frac{\lambda D}{d} = 80\,\mu m.$$

The slits are visible with the naked eye, and the interference pattern is macroscopic. Again, the theoretical calculation taking into account the various parameters of the experiment is in excellent agreement with the experimental interference pattern (Fig. 1.9).

However, there is a crucial difference from an experiment on optical interference: the interference pattern is made up of impacts of isolated neutrons and it is reconstructed afterwards, when the experiment is completed. Actually, the counter is moved along the screen (or an array of identical counters covers the screen), and the neutrons arriving in the neighborhood of each point of the screen are recorded during identical time intervals. Let $N(x)\Delta x$ be the number of neutrons detected per second in the interval

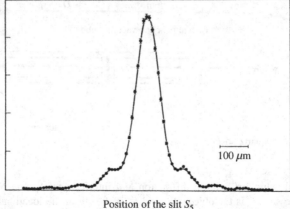

Fig. 1.8. Neutron diffraction by a slit. The full line is the theoretical prediction. From A. Zeilinger *et al.*, *Rev. Mod. Phys.* **60**, 1067 (1988).

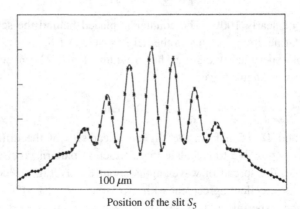

Fig. 1.9. Young's slit experiment using neutrons. The full line is the theoretical prediction. From A. Zeilinger *et al.*, *Rev. Mod. Phys.* **60**, 1067 (1988).

$[x - \Delta x/2, x + \Delta x/2]$, where x is the abscissa of a point on the screen. The intensity $\mathcal{J}(x)$ can be defined as being equal to $N(x)$, and the number of neutrons arriving in the neighborhood of a point of the screen is proportional to the intensity $\mathcal{J}(x)$ of the interference pattern, with statistical fluctuations of order $\sqrt{N}$ about the average value. The isolated impacts are illustrated in Fig. 1.10 for an experiment performed using not neutrons, but cold atoms (see Section 14.4) which were allowed to fall through Young slits. The impacts of the atoms that hit the screen were recorded, giving the pattern in Fig. 1.10.

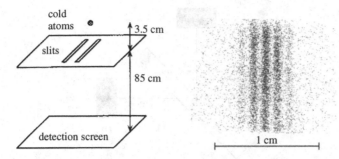

Fig. 1.10. Interference using cold atoms. From Basdevant and Dalibard [2002].

1.4.3 Interpretation of the experiments

In addition to cold neutrons and atoms, other types of particle have been used in diffraction and interference experiments:

- photons, with the light intensity reduced such that the photons arrive at the screen one by one. Nevertheless, an experiment performed under these conditions is not entirely convincing, because it can be explained semiclassically taking into account the quantum nature of the detector; see Footnote 26. However, it is now known how to construct sources that provide truly isolated photons, and experiments using such photons unarguably demonstrate interference produced by one photon at a time[28]
- electrons
- light molecules (Na_2)
- fullerenes C_{60} (Exercise 1.6.1).

There is every reason to assume that the results are universal, independent of the type of particle – atoms, molecules, virus particles, etc.[29] However, a difficulty of principle seems to arise in interpreting these experimental results. In a classical Young's slit interference experiment realized using waves, the incident wave is split into two waves which recombine and interfere, a phenomenon which is visible to the naked eye in, for example, the case of waves on the surface of water. In the case of neutrons, each neutron arrives separately, and the interval between the arrivals of two successive neutrons is such that when a neutron is detected on the screen, the next one is still in the reactor confined inside a uranium atom. Can we imagine that a neutron is split in two, with each half passing through a slit? It is easy to convince ourselves that this hypothesis is absurd: a counter always detects an entire neutron, never a fraction of one. The same situation occurs if a semi-transparent mirror is used to split a light wave of intensity

---

[28] A. Aspect, P. Grangier, and G. Roger, Dualité onde–corpuscule pour un photon unique, *J. Optics (Paris)* **20**, 119 (1989).

[29] However, wave effects become more and more difficult to observe for larger particles, in practice because the wavelength becomes shorter and shorter, and more fundamentally because decoherence effects (Section 15.4.5) become more and more important as an object becomes larger. See M. Arndt, K. Hornberger, and A. Zeilinger, Probing the limits of quantum worlds, *Physics World* **18** (3), 35 (2005).

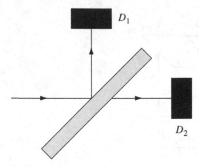

Fig. 1.11. Beam-splitting plate and photon counting by photodetectors D_1 and D_2.

reduced enough to permit the detection of individual photons. The photodetectors D_1 and D_2 always detect an entire photon, never a fraction of one (Fig. 1.11). The photon, like the neutron, is indivisible, at least in a vacuum (though by interaction with a nonlinear medium a photon can be split into two of lower energy; see Section 6.3.2).

We therefore must assume that a quantum particle possesses wave and particle properties *simultaneously*. It is an entirely new and strange object, at least to our intuition based on experience with macroscopic objects. As Lévy-Leblond and Balibar, paraphrasing Feynman, have written, "quantum objects are completely crazy." However, they add "at least they are all crazy in the same way." Photons, electrons, neutrons, atoms, molecules – all behave the same way, like waves and particles at the same time. In order to emphasize this unity of quantum behavior, some authors have proposed the term "quanton" to refer to such an object. Here we shall continue to use "quantum particle" or simply "particle," because the particles we shall consider in this book generally display quantum behavior. We will specify "classical particle" when we need to refer to particles that behave like little billiard balls.

If the neutron is indivisible, is it possible to know which slit it has passed through? If one slit is closed, we observe on the screen the diffraction pattern corresponding to the other slit and vice versa. If the experimental situation is such that it is possible to tell which slit the neutron has passed through, then we observe on the screen the superposition of the *intensities* of the diffraction patterns of each slit: the neutrons can effectively be divided into two groups, those that passed through the upper slit and for which the lower slit could have been closed without changing the result, and those that passed through the lower slit. We observe an interference pattern only if the experimental apparatus is such that we cannot know, *even in principle*, which slit a neutron has passed through. Summarizing:

(i) If the experimental apparatus does not permit knowledge of which slit a neutron passed through, an interference pattern is observed.

(ii) If the apparatus permits us in principle to determine which of the two slits a neutron passed through, the interference will be destroyed independently of whether we actually bother to determine which slit it was.

A fundamental point to note is that we cannot know a priori at which point of the screen a given neutron will arrive. We can only state that the *probability* of arriving at the screen is large at a point of an interference maximum and small at a point of an interference minimum. More precisely, the probability of arriving at an abscissa x is proportional to the intensity $\mathcal{J}(x)$ of the interference pattern at this point. Likewise, in the experiment of Fig. 1.11 each photomultiplier has a probability of 1/2 of being triggered by a given photon, but it is impossible to know in advance which of the two detectors will be triggered.

Let us try to make the preceding discussion quantitative. First of all, by analogy with waves, we shall introduce a complex function of x, $a_1(x)$ $[a_2(x)]$, associated with the passage through the upper slit [lower slit] of a neutron that reaches a point x on the screen. For reasons to be explained below, this function will be called the probability amplitude. The squared modulus of the probability amplitude gives the intensity: if slit 2 is closed $\mathcal{J}_1(x) = |a_1(x)|^2$, and, conversely, if slit 1 is closed $\mathcal{J}_2(x) = |a_2(x)|^2$. In case (i) above we add the *amplitudes* before calculating the intensity:

$$\mathcal{J}(x) \propto |a_1(x) + a_2(x)|^2, \tag{1.27}$$

while in case (ii) we add the *intensities*

$$\mathcal{J}(x) \propto |a_1(x)|^2 + |a_2(x)|^2 = \mathcal{J}_1(x) + \mathcal{J}_2(x). \tag{1.28}$$

As above, the intensity can be defined as the number of neutrons arriving per second per unit length of the screen. To take into account the probabilistic nature of the neutron point of impact, the amplitudes a_1 and a_2 will not be wave amplitudes measuring the amplitude of a vibration, but *probability amplitudes*, with the squared modulus being the probability of arriving at a point x on the screen. The concept of probability amplitude in quantum physics will be developed and given mathematical status in Chapter 3.

A more general statement of (1.27) and (1.28) is the following. Let us suppose that starting from an initial state i we arrive at a final state f. To find the probability $p_{i \to f}$ of observing the final state f, we must add all the amplitudes that lead to the result f starting from i:

$$a_{i \to f} = a_{i \to f}^{(1)} + a_{i \to f}^{(2)} + \cdots + a_{i \to f}^{(n)},$$

and then $p_{i \to f} = |a_{i \to f}|^2$. It should be understood that the states i and f are specified uniquely by the parameters that define the initial and final states of the full ensemble of the experimental apparatus. If, for example, we desire information about the passage of a neutron through a given slit, we can obtain it by integrating the Young's slits into a larger apparatus. Then the final state of this larger apparatus, which will be a function of other parameters in addition to the neutron point of impact, is capable of informing us whether the neutron has passed through the given slit. Just what is the final state of this larger apparatus will depend on which slit the neutron passed through.

In summary, we must sum the amplitudes for identical final states and the probabilities for different final states, even if these final states differ only by physical parameters other than those of interest. It is sufficient that these other parameters be accessible in principle, even if they are not actually observed, for us to consider the final states as being different. We shall illustrate this point by a concrete example in the following paragraph. Another way of saying this which is easier to visualize is the following: identical final states are associated with *indistinguishable paths*, and it is necessary to sum the amplitudes corresponding to all indistinguishable paths.

1.4.4 Heisenberg inequalities I

Let us return to the neutron diffraction experiment in order to extract from it a fundamental relation called the *Heisenberg inequality*, or, more commonly but ambiguously, the *Heisenberg uncertainty principle*. If the slit width is a and if we orient the x axis along the slit, perpendicular to the direction through the slit, the neutron position relative to this axis immediately on leaving the slit is known to within $\Delta x = a$. Because the angular width of the diffraction maximum is $\sim \lambda/\Delta x$, the x component of the neutron momentum is $\Delta p_x \approx (\lambda/\Delta x)p = h\Delta p_x$, where p is the neutron momentum (we assume that $p \gg \Delta p_x$). We then obtain the relation

$$\Delta p_x \Delta x \sim h. \tag{1.29}$$

In Chapter 9 we shall discuss a more accurate version of Eq. (1.29) involving the standard deviations, which we shall call simply the dispersions, of momentum and position Δp_i and Δx_i for identical values of $i = x, y, z$:

$$\Delta p_i \Delta x_i \geq \frac{1}{2}\hbar. \tag{1.30}$$

There are no inequalities relating different components of momentum and position, for example Δp_x and Δy. When interpreting a diffraction experiment it is often said that the passage of a neutron through a slit of width Δx allows the neutron's x coordinate to be measured with a precision Δx, and that this measurement perturbs the neutron's momentum by an amount $\Delta p_x \approx h/\Delta x$. We shall see in Section 4.2.4 that the inequalities (1.30) in fact have nothing to do with the experimental measurement of position or momentum, but instead arise from the mathematical description of a quantum particle as a wave packet, and we shall also elaborate on the precise meaning of these relations.

We are now going to use (1.29) to discuss the question of observing trajectories in a neutron interference experiment. Einstein proposed the apparatus of Fig. 1.12 for determining the neutron trajectory, i.e., for determining whether the neutron passes through the upper or the lower slit. When the neutron passes through the first slit S_0, owing to momentum conservation it transfers a downward momentum to the screen E_0 if it passes through the upper slit S_1 and an upward momentum to the screen if it passes through the lower slit S_2. It is then possible to determine which slit the neutron has passed through. Bohr's response was the following. If the screen E_0 receives a momentum δp_x which can be measured, this means that the *initial* momentum Δp_x of the screen was much less than δp_x, and the initial position is determined with an uncertainty at least of order

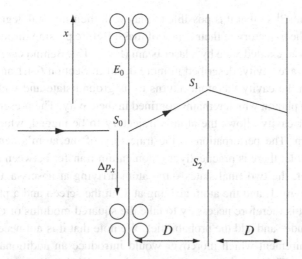

Fig. 1.12. The Bohr–Einstein controversy. Slits S_1 and S_2 are Young's slits. Slit S_0 is located in a screen which can move vertically.

$h/\Delta p_x$. Such an inaccuracy in the position of the source is sufficient to make the interference pattern disappear (Exercise 1.6.3). All the various types of apparatus that can be imagined for determining the neutron trajectory are either efficient, in which case there is no interference pattern, or inefficient, in which case there is an interference pattern, but the slit through which the neutron has passed cannot be known. The interference pattern becomes more and more fuzzy as the apparatus becomes more and more efficient.

The above discussion is completely correct, but one should not conclude that it is the perturbation of the neutron trajectory on hitting the first screen that spoils the interference pattern.[30] The crucial point is the possibility of tagging the trajectory. It is possible to imagine and even experimentally construct an apparatus that tags trajectories without disturbing the observed degrees of freedom at all, and yet this tagging is sufficient to destroy the interference pattern. Let us briefly describe an apparatus which has not yet been realized experimentally, but may become feasible when technology has evolved further. Other types of apparatus that tag trajectories without perturbing them have been *effectively* realized and are discussed in Exercise 3.3.9, Section 6.3.2, and Appendix B. However, the principle governing such devices is based on ideas which we have not yet introduced, and so for now we shall return to the familiar example of Young's slits. The proposed

[30] The same remark applies to the apparatus imagined by Feynman for a Young's slit experiment using electrons (Feynman *et al.* [1965], Vol. III, Chapter 1). A photon source placed behind the slits makes it possible in theory to observe the electron passage. When short-wavelength photons are used the electron–photon collisions permit the two slits to be distinguished, but the collisions perturb the trajectories enough to spoil the interference pattern. If the photon wavelength is increased, the impacts are less violent, but the resolving power of the photons decreases. The interference fringes reappear when the resolution becomes such that it is no longer possible to distinguish between the slits.

apparatus uses atoms,[31] so that it is possible to play with their internal degrees of freedom without affecting the trajectory of their center of mass. Before passing through the slits, the atoms are raised to an excited state by a laser beam (Fig. 1.13). Behind each slit is a super-conducting microwave cavity, described in more detail in Section 6.4.1 and Appendix B. In passing through the cavity the atom returns to its ground state and with nearly 100% probability emits a photon which remains confined in the cavity. The presence of a photon in one or the other cavity allows the atom's trajectory to be tagged, which destroys the interference pattern. The perturbation to the trajectory of the atom's center of mass is completely negligible: there is practically no momentum transfer between the photon and the atom. However, the two final states – the atom arriving at abscissa x on the screen and a photon in cavity 1, and the atom arriving at x on the screen and a photon in cavity 2 – are different. It is therefore necessary to take the squared modulus of each of the corresponding amplitudes and add the probabilities. We note that it is not necessary to detect the photon, a requirement which moreover would introduce an additional experimental complication. It is sufficient to know that the atom has emitted a photon in a quasi-certain way in its passage through the cavity. As we have already emphasized, it is not at all necessary that the final state is effectively observed, it is only necessary that it can be observed in principle, even if the present or future state of technology does not permit such observation. In the terminology to be defined in Chapters 6 and 15, we can say that interference is destroyed if "which path" information is encoded in the environment. We shall return to this subject in Appendix B.1, where we will discuss it in a mathematical context.

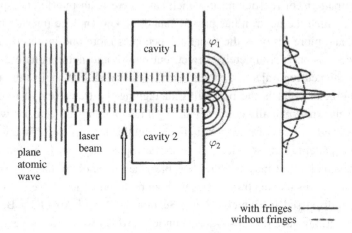

Fig. 1.13. Tagging of trajectories in Young's slit experiments. Taken from B. Englert, M. Scully, and H. Walther, Origin of quantum mechanical complementarity probed by a "which way" experiment in an atom interferometer. *Nature* **351**, 111 (1991).

---

[31] This has been imagined by B. Englert, M. Scully, and H. Walther, Quantum optical tests of complementarity, *Nature* **351**, 111 (1991), and they present a popularized description of it in *Scientific American* **271**, 86 (December 1994). The atoms are assumed to be in Rydberg states (cf. Exercise 14.5.4). A related experiment based on the same principle but with a more complicated realization has been performed by S. Dürr, T. Nonn, and G. Rempe, Origin of quantum mechanical complementarity probed by a "which way" experiment in an atom interferometer, *Nature* **395**, 33 (1998). See also P. Bertet *et al.*, A complementarity experiment with an interferometer at the quantum–classical boundary, *Nature* **411**, 166–170 (2001).

1.5 Energy levels

The goal of this section is to define the concept of energy level, first on the basis of the classical notion. Taking as an example the Bohr atom, we can then proceed in a simple way to the quantum notion, after which we shall examine radiative transitions between levels.

1.5.1 Energy levels in classical mechanics and classical models of the atom

Let us imagine a classical particle which we take, for the sake of simplicity, to be moving along the x axis and which has potential energy $U(x)$. In quantum mechanics, $U(x)$ is referred to in general as the *potential*. It is well known that the mechanical energy E, the sum of the kinetic energy K and the potential energy U, is constant: $E = K + U =$ const. Let us assume that the potential energy has the form shown in Fig. 1.14, that of a "potential well" which tends to the same constant value for $x \to \pm\infty$. It will be convenient to fix the zero of the energy such that $E = 0$ for a particle of kinetic energy that vanishes at infinity.

There are two possible situations.

(i) The particle has energy $E > 0$. Then if, for example, it leaves from $x = -\infty$, it is first accelerated and then decelerated in passing through the potential well, and at $x = +\infty$ it reaches a final velocity equal to the initial one. Such a particle is said to be in a *scattering state*.

(ii) The particle has negative energy $U_0 < E < 0$. Then the particle cannot escape from the well, but travels back and forth inside it between the points x_1 and x_2 satisfying $E = U(x_{1,2})$. It is confined inside a finite region of the x axis, $x_1 \le x \le x_2$, and is said to be in a *bound state*.

When the potential energy is positive (Fig. 1.15) we have the case of a "potential barrier."[32] In this case $E > 0$ and only scattering states are observed. If $E < U_0$, a particle leaving from $x = -\infty$ is at first decelerated, and when it arrives at the point x_1 satisfying $U(x_1) = E$ it is reflected by the potential barrier. If $E > U_0$ the particle passes over the potential barrier and reaches $x = +\infty$ with its initial velocity.

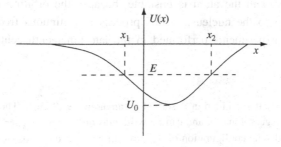

Fig. 1.14. A potential well.

---

[32] Naturally, situations more complex than the ones in these figures can be imagined, for example a double well. Here we shall discuss only the simplest cases.

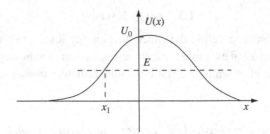

Fig. 1.15. A potential barrier.

In classical mechanics the energy of a bound state can take all possible values between U_0 and 0. In quantum mechanics, we shall see in Chapter 9 that it can take only discrete values. On the other hand, as in classical mechanics, the energy of a scattering state is arbitrary. However, there are still notable differences (Sections 9.3 and 9.4) from the case of classical mechanics. For example, the particle can pass over a potential barrier even if $E < U_0$. This is called "tunneling." Moreover, the particle can be reflected even if $E > U_0$.

Let us apply these ideas from classical mechanics to atoms. The first atomic model was proposed by Thomson (Fig. 1.16a). Here the atom is represented as a sphere of uniform positive charge, with electrons moving around inside this charge distribution. It is a result of elementary electrostatics that the electrons here experience a harmonic potential, and their ground (stable) energy level is the state in which they are at rest at the bottom of the potential well. Excited states correspond to vibrations about the equilibrium position. This model was ruled out by the experiments of Geiger and Marsden, who showed that α-particle (^{4}He nucleus) scattering by atoms is incompatible with it.[33] Rutherford deduced from his experiments the existence of an atomic nucleus of size less than 10 F, and proposed a planetary model of the atom (Fig. 1.16b): the electrons orbit the nucleus like the planets orbit the Sun, with the Coulomb interaction playing the role of gravitational attraction. This model possesses two major, related shortcomings: there is no scale which fixes the atomic size, and the atom is unstable, because the orbiting electrons radiate and end up falling onto the nucleus. In this process a continuous frequency spectrum is emitted, whereas experiments performed in the late nineteenth century showed that (Fig. 1.17)

- the frequencies of radiation emitted or absorbed by an atom are discrete. They are expressed as a function of two integers n and m and can be written as differences, $\omega_{nm} = A_n - A_m$;
- there exists a ground-state configuration of the atom in which it does not radiate.

---

[33] Though atomic physicists still often make use of it …

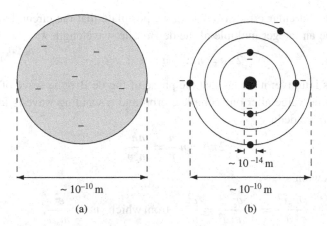

Fig. 1.16. Models of the atom. (a) Thomson: the electrons are located inside a uniform distribution of positive charge. (b) Rutherford: the electrons orbit a nucleus.

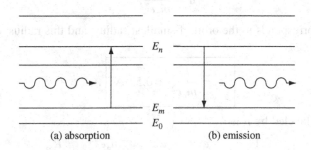

Fig. 1.17. Emission and absorption of radiation between two levels E_n and E_m.

These results suggest that the atom emits or absorbs a photon in passing from one level to another, with the photon frequency ω_{nm} given by $(E_n > E_m)$

$$\hbar\omega_{nm} = E_n - E_m. \tag{1.31}$$

The frequencies ω_{nm} are called the Bohr frequencies. According to these arguments, only certain levels labeled by a discrete index can exist. This is referred to as the quantization of energy levels.

1.5.2 The Bohr atom

In order to explain this quantization, Bohr imposed an ad hoc quantization rule on classical mechanics and the Rutherford atom. We shall follow an argument slightly different from his original one. Taking for simplicity the hydrogen atom with an electron of mass m_e

and charge q_e in a circular orbit of radius a, we postulate that the circumference $2\pi a$ of the orbit must be an integer multiple of the de Broglie wavelength λ:

$$2\pi a = n\lambda, \quad n = 1, 2, \ldots \tag{1.32}$$

This postulate is intuitive; it means that the phase of the de Broglie wave of the electron returns to its initial value after one complete orbit and a standing wave is formed. From (1.32) and (1.26) we deduce

$$2\pi a = n\frac{h}{p} = \frac{nh}{m_e v}.$$

According to Newton's law,

$$\frac{m_e v^2}{a} = \frac{q_e^2}{4\pi\varepsilon_0 a^2} = \frac{e^2}{a^2}, \quad \text{from which} \quad v^2 = \frac{e^2}{m_e a},$$

where we have defined the quantity $e^2 = q_e^2/4\pi\varepsilon_0$. Eliminating the speed v between the two equations, we obtain the orbital radius:

$$a = \frac{n^2\hbar^2}{m_e e^2}. \tag{1.33}$$

The case $n = 1$ corresponds to the orbit of smallest radius, and this radius, denoted a_0, is called the *Bohr radius*:

$$\boxed{a_0 = \frac{\hbar^2}{m_e e^2} \simeq 0.53 \text{ Å}}. \tag{1.34}$$

The energy level labeled by n is

$$E_n = \frac{1}{2}m_e v^2 - \frac{e^2}{a} = -\frac{e^2}{2a} = -\frac{m_e e^4}{2n^2\hbar^2} = -\frac{R_\infty}{n^2}.$$

The energy levels E_n are expressed as a function of the *Rydberg constant* R_∞,[34]

$$\boxed{R_\infty = \frac{m_e e^4}{2\hbar^2} \simeq 13.6 \text{ eV}} \tag{1.35}$$

as

$$\boxed{E_n = -\frac{R_\infty}{n^2}}. \tag{1.36}$$

This formula gives the *level spectrum* of the hydrogen atom. The ground state corresponds to $n = 1$ and the ionization energy of the hydrogen atom is R_∞. The photons emitted by the hydrogen atom have frequencies

$$\hbar\omega_{nm} = -R_\infty\left(\frac{1}{n^2} - \frac{1}{m^2}\right), n > m, \tag{1.37}$$

---

[34] The subscript ∞ is used because the theory described here assumes that the proton is infinitely heavy. When the finite mass m_p of the proton is taken into account, R_∞ is changed to $R_\infty[1/(1 + m_e/m_p)]$; cf. Exercise 1.6.5.

in perfect agreement with the spectroscopic data for hydrogen. However, the simplicity with which the spectrum of the hydrogen atom can be calculated using the Bohr theory should not be allowed to mask the artificial nature of this theory.

Sommerfeld's generalization of the Bohr theory consists of the postulate

$$\int p_i \, dq_i = nh, \tag{1.38}$$

where q_i and p_i are coordinates and momenta conjugate in the sense of classical mechanics and n is an integer ≥ 1. However, we now know that the conditions (1.38) are valid only for certain very special systems and for large n, with some exceptions. The Bohr–Sommerfeld theory cannot describe atoms with many electrons, or scattering states. The success of the Bohr theory in the case of the hydrogen atom is only a happy accident.

1.5.3 Orders of magnitude in atomic physics

Metre/Kilogram/Second units, which are adapted to measuring things at the human scale, are not convenient in atomic physics. A priori, a convenient system of units should feature the fundamental constants $\hbar$ and c, as well as the electron mass m_e. The proton can be considered infinitely heavy, or, more precisely, the electron mass can be replaced by the reduced mass (cf. Footnote 34). Let us recall the values of these constants with an accuracy of $\sim 10^{-3}$ sufficient for the numerical applications in this book:

$$\hbar = 1.054 \times 10^{-34} \, \text{J s},$$

$$c = 3 \times 10^8 \, \text{m s}^{-1},$$

$$m_e = 0.911 \times 10^{-30} \, \text{kg}.$$

From these constants we can form the following natural units:

- The unit of length:[35] $\dfrac{\hbar}{m_e c} = 3.86 \times 10^{-13} \, \text{m}$;
- The unit of time: $\dfrac{\hbar}{m_e c^2} = 1.29 \times 10^{-21} \, \text{s}$;
- The unit of energy: $m_e c^2 = 5.11 \times 10^5 \, \text{eV}$.

These units are much closer than MKS units to the orders of magnitude characteristic of atomic physics, though a few orders of magnitude are still lacking. This is fixed by introducing a quantity which measures the strength of the electromagnetic force, the

---

[35] Called the *Compton wavelength* of the electron.

coupling constant $e^2 = q_e^2/4\pi\varepsilon_0$. From $\hbar$, c, and e^2 we can form a dimensionless quantity called the *fine-structure constant* α:[36]

$$\alpha = \frac{e^2}{\hbar c} = \frac{q_e^2}{4\pi\varepsilon_0 \hbar c} \simeq \frac{1}{137}. \tag{1.39}$$

The relations between atomic units and natural units are now easy to find. For the Bohr radius, the natural unit of length in atomic physics, we obtain

$$a_0 = \frac{\hbar^2}{m_e e^2} = \frac{\hbar c}{e^2} \frac{\hbar}{m_e c} = \frac{1}{\alpha} \frac{\hbar}{m_e c} \approx 0.53 \text{ Å}. \tag{1.40}$$

The Rydberg, the natural unit of energy in atomic physics, is related to $m_e c^2$ as

$$R_\infty = \frac{1}{2} \frac{m_e e^4}{\hbar^2} = \frac{1}{2} \left(\frac{e^2}{\hbar c}\right)^2 m_e c^2 = \frac{1}{2} \alpha^2 m_e c^2 \approx 13.6 \text{ eV}. \tag{1.41}$$

The speed of the electron in the ground state is $v = \alpha c = e^2/\hbar$, and the period of this orbit, which is the atomic unit of time, is

$$T = \frac{2\pi a_0}{v} = 2\pi \frac{1}{\alpha} \frac{\hbar}{m_e c} \frac{1}{\alpha c} = \frac{2\pi}{\alpha^2} \frac{\hbar}{m_e c^2} \approx 1.5 \times 10^{-16} \text{ s}. \tag{1.42}$$

Equations (1.40)–(1.42) show that the natural units and atomic units are related by powers of α.

As a final example, let us estimate the average lifetime of an electron in an excited state. We shall use a classical picture, viewing the electron as traveling in an orbit of radius a.[37] We shall push this picture until it breaks down, and then we shall attempt to correct it by taking into account quantum considerations; this is called *semiclassical reasoning*. A calculation in classical electromagnetism shows that an electron in a circular orbit which moves with speed $v = \omega a \ll c$ radiates a power

$$P = \frac{2}{3c^3} e^2 a^2 \omega^4 = \frac{2}{3} \left(\frac{e^2}{\hbar c}\right) \frac{a^2 \hbar \omega^4}{c^2} \sim \alpha \omega^2 \hbar \left(\frac{a\omega}{c}\right)^2. \tag{1.43}$$

In a purely classical picture, the electron will lose energy in a continuous fashion by emitting electromagnetic radiation. This is where an admittedly ad hoc quantum argument

---

[36] This terminology arose for historical reasons and is somewhat confusing; it would be better to say "atomic constant" α. This is the *coupling constant of electrodynamics*, although it is not really constant owing to subtleties of quantum field theory. The quantum fluctuations of the electron–positron field have the effect of screening electric charges: owing to (virtual) electron–positron pair production, the charge of a particle measured far from the particle is smaller than the charge measured close to it. Owing to the Heisenberg inequality (1.30), short distance implies large momentum and therefore high energy, i.e., particles of high energy must be used to explore short distances. It can therefore be concluded that the fine-structure constant is an increasing function of energy, and in fact at energies of the order of the Z^0 boson rest energy, $m_Z c^2 \approx 90$ GeV, we have $\alpha \approx 1/129$ instead of the low-energy value $\alpha \approx 1/137$. The renormalization procedure of eliminating infinities allows us to choose an arbitrary energy (or distance) scale for defining α. In sum, α depends on the energy scale characteristic of the process under study, and also on details of the renormalization procedure (cf. Footnote 13). This energy dependence of α has been observed for several years now in precision experiments in high-energy physics. See also Exercise 14.6.3.

[37] One can also view an atom as a dipole oscillating with frequency ω, as in the Thomson model. The only difference is that the factor of 2/3 in (1.43) becomes 1/3, which has no effect on the orders of magnitude.

enters: the atom emits a photon when it has accumulated an energy $\sim \hbar\omega$, which takes a time τ corresponding to the *lifetime of the excited state*:

$$\frac{1}{\tau} \sim \frac{P}{\hbar\omega} \sim a\omega \left(\frac{a\omega}{c}\right)^2. \tag{1.44}$$

However, we have seen that $a\omega/c = v/c \sim \alpha$, and the relation between the period T and the average lifetime τ is

$$\frac{T}{\tau} \sim \frac{1}{\tau\omega} \sim \alpha^3 \sim 10^{-6}. \tag{1.45}$$

The electron orbits about a million times before emitting a photon, and so an excited state is well defined. For the ground state of the hydrogen atom where the energy is $\sim 10\,\text{eV}$ we have seen that $T \sim 10^{-16}\,\text{s}$, while for an outer-shell electron of an alkaline atom with energy $\sim 1\,\text{eV}$ we have instead $T \sim 10^{-15}$ s and the order of magnitude of the lifetime of an excited state is $\sim 10^{-7}$–10^{-9} s. For example, the first excited state of rubidium (D_2 line) has an average lifetime of 2.7×10^{-8} s.

The reasoning we have followed in this section has the merit of simplicity, but it is not satisfying. We had to impose a somewhat ad hoc quantum constraint on the classical arguments when they became untenable, and the reader can justly fail to be convinced by this sort of reasoning. It is therefore necessary to develop an entirely new theory which is no longer guided by classical physics, but instead develops in an autonomous fashion, *without reference to classical physics*.

1.6 Exercises

1.6.1 Orders of magnitude

1. We would like to explore distances at the atomic scale, that is, 1 Å, using photons, neutrons, or electrons. What should the order of magnitude of the energy of these particles be in eV?
2. When the wavelength λ of a sound wave is large compared with the lattice spacing of the crystal in which the vibration propagates, the frequency ω of the wave is linear in the wave vector $k = 2\pi/\lambda$: $\omega = c_s k$, where c_s is the speed of sound (cf. Section 11.3.1). In the case of steel $c_s \simeq 5 \times 10^3\,\text{m s}^{-1}$. What is the energy $\hbar\omega$ of a sound wave for $k = 1\,\text{nm}^{-1}$? The particle analogous to the photon in the case of sound waves is called the *phonon* (see Section 11.3.1), and $\hbar\omega$ is the phonon energy. Using the fact that a phonon can be created in an inelastic collision with a crystal, should neutrons or photons be used to study phonons?
3. In an interference experiment using fullerenes C_{60}, which are at present the largest objects for which wave behavior has been verified experimentally,[38] the average speed of the molecules is about 220 m s^{-1}. What is their de Broglie wavelength? How does it compare with the size of the molecule?
4. A diatomic molecule is composed of two atoms of masses M_1 and M_2 and has the form of a dumb-bell. The two nuclei are located a distance $r_0 = ba_0$ apart, where a_0 is the Bohr radius (1.34)

[38] M. Arndt, O. Nairz, J. Vos-Andreae, C. Keller, G. van der Zouw, and A. Zeilinger, Wave–particle duality of C_{60} molecules, *Nature* **401**, 680 (1999). For more recent results see M. Arndt, K. Hornberger, and A. Zeilinger, *Physics World* **18**(3), 35 (2005).

and b is a numerical coefficient ~ 1. It is assumed that the molecule rotates about its center of inertia, through which passes the axis perpendicular to the line joining the nuclei, referred to as the nuclear axis. Show that the moment of inertia is $I = \mu r_0^2$, where $\mu = M_1 M_2/(M_1 + M_2)$ is the reduced mass. If we assume that the angular momentum is $\hbar$, what is the angular speed of rotation and the corresponding energy ε_{rot}? Show that this energy is proportional to $(m_e/\mu)R_\infty$, where m_e is the electron mass and $R_\infty = m_e e^4/(2\hbar^2) = e^2/(2a_0)$.

5. The molecule can also vibrate along the nuclear axis about the equilibrium position $r = r_0$, where the restoring force has the form $-K(r - r_0)$, with $Kr_0^2 = dR_\infty$ and d a numerical coefficient ~ 1. What are the vibrational frequency ω_v and the corresponding energy $\hbar\omega_v$? Show that this energy is proportional to $\sqrt{m_e/\mu}\, R_\infty$. An example is the $H^{35}Cl$ molecule, for which the experimental values are $r_0 = 1.27\,\text{Å}$, $\varepsilon_{rot} = 1.3 \times 10^{-3}\,\text{eV}$, and $\hbar\omega_v = 0.36\,\text{eV}$. Calculate the numerical values of b and d. What will the wavelengths of photons of energy ε_{rot} and $\hbar\omega_v$ be? In which regions do these wavelengths lie?

6. The absence of a quantum theory of gravity makes it necessary to restrict all theories to energies lower than E_P, the Planck energy. Use a dimensional argument to construct E_P as a function of the gravitational constant G (Eq. (1.5)), $\hbar$, and c and find its numerical value. What is the corresponding wavelength (or Planck length) l_P?

1.6.2 The black body

1. Prove the following equation (Footnote 21):

$$\int \mathrm{d}x \mathrm{d}p\, \delta \left(E - \frac{p^2}{2m} - \frac{1}{2} m\omega^2 x^2 \right) f(E) = \frac{2\pi}{\omega} f(E).$$

2. We want to relate the energy density per unit frequency $\epsilon(\omega, T)$ to the emitted power $u(\omega, T)$, Eq. (1.15). We consider a cavity maintained at temperature T (Fig. 1.4). Let $\tilde{\epsilon}(k, T)\mathrm{d}^3 k$ be the energy density in a volume $\mathrm{d}^3 k$ about $\vec{k}$, which depends only on $k = |\vec{k}|$. Show that

$$\tilde{\epsilon}(k, T) = \frac{c}{4\pi k^2} \epsilon(\omega, T).$$

The Poynting vector of a wave with wave vector $\vec{k}$ escaping from the cavity is $c\tilde{\epsilon}(k, T)\hat{k}$. Show that the flux of the Poynting vector through an opening of area S is

$$\Phi = \frac{1}{4} cS \int_0^\infty \epsilon(\omega, T)\mathrm{d}\omega$$

and derive (1.15).

3. Show by dimensional analysis that in classical physics the energy density of a black body is given by

$$\epsilon(T) = A(k_B T)c^{-3} \int_0^\infty \omega^2\, \mathrm{d}\omega,$$

where A is a numerical coefficient.

4. Each mode $\vec{k}$ of the electromagnetic field inside the cavity is a harmonic oscillator. In classical statistical mechanics the energy of such a mode is $2k_B T$ (where does the factor of 2 come from?). Show that the energy density inside the cavity is

$$\epsilon(T) = \frac{1}{\pi^2} (k_B T)c^{-3} \int_0^\infty \omega^2\, \mathrm{d}\omega$$

and compute A.

5. Demonstrate (1.22) and show that the classical expression is recovered for $\hbar\omega \ll k_B T$, that is, for a sufficiently high temperature with ω fixed. This is a very general result: *the classical approximation is valid at high temperature.*

1.6.3 Heisenberg inequalities

In the thought experiment of Fig. 1.12, show that the momentum δp_x transferred to the screen must be $pa/(2D)$, where a is the spacing between the slits S_1 and S_2 (Fig. 1.12) and p is the neutron momentum. Determination of the trajectory implies that $\Delta p_x \ll \delta p_x$, where Δp_x is the spread in the *initial* momentum of the screen. What is the dispersion Δx at the location of S_0? Show that in this case the interference pattern is destroyed.[39]

1.6.4 Neutron diffraction by a crystal

Neutron diffraction is one of the principal techniques used to analyze crystal structure. For simplicity, let us consider a two-dimensional crystal composed of identical atoms with wave vectors lying in the plane of the crystal.[40] The atoms of the crystal are located at the lattice sites (Fig. 1.18)

$$\vec{r}_i = na\hat{x} + mb\hat{y}, \quad n = 0, 1, \ldots, N-1, \quad m = 0, 1, \ldots, M-1.$$

The neutrons interact with the atomic nuclei via the nuclear interaction.[41] We use $f(\theta)$ to denote the probability amplitude that a neutron of momentum $\hbar\vec{k}$ is scattered in the direction $\hat{k}'$ by an atom located at the origin, where θ is the angle between $\hat{k}$ and $\hat{k}'$. Since

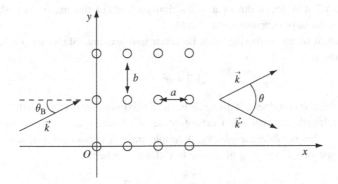

Fig. 1.18. Neutron diffraction by a crystal. The incident neutron has momentum $\hbar\vec{k}$ and the scattered neutron $\hbar\vec{k}'$. The Bragg angle θ_B is defined in question 4.

[39] See W. Wootters and W. Zurek, Complementarity in the double slit experiment: quantum nonseparability and a quantitative statement of Bohr's principle, *Phys. Rev.* **D19**, 473–484 (1979).

[40] One can also imagine 3D scattering by a 2D crystal; cf. Wichman [1974], Chapter 5, where a model for diffraction by the surface of a crystal is presented.

[41] There is also an interaction between the neutron magnetic moment and the atomic magnetism. It plays a very important role in studies of magnetism, but is not relevant to the present discussion.

the neutron energy is very low, $\sim 0.01\,\text{eV}$, $f(\theta)$ is independent of θ (Section 12.2.4): $f(\theta) = f$. The collision between a neutron and an atomic nucleus is elastic and leaves the state of the crystal unchanged: it is impossible to know which atom has scattered the neutron.

1. Show that the amplitude for scattering by an atom located at a site $\vec{r}_i$ is

$$f_i = f\,e^{i(\vec{k} - \vec{k}')\cdot\vec{r}_i} = f\,e^{-i\vec{q}\cdot\vec{r}_i},$$

with $\vec{q} = \vec{k}' - \vec{k}$.

2. Show that the amplitude f_{tot} for scattering by a crystal has the form

$$f_{\text{tot}} = fF(aq_x, bq_y),$$

with the function $F(aq_x, bq_y)$ given by

$$F(aq_x, bq_y) = \exp\left(-i\frac{aq_x(N-1)}{2}\right)\exp\left(-i\frac{bq_y(M-1)}{2}\right)$$

$$\times\left[\frac{\sin(aq_x N/2)}{\sin(aq_x/2)}\right]\left[\frac{\sin(bq_y M/2)}{\sin(bq_y/2)}\right].$$

3. Show that for $N, M \gg 1$ the scattering probability is proportional to $(NM)^2$ when $\vec{q}$ has components

$$q_x = \frac{2\pi n_x}{a}, \quad q_y = \frac{2\pi n_y}{b}$$

n_x and n_y being integers. When the components of $\vec{q}$ are of this form, it is said that $\vec{q}$ belongs to the *reciprocal lattice* of the crystal lattice. Diffraction maxima are obtained if $\vec{q}$ is a reciprocal lattice vector. What is the width of a diffraction peak about the maximum? Show that the intensity inside the peak is proportional to NM.

4. The elastic nature of the scattering must be taken into account. Show that the condition for elastic scattering is

$$2\vec{k}\cdot\vec{q} + q^2 = 0.$$

A reciprocal lattice vector does not give a diffraction maximum unless this condition is satisfied. For fixed wavelength, this condition cannot be satisfied unless the angle of incidence takes special values, called the *Bragg angles* θ_{B}. A simple analysis is possible if $n_x = 0$. Show that in this case an angle of incidence θ_{B} gives rise to diffraction when

$$\sin\theta_{\text{B}} = \frac{\pi n}{bk}, \quad n = 1, 2, \ldots$$

In general, it is convenient to interpret the Bragg condition geometrically: the tip of the vector $\vec{k}$ is located at a point of the reciprocal lattice and traces a circle of radius k. If this circle passes through another point of the reciprocal lattice a diffraction maximum is obtained. In general, a beam of neutrons incident on a crystal will not give rise to a diffraction peak. The angle of incidence and/or wavelength must be chosen appropriately. Why doesn't this phenomenon occur in diffraction by a one-dimensional lattice? What happens if only the first vertical column of atoms on the line $y = 0$ is present?

5. Now let us assume that the crystal is composed of atoms of two types. The basic crystal pattern, or *cell*, is formed as follows. Two atoms of type 1 are respectively located at

$$\vec{r}_1 = 0 \quad \text{and} \quad \vec{r}_1' = a\hat{x} + b\hat{y},$$

and two atoms of type 2 at

$$\vec{r}_2 = a\hat{x} \quad \text{and} \quad \vec{r}_2' = b\hat{y}.$$

The pattern is repeated with periodicity $2a$ in the x direction and $2b$ in the y direction. Let f_1 [f_2] be the amplitude for neutron scattering by an atom of type 1 [2] located at the origin; these amplitudes can be taken to be real. If NM is the number of cells, show that the amplitude for scattering by the crystal is proportional to $F(2aq_x, 2bq_y)$. Find the proportionality factor as a function of f_1 and f_2. Show that if q_x and q_y correspond to a diffraction maximum, this proportionality factor must be

$$f_1 \left[1 + (-1)^{n_x + n_y} \right] + f_2 \left[(-1)^{n_x} + (-1)^{n_y} \right].$$

Discuss the result as a function of the parity of n_x and n_y.

6. The atoms 1 and 2 form an alloy.[42] At low temperatures the atoms are in the configuration described in question **5** above, but above a certain temperature each atom has a 50% probability of occupying any site, and all sites are equivalent. How will the diffraction picture change?

1.6.5 Hydrogen-like atoms

Calculate, as a function of R_∞, the ground-state energy of the ordinary hydrogen atom, the deuterium atom, and the singly ionized helium atom taking into account the fact that nucleons have finite mass. Hint: what are the reduced masses?

1.6.6 The Mach–Zehnder interferometer

In a Mach–Zehnder interferometer (Fig. 1.19), a light beam arrives at the first beam splitter BS_1. The two resulting beams are then reflected by two mirrors and recombined

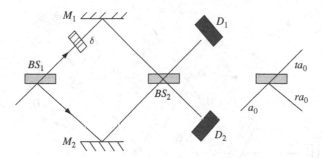

Fig. 1.19. The Mach–Zehnder interferometer.

[42] An example of the phenomenon described in this exercise is brass with composition 50% copper and 50% zinc.

by a second beam splitter BS_2. The intensity of the incident light is reduced to the level at which the photons arrive one by one. More precisely, the time between the arrival of two successive photons is very large compared with the resolution times of the photodetectors D_1 and D_2. If a photon arrives at a beam splitter with probability amplitude a_0, it will be transmitted with an amplitude ta_0 and reflected with an amplitude ra_0, where t and r are complex numbers

$$t = |t|e^{i\alpha}, \quad r = |r|e^{i\beta}$$

and $|t| = |r| = 1/\sqrt{2}$. A phase shift δ can be introduced into, for example, the upper path of the interferometer by means of a plate with parallel faces of variable thickness. In the absence of this plate $\delta = \delta_0 \neq 0$ because the two beam paths in the interferometer are never exactly equal. Let p_1 and p_2 denote the probabilities of detecting a photon by D_1 and D_2.

1. Calculate p_1 and p_2 as functions of α, β, and δ. What is observed when δ is varied?
2. What is the relation between p_1 and p_2? Derive the expression

$$\alpha - \beta = \frac{\pi}{2} \pm n\pi, \quad \text{integer } n.$$

1.6.7 Neutron interferometry and gravity

A neutron interferometer is realized in the following way (Fig. 1.20). A monochromatic (i.e., fixed wavelength) incident beam arrives at the first crystal at point A, with the angle of incidence and wavelength chosen such that a diffraction maximum is obtained (see Exercise 1.6.4, question 4); this angle of incidence is the Bragg angle θ_B. Part of the beam is transmitted as beam I with probability amplitude t and the rest is refracted as beam II

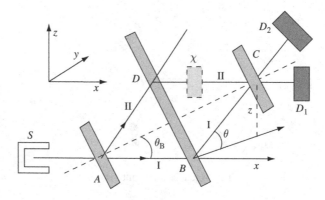

Fig. 1.20. Neutron interferometry.

with probability amplitude r. These amplitudes satisfy $|t|^2 + |r|^2 = 1$. Beams I and II arrive at a second crystal at points B and D, respectively, and the refracted parts of I and II are recombined by a third crystal at point C. The neutrons are detected by the two counters D_1 and D_2. On trajectory II the neutrons undergo a phase shift χ which can have various origins (a difference between the lengths of the trajectories, gravity, passage through a magnetic field, etc.), and the objective of neutron interferometry is to measure this phase shift.

1. Show that the probability amplitude a_1 for a neutron to arrive at D_1 is

$$a_1 = a_0(e^{i\chi}\, trr + rrt),$$

and that the probability of detection by D_1 is

$$p_1 = 2|a_0|^2|t|^2|r|^4(1+\cos\chi) = A(1+\cos\chi),$$

where a_0 is the amplitude incident on the first crystal.

2. What is the amplitude a_2 for a neutron to reach detector D_2 as a function of r, t, and a_0, and the corresponding probability p_2? Why must we have $p_1 + p_2 = $ constant? Show that

$$p_2 = B - A\cos\chi.$$

What is B as a function of t, r, and a_0? Letting

$$t = |t|e^{i\alpha}, \quad r = |r|e^{i\beta},$$

show that

$$\alpha - \beta = \frac{\pi}{2} \pm n\pi, \quad n = 0, 1, 2, \ldots$$

3. We now take gravity into account. How does the wave vector $k = 2\pi/\lambda$ of a neutron vary with height z when the neutron is located in a gravitational field with gravitational acceleration g? Compare the numerical values of the neutron kinetic energy and gravitational energy[43] $m_n gz$ (where m_n is the neutron mass), and derive an approximation for k. Assuming that the plane $ABCD$ is initially horizontal, it can be rotated about the axis AB such that it becomes vertical. Show that such a rotation induces the following phase difference between the two trajectories:

$$\Delta\phi = \frac{m_n^2 g \mathcal{S}}{\hbar^2 k} = \frac{2\pi m_n^2 g \mathcal{S}\lambda}{h^2},$$

where $\mathcal{S}$ is the area of the rhombus $ABCD$.

---

[43] The energy is defined up to an additive constant, with the zero of energy fixed according to the following convention: a neutron of zero velocity and height $z = 0$ has zero energy.

4. If the plane $ABDC$ lies at a variable angle θ with respect to the vertical direction, give a qualitative discussion of the variation of the neutron detection probability as a function of θ. Numerical data:[44] $\lambda = 1.44\,\text{Å}$, $\mathcal{S} = 10.1$ cm^2.

1.6.8 Coherent and incoherent neutron scattering by a crystal

We want to study neutron scattering by a crystal composed of two types of nucleus. A given lattice site is occupied by a nucleus of type 1 with probability p_1 or by a nucleus of type 2 with probability $p_2 = 1 - p_1$. The total number of nuclei is $\mathcal{N}$, and so there are $p_1 \mathcal{N}$ nuclei of type 1 and $p_2 \mathcal{N}$ nuclei of type 2 in the crystal. With a site i, $i = 1, \dots, \mathcal{N}$, we associate a number α_i which takes the value 1 if the site is occupied by a nucleus of type 1 and 0 if it is occupied by a nucleus of type 2. The ensemble $\{\alpha_i\}$ of the α_i, with $\sum_i \alpha_i = p_1 \mathcal{N}$, defines a configuration of the crystal. The amplitude of neutron scattering by the crystal in a configuration $\{\alpha_i\}$ is (cf. Exercise 1.6.4)

$$f_{\text{tot}} = \sum_{i=1}^{\mathcal{N}} (\alpha_i f_1 + (1 - \alpha_i) f_2)\, e^{i\vec{q}\cdot\vec{r}_i},$$

where f_1 (f_2) is the amplitude for neutron scattering by a nucleus of type 1 (2).

1. We shall use brackets $\langle \bullet \rangle$ to denote the average over all possible configurations of the crystal, assuming that the occupation numbers of the sites are not correlated (for example, the occupation of a site by a nucleus of type 1 does not increase the probability that a nearest-neighbor site is also occupied by a nucleus of type 1). Prove the identities

$$\langle \alpha_i \alpha_j \rangle = p_1^2 + p_1 p_2 \delta_{ij}, \quad \langle \alpha_i (1 - \alpha_j) \rangle = p_1 p_2 (1 - \delta_{ij}).$$

2. Use these identities to derive the average of $|f_{\text{tot}}|^2$ over configurations:

$$\langle |f_{\text{tot}}|^2 \rangle = (p_1 f_1 + p_2 f_2)^2 \sum_{i,j} e^{i\vec{q}\cdot(\vec{r}_i - \vec{r}_j)} + \mathcal{N} p_1 p_2 (f_1 - f_2)^2.$$

The first term describes *coherent scattering* and gives rise to diffraction peaks. The second term is proportional to the number of sites and independent of angles; it corresponds to *incoherent scattering*.

1.7 Further reading

The introductory Chapters 1–3 of Feynman *et al.* [1965], vol. III, and Chapters 1–5 of Wichman [1967] are strongly recommended as an elementary introduction to quantum physics. Another source is Chapters 1–3 of Lévy-Leblond and Balibar [1990]. For a pedagogical discussion of elementary particle physics see D. Perkins, *An Introduction to High Energy Physics*, 4th edn, Cambridge: Cambridge University Press (2000). A detailed

---

[44] R. Colella, A. Overhauser, and S. Werner, Observation of gravitationally induced quantum interference, *Phys. Rev. Lett.* **34**, 1472–1474 (1975).

discussion of black-body radiation can be found in, for example, Le Bellac *et al.* [2004], Chapter 4. Interference and diffraction experiments using cold neutrons have been performed by A. Zeilinger, R. Gähler, C. Shull, W. Treimer, and W. Mampe, Single and double-slit diffraction of neutrons, *Rev. Mod. Phys.* **60**, 1067 (1988), and interference experiments using cold atoms by F. Shimizu, K. Shimizu, and H. Takuma, Double-slit interference with ultracold metastable neon atoms, *Phys. Rev.* **A46**, R17 (1992). Neutron diffraction by a crystal is discussed by Kittel [1996], Chapter 2. A recent book on neutron interferometry is that by H. Rauch and S. Werner, *Neutron Interferometry*, Oxford: Clarendon Press (2000).

2

The mathematics of quantum mechanics I: finite dimension

The superposition principle is a founding principle of quantum mechanics; we have already made use of it in interpreting the Young's slit experiment. Quantum mechanics is a linear theory, and so it is natural that vector spaces play an important role in it. We shall see that a physical state is represented mathematically by a vector in a space whose characteristics we shall define; this is called the *space of states*. A second founding principle, which can also be deduced from the Young's slit experiment, is the existence of probability amplitudes. These probability amplitudes will be represented mathematically by scalar products defined on the space of states. In the theory of waves, the use of complex numbers is just a convenience, but in quantum mechanics the probability amplitudes are fundamentally complex numbers – the scalar product will a priori be a complex number. Physical properties like momentum, position, energy, and so on will be represented by operators acting in the space of states. In this chapter we shall introduce the essential properties of Hilbert spaces, that is, vector spaces on which a positive-definite scalar product is defined, and we shall limit ourselves to the case of finite dimension. This restriction will be lifted later on, because the space of states is in general of infinite dimension. The mathematical theory of Hilbert spaces of infinite dimension is much more complicated than that of spaces of finite dimension, and we shall put off studying them until Chapter 7. The reader familiar with vector spaces of finite dimension and operators in such spaces can proceed directly to Chapter 3 after reviewing the notation.

2.1 Hilbert spaces of finite dimension

Let $\mathcal{H}$ be a vector space of dimension N over complex numbers. We shall use $|\varphi\rangle$, $|\chi\rangle$, ... to denote the elements (vectors) of $\mathcal{H}$. If λ, μ, ... are complex numbers and if $|\varphi\rangle$ and $|\chi\rangle \in \mathcal{H}$, linearity implies that $\lambda|\varphi\rangle \equiv |\lambda\varphi\rangle \in \mathcal{H}$ and that $(|\varphi\rangle + \lambda|\chi\rangle) \in \mathcal{H}$.

The space $\mathcal{H}$ is endowed with a positive-definite scalar product, which makes it a *Hilbert space*. The scalar product[1] of two vectors $|\varphi\rangle$ and $|\chi\rangle$ will be denoted $\langle\chi|\varphi\rangle$; it is linear in $|\varphi\rangle$,

$$\langle\chi|(\varphi_1 + \lambda\varphi_2)\rangle = \langle\chi|\varphi_1\rangle + \lambda\langle\chi|\varphi_2\rangle, \tag{2.1}$$

---

[1] We could use the mathematicians' notation $(\chi, \varphi) \equiv \langle\chi|\varphi\rangle$ for the scalar product. However, it should be noted that for mathematicians the scalar product (χ, φ) is linear in χ!

and it possesses the property of complex conjugation

$$\langle \chi | \varphi \rangle = \langle \varphi | \chi \rangle^*, \tag{2.2}$$

which implies that $\langle \varphi | \varphi \rangle$ is a real number. From (2.1) and (2.2) we deduce the fact that the scalar product $\langle \chi | \varphi \rangle$ is antilinear in $|\chi\rangle$:

$$\langle (\chi_1 + \lambda \chi_2) | \varphi \rangle = \langle \chi_1 | \varphi \rangle + \lambda^* \langle \chi_2 | \varphi \rangle. \tag{2.3}$$

Finally, the scalar product is positive-definite:

$$\langle \varphi | \varphi \rangle = 0 \Longleftrightarrow |\varphi\rangle = 0. \tag{2.4}$$

It will be convenient to choose an orthonormal basis in $\mathcal{H}$ of N vectors $\{|n\rangle\} \equiv \{|1\rangle, |2\rangle, \dots, |n\rangle, \dots, |N\rangle\}$

$$\langle n | m \rangle = \delta_{nm}. \tag{2.5}$$

Any vector $|\varphi\rangle$ can be decomposed on this basis with coefficients c_n which are the components of $|\varphi\rangle$ in this basis:

$$|\varphi\rangle = \sum_{n=1}^{N} c_n |n\rangle. \tag{2.6}$$

Taking the scalar product of (2.6) with the basis vector $|m\rangle$, we find the following for the c_m:

$$c_m = \langle m | \varphi \rangle. \tag{2.7}$$

If a vector $|\chi\rangle$ is decomposed on this basis as $|\chi\rangle = \sum d_n |n\rangle$, the scalar product $\langle \chi | \varphi \rangle$ is written as follows using (2.5):

$$\langle \chi | \varphi \rangle = \sum_{n,m=1}^{N} d_m^* c_n \langle m | n \rangle = \sum_{n=1}^{N} d_n^* c_n. \tag{2.8}$$

The *norm* of $|\varphi\rangle$, denoted $||\varphi||$, is defined using the scalar product:

$$||\varphi||^2 = \langle \varphi | \varphi \rangle = \sum_{n=1}^{N} |c_n|^2 \geq 0. \tag{2.9}$$

An important property of the scalar product is the Schwarz inequality:

$$\boxed{|\langle \chi | \varphi \rangle|^2 \leq \langle \chi | \chi \rangle \langle \varphi | \varphi \rangle = ||\chi||^2 \, ||\varphi||^2} \,. \tag{2.10}$$

The equality holds if and only if $|\varphi\rangle$ and $|\chi\rangle$ are proportional to each other: $|\chi\rangle = \lambda|\varphi\rangle$.

Proof.[2] The theorem is proved if $\langle\chi|\varphi\rangle = 0$. We can then assume that $\langle\chi|\varphi\rangle \neq 0$ so that $|\varphi\rangle \neq 0$ and $|\chi\rangle \neq 0$. From the positivity (2.9) of the norm we have

$$\langle(\varphi - \lambda\chi)|(\varphi - \lambda\chi)\rangle = ||\varphi||^2 - \lambda^*\langle\chi|\varphi\rangle - \lambda\langle\varphi|\chi\rangle + |\lambda|^2||\chi||^2 \geq 0.$$

Choosing

$$\lambda = \frac{||\varphi||^2}{\langle\varphi|\chi\rangle}, \quad \lambda^* = \frac{||\varphi||^2}{\langle\chi|\varphi\rangle},$$

we obtain

$$||\varphi||^2 - 2||\varphi||^2 + \frac{||\varphi||^4||\chi||^2}{|\langle\chi|\varphi\rangle|^2} \geq 0,$$

from which (2.10) follows immediately. According to (2.4), the equality can hold only if $|\varphi\rangle = \lambda|\chi\rangle$ and vice versa.

2.2 Linear operators on $\mathcal{H}$

2.2.1 Linear, Hermitian, unitary operators

A *linear operator* A establishes a linear correspondence between a vector $|\varphi\rangle$ and a vector $|A\varphi\rangle$:

$$|A(\varphi + \lambda\chi)\rangle = |A\varphi\rangle + \lambda|A\chi\rangle. \tag{2.11}$$

This operator is represented in a given basis $\{|n\rangle\}$ by a matrix with elements A_{mn}.[3] Using the property of linearity and the decomposition (2.6)

$$|A\varphi\rangle = \sum_{n=1}^{N} c_n|An\rangle,$$

we obtain the components d_m of $|A\varphi\rangle = \sum_m d_m|m\rangle$:

$$d_m = \langle m|A\varphi\rangle = \sum_{n=1}^{N} c_n\langle m|An\rangle = \sum_{n=1}^{N} A_{mn}c_n. \tag{2.12}$$

An element A_{mn} of the matrix is then given by

$$A_{mn} = \langle m|An\rangle. \tag{2.13}$$

The *Hermitian conjugate* (or adjoint) of A, $A^\dagger$, is defined as

$$\langle\chi|A^\dagger\varphi\rangle = \langle A\chi|\varphi\rangle = \langle\varphi|A\chi\rangle^* \tag{2.14}$$

---

[2] This proof can be carried over directly to spaces of infinite dimension.
[3] We note that physicists often casually use the terms operator and matrix interchangeably, the latter referring to the matrix representing the operator in a given basis.

for every pair of vectors $|\varphi\rangle$, $|\chi\rangle$. It can easily be shown that $A^\dagger$ is also a linear operator. Its matrix elements in the basis $\{|n\rangle\}$ are obtained by taking $|\varphi\rangle$ and $|\chi\rangle$ to be the basis vectors, and $(A^\dagger)_{mn}$ satisfies

$$(A^\dagger)_{mn} = A^*_{nm}. \tag{2.15}$$

The Hermitian conjugate of the product AB of two operators is $B^\dagger A^\dagger$:

$$\langle\chi|(AB)^\dagger\varphi\rangle = \langle AB\chi|\varphi\rangle = \langle B\chi|A^\dagger\varphi\rangle = \langle\chi|B^\dagger A^\dagger\varphi\rangle.$$

An operator satisfying $A = A^\dagger$ is termed *Hermitian* or *self-adjoint*. The two terms are equivalent for finite-dimensional spaces, but not for infinite-dimensional ones.

An operator that satisfies $UU^\dagger = U^\dagger U = I$ or, equivalently, $U^{-1} = U^\dagger$, is called a *unitary* operator. Throughout this book we shall use I to denote the identity operator of the Hilbert space. In a finite-dimensional space the necessary and sufficient condition for an operator U to be unitary is that it leave unchanged the norm

$$||U\varphi||^2 = ||\varphi||^2 \quad \text{or} \quad \langle U\varphi|U\varphi\rangle = \langle\varphi|\varphi\rangle \ \ \forall\varphi \in \mathcal{H}. \tag{2.16}$$

Proof. Let us calculate the squared norm of $|U(\varphi + \lambda\chi)\rangle$, which by hypothesis is equal to the squared norm of $|\varphi + \lambda\chi\rangle$:

$$\langle\varphi + \lambda\chi|\varphi + \lambda\chi\rangle = \langle\varphi|\varphi\rangle + |\lambda|^2\langle\chi|\chi\rangle + 2\mathrm{Re}\,(\lambda\langle\varphi|\chi\rangle),$$

while

$$\langle U(\varphi + \lambda\chi)|U(\varphi + \lambda\chi)\rangle = \langle U\varphi|U\varphi\rangle + |\lambda|^2\langle U\chi|U\chi\rangle + 2\mathrm{Re}\,(\lambda\langle U\varphi|U\chi\rangle).$$

Subtracting the second of these equations from the first gives

$$\mathrm{Re}\,(\lambda\langle\varphi|\chi\rangle) = \mathrm{Re}\,(\lambda\langle U\varphi|U\chi\rangle),$$

and choosing $\lambda = 1$ and then $\lambda = \mathrm{i}$ we find

$$\langle U\varphi|U\chi\rangle = \langle\varphi|\chi\rangle \Rightarrow U^\dagger U = I.$$

In a vector space of finite dimension the existence of a left inverse implies the existence of a right inverse, and so we also have $UU^\dagger = I$. An operator that preserves the norm is an *isometry*. In a space of finite dimension an isometry is a unitary operator.

Unitary operators perform changes of orthonormal basis in $\mathcal{H}$. Let $|n'\rangle = |Un\rangle$. Then

$$\langle m'|n'\rangle = \langle Um|Un\rangle = \langle m|n\rangle = \delta_{mn} = \delta_{m'n'}$$

and the ensemble of vectors $\{|n'\rangle\}$ forms an orthonormal basis. It should be noted that the components c_n of a vector are transformed using $U^\dagger$ (or U^{-1})

$$c'_n = \langle n'|\varphi\rangle = \langle Un|\varphi\rangle = \langle n|U^\dagger\varphi\rangle = \sum_{m=1}^{N} U^\dagger_{nm} c_m. \tag{2.17}$$

We also note the transformation law of the matrix elements:

$$A'_{mn} = \langle m'|An'\rangle = \langle Um|AUn\rangle = \langle m|U^\dagger AUn\rangle = \sum_{k,l=1}^{N} U^\dagger_{mk} A_{kl} U_{ln}. \tag{2.18}$$

2.2.2 Projection operators and Dirac notation

We shall frequently use *projection operators* (projectors). Let $\mathcal{H}_1$ be a subspace of $\mathcal{H}$ and $\mathcal{H}_2$ be the orthogonal subspace. Any vector $|\varphi\rangle$ can be decomposed uniquely into a vector $|\varphi_1\rangle$ belonging to $\mathcal{H}_1$ and a vector $|\varphi_2\rangle$ belonging to $\mathcal{H}_2$:

$$|\varphi\rangle = |\varphi_1\rangle + |\varphi_2\rangle, \quad |\varphi_1\rangle \in \mathcal{H}_1, \quad |\varphi_2\rangle \in \mathcal{H}_2, \quad \langle\varphi_1|\varphi_2\rangle = 0.$$

The projector $\mathcal{P}_1$ onto $\mathcal{H}_1$ is defined by its action on an arbitrary vector $|\varphi\rangle$:

$$|\mathcal{P}_1\varphi\rangle = |\varphi_1\rangle. \tag{2.19}$$

$\mathcal{P}_1$ is obviously a linear operator, and it is also a Hermitian operator because if the decomposition of $|\chi\rangle$ into vectors belonging to $\mathcal{H}_1$ and $\mathcal{H}_2$ is $|\chi\rangle = |\chi_1\rangle + |\chi_2\rangle$, then

$$\langle\chi|\mathcal{P}_1\varphi\rangle = \langle\chi|\varphi_1\rangle = \langle\chi_1|\varphi_1\rangle,$$
$$\langle\chi|\mathcal{P}_1^\dagger\varphi\rangle = \langle\mathcal{P}_1\chi|\varphi\rangle = \langle\chi_1|\varphi\rangle = \langle\chi_1|\varphi_1\rangle.$$

It should also be noted that

$$|\mathcal{P}_1^2\varphi\rangle = |\mathcal{P}_1\varphi_1\rangle = |\varphi_1\rangle \Rightarrow \mathcal{P}_1^2 = \mathcal{P}_1.$$

Conversely, every linear operator satisfying $\mathcal{P}_1^\dagger\mathcal{P}_1 = \mathcal{P}_1$ is a projector.

Proof. First we notice that $\mathcal{P}_1^\dagger = \mathcal{P}_1$, and then that vectors of the form $|\mathcal{P}_1\varphi\rangle$ form a vector subspace $\mathcal{H}_1$ of $\mathcal{H}$. If we write

$$|\varphi\rangle = |\mathcal{P}_1\varphi\rangle + (|\varphi\rangle - |\mathcal{P}_1\varphi\rangle) = |\mathcal{P}_1\varphi\rangle + |\varphi_2\rangle,$$

then $|\varphi_2\rangle$ is orthogonal to every vector $|\mathcal{P}_1\chi\rangle$:

$$\langle\varphi - \mathcal{P}_1\varphi|\mathcal{P}_1\chi\rangle = \langle\mathcal{P}_1\varphi - \mathcal{P}_1^2\varphi|\chi\rangle = 0.$$

We have in fact decomposed $|\varphi\rangle$ into $|\mathcal{P}_1\varphi\rangle$ and a vector of the subspace orthogonal to $\mathcal{H}_1$.

The property $\mathcal{P}_1^2 = \mathcal{P}_1$ demonstrates that the eigenvalues of a projector are 0 or 1, and $\mathrm{Tr}\,\mathcal{P}_1$ (see 2.23) is the dimension of the projection space, as is easily seen by writing $\mathcal{P}_1$ in a basis in which it is diagonal: as we shall see in the next section, such a basis always exists because $\mathcal{P}_1$ is Hermitian. Furthermore, we can prove the following properties (Exercise 2.4.6):

- If $\mathcal{P}_1$ and $\mathcal{P}_1'$ are projectors onto $\mathcal{H}_1$ and $\mathcal{H}_1'$, respectively, $\mathcal{P}_1\mathcal{P}_1'$ is a projector if and only if $\mathcal{P}_1\mathcal{P}_1' = \mathcal{P}_1'\mathcal{P}_1$. Then $\mathcal{P}_1\mathcal{P}_1'$ projects onto the intersection $\mathcal{H}_1 \cap \mathcal{H}_1'$.
- $\mathcal{P}_1 + \mathcal{P}_1'$ is a projector if and only if $\mathcal{P}_1\mathcal{P}_1' = 0$. In this case $\mathcal{H}_1$ and $\mathcal{H}_1'$ are orthogonal and $\mathcal{P}_1 + \mathcal{P}_1'$ projects onto the direct sum $\mathcal{H}_1 \oplus \mathcal{H}_1'$.
- If $\mathcal{P}_1\mathcal{P}_1' = \mathcal{P}_1'\mathcal{P}_1$, then $\mathcal{P}_1 + \mathcal{P}_1' - \mathcal{P}_1\mathcal{P}_1'$ projects onto the union $\mathcal{H}_1 \cup \mathcal{H}_1'$. The second property is a special case of this one.

Dirac notation. Instead of writing $|A\varphi\rangle$, from now on we shall use the notation $A|\varphi\rangle$ introduced by Dirac.[4] The scalar product $\langle\chi|A\varphi\rangle$ is written as $\langle\chi|A|\varphi\rangle$ in Dirac notation. The vectors $|\varphi\rangle$ of $\mathcal{H}$ are called "kets," and the vectors $\langle\chi|$ of the dual space are called "bras." The bra associated with the ket $|\lambda\varphi\rangle$ is $\lambda^*\langle\varphi|$; indeed,

$$\langle\lambda\varphi|\chi\rangle = \lambda^*\langle\varphi|\chi\rangle.$$

In $\langle\chi|A|\varphi\rangle$, A acts on $|\varphi\rangle$ from the right: $\langle\chi|A|\varphi\rangle = \langle\chi|(A|\varphi\rangle)$ and not $\langle A\chi|\varphi\rangle$. Since $(A|\varphi\rangle)^\dagger = \langle\varphi|A^\dagger$, there are no ambiguities if A is Hermitian. The main virtue of the Dirac notation is that it allows us to write projectors in a very simple way. Let $|\varphi\rangle$ be a normalized vector: $\langle\varphi|\varphi\rangle = 1$. The decomposition of $|\chi\rangle$ into $|\varphi\rangle$ and a vector $|\chi_\perp\rangle$ orthogonal to $|\varphi\rangle$ is

$$|\chi\rangle = |\varphi\rangle\langle\varphi|\chi\rangle + (|\chi\rangle - |\varphi\rangle\langle\varphi|\chi\rangle) = |\varphi\rangle\langle\varphi|\chi\rangle + |\chi_\perp\rangle = \mathcal{P}_\varphi|\chi\rangle + |\chi_\perp\rangle.$$

We can then write[5]

$$\boxed{\mathcal{P}_\varphi = |\varphi\rangle\langle\varphi|} \,. \tag{2.20}$$

If the vectors $\{|1\rangle, \ldots, |M\rangle\}$, $M \leq N$, form an orthonormal basis of the subspace $\mathcal{H}_1$, then $\mathcal{P}_1$ can be written as

$$\mathcal{P}_1 = \sum_{n=1}^{M} |n\rangle\langle n|. \tag{2.21}$$

If $M = N$ we obtain the decomposition of the identity operator:

$$\boxed{I = \sum_{n=1}^{N} |n\rangle\langle n|} \,. \tag{2.22}$$

---

[4] This notation is convenient and very widely used, but it is not free of ambiguities. For example, it is not wise to use it when dealing with time reversal: see Appendix A.
[5] If $\||\varphi\|^2 \neq 1$, then $P_\varphi = |\varphi\rangle\langle\varphi|/\||\varphi\|^2$.

This relation is called the *completeness relation*. It often proves very useful in calculations. For example, it provides a simple proof of the matrix multiplication law:

$$(AB)_{nm} = \langle n|AB|m \rangle = \langle n|AIB|m \rangle = \sum_{l=1}^{N} \langle n|A|l \rangle \langle l|B|m \rangle = \sum_{l=1}^{N} A_{nl} B_{lm}.$$

Finally, let us give an important definition. The *trace* of an operator is the sum of its diagonal elements:

$$\boxed{\operatorname{Tr} A = \sum_{n=1}^{N} A_{nn}} .$$

(2.23)

It is easily shown (Exercise 2.4.2) that the trace is invariant under a change of basis and that

$$\operatorname{Tr} AB = \operatorname{Tr} BA.$$

(2.24)

2.3 Spectral decomposition of Hermitian operators

2.3.1 Diagonalization of a Hermitian operator

Let A be a linear operator. If there exists a vector $|\varphi\rangle$ and a complex number a such that

$$A|\varphi\rangle = a|\varphi\rangle,$$

(2.25)

then $|\varphi\rangle$ is called an *eigenvector* and a an *eigenvalue* of A. The eigenvalues are found by solving the equation for a:

$$\det(A - aI) = 0.$$

(2.26)

The eigenvectors and eigenvalues of Hermitian operators possess remarkable properties.

Theorem. The eigenvalues of a Hermitian operator are real and the eigenvectors corresponding to two different eigenvalues are orthogonal.

The proof is simple. It is sufficient to consider the scalar product $\langle \varphi|A|\varphi \rangle$, where $|\varphi\rangle$ satisfies (2.25):

$$\langle \varphi|A|\varphi \rangle = \langle \varphi|a\varphi \rangle = a||\varphi||^2$$
$$= \langle A\varphi|\varphi \rangle = \langle a\varphi|\varphi \rangle = a^*||\varphi||^2,$$

which gives $a = a^*$; on the other hand, if $A|\varphi\rangle = a|\varphi\rangle$ and $A|\chi\rangle = b|\chi\rangle$, then

$$\langle \chi|A\varphi \rangle = a\langle \chi|\varphi \rangle = \langle A\chi|\varphi \rangle = b\langle \chi|\varphi \rangle,$$

from which we find $\langle \chi|\varphi \rangle = 0$ if $a \neq b$. An immediate consequence of this result is that the eigenvectors of a Hermitian operator normalized to unity form an orthonormal basis of $\mathcal{H}$ if the eigenvalues are all distinct, that is, if the roots of Eq. (2.26) are all different. However, it may happen that one (or more) of the roots of (2.26) are the same, that is, one finds multiple roots. Let a_n be a multiple root: the eigenvalue a_n is then said to be *degenerate*. Again in this case it is possible to use the eigenvectors of A to construct an orthonormal basis of $\mathcal{H}$. Indeed, we have at our disposal the following theorem, which we state without proof.

Theorem. If an operator A is Hermitian, it is always possible to find a (nonunique) unitary matrix U such that $U^{-1}AU$ is a diagonal matrix, where the diagonal elements are the eigenvalues of A, each of which appears a number of times equal to its multiplicity:

$$U^{-1}AU = \begin{pmatrix} a_1 & 0 & 0 & \ldots & 0 \\ 0 & a_2 & 0 & \ldots & \vdots \\ 0 & 0 & a_3 & 0 & \vdots \\ \vdots & \vdots & \ddots & \ddots & 0 \\ 0 & \ldots & \ldots & 0 & a_N \end{pmatrix}. \tag{2.27}$$

Let a_n be a degenerate eigenvalue and let $G(n)$ be its multiplicity in (2.26); it is also said that a_n is $G(n)$ times degenerate. Then there exist $G(n)$ independent eigenvectors corresponding to this eigenvalue. These $G(n)$ eigenvectors span a vector subspace of dimension $G(n)$ called the *subspace of the eigenvalue a_n*, in which we can find a (nonunique) orthonormal basis $|n, r\rangle$, $r = 1, \ldots, G(n)$:

$$A|n, r\rangle = a_n|n, r\rangle. \tag{2.28}$$

Using (2.21), we can write the projector $\mathcal{P}_n$ onto this vector subspace as

$$\mathcal{P}_n = \sum_{r=1}^{G(n)} |n, r\rangle\langle n, r|. \tag{2.29}$$

The sum of the $\mathcal{P}_n$ gives the identity operator since the set of vectors $|n, r\rangle$ forms a basis of $\mathcal{H}$, and we obtain the completeness relation (2.22):

$$\boxed{\sum_n \mathcal{P}_n = \sum_n \sum_{r=1}^{G(n)} |n, r\rangle\langle n, r| = I}. \tag{2.30}$$

Let $|\varphi\rangle$ be some vector of $\mathcal{H}$:

$$A|\varphi\rangle = \sum_n A\mathcal{P}_n|\varphi\rangle = \sum_n a_n \mathcal{P}_n|\varphi\rangle,$$

since $\mathcal{P}_n|\varphi\rangle$ belongs to the subspace of the eigenvalue a_n. We can then cast A in the form

$$\boxed{A = \sum_n a_n \mathcal{P}_n = \sum_n \sum_{r=1}^{G(n)} |n, r\rangle a_n \langle n, r|}. \tag{2.31}$$

This fundamental relation is called the *spectral decomposition* of A. Reciprocally, an operator of the form $\sum_n a_n \mathcal{P}_n$ is Hermitian with eigenvalues a_n if $a_n = a_n^*$ and if $\mathcal{P}_n \mathcal{P}_m = \delta_{nm} \mathcal{P}_n$, namely, if the $\mathcal{P}_n$ are pairwise orthogonal.

2.3.2 Diagonalization of a 2×2 Hermitian matrix

We shall often need to diagonalize 2×2 Hermitian matrices. The most general form of such a matrix in a $\{|1\rangle, |2\rangle\}$ basis,

$$|1\rangle = \begin{pmatrix} 1 \\ 0 \end{pmatrix}, \quad |2\rangle = \begin{pmatrix} 0 \\ 1 \end{pmatrix},$$

is

$$A = \begin{pmatrix} A_{11} & A_{12} \\ A_{21} & A_{22} \end{pmatrix} = \begin{pmatrix} a & b \\ b^* & a' \end{pmatrix},$$

where a and a' are real numbers and b is a priori complex. However, we shall see that in quantum mechanics it is always possible to redefine the phase of the basis vectors:

$$|1\rangle \to |1'\rangle = e^{i\alpha}|1\rangle, \quad |2\rangle \to |2'\rangle = e^{i\beta}|2\rangle.$$

In this new basis the matrix element A'_{12} of the operator A is

$$A'_{12} = \langle 1'|A|2'\rangle = e^{i(\beta-\alpha)} \langle 1|A|2\rangle = e^{i(\beta-\alpha)} A_{12} = e^{i(\beta-\alpha)} b.$$

If $b = |b|\exp(i\delta)$, it is sufficient to take $(\alpha - \beta) = \delta$ to eliminate the phase of b, which can then be chosen to be real. The simplest case is that where $a = a'$:

$$A = \begin{pmatrix} a & b \\ b & a \end{pmatrix}. \tag{2.32}$$

In this case we immediately verify that the two vectors $|\chi_+\rangle$ and $|\chi_-\rangle$

$$|\chi_+\rangle = \frac{1}{\sqrt{2}} \begin{pmatrix} 1 \\ 1 \end{pmatrix}, \quad |\chi_-\rangle = \frac{1}{\sqrt{2}} \begin{pmatrix} -1 \\ 1 \end{pmatrix}, \tag{2.33}$$

are eigenvectors of A with eigenvalues $(a+b)$ and $(a-b)$, respectively. This very simple result has an interesting origin. Let U_P be a unitary operator which performs a permutation of the basis vectors $|1\rangle$ and $|2\rangle$:

$$U_P = \begin{pmatrix} 0 & 1 \\ 1 & 0 \end{pmatrix}.$$

The operator U_P has unit square: $U_P^2 = I$, and its eigenvalues then are ± 1. The corresponding eigenvectors are $|\chi_+\rangle$ and $\chi_-\rangle$. We observe that A can be written in the form

$$A = aI + bU_P,$$

which shows that A and U_P commute: $AU_P = U_P A$. Then, as we shall see in the following subsection, we can find a basis constructed from eigenvectors common to A and U_P. It is easy to diagonalize A because A commutes with a symmetry operation, a property which we shall often use in this book.

In the general case $a \neq a'$, the symmetry property does not hold and the diagonalization is not so simple. It is convenient to write A in the form

$$A = \begin{pmatrix} a+c & b \\ b & a-c \end{pmatrix} = aI + \sqrt{b^2+c^2} \begin{pmatrix} \cos\theta & \sin\theta \\ \sin\theta & -\cos\theta \end{pmatrix}, \qquad (2.34)$$

where the angle θ is defined by

$$c = \sqrt{b^2+c^2} \cos\theta,$$
$$b = \sqrt{b^2+c^2} \sin\theta.$$

We note that $\tan\theta = b/c$, and that care must be taken to choose a correct definition of θ in $[0, 2\pi]$. We then verify that the eigenvectors are

$$|\chi_+\rangle = \begin{pmatrix} \cos\theta/2 \\ \sin\theta/2 \end{pmatrix}, \quad |\chi_-\rangle = \begin{pmatrix} -\sin\theta/2 \\ \cos\theta/2 \end{pmatrix}, \qquad (2.35)$$

corresponding to the eigenvalues $a+\sqrt{b^2+c^2}$ and $a-\sqrt{b^2+c^2}$, respectively. We recover the preceding case for $c=0$, which corresponds to $\theta = \pm\pi/2$.

2.3.3 Complete sets of compatible operators

By definition, two operators A and B commute if $AB = BA$, and in this case their *commutator* $[A, B]$ defined as

$$[A, B] = AB - BA \qquad (2.36)$$

vanishes. Let A and B be two Hermitian operators that commute. We can then prove the following theorem.

Theorem. Let A and B be two Hermitian operators such that $[A, B] = 0$. We can then find a basis of $\mathcal{H}$ constructed from eigenvectors common to A and B.

Proof. Let a_n be the eigenvalues of A and $|n, r\rangle$ be a basis of $\mathcal{H}$ constructed using the corresponding eigenvectors. We multiply the two sides of (2.28) by B and use the commutation relation

$$BA|n, r\rangle = A(B|n, r\rangle) = a_n(B|n, r\rangle),$$

which implies that the vector $B|n, r\rangle$ belongs to the subspace of the eigenvalue a_n. If a_n is nondegenerate, this subspace has dimension one, and $B|n, r\rangle$ is necessarily proportional to $|n, r\rangle$ which then is also an eigenvector of B. If a_n is degenerate, we can only deduce that $B|n, r\rangle$ is necessarily orthogonal to every eigenvector $|m, s\rangle$ of A with $m \neq n$:

$$\langle m, s|B|n, r\rangle = \delta_{nm} B_{sr}^{(n)},$$

which implies that in the basis $|n, r\rangle$ the matrix representation of B is block-diagonal:

$$B = \begin{pmatrix} B^{(1)} & 0 & 0 \\ 0 & B^{(2)} & 0 \\ 0 & 0 & B^{(3)} \end{pmatrix}.$$

Each block $B^{(k)}$ can be diagonalized separately by a change of basis which acts only in each subspace without affecting the diagonalization of A as a whole, since inside each subspace A is represented by a diagonal matrix.

Reciprocally, let us suppose that we have found a basis $|(n, p)r\rangle$ of $\mathcal{H}$ constructed from eigenvectors common to A and B:

$$A|(n, p)r\rangle = a_n|(n, p)r\rangle, \quad B|(n, p)r\rangle = b_p|(n, p)r\rangle.$$

It is then obvious that

$$[A, B]|(n, p)r\rangle = 0,$$

and since the vectors $|(n, p)r\rangle$ form a basis, $[A, B] = 0$. If $[A, B] = 0$, it may happen that given only the eigenvalues a_n and b_p, the basis vectors can be specified uniquely up to a multiplicative constant of modulus unity; there exists one and only one vector $|(n, p)\rangle$ such that

$$A|(n, p)\rangle = a_n|(n, p)\rangle, \quad B|(n, p)\rangle = b_p|(n, p)\rangle. \tag{2.37}$$

It is then said that A and B form a complete set of compatible operators. If there is still some indeterminacy, that is, if there exists more than one linearly independent vector satisfying (2.37), it can happen that knowing the eigenvalues of a third operator C commuting with A and B lifts the indeterminacy. An ensemble of Hermitian operators $A_1, \ldots, A_M$ that commute pairwise and whose eigenvalues unambiguously define the vectors of a basis of $\mathcal{H}$ is called a *complete set of compatible operators* (or a complete set of commuting operators).

2.3.4 Unitary operators and Hermitian operators

The properties of unitary operators $U^\dagger = U^{-1}$ are intimately related to those of Hermitian operators. In particular, such operators can always be diagonalized. The basic theorem for unitary operators is stated as follows.

Theorem. (a) *The eigenvalues a_n of a unitary operator have modulus unity: $a_n = \exp(i\alpha_n)$, α_n real.* (b) *The eigenvectors corresponding to two different eigenvalues are orthogonal.* (c) *The spectral decomposition of a unitary operator is written as a function of pairwise orthogonal projectors $\mathcal{P}_n$ as*

$$U = \sum_n a_n \mathcal{P}_n = \sum_n e^{i\alpha_n} \mathcal{P}_n \quad \text{with} \quad \sum_n \mathcal{P}_n = I. \tag{2.38}$$

The proof of (a) and (b) is trivial. To obtain (c) we write

$$U = \frac{1}{2}(U + U^{\dagger}) + i\frac{1}{2i}(U - U^{\dagger}) = A + iB. \tag{2.39}$$

The operators A and B are Hermitian and $[A, B] = 0$, so that the operators A and B can be diagonalized simultaneously, and the eigenvectors common to A and B are also eigenvectors of U. The eigenvalues of A and B are $\cos\alpha_n$ and $\sin\alpha_n$, respectively. Equation (2.39) generalizes to unitary operators the decomposition of a complex number into real and imaginary parts, with Hermitian operators playing the role of real numbers. The operator C

$$C = \sum_n \alpha_n \mathcal{P}_n$$

is a Hermitian operator and $U = \exp(iC)$. Inversely, let $A = \sum_n a_n \mathcal{P}_n$ be a Hermitian operator. The operator

$$U = \sum_n e^{i\alpha a_n} \mathcal{P}_n = e^{i\alpha A} \tag{2.40}$$

is manifestly a unitary operator. This notation generalizes the representation $\exp(i\alpha)$ of a complex number of unit modulus to unitary operators.

2.3.5 *Operator-valued functions*

In writing down (2.40) we have introduced the exponential of an operator. More generally, it is useful to know how to construct a function $f(A)$ of an operator. The construction is obvious if the operator A can be diagonalized: $A = XDX^{-1}$, where D is a diagonal matrix whose elements are d_n. Let us assume that a function f is defined by a Taylor series which converges in a certain region of the complex plane $|z| < R$:

$$f(z) = \sum_{p=0}^{\infty} c_p z^p.$$

The operator-valued function $f(A)$ will be given by

$$f(A) = \sum_{p=0}^{\infty} c_p A^p = \sum_{p=0}^{\infty} c_p X D^p X^{-1} = X \left[\sum_{p=0}^{\infty} c_p D^p \right] X^{-1}. \tag{2.41}$$

The expression inside the square brackets is just a diagonal matrix with elements $f(d_n)$ well defined if $|d_n| < R$ for any n. In general, it is possible to find an analytic continuation for $f(A)$ even if some eigenvalues d_n lie outside the region of convergence of the Taylor series, just as it is possible to analytically continue

$$\sum_{p=0}^{\infty} z^p = \frac{1}{1-z}$$

outside the region of convergence $|z| < 1$ for any value of z different from unity. A particularly important case is that of the exponential of an operator:

$$\exp A = \sum_{p=0}^{\infty} \frac{A^p}{p!}. \tag{2.42}$$

Since the radius of convergence of an exponential is infinite, the above argument implies that $\exp A$ is well defined by the series (2.42) if A is diagonalizable (in fact, it is easy to show directly that the series (2.42) is convergent in any case). Care must be taken of the fact that, in general,

$$\exp A \exp B \neq \exp B \exp A;$$

a sufficient (but not necessary!) condition for the equality to hold is that A and B commute (Exercise 3.3.6).

In summary, given a Hermitian operator A whose spectral decomposition is given by (2.31), it is straightforward to define any function of A by

$$f(A) = \sum_n f(a_n)\mathcal{P}_n, \tag{2.43}$$

for example, the exponential $\exp A$, the logarithm $\ln A$, or the resolvent $R(z, A)$:

$$e^{i\alpha A} = \sum_n e^{i\alpha a_n}\mathcal{P}_n, \tag{2.44}$$

$$\ln A = \sum_n (\ln a_n)\mathcal{P}_n, \tag{2.45}$$

$$R(z, A) = (zI - A)^{-1} = \sum_n \frac{1}{z - a_n}\mathcal{P}_n. \tag{2.46}$$

The resolvent $R(z, A)$ is of course defined only for $z \neq a_n$ for any n, and the logarithm is defined only if none of the eigenvalues a_n is zero.

2.4 Exercises

2.4.1 The scalar product and the norm

Let us take a norm $||\varphi||$ derived from a scalar product: $||\varphi||^2 = (\varphi, \varphi)$.

1. Show that this norm satisfies the triangle inequality

$$||\chi + \varphi|| \leq ||\chi|| + ||\varphi||,$$

as well as

$$|\,||\chi|| - ||\varphi||\,| \leq ||\chi + \varphi||.$$

2. Show also that

$$||\chi+\varphi||^2+||\chi-\varphi||^2=2(||\chi||^2+||\varphi||^2).$$

What is the interpretation of this equality in the real plane $\mathbb{R}^2$? Conversely, if a norm possesses this property in a *real* vector space, show that

$$(\varphi,\chi)=(\chi,\varphi)=\frac{1}{4}\left(||\chi+\varphi||^2-||\chi-\varphi||^2\right)$$

defines a scalar product. This scalar product must satisfy

$$(\chi,\varphi_1+\varphi_2)=(\chi,\varphi_1)+(\chi,\varphi_2),\quad(\chi,\lambda\varphi)=\lambda(\chi,\varphi).$$

In the case of a complex vector space, show that

$$(\chi,\varphi)=\frac{1}{4}\left[\left(||\chi+\varphi||^2-||\chi-\varphi||^2\right)-i\left(||\chi+i\varphi||^2-||\chi-i\varphi||^2\right)\right].$$

2.4.2 Commutators and traces

1. Show that

$$[A,BC]=B[A,C]+[A,B]C. \tag{2.47}$$

2. The trace of an operator is the sum of the diagonal elements of its representation matrix in a given basis:

$$\mathrm{Tr}\,A=\sum_n A_{nn}. \tag{2.48}$$

Show that

$$\mathrm{Tr}\,AB=\mathrm{Tr}\,BA, \tag{2.49}$$

and deduce that the trace is invariant under a change of basis $A\to A'=SAS^{-1}$. The trace of an operator is (fortunately) independent of the basis.

3. Show that the trace is invariant under cyclic permutations:

$$\mathrm{Tr}\,ABC=\mathrm{Tr}\,BCA=\mathrm{Tr}\,CAB. \tag{2.50}$$

2.4.3 The determinant and the trace

1. Let a matrix $A(t)$ depending on a parameter t satisfy

$$\frac{dA(t)}{dt}=A(t)\,B.$$

Show that $A(t)=A(0)\exp(Bt)$. What is the solution of

$$\frac{dA(t)}{dt}=BA(t)\,?$$

2. Show that

$$\det e^{At_1} \times \det e^{At_2} = \det e^{A(t_1+t_2)}.$$

Then derive the relation

$$\det e^A = e^{\operatorname{Tr} A},$$

or, equivalently,

$$\det B = e^{\operatorname{Tr}\ln B}. \tag{2.51}$$

Hint: Find a differential equation for the function $g(t) = \det[\exp(At)]$. The results are obvious if A is diagonalizable.

2.4.4 A projector in $\mathbb{R}^3$

1. Let us take two vectors $\vec{u}_1$ and $\vec{u}_2$ in real three-dimensional space $\mathbb{R}^3$ which are linearly independent but not necessarily orthogonal and which have any norm. Let $\mathcal{P}$ be the projector onto the plane defined by these two vectors. Show that the action of $\mathcal{P}$ on a vector $\vec{V}$ can be written as

$$\mathcal{P}\vec{V} = \sum_{i,j=1}^{2} C_{ij}^{-1}(\vec{V}\cdot\vec{u}_i)\vec{u}_j, \tag{2.52}$$

where the 2×2 matrix $C_{ij} = \vec{u}_i\cdot\vec{u}_j$.
2. Generalization: assume that we have p linearly independent vectors $\vec{u}_1,\ldots,\vec{u}_p$ in $\mathbb{R}^N$, $p < N$. Write down the projector onto the vector space generated by these p vectors.

2.4.5 The projection theorem

Let $\mathcal{H}_1$ be a vector subspace of $\mathcal{H}$ and $|\varphi\rangle \in \mathcal{H}$. Show that then there exists a unique element $|\varphi_1\rangle$ of $\mathcal{H}_1$ such that the norm $||\varphi_1 - \varphi||$ is a minimum: $||\varphi_1 - \varphi||$ is the distance from $|\varphi\rangle$ to $\mathcal{H}_1$. Find $|\varphi_1\rangle$.

2.4.6 Properties of projectors

Show the following properties of projectors.

Property 1. If $\mathcal{P}_1$ and $\mathcal{P}_1'$ are projectors onto $\mathcal{H}_1$ and $\mathcal{H}_1'$, respectively, then $\mathcal{P}_1\mathcal{P}_1'$ is a projector if and only if $\mathcal{P}_1\mathcal{P}_1' = \mathcal{P}_1'\mathcal{P}_1$. Then $\mathcal{P}_1\mathcal{P}_1'$ projects onto the intersection $\mathcal{H}_1 \cap \mathcal{H}_1'$.

Property 2. $\mathcal{P}_1 + \mathcal{P}_1'$ is a projector if and only if $\mathcal{P}_1\mathcal{P}_1' = 0$. In this case $\mathcal{H}_1$ and $\mathcal{H}_1'$ are orthogonal and $\mathcal{P}_1 + \mathcal{P}_1'$ projects onto the direct sum $\mathcal{H}_1 \oplus \mathcal{H}_1'$.

Property 3. If $\mathcal{P}_1\mathcal{P}_1' = \mathcal{P}_1'\mathcal{P}_1$, then $\mathcal{P}_1 + \mathcal{P}_1' - \mathcal{P}_1\mathcal{P}_1'$ projects onto the union $\mathcal{H}_1 \cup \mathcal{H}_1'$. The property 2 is a special case of this result.

Property 4. Assume that we have an operator Ω such that $\Omega^\dagger \Omega$ is a projector:

$$\Omega^\dagger \Omega = \mathcal{P}.$$

Show that $\Omega \Omega^\dagger$ is also a projector. Hint: show that

$$\Omega |\varphi\rangle = 0 \iff \mathcal{P} |\varphi\rangle = 0.$$

2.4.7 The Gaussian integral

Let A be a real $N \times N$ matrix which is symmetric and strictly positive (cf. Exercise 2.4.10). Show that the multiple integral

$$I(b) = \int \prod_{i=1}^{N} \mathrm{d}x_i \exp\left(-\frac{1}{2} \sum_{jk} x_j A_{jk} x_k + b_j x_j \right)$$

becomes

$$I(b) = \frac{(2\pi)^{N/2}}{\sqrt{\det A}} \exp\left(\frac{1}{2} \sum_{jk} b_j A_{jk}^{-1} b_k \right). \tag{2.53}$$

Hint: write

$$\sum_{jk} x_j A_{jk} x_k = x^T A x = \langle x | A | x \rangle,$$

where x is a column vector and x^T is a row vector, and make the change of variable

$$x' = x - A^{-1} b.$$

These Gaussian integrals are fundamental in probability theory and arise in many physics problems.

2.4.8 Commutators and a degenerate eigenvalue

Let us take three $N \times N$ matrices A, B, and C satisfying

$$[A, B] = 0, \quad [A, C] = 0, \quad [B, C] \neq 0.$$

Show that at least one eigenvalue of A is degenerate.

2.4.9 Normal matrices

A matrix C is termed *normal* if it commutes with its Hermitian conjugate:

$$C^\dagger C = C C^\dagger.$$

Writing

$$C = \frac{1}{2}(C + C^\dagger) + i\frac{1}{2i}(C - C^\dagger) = A + iB,$$

show that C is diagonalizable.

2.4.10 Positive matrices

A matrix A is termed *positive* (or non-negative by some authors) if for any vector $|\varphi\rangle \neq 0$ the average value is real and positive: $\langle\varphi|A|\varphi\rangle \geq 0$. It is termed *strictly positive* if $\langle\varphi|A|\varphi\rangle > 0$.

1. Show that any positive matrix is Hermitian and that a necessary and sufficient condition for a matrix to be positive is that its eigenvalues are all ≥ 0.
2. Show that in a real Hilbert space, where a Hermitian matrix is symmetric ($A = A^T$), a positive matrix is not in general symmetric.

2.4.11 Operator identities

1. Let an operator $f(t)$ be a function of a parameter t such that

$$f(t) = e^{tA} B e^{-tA},$$

where the operators A and B are represented by $N \times N$ matrices. Show that

$$\frac{df}{dt} = [A, f(t)], \qquad \frac{d^2 f}{dt^2} = [A, [A, f(t)]], \text{ etc.}$$

Derive the expression

$$e^{tA} B e^{-tA} = B + \frac{t}{1!}[A, B] + \frac{t^2}{2!}[A, [A, B]] + \cdots \tag{2.54}$$

Application: let three operators A, B, and C obey

$$[A, B] = iC, \quad [B, C] = iA.$$

Show that

$$e^{iBt} A e^{-iBt} = A \cos t + C \sin t.$$

An example is provided by the angular momentum operators J_x, J_y, J_z (see Chapter 10).

2. Let us assume that A and B both commute with their commutator $[A, B]$. Write down a differential equation for the operator

$$g(t) = e^{At} e^{Bt}$$

and derive the expression

$$e^{A+B} = e^A e^B e^{-\frac{1}{2}[A,B]}. \tag{2.55}$$

Careful! This identity is not valid in general. It is guaranteed to hold only when $[A, [A, B]] = [B, [A, B]] = 0$. Using the same assumptions, show also that

$$e^A e^B = e^B e^A e^{[A,B]}. \tag{2.56}$$

2.4.12 A beam splitter

Let us consider a beam splitter (a mirror which is semi-transparent to a light wave, a crystal aligned at a Bragg angle for a neutron, etc.) which we *assume to be nonabsorbing*. Waves arrive at the same angle of incidence on the left and right sides of the beam splitter with amplitudes A_L and A_R, respectively (see Fig. 2.1). The amplitudes B_L and B_R of the outgoing waves, which are made up of both reflected and transmitted waves, are linearly related to the amplitudes of the incoming waves as[6]

$$\begin{pmatrix} B_R \\ B_L \end{pmatrix} = M' \begin{pmatrix} A_R \\ A_L \end{pmatrix}, \qquad M' = \begin{pmatrix} a & b \\ c & d \end{pmatrix}.$$

1. Show that M' is unitary and that $\det M' = \exp(i\theta)$.
2. Since we are interested in experiments where the outgoing waves interfere, a global phase factor has no physical consequences and M' can be replaced by $M = \exp(-i\theta/2)M'$ with $\det M = -1$. Derive the general form of M:

$$M = \begin{pmatrix} r & t^* \\ t & -r^* \end{pmatrix}, \qquad |r|^2 + |t|^2 = 1.$$

3. Show that M can be written as

$$M = \begin{pmatrix} |r|e^{i\chi} & |t|e^{-i\phi} \\ |t|e^{i\phi} & -|r|e^{-i\chi} \end{pmatrix}.$$

Let δ_R be the difference of the phases of the reflected and transmitted waves for the wave incident from the right ($A_R = 1$, $A_L = 0$), and let δ_L be the same phase difference for the wave incident from the left ($A_R = 0$, $A_L = 1$). Show that

$$\delta_R + \delta_L = \pi \pm 2n\pi, \qquad n = 0, 1, 2, \dots$$

This result generalizes that obtained using the Mach–Zehnder interferometer in Exercise 1.6.5 to the case where the beam splitter is not symmetric. If it is symmetric

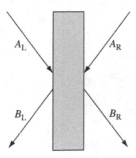

Fig. 2.1. A beam splitter.

[6] A. Zeilinger, General properties of lossless beam splitters in interferometry, *Am. J. Phys.* **49**, 882 (1981).

$\delta_R = \delta_L = \pi/2$. What is the form of M in the symmetric case? Rederive the results of Exercise 1.6.5 and show that for suitably chosen phases we can write the following in the symmetric case:

$$M = \frac{1}{\sqrt{2}} \begin{pmatrix} i & 1 \\ 1 & i \end{pmatrix}, \quad \text{or} \quad M = H = \frac{1}{\sqrt{2}} \begin{pmatrix} 1 & 1 \\ 1 & -1 \end{pmatrix}.$$

The matrix H is called the *Hadamard matrix* (or gate) and is widely used in quantum computing (Section 6.4.2).

2.5 Further reading

The results on finite-dimensional vector spaces and operators can be found in any undergraduate linear algebra text. In addition, the reader can consult Isham [1995], Chapters 2 and 3, or Nielsen and Chuang [2000], Chapter 2, which gives an elegant demonstration of the spectral decomposition theorem for a Hermitian operator.

3

Polarization: photons and spin-1/2 particles

In this chapter we build up the basic concepts of quantum mechanics using two simple examples, following a heuristic approach which is more inductive than deductive. We start with a familiar phenomenon, that of the polarization of light, which will allow us to introduce the necessary mathematical formalism. We show that the description of polarization leads naturally to the need for a two-dimensional complex vector space, and we establish the correspondence between a polarization state and a vector in this space, referred to as the space of polarization states. We then move on to the quantum description of photon polarization and illustrate the construction of probability amplitudes as scalar products in this space. The second example will be that of spin 1/2, where the space of states is again two-dimensional. We construct the most general states of spin 1/2 using rotational invariance. Finally, we introduce dynamics, which allows us to follow the time evolution of a state vector.

The analogy with the polarization of light will serve as a guide to constructing the quantum theory of photon polarization, but no such classical analog is available for constructing the quantum theory for spin 1/2. In this case the quantum theory will be constructed without reference to any classical theory, using an assumption about the dimension of the space of states and symmetry principles.

3.1 The polarization of light and photon polarization

3.1.1 The polarization of an electromagnetic wave

The polarization of light or, more generally, of an electromagnetic wave, is a familiar phenomenon related to the vector nature of the electromagnetic field. Let us consider a plane wave of monochromatic light of frequency ω propagating in the positive z direction. The electric field $\vec{E}(t)$ at a given point is a vector orthogonal to the direction of propagation. It therefore lies in the xOy plane and has components $\{E_x(t), E_y(t), E_z(t) = 0\}$ (Fig. 3.1). The most general case is that of elliptical polarization, where the electric field has the form

$$\vec{E}(t) = \begin{cases} E_x(t) = E_{0x} \cos(\omega t - \delta_x) \\ E_y(t) = E_{0y} \cos(\omega t - \delta_y) \end{cases}. \tag{3.1}$$

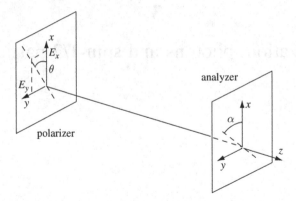

Fig. 3.1. A polarizer–analyzer ensemble.

We have not made the z dependence explicit because we are only interested in the field in a plane $z = \text{constant}$. By a suitable choice of the origin of time, it is always possible to choose $\delta_x = 0$, $\delta_y = \delta$. The intensity $\mathcal{I}$ of the light wave is proportional to the square of the electric field:

$$\mathcal{I} = \mathcal{I}_x + \mathcal{I}_y = k(E_{0x}^2 + E_{0y}^2) = kE_0^2, \tag{3.2}$$

where k is a proportionality constant which need not be specified here. When $\delta = 0$ or π, the polarization is *linear*: if we take $E_{0x} = E_0 \cos\theta$, $E_{0y} = E_0 \sin\theta$, Eq. (3.1) for $\delta_x = \delta_y = 0$ shows that the electric field oscillates in the $\hat{n}_\theta$ direction of the xOy plane, making an angle θ with the Ox axis. Such a light wave can be obtained using a linear polarizer whose axis is parallel to $\hat{n}_\theta$.

When we are interested only in the polarization of this light wave, the relevant parameters are the ratios $E_{0x}/E_0 = \cos\theta$ and $E_{0y}/E_0 = \sin\theta$, where θ can be chosen to lie in the range $[0, \pi]$. Here E_0 is a simple proportionality factor which plays no role in the description of the polarization. We can establish a correspondence between waves linearly polarized in the Ox and Oy directions and orthogonal unit vectors $|x\rangle$ and $|y\rangle$ in the xOy plane forming an orthonormal basis in this plane. The most general state of *linear* polarization in the $\hat{n}_\theta$ direction will correspond to the vector $|\theta\rangle$ in the xOy plane:

$$|\theta\rangle = \cos\theta|x\rangle + \sin\theta|y\rangle, \tag{3.3}$$

which also has unit norm:

$$\langle\theta|\theta\rangle = \cos^2\theta + \sin^2\theta = 1.$$

The fundamental reason for using a vector space to describe polarization is the *superposition principle*: a polarization state can be decomposed into two (or more) other states, or, conversely, two polarization states can be added together vectorially. To illustrate decomposition, let us imagine that a wave polarized in the $\hat{n}_\theta$ direction passes through a second polarizer, called an analyzer, oriented in the $\hat{n}_\alpha$ direction of the xOy plane making an angle α with Ox (Fig. 3.1). Only the component of the electric field in the

$\hat{n}_\alpha$ direction, that is, the projection of the field on $\hat{n}_\alpha$, will be transmitted. The amplitude of the electric field will be multiplied by a factor $\cos(\theta - \alpha)$ and the light intensity at the exit from the analyzer will be reduced by a factor $\cos^2(\theta - \alpha)$. We shall use $\overline{a}(\theta \to \alpha)$ to denote the projection factor, which we refer to as the *amplitude of the $\hat{n}_\theta$ polarization in the $\hat{n}_\alpha$ direction*. This amplitude is just the scalar product of the vectors $|\theta\rangle$ and $|\alpha\rangle$:

$$\overline{a}(\theta \to \alpha) = \langle\alpha|\theta\rangle = \cos(\theta - \alpha) = \hat{n}_\alpha \cdot \hat{n}_\theta. \tag{3.4}$$

The intensity at the exit of the analyzer is given by the Malus law:

$$\mathcal{I} = \mathcal{I}_0 |\overline{a}(\theta \to \alpha)|^2 = \mathcal{I}_0 |\langle\alpha|\theta\rangle|^2 = \mathcal{I}_0 \cos^2(\theta - \alpha) \tag{3.5}$$

if $\mathcal{I}_0$ is the intensity at the exit of the polarizer. Another illustration of decomposition is given by the apparatus of Fig. 3.2. Using a uniaxial birefringent plate perpendicular to the direction of propagation and with optical axis lying in the xOz plane, a light beam can be decomposed into a wave polarized in the Ox direction and a wave polarized in the Oy direction. The wave polarized in the Ox direction propagates in the direction of the extraordinary ray refracted at the entrance and exit of the plate, and the wave polarized in the Oy direction follows the ordinary ray propagating in a straight line.

The addition of two polarization states can be illustrated using the apparatus of Fig. 3.3. The two beams are recombined by a second birefringent plate, symmetrically located relative to the first with respect to a vertical plane, before the beam passes through the analyzer.[1] In order to simplify the arguments, we shall neglect the phase difference

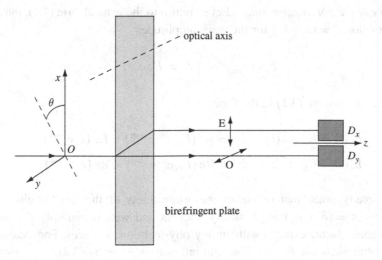

optical axis

Fig. 3.2. Decomposition of the polarization by a birefringent plate. The ordinary ray O is polarized horizontally, and the extraordinary ray E is polarized vertically.

---

[1] This recombination of amplitudes is possible because two beams from the same source are coherent. Of course, it would be impossible to add the amplitudes of two polarized beams from different sources; the situation is identical to that in the case of interference.

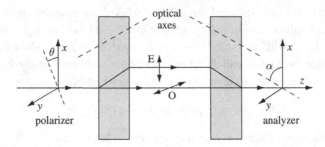

Fig. 3.3. Decomposition and recombination of polarizations using birefringent plates.

originating from the difference between the ordinary and extraordinary indices in the birefringent plates (equivalently, we can imagine that this difference is cancelled by an intermediate birefringent plate which is oriented appropriately; see Exercise 3.3.1). Under these conditions the light wave at the exit of the second birefringent plate is polarized in the $\hat{n}_\theta$ direction. The recombination of the two x and y beams gives the initial light beam polarized in the $\hat{n}_\theta$ direction, and the intensity at the exit of the analyzer is reduced as before by a factor $\cos^2(\theta - \alpha)$.

If we limit ourselves to linear polarization states, we can describe any polarization state as a real unit vector in the xOy plane, in which a possible orthonormal basis is constructed from the vectors $|x\rangle$ and $|y\rangle$. However, if we want to describe an arbitrary polarization, we need to introduce a two-dimensional *complex* vector space $\mathcal{H}$. This space will be the *vector space of the polarization states*. Let us return to the general case (3.1), introducing complex notation $\vec{\mathcal{E}} = (\mathcal{E}_x, \ \mathcal{E}_y)$ for the wave amplitudes:

$$\mathcal{E}_x = E_{0x} e^{i\delta_x}, \quad \mathcal{E}_y = E_{0y} e^{i\delta_y}, \tag{3.6}$$

which allows us to write (3.1) in the form

$$
\begin{aligned}
E_x(t) &= E_{0x}\cos(\omega t - \delta_x) = \mathrm{Re}\left(E_{0x}e^{i\delta_x}e^{-i\omega t}\right) = \mathrm{Re}\left(\mathcal{E}_x e^{-i\omega t}\right), \\
E_y(t) &= E_{0y}\cos(\omega t - \delta_y) = \mathrm{Re}\left(E_{0y}e^{i\delta_y}e^{-i\omega t}\right) = \mathrm{Re}\left(\mathcal{E}_y e^{-i\omega t}\right).
\end{aligned}
\tag{3.7}
$$

We have already noted that owing to the arbitrariness of the time origin, only the relative phase $\delta = (\delta_y - \delta_x)$ is physically relevant and we can multiply $\mathcal{E}_x$ and $\mathcal{E}_y$ by a common phase factor $\exp(i\beta)$ without any physical consequences. For example, it is always possible to choose $\delta_x = 0$. The light intensity is given by (3.2):

$$\mathcal{I} = k(|\mathcal{E}_x|^2 + |\mathcal{E}_y|^2) = k|\vec{\mathcal{E}}|^2 = kE_0^2. \tag{3.8}$$

An important special case of (3.7) is that of *circular polarization*, where $E_{0x} = E_{0y} = E_0/\sqrt{2}$ and $\delta_y = \pm\pi/2$ (we have conventionally chosen $\delta_x = 0$). If $\delta_y = +\pi/2$, the tip of the electric field vector traces a circle in the xOy plane in the counterclockwise sense. The components $E_x(t)$ and $E_y(t)$ are given by

$$E_x(t) = \mathrm{Re}\left(\frac{E_0}{\sqrt{2}}e^{-i\omega t}\right) = \frac{E_0}{\sqrt{2}}\cos\omega t,$$

$$E_y(t) = \mathrm{Re}\left(\frac{E_0}{\sqrt{2}}e^{-i\omega t}e^{i\pi/2}\right) = \frac{E_0}{\sqrt{2}}\cos(\omega t - \pi/2) = \frac{E_0}{\sqrt{2}}\sin\omega t. \tag{3.9}$$

An observer at whom the light wave arrives sees the tip of the electric field vector tracing a circle of radius $E_0/\sqrt{2}$ counterclockwise in the xOy plane. The corresponding polarization is termed *right-handed circular polarization*.[2] When $\delta_y = -\pi/2$, we obtain *left-handed circular polarization* – the circle is traced in the clockwise sense:

$$E_x(t) = \mathrm{Re}\left(\frac{E_0}{\sqrt{2}}e^{-i\omega t}\right) = \frac{E_0}{\sqrt{2}}\cos\omega t,$$

$$E_y(t) = \mathrm{Re}\left(\frac{E_0}{\sqrt{2}}e^{-i\omega t}e^{-i\pi/2}\right) = \frac{E_0}{\sqrt{2}}\cos(\omega t + \pi/2) = -\frac{E_0}{\sqrt{2}}\sin\omega t. \tag{3.10}$$

These right- and left-handed circular polarization states are obtained experimentally starting from linear polarization at an angle of 45° to the axes and then introducing a phase shift $\pm\pi/2$ of the field in the Ox or Oy direction by means of a quarter-wave plate.

In complex notation the fields $\mathcal{E}_x$ and $\mathcal{E}_y$ are written as

$$\mathcal{E}_x = \frac{1}{\sqrt{2}}E_0, \quad \mathcal{E}_y = \frac{1}{\sqrt{2}}E_0 e^{\pm i\pi/2} = \frac{\pm i}{\sqrt{2}}E_0,$$

where the $+$ sign corresponds to right-handed circular polarization and the $-$ to left-handed. The proportionality factor E_0 common to $\mathcal{E}_x$ and $\mathcal{E}_y$ defines the intensity of the light wave and plays no role in describing the polarization, which is characterized by the normalized vectors

$$\boxed{|R\rangle = -\frac{1}{\sqrt{2}}(|x\rangle + i|y\rangle), \quad |L\rangle = \frac{1}{\sqrt{2}}(|x\rangle - i|y\rangle)} . \tag{3.11}$$

The overall minus sign in the definition of $|R\rangle$ has been introduced to be consistent with the conventions of Chapter 10. Equation (3.11) shows that the mathematical description of polarization leads naturally to the use of unit vectors in a complex two-dimensional vector space $\mathcal{H}$, in which the vectors $|x\rangle$ and $|y\rangle$ form one possible orthonormal basis.

---

[2] See Fig. 10.8. Our definition of right- and left-handed circular polarization is the one used in elementary particle physics. With this definition, right- (left-) handed circular polarization corresponds to positive (negative) helicity, that is, to projection of the photon spin on the direction of propagation equal to $+\hbar$ $(-\hbar)$. However, this definition is not universal; optical physicists often use the opposite, but, as one of them has remarked (E. Hecht, *Optics*, New York: Addison-Wesley (1987), Chapter 8): "This choice of terminology is admittedly a bit awkward. Yet its use in optics is fairly common, even though it is completely antithetic to the more reasonable convention adopted in elementary particle physics."

Above we have established the correspondence between linear polarization in the $\hat{n}_\theta$ direction and the unit vector $|\theta\rangle$ of $\mathcal{H}$, as well as the correspondence between the two circular polarizations and the two vectors (3.11) of $\mathcal{H}$. We are now going to generalize this correspondence by constructing the polarization corresponding to the most general normalized vector $|\Phi\rangle$ of $\mathcal{H}$:[3]

$$|\Phi\rangle = \lambda|x\rangle + \mu|y\rangle, \quad |\lambda|^2 + |\mu|^2 = 1. \tag{3.12}$$

It is always possible to choose λ to be real (in Exercise 3.3.2 we show that the physics is unaffected if λ is complex). The numbers λ and μ can then be parametrized by two angles θ and η:

$$\lambda = \cos\theta, \quad \mu = \sin\theta e^{i\eta}.$$

We shall imagine a device containing two birefringent plates and a linear polarizer, on which an electromagnetic wave (3.7) is incident. This device will be called a (λ, μ) *polarizer*.

- The first birefringent plate changes the phase of $\mathcal{E}_y$ by $-\eta$ while leaving $\mathcal{E}_x$ unchanged:

$$\mathcal{E}_x \to \mathcal{E}_x^{(1)} = \mathcal{E}_x, \quad \mathcal{E}_y \to \mathcal{E}_y^{(1)} = \mathcal{E}_y e^{-i\eta}.$$

- The linear polarizer projects on the $\hat{n}_\theta$ direction:

$$\vec{\mathcal{E}}^{(1)} \to \vec{\mathcal{E}}^{(2)} = \left(\mathcal{E}_x^{(1)} \cos\theta + \mathcal{E}_y^{(1)} \sin\theta\right) \hat{n}_\theta$$
$$= \left(\mathcal{E}_x \cos\theta + \mathcal{E}_y \sin\theta e^{-i\eta}\right) \hat{n}_\theta.$$

- The second birefringent plate leaves $\mathcal{E}_x^{(2)}$ unchanged and shifts the phase of $\mathcal{E}_y^{(2)}$ by η:

$$\mathcal{E}_x^{(2)} \to \mathcal{E}_x' = \mathcal{E}_x^{(2)}, \quad \mathcal{E}_y^{(2)} \to \mathcal{E}_y' = \mathcal{E}_y^{(2)} e^{i\eta}.$$

The combination of the three operations is represented by the transformation $\vec{\mathcal{E}} \to \vec{\mathcal{E}}'$ which can be written in terms of components:

$$\mathcal{E}_x' = \mathcal{E}_x \cos^2\theta + \mathcal{E}_y \sin\theta \cos\theta e^{-i\eta} = |\lambda|^2 \mathcal{E}_x + \lambda\mu^* \mathcal{E}_y,$$
$$\mathcal{E}_y' = \mathcal{E}_x \sin\theta \cos\theta e^{i\eta} + \mathcal{E}_y \sin^2\theta = \lambda^*\mu \mathcal{E}_x + |\mu|^2 \mathcal{E}_y. \tag{3.13}$$

The operation (3.13) amounts to *projection* on $|\Phi\rangle$. In fact, if we choose to write the vectors $|x\rangle$ and $|y\rangle$ as column vectors

$$|x\rangle = \begin{pmatrix} 1 \\ 0 \end{pmatrix}, \quad |y\rangle = \begin{pmatrix} 0 \\ 1 \end{pmatrix}, \tag{3.14}$$

then the projector $\mathcal{P}_\Phi$

$$\mathcal{P}_\Phi = |\Phi\rangle\langle\Phi| = \left(\lambda|x\rangle + \mu|y\rangle\right)\left(\lambda^*\langle x| + \mu^*\langle y|\right)$$

---

[3] We shall use upper-case letters $|\Phi\rangle$ or $|\Psi\rangle$ for generic vectors of $\mathcal{H}$ of the form (3.12) or (3.16), to avoid any confusion with an angle, as for $|\theta\rangle$ or $|\alpha\rangle$.

is represented by the matrix

$$\mathcal{P}_\Phi = \begin{pmatrix} |\lambda|^2 & \lambda\mu^* \\ \lambda^*\mu & |\mu|^2 \end{pmatrix}. \tag{3.15}$$

We can put the incident field $\vec{\mathcal{E}}$ (3.7) in correspondence with a (non-normalized) vector $|\mathcal{E}\rangle$ of $\mathcal{H}$ with the complex components $\mathcal{E}_x$ and $\mathcal{E}_y$:

$$|\mathcal{E}\rangle = \mathcal{E}_x|x\rangle + \mathcal{E}_y|y\rangle.$$

Using $|\mathcal{E}\rangle$ we can define a vector $|\Psi\rangle$ normalized to unity by $|\mathcal{E}\rangle = E_0|\Psi\rangle$:

$$|\Psi\rangle = \nu|x\rangle + \sigma|y\rangle, \quad |\nu|^2 + |\sigma|^2 = 1, \tag{3.16}$$

where

$$\nu = \frac{\mathcal{E}_x}{E_0}, \quad \sigma = \frac{\mathcal{E}_y}{E_0}.$$

The normalized vector $|\Psi\rangle$ which describes the polarization of the wave (3.7) is called the *Jones vector*. According to (3.13) and (3.15), the electric field at the exit of the (λ, μ) polarizer will be

$$|\mathcal{E}'\rangle = \mathcal{P}_\Phi|\mathcal{E}\rangle = E_0\mathcal{P}_\Phi|\Psi\rangle = E_0|\Phi\rangle\langle\Phi|\Psi\rangle. \tag{3.17}$$

Now let us generalize everything we have obtained for the linear polarizer to the (λ, μ) polarizer. The latter projects the polarization state $|\Psi\rangle$ onto $|\Phi\rangle$ with amplitude equal to $\langle\Phi|\Psi\rangle$:

$$\bar{a}(\Psi \to \Phi) = \langle\Phi|\Psi\rangle. \tag{3.18}$$

At the exit of the polarizer the intensity is reduced by a factor $|\bar{a}(\Psi \to \Phi)|^2 = |\langle\Phi|\Psi\rangle|^2$. If the polarization state is described by the unit vector $|\Phi\rangle$ (3.12), then the transmission through the (λ, μ) polarizer is 100%. On the other hand, the polarization state

$$|\Phi_\perp\rangle = -\mu^*|x\rangle + \lambda^*|y\rangle \tag{3.19}$$

is completely stopped by the (λ, μ) polarizer. The polarization state (3.16) is in general an elliptic polarization. It is easy to determine the characteristics of the corresponding ellipse and the direction in which it is traced (Exercise 3.3.2).

The states $|\Phi\rangle$ and $|\Phi_\perp\rangle$ form an orthonormal basis of $\mathcal{H}$ obtained from the $\{|x\rangle, |y\rangle\}$ basis by a unitary transformation U:

$$U = \begin{pmatrix} \lambda & \mu \\ -\mu^* & \lambda^* \end{pmatrix}.$$

In summary, we have shown that any polarization state can be put into correspondence with a normalized vector $|\Phi\rangle$ of a two-dimensional complex space $\mathcal{H}$. The vectors $|\Phi\rangle$ and $\exp(i\beta)|\Phi\rangle$ represent the same polarization state. Stated more precisely, a polarization state can be put into correspondence with a vector up to a phase.

3.1.2 The photon polarization

Now we shall show that the mathematical formalism used above to describe the polarization of a light wave can be carried over without modification to the description of the polarization of a photon. However, the fact that the mathematical formalism is identical in the two cases should not obscure the fact that the physical interpretation is radically modified. We shall return to the experiment of Fig. 3.2 and reduce the light intensity such that individual photons are registered by the photomultipliers D_x and D_y, which respectively detect photons polarized in the Ox and Oy directions. We then observe the following:

- only one of the two photomultipliers is triggered by a photon incident on the plate. Like the neutrons of Chapter 1, the photons arrive in lumps: they are never split.
- the probability p_x (p_y) of D_x (D_y) being triggered by a photon incident on the plate is $\mathsf{p}_x = \cos^2 \theta$ ($\mathsf{p}_y = \sin^2 \theta$).

This result must hold true if we want to recover classical optics in the limit where the number N of photons is large. In fact, if N_x and N_y are the numbers of photons detected by D_x and D_y, we must have

$$\mathsf{p}_x = \lim_{N\to\infty} \frac{N_x}{N}, \quad \mathsf{p}_y = \lim_{N\to\infty} \frac{N_y}{N}$$

and $\mathcal{I}_x \propto N_x = N\cos^2 \theta$, $\mathcal{I}_y \propto N_y = N\sin^2 \theta$ in the limit $N \to \infty$. However, the fate of an individual photon cannot be predicted. We can only know its *probability* of detection by D_x or D_y. The need to resort to probabilities is an intrinsic feature of quantum physics, whereas in classical physics resorting to probabilities is only a way to take into account the complexity of a phenomenon whose details we cannot (or do not want to) know. For example, when flipping a coin, complete knowledge of the initial conditions under which the coin is thrown and inclusion of the air resistance, the state of the ground on which the coin lands, etc. permit us in principle to predict the result. Some physicists[4] have suggested that the probabilistic nature of quantum mechanics has an analogous origin: if we had access to additional variables which at present we do not know, the so-called *hidden variables*, we would be able to predict with certainty the fate of each individual photon. This hidden variable hypothesis has some utility in discussions of the foundations of quantum physics. Nevertheless, in Chapter 6 we shall see that, given very plausible hypotheses, such variables are excluded by experiment.

However, probabilities alone provide only a very incomplete description of the photon polarization. A complete description requires also the introduction of *probability amplitudes*. Probability amplitudes, which we denote a (the difference between the wave amplitudes of the preceding subsection and probability amplitudes is emphasized by using different notation: a instead of $\bar{a}$), are complex numbers, and probabilities correspond to their squared modulus $|a|^2$. To make manifest the incomplete nature of probabilities

---

[4] Including de Broglie and Bohm.

alone, let us again consider the apparatus of Fig. 3.3. Between the two plates a photon follows either the trajectory of an extraordinary ray polarized in the Ox direction, called an x trajectory, or the trajectory of an ordinary ray polarized in the Oy direction, called a y trajectory. According to purely probabilistic reasoning, a photon following an x trajectory has probability $\cos^2\theta\cos^2\alpha$ of being transmitted by the analyzer, and a photon following a y trajectory has the corresponding probability $\sin^2\theta\sin^2\alpha$. The total probability for a photon to be transmitted by the analyzer is therefore

$$p_{\text{tot}} = \cos^2\theta\cos^2\alpha + \sin^2\theta\sin^2\alpha. \tag{3.20}$$

This is not what is found from experiment, which confirms the result obtained earlier using wave arguments:

$$p_{\text{tot}} = \cos^2(\theta - \alpha).$$

A correct reasoning must be based on probability amplitudes, just as before we used wave amplitudes. Probability amplitudes obey the same rules as wave amplitudes, which guarantees that the results of optics are reproduced when the number of photons $N \to \infty$. The probability amplitude for a photon linearly polarized in the $\hat{n}_\theta$ direction to be polarized in the $\hat{n}_\alpha$ direction is given by (3.4): $a(\theta \to \alpha) = \cos(\theta - \alpha) = \hat{n}_\theta \cdot \hat{n}_\alpha$. We obtain the following table of probability amplitudes for the experiment of Fig. 3.3:

$$a(\theta \to x) = \cos\theta, \quad a(x \to \alpha) = \cos\alpha,$$

$$a(\theta \to y) = \sin\theta, \quad a(y \to \alpha) = \sin\alpha.$$

This example provides an illustration of the rules governing the combination of probability amplitudes. The probability amplitude a_x for an incident photon following an x trajectory to be transmitted by the analyzer is

$$a_x = a(\theta \to x)a(x \to \alpha) = \cos\theta\cos\alpha.$$

This expression suggests the *factorization* rule for amplitudes: a_x is the product of the amplitudes $a(\theta \to x)$ and $a(x \to \alpha)$. This factorization rule guarantees that the corresponding rule for the probabilities holds. We also have

$$a_y = a(\theta \to y)a(y \to \alpha) = \sin\theta\sin\alpha.$$

If the experimental setup does not allow us to know which trajectory a photon has followed, the amplitudes must be added. The total probability amplitude for a photon to be transmitted by the analyzer is then

$$a_{\text{tot}} = a_x + a_y = \cos\theta\cos\alpha + \sin\theta\sin\alpha = \cos(\theta - \alpha), \tag{3.21}$$

and the corresponding probability is $\cos^2(\theta - \alpha)$, in agreement with the result (3.5) of classical optics. If there is a way to distinguish between the two trajectories, the interference is destroyed and the probabilities must be added as in (3.20).

Since the rules for combining probability amplitudes are the same as those for wave amplitudes, these rules will apply if the polarization state of a photon is described by a normalized vector in a two-dimensional vector space $\mathcal{H}$, called the *space of states*. In the present case this is the space of polarization states. When a photon is linearly polarized in the Ox (Oy) direction, we can put this polarization state in correspondence with a vector $|x\rangle$ $(|y\rangle)$ of this space. Such a polarization state is obtained by allowing a photon to pass through a linear polarizer oriented in the Ox (Oy) direction. The probability that a photon polarized in the Ox direction will be transmitted by an analyzer oriented in the Oy direction is zero: the probability amplitude $a(x \to y) = 0$. Conversely, the probability that a photon polarized in the Ox or Oy direction will be transmitted by an analyzer oriented in the same direction is equal to unity, and so

$$|a(x \to x)| = |a(y \to y)| = 1, \quad a(x \to y) = a(y \to x) = 0.$$

These relations are satisfied if $|x\rangle$ and $|y\rangle$ form an orthonormal basis of $\mathcal{H}$ and if we identify the probability amplitudes as scalar products:

$$a(x \to x) = \langle x|x \rangle = 1, \quad a(y \to y) = \langle y|y \rangle = 1, \quad a(y \to x) = \langle x|y \rangle = 0. \quad (3.22)$$

The most general *linear* polarization state is the state in which the polarization makes an angle θ with Ox. This state will be represented by the vector

$$|\theta\rangle = \cos\theta|x\rangle + \sin\theta|y\rangle. \quad (3.23)$$

Equations (3.22) and (3.23) ensure that the probability amplitudes listed above are correctly given by the scalar products, for example,

$$a(\theta \to x) = \langle x|\theta \rangle = \cos\theta,$$

or, in general, if $|\alpha\rangle$ is a state of linear polarization,

$$a(\theta \to \alpha) = \langle \alpha|\theta \rangle = \cos(\theta - \alpha).$$

The most general polarization state will be described by a normalized vector called a *state vector*:

$$|\Phi\rangle = \lambda|x\rangle + \mu|y\rangle, \quad |\lambda|^2 + |\mu|^2 = 1.$$

As in the wave case, the vectors $|\Phi\rangle$ and $\exp(i\beta)|\Phi\rangle$ represent the same physical state: a physical state is represented by a vector up to a phase in the space of states. The *probability amplitude for finding a polarization state* $|\Psi\rangle$ *in* $|\Phi\rangle$ will be given by the scalar product $\langle \Phi|\Psi \rangle$, and the projection onto a given polarization state will be realized by the (λ, μ) polarizer described in the preceding subsection. In summary, we have used a specific example, that of the polarization of a photon, to illustrate the construction of the Hilbert space of states.

The photon polarization along some (complex) direction is an example of a *quantum physical property*. The interpretation of a quantum physical property differs radically from that of a classical physical property. We shall illustrate this by examining the photon polarization. At first we limit ourselves to the simplest case, that of a linear polarization state. Using a linear polarizer oriented in the Ox direction, we prepare an ensemble of photons all in the state $|x\rangle$. The photons arrive one by one at the polarizer, and all the photons which are transmitted by the polarizer are in the state $|x\rangle$. This is the stage of *preparation of the quantum system*, where one only keeps the photons which have passed through the polarizer aligned in the Ox direction. The next stage, the *test stage*, consists of testing this polarization by allowing the photons to pass through a linear analyzer. If the analyzer is parallel to Ox the photons are transmitted with unit probability and if it is parallel to Oy they are transmitted with zero probability. In both cases the result of the test can be predicted with certainty. The physical property "polarization of a photon prepared in the state $|x\rangle$" takes well-defined values if the basis $\{|x\rangle, |y\rangle\}$ is chosen for the test. On the other hand, if we use analyzers oriented in the direction $\hat{n}_\theta$ corresponding to the state $|\theta\rangle$ (3.23) and in the perpendicular direction $\hat{n}_{\theta_\perp}$ corresponding to the state

$$|\theta_\perp\rangle = -\sin\theta|x\rangle + \cos\theta|y\rangle, \tag{3.24}$$

we can predict only the transmission probability $|\langle\theta|x\rangle|^2 = \cos^2\theta$ in the first case and $|\langle\theta_\perp|x\rangle|^2 = \sin^2\theta$ in the second. The physical property "polarization of the photon in the state $|x\rangle$" has no well-defined value in the basis $\{|\theta\rangle, |\theta_\perp\rangle\}$. In other words, the physical property "polarization" is associated with a given basis, and the two bases $\{|x\rangle, |y\rangle\}$ and $\{|\theta\rangle, |\theta_\perp\rangle\}$ are termed *incompatible* (except when $\theta = 0$ and $\theta = \pi/2$). *Complementary bases* are a special case of incompatible ones: in a Hilbert space of dimension N, two bases $\{|m\rangle\}$ and $\{|\mu\rangle\}$ are termed complementary if $|\langle m|\mu\rangle|^2 = 1/N$ for all m and μ.

The preceding discussion should be made more precise in two respects. First, it is clearly impossible to test the polarization of an isolated photon. The polarization test requires that we are provided with a number $N \gg 1$ of photons prepared under identical conditions. Let us then suppose that N photons have been prepared in a certain polarization state and that they are tested by a linear analyzer oriented in the Ox direction. If we find – within the experimental accuracy of the apparatus – that the photons pass through the analyzer with a probability of 100%, we can deduce that the photons have been prepared in the state $|x\rangle$. The observation of a single photon obviously does not allow us to arrive at this conclusion, unless we know beforehand in which basis it was prepared. The second point is that even if the photons are transmitted with a probability $\cos^2\theta$, we cannot deduce that they have been prepared in the linear polarization state (3.23). In fact, we will observe the same transmission probability if the photons have been prepared in an elliptic polarization state (3.12) with

$$\lambda = \cos\theta e^{i\delta_x}, \quad \mu = \sin\theta e^{i\delta_y}.$$

Only a test whose results have probability 0 or 1 allows the photon polarization state to be determined unambiguously with one orientation of the analyzer. Otherwise, a second orientation will be necessary to determine the phases.

In the representation (3.14) of the basis vectors of $\mathcal{H}$, the projectors $\mathcal{P}_x$ and $\mathcal{P}_y$ onto the states $|x\rangle$ and $|y\rangle$ are represented by matrices

$$\mathcal{P}_x = \begin{pmatrix} 1 & 0 \\ 0 & 0 \end{pmatrix}, \quad \mathcal{P}_y = \begin{pmatrix} 0 & 0 \\ 0 & 1 \end{pmatrix}$$

which commute: $[\mathcal{P}_x, \mathcal{P}_y] = 0$. The two operators are compatible according to the definition of Section 2.3.3. The projectors $\mathcal{P}_\theta$ and $\mathcal{P}_{\theta_\perp}$ can be calculated directly from (3.15):

$$\mathcal{P}_\theta = \begin{pmatrix} \cos^2\theta & \sin\theta\cos\theta \\ \sin\theta\cos\theta & \sin^2\theta \end{pmatrix}, \quad \mathcal{P}_{\theta_\perp} = \begin{pmatrix} \sin^2\theta & -\sin\theta\cos\theta \\ -\sin\theta\cos\theta & \cos^2\theta \end{pmatrix}.$$

They commute with each other, but not with either $\mathcal{P}_x$ or $\mathcal{P}_y$: $\mathcal{P}_x$ and $\mathcal{P}_\theta$, for example, are incompatible. The commutation (or noncommutation) of operators is the mathematical translation of the compatibility (or incompatibility) of physical properties.

As another choice of basis we can use the right- and left-handed circular polarization states $|R\rangle$ and $|L\rangle$ of (3.11). The basis $\{|R\rangle, |L\rangle\}$ is incompatible with any basis constructed using linear polarization states, and in fact complementary to any such basis. The projectors $\mathcal{P}_R$ and $\mathcal{P}_L$ onto these circular polarization states are

$$\mathcal{P}_R = \frac{1}{2}\begin{pmatrix} 1 & -i \\ i & 1 \end{pmatrix}, \quad \mathcal{P}_L = \frac{1}{2}\begin{pmatrix} 1 & i \\ -i & 1 \end{pmatrix}. \tag{3.25}$$

We can use $\mathcal{P}_R$ and $\mathcal{P}_L$ to construct the remarkable Hermitian operator Σ_z:

$$\boxed{\Sigma_z = \mathcal{P}_R - \mathcal{P}_L = \begin{pmatrix} 0 & -i \\ i & 0 \end{pmatrix}.} \tag{3.26}$$

This operator has the states $|R\rangle$ and $|L\rangle$ as its eigenvectors, and their respective eigenvalues are $+1$ and -1:

$$\Sigma_z|R\rangle = |R\rangle, \quad \Sigma_z|L\rangle = -|L\rangle. \tag{3.27}$$

This result suggests that the Hermitian operator Σ_z with eigenvectors $|R\rangle$ and $|L\rangle$ is associated with the physical property called "circular polarization." We shall see in Chapter 10 that $\hbar\Sigma_z = J_z$ is the operator representing the physical property called "z component of the photon angular momentum (or spin)." We also observe that $\exp(-i\theta\Sigma_z)$ is an operator which performs rotations by an angle θ about the Oz axis, as can be seen from a simple calculation (Exercise 3.3.3)

$$\exp(-i\theta\Sigma_z) = \begin{pmatrix} \cos\theta & -\sin\theta \\ \sin\theta & \cos\theta \end{pmatrix}, \tag{3.28}$$

and $\exp(-i\theta\Sigma_z)$ transforms the state $|x\rangle$ into the state $|\theta\rangle$ and $|y\rangle$ into $|\theta_\perp\rangle$:

$$\exp(-i\theta\Sigma_z)|x\rangle = |\theta\rangle, \quad \exp(-i\theta\Sigma_z)|y\rangle = |\theta_\perp\rangle. \tag{3.29}$$

3.1.3 Quantum cryptography

Quantum cryptography is a recent invention based on the incompatibility of two different bases of linear polarization states. Ordinary cryptography makes use of an encryption key known only to the transmitter and receiver. This is called *secret-key* cryptography. It is in principle very secure,[5] but it is necessary that the transmitter and receiver be able to exchange the key without its being intercepted by a spy. The key must be changed often, because a set of messages encoded using the same key can reveal regularities which permit decipherment by a third party. The process of transmitting a secret key is risky, and for this reason it is preferable to use systems based on a different principle, the so-called *public-key* systems, where the key is made public, for example via the Internet. A public-key system currently in use is based on the difficulty of factoring a very large number N into primes,[6] whereas the reverse operation is straightforward: without a calculator one can obtain $137 \times 53 = 7261$ in a few seconds, but given 7261 it would take some time to factor it into primes. The number of instructions needed for a computer using the best modern algorithms to factor a number N into primes grows with N roughly as $\exp[(\ln N)^{1/3}]$.[7] In a public-key system, the receiver, conventionally named Bob, publicly sends to the transmitter, conventionally named Alice, a very large number $N = pq$ which is the product of two primes p and q, as well a number c having no common factor with $(p-1)(q-1)$. Knowledge of N and c is sufficient for Alice to encrypt the message, but decipherment requires knowing the numbers p and q. Of course, a spy, conventionally named Eve, possessing a sufficiently powerful computer and enough time can manage to crack the code, but in general one can count on keeping the contents of the message secret for a limited period of time. However, it is not impossible that eventually very powerful algorithms will be found for factoring a number into primes, and, moreover, if quantum computers (Section 6.4.2) ever see the light of day, they will push the limits of factorization very far. Fortunately, thanks to quantum mechanics we are nearly at the point of being able to counteract the efforts of spies.

"Quantum cryptography" is a catchy phrase, but somewhat inaccurate. The point is not that a message is encrypted using quantum physics, but rather that quantum physics is used to ensure that the key has been transmitted securely: a more accurate terminology is thus "quantum key distribution" (QKD). A message, encrypted or not, can be transmitted using the two orthogonal linear polarization states of a photon, for example, $|x\rangle$ and $|y\rangle$. We can adopt the convention of assigning the value 1 to the polarization $|x\rangle$ and 0 to the polarization $|y\rangle$; then each photon transports a bit of information. The entire message, encrypted or not, can be written in binary code, that is, as a series of ones and zeros, and the message 1001110 can be encoded by Alice using the photon sequence *xyyxxxy* and then sent to Bob via, for example, an optical fiber. Using a birefringent plate, Bob

---

[5] An absolutely secure encryption was discovered by Vernam in 1917. However, absolute security requires that the key be as long as the message and that it be used only a single time!

[6] Called RSA encryption, discovered by Rivest, Shamir, and Adleman in 1977.

[7] At present the best factorization algorithm requires a number of operations $\sim \exp[1.9(\ln N)^{1/3}(\ln \ln N)^{2/3}]$. One cannot hope to factor numbers with more than 180 figures ($\sim 10^{20}$ instructions) in a reasonable amount of time.

will separate the photons of vertical and horizontal polarization as in Fig. 3.2, and two detectors located behind the plate will permit him to decide if a photon was horizontally or vertically polarized. In this way he can reconstruct the message. If this were an ordinary message, there would of course be much simpler and more efficient methods of sending it! At this point, let us just note that if Eve eavesdrops on the fiber, detects the photons and their polarization, and then sends to Bob other photons with the same polarization as the ones sent by Alice, Bob is none the wiser. The situation would be the same for any device functioning in a classical manner, that is, any device that does not use the superposition principle: if the spy takes sufficient precautions, the spying is undetectable, because she can send a signal that is arbitrarily close to the original one.

This is where quantum mechanics and the superposition principle come to the aid of Alice and Bob, allowing them to be sure that their message has not been intercepted. The message need not be long (the method of transmission via polarization is not very efficient). The idea in general is to transmit the key permiting encryption of a later message, a key which can be replaced when necessary. Alice sends Bob four types of photon: photons polarized along Ox ($\updownarrow$) and Oy ($\leftrightarrow$) as before, and photons polarized along axes rotated by $\pm 45°$, that is, Ox' ($\nwarrow$) and Oy' ($\nearrow$), respectively corresponding to bits 1 and 0. Again Bob analyzes the photons sent by Alice, now using analyzers oriented in four directions, vertical/horizontal and $\pm 45°$. One possibility is to use a birefringent crystal randomly oriented vertically or at 45° from the vertical and to detect the photons leaving this crystal as in Fig. 3.3. However, instead of rotating the crystal+detector ensemble, it is easier to use a Pockels cell, which allows a given polarization to be transformed into one of arbitrary orientation while keeping the crystal+detector ensemble fixed (Fig. 3.4). Bob records 1 if the photon has polarization $\updownarrow$ or $\nwarrow$, and 0 if it has polarization $\leftrightarrow$ or $\nearrow$. After recording a sufficient number of photons, Bob announces publicly the analyzer sequence he has used, but not his results. Alice compares her polarizer sequence to that of Bob and also publicly gives him the list of polarizers compatible with his analyzers. The bits corresponding to incompatible analyzers and polarizers are rejected ($-$), and, for the other bits, Alice and Bob are certain that their values are the same. It is these bits which will serve to construct the key, and they are known only to Bob and Alice, because an outsider knows only the list of orientations and not the results. An example of photon exchanges between Alice and Bob is given in Fig. 3.5.

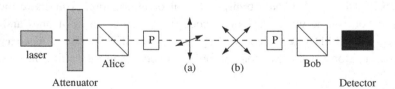

Fig. 3.4. The BB84 protocol. An attenuted laser beam allows Alice to send individual photons. A birefringent crystal selects a given linear polarization, which can be rotated thanks to a Pockels cell P. The photons are polarized, either vertically/horizontally (a), or to $\pm 45°$ (b).

Alice's polarizers	↕	↔	↗	↕	↗	↗	↘	↕	↘
sequence of bits	1	0	0	1	0	0	1	1	1
Bob's analyzers	↕	✕	↕	↕	✕	✕	↕	↕	✕
Bob's measurements	1	1	0	1	0	0	1	1	1
retained bits	1	–	–	1	0	0	–	1	1

Fig. 3.5. Quantum cryptography: transmission of polarized photons between Bob and Alice.

The only thing left is to ensure that the message has not been intercepted and that the key it contains can be used without risk. Alice and Bob randomly choose a subset of their key and compare it publicly. If Eve has intercepted the photons, this will result in a reduction of the correlation between the values of their bits. Suppose, for example, that Alice sends a photon polarized in the Ox direction. If Eve intercepts it using a polarizer oriented in the Ox' direction, and if the photon is transmitted by her analyzer, she does not know that this photon was initially polarized along the Ox direction, and so she resends Bob a photon polarized in the Ox' direction, and in 50% of cases Bob will not obtain the right result. Since Eve has one chance in two of orienting her analyzer in the right direction, Alice and Bob will register a difference in 25% of cases and conclude that the message has been intercepted. The use of two complementary bases maximizes the security of the BB84 protocol. Of course, this discussion is greatly simplified. It does not take into account the possibilities of errors which must be corrected, and moreover it is based on recording impacts of isolated photons, while in practice one sends packets of coherent states with a small ($\langle n \rangle \sim 0.1$) average number of photons by using an attenuated laser beam.[8] Nevertheless, the method is correct in principle, and, to this day, two devices capable of realizing transmissions over several tens of kilometers are available on the market.

3.2 Spin 1/2

3.2.1 Angular momentum and magnetic moment in classical physics

Our second example of an elementary quantum system will be that of spin 1/2. Since for such a system there is no classical wave limit as there is in the case of the photon, our classical discussion will be much shorter than that of the preceding section. We consider a particle of mass m and charge q describing a closed orbit in the field of a central force (Fig. 3.6). We denote the position and momentum of this particle as $\vec{r}(t)$ and $\vec{p}(t)$.

---

[8] In the case of the transmission of isolated photons, the theorem of quantum cloning (Section 6.4.2) guarantees that it is impossible for Eve to fool Bob. However, Eve can slightly reduce her error rate by using a more sophisticated method: see Exercise 15.5.3.

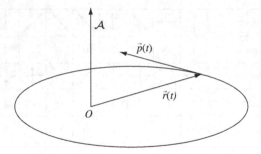

Fig. 3.6. The gyromagnetic ratio.

Let $\mathrm{d}\vec{A}$ be the oriented element of area swept out by the radius vector. It satisfies the relation

$$\frac{\mathrm{d}\vec{A}}{\mathrm{d}t} = \frac{1}{2m}\vec{r} \times \vec{p} = \frac{1}{2m}\vec{j},$$

where $\vec{j}$ is the angular momentum. We recall that for motion in a central force field, the angular momentum is a fixed vector perpendicular to the orbital plane. Integrating over a period, we can relate the total oriented area of the orbit $\vec{A}$ to $\vec{j}$ and to the period T:

$$\vec{A} = \frac{T}{2m}\vec{j}.$$

The current induced by the charge is $I = q/T$ because the charge q passes a given point $1/T$ times per second, and the magnetic moment $\vec{\mu}$ induced by this current will be

$$\boxed{\vec{\mu} = I\vec{A} = \frac{q}{2m}\vec{j} = \gamma\vec{j}} \; . \tag{3.30}$$

The *gyromagnetic ratio* γ defined by (3.30) is $q/2m$. The motion of the electrons inside an atom gives rise to atomic magnetism and the motion of protons inside atomic nuclei gives rise to nuclear magnetism. However, the motion of the charges cannot quantitatively explain either atomic magnetism or nuclear magnetism. It must be assumed that particles have an intrinsic magnetism. Experiment shows that elementary particles of nonzero spin carry a magnetic moment associated with an intrinsic angular momentum, called the *spin* of the particle, which we denote as $\vec{s}$. We can try to represent this angular momentum intuitively as arising from rotation of the particle about its axis. Such a picture may be useful, but it should not be taken very seriously, as it leads to insurmountable contradictions if pushed too far. Only quantum mechanics can give a correct description of spin. Experiments show that the electron, the proton, and the neutron have spin $\frac{1}{2}\hbar$. The factor $\hbar$ is often omitted, and it is simply said that the electron, proton, and neutron are spin-1/2 particles. The gyromagnetic ratio associated with spin is different from (3.30). For example, for the electron[9] and the proton we have

$$\text{electron}: \; \gamma_e = 2\frac{q_e}{2m_e}, \quad \text{proton}: \; \gamma_p = 5.59\frac{q_p}{2m_p},$$

where $(q_e, q_p = -q_e)$ and (m_e, m_p) are the charges and masses of the electron and proton. Moreover, even though its charge is zero, the neutron possesses a magnetic moment. Its gyromagnetic ratio is given by

$$\gamma_n = -3.83\frac{q_p}{2m_p}.$$

Atomic magnetism arises from the electron motion (orbital magnetism) combined with the magnetism associated with the electron spin. The magnetism of atomic nuclei arises from the proton motion and the magnetism associated with the spins of the neutrons and protons. Equation (3.30) shows that the gyromagnetic ratio is inversely proportional to the mass: magnetism of nuclear origin is weaker than that of electron origin by a factor $\sim m_e/m_p \approx 1/1000$. In spite of this suppression, nuclear magnetism is of great practical importance as it lies at the basis of nuclear magnetic resonance (NMR; see Section 5.2.3) and derived technologies such as magnetic resonance imaging (MRI).

Let us use classical physics to study the motion of a magnetic moment $\vec{\mu}$ in a constant magnetic field $\vec{B}$. This magnetic moment is subject to a torque $\vec{\Gamma} = \vec{\mu} \times \vec{B}$, and the equation of motion is

$$\frac{d\vec{s}}{dt} = \vec{\mu} \times \vec{B} = \frac{q}{2m}\vec{s} \times \vec{B} = -\frac{qB}{2m}\hat{B} \times \vec{s}. \tag{3.31}$$

This equation implies that $\vec{s}$ and $\vec{\mu}$ rotate about $\vec{B}$ with constant angular speed $\omega = -qB/2m$ called the *Larmor frequency*. It is convenient to assign an algebraic value to ω: the rotation occurs in the counterclockwise sense for $q < 0$ ($\omega > 0$). This rotational motion is called Larmor precession (Fig. 3.7).

3.2.2 The Stern–Gerlach experiment and Stern–Gerlach filters

The experiment performed by Stern and Gerlach in 1921 is shown schematically in Fig. 3.8. A beam of silver atoms leaves an oven and is collimated by two slits, then passes between the poles of a magnet with the magnetic field pointing in the Oz direction.[10] The magnetic field is nonuniform: B_z is a function of z. A silver atom possesses a magnetic moment due to that of its valence electron. From the point of view of the magnetic forces, it is just as though an electron were passing through the magnet gap. However, the dynamics is simplified owing to the absence of the Lorentz force, as the silver atom is electrically neutral; moreover, the electron mass is replaced by the atomic mass. The

---

[9] Up to corrections of order 0.1%, which can be calculated using quantum electrodynamics.
[10] The reader will note that the orientation of the axes is different from that in the preceding section; the direction of propagation is now the Oy direction. This new choice is made in order to conform with the usual conventions.

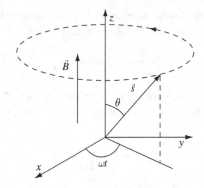

Fig. 3.7. Larmor precession: the spin $\vec{s}$ precesses about $\vec{B}$ with angular frequency ω.

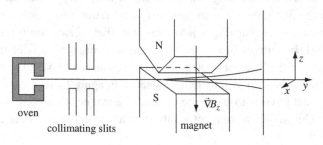

Fig. 3.8. The Stern–Gerlach experiment.

potential energy U of a magnetic moment in $\vec{B}$ is $U = -\vec{\mu} \cdot \vec{B}$, and the corresponding force is

$$\vec{F} = -\vec{\nabla}U, \quad F_z = \mu_z \frac{\partial B_z}{\partial z}. \tag{3.32}$$

In reality, $\vec{B}$ cannot be strictly parallel to Oz; if $\vec{B} = (0, 0, B)$, $\partial B/\partial z \neq 0$ is incompatible with the Maxwell equation $\vec{\nabla} \cdot \vec{B} = 0$. A complete justification of (3.32) can be found in Exercise 9.7.13, where it is shown that this expression gives the effective force on an atom. When the magnetic field is zero, the atoms arrive in the vicinity of a point on the screen and form a spot of finite size owing to their velocity spread, as they are not perfectly collimated. The orientation of the magnetic moments at the exit of the oven is a priori random, and when a magnetic field is present we would expect the spot to be larger: the atoms with magnetic moment $\vec{\mu}$ antiparallel to Oz should undergo maximal upward deflection for $(\partial B_z/\partial z) < 0$, while those with $\vec{\mu}$ parallel to Oz should undergo maximal downward deflection, with all intermediate deflections being possible. But in fact it is observed experimentally that there are *two* spots symmetrically located about the point of arrival in the absence of a magnetic field. It is as though μ_z, and thus s_z, could take two and only two values, and we find[11] that they correspond to $s_z = \pm \hbar/2$, i.e., s_z is *quantized*. We note that since the gyromagnetic ratio is negative ($\gamma < 0$), upward (downward) deflection corresponds to $s_z > 0$ (< 0). The Stern–Gerlach apparatus acts like the birefringent plate of Fig. 3.2: at the exit of the device the atom follows a trajectory[12] on which its spin points either up, $s_z = +\hbar/2$, or down, $s_z = -\hbar/2$. The analogy with photon polarization suggests that the space of spin-1/2 states is a two-dimensional vector space, which is in fact the case. A possible basis in this space is formed by the two vectors $|+\rangle$ and $|-\rangle$ describing the physical states obtained by selecting atoms deflected upward or downward by the Stern–Gerlach device and respectively corresponding to $s_z = +\hbar/2$ and $-\hbar/2$. The states $|+\rangle$ and $|-\rangle$ are called "spin up" and "spin down." These spin states are the analog of the two orthogonal polarization states $|\Phi\rangle$ and $|\Phi_\perp\rangle$ in the case of photons.[13]

The apparatus shown schematically in Fig. 3.9 can be used to recombine atoms deflected upward or downward along a single trajectory, just as the set of two birefringent plates of Fig. 3.3 allows the trajectories of photons polarized in the Ox and Oy directions to be recombined. This apparatus, which we shall refer to as a Stern–Gerlach filter, was not actually realized experimentally by Stern and Gerlach. It was imagined 40 years later by Wigner, and it allows us to illustrate the following theoretical argument. If two Stern–Gerlach filters are located one after the other with the same orientation of $\vec{B}$ and, for example, the two lower paths are blocked (Fig. 3.10(a)), then it can be stated that 100% of the atoms that pass through the first filter will also be transmitted by the second, just as a photon selected by a polarizer oriented in the Ox direction is transmitted with 100% probability by an analyzer of the same orientation. If, on the other hand, the lower path is blocked in the first filter and the upper one in the second filter (Fig. 3.10(b)), then not a single atom is transmitted, just as no photons are transmitted if the analyzer and polarizer are orthogonal. As in the preceding section, these results can be expressed by

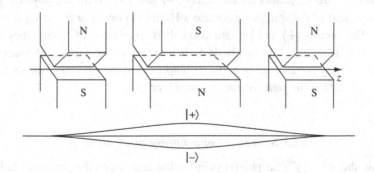

Fig. 3.9. A Stern–Gerlach filter.

---

[11] Knowledge of $\partial B_z / \partial z$ and γ makes it possible in principle to obtain s_z from the deflection; see Exercise 9.7.13.

[12] It can be shown (Exercise 9.7.13) that the trajectories can be treated classically.

[13] This analogy should not be pushed too far; as we shall see in Chapter 10, the photon has spin $\hbar$, not $\hbar/2$. Spin $\hbar$ normally has three possible polarization states. However, in the case of the photon there are only two because the photon is massless.

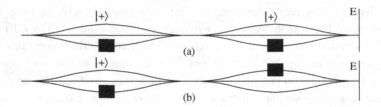

Fig. 3.10. Stern–Gerlach filters in series.

writing the probability amplitudes $a(+ \rightarrow +)$ and $a(+ \rightarrow -)$ as scalar products of the basis vectors:[14]

$$a(+ \rightarrow +) = \langle +|+\rangle = 1, \quad a(- \rightarrow -) = \langle -|-\rangle = 1, \quad a(+ \rightarrow -) = \langle -|+\rangle = 0. \quad (3.33)$$

If the vectors $|+\rangle$ and $|-\rangle$ are represented as column vectors

$$|+\rangle = \begin{pmatrix} 1 \\ 0 \end{pmatrix}, \quad |-\rangle = \begin{pmatrix} 0 \\ 1 \end{pmatrix}, \quad (3.34)$$

the most general (normalized) state vector $|\chi\rangle \in \mathcal{H}$ can be written as

$$|\chi\rangle = \lambda|+\rangle + \mu|-\rangle \quad \text{or} \quad |\chi\rangle = \begin{pmatrix} \lambda \\ \mu \end{pmatrix}. \quad (3.35)$$

The vectors $|+\rangle$ and $|-\rangle$ can be used to construct a Hermitian operator S_z such that these vectors are eigenvectors of S_z with eigenvalues $\pm\hbar/2$:

$$S_z = \frac{1}{2}\hbar\left(|+\rangle\langle +| - |-\rangle\langle -|\right) = \frac{1}{2}\hbar\left(\mathcal{P}_+ - \mathcal{P}_-\right) = \frac{1}{2}\hbar\begin{pmatrix} 1 & 0 \\ 0 & -1 \end{pmatrix}, \quad (3.36)$$

where $\mathcal{P}_+$ and $\mathcal{P}_-$ are projectors on the states $|+\rangle$ and $|-\rangle$. With the physical property $\mathcal{S}_z$, the z component of the spin, we associate a Hermitian operator S_z acting in the space of states $\mathcal{H}$. The vectors $|+\rangle$ and $|-\rangle$ are also called *eigenstates of S_z*, and they form the basis in which S_z is diagonal. In this basis S_z is represented by a diagonal matrix (3.36). The physical property corresponding to the z component of the spin takes the well-defined value $+\hbar/2$ or $-\hbar/2$ if the state vector $|\chi\rangle$ is $|+\rangle$ or $|-\rangle$.

3.2.3 Spin states of arbitrary orientation

Let us pursue the analogy with photon polarization and rotate the magnetic field in the Stern–Gerlach filter so that it points in the $\hat{n}$ direction. Then only the magnetic field component $B_{\hat{n}} = \vec{B} \cdot \hat{n}$ is nonzero. With this new orientation the Stern–Gerlach filter will produce states denoted as $|+, \hat{n}\rangle$ and $|-, \hat{n}\rangle$ which are obtained by selecting atoms

---

[14] More rigorously, we know only that $|a(+ \rightarrow +)| = |a(- \rightarrow -)| = 1$, but a suitable choice of phase always leads to (3.33).

deflected respectively in the direction of $\hat{n}$ and opposite to it.[15] By analogy with the case of photons, we say that the spin 1/2 is *polarized* in the direction $+\hat{n}$ or $-\hat{n}$. We proceed as in the discussion of photon polarization, with the first Stern–Gerlach filter acting as the polarizer; its magnetic field is oriented in the Oz direction and selects spins in the state $|+\rangle$. The second filter has its magnetic field oriented in the $\hat{n}$ direction and acts as the analyzer. It allows experimental measurement of the probabilities $\mathsf{p}(+ \to [+, \hat{n}]) = |\langle +, \hat{n}|+\rangle|^2$ and $\mathsf{p}(+ \to [-, \hat{n}]) = |\langle -, \hat{n}|+\rangle|^2$; as in the preceding section, we assume that these probabilities are given by the squared modulus of a scalar product. Like the states[16] $|+\rangle$ and $|-\rangle$, the states $|+, \hat{n}\rangle$ and $|-, \hat{n}\rangle$ are orthogonal: $\langle +, \hat{n}|-, \hat{n}\rangle = 0$. If the polarizer and analyzer are oriented in the same direction, a state prepared by the polarizer is transmitted with 100% probability by the analyzer. If their orientations are opposite[17] there is 0% transmission probability. The result of testing the polarization is certain. If the directions are not the same, we observe only a certain transmission probability. Just as the bases of photon polarization states $\{|x\rangle, |y\rangle\}$ and $\{|\theta\rangle, |\theta_\perp\rangle\}$ are incompatible (Section 3.1.2), the bases $\{|+\rangle, |-\rangle\}$ and $\{|+, \hat{n}\rangle, |-, \hat{n}\rangle\}$ are incompatible for states of spin 1/2.

Now let us determine the transmission probabilities using the invariance under rotation, i.e., the fact that the physics of the problem cannot depend on the orientation of the axes. The first consequence of this invariance is that the Oz direction is in no way special, and so there must exist a Hermitian operator $S_{\hat{n}} = \vec{S} \cdot \hat{n}$, the spin projection on the $\hat{n}$ axis, which has eigenvalues $\hbar/2$ and $-\hbar/2$ and takes the form (3.36) in a basis $\{|+, \hat{n}\rangle, |-, \hat{n}\rangle\}$ which we must determine. The operator $S_{\hat{n}}$ is written as a function of its eigenvalues and eigenvectors as

$$S_{\hat{n}} = \frac{1}{2}\hbar\Big(|+, \hat{n}\rangle\langle +, \hat{n}| - |-, \hat{n}\rangle\langle -, \hat{n}|\Big). \tag{3.37}$$

We introduce the concept of the *expectation value* of the spin component in the $\hat{n}$ direction, which we denote $\langle S_{\hat{n}}\rangle$. Since deflection in the direction $\pm\hat{n}$ corresponds to a value $s_{\hat{n}} = \pm\hbar/2$ when the spin is in an arbitrary state $|\chi\rangle$, this expectation value, denoted $\langle S_{\hat{n}}\rangle$, will be given by

$$\begin{aligned}
\langle S_{\hat{n}}\rangle &= \frac{1}{2}\hbar\Big(\mathsf{p}(\chi \to [+, \hat{n}]) - \mathsf{p}(\chi \to [-, \hat{n}])\Big) \\
&= \frac{1}{2}\hbar\Big(\langle \chi|+, \hat{n}\rangle\langle +, \hat{n}|\chi\rangle - \langle \chi|-, \hat{n}\rangle\langle -, \hat{n}|\chi\rangle\Big) \\
&= \langle \chi|\frac{1}{2}\hbar\Big(|+, \hat{n}\rangle\langle +, \hat{n}| - |-, \hat{n}\rangle\langle -, \hat{n}|\Big)|\chi\rangle \\
&= \langle \chi|S_{\hat{n}}|\chi\rangle. \tag{3.38}
\end{aligned}$$

---

[15] This presupposes that we know how to change the electron propagation direction to make it orthogonal to $\hat{n}$. Since we are discussing a "thought experiment," we shall not dwell on how this can be done in practice.

[16] Thus $|+\rangle$ and $|-\rangle$ are shorthand notations for $|+, \hat{z}\rangle$ and $|-, \hat{z}\rangle$.

[17] And not orthogonal as in the case of photons!

The matrix representing $S_{\hat{n}}$ in the basis (3.34) in which S_z is diagonal is a priori given by the most general Hermitian 2×2 matrix with eigenvalues $\pm\hbar/2$:

$$S_{\hat{n}} = \frac{1}{2}\hbar \begin{pmatrix} a & b \\ b^* & c \end{pmatrix} = \frac{1}{2}\hbar A, \tag{3.39}$$

where a and c are real numbers. The equation for the eigenvalues $\lambda_\pm$ of the matrix A is

$$\lambda^2 - (a+c)\lambda + ac - |b|^2 = 0.$$

We must have $\lambda_+ + \lambda_- = 0$ and $\lambda_+\lambda_- = -1$, and so

$$a + c = 0, \quad ac - |b|^2 = -1 \Rightarrow a^2 + |b|^2 = 1.$$

We parametrize a and b using the two angles α and β: $a = \cos\beta$ and $b = \exp(-i\alpha)\sin\beta$. Then for $S_{\hat{n}}$ we find

$$S_{\hat{n}} = \frac{1}{2}\hbar \begin{pmatrix} \cos\beta & e^{-i\alpha}\sin\beta \\ e^{i\alpha}\sin\beta & -\cos\beta \end{pmatrix}, \tag{3.40}$$

where the eigenvectors up to a phase are (cf. (2.35))

$$|+, \hat{n}\rangle = \begin{pmatrix} e^{-i\alpha/2}\cos\beta/2 \\ e^{i\alpha/2}\sin\beta/2 \end{pmatrix}, \quad |-, \hat{n}\rangle = \begin{pmatrix} -e^{-i\alpha/2}\sin\beta/2 \\ e^{i\alpha/2}\cos\beta/2 \end{pmatrix}. \tag{3.41}$$

3.2.4 Rotation of spin 1/2

We still need to find a geometrical interpretation for the angles α and β. We shall hypothesize that the expectation value $\langle \vec{S} \rangle$, which has components $(\langle S_x \rangle, \langle S_y \rangle, \langle S_z \rangle)$, transforms under rotation as a vector in a three-dimensional space, that is, as the corresponding classical object $\vec{s}$. Again we use the polarizer/analyzer experiment. First we have the magnetic fields of the polarizer and the analyzer point in the Oz direction. We know that in 100% of cases the spins pass through the analyzer. If the field of the analyzer is oriented antiparallel to Oz none of the spins is transmitted. We can express this result as follows. At the exit of the polarizer the expectation value of S_z, that is, $\langle S_z \rangle$, is equal to $\hbar/2$. Now we orient the magnetic field of the analyzer in the Ox direction. It can be verified experimentally that the spins now have one chance in two of being deflected toward positive x and one chance in two of being deflected toward negative x, which corresponds to expectation value of S_x equal to zero: $\langle S_x \rangle = 0$. This result is not unexpected. One argument for it is based on classical reasoning: a classical spin parallel to Oz is not deflected by a field gradient in the Ox direction. A second, more general argument is based on rotational invariance.[18] In our problem the spin variables are decoupled from the spatial variables associated with the propagation of the atom and, for spin rotations,

---

[18] It is also possible to invoke parity invariance without resorting to the decoupling of the spin and spatial variables; see Exercise 9.7.13.

the system is invariant under rotations about the Oz direction: in the absence of a privileged direction in the xOy plane, $\langle S_x \rangle = \langle S_y \rangle = 0$. The vector $\langle \vec{S} \rangle$ then has components $(0, 0, \hbar/2)$.

Let us now suppose that the experimentalist decides to use the set of axes $x'Oz'$ obtained from xOz by a rotation of angle $-\theta$ about Oy (Fig. 3.11(a)). If $\langle \vec{S} \rangle$ is a vector, its components in the new set of axes will be $\frac{1}{2}\hbar(\sin\theta, 0, \cos\theta)$. An equivalent physical situation is obtained by keeping the original set of axes and orienting the magnetic field gradient of the polarizer in the direction making an angle θ with Oz (Fig. 3.11(b)).[19] The polarizer then prepares the spins in a state which we denote $|+, \hat{n}_\theta\rangle$. The expectation values become

$$\langle S_x \rangle = \langle +, \hat{n}_\theta | S_x | +, \hat{n}_\theta \rangle = \frac{\hbar}{2}\sin\theta, \quad \langle S_z \rangle = \langle +, \hat{n}_\theta | S_z | +, \hat{n}_\theta \rangle = \frac{\hbar}{2}\cos\theta. \quad (3.42)$$

In general, the magnetic field $\vec{B}$ of the polarizer can be oriented in any direction $\hat{n}$: the polarizer prepares the spins in the state $|+, \hat{n}\rangle$. Let θ and ϕ be the polar and azimuthal angles defining the direction of $\hat{n}$ (Fig. 3.12). Direct generalization of the preceding argument shows that the expectation values of $\vec{S}$ then become

$$\langle S_x \rangle = \langle +, \hat{n} | S_x | +, \hat{n} \rangle = \frac{\hbar}{2}\sin\theta\cos\phi = \frac{\hbar}{2}n_x,$$

$$\langle S_y \rangle = \langle +, \hat{n} | S_y | +, \hat{n} \rangle = \frac{\hbar}{2}\sin\theta\sin\phi = \frac{\hbar}{2}n_y, \quad (3.43)$$

$$\langle S_z \rangle = \langle +, \hat{n} | S_z | +, \hat{n} \rangle = \frac{\hbar}{2}\cos\theta = \frac{\hbar}{2}n_z,$$

or, in vector notation,

$$\langle \vec{S} \rangle = \langle +, \hat{n} | \vec{S} | +, \hat{n} \rangle = \frac{\hbar}{2}\hat{n}. \quad (3.44)$$

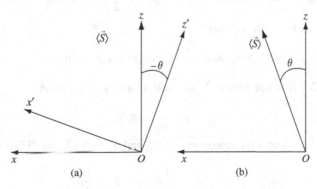

Fig. 3.11. (a) $\langle \vec{S} \rangle$ in two sets of axes. (b) Rotation of $\langle \vec{S} \rangle$.

[19] We shall see in Section 8.1.1 that this amounts to going from a passive to an active point of view for a symmetry operation.

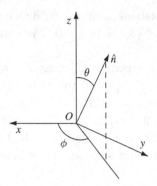

Fig. 3.12. Orientation of $\hat{n}$.

We went through a rather detailed and lengthy argument leading to (3.44), but we could have taken a shortcut by noting that the only vector at our disposal is $\hat{n}$, and $\langle \vec{S} \rangle$ is necessarily parallel to $\hat{n}$, whence (3.44). Let us now calculate the expectation values taking into account (3.41):

$$\langle S_z \rangle = \frac{\hbar}{2} \left(\cos^2 \beta/2 - \sin^2 \beta/2 \right) = \frac{\hbar}{2} \cos \beta.$$

We must therefore have $\beta = \pm\theta$. We choose the solution $\beta = \theta$ and calculate the matrices representing S_x and S_y in the basis (3.34). Since $\theta = \beta = \pi/2$ in both cases, (3.40) becomes

$$S_x = \frac{1}{2}\hbar \begin{pmatrix} 0 & e^{-i\alpha_x} \\ e^{i\alpha_x} & 0 \end{pmatrix}, \quad S_y = \frac{1}{2}\hbar \begin{pmatrix} 0 & e^{-i\alpha_y} \\ e^{i\alpha_y} & 0 \end{pmatrix}.$$

This gives the expectation values

$$\langle S_x \rangle = \frac{1}{2}\hbar \sin \theta \cos(\alpha - \alpha_x), \quad \langle S_y \rangle = \frac{1}{2}\hbar \sin \theta \cos(\alpha - \alpha_y).$$

By identification with (3.43) we obtain

$$\cos(\alpha - \alpha_x) = \cos \phi, \quad \cos(\alpha - \alpha_y) = \sin \phi. \tag{3.45}$$

The solution of (3.45) is not unique;[20] we shall adopt by convention

$$\alpha_x = 0, \quad \alpha_y = \pi/2.$$

With this choice $\alpha = \phi$ and the operators S_x, S_y, and S_z in the basis (3.34) take the form

$$S_x = \frac{1}{2}\hbar\sigma_x, \quad S_y = \frac{1}{2}\hbar\sigma_y, \quad S_z = \frac{1}{2}\hbar\sigma_z. \tag{3.46}$$

---

[20] The other solutions correspond to the set of axes obtained by rotating the Ox and Oy axes about Oz, or to the set of axes obtained by inversion of Oy; cf. Exercise 3.3.4.

The matrices σ_x, σ_y, and σ_z are called the *Pauli matrices*:

$$\sigma_x = \begin{pmatrix} 0 & 1 \\ 1 & 0 \end{pmatrix}, \quad \sigma_y = \begin{pmatrix} 0 & -i \\ i & 0 \end{pmatrix}, \quad \sigma_z = \begin{pmatrix} 1 & 0 \\ 0 & -1 \end{pmatrix}. \tag{3.47}$$

These matrices satisfy the following important, frequently used relations:

$$\sigma_x^2 = \sigma_y^2 = \sigma_z^2 = I, \quad \sigma_x\sigma_y = i\sigma_z \text{ and permutations,} \tag{3.48}$$

which can be written compactly as

$$\sigma_i\sigma_j = \delta_{ij} + i\sum_k \varepsilon_{ijk}\sigma_k, \tag{3.49}$$

where the indices (i, j, k) take the values (x, y, z), and ε_{ijk} is the completely antisymmetric tensor, equal to $+1$ if (ijk) is a cyclic permutation of (xyz), -1 for a noncyclic permutation, and zero otherwise.[21] An equivalent form of (3.49) is the following: if $\vec{a}$ and $\vec{b}$ are two vectors, then

$$(\vec{\sigma} \cdot \vec{a})(\vec{\sigma} \cdot \vec{b}) = \vec{a} \cdot \vec{b} + i\vec{\sigma} \cdot (\vec{a} \times \vec{b}), \tag{3.50}$$

which is readily deduced from the form of the vector product

$$(\vec{a} \times \vec{b})_i = \sum_{j,k} \varepsilon_{ijk} a_j b_k. \tag{3.51}$$

Equation (3.49) also implies the commutation relations[22]

$$[\sigma_i, \sigma_j] = 2i\sum_k \varepsilon_{ijk}\sigma_k, \tag{3.52}$$

or equivalently for the spin components

$$[S_i, S_j] = i\hbar \sum_k \varepsilon_{ijk} S_k. \tag{3.53}$$

The Pauli matrices together with the identity matrix I form a basis for the vector space of matrices on $\mathcal{H}$. Any 2×2 matrix can be written as

$$A = \lambda_0 I + \sum_i \lambda_i \sigma_i, \tag{3.54}$$

where the coefficients λ_0 and λ_i are real for a Hermitian matrix $A = A^\dagger$ and are given by (Exercise 3.3.5)

$$\lambda_0 = \frac{1}{2}\mathrm{Tr}\,A, \quad \lambda_i = \frac{1}{2}\mathrm{Tr}\,A\sigma_i. \tag{3.55}$$

---

[21] For example, $\varepsilon_{yzx} = 1$, $\varepsilon_{yxz} = -1$, and $\varepsilon_{xxz} = 0$.

[22] If the indices are written out explicitly, we have $[\sigma_x, \sigma_y] = 2i\sigma_z$ along with the two other relations obtained by cyclic permutation of the indices (x, y, z).

Since the Pauli matrices form a basis for the matrices acting in any two-dimensional Hilbert space, they are often used in problems where the space of states is two-dimensional, even if the physical situation has nothing to do with spin 1/2. For example, they are very useful for dealing with a common model in atomic physics, that of the "two-level atom" (see Sections 5.4 and 14.4.1).

The eigenvectors $|+, \hat{n}\rangle$ and $|-, \hat{n}\rangle$ of $S_{\hat{n}} = \frac{1}{2}\hbar\vec{\sigma} \cdot \hat{n}$ are derived from (3.41) with $\beta = \theta$ and $\alpha = \varphi$:

$$|+, \hat{n}\rangle = \begin{pmatrix} e^{-i\phi/2}\cos\theta/2 \\ e^{i\phi/2}\sin\theta/2 \end{pmatrix}, \quad |-, \hat{n}\rangle = \begin{pmatrix} -e^{-i\phi/2}\sin\theta/2 \\ e^{i\phi/2}\cos\theta/2 \end{pmatrix}. \tag{3.56}$$

The states $|+, \hat{n}\rangle$ and $|-, \hat{n}\rangle$ are obtained by transforming $|+\rangle$ and $|-\rangle$ by a rotation that aligns the Oz azis with $\hat{n}$. A possible choice which is consistent with that which will be made in Chapter 10 is to rotate first by an angle θ about Oy, then rotate by an angle ϕ about Oz. Then (3.56) can be written as

$$\begin{aligned} |+, \hat{n}\rangle &= D_{++}^{(1/2)}(\theta, \phi)|+\rangle + D_{-+}^{(1/2)}(\theta, \phi)|-\rangle, \\ |-, \hat{n}\rangle &= D_{+-}^{(1/2)}(\theta, \phi)|+\rangle + D_{--}^{(1/2)}(\theta, \phi)|-\rangle. \end{aligned} \tag{3.57}$$

This equation defines a matrix $D^{(1/2)}(\theta, \phi)$, called the *rotation matrix* for spin 1/2:[23]

$$D^{(1/2)}(\theta, \phi) = \begin{pmatrix} e^{-i\phi/2}\cos\theta/2 & -e^{-i\phi/2}\sin\theta/2 \\ e^{i\phi/2}\sin\theta/2 & e^{i\phi/2}\cos\theta/2 \end{pmatrix}. \tag{3.58}$$

This matrix is unitary because it performs a change of basis in $\mathcal{H}$. We can also check that it has determinant 1, and so it is a matrix belonging to the group $SU(2)$ (cf. Exercise 8.5.2). It is interesting to consider rotations by 2π, which return the physical system to its initial position. We have, for example, $D^{1/2}(\theta = 2\pi, \phi = 0) = -I$. Under a rotation by 2π about Oy, the state vector $|\chi\rangle \rightarrow -|\chi\rangle$! However, there is no paradox: the vectors $|\chi\rangle$ and $-|\chi\rangle$ represent the same physical state, and, as must be the case, a rotation by 2π does not change this state. This behavior of spin 1/2 contrasts with that of photons. According to (3.28), $\exp(-2i\pi\Sigma_z) = +I$ and the state vector is unchanged under a rotation by 2π. Here we see a remarkable difference between integer and half-integer spins, to which we shall return in Chapter 10.

The form (3.56) of the eigenvectors of $S_{\hat{n}}$ allows the probability amplitudes to be calculated:

$$\begin{aligned} a(+ \rightarrow [+, \hat{n}]) &= \langle +, \hat{n}|+\rangle = e^{i\phi/2}\cos\theta/2 , \\ a(+ \rightarrow [-, \hat{n}]) &= \langle -, \hat{n}|+\rangle = -e^{i\phi/2}\sin\theta/2, \end{aligned} \tag{3.59}$$

---

[23] It should be noted that this matrix is a function of $\theta/2$ and not θ as in the photon case (3.28): the photon has spin 1 rather than 1/2!

along with the corresponding probabilities:

$$p(+ \to [+, \hat{n}]) = |\langle +, \hat{n}|+\rangle|^2 = \cos^2 \theta/2,$$
$$p(+ \to [-, \hat{n}]) = |\langle -, \hat{n}|+\rangle|^2 = \sin^2 \theta/2. \tag{3.60}$$

We have obtained the essential properties of spin 1/2 on the basis of only three hypotheses, with the first two following from invariance under rotation:

• The expectation value $\langle \vec{S} \rangle$ transforms like a vector under rotations.
• The eigenvalues of $\vec{S} \cdot \hat{n}$ are independent of $\hat{n}$.
• The space of states is two-dimensional.

Some of these properties, like the commutation relations (3.53) or the existence of rotation matrices, can be carried over to any angular momentum $\vec{J}$ (Chapter 10). However, other properties are specific to spin 1/2; for example, it is only in this case that any state of $\mathcal{H}$ can be written as an eigenvector of $\vec{J} \cdot \hat{n} = \vec{S} \cdot \hat{n}$ for some $\hat{n}$.

3.2.5 Dynamics and time evolution

Let us return to the problem of a spin placed in a uniform constant magnetic field $\vec{B}$, which we assume to be oriented along the z axis. Our classical study of Section 3.2.1 revealed the phenomenon of Larmor precession. In classical physics, the energy is a number

$$U = -\vec{\mu} \cdot \vec{B} = -\gamma \vec{s} \cdot \vec{B} = -\gamma s_z B = \omega s_z, \tag{3.61}$$

where $\omega = -\gamma B$ is the Larmor frequency. In quantum physics the energy becomes a Hermitian operator called the *Hamiltonian* and denoted H which acts in the space of states. Since this space is two-dimensional, the Hamiltonian will be represented by a 2×2 matrix. We *assume*[24] that in quantum mechanics the Hamiltonian formally remains of the form (3.61), with the condition that the classical quantity s_z is replaced by the operator S_z, the projection on Oz of the spin operator $\vec{S}$:

$$H = \omega S_z = \frac{\omega}{2} \hbar \begin{pmatrix} 1 & 0 \\ 0 & -1 \end{pmatrix}. \tag{3.62}$$

Here the second form of H is its matrix representation in a basis in which S_z is diagonal. The eigenvalues of H are $+\hbar\omega/2$ and $-\hbar\omega/2$. These are the two possible values of the energy, and the corresponding eigenvectors are of course those of S_z: $|+\rangle$ and $|-\rangle$. The energy-level scheme is given in Fig. 3.13 for $\omega > 0$, and the two levels are called the *Zeeman levels* of a spin 1/2 in a magnetic field.

Let us assume that at time $t = 0$ the spin is found in the eigenstate $|+, \hat{n}\rangle$. We can then ask the following question: what will the spin state be at a later time t? To answer this question we need an additional postulate. This postulate, whose details will be made

---

[24] In the end, the expression for the Hamiltonian will be justified by agreement with experiment.

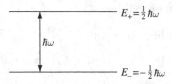

Fig. 3.13. Spectrum of the Hamiltonian (3.62), or Zeeman levels of a spin 1/2 in a magnetic field.

more explicit in the following chapter, stipulates that the state vector $|\chi(t)\rangle$ at time t is derived from the state vector at time $t = 0$, $|\chi(t = 0)\rangle$, as follows:

$$|\chi(t)\rangle = \exp\left(-\frac{iHt}{\hbar}\right)|\chi(0)\rangle. \tag{3.63}$$

This evolution law is particularly simple for eigenvectors of H, which are called *stationary states*:

$$|+\rangle \rightarrow \exp\left(-\frac{i\omega t}{2}\right)|+\rangle, \quad |-\rangle \rightarrow \exp\left(\frac{i\omega t}{2}\right)|-\rangle.$$

If $|\psi\rangle$ is an arbitrary state, the probability of finding a stationary state in $|\psi\rangle$ is independent of time. For example,

$$\left|\langle\psi|\exp\left(-\frac{iHt}{\hbar}\right)|+\rangle\right|^2 = |\langle\psi|+\rangle|^2.$$

Let us suppose that a spin points in the direction $\hat{n}$ at time $t = 0$:

$$|\chi(0)\rangle = \cos\frac{1}{2}\theta\exp(-i\phi/2)|+\rangle + \sin\frac{1}{2}\theta\exp(i\phi/2)|-\rangle.$$

At time t we have

$$|\chi(t)\rangle = \cos\frac{1}{2}\theta\exp[-i(\phi+\omega t)/2]|+\rangle + \sin\frac{1}{2}\theta\exp[i(\phi+\omega t)/2]|-\rangle. \tag{3.64}$$

If at time $t = 0$ the spin points in a direction $\hat{n}$ defined by the angles θ and ϕ, $\langle\vec{S}\rangle = \frac{1}{2}\hbar\hat{n}$, at time t the spin will point in the direction $(\theta, \phi + \omega t)$. The rotation is in the counterclockwise sense for $q < 0$ and, of course, coincides with that of the classical spin. The expectation value of the spin precesses about $\vec{B}$ with the Larmor frequency.

The evolution law (3.64) allows us to introduce a relation between the energy spread ΔE and the characteristic evolution time of a quantum system, which will be written in the general form of a *temporal Heisenberg inequality* in Section 4.2.4. We rewrite (3.64) using the notation c_+ and c_- for the components of $|\chi(0)\rangle$ in the basis $\{|+\rangle, |-\rangle\}$:

$$c_+ = \cos\frac{1}{2}\theta\exp(-i\phi/2), \quad c_- = \sin\frac{1}{2}\theta\exp(i\phi/2),$$

and we define the frequencies $\omega_\pm$ as

$$\omega_+ = \frac{E_+}{\hbar} = +\frac{1}{2}\omega, \quad \omega_- = \frac{E_-}{\hbar} = -\frac{1}{2}\omega,$$

so that for $|\chi(t)\rangle$ we have

$$|\chi(t)\rangle = c_+ \exp(-i\omega_+ t)|+\rangle + c_- \exp(-i\omega_- t)|-\rangle.$$

Let us calculate the probability of finding the state vector $|\chi(t)\rangle$ in an arbitrary state $|\psi\rangle$:

$$|\langle\psi|\chi(t)\rangle|^2 = |c_+|^2 |\langle\psi|+\rangle|^2 + |c_-|^2 |\langle\psi|-\rangle|^2$$
$$+ 2\mathrm{Re}\Big[c_+^* c_- \exp[i(\omega_+ - \omega_-)t]\langle +|\psi\rangle\langle\psi|-\rangle\Big]. \qquad (3.65)$$

The first two terms of (3.65) are independent of time and the third oscillates with frequency

$$\omega_+ - \omega_- = \frac{E_+ - E_-}{\hbar} = \frac{\Delta E}{\hbar},$$

where ΔE is the energy spread. The energy of the system does not have a well-defined value because the system evolves from one level to another in a characteristic time $\Delta t \simeq \hbar/\Delta E$. We can express this as a relation between the energy spread and the characteristic evolution time:

$$\Delta E \, \Delta t \simeq \hbar. \qquad (3.66)$$

This expression, which we shall write as an inequality using the more general method of Section 4.2.4, is an example of a temporal Heisenberg inequality.

3.3 Exercises

3.3.1 Decomposition and recombination of polarizations

Figure 3.3 illustrates an experiment in which a birefringent plate decomposes a linear polarization into polarizations in the Ox and Oy directions, with the two polarizations corresponding to distinct light rays. This decomposition is followed by a recombination of the two polarizations by a second plate which restores the initial polarization. In fact, the scheme shown in Fig. 3.3 does not lead to the advertised result, because the indices of refraction of the ordinary ray and the extraordinary ray are different, which leads to a difference in the optical paths of the two rays. It is necessary to compensate for this difference if we wish to recombine the two polarizations. We recall that the extraordinary ray is always polarized in the plane containing the optical axis, while the ordinary ray is polarized in the plane perpendicular to it. The two birefringent plates are assumed to be identical; they are cut from calcite crystals and have thickness a.

1. The extraordinary ray in the calcite plate makes an angle $\alpha = 6.20°$ (0.1082 rad) to the normal. The thickness of the plate is 10 mm and the ordinary and extraordinary indices are

$$n_O = 1.65567, \quad n_E' = 1.55405,$$

respectively.[25] The incident light beam is produced by a helium–neon laser of wavelength $\lambda = 632.8$ nm, and the beam diameter is 250 μm.[26] Are the two rays well separated at the exit of the first plate? What is the difference between the optical paths of the ordinary and extraordinary rays?

2. We want to compensate for this difference in the optical paths, as well as for that induced by the second plate, by inserting an intermediate calcite plate (a compensating plate) with optical axis perpendicular to the plane of Fig. 3.14. In this plate ray x propagates like an ordinary ray and ray y like an extraordinary ray with index $n_E = 1.48465$. What thickness D must this intermediate plate have if we wish to compensate for the difference of the optical paths so as to be able to recombine the two polarizations at the exit of the second plate?

3. Show that a precision of 10^{-5} for the indices is sufficient for determining the thickness of the compensating plate. Compare this with the precision required for the indices if we want to avoid using a compensating plate and instead fix the thicknesses of the entrance and exit plates such that the difference induced in the optical path by the two plates is an integer multiple of the wavelength. In order to simplify the discussion, neglect the difference between n'_E and n_E in the calculation of the error.

4. The apparatus is very sensitive to temperature variations owing to expansion of the calcite and variation of the indices. In order to simplify the discussion, we shall limit ourselves to the effects of variation of the indices, which are

$$\delta n_O = 2.1 \times 10^{-6}\,\mathrm{K}^{-1}, \quad \delta n_E = 11.9 \times 10^{-6}\,\mathrm{K}^{-1}.$$

We assume that the compensation is perfect at a particular temperature T. Then what will be the total difference in the optical paths (induced by the three plates) if the temperature varies by 1 degree? What will happen if a compensating plate is not used?

5. Now let the first plate have a thickness of 2 mm. Describe the polarization at the exit of this plate.

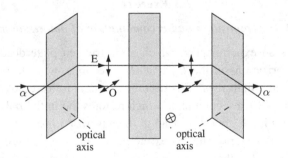

Fig. 3.14. Compensation of the phase shift by an intermediate plate. The optical axis of the intermediate plate is perpendicular to the plane of the figure.

---

[25] The value of n'_E has been calculated using the ellipsoid of indices.
[26] In fact, this diameter $w(z)$ is not constant, but varies as

$$w(z) = w_0\sqrt{1 + \left(\frac{z}{z_R}\right)^2},$$

where $z_R \simeq 0.31$ m and w_0 is the minimum diameter or waist of the beam. If the entire apparatus is about 10 cm long, this variation in diameter is negligible if the waist is located at the center of the apparatus.

3.3.2 Elliptical polarization

1. Determine the axes of the ellipse and the direction in which it is traced for a polarization state (3.12):

$$|\Phi\rangle = \lambda|x\rangle + \mu|y\rangle, \quad |\lambda|^2 + |\mu|^2 = 1.$$

2. Show that the state $|\Phi_\perp\rangle$ (3.19) orthogonal to $|\Phi\rangle$,

$$|\Phi_\perp\rangle = -\mu^*|x\rangle + \lambda^*|y\rangle,$$

   is not transmitted by the linear polarizer of the (λ, μ) polarizer.

3. Show that the physical properties of the (λ, μ) polarizer are unchanged if a general parametrization with complex λ and μ is used:

$$\lambda = \cos\theta e^{i\eta_x}, \quad \mu = \sin\theta e^{i\eta_y},$$

   with $\eta = \eta_y - \eta_x$. Recover the expression for $\mathcal{P}_\Phi$.

3.3.3 Rotation operator for the photon spin

Prove (3.28). Hint: expand $\exp(-i\theta\Sigma_z)$ in a series. What is $(\Sigma_z)^2$?

3.3.4 Other solutions of (3.45)

1. In the space of spin-1/2 states, the unitary matrix $D^{(1/2)}(\theta, \psi)$ transforms the state $|+\rangle$ into the state $|+, \hat{n}\rangle$, where the unit vector $\hat{n}$ is given by $\hat{n} = (\sin\theta\cos\psi, \sin\theta\sin\psi, \cos\theta)$. If the rotation is performed about the z axis, $\theta = 0$ in (3.58) and

$$D^{(1/2)}(\theta = 0, \psi) = U = \begin{pmatrix} e^{-i\psi/2} & 0 \\ 0 & e^{i\psi/2} \end{pmatrix}.$$

   Discuss what action U has on the states $|+\rangle$ and $|-\rangle$.

2. The operator U can be considered a change of basis in which an operator A is transformed according to (2.18) into

$$A \rightarrow A' = U^\dagger A U.$$

   What are the transforms of σ_x, σ_y, and σ_z?

3. The conditions (3.45) have the solution (1) $\alpha - \alpha_x = \phi$ or (2) $\alpha - \alpha_x = -\phi$. Show that in case (1), σ_x and σ_y are given by

$$\sigma_x = \begin{pmatrix} 0 & e^{-i\alpha_x} \\ e^{-i\alpha_x} & 0 \end{pmatrix}, \quad \sigma_y = \begin{pmatrix} 0 & -ie^{-i\alpha_x} \\ ie^{-i\alpha_x} & 0 \end{pmatrix},$$

   and that with reference to the standard solution (3.47) this solution corresponds to a simple rotation of the axes about Oz.

4. Show that if we choose $\alpha - \alpha_x = -\phi$ the standard solution is

$$\sigma_x = \begin{pmatrix} 0 & 1 \\ 1 & 0 \end{pmatrix}, \quad \sigma_y = \begin{pmatrix} 0 & i \\ -i & 0 \end{pmatrix}.$$

   What is the interpretation of this result?

3.3.5 Decomposition of a 2×2 matrix

1. We introduce the notation

$$\hat{\sigma}_0 = I, \quad \hat{\sigma}_i = \sigma_i, \quad i = 1, 2, 3.$$

   Show that if a 2×2 matrix A satisfies $\mathrm{Tr}(\hat{\sigma}_i A) = 0 \forall i = 0, \ldots, 3$, then $A = 0$.
2. Let us write a 2×2 matrix as

$$A = \lambda_0 I + \sum_{i=1}^{3} \lambda_i \sigma_i = \sum_{i=0}^{3} \lambda_i \hat{\sigma}_i.$$

   Show that

$$\lambda_i = \frac{1}{2} \mathrm{Tr}(A \hat{\sigma}_i).$$

   Show that any 2×2 matrix can always be written as

$$A = \sum_{i=0}^{3} \lambda_i \hat{\sigma}_i.$$

   What condition must the coefficients λ_i obey when A is Hermitian, $A = A^\dagger$?

3.3.6 Exponentials of Pauli matrices and rotation operators

1. Show that

$$\boxed{\exp\left(-i\frac{\theta}{2}\vec{\sigma} \cdot \hat{p}\right) = I \cos\frac{\theta}{2} - i(\vec{\sigma} \cdot \hat{p}) \sin\frac{\theta}{2}}, \tag{3.67}$$

   where $\hat{p}$ is a unit vector. Hint: calculate $(\vec{\sigma} \cdot \hat{p})^2$. The operator $\exp(-i\theta\vec{\sigma} \cdot \hat{p}/2)$ is the rotation operator $U[\mathcal{R}_{\hat{p}}(\theta)]$ of an angle θ around the $\hat{p}$ axis. To see it, show that in order to rotate the state $|\pm\rangle$ into $|\pm, \hat{n}\rangle$, as in (3.57), one can use as a rotation axis $\hat{p} = (-\sin\phi, \cos\phi, 0)$. Compare with (3.57) and show that $\exp(-i\theta\vec{\sigma} \cdot \hat{p}/2)|\pm\rangle$ gives the correct result, up to an overall, physically irrelevant, phase factor. Compute the operator $U[\mathcal{R}_x(\theta)]$ and give its explicit matrix form.
2. Show that any 2×2 matrix U which is unitary and has unit determinant can be written in the form in question 1 above. Hint: show that U has the form

$$\begin{pmatrix} a & b \\ -b^* & a^* \end{pmatrix}$$

   and write $a = a_1 + ia_2$, $b = b_1 + ib_2$. Show that $a_1 = \cos\theta/2$.
3. Find two 2×2 matrices A and B such that

$$e^A e^B = e^{(A+B)} \quad \text{with} \quad [A, B] \neq 0.$$

3.3.7 The tensor ε_{ijk}

1. Prove the identity

$$\sum_k \varepsilon_{ijk}\varepsilon_{lmk} = \delta_{il}\delta_{jm} - \delta_{im}\delta_{jl}.$$

Use this identity to derive

$$\vec{a} \times (\vec{b} \times \vec{c}) = (a \cdot c)\vec{b} - (\vec{a} \cdot \vec{b})\vec{c}.$$

What is the result for

$$\sum_{jk} \varepsilon_{ijk}\varepsilon_{ljk} \ ?$$

2. The ith component of the curl of a vector $\vec{A}$ can be written as

$$(\vec{\nabla} \times \vec{A})_i = \sum_{i,j} \varepsilon_{ijk}\partial_j A_k,$$

with $\partial_j = \partial/\partial x_j$. Use the identity of question 1 to show that

$$\vec{\nabla} \times \vec{\nabla} \times \vec{A} = \vec{\nabla}(\vec{\nabla} \cdot \vec{A}) - \nabla^2 \vec{A}.$$

3.3.8 A 2π rotation of spin 1/2

Let us return to the neutron interferometer of Exercise 1.6.7, where the plane $ABDC$ is horizontal and θ_B is a Bragg angle. A variable phase shift χ is obtained by having the neutrons of beam I pass through a uniform constant magnetic field $\vec{B}$ over a distance l, where the magnetic field is perpendicular to the plane of the figure (Fig. 3.15).[27] The neutrons are assumed to be polarized parallel to the plane of the figure. Determine the rotation angle of the neutron spin at the exit of the magnetic field as a function of l, the (known) speed v of the neutron, and the neutron gyromagnetic ratio γ_n. Show that

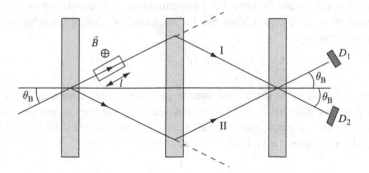

Fig. 3.15. Experimental demonstration of a 2π rotation of spin 1/2.

[27] S. Werner, R. Colella, A. Overhauser, and C. Eagen, Observation of the phase shift of a neutron due to precession in a magnetic field, *Phys. Rev. Lett.* **35**, 1053–1055 (1975).

the counting rates of the detectors D_1 and D_2 depend sinusoidally on B. Show that from these oscillations we can deduce that the spin state vector is multiplied by -1 in a single rotation by 2π.

3.3.9 Neutron scattering by a crystal: spin-1/2 nuclei

Let us revisit the experiment described in Exercise 1.6.4 on neutron diffraction by a crystal, assuming that the atomic nuclei have spin 1/2 (some examples are ^{1}H, ^{13}C, ^{19}F, and so on). We shall limit ourselves at first (questions 1 and 2) to the case where the neutrons have spin up ($\uparrow$) and the nuclei have spin down ($\downarrow$): the neutrons and nuclei are polarized. Under these conditions there are two possible scattering amplitudes, because it can be shown (Chapter 12) that the z component of the total spin is conserved in the neutron–nucleus scattering. These two amplitudes are

- The amplitude f_a where the scattering occurs without change of the spin state:

$$\text{neutron} \uparrow + \text{nucleus} \downarrow \rightarrow \text{neutron} \uparrow + \text{nucleus} \downarrow .$$

- The amplitude f_b where the scattering occurs with spin flip:

$$\text{neutron} \uparrow + \text{nucleus} \downarrow \rightarrow \text{neutron} \downarrow + \text{nucleus} \uparrow .$$

1. Show that in the first case we obtain the same results as in scattering without spin.
2. Show that in the second case there are no diffraction peaks as the scattering probability is independent of $\vec{q}$.
3. In general, nuclei are not polarized, and so they have one chance in two of having spin up and one chance in two of having spin down. It becomes necessary to take into account a third amplitude f_c corresponding to the scattering

$$\text{neutron} \uparrow + \text{nucleus} \uparrow \rightarrow \text{neutron} \uparrow + \text{nucleus} \uparrow .$$

Following the method used in Exercise 1.6.8, we introduce a number α_i that takes the value 0 if the nucleus i has spin up and the value 1 if it has spin down. The ensemble of $\{\alpha_i\}$ characterizes a spin configuration of the crystal. Show that the amplitude for neutron scattering by the crystal in the configuration $\{\alpha_i\}$ is

$$\sum_i (\alpha_i f_a + (1 - \alpha_i) f_c) \, e^{i\vec{q} \cdot \vec{r}_i} + \sum_i \alpha_i f_b e^{i\vec{q} \cdot \vec{r}_i} .$$

What would the intensity be if the configuration $\{\alpha_i\}$ were fixed? Care must be taken to add the probabilities for different final states. In addition, it is necessary to use the average over different crystal configurations, with the spin of each nucleus assumed to be independent of the other spins. If $\langle \bullet \rangle$ denotes the average over configurations, show that

$$\langle \alpha_i \alpha_j \rangle = \frac{1}{4} + \frac{1}{4} \delta_{ij} .$$

Show that the scattering probability is proportional to

$$\mathcal{I} = \frac{1}{4} (f_a + f_c)^2 \sum_{i,j} e^{i\vec{q} \cdot (\vec{r}_i - \vec{r}_j)} + \frac{\mathcal{N}}{4} [(f_a - f_c)^2 + 2 f_b^2] ,$$

where $\mathcal{N}$ is the number of nuclei. In reality, the three amplitudes f_a, f_b, and f_c are not independent. In Exercise 12.5.5 we shall see that

$$-f_a = \frac{1}{2}(a_t + a_s), \quad -f_b = \frac{1}{2}(a_t - a_s), \quad -f_c = a_t,$$

where a_t and a_s are the scattering lengths in the triplet and singlet states.
4. What happens if the neutrons are not polarized, as is usually the case in practice?

3.4 Further reading

The polarization of light and its propagation in anisotropic media are explained in detail in, for example, E. Hecht, *Optics*, New York: Addison-Wesley (1987), Chapter 8. As a complement to the discussion of photon polarization, one can consult Lévy-Leblond and Balibar [1990], Chapter 4, or G. Baym, *Lectures on Quantum Mechanics*, Reading: Benjamin (1969), Chapter 1. A recent journal article on quantum cryptography with numerous references to previous studies is the review by N. Gisin, G. Ribordy, W. Tittel, and H. Zbinden, Quantum cryptography, *Rev. Mod. Phys.* **74**, 145 (2002); a popularized account of quantum cryptography can be found in C. Bennett, G. Brassard, and A. Ekert, Quantum cryptography, *Scientific American*, **26** (October 1992). The Stern–Gerlach experiment is discussed by Feynman *et al.* [1965], vol. III, Chapter 5; by Cohen-Tannoudji *et al.* [1977], Chapter IV; and by Peres [1993], Chapter 1.

4

Postulates of quantum physics

In this chapter we shall present the basic postulates of quantum physics, generalizing the results obtained in the preceding chapter for the two special cases of photon polarization and spin 1/2. In general, the space of states will a priori have any dimension N, which may even be infinite, rather than only two dimensions. The postulates which we present in this chapter fix the general conceptual framework of quantum mechanics and do not directly provide the tools necessary for solving specific problems. The solution of a specific physical problem always involves a modeling stage, where the system to be studied is simplified, the approximations to be used are defined, and so on, and this modeling stage inevitably rests on more or less heuristic arguments which cannot be derived within the general framework of quantum physics.[1] In Section 3.2.5 we gave an example of a heuristic procedure leading to the solution of a specific problem, that of the motion of a spin 1/2 in a magnetic field.

Other sets of postulates can be used. For example, another approach is to state the postulates of quantum mechanics in terms of path integrals.[2] As is often the case, the same physical theory can be dressed in various different mathematical clothes. Finally, it should be emphasized that the postulates of quantum physics give rise to some difficult epistemological problems which are still largely under debate and which we do not discuss in this book. The interested reader may consult, for example, the book by Isham [1995].

4.1 State vectors and physical properties

4.1.1 The superposition principle

In Chapter 3 we learned how to characterize the polarization state of a photon or of a spin-1/2 particle by means of a vector belonging to a complex Hilbert space, the space of states. Postulate **I** generalizes the ideas of state vector and space of states to any quantum system.

---

[1] This procedure does not differ fundamentally from that followed in classical physics. For example, the three laws of Newton fix the conceptual framework of classical mechanics, but the solution of a specific problem always requires some modeling: simplification of the posed problem, approximations for the forces, and so on.

[2] See, for example, L. S. Schulman, *Techniques and Applications of Path Integration*, New York: Wiley (1981).

Postulate I: the space of states

The properties of a quantum system are completely defined by specification of its *state vector* $|\varphi\rangle$, which fixes the mathematical representation of the physical state of the system.[3] The state vector is an element of a complex Hilbert space $\mathcal{H}$ called the *space of states*. It will be convenient to choose $|\varphi\rangle$ to be normalized that is, to have unit norm: $||\varphi||^2 = \langle \varphi|\varphi\rangle = 1$.

The fact that a physical state is represented by a vector implies, under certain conditions, the *superposition principle* characteristic of the *linearity* of the theory: if $|\varphi\rangle$ and $|\chi\rangle$ are vectors of $\mathcal{H}$ representing physical states, the normalized vector

$$|\psi\rangle = \frac{\lambda|\varphi\rangle + \mu|\chi\rangle}{||\lambda|\varphi\rangle + \mu|\chi\rangle||},$$ (4.1)

where λ and μ are complex numbers, is a vector of $\mathcal{H}$ and also represents a physical state.

In the preceding chapter we defined probability amplitudes as scalar products of vectors belonging to the space of states. For example, if $|\varphi\rangle$ represents the state of a photon linearly polarized in the Ox direction, $|\varphi\rangle = |x\rangle$, and $|\chi\rangle$ the state of a photon linearly polarized in the $\hat{n}_\theta$ direction (3.3), $|\chi\rangle = |\theta\rangle$, the probability amplitude $a(x \rightarrow \theta) = \langle\theta|x\rangle = \cos\theta$. We also showed that the squared modulus of this amplitude possesses a remarkable physical interpretation: if we test the polarization by having the photon $|x\rangle$ pass through a linear analyzer oriented in the $\hat{n}_\theta$ direction, we obtain the transmission probability

$$p(x \rightarrow \theta) = |a(x \rightarrow \theta)|^2 = |\langle\theta|x\rangle|^2 = \cos^2\theta,$$

which is the *probability for the photon in the state* $|x\rangle$ *to pass the* $|\theta\rangle$ *test*. We shall generalize the ideas of probability amplitude and testing as postulate **II**.

Postulate II: probability amplitudes and probabilities

If $|\varphi\rangle$ is the vector representing the state of a system and if $|\chi\rangle$ represents another physical state, there exists a *probability amplitude* $a(\varphi \rightarrow \chi)$ *of finding* $|\varphi\rangle$ *in state* $|\chi\rangle$, which is given by a scalar product on $\mathcal{H}$: $a(\varphi \rightarrow \chi) = \langle\chi|\varphi\rangle$. The probability $p(\varphi \rightarrow \chi)$ for the state $|\varphi\rangle$ to pass the $|\chi\rangle$ test is obtained by taking the squared modulus $|\langle\chi|\varphi\rangle|^2$ of this amplitude:[4]

$$\boxed{p(\varphi \rightarrow \chi) = |a(\varphi \rightarrow \chi)|^2 = |\langle\chi|\varphi\rangle|^2}.$$ (4.2)

This postulate is often called the *Born rule*.

---

[3] The viewpoint of the present author is that the state vector describes the physical reality of an *individual* quantum system. This point of view is far from universally shared, and the reader can easily find other interpretations, for example: "the state vector describes the available information on a quantum system," or "the state vector is not a property of an individual physical system, but simply a protocol for preparing a set of such states," or even "quantum mechanics is a set of rules which allow the probability of an experimental result to be calculated." This diversity of viewpoints has no effect on the practical application of quantum mechanics.

[4] To make the order of the factors correspond to that of the scalar product, it is sometimes useful to denote probability amplitudes as $a(\chi \leftarrow \varphi)$ and probabilities as $p(\chi \leftarrow \varphi)$. We also note that although (4.2) is not intuitive, it is at least consistent: the probability of finding a state in itself is unity, and according to the Schwarz inequality $0 \leq |\langle\chi|\varphi\rangle|^2 \leq 1$.

Let us add a few remarks to complete our statement of the first two postulates.

- Unless the contrary is explicitly stated, we assume that state vectors have unit norm. If this is not the case, care must be taken to divide by the norm. For example, Eq. (4.2) becomes

$$p(\varphi \to \chi) = \frac{|\langle \chi | \varphi \rangle|^2}{||\chi||^2 ||\varphi||^2}.$$

- The vectors $|\varphi\rangle$ and $|\varphi'\rangle = \exp(i\beta)|\varphi\rangle$ represent the same physical state. Actually, we know only how to measure probabilities, and

$$|\langle \chi | \varphi \rangle|^2 = |\langle \chi | \varphi' \rangle|^2 \; \forall \; |\chi\rangle \in \mathcal{H}.$$

It is therefore impossible to distinguish between $|\varphi\rangle$ and $|\varphi'\rangle$, which differ by a *phase factor*. To be rigorous, a physical state is represented by a *ray*, or a vector up to a phase, in the Hilbert space. However, the superposition $\lambda|\varphi\rangle + \mu|\chi\rangle$ represents a physical state that is different from $\lambda|\varphi'\rangle + \mu|\chi\rangle$. The answer to the question "Which are the arbitrary phases and which are the physically relevant ones?" may be tricky in some cases.

- We limit ourselves to physical systems called *pure states*, where there is maximal information about the physical state. In cases where the available information is incomplete, we must resort to the state (or density) operator formalism, which will be described in Section 6.2.

- We have taken great care to use the term "quantum system" rather than "quantum particle," which is a special case of the former. In fact, we shall see in Chapter 6 that for a system of two or more particles it is in general impossible to attribute an individual state vector to each particle; a state vector can be associated only with the ensemble of particles, that is, with the whole quantum system. This point will be developed and illustrated in Section 6.3.

- There exist restrictions on the superposition principle called "superselection rules",[5] which we shall not consider in this book.

4.1.2 Physical properties and measurement

In Chapter 3 we showed that the physical property "spin component along the $\hat{n}$ axis" can be put into correspondence with a Hermitian operator $\vec{S} \cdot \hat{n}$ acting in the space of states. Postulate **III** generalizes this result to any physical property.

Postulate III: physical properties and operators

With every physical property $\mathcal{A}$ (energy, position, momentum, angular momentum, and so on) there exists an associated Hermitian operator A which acts in the space of states $\mathcal{H}$: A fixes the mathematical representation of $\mathcal{A}$.

---

[5] It is generally agreed that a state of spin 1/2, $|\chi\rangle_{1/2}$, and a state of spin 1, $|\varphi\rangle_1$, cannot be superposed. This impossibility is an example of a superselection rule. As we have seen in Chapter 3 (and this observation will be generalized in Chapter 10), the state vector of a spin-1/2 particle is multiplied by -1 in a rotation by 2π, while that of a spin-1 particle is multiplied by $+1$. In a rotation by 2π which takes the system back to its original situation, if the state vector is of the form $|\psi\rangle = \lambda|\varphi\rangle_1 + \mu|\chi\rangle_{1/2}$ it is transformed by a 2π rotation into $|\psi'\rangle = \lambda|\varphi\rangle_1 - \mu|\chi\rangle_{1/2} \neq |\psi\rangle$. In contrast, the fact that $|\chi\rangle_{1/2}$ is transformed into $-|\chi\rangle_{1/2}$ does not present any problem, because the two vectors differ by only a phase factor. Another example is the superselection rule on the mass in the case of Galilean invariance. For a critical view of superselection rules, see Weinberg [1995], Chapter 2.

To simplify our discussion, let us start by considering a physical property $\mathcal{A}$ represented by a Hermitian operator A whose eigenvalues a_n are nondegenerate: $A|n\rangle = a_n|n\rangle$. We can then write down the spectral decomposition

$$A = \sum_n |n\rangle a_n \langle n|.$$

If the quantum system is in a state $|\varphi\rangle \equiv |n\rangle$, the value of the operator A in this state is a_n, that is, the physical property $\mathcal{A}$ takes the exact numerical value a_n. If $|\varphi\rangle$ is not an eigenstate (or eigenvector) of A, we know from postulate **II** that the probability $\mathsf{p}_n \equiv \mathsf{p}(a_n)$ of finding $|\varphi\rangle$ in $|n\rangle$, and therefore of measuring the value a_n of $\mathcal{A}$, is $\mathsf{p}_n = |\langle n|\varphi\rangle|^2$. To determine if the quantum system is in the state $|n\rangle$, $n = 1, \ldots, N$, we can imagine a generalization of the Stern–Gerlach experiment with N exit channels instead of the two channels $|+\rangle$ and $|-\rangle$, with a detector associated with each channel. Let us carry out a series of tests on a set of quantum systems that are all in the state $|\varphi\rangle$. It is said that these systems have been *prepared in the state* $|\varphi\rangle$; we have already encountered the idea of preparing a quantum system in the case of photon polarization, and we shall return to it again below. If the number of tests $\mathcal{N}$ is very large, one can obtain experimentally an accurate estimate of *the expectation value of the physical property $\mathcal{A}$ in the state $|\varphi\rangle$*, denoted $\langle A \rangle_\varphi$:

$$\langle A \rangle_\varphi = \lim_{\mathcal{N} \to \infty} \frac{1}{\mathcal{N}} \sum_{p=1}^{\mathcal{N}} \mathcal{A}_p, \tag{4.3}$$

where $\mathcal{A}_p$ is the result of the pth measurement. This result varies from one test to another, but it always takes one of the eigenvalues a_n. The expectation value is given as a function of A and $|\varphi\rangle$ by

$$\langle A \rangle_\varphi = \sum_n \mathsf{p}_n a_n = \sum_n \langle \varphi|n\rangle a_n \langle n|\varphi\rangle = \langle \varphi|A|\varphi\rangle.$$

We have already encountered a special case of this relation in (3.38). It is not difficult to generalize to the case of degenerate eigenvalues. If the system is in some state $|\varphi\rangle$, we can decompose $|\varphi\rangle$ on the basis formed by the eigenvectors of A using the completeness relation (2.30)

$$|\varphi\rangle = \sum_{n,r} |n, r\rangle \langle n, r|\varphi\rangle = \sum_{n,r} c_{nr}|n, r\rangle.$$

To find the probability $\mathsf{p}(a_n)$ of observing the eigenvalue a_n, we now need to sum all the probabilities of finding $|\varphi\rangle$ in any state $|n, r\rangle$ over the index r with n fixed:

$$\mathsf{p}(a_n) = \sum_r |c_{nr}|^2 = \sum_r \langle \varphi|n, r\rangle \langle n, r|\varphi\rangle$$

$$= \langle \varphi|\mathcal{P}_n|\varphi\rangle, \tag{4.4}$$

where $\mathcal{P}_n$ is the projector on the subspace of the eigenvalue a_n (cf. (2.29)):

$$\mathcal{P}_n = \sum_r |n, r\rangle \langle n, r|. \tag{4.5}$$

As above, by carrying out a large number of measurements on quantum systems prepared under identical conditions, we can obtain the expectation value $\langle A \rangle_\varphi$ of A in the state $|\varphi\rangle$:

$$\langle A \rangle_\varphi = \sum_n a_n \mathsf{p}(a_n) = \sum_{n,r} \langle \varphi | n, r \rangle \, a_n \, \langle n, r | \varphi \rangle,$$

and then, using (2.31), we find

$$\boxed{\langle A \rangle_\varphi = \langle \varphi | A | \varphi \rangle} \,, \tag{4.6}$$

which generalizes the preceding result. The operators representing physical properties are often called "observables" in the literature. We shall avoid this terminology, as it does not seem to provide further insight into quantum physics.[6]

The simplest Hermitian operator is the projector on a vector of $\mathcal{H}$, and subjecting a quantum system to a $|\chi\rangle$ test is equivalent to measuring the projector $\mathcal{P}_\chi = |\chi\rangle\langle\chi|$, with result 1 if the system passes the $|\chi\rangle$ test and 0 if it fails. Viewing the spectral decomposition of a Hermitian operator as the sum of projectors, we see that the ideas of testing and measuring a physical property are closely related. We shall emphasize the measurement aspect if we are interested in the eigenvalues of A, and the test aspect if we are interested in the probability of finding the system in an eigenstate of A.[7] Let us illustrate this using the Stern–Gerlach experiment of Section 3.2.2. In the spin-measurement interpretation the Stern–Gerlach apparatus measures the z component of the spin from the upward or downward deflection of the beam of silver atoms; detection of an atom on the screen at the exit of the device makes it possible to distinguish between the values $+\hbar/2$ and $-\hbar/2$ of the physical property S_z, the spin component on the Oz axis. Equivalently, we can say that we have subjected the atoms to $|+\rangle$ and $|-\rangle$ tests. The probability of upward (downward) deflection is $|\langle +|\varphi\rangle|^2$ ($|\langle -|\varphi\rangle|^2$).

However, the measurements, or tests, described in Section 3.2.2 have a major drawback: the measurement is not complete until the atoms are absorbed by the screen, and then they are no longer available for further experiments. In an *ideal measurement* (or *ideal test*) it is assumed that the physical system is not destroyed by the measurement.[8] From postulate **II**, if before the measurement of A the state vector is $|\varphi\rangle = \sum_n c_n |n\rangle$, the probability that the system *after* the measurement will be in the state $|n\rangle$ is $|c_n|^2$. It is

---

[6] This terminology goes back to a seminal article of Heisenberg containing the following statement: "The present paper seeks to establish a basis for theoretical quantum mechanics founded exclusively upon relationships between quantities which are in principle observable." Limiting ourselves to this approach is somewhat restrictive, and Heisenberg himself did not follow it in practice!

[7] We can view the photon polarization test in, for example, the basis $\{|x\rangle, |y\rangle\}$ as a measurement by introducing the physical property A_x represented by the operator

$$A_x = |x\rangle\langle x| - |y\rangle\langle y|,$$

which takes the value $+1$ if the photon is polarized in the Ox direction and -1 if it is polarized in the Oy direction.

[8] If the same ideal measurement could be repeated a number of times, one would have a "quantum nondemolition (QND) measurement." See, for example, C. Caves *et al.*, On the measurement of a weak classical force coupled to a quantum mechanical oscillator, *Rev. Mod. Phys.* **52**, 341–392 (1980) or V. Braginsky, Y. Vorontsov, and K. Thorne, Quantum non-demolition measurements, *Science* **209**, 547–557 (1980).

possible to think up a way to perform an ideal measurement[9] of the spin (but completely beyond present technology!) using a Stern–Gerlach filter modified in the spirit of the apparatus described in Section 1.1.4. Taking as our starting point the filter of Fig. 3.8, the atom entering the filter is illuminated by a suitable laser beam so as to induce a transition to one of its excited levels. When the two trajectories inside the filter are maximally separated, they pass through two different resonant cavities in which the atom returns to its ground state by emitting a photon with near 100% probability (Fig. 4.1). This photon is detected in one of the two cavities, and it is thus possible to tag the trajectory inside the filter without disturbing whatever spin state it is in, assuming that the transition is of the electric dipole kind. Such a measurement involves a profound modification in the description of the spin state. Assume, for example, that the spin state at the entrance to the filter is the eigenstate $|+, \hat{x}\rangle$ of S_x. When no measurement is made the coherence of the two trajectories will be preserved, and they can be recombined at the exit of the filter to reconstruct the state $|+, \hat{x}\rangle$. The filter contains a *coherent* superposition of the eigenstates of S_z, $|+\rangle$ and $|-\rangle$, with amplitude $1/\sqrt{2}$:

$$|+, \hat{x}\rangle = \frac{1}{\sqrt{2}}(|+\rangle + |-\rangle).$$

In contrast, when a measurement is made, the spin is projected onto one of the states $|+\rangle$ or $|-\rangle$ with 50% probability, and it is impossible to go backward and reconstruct the state $|+, \hat{x}\rangle$. Later on we shall return to this point of the irreversible nature of a measurement. As we shall see in more detail in Chapter 6 and Appendix B, the measurement has transformed the coherent superposition $|+, \hat{x}\rangle$ into a classical statistical ensemble of 50% spins up and 50% spins down, *but an experiment performed on an individual atom always gives a unique result.*

If a measurement of S_z has given the result $+\hbar/2$ and if this measurement is repeated, the result will always be $+\hbar/2$: immediately after a measurement of S_z that has given

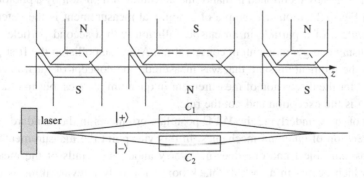

Fig. 4.1. An ideal measurement of the spin.

[9] Another thought experiment has been suggested by M. Scully, B. Englert, and J. Schwinger, Spin coherence and Humpty-Dumpty III. The effect of observation, *Phys. Rev. A* **40**, 1775–1784 (1989).

the result $+\hbar/2$, the spin is in the state $|+\rangle$. In general, a quantum system that passes the $|\chi\rangle$ test will be found in the state $|\chi\rangle$ immediately after the test:

$$|\varphi\rangle \rightarrow \frac{\mathcal{P}_\chi|\varphi\rangle}{||\mathcal{P}_\chi|\varphi\rangle||}.$$

The system has undergone an irreversible evolution which has projected it onto the state $|\chi\rangle$. The general statement is the contents of a supplementary postulate called wave-function collapse (WFC), which complements postulate **II**.

The **WFC** postulate

If a system is initially in a state $|\varphi\rangle$, and if the result of an ideal measurement of $\mathcal{A}$ is a_n, then immediately after this measurement the system is in the state projected on the subspace of the eigenvalue a_n:

$$|\varphi\rangle \rightarrow |\psi\rangle = \frac{\mathcal{P}_n|\varphi\rangle}{(\langle\varphi|\mathcal{P}_n|\varphi\rangle)^{1/2}}. \tag{4.7}$$

The vector $|\psi\rangle$ in (4.7) is normalized because

$$||\mathcal{P}_n|\varphi\rangle||^2 = \langle\varphi|\mathcal{P}_n^\dagger\mathcal{P}_n|\varphi\rangle = \langle\varphi|\mathcal{P}_n|\varphi\rangle$$

owing to the properties of projectors. The **WFC** postulate presupposes that the measurement is ideal, that is, nondestructive, so that the tests can be repeated. From a purely pragmatic viewpoint, this postulate is only interesting if at least two consecutive measurements are made. Above we have given the example of an ideal measurement of the spin of a silver atom (Fig. 4.1). At the exit of the filter we know the spin state of the atom, which is now available for further tests. A repetition of the measurement of $\mathcal{S}_z$ will again give $+\hbar/2$ for atoms that have emitted a photon in C_1 and $-\hbar/2$ for those that have emitted a photon in C_2. It should be noted that an ideal measurement is rarely possible in practice. In general, detection destroys the system under observation.[10] An example which we have already mentioned is that of the detection of a photon by a photomultiplier D_x or D_y in Fig. 3.2. Another example of a nonideal measurement is the determination of the momentum of a particle in an elastic collision with a second particle of known momentum using energy–momentum conservation. After the collision the first particle is no longer in the momentum state that was measured. The concept of ideal measurement is convenient for the discussion of measurement in quantum physics, but in practice ideal measurement is the exception and not the rule.

The point of view underlying the **WFC** postulate originates in the standard, or "orthodox" interpretation of quantum mechanics. In this viewpoint the measurement apparatus acts as a classical object and one does not worry about the details of the measurement procedure, which occurs in a sort of "black box." The only relevant thing is the result, which is read from a classical measurement such as the position of a needle on a meter. In

---

[10] It is now known how to make nondestructive measurements on a photon; see G. Nogues *et al.*, Seeing a single photon without observing it, *Nature* **400**, 239–242 (1999).

Section 6.4.1 and Appendix B we shall return to the topic of measurement procedure in quantum mechanics and try to go beyond this viewpoint. A complete analysis of the measurement procedure including the quantum interactions with the two devices performing consecutive measurements, as well as the interactions with the environment, shows that the **WFC** postulate is a consequence of postulate **II** and of the time evolution postulate **IV** stated below in Eq. (4.11), and is thus not independent of the other postulates. However, the standard viewpoint is perfectly operational in all current applications of quantum mechanics, and from now on we shall use it without further comment.

When we try to completely determine the state vector $|\varphi\rangle$ of a physical system, it can happen that an ideal measurement of a physical property $\mathcal{A}$ gives the result a, where the eigenvalue a of A is nondegenerate. Immediately after the measurement the state vector is then the eigenvector $|a\rangle$ of A. If the eigenvalue is degenerate, it is necessary to find a second physical property $\mathcal{B}$ compatible with $\mathcal{A}$: $[A, B] = 0$. In this case it is possible that the known eigenvalues a and b completely specify the state vector. If this is not yet so, it is necessary to find a third physical property $\mathcal{C}$ compatible with $\mathcal{A}$ and $\mathcal{B}$, and so on. When the known eigenvalues $\{a, b, c \ldots\}$ of the compatible operators $\{A, B, C \ldots\}$ entirely specify the state vector we say, following the terminology introduced in Section 2.3.3, that these operators (or the physical properties which they represent) form a *complete set of compatible operators* (*or compatible physical properties*). The simultaneous measurement of the complete set of compatible physical properties $\{\mathcal{A}, \mathcal{B}, \mathcal{C} \ldots\}$ constitutes a *maximal test* of a state vector. If the space of states has dimension N, the maximal test must have N different mutually exclusive outcomes. When an ideal maximal test has been carried out on a quantum system the state vector of the latter is known exactly, and in this way the quantum system has been *prepared* in a determined state. The stage corresponding to preparation of the system has been completed. However, the preparation stage need not (and in general does not) involve a measurement: for example, the left filter of Fig. 3.10 prepares the spin in the $|+\rangle$ state without measuring it.

To illustrate these ideas, let us suppose that two known eigenvalues a_r and b_s of two compatible operators A and B completely specify a vector $|r, s\rangle$ of $\mathcal{H}$:

$$A|r, s\rangle = a_r |r, s\rangle, \quad B|r, s\rangle = b_s |r, s\rangle.$$

The simultaneous measurement of the physical properties $\mathcal{A}$ and $\mathcal{B}$ is then a maximal test and the N possible results are labeled by the set (r, s). An example of a device that performs a maximal test is the Stern–Gerlach apparatus of Fig. 3.7. This apparatus separates the spin states $|+\rangle$ and $|-\rangle$, giving two different spots on the screen because the space of states has dimension 2: $N = 2$. In the general case, the measurement of $\mathcal{A}$ and $\mathcal{B}$ allows the system to be prepared in the state $|r, s\rangle$ by selecting the systems that have given the result (a_r, b_s). If the selected quantum systems in the state $|r, s\rangle$ are again subjected to simultaneous measurement of $\mathcal{A}$ and $\mathcal{B}$, the result of this new measurement will be (a_r, b_s) with 100% probability. When a physical system is described by a state vector, there must exist, at least in principle, a maximal test one of whose possible results has 100% probability. For a spin 1/2 in the state $|+\rangle$, one such maximal test is that performed using a Stern–Gerlach apparatus with magnetic field in the Oz direction.

It is also instructive to study the case of a physical property $\mathcal{A}$ which is compatible with $\mathcal{B}$ and $\mathcal{C}$, $[A, B] = [A, C] = 0$, while $\mathcal{B}$ and $\mathcal{C}$ are incompatible: $[B, C] \neq 0$. In this case the result of a measurement of $\mathcal{A}$ depends on whether $\mathcal{B}$ or $\mathcal{C}$ is measured simultaneously. This property is called *contextuality*, and an example of it will be given in Section 6.3.3.

By now the reader will have realized that measurement in quantum physics is fundamentally different from that in classical physics. In classical physics, a measurement reveals a *pre-existing* property of the physical system that is tested. If a car is driving at 180 km h^{-1} on the highway, the measurement of its speed by radar determines a property that exists prior to the measurement, which gives the police the legitimacy to give a ticket to the driver. On the contrary, the measurement of the S_x component of a spin-1/2 particle in the state $|+\rangle$ does not reveal a value of S_x existing before the measurement. The spread in the results of measuring S_x in this case is sometimes attributed to "uncontrollable perturbation of the spin due to the measurement process," but the value of S_x does not exist before the measurement, and that which does not exist cannot be perturbed. We shall return to this point in Section 6.4.1.

4.1.3 Heisenberg inequalities II

In the preceding chapter we introduced the idea of incompatible physical properties. We shall now discuss this idea and its consequences for measurement in a more quantitative way. Two physical properties $\mathcal{A}$ and $\mathcal{B}$ are *incompatible* if the commutator of the operators A and B representing them is nonzero: $[A, B] \neq 0$. Let us assume that the first measurement of $\mathcal{A}$ has given the result a and has projected the initial state vector onto the eigenvector $|a\rangle$ of $A: A|a\rangle = a|a\rangle$. If $\mathcal{B}$ is measured immediately after $\mathcal{A}$, in general the vector $|a\rangle$ will not be an eigenvector of B and the result of the measurement will only be known with a certain probability. For example, if b is a nondegenerate eigenvalue of B corresponding to eigenvector $|b\rangle$, $B|b\rangle = b|b\rangle$, then the probability of measuring b will be $\mathsf{p}(a \to b) = |\langle b|a\rangle|^2$. In general, it will not be possible to find states for which the values of $\mathcal{A}$ and $\mathcal{B}$ are both known exactly. Let us derive an important result on the dispersion (or standard deviation) of measurements performed starting from an arbitrary initial state $|\varphi\rangle$. It is convenient to define the *dispersions* $\Delta_\varphi A$ and $\Delta_\varphi B$ in the state $|\varphi\rangle$ as

$$(\Delta_\varphi A)^2 = \langle A^2 \rangle_\varphi - (\langle A \rangle_\varphi)^2 = \langle (A - \langle A \rangle_\varphi I)^2 \rangle_\varphi,$$
$$(\Delta_\varphi B)^2 = \langle B^2 \rangle_\varphi - (\langle B \rangle_\varphi)^2 = \langle (B - \langle B \rangle_\varphi I)^2 \rangle_\varphi. \tag{4.8}$$

The commutator of A and B is of the form iC, where C is a Hermitian operator because

$$[A, B]^\dagger = [B^\dagger, A^\dagger] = [B, A] = -[A, B].$$

We can then write

$$[A, B] = iC, \quad C = C^\dagger. \tag{4.9}$$

Let us define the Hermitian operators of zero expectation value (a priori specific to the state $|\varphi\rangle$):

$$A_0 = A - \langle A \rangle_\varphi I, \quad B_0 = B - \langle B \rangle_\varphi I.$$

Their commutator is also iC, $[A_0, B_0] = iC$, because $\langle A \rangle_\varphi$ and $\langle B \rangle_\varphi$ are numbers. The squared norm of the vector

$$(A_0 + i\lambda B_0)|\varphi\rangle,$$

where λ is chosen to be real, must be positive:

$$||(A_0 + i\lambda B_0)|\varphi\rangle||^2 = ||A_0|\varphi\rangle||^2 + i\lambda\langle\varphi|A_0B_0|\varphi\rangle - i\lambda\langle\varphi|B_0A_0|\varphi\rangle + \lambda^2||B_0|\varphi\rangle||^2$$

$$= \langle A_0^2 \rangle_\varphi - \lambda\langle C \rangle_\varphi + \lambda^2\langle B_0^2 \rangle_\varphi \geq 0.$$

The second-degree polynomial in λ must be positive for any λ, which implies

$$\langle C \rangle_\varphi^2 - 4\langle A_0^2 \rangle_\varphi \langle B_0^2 \rangle_\varphi \leq 0.$$

This demonstrates the *Heisenberg inequality*

$$\boxed{(\Delta_\varphi A)(\Delta_\varphi B) \geq \frac{1}{2}|\langle C \rangle_\varphi|.} \tag{4.10}$$

This is the desired relation constraining the dispersions in the measurements of $\mathcal{A}$ and $\mathcal{B}$: the product of the dispersions in the measurements is greater than or equal to half the modulus of the expectation value of the commutator of A and B. It is easy to show (Exercise 4.4.1) that a necessary and sufficient condition for $\Delta_\varphi A = 0$ is that $|\varphi\rangle$ be an eigenvector of A. In a vector space of *finite dimension* we then have $\langle C \rangle_\varphi = 0$. It is important to stress the correct interpretation of (4.10): when, as in (4.3), a large number $\mathcal{N}$ of measurements of $\mathcal{A}$ are performed on systems all prepared in the same state $|\varphi\rangle$, and similarly for $\mathcal{B}$ and $\mathcal{C}$, we can obtain accurate experimental estimates for the dispersions $\Delta_\varphi A$ and $\Delta_\varphi B$ as well as the expectation value $\langle C \rangle_\varphi$, which then obey (4.10). We emphasize that $\mathcal{A}$, $\mathcal{B}$, and $\mathcal{C}$ are of course measured in different experiments: they cannot be measured simultaneously if A, B, and C do not commute. Furthermore, $\Delta_\varphi A$ and $\Delta_\varphi B$ are in no way related to errors of measurement. If, for example, δA is the experimental resolution for the measurement of $\mathcal{A}$, we must have $\delta A \ll \Delta_\varphi A$ for an accurate determination of the dispersion. The error on $\langle A \rangle$ is governed by the experimental resolution, and not at all by $\Delta_\varphi A$, and $\langle A \rangle_\varphi$ may be determined with an accuracy much better than $\Delta_\varphi A$.

4.2 Time evolution

4.2.1 The evolution equation

So far we have considered a physical system at a certain instant of time, or during the time interval necessary to perform the measurement, which is assumed to be very short.

We shall now take into account the time evolution of the state vector, which will be written as explicitly dependent on the time t: $|\varphi(t)\rangle$.

Postulate IV: the evolution equation

The time evolution of the state vector $|\varphi(t)\rangle$ of a quantum system is governed by the evolution equation

$$\boxed{i\hbar\frac{d|\varphi(t)\rangle}{dt} = H(t)|\varphi(t)\rangle}. \tag{4.11}$$

The Hermitian operator $H(t)$ is called the *Hamiltonian*.

Let us be precise on the conditions under which Eq. (4.11) applies. It holds for a *closed quantum system*, and this statement should be understood as follows: the quantum system under consideration must not be part of a larger quantum system, a situation dealt with at length in Chapter 15. However, (4.11) is valid if the quantum system interacts with a *classical* system, which means that it is not necessarily isolated. It is valid, for example, in the case of a spin 1/2 submitted to a time-dependent magnetic field (Section 5.2), or for a two-level atom submitted to a classical electromagnetic field (Sections 14.3.1 to 14.3.3), but not for an atom interacting with a quantized electromagnetic field (Section 14.4). In the latter case, the time evolution of the state vector (or more accurately of the state operator) of the atom is not governed by a Hamiltonian. A Hamiltonian evolution holds only for the *atom + field* system.

The operator H has the dimensions of energy, and we do identify H later on as the Hermitian operator representing the physical property of energy (Eq. (4.23)). Equation (4.11) is of first order in time, and the evolution is deterministic: given an initial condition $|\varphi(t_0)\rangle$ for the state vector at time $t = t_0$, the evolution (4.11) determines $|\varphi(t)\rangle$ at any later time $t > t_0$, provided of course that the Hamiltonian is known. In fact, the restriction to $t > t_0$ is unnecessary: the evolution (4.11) is reversible and we can perfectly well go backwards in time. A schematic view of a typical experiment is given in Fig. 4.2. The system is prepared at time $t = t_0$ by an ideal measurement of an ensemble of compatible physical properties, which determines the state vector $|\varphi(t_0)\rangle$. The state vector then evolves until time t according to (4.11), and a second measurement of one or a set of physical properties (either the same ones as in the first measurement, or different ones)

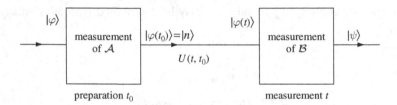

Fig. 4.2. Preparation and measurement. Measurement of $\mathcal{A}$ at time t_0 gives the result a_n. The state vector evolves between t_0 and t as $|\varphi(t)\rangle = U(t, t_0)|\varphi(t_0)\rangle$ (4.14). Then $\mathcal{B}$ is measured at time t.

is made at time t. Note that the duration of the measurements is assumed to be very short with respect to the characteristic evolution time of the Schrödinger equation. This second measurement permits the complete or partial determination of $|\varphi(t)\rangle$ from which we may infer, for example, the properties of H. For (4.11) to hold between the two measurements it is of course necessary that the quantum system be closed, as defined above, during the corresponding time interval.

The (necessary) conservation of the norm of the state vector is assured by the Hermiticity of H. We have

$$\frac{d}{dt}\|\varphi(t)\|^2 = \frac{d}{dt}\langle\varphi(t)|\varphi(t)\rangle$$

$$= \langle\varphi(t)|\left(\frac{1}{i\hbar}H\right)^\dagger|\varphi(t)\rangle + \langle\varphi(t)|\left(\frac{1}{i\hbar}H\right)|\varphi(t)\rangle$$

$$= \frac{1}{i\hbar}\langle\varphi(t)|(H-H^\dagger)|\varphi(t)\rangle = 0 \qquad (4.12)$$

because $H = H^\dagger$. If $|\varphi(t)\rangle$ is decomposed on a basis $|n, r\rangle$

$$|\varphi(t)\rangle = \sum_{n,r}|n, r\rangle\langle n, r|\varphi(t)\rangle = \sum_{n,r}c_{nr}(t)|n, r\rangle,$$

the components $c_{nr}(t)$ satisfy

$$\frac{d}{dt}\left(\sum_{n,r}|c_{nr}(t)|^2\right) = \frac{d}{dt}\left(\sum_n \mathsf{p}(a_n, t)\right) = 0.$$

The sum of the probabilities $\mathsf{p}(a_n, t)$ must always be unity.

The matrix form of the evolution equation (4.11) is obtained in an arbitrary basis $\{|\alpha\rangle\}$ of $\mathcal{H}$ by multiplying (4.11) on the left by $\langle\alpha|$ and using the completeness relation:

$$i\hbar\frac{d}{dt}\langle\alpha|\varphi(t)\rangle = \langle\alpha|H(t)|\varphi(t)\rangle = \sum_\beta\langle\alpha|H(t)|\beta\rangle\langle\beta|\varphi(t)\rangle,$$

which gives

$$i\hbar\dot{c}_\alpha(t) = \sum_\beta H_{\alpha\beta}(t)\,c_\beta(t). \qquad (4.13)$$

We have emphasized the reversible and unitary nature of the evolution (4.11). This should be contrasted with the nature of the evolution in a measurement, which is nonunitary and irreversible. The projection of the initial state vector on the eigenvector of the measured physical property is not unitary – the norm is not conserved, and the result $\mathcal{P}_n|\varphi\rangle$ of the projection (cf. the denominator in (4.7)) must be normalized. Moreover, it is impossible to reconstruct the initial state vector once the measurement has been made. From the orthodox point of view this implies that there are two types of evolution: one reversible (4.11) and one irreversible (4.7). This is not a very satisfying state of affairs, and we shall examine this problem in Appendix B.

4.2.2 The evolution operator

In (4.11) we gave the differential form of the evolution equation. There exists an integral formulation of this equation involving the evolution operator $U(t, t_0)$. In this formulation postulate **IV** becomes the following.

*Postulate **IV'**: the evolution operator*

The state vector $|\varphi(t)\rangle$ at time t is derived from the state vector $|\varphi(t_0)\rangle$ at time t_0 by applying a unitary operator $U(t, t_0)$, called the *evolution operator*:

$$\boxed{|\varphi(t)\rangle = U(t, t_0)|\varphi(t_0)\rangle} . \tag{4.14}$$

The unitarity of U, $U^\dagger U = UU^\dagger = I$, ensures conservation of the norm (4.12):

$$\langle\varphi(t)|\varphi(t)\rangle = \langle\varphi(t_0)|U^\dagger(t, t_0)U(t, t_0)|\varphi(t_0)\rangle = \langle\varphi(t_0)|\varphi(t_0)\rangle = 1.$$

Inversely, we can start from conservation of the norm and show that $U^\dagger U = I$. In a vector space of finite dimension this is sufficient to ensure that $UU^\dagger = I$ (cf. Section 2.2.1), but this is not necessarily true in a space of infinite dimension. The evolution operator also satisfies the *group property*:

$$U(t, t_1)U(t_1, t_0) = U(t, t_0), \quad t_0 \le t_1 \le t. \tag{4.15}$$

In effect, going directly from t_0 to t is equivalent to going first from t_0 to t_1 and then from t_1 to t:

$$|\varphi(t)\rangle = U(t, t_0)|\varphi(t_0)\rangle$$
$$= U(t, t_1)|\varphi(t_1)\rangle = U(t, t_1)U(t_1, t_0)|\varphi(t_0)\rangle.$$

As before, the restriction $t_0 < t_1 < t$ is unnecessary: t_1 can take any value. Obviously $U(t_0, t_0) = I$, and the group property together with the unitarity of U implies

$$U(t, t_0) = U^{-1}(t_0, t) = U^\dagger(t_0, t). \tag{4.16}$$

Of course, the temporal evolution postulates **IV** and **IV'** are not independent. In fact, it is easy to write down a differential equation for $U(t, t_0)$ starting from (4.11). Differentiating (4.14) with respect to time

$$i\hbar\frac{d}{dt}|\varphi(t)\rangle = i\hbar\left[\frac{d}{dt}U(t, t_0)\right]|\varphi(t_0)\rangle$$

and comparing the result with (4.11), we obtain

$$i\hbar\left[\frac{d}{dt}U(t, t_0)\right]|\varphi(t_0)\rangle = H(t)U(t, t_0)|\varphi(t_0)\rangle.$$

Since this equation must hold for any $|\varphi(t_0)\rangle$, we can derive from it a differential equation for $U(t, t_0)$:

$$\boxed{i\hbar \frac{d}{dt} U(t, t_0) = H(t)U(t, t_0)} \;, \tag{4.17}$$

which leads to

$$H(t_0) = i\hbar \frac{d}{dt} U(t, t_0)\Big|_{t=t_0} \tag{4.18}$$

by taking the limit $t \to t_0$. Then it is easy to pass from the integral formulation (4.14) to the differential formulation (4.11). The reverse is more complicated. If $H(t)$ were a number, it would be possible to integrate (4.17) immediately; however, $H(t)$ is an operator and in general

$$U(t, t_0) \neq \exp\left(-\frac{i}{\hbar} \int_{t_0}^{t} H(t')\, dt'\right), \tag{4.19}$$

because there is no reason to have $[H(t'), H(t'')] = 0$. However, there exists a general expression[11] for calculating $U(t, t_0)$ from $H(t)$, and postulates **IV** and **IV'** are strictly equivalent.[12]

4.2.3 Stationary states

A very important special case is that of a system that is isolated from any kind of environment, be it quantum or classical. The evolution operator of such a system cannot depend on the choice of time origin – it is of no importance if we choose to describe a system isolated from all external influences using the time of London or that of New York, which, as is well known, differ by $\tau = 5$ hours:

$$t_{\text{NewYork}} = t_{\text{London}} - \tau.$$

Whatever τ is, we must have

$$U(t - \tau, t_0 - \tau) = U(t, t_0). \tag{4.20}$$

This implies that U can only depend on the *difference* $(t - t_0)$. Equation (4.18) then shows that the Hamiltonian is independent of time, because the choice of t_0 is arbitrary. Naturally, it can perfectly well happen that the Hamiltonian is independent of time even for a system that is not isolated, for example, if the system is exposed to a time-independent magnetic field like the spin-1/2 particle of Section 3.2.5. On the other hand, if a magnetic field is switched on between 12:00 and 12:10 London time, the choice of time origin will matter!

---

[11] See, for example, Messiah [1999], Chapter XVII.
[12] To be completely accurate, it is possible to find exceptions where U is defined but H is not; see Peres [1993], 85.

Since the Hamiltonian is independent of time, the differential equation (4.17) can easily be integrated and we find

$$U(t, t_0) = \exp\left(-\frac{\mathrm{i}(t - t_0)}{\hbar} H\right), \tag{4.21}$$

which depends only on $(t - t_0)$.

The operator $U(t - t_0)$ (4.21) is obtained by exponentiating the Hermitian operator H; $U(t - t_0)$ performs a time-translation $(t - t_0)$ on the state vector, and if $(t - t_0)$ is infinitesimal

$$U(t - t_0) \simeq I - \frac{\mathrm{i}(t - t_0)}{\hbar} H. \tag{4.22}$$

This equation can be interpreted as follows: H is the *infinitesimal generator of time-translations*, and, for an isolated system, the most general definition of the Hamiltonian is precisely that of an infinitesimal time-translation generator. The concept of infinitesimal generator will be extended to other transformations in Chapter 8.

Let us consider an isolated physical system which can to a good approximation be described by a state vector of a Hilbert space of dimension 1. This might be a stable elementary particle, an atom in its ground state, and so on. The state vector is a complex number $\varphi(t)$ and H is a real number, $H = E$. The evolution law (4.13) becomes, taking into account (4.20),

$$\varphi(t) = \exp\left(-\frac{\mathrm{i}}{\hbar} E(t - t_0)\right) \varphi(t_0) = \exp(-\mathrm{i}\omega(t - t_0))\varphi(t_0), \tag{4.23}$$

where we have defined $E = \hbar\omega$. According to the Planck–Einstein relation $E = \hbar\omega$, it is natural to identify E as the energy.

Now let us consider a less trivial case. Let $|n, r\rangle$ be an eigenvector of H corresponding to the eigenvalue E_n: $H|n, r\rangle = E_n|n, r\rangle$. Its time evolution is particularly simple. If $|\varphi(t_0)\rangle = |n, r\rangle$, then

$$|\varphi(t)\rangle = \exp\left(-\frac{\mathrm{i}(t - t_0)}{\hbar} H\right) |n, r\rangle = \exp\left(-\frac{\mathrm{i}}{\hbar} E_n(t - t_0)\right) |n, r\rangle. \tag{4.24}$$

The probability of finding $|\varphi(t)\rangle$ in any state $|\chi\rangle$ is independent of time:

$$|\langle\chi|\varphi(t)\rangle|^2 = \left|\langle\chi| \exp\left(-\frac{\mathrm{i}}{\hbar} E_n(t - t_0)\right) |\varphi(t_0)\rangle\right|^2 = |\langle\chi|\varphi(t_0)\rangle|^2.$$

For this reason an eigenstate of H is called a *stationary state*.

Sometimes it is useful to write the time-evolution law in component form. Let us write down the decomposition of an arbitrary state vector $|\varphi(t_0)\rangle$ at time $t = t_0$ on the basis $\{|n, r\rangle\}$ of eigenvectors of H:

$$|\varphi(t_0)\rangle = \sum_{n,r} c_{nr}(t_0)|n, r\rangle, \quad c_{nr}(t_0) = \langle n, r|\varphi(t_0)\rangle.$$

We then find

$$|\varphi(t)\rangle = \sum_{n,r} c_{nr}(t_0) \exp\left(-\frac{\mathrm{i}(t-t_0)}{\hbar} H\right) |n, r\rangle = \sum_{n,r} c_{nr}(t_0) \exp\left(-\frac{\mathrm{i}}{\hbar} E_n(t-t_0)\right) |n, r\rangle,$$

which gives the variation of the coefficients c_{nr} as a function of t:

$$c_{nr}(t) = \exp\left(-\frac{\mathrm{i}}{\hbar} E_n(t-t_0)\right) c_{nr}(t_0). \tag{4.25}$$

4.2.4 The temporal Heisenberg inequality

In Section 3.2.5 we gave an elementary explanation of the relation between a characteristic evolution time Δt and an energy spread ΔE. Now we shall give a general derivation of an inequality for the product $\Delta E \, \Delta t$, the *temporal Heisenberg inequality*. First we write down the evolution equation for the expectation value $\langle A \rangle_\varphi(t) = \langle \varphi(t)|A|\varphi(t)\rangle$ of an operator A representing a physical property $\mathcal{A}$, assumed to be independent of time:

$$\frac{\mathrm{d}}{\mathrm{d}t} \langle \varphi(t)|A|\varphi(t)\rangle = \frac{1}{\mathrm{i}\hbar} \left[-\langle \varphi(t)|HA|\varphi(t)\rangle + \langle \varphi(t)|AH|\varphi(t)\rangle\right]$$

$$= \frac{1}{\mathrm{i}\hbar} \langle \varphi(t)|AH - HA|\varphi(t)\rangle,$$

which gives the *Ehrenfest theorem*:

$$\boxed{\frac{\mathrm{d}}{\mathrm{d}t} \langle A \rangle_\varphi(t) = \frac{1}{\mathrm{i}\hbar} \langle \varphi(t)|[A, H]|\varphi(t)\rangle = \frac{1}{\mathrm{i}\hbar} \langle [A, H] \rangle_\varphi} \quad . \tag{4.26}$$

Now we use (4.10), replacing B by H:

$$\boxed{\Delta_\varphi H \, \Delta_\varphi A \geq \frac{1}{2} |\langle [A, H] \rangle_\varphi| = \frac{1}{2} \hbar \left|\frac{\mathrm{d}}{\mathrm{d}t} \langle A \rangle_\varphi(t)\right|,} \tag{4.27}$$

and define the time $\tau_\varphi(A)$ as

$$\frac{1}{\tau_\varphi(A)} = \left|\frac{\mathrm{d}\langle A \rangle_\varphi(t)}{\mathrm{d}t}\right| \frac{1}{\Delta_\varphi A}.$$

The time $\tau_\varphi(A)$ is the characteristic time for the expectation value of A to change by $\Delta_\varphi A$, that is, by an amount of the order of the dispersion. The preceding inequality becomes

$$\Delta_\varphi H \, \tau_\varphi(A) \geq \frac{1}{2} \hbar, \tag{4.28}$$

which is the rigorous form of the temporal Heisenberg inequality. This inequality is often written as

$$\Delta E \, \Delta t \gtrsim \frac{1}{2} \hbar \tag{4.29}$$

where ΔE represents the energy spread and Δt the characteristic evolution time.[13] This equation has great heuristic value, but the meaning of ΔE may be ambiguous, as explained below. The value of the energy can be fixed exactly only when the spread ΔE is zero, which implies that the characteristic time must be infinite. This is not possible unless the system is in a stationary state, which occurs, for example, for a stable elementary particle or an atom in its ground state in the absence of external perturbations. However, an atom or a nucleus raised to an excited state is not in a stationary state. Owing to the coupling with the vacuum fluctuations of the electromagnetic field (cf. Section 14.3.4), the atom, or the nucleus, emits a photon after an average time τ, called the *lifetime of the excited state* (cf. Section 1.5.3). The energy of the final photon has a spread ΔE called the *width* of the state and often denoted as $\hbar\Gamma$; an example is given in Appendix C, Fig. C.1. The decay law of the excited state is generally very nearly exponential: the survival probability $\mathsf{p}(t)$ of the excited state is given by $\mathsf{p}(t) = \exp(-t/\tau)$. The width ΔE of the state and the lifetime τ are related by Fourier transformation and one can show that $\tau\Delta E \simeq \hbar$, so that, from $\Delta E = \hbar\Gamma$, one has

$$\Gamma\tau \simeq 1. \tag{4.30}$$

However, ΔE is not the same thing as the dispersion ΔH of the Hamiltonian *computed in the excited state*. It fact, it can be shown that $\hbar\Gamma = \Delta E \ll \Delta H$ for the exponential decay law to be valid; see Exercise 4.4.5 and Appendix C for more details.[14]

Let us look at orders of magnitude for a typical system in atomic physics, the first excited state of the rubidium atom. An atom in this state returns to its ground state by emitting a photon of wavelength $\lambda = 0.78\,\mu\mathrm{m}$ corresponding to energy $\varepsilon = 1.6$ eV. The width and lifetime of the state are $\hbar\Gamma = 2.4 \times 10^{-8}$ eV and $\tau \simeq 1/\Gamma = 2.7 \times 10^{-8}$ s. The energy spread of the excited state is therefore very small compared with the difference between the energies of the ground and excited states: $\hbar\Gamma/\varepsilon \simeq 10^{-8}$, which means that the energy of the excited state is very precisely defined. The relation (4.30) can be generalized to any particle decay, for example, a two-body decay $C \rightarrow A + B$.

As in the case of the Heisenberg inequality (4.10), the dispersion ΔE is in no way related to the accuracy with which the energy can be measured. It is of course possible to measure an energy with a precision better than ΔE. Let us take as an example the energy E of the Z^0 boson, a carrier of the weak interaction (cf. Section 1.1.4); in the Z^0 rest frame $E = m_Z c^2$, where m_Z is the Z^0 mass. The Z^0 boson is unstable and therefore has a width, which has been measured very precisely: $\hbar\Gamma_Z = 2.4952 \pm 0.0023$ GeV. However, the Z^0 mass has actually been measured more precisely than Γ_Z! The best measurement gives $m_Z c^2 = 91.1875 \pm 0.0021$ GeV (Fig. 4.3). In other words, it is possible to locate the center of the peak with an accuracy much better than its spread.

---

[13] The status of the inequality $\Delta E\,\Delta t \gtrsim \hbar$ is different from that of (4.10) in that, as shown by Pauli, there is no operator T which obeys the commutation relation $[T, H] = \mathrm{i}\hbar$. The quantity Δt is often incorrectly interpreted as the time necessary to measure the energy. Also, one cannot invoke the time–frequency inequality for a signal, $\Delta t \Delta \omega \geq 1/2$, because we do not have $E = \hbar\omega$, but rather $\hbar\omega = (E_1 - E_2)$, at least in nonrelativistic quantum mechanics.

[14] The conditions of validity of the exponential decay law are examined by A. Peres, Nonexponential decay law, *Ann. Phys. (NY)*, **129**, 33 (1980).

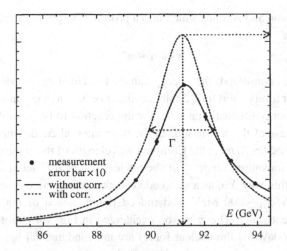

Fig. 4.3. Mass spectrum of the Z^0 boson. The solid line shows the raw experimental data. This result must be corrected taking into account radiative corrections (photon emission), which can be calculated with extremely high accuracy. The dotted line shows the Z^0 mass spectrum. From the LEP collaboration, CERN Preprint EP-2000-13 (2000).

The relation (4.29) also leads to the idea of "virtual particles." It is possible to interpret processes in quantum field theory in terms of virtual particle exchange. For example, the Coulomb interaction in the hydrogen atom corresponds to the exchange of virtual photons between the proton and the electron. Virtual exchange does not correspond to an observable reaction between the particles, because virtual particles cannot satisfy energy–momentum conservation together with the condition relating the energy to the momentum and the mass $E^2 = \vec{p}^{\,2}c^2 + m^2c^4$. Let us take the example of interactions between nucleons, or strong interactions (cf. Section 1.1.4). In 1935 Yukawa imagined that these interactions arose from the exchange of a then-unknown particle which today we call the π meson. This exchange is represented in Fig. 4.4 by a "Feynman graph." The proton on the left (p) emits a π^+ meson and is transformed into a neutron (n), while the neutron on the right

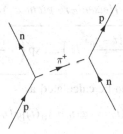

Fig. 4.4. Feynman diagram for π-meson exchange.

absorbs this π^+ meson and is transformed into a proton. Energy–momentum conservation forbids the reaction

$$p \to n + \pi^+.$$

If the momentum is conserved, the energy cannot be. However, if we assume that the reaction occurs over a very short time Δt, it becomes possible to have an energy fluctuation $\Delta E \simeq \hbar/\Delta t$. The energy fluctuation needed for the reaction to be possible is $\Delta E \sim m_\pi c^2$, where m_π is the mass of the π^+ meson. In the time interval Δt the meson can travel at most a distance[15] $\sim c\Delta t \sim \hbar/m_\pi c$, the Compton wavelength of the π meson. This distance corresponds to the maximum range r_0 of the nuclear forces (cf. Section 1.1.4), which is of order 1 fm. In this way Yukawa succeeded in predicting the existence of a particle of mass of order $\hbar/cr_0 \sim 200$ MeV, and indeed the π meson of mass 140 MeV was discovered some years later. The π meson exchanged in Fig. 4.4 is not observable: it is virtual. We know today that the nuclear forces are not fundamental but are derived from the fundamental forces between quarks. Nevertheless, the argument of Yukawa remains valid, because it is possible to write down an effective theory of nuclear forces involving meson exchange, where the maximum range of the forces is determined by the lightest meson, the π meson. Since the photon has zero mass, the range of electromagnetic forces is infinite. Indeed, we have seen in Section 1.1.4 that the Coulomb potential is long-range.

4.2.5 The Schrödinger and Heisenberg pictures

The point of view adopted above, in which the state vector evolves with time while the operators are independent of time, is called the *Schrödinger picture*. An equivalent viewpoint as regards physical results is that of Heisenberg, where the state vectors are independent of time and the operators depend on time. To simplify the discussion, we shall consider the case of a Hamiltonian H and an operator A which are time-independent. This is not the most general situation, because it may happen that even in the Schrödinger picture an operator A has an explicit time dependence, or that H depends on time. We shall assume that this is not so here, and leave the general case to Exercise 4.4.7. The expectation value of A at time t is

$$\langle A \rangle_\varphi(t) = \langle \varphi(t_0) | \exp\left(\frac{\mathrm{i}(t-t_0)}{\hbar}H\right) A \exp\left(-\frac{\mathrm{i}(t-t_0)}{\hbar}H\right) | \varphi(t_0) \rangle.$$

If we define the *operator A in the Heisenberg picture* $A_H(t)$ as

$$\boxed{A_H(t) = \exp\left(\frac{\mathrm{i}(t-t_0)}{\hbar}H\right) A \exp\left(-\frac{\mathrm{i}(t-t_0)}{\hbar}H\right)}, \qquad (4.31)$$

then the expectation value of A can be calculated as

$$\langle A \rangle_\varphi(t) = \langle \varphi(t_0) | A_H(t) | \varphi(t_0) \rangle. \qquad (4.32)$$

---

[15] For simplicity we neglect time dilation.

The time dependence is incorporated in the operator, leaving the state vector independent of t.

4.3 Approximations and modeling

We have now stated the general principles that determine the universal framework of quantum theory. However, we are not yet ready to take on a physical problem. In order to solve a specific problem, for example that of calculating the energy levels of the hydrogen atom, we need to fix the space of states and the Hamiltonian appropriately according to the degree of precision with which we hope to solve the problem. Choosing the space of states and Hamiltonian always implies that we are using a certain approximation, and this approximation (model) should not be confused with the fundamental principles. For example, as we shall show immediately below, the space of states is always initially of infinite dimension, but it may turn out that it is possible to find an approximation framework where it reduces to a space of finite dimension, and maybe even of small dimensions. The dimension N of this space is called the *number of levels* of the approximation. We have already seen an example in our study of spin 1/2. In the first approximation the spin degrees of freedom are decoupled from the spatial degrees of freedom, which is what allowed us to consider a two-dimensional space and ignore the spatial degrees of freedom. Another example is that of a two-level atom, a standard model in atomic physics. When we are interested in the interaction between an atom and an electromagnetic field of frequency ω (in practice, the field of a laser), and if the spacing of two energy levels is $\hbar\omega_0 \simeq \hbar\omega$, we can limit ourselves to these two energy levels. They form a basis for a two-dimensional space of states, and then we can write down a Hamiltonian for the interaction with the laser field acting in this space; cf. Sections 5.4 and 14.1.1. This approach provides an excellent approximation for the laser–atom interaction and can easily be refined, for example, by taking into account the effects of level splitting due to the spins.

Unfortunately, the situation is not always so simple. As we shall see in Chapter 9, spatial degrees of freedom can be dealt with using the *correspondence principle*. According to this principle, the physical properties corresponding to position and momentum are represented by operators $\vec{R}$ and $\vec{P}$ with components X_i and P_j, $i, j = (x, y, z)$, satisfying commutation relations called *canonical commutation relations*:

$$[X_i, P_j] = i\hbar\delta_{ij}I. \tag{4.33}$$

Taking the trace of the two sides, we see that it is impossible to satisfy these relations in a space of finite dimension: the trace of the quantity on the left is zero (the trace of a commutator is always zero), while that of the quantity on the right is $i\hbar N$, where N is the dimension of $\mathcal{H}$. Once this feature is recognized, the rest of the procedure (which itself is not always unambiguous) consists of replacing the positions and momenta $\vec{r}$ and $\vec{p}$ in the classical expression for the energy E by the operators $\vec{R}$ and $\vec{P}$, thus obtaining the quantum Hamiltonian of a particle of mass m with potential energy $V(\vec{r})$. The correspondence principle therefore gives the transformation $E \to H$:

$$E = \frac{\vec{p}^2}{2m} + V(\vec{r}) \to H = \frac{\vec{P}^2}{2m} + V(\vec{R}). \qquad (4.34)$$

In the case of the hydrogen atom, (4.34) provides a very good approximation if the Coulomb potential corresponding to the force law (1.3) is used for $V(\vec{r})$ and the space of states is taken to be that of the electron. The effect of the finite proton mass is taken into account by using the reduced mass. It should be clear that (4.33) and (4.34) represent a choice for the space of states and the Hamiltonian, and that approximations have been made. In particular, we have neglected relativistic effects, the inclusion of which would greatly complicate the problem. As a first step, one could try to generalize the expression for the Hamiltonian (which leads to the Dirac equation), but a theory that is truly quantum and relativistic requires the introduction of quantized electron–positron and electromagnetic fields. This theory is called *quantum electrodynamics* (QED). Under these conditions, the correspondence principle in the form (4.33) is no longer valid;[16] in fact, there is no longer a position operator. Moreover, quantum electrodynamics itself is very likely just an approximation to a more comprehensive theory, and so on. It is therefore necessary to distinguish carefully between fundamental principles and the approximations needed to solve a specific physical problem. As Isham [1995] has emphasized, the standard procedure of "quantizing a classical theory" using the correspondence principle has only heuristic value; in the end, the approximations based on this principle or any other heuristic approach must be validated by confrontation with the experimental results.

Up to now we have used different notation for a physical property ($\mathcal{A}$) and the associated Hermitian operator (A). Now we shall abandon this distinction and, unless explicitly stated otherwise, denote both the property and the operator by upper-case letters: the Hamiltonian H, position $\vec{R}$, momentum $\vec{P}$, angular momentum $\vec{J}$, and so on. Eigenvalues will be denoted by the corresponding lower-case letter: $\vec{r}$, $\vec{p}$, $\vec{j}$, ... , with the exception of the energy for which we use two different letters: the eigenvalues of H will be denoted by E.

4.4 Exercises

4.4.1 Dispersion and eigenvectors

Show that a necessary and sufficient condition for $|\varphi\rangle$ to be an eigenvector of a Hermitian operator A is that the dispersion (4.8) $\Delta_\varphi A = 0$.

---

[16] It is replaced by canonical commutation relations between the fields and their conjugate momenta, which lead to complicated mathematical objects called operator-valued distributions. But there is still such a long way to go (gauge invariance, renormalization) before calculating a physical quantity that the correspondence principle appears of rather secondary importance, and anyway in practice it is nowadays replaced by the Feynman path integral approach.

4.4.2 The variational method

1. Let $|\varphi\rangle$ be a vector (not normalized) in the Hilbert space of states and H be a Hamiltonian. The expectation value $\langle H\rangle_\varphi$ is

$$\langle H\rangle_\varphi = \frac{\langle\varphi|H|\varphi\rangle}{\langle\varphi|\varphi\rangle}.$$

Show that if the minimum of this expectation value is obtained for $|\varphi\rangle = |\varphi_m\rangle$ and the maximum for $|\varphi\rangle = |\varphi_M\rangle$, then

$$H|\varphi_m\rangle = E_m|\varphi_m\rangle \quad \text{and} \quad H|\varphi_M\rangle = E_M|\varphi_M\rangle,$$

where E_m and E_M are the smallest and largest eigenvalues.

2. We assume that the vector $|\varphi\rangle$ depends on a parameter α: $|\varphi\rangle = |\varphi(\alpha)\rangle$. Show that if

$$\left.\frac{\partial\langle H\rangle_{\varphi(\alpha)}}{\partial\alpha}\right|_{\alpha=\alpha_0} = 0,$$

then $E_m \leq \langle H\rangle_{\varphi(\alpha_0)}$ if α_0 corresponds to a minimum of $\langle H\rangle_{\varphi(\alpha)}$, and $\langle H\rangle_{\varphi(\alpha_0)} \leq E_M$ if α_0 corresponds to a maximum. This result forms the basis of an approximation method called the variational method (Section 14.1.4).

3. If H acts in a two-dimensional space, its most general form is

$$H = \begin{pmatrix} a+c & b \\ b & a-c \end{pmatrix},$$

where b can always be chosen to be real. Parametrizing $|\varphi(\alpha)\rangle$ as

$$|\varphi(\alpha)\rangle = \begin{pmatrix} \cos\alpha/2 \\ \sin\alpha/2 \end{pmatrix},$$

find the values of α_0 by seeking the extrema of $\langle\varphi(\alpha)|H|\varphi(\alpha)\rangle$. Rederive (2.35).

4.4.3 The Feynman–Hellmann theorem

Let a Hamiltonian H depend on a parameter λ: $H = H(\lambda)$. Let $E(\lambda)$ be a nondegenerate eigenvalue and $|\varphi(\lambda)\rangle$ be the corresponding normalized eigenvector ($\||\varphi(\lambda)\rangle\|^2 = 1$):

$$H(\lambda)|\varphi(\lambda)\rangle = E(\lambda)|\varphi(\lambda)\rangle.$$

Demonstrate the Feynman–Hellmann theorem:

$$\frac{\partial E}{\partial\lambda} = \langle\varphi(\lambda)\Big|\frac{\partial H}{\partial\lambda}\Big|\varphi(\lambda)\rangle. \tag{4.35}$$

4.4.4 Time evolution of a two-level system

We consider a two-level system with Hamiltonian H represented by the matrix

$$H = \hbar \begin{pmatrix} A & B \\ B & -A \end{pmatrix}$$

in the basis

$$|+\rangle = \begin{pmatrix} 1 \\ 0 \end{pmatrix}, \quad |-\rangle = \begin{pmatrix} 0 \\ 1 \end{pmatrix}.$$

According to (2.35), the eigenvalues and eigenvectors of H are

$$E_+ = \hbar\sqrt{A^2 + B^2}, \quad |\chi_+\rangle = \cos\frac{\theta}{2}|+\rangle + \sin\frac{\theta}{2}|-\rangle$$

$$E_- = -\hbar\sqrt{A^2 + B^2}, \quad |\chi_-\rangle = -\sin\frac{\theta}{2}|+\rangle + \cos\frac{\theta}{2}|-\rangle$$

with

$$A = \sqrt{A^2 + B^2}\cos\theta, \quad B = \sqrt{A^2 + B^2}\sin\theta, \quad \tan\theta = \frac{B}{A}.$$

1. The state vector $|\varphi(t)\rangle$ at time t can be decomposed on the $\{|+\rangle, |-\rangle\}$ basis:

$$|\varphi(t)\rangle = c_+(t)|+\rangle + c_-(t)|-\rangle.$$

Write down the system of coupled differential equations which the components $c_+(t)$ and $c_-(t)$ satisfy.

2. Let $|\varphi(t=0)\rangle$ be decomposed on the $\{|\chi_+\rangle, |\chi_-\rangle\}$ basis:

$$|\varphi(t=0)\rangle = |\varphi(0)\rangle = \lambda|\chi_+\rangle + \mu|\chi_-\rangle, \quad |\lambda|^2 + |\mu|^2 = 1.$$

Show that $c_+(t) = \langle +|\varphi(t)\rangle$ is written as

$$c_+(t) = \lambda e^{-i\Omega t/2}\cos\frac{\theta}{2} - \mu e^{i\Omega t/2}\sin\frac{\theta}{2}$$

with $\Omega = 2\sqrt{A^2 + B^2}$. Here $\hbar\Omega$ is the energy difference of the two levels. Show that $c_+(t)$ (as well as $c_-(t)$) satisfies the differential equation

$$\ddot{c}_+(t) + \left(\frac{\Omega}{2}\right)^2 c_+(t) = 0.$$

3. We assume that $c_+(0) = 0$. Find λ and μ up to a phase as well as $c_+(t)$. Show that the probability of finding the system in the state $|+\rangle$ at time t is

$$p_+(t) = \sin^2\theta \sin^2\left(\frac{\Omega t}{2}\right) = \frac{B^2}{A^2 + B^2}\sin^2\left(\frac{\Omega t}{2}\right).$$

4. Show that if $c_+(t=0) = 1$, then

$$c_+(t) = \cos\frac{\Omega t}{2} - i\cos\theta\sin\frac{\Omega t}{2}.$$

Find $p_+(t)$ and $p_-(t)$, and verify that the result is compatible with that of the preceding question.

4.4.5 Unstable states

Let $|\varphi(0)\rangle$ represent the state vector at time $t = 0$ of an unstable particle, or more generally that of an unstable quantum state such as an atom in an excited state, and let $\mathsf{p}(t)$ be the probability (survival probability) that it has not decayed at time t. The particle is assumed to be isolated from external influences (but not from quantized fields), so that the Hamiltonian H that governs the decay is time-independent. Let $|\Psi(t)\rangle$ be the state vector at time t of the full quantum system

$$|\Psi(t)\rangle = \exp\left(-\frac{iHt}{\hbar}\right)|\varphi(0)\rangle.$$

The probability amplitude for finding the state of the quantum system at time t in $|\varphi(0)\rangle$ is

$$c(t) = \langle\varphi(0)|\Psi(t)\rangle = \langle\varphi(0)|\exp\left(-\frac{iHt}{\hbar}\right)|\varphi(0)\rangle,$$

and the survival probability is

$$\mathsf{p}(t) = |c(t)|^2 = |\langle\Psi(t)|\varphi(0)\rangle|^2 = \langle\Psi(t)|\mathcal{P}|\Psi(t)\rangle,$$

where $\mathcal{P} = |\varphi(0)\rangle\langle\varphi(0)|$ is the projector on the initial state.

1. Let us first restrict ourselves to very short times. Show that for $t \to 0$

$$\mathsf{p}(t) \simeq 1 - \frac{(\Delta H)^2}{\hbar^2} t^2,$$

so that, for very short times, the decay law is certainly not exponential. The expectation values of H and H^2 are computed in the state $|\varphi(0)\rangle$. Note that ΔH must be finite, otherwise $|\varphi(0)\rangle$ would not belong to the domain of H^2, which would be difficult to imagine physically (see Chapter 7 for the definition of the domain of an operator).

2. A more general result is obtained as follows. Show first that

$$\Delta\mathcal{P}^2 = \langle\mathcal{P}\rangle - \langle\mathcal{P}\rangle^2$$

and use (4.27) to deduce the inequality ($\Delta H = (\langle H^2\rangle - \langle H\rangle^2)^{1/2}$)

$$\left|\frac{d\mathsf{p}(t)}{dt}\right| \leq \frac{2\Delta H}{\hbar}\sqrt{\mathsf{p}(1-\mathsf{p})}.$$

Integrating this differential equation, derive

$$\mathsf{p}(t) \geq \cos^2\left(\frac{t\Delta H}{\hbar}\right) \quad 0 \leq t \leq \frac{\pi\hbar}{2\Delta H}.$$

3. Let $|n\rangle$ be a complete set of eigenstates of the Hamiltonian

$$H|n\rangle = E_n|n\rangle.$$

Show that $c(t)$ is given by the Fourier transform of a spectral function $w(E)$

$$w(E) = \sum_n |\langle n|\varphi(0)\rangle|^2 \, \delta(E - E_n).$$

Set $E_0 = \langle H\rangle$ and give the expression of $(\Delta H)^2$ in terms of $w(E)$ and E_0.

4. If $w(E)$ has a Lorentzian shape

$$w(E) = \frac{\Gamma\hbar}{2\pi} \frac{1}{(E-E_0)^2 + \hbar^2\Gamma^2/4},$$

show that

$$c(t) = \mathrm{e}^{-iE_0 t/\hbar}\,\mathrm{e}^{-\Gamma t/2}$$

and that the decay law is an exponential. The width of $w(E)$ is $\hbar\Gamma$, but ΔH is infinite, Thus ΔH is a rather poor measure of energy spread, and the width $\hbar\Gamma = \Delta E$ is the physically relevant quantity.

4.4.6 The solar neutrino puzzle

The nuclear reactions occurring in the interior of the Sun produce an abundance of electron neutrinos ν_e; 95% of these are produced in the reaction

$$\mathrm{p+p} \rightarrow {}^2\mathrm{H} + \mathrm{e}^+ + \nu_e.$$

The Earth receives 6.5×10^{14} neutrinos per second and per square metre from the Sun. For about thirty years several experiments sought to detect these neutrinos, but all of them concluded that the measured neutrino flux is only about half the flux calculated using the standard solar model. Now this model is considered to be quite reliable,[17] in particular owing to recent results from helioseismology. In any case, the uncertainties in the solar model cannot explain this "solar neutrino deficit." The combined results of three experiments (see Footnote 4, Chapter 1) have now shown with no possible doubt that this neutrino deficit is due to the transformation of ν_e neutrinos into other types of neutrino during the passage from the Sun to the Earth. These experiments show that the total neutrino flux predicted by the solar model is correct, but that the measured *electron neutrino* flux is too small. We shall construct a simplified theory which gives the essential physics. We assume that

- there exist only two types of neutrino, the electron neutrino ν_e and the muon neutrino ν_μ (in fact, there is also a third type, the τ neutrino ν_τ);
- the entire phenomenon takes place in a vacuum during the propagation from the Sun to the Earth (the propagation inside the Sun actually plays an important role).[18]

It has long been thought that neutrinos have zero mass. If, on the contrary, they are massive, we can place them in their rest frame and write down the Hamiltonian in the $\{|\nu_e\rangle, |\nu_\mu\rangle\}$ basis:

$$|\nu_e\rangle = \begin{pmatrix} 1 \\ 0 \end{pmatrix}, \quad |\nu_\mu\rangle = \begin{pmatrix} 0 \\ 1 \end{pmatrix}, \quad H = c^2 \begin{pmatrix} m_e & m \\ m & m_\mu \end{pmatrix}.$$

---

[17] It is often said that the interior of the Sun is much better understood than that of the Earth.
[18] See E. Abers, *Quantum Mechanics*, New Jersey: Pearsons Education (2004), Chapter 6, for an elementary discussion.

The off-diagonal element m makes transitions between electron neutrinos and muon neutrinos possible.

1. Show that the states of definite mass are $|v_1\rangle$ and $|v_2\rangle$:

$$|v_1\rangle = \cos\frac{\theta}{2}\,|v_e\rangle + \sin\frac{\theta}{2}\,|v_\mu\rangle,$$

$$|v_2\rangle = -\sin\frac{\theta}{2}\,|v_e\rangle + \cos\frac{\theta}{2}\,|v_\mu\rangle,$$

with

$$\tan\theta = \frac{2m}{m_e - m_\mu},$$

and that the masses m_1 and m_2 are

$$m_1 = \frac{m_e + m_\mu}{2} + \sqrt{m^2 + \left(\frac{m_e - m_\mu}{2}\right)^2},$$

$$m_2 = \frac{m_e + m_\mu}{2} - \sqrt{m^2 + \left(\frac{m_e - m_\mu}{2}\right)^2}.$$

2. Neutrinos propagate with a speed close to that of light; their energy is very high compared with $\langle m\rangle c^2$, where $\langle m\rangle$ is the typical mass in H. Show that if an electron neutrino is produced inside the Sun at time $t = 0$ with state vector

$$|\varphi(t = 0)\rangle = |v_e\rangle = \cos\frac{\theta}{2}\,|v_1\rangle - \sin\frac{\theta}{2}\,|v_2\rangle,$$

the state vector at time t has component on $|v_e\rangle$ given by

$$\langle v_e|\varphi(t)\rangle = e^{-iE_1 t/\hbar}\left(\cos^2\frac{\theta}{2} + \sin^2\frac{\theta}{2}\,e^{-i\Delta E t/\hbar}\right),$$

where $\Delta E = E_2 - E_1$. Show that the probability of finding a neutrino v_e at time t is

$$p_e(t) = 1 - \sin^2\theta\,\sin^2\left(\frac{\Delta E\, t}{2\hbar}\right).$$

This transformation phenomenon is called *neutrino oscillation*.

3. If $p \gg \langle m\rangle c$ is the neutrino momentum, show that ΔE, as measured in the Sun rest frame, is

$$\Delta E = \frac{(m_2^2 - m_1^2)c^3}{2p} = \frac{\Delta m^2 c^3}{2p}$$

with $\Delta m^2 = m_2^2 - m_1^2$. Then t must also be measured in the Sun rest frame, and not in the neutrino rest frame!

4. Assuming that half an oscillation occurs during the trip from the Sun to the Earth (that is, $\Delta E\, t/\hbar = \pi$) for neutrinos of energy 8 MeV, what is the order of magnitude of the difference of the squared masses Δm^2? The Earth–Sun separation is 150 million kilometers.

4.4.7 The Schrödinger and Heisenberg pictures

Let a Hermitian operator A be time-dependent in the Schrödinger picture: $A = A(t)$. The Hamiltonian H is also assumed to be time-dependent. Show that

$$A_H(t) = U^{-1}(t, t_0)A(t)U(t, t_0)$$

satisfies

$$i\hbar\frac{\mathrm{d}A_H}{\mathrm{d}t} = [A_H(t), H_H(t)] + i\hbar\left(\frac{\partial A(t)}{\partial t}\right)_H,$$

where $H_H(t)$ and $(\partial A/\partial t)_H$ are obtained from $H(t)$ and $(\partial A(t)/\partial t)$ by the transformation law used for A.

4.4.8 The system of neutral K mesons

Let us suppose that at time $t = 0$ an unstable particle A of mass m is created whose state vector at time $t = 0$ is $|\varphi(0)\rangle$. If the particle A were stable, $c(t)$ would simply be given by

$$c(t) = \exp\left(-i\frac{Et}{\hbar}\right) = \exp\left(-i\frac{mc^2 t}{\hbar}\right)$$

in the particle rest frame, where its energy is $E = mc^2$, and we would have $|c(t)|^2 = 1$ for all times t, as the probability that the particle exists at any time t would always be unity. Now let us suppose that the particle is unstable and that its decay follows an exponential law. Then, from Exercise 4.4.5,

$$c(t) = \exp\left(-i\frac{mc^2 t}{\hbar}\right)\exp\left(-\frac{t}{2\tau}\right).$$

We would like to adapt this description of particle decay to a two-level system, the system of neutral K mesons, by generalizing the differential equation obeyed by $c(t)$ ($\tau = 1/\Gamma$)

$$i\hbar\dot{c}(t) = \left(mc^2 - i\frac{\hbar\Gamma}{2}\right)c(t).$$

There exist two types of neutral K meson,[19] the K^0 formed from the down quark d and the strange antiquark $\bar{s}$, and the $\overline{K^0}$ formed from the $\bar{d}$ and the s. We recall that the charges of the u, d, and s quarks are respectively 2/3, $-1/3$, and $-1/3$ in units of the proton charge. These mesons are produced by the strong interaction, for which there is a conservation law analogous to that for electric charge: the number of strange quarks minus the number of strange antiquarks is conserved (just as in a reaction involving only electrons and positrons the number of electrons minus the number of positrons is conserved owing to electric charge conservation). Let us give some examples. The π^+

---

[19] There also exist two charged K mesons, the $K^+(u\bar{s})$ and the $K^-(\bar{u}s)$.

meson is the combination (u $\overline{\text{d}}$), the π^- meson is the combination ($\overline{\text{u}}$ d), and the Λ^0 is the combination (uds). The reactions

$$\pi^- (\overline{\text{u}}\text{d}) + \text{proton (uud)} \rightarrow \text{K}^0(\text{d}\,\overline{\text{s}}) + \Lambda^0 \text{ (uds)}$$

and

$$\overline{\text{K}^0} (\overline{\text{d}}\text{s}) + \text{proton (uud)} \rightarrow \pi^+ (\text{u}\,\overline{\text{d}}) + \Lambda^0 \text{ (uds)}$$

are allowed, while

$$\pi^- (\overline{\text{u}}\text{d}) + \text{proton (uud)} \rightarrow \overline{\text{K}^0}(\overline{\text{d}}\text{s}) + \Lambda^0 \text{ (uds)}$$

and

$$\text{K}^0 (\text{d}\,\overline{\text{s}}) + \text{proton (uud)} \rightarrow \pi^+(\text{u}\,\overline{\text{d}}) + \Lambda^0 \text{ (uds)}$$

are forbidden.

1. The $(\text{K}^0, \overline{\text{K}^0})$ system is a two-level system and its state vector $|\varphi(t)\rangle$ can be written as

$$|\varphi(t)\rangle = c(t)|\text{K}^0\rangle + \overline{c}(t)|\overline{\text{K}^0}\rangle$$

   in the $\{|\text{K}^0\rangle, |\overline{\text{K}^0}\rangle\}$ basis. The components of the vector $|\varphi(t)\rangle$ satisfy an evolution equation

$$i\hbar \begin{pmatrix} \dot{c}(t) \\ \dot{\overline{c}}(t) \end{pmatrix} = M \begin{pmatrix} c(t) \\ \overline{c}(t) \end{pmatrix},$$

   where M is a 2×2 matrix. Let $\mathcal{C}$ be the "charge conjugation operator" which exchanges particles and antiparticles:[20]

$$\mathcal{C}|\text{K}^0\rangle = |\overline{\text{K}^0}\rangle, \quad \mathcal{C}|\overline{\text{K}^0}\rangle = |\text{K}^0\rangle.$$

   Show that if M commutes with $\mathcal{C}$, its most general form is

$$M = \begin{pmatrix} A & B \\ B & A \end{pmatrix},$$

   where A and B are a priori complex numbers, because the matrix M is not Hermitian.

2. What are the eigenvectors $|K_1\rangle$ and $|K_2\rangle$ of M? Show that it is these two states which have well-defined energy and lifetime. If $|\varphi(t)\rangle$ has components $c(0)$ and $\overline{c}(0)$ at time $t = 0$, calculate $c(t)$ and $\overline{c}(t)$. We can write

$$A = \frac{1}{2}\left[(E_1 + E_2) - \frac{i\hbar}{2}(\Gamma_1 + \Gamma_2)\right],$$

$$B = \frac{1}{2}\left[(E_1 - E_2) - \frac{i\hbar}{2}(\Gamma_1 - \Gamma_2)\right].$$

3. Imagine that at time $t = 0$ a K^0 meson is produced in the reaction

$$\pi^- (\overline{\text{u}}\text{d}) + \text{proton (uud)} \rightarrow \text{K}^0(\text{d}\overline{\text{s}}) + \Lambda^0 \text{ (uds)}.$$

---

[20] We can generalize the argument using not $\mathcal{C}$ but the product $\mathcal{CP}$, where $\mathcal{P}$ is the parity operator. In fact, experiment shows that $[M, \mathcal{CP}] \neq 0$, but the corrections are very small.

What is the probability of finding a $\overline{K^0}$ meson at time t?[21] Assuming that $\Gamma_1 \gg \Gamma_2$, show that the probability of observing the reaction

$$\overline{K^0}\,(\overline{d}\,s) + \text{proton (uud)} \rightarrow \pi^+\,(u\,\overline{d}) + \Lambda^0\,(\text{uds})$$

for $t \sim \tau_1 = 1/\Gamma_1$ is proportional to

$$p(t) = 1 - 2\exp\left(-\frac{\Gamma_1 t}{2}\right)\cos\frac{(E_1 - E_2)t}{\hbar} + \exp\left(-\Gamma_1 t\right).$$

Plot the curve representing $p(t)$. What can be said about the order of magnitude of $(E_1 - E_2)$ versus that of E_1 or E_2? How can $(E_1 - E_2)$ be measured? The numerical values are $\tau_1 \simeq 10^{-10}$ s, $\tau_2 \simeq 10^{-7}$ s, and $E_1 \simeq E_2 \simeq 500\,\text{MeV}$.

4.5 Further reading

Our presentation of the postulates of quantum mechanics essentially follows the classical expositions of, for example, Messiah [1999], Chapter VIII, Cohen-Tannoudji *et al.* [1977], Chapter III, and Basdevant and Dalibard [2002], Chapter 5. The reader can also consult Peres [1993], Chapter 2; Isham [1995], Chapter 5; Ballentine [1998], Chapters 8 and 9; and Omnès [1999]. A qualitative discussion of the Heisenberg inequalities can be found in Lévy-Leblond and Balibar [1990], Chapter 3. Ballentine [1998], Chapter 12, and Peres [1993], Chapter 12, give particularly lucid discussions of the temporal Heisenberg inequality. A recent book on epistemological problems in quantum mechanics is J. Baggot, *Beyond Measure*, Oxford: Oxford University Press (2004).

---

[21] In practice, the K mesons travel in a straight line from their production point with a speed close to the speed of light, and the detector is located a distance $l \simeq ct(1 - v^2/c^2)^{-1/2}$ from the production point.

5

Systems with a finite number of levels

In this chapter we examine some simple applications of quantum mechanics in situations where it is possible to model quantum systems accurately by restricting ourselves to a space of states of finite dimension. If each energy level, including degenerate ones, is counted once, the dimension of $\mathcal{H}$ is equal to the number of levels, and this is why we use the term *system with a finite number of levels*. The first two examples (Section 5.1) are taken from quantum chemistry and allow us to study a stationary situation where the Hamiltonian is time-independent. But the most important point in this chapter is the introduction of time dependence, which will be implemented by coupling a two-level system to an external periodic classical field. This will be illustrated by three examples of great practical importance: nuclear magnetic resonance (Section 5.2), the ammonia molecule (Section 5.3), and the two-level atom (Section 5.4).

5.1 Elementary quantum chemistry

5.1.1 The ethylene molecule

The ethylene molecule C_2H_4 will serve as an introduction to the subject. The "skeleton" of this molecule is formed by the so-called σ bonds, pairs of σ *electrons* of opposite spin common to two carbon atoms or to a carbon and a hydrogen atom, thus forming the $(C_2H_4)^{++}$ ion (Fig. 5.1). The remaining two electrons, called π *electrons*, are mobile – they can jump from one carbon atom to another. It is said that they are *delocalized*. The separate treatment of the π and σ electrons is, of course, an approximation, but one that plays an important role in the theory of chemical bonding. Let us begin by putting the first π electron in place. It can be localized near carbon atom 1; we shall denote the corresponding quantum state as $|\varphi_1\rangle$.[1] It can also be localized near carbon atom 2, and the corresponding quantum state will be denoted as $|\varphi_2\rangle$ (Fig. 5.2). The energy E_0 of this electron when localized near atom 1 or atom 2 is the same owing to the symmetry between the two atoms. We shall approximate the space of states as a two-dimensional

---

[1] Dirac notation is superfluous in this chapter. We use it for coherence, but the reader can dispense with it if desired.

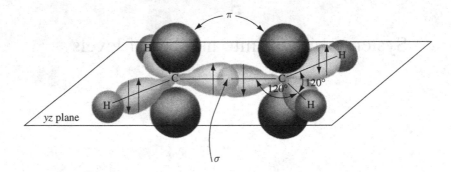

Fig. 5.1. The ethylene molecule.

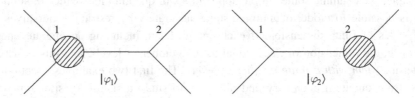

Fig. 5.2. The two possible states of a π electron, localized near atom 1 or near atom 2.

space $\mathcal{H}$ in which the basis vectors are $\{|\varphi_1\rangle, |\varphi_2\rangle\}$. In this basis the Hamiltonian can be written provisionally as

$$H_0 = \begin{pmatrix} E_0 & 0 \\ 0 & E_0 \end{pmatrix}, \quad H|\varphi_{1,2}\rangle = E_0|\varphi_{1,2}\rangle. \tag{5.1}$$

However, this Hamiltonian is incomplete, because we have neglected the possibility of the electron jumping from one carbon atom to another. Within our approximations, which are those of *Hückel's theory of molecular orbitals*, the most general form of H is

$$H = \begin{pmatrix} E_0 & -A \\ -A & E_0 \end{pmatrix}, \tag{5.2}$$

and the off-diagonal element $-A$ is precisely what gives rise to transitions between $|\varphi_1\rangle$ and $|\varphi_2\rangle$. By suitable choice of the phase of the basis vectors we can take A to be real; cf. Section 2.3.2. We have written A with a minus sign, which is significant because it can be shown that $A > 0$.

If $A \neq 0$, the states $|\varphi_1\rangle$ and $|\varphi_2\rangle$ will no longer be stationary states. As we have seen in Section 2.3.2, the eigenvectors of H are now

$$|\chi_+\rangle = \frac{1}{\sqrt{2}} \left(|\varphi_1\rangle + |\varphi_2\rangle \right) = \frac{1}{\sqrt{2}} \begin{pmatrix} 1 \\ 1 \end{pmatrix}, \tag{5.3}$$

$$|\chi_-\rangle = \frac{1}{\sqrt{2}} \left(|\varphi_1\rangle - |\varphi_2\rangle \right) = \frac{1}{\sqrt{2}} \begin{pmatrix} 1 \\ -1 \end{pmatrix}, \tag{5.4}$$

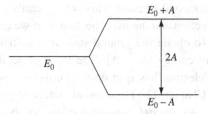

Fig. 5.3. Energy levels of a π electron.

with

$$H|\chi_+\rangle = (E_0 - A)|\chi_+\rangle, \quad H|\chi_-\rangle = (E_0 + A)|\chi_-\rangle. \tag{5.5}$$

Since $A > 0$, the symmetric state $|\chi_+\rangle$ is the state of lowest energy. The spectrum of the Hamiltonian is shown in Fig. 5.3, where we see that the ground state is the state $|\chi_+\rangle$ of energy $(E_0 - A)$. These results can be interpreted spatially by studying the localization of the electron on the line joining the two carbon atoms, which we take to be the x axis, with the origin located at the center of the line. As we shall see in detail in Chapter 9, if $|x\rangle$ is an eigenvector of the position operator, the quantity $\langle x|\varphi_1\rangle$ is the probability amplitude for finding the electron in the state $|\varphi_1\rangle$ at point x. In Chapter 9 we shall call this probability amplitude the *wave function* of the electron. The squared modulus of this probability amplitude gives the probability of finding the electron at point x,[2] also called the *probability density for the electron at point x*. This interpretation allows us to qualitatively represent the probability amplitudes $\chi_\pm(x) = \langle x|\chi_\pm\rangle$ corresponding to the states $|\chi_\pm\rangle$ as in Fig. 5.4. This probability vanishes at the origin in the antisymmetric case $|\chi_-\rangle$, but not in the symmetric one $|\chi_+\rangle$. The symmetric or antisymmetric nature of the ground-state wave function is related to the sign of A. Most of the time, ground states are symmetric, which corresponds to $A > 0$.

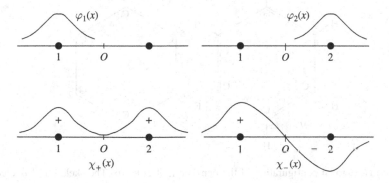

Fig. 5.4. Probability amplitudes for finding a π electron at a point x.

---

[2] More precisely, the probability per unit length: $|\langle x|\varphi\rangle|^2 dx$ is the probability of finding the particle in the range $[x, x+dx]$; see Section 9.1.2.

We still need to place the second electron. This is very easily done if we can ignore the interactions between this electron and the first one, that is, if we can use the approximation of independent electrons. To obtain the ground state it is sufficient to place the second electron in the state $|\chi_+\rangle$ of energy $(E_0 - A)$. The Pauli principle (Chapter 13) restricts the spin states: if the first electron has spin up $(|+\rangle)$, the second must have spin down $(|-\rangle)$, as we shall see in Chapter 13. The ground-state energy of the π bond then is $2(E_0 - A)$, where $-2A$ is called the *delocalization energy* of the π electrons. The crucial role played by the *independent particle approximation* should be emphasized. We have assumed that the π electrons do not interact with the σ electrons or with each other. It is difficult to justify this model on the basis of fundamental principles or from what are now termed *ab initio* calculations, but nevertheless it is of considerable practical importance.

5.1.2 The benzene molecule

In the benzene molecule the σ skeleton of the $(C_6H_6)^{6+}$ ion forms a hexagon. If we again add the six π electrons so as to form three double bonds we obtain the Kékulé formula (Fig. 5.5a) and the prediction $6(E_0 - A)$ for the ground-state energy. It is known from chemistry that the Kékulé formula cannot be completely correct,[3] and we shall see that taking into account the delocalization of the π electrons along the entire hexagonal chain leads to an energy lower than $6(E_0 - A)$. Therefore, the Kékulé formula does not give the correct ground-state energy. Let us begin by considering the addition of a single electron, assigning the numbers 0 to 5 to the carbon atoms along the hexagonal chain starting from an arbitrary origin (Fig. 5.5b).[4] For example, we use $|\varphi_3\rangle$ to denote the state where the electron is localized near atom 3. Since it is just as easy to deal with

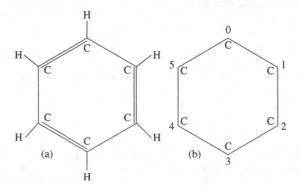

Fig. 5.5. (a) Hexagonal configuration of the benzene molecule. (b) The skeleton of σ electrons.

---

[3] For example, there exists a single form of orthodibromobenzene, whereas the Kékulé formula predicts two different ones. Moreover, the length of the bond between two carbon atoms in benzene (1.40 Å) is intermediate between the lengths of a simple (1.54 Å) and a double (1.35 Å) bond.

[4] As we shall soon see, it is much more convenient to number from 0 to 5 rather than from 1 to 6!

any number N of carbon atoms forming a closed chain, that is, a regular polygon of N sides, we shall use $|\varphi_n\rangle$ to denote the state where the electron is localized near the nth atom, $n = 0, 1, \ldots, N-1$, with $N = 6$ for benzene. Atoms n and $n+N$ are identical: $n \equiv n+N$. The space of states has N dimensions, and the Hamiltonian is defined by its action on $|\varphi_n\rangle$:

$$H|\varphi_n\rangle = E_0|\varphi_n\rangle - A(|\varphi_{n-1}\rangle + |\varphi_{n+1}\rangle). \qquad (5.6)$$

We shall use the symmetry of the problem under circular permutations of the N atoms of the chain to find the eigenvalues and eigenvectors of H. Let U_P be the unitary operator performing a circular permutation of the atoms in the direction $n \to (n-1)$:

$$U_P|\varphi_n\rangle = |\varphi_{n-1}\rangle, \qquad U_P^\dagger|\varphi_n\rangle = U_P^{-1}|\varphi_n\rangle = |\varphi_{n+1}\rangle. \qquad (5.7)$$

According to (5.6) and (5.7), we can write the Hamiltonian as

$$H = E_0 I - A(U_P + U_P^\dagger), \qquad (5.8)$$

which implies that H and U_P commute:

$$[H, U_P] = 0, \qquad (5.9)$$

and therefore have a basis of common eigenvectors. Let us look for the eigenvectors and eigenvalues of U_P, as this operator is a priori simpler than H. Since U_P is unitary, its eigenvalues have the form $\exp(i\delta)$ (see Section 2.3.4). From $(U_P)^N = I$, we deduce $\exp(iN\delta) = 1$, and so the eigenvalues can be classified by an integer index s:

$$\delta = \delta_s = \frac{2\pi s}{N}, \qquad s = 0, 1, \ldots, N-1. \qquad (5.10)$$

We have therefore determined the N distinct eigenvalues of U_P. Since the latter acts in a space of dimension N, the corresponding eigenvectors are orthogonal and form a basis of $\mathcal{H}$. Let us write a normalized eigenvector $|\chi_s\rangle$ in the form

$$|\chi_s\rangle = \sum_{n=0}^{N-1} c_n|\varphi_n\rangle, \qquad \sum_{n=0}^{N-1} |c_n|^2 = 1. \qquad (5.11)$$

On the one hand we have

$$U_P|\chi_s\rangle = \sum_{n=0}^{N-1} c_n|\varphi_{n-1}\rangle = \sum_{n=0}^{N-1} c_{n+1}|\varphi_n\rangle,$$

while on the other

$$U_P|\chi_s\rangle = e^{i\delta_s}|\chi_s\rangle = \sum_{n=0}^{N-1} e^{i\delta_s} c_n|\varphi_n\rangle.$$

Equating the coefficients of $|\varphi_n\rangle$ in these two equations leads to

$$c_{n+1} = e^{i\delta_s} c_n \quad \text{or} \quad c_n = e^{in\delta_s} c_0.$$

The eigenvector corresponding to the eigenvalue $\exp(i\delta_s)$ then is

$$|\chi_s\rangle = \frac{1}{\sqrt{N}} \sum_{n=0}^{N-1} e^{in\delta_s} |\varphi_n\rangle. \tag{5.12}$$

The choice $c_0 = 1/\sqrt{N}$ ensures that $|\chi_s\rangle$ is normalized. The bases $|\varphi_n\rangle$ and $|\chi_s\rangle$ are complementary according to the definition given in Section 3.1.2. Taking into account the expression (5.8) for H, the eigenvalue E_s is given by

$$E_s = E_0 - A\left(e^{i\delta_s} + e^{-i\delta_s}\right) = E_0 - 2A\cos\delta_s$$

or (Fig. 5.6)

$$E_s = E_0 - 2A\cos\frac{2\pi s}{N}. \tag{5.13}$$

We could have obtained (5.13) directly without the intermediary of the circular permutation operator U_P. However, our use of U_P illustrates a general strategy and is not just a computational trick. We shall often use this strategy, as it simplifies, sometimes greatly, the diagonalization of the Hamiltonian: instead of diagonalizing H directly, we first diagonalize the unitary symmetry operators which commute with H, when such operators exist owing to some symmetry of the physical problem.

It should be noted that the values s and $\tilde{s} = N - s$ give the same value of the energy; aside from $s = 0$ and $s = N - 1$ (for N even), the energy levels are doubly degenerate. It is

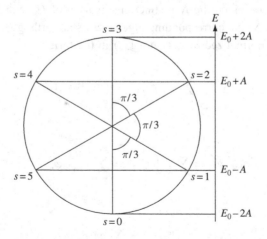

Fig. 5.6. Energy levels of a π electron of the benzene molecule.

possible to obtain eigenvectors of H with real components by forming linear combinations of $|\chi_s\rangle$ and $|\chi_{\tilde{s}}\rangle$:

$$|\chi_s^+\rangle = \frac{1}{\sqrt{2}}(|\chi_s\rangle + |\chi_{\tilde{s}}\rangle) = \sqrt{\frac{2}{N}}\sum_{n=0}^{N-1}\cos\frac{2\pi ns}{N}|\varphi_n\rangle, \qquad (5.14)$$

$$|\chi_s^-\rangle = \frac{1}{i\sqrt{2}}(|\chi_s\rangle - |\chi_{\tilde{s}}\rangle) = \sqrt{\frac{2}{N}}\sum_{n=0}^{N-1}\sin\frac{2\pi ns}{N}|\varphi_n\rangle. \qquad (5.15)$$

Now we can write down the results for the eigenvalues of H and the corresponding eigenvectors in the case of benzene, where $N = 6$, $\cos(2\pi/6) = 1/2$, and $\sin(2\pi/6) = \sqrt{3}/2$ (Fig. 5.6):

$$s = 0 \quad E = E_0 - 2A$$

$$|\chi_0\rangle = \frac{1}{\sqrt{6}}(1, 1, 1, 1, 1, 1);$$

$$s = 1, \ \tilde{s} = 5 \quad E = E_0 - A$$

$$|\chi_1^+\rangle = \frac{1}{\sqrt{3}}\left(1, \frac{1}{2}, -\frac{1}{2}, -1, -\frac{1}{2}, \frac{1}{2}\right), \qquad |\chi_1^-\rangle = \left(0, \frac{1}{2}, \frac{1}{2}, 0, -\frac{1}{2}, -\frac{1}{2}\right);$$

$$s = 2, \ \tilde{s} = 4 \quad E = E_0 + A$$

$$|\chi_2^+\rangle = \frac{1}{\sqrt{3}}\left(1, -\frac{1}{2}, -\frac{1}{2}, 1, -\frac{1}{2}, -\frac{1}{2}\right), \qquad |\chi_2^-\rangle = \left(0, \frac{1}{2}, -\frac{1}{2}, 0, \frac{1}{2}, -\frac{1}{2}\right);$$

$$s = \tilde{s} = 3 \quad E = E_0 + 2A$$

$$|\chi_3\rangle = \frac{1}{\sqrt{6}}(1, -1, 1, -1, 1, -1). \qquad (5.16)$$

Let us now find the ground state, that is, the state of lowest energy of the benzene molecule, by placing the six delocalized π electrons. In the approximation where the electrons are independent, this state will be obtained by first putting two electrons of opposite spins in the level $E_0 - 2A$. The Pauli principle (Chapter 13) forbids any more electrons in this level. As the level $(E_0 - A)$ is doubly degenerate, we can put four electrons in it (two pairs of electrons with opposite spins). This gives the total energy

$$E = 2(E_0 - 2A) + 4(E_0 - A) = 6E_0 - 8A. \qquad (5.17)$$

This energy is lower by $2A$ than the energy in the Kékulé formula $(6E_0 - 6A)$. The π electrons of benzene are not localized on the double bonds, but are delocalized along the entire hexagonal chain, and this form of delocalization decreases the energy by $2A$.

By comparing the heat of hydrogenation[5] of benzene into cyclohexane

$$C_6H_6 + 3H_2 \rightarrow C_6H_{12} - 49.8 \text{ kcal mol}^{-1}$$

---

[5] For purists: this is in fact a variation of the enthalpy, but the difference is negligible.

with that of cyclohexene, which contains a single double bond,

$$C_6H_{10} + H_2 \rightarrow C_6H_{12} - 28.6 \text{ kcal mol}^{-1},$$

we can estimate $2A$: $2A = 3 \times 28.6 - 49.8 = 36$ kcal mol$^{-1} \simeq 1.6$ eV. However, this estimate is at best an order of magnitude, because it involves uncertainties which are difficult to evaluate. They arise mainly from the approximation of independent electrons, which is poorly controlled.

5.2 Nuclear magnetic resonance (NMR)

In Section 5.1 we studied the energy levels of time-independent Hamiltonians. In the next three sections we introduce a time-dependent interaction for a two-level system by placing it in an external *classical* field which is periodic with frequency ω. Under these conditions it is clear that stationary states no longer exist, and the interesting problem is now the study of transitions from one level to another induced by the external field. We shall find the following fundamental result: if $\omega \simeq \omega_0$, where $\hbar\omega_0$ is the energy difference between the two levels, a remarkable resonance phenomenon occurs. We are going to give three examples of great practical importance: nuclear magnetic resonance in the present section, the ammonia molecule in Section 5.3, and the two-level atom in Section 5.4.

5.2.1 A spin 1/2 in a periodic magnetic field

Nuclear magnetic resonance (NMR) rests on the fact that an atomic nucleus with nonzero spin possesses a magnetic moment. We shall limit ourselves to spin-1/2 nuclei (^{1}H, ^{13}C, ^{19}F, etc.), for which the magnetic moment, which is an operator in quantum mechanics, is given by

$$\vec{\mu} = \gamma \vec{S} = \frac{1}{2}\gamma\hbar\vec{\sigma}, \tag{5.18}$$

where $\vec{S}$ is the spin operator defined in Section 3.2 and γ is the gyromagnetic ratio:

$$\gamma = \overline{\gamma}\frac{q_p}{2m_p}; \tag{5.19}$$

$\overline{\gamma} = 5.59$ for the proton, 1.40 for ^{13}C, 5.26 for ^{19}F, and so on. The nuclear spin is placed in a magnetic field $\vec{B}_0$ pointing in the Oz direction. Following (3.61), we can write the Hamiltonian H_0 of the nuclear spin as

$$H_0 = -\vec{\mu}\cdot\vec{B}_0 = -\frac{1}{2}\gamma\hbar B_0\sigma_z = -\frac{1}{2}\hbar\omega_0\sigma_z, \tag{5.20}$$

with $\omega_0 = \gamma B_0$, or in matrix form in the basis in which σ_z is diagonal:

$$H_0 = -\frac{1}{2}\hbar\begin{pmatrix} \omega_0 & 0 \\ 0 & -\omega_0 \end{pmatrix}. \tag{5.21}$$

We note that since the proton charge q_p is positive there is no minus sign in the definition of ω_0, in contrast to the case of Section 3.2.5 for the electron. Here ω_0 is the Larmor frequency, the frequency with which the classical magnetic moment precesses about $\vec{B}_0$ (Fig. 3.7). In the case of the proton the Larmor precession is in the clockwise direction. The state $|+\rangle$ has energy $-\hbar\omega_0/2$, and the state $|-\rangle$ has energy $\hbar\omega_0/2$. We therefore have a two-level system, the two Zeeman levels of a spin 1/2 in a magnetic field, with the energy difference of the levels being $\hbar\omega_0$.

Now let us add to the constant field $\vec{B}_0$ a periodic radiofrequency field $\vec{B}_1(t)$ parallel to the xOy plane and rotating in the clockwise direction,[6] that is, in the same direction as the Larmor precession, with angular speed ω:

$$\vec{B}_1(t) = B_1(\hat{x}\cos\omega t - \hat{y}\sin\omega t). \tag{5.22}$$

In practice, such a field can be obtained by means of two coils placed along the Ox and Oy axes and fed by an alternating current of frequency ω. The contribution to the Hamiltonian due to the field $\vec{B}_1(t)$ is

$$H_1(t) = -\vec{\mu}\cdot\vec{B}_1(t) = -\frac{1}{2}\hbar\omega_1(\sigma_x\cos\omega t - \sigma_y\sin\omega t),$$

where $\omega_1 = \gamma B_1$ is the *Rabi frequency*, often called the *nutation frequency* ω_{nut} in NMR. The total time-dependent Hamiltonian $H(t)$ in matrix form is then

$$H(t) = H_0 + H_1(t) = -\frac{1}{2}\hbar\begin{pmatrix} \omega_0 & \omega_1 e^{i\omega t} \\ \omega_1 e^{-i\omega t} & -\omega_0 \end{pmatrix}, \tag{5.23}$$

where we have used the expressions (3.49) for σ_x and σ_y. It is now easy to write down the Schrödinger equation in matrix form (4.13), decomposing the state vector $|\psi(t)\rangle$ onto the basis vectors $|+\rangle$ and $|-\rangle$:

$$|\psi(t)\rangle = c_+(t)|+\rangle + c_-(t)|-\rangle. \tag{5.24}$$

We obtain the following system of differential equations for $c_\pm(t)$:

$$i\frac{dc_\pm}{dt} = \mp\frac{1}{2}\omega_0 c_\pm - \frac{1}{2}\omega_1 e^{\pm i\omega t}c_\mp. \tag{5.25}$$

5.2.2 Rabi oscillations

To solve the system of differential equations (5.25), we define the coefficients $\gamma_\pm(t)$ as

$$c_\pm(t) = \gamma_\pm(t)\,e^{\pm i\omega_0 t/2}. \tag{5.26}$$

This definition has an interesting geometrical interpretation. When $\vec{B}_1 = 0$ the spin simply performs Larmor precession (Fig. 3.7) about $\vec{B}_0$ in the clockwise direction with

---

[6] We could also use a field $\vec{B}_1(t)$ parallel to Ox; see Exercise 5.5.6.

frequency ω_0. Instead of using the laboratory frame to measure the x and y components of the spin, we can use the reference frame rotating around Oz with the Larmor frequency ω_0, in which $|\psi(t)\rangle$ becomes $|\psi'(t)\rangle$:[7]

$$|\psi(t)\rangle \to |\psi'(t)\rangle = \mathrm{e}^{-\mathrm{i}\omega_0\sigma_z t/2}|\psi(t)\rangle$$

$$= c_+(t)\,\mathrm{e}^{-\mathrm{i}\omega_0 t/2}\,|+\rangle + c_-(t)\,\mathrm{e}^{\mathrm{i}\omega_0 t/2}|-\rangle. \tag{5.27}$$

The operator which performs a rotation by an angle θ about Oz is $\exp(-\mathrm{i}\theta\sigma_z/2)$, so that the coefficients $\gamma_\pm(t)$ are just the components of the state vector in the rotating reference frame. Another way of interpreting the transformation (5.27) is to note that if $\bar{B}_1 = 0$, then

$$c_\pm(t) = \mathrm{e}^{\pm\mathrm{i}\omega_0 t/2}\,c_\pm(0) \quad \gamma_\pm(t) = \mathrm{const.}$$

and the transformation (5.26) allows us to eliminate the trivial time dependence due to H_0. Using

$$\mathrm{i}\frac{\mathrm{d}c_\pm}{\mathrm{d}t} = \left(\mp\frac{1}{2}\omega_0\gamma_\pm + \mathrm{i}\frac{\mathrm{d}\gamma_\pm}{\mathrm{d}t}\right)\mathrm{e}^{\pm\mathrm{i}\omega_0 t/2},$$

we can transform (5.25) into

$$\mathrm{i}\frac{\mathrm{d}\gamma_\pm}{\mathrm{d}t} = -\frac{1}{2}\,\omega_1\,\mathrm{e}^{\pm\mathrm{i}(\omega-\omega_0)t}\,\gamma_\mp(t) = -\frac{1}{2}\,\omega_1\,\mathrm{e}^{\pm\mathrm{i}\delta t}\,\gamma_\mp(t). \tag{5.28}$$

The difference $\delta = (\omega - \omega_0)$ between the frequency of the external field and the Larmor frequency is called the *detuning*, and the *offset frequency* by NMR practitioners. It is particularly easy to solve (5.28) in the case of resonance, $\delta = 0$ (we shall see shortly the reason for this terminology):

$$\mathrm{i}\frac{\mathrm{d}\gamma_\pm}{\mathrm{d}t} = -\frac{1}{2}\,\omega_1\gamma_\mp(t). \tag{5.29}$$

Differentiating one of the equations with respect to time and using the second equation, we obtain

$$\frac{\mathrm{d}^2\gamma_\pm}{\mathrm{d}t^2} = -\frac{1}{4}\,\omega_1^2\gamma_\pm(t). \tag{5.30}$$

This equation can be integrated immediately. The solution depends on two constants a and b, $|a|^2 + |b|^2 = 1$, which are related to the initial conditions:

$$\gamma_+(t) = a\cos\left(\frac{\omega_1 t}{2}\right) + b\sin\left(\frac{\omega_1 t}{2}\right),$$

$$\gamma_-(t) = \mathrm{i}a\sin\left(\frac{\omega_1 t}{2}\right) - \mathrm{i}b\cos\left(\frac{\omega_1 t}{2}\right). \tag{5.31}$$

Equation (5.31) can be given a very interesting geometrical interpretation in the rotating reference frame. If the angle θ is defined as $\omega_1 t = \theta$, the operation (5.31) amounts to

---

[7] Another method of solving (5.25) is to use a reference frame rotating with frequency ω.

rotating the spin by an angle $-\theta$ about the Ox axis. This can be seen using the expression for the operator that performs a rotation by an angle $-\theta$ about the Ox axis:[8]

$$U[\mathcal{R}_x(-\theta)] = \exp\left(i\frac{\theta}{2}\,\sigma_x\right).$$

We then have

$$\begin{pmatrix} \gamma_+(t) \\ \gamma_-(t) \end{pmatrix} = e^{i\theta\sigma_x/2} \begin{pmatrix} \gamma_+(0) \\ \gamma_-(0) \end{pmatrix} = \begin{pmatrix} \cos\frac{\theta}{2} & i\sin\frac{\theta}{2} \\ i\sin\frac{\theta}{2} & \cos\frac{\theta}{2} \end{pmatrix} \begin{pmatrix} a \\ -ib \end{pmatrix}, \tag{5.32}$$

in agreement with (5.31). The classical picture of the rotation is also interesting. In the rotating frame, the spin sees a time-independent field $\vec{B}_1$, which is aligned along Ox. Thus (5.31) is nothing other than the Larmor precession about $\vec{B}_1$ with an angular frequency ω_1. To illustrate this rotation, let us suppose that at time $t = 0$ the spin is in the state $|+\rangle$, which has the lowest energy $-\hbar\omega_0/2$: $a = 1$, $b = 0$. At time t the probability $\mathsf{p}_\pm$ of finding the spin in the state $|\pm\rangle$ will be

$$\mathsf{p}_+(t) = |\langle+|\psi(t)\rangle|^2 = |\gamma_+(t)|^2 = \cos^2\left(\frac{\omega_1 t}{2}\right),$$
$$\mathsf{p}_-(t) = |\langle-|\psi(t)\rangle|^2 = |\gamma_-(t)|^2 = \sin^2\left(\frac{\omega_1 t}{2}\right). \tag{5.33}$$

The oscillations between the two levels are called *Rabi oscillations*. A spin which is initially in the state $|+\rangle$ will be found in the state $|-\rangle$ at times t given by

$$\frac{\omega_1 t}{2} = \left(n + \frac{1}{2}\right)\pi, \quad n = 0, 1, 2, 3, \ldots \tag{5.34}$$

If the radiofrequency field $\vec{B}_1(t)$ is applied during a time interval $[0, t]$ satisfying (5.34), in general with $n = 0$, it is said that a π *pulse* has been applied. When

$$\frac{\omega_1 t}{2} = \left(n + \frac{1}{2}\right)\frac{\pi}{2}, \quad n = 0, 1, 2, 3, \ldots, \tag{5.35}$$

we say that a $\pi/2$ *pulse* has been applied. The spin is then in a linear combination of the states $|+\rangle$ and $|-\rangle$ with equal weights.

In the off-resonance case, starting from (5.28) we obtain a second-order differential equation for γ_+:

$$\frac{2}{\omega_1}\frac{d^2\gamma_+}{dt^2} - \frac{2i}{\omega_1}\delta\frac{d\gamma_+}{dt} + \frac{1}{2}\omega_1\gamma_+ = 0, \tag{5.36}$$

---

[8] This expression is derived from Exercise 3.3.6, eq. (3.67), by taking the unit vector $\hat{p}$ parallel to Ox.

the solutions of which we seek in the form

$$\gamma_+(t) = e^{i\Omega_\pm t}.$$

The values of $\Omega_\pm$ are the roots of a second-order equation given as a function of the frequency $\Omega = (\omega_1^2 + \delta^2)^{1/2}$ by

$$\Omega_\pm = \frac{1}{2}[\delta \pm \Omega]. \tag{5.37}$$

The solution of (5.36) for γ_+ is a linear combination of $\exp(i\Omega_+ t)$ and $\exp(i\Omega_- t)$:

$$\gamma_+(t) = \lambda \exp(i\Omega_+ t) + \mu \exp(i\Omega_- t).$$

Let us choose the initial conditions $\gamma_+(0) = 1$, $\gamma_-(0) = 0$. Since $\gamma_-(0) \propto \dot{\gamma}_+(0)$, these initial conditions are equivalent to

$$\lambda + \mu = 1 \quad \text{and} \quad \lambda\Omega_+ + \mu\Omega_- = 0,$$

and so

$$\lambda = -\frac{\Omega_-}{\Omega}, \quad \mu = \frac{\Omega_+}{\Omega}.$$

The final result can be written as

$$\gamma_+(t) = \frac{e^{i\delta t/2}}{\Omega}\left[\Omega \cos\frac{\Omega t}{2} - i\delta \sin\frac{\Omega t}{2}\right], \tag{5.38}$$

$$\gamma_-(t) = \frac{i\omega_1}{\Omega}e^{-i\delta t/2}\sin\frac{\Omega t}{2}, \tag{5.39}$$

which reduces to (5.31) when $\delta = 0$. The factor $\exp(\pm i\delta t/2)$ arises because δ is the Larmor frequency in the rotating reference frame. Equation (5.39) is particularly interesting. It shows that if we start from the state $|+\rangle$ at $t = 0$, the probability of finding the spin in the state $|-\rangle$ at time t is

$$p_-(t) = \frac{\omega_1^2}{\Omega^2}\sin^2\left(\frac{\Omega t}{2}\right). \tag{5.40}$$

We see that the maximum probability of making a transition from the state $|+\rangle$ to the state $|-\rangle$ for $\Omega t/2 = \pi/2$ is given by a *resonance curve* of width δ:

$$p_-^{\text{max}} = \frac{\omega_1^2}{\Omega^2} = \frac{\omega_1^2}{\omega_1^2 + \delta^2} = \frac{\omega_1^2}{\omega_1^2 + (\omega - \omega_0)^2}. \tag{5.41}$$

As shown in Fig. 5.7, the Rabi oscillations are maximal at resonance and decrease rapidly in amplitude with growing δ. This has a clear intuitive interpretation: the influence of the radiofrequency (RF) field $\vec{B}_1$ is maximal when it rotates with the same speed as the spin undergoing Larmor precession about Oz, so that the spin sees a constant field $\vec{B}_1$ instead of a periodic one.

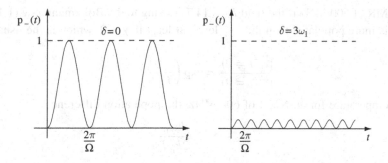

Fig. 5.7. Rabi oscillations. (a) $\delta = 0$, (b) $\delta = 3\omega_1$. In case (b) the maximum value of $p_-(t)$ is 1/10.

5.2.3 Principles of NMR and MRI

NMR is principally used to determine the structure of molecules in chemistry or biology, and for studying condensed matter in the solid or liquid state. A detailed description of how NMR works would take us too far afield, and so we shall only touch upon the subject. The sample under study is placed in a uniform field $\vec{B}_0$ of several teslas, the maximum strength attainable at present being about 20 T (Fig. 5.8). An NMR is usually characterized by specifying the resonance frequency[9] $\nu_0 = \omega_0/(2\pi) = \gamma B_0/(2\pi)$ for a proton: a field of 1 T corresponds to a frequency of about 42.5 MHz, and so we

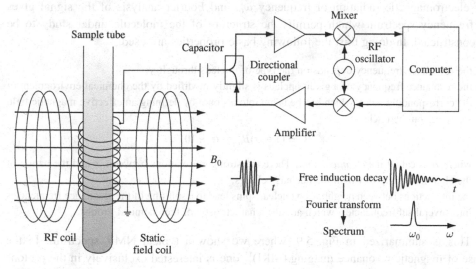

Fig. 5.8. Schematic depiction of an NMR. The static field $\vec{B}_0$ is horizontal and the RF field is generated by the vertical solenoid, which is also used for signal detection. The RF pulse and the signal are drawn on the bottom right of the figure. One notices the exponential decay of the signal and the peak of its Fourier transform at $\omega = \omega_0$. After Nielsen and Chuang [2000].

---

[9] See Footnote 23 of Chapter 1.

have an NMR of 600 MHz if the field B_0 is 14 T. Owing to the Boltzmann law (1.12), the $|+\rangle$ level is more populated than the $|-\rangle$ level, at least if $\gamma > 0$, which is the usual case:

$$\frac{\mathsf{p}_+(t=0)}{\mathsf{p}_-(t=0)} = \exp\left(\frac{\hbar\omega_0}{k_\mathrm{B}T}\right). \tag{5.42}$$

At room temperature for an NMR of 600 MHz, the population difference

$$\mathsf{p}_+ - \mathsf{p}_- \simeq \frac{\hbar\omega_0}{2k_\mathrm{B}T}$$

between the levels $|+\rangle$ and $|-\rangle$ is $\sim 5 \times 10^{-5}$.

The application of a radiofrequency field $\vec{B}_1(t)$ near resonance during a time t such that $\omega_1 t = \pi$, or a π-pulse (see (5.34)), causes the spins in the state $|+\rangle$ to flip to the state $|-\rangle$, thus inducing a *population inversion* relative to the equilibrium situation, so that the sample is no longer in equilibrium. The return to equilibrium is governed by a relaxation time T_1,[10] the *longitudinal relaxation time*. For reasons which will be explained in Section 6.2.4, a $\pi/2$ pulse is generally used, and so $\omega_1 t = \pi/2$. This corresponds geometrically to rotating the spin by an angle $\pi/2$ about an axis in the xOy plane (cf. (5.32)); if the spin is initially parallel to $\vec{B}_0$, it ends up in a plane perpendicular to $\vec{B}_0$, a transverse plane (whereas a π-pulse aligns the spin in the longitudinal direction $-\vec{B}_0$). The return to equilibrium is then governed by a relaxation time T_2, the *transverse relaxation time*. In any case, the return to equilibrium is accompanied by the emission of electromagnetic radiation of frequency ω_0, and Fourier analysis of the signal gives a frequency spectrum which permits the structure of the molecule under study to be reconstructed. In doing this, the following basic properties are used:

- the resonance frequency depends on the type of nucleus through γ;
- the resonance frequency of a given nucleus is slightly modified by the chemical environment of the corresponding atom, which can be taken into account by defining an effective magnetic field B_0' acting on the nucleus:

$$B_0' = (1 - \sigma)B_0, \quad \sigma \sim 10^{-6},$$

where σ is called the *chemical shift*. There are strong correlations between σ and the nature of the chemical group to which the nucleus belongs;
- the interactions between neighboring nuclear spins lead to a splitting of the resonance frequencies into several subfrequencies, which are also characteristic of the chemical groups.

This is summarized in Fig. 5.9, where we show a typical NMR spectrum. In the case of magnetic resonance imaging (MRI)[11] one is interested exclusively in the protons contained in water and fats. The sample is placed in a nonuniform field $\vec{B}_0$, which makes the resonance frequency spatially dependent. Since the signal amplitude is directly proportional to the spin density, and thus to the proton density, it is possible to obtain a

---

[10] When a field $\vec{B}_0$ is applied, thermodynamical equilibrium (5.42) is not established instantly, but only after a time $\sim T_1$.
[11] The adjective "nuclear" was dropped in order not to frighten the public!

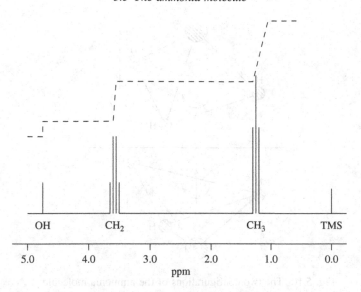

Fig. 5.9. NMR spectrum of protons of ethanol CH_3CH_2OH, obtained using an NMR of 200 MHz. The observed peaks are associated with the three groups OH, CH_3, and CH_2. The dashed line represents the integrated area of the signals, and the peak splitting is explained in Exercise 6.5.6. The TMS (tetramethyl silane) is a reference signal.

three-dimensional image of the density of water in biological tissues by means of complex computer calculations. The spatial resolution is of the order of a millimeter, and an image can be made in 0.1 s. This has permitted the development of functional MRI (fMRI), which can be used, for example, to watch the brain in action by measurement of local variations in the blood flow. The longitudinal and transverse relaxation times T_1 and T_2 play an important role in obtaining and interpreting MRI signals.

Although we shall meet the Rabi oscillations between two levels again in the next two sections, there are important differences of principle between NMR and the problems of molecular and atomic physics of these sections, on which we shall comment at the end of Section 5.4.

5.3 The ammonia molecule

The ammonia molecule will serve as the second example of a two-level system which can be coupled to an external periodic field.

5.3.1 The ammonia molecule as a two-level system

The ammonia molecule has the form of a pyramid with the nitrogen atom at the summit and the three hydrogen atoms forming an equilateral triangle which is the base (Fig. 5.10). There are a great many possible motions of this molecule. It can undergo translations

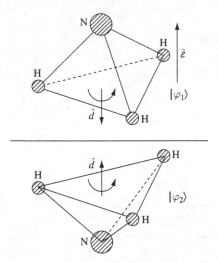

Fig. 5.10. The two configurations of the ammonia molecule.

and rotations in space, the atoms can oscillate about their equilibrium position, and the electrons can be in excited states. Once the degrees of freedom corresponding to the translation, rotation, and vibration of the molecule in its electronic ground state are fixed, there are still two possible configurations for the molecule rotating about its symmetry axis.[12] These two configurations are reflection-symmetric, one being the reflection of the other in a plane (Fig. 5.10). To go from one configuration to the other, the nitrogen atom must cross the plane formed by the hydrogen atoms. This is possible owing to the tunnel effect, which we shall explain in Section 9.4.2. Here we shall focus exclusively on these two configurations, which is justified by the energies involved.[13] As in the case of the ethylene molecule, we shall use a two-dimensional space to describe these two configurations. The molecule in state 1 (2) of Fig. 5.10 will be described by the basis vector $|\varphi_1\rangle$ ($|\varphi_2\rangle$). If the nitrogen atom were unable to cross the plane of the hydrogen atoms, the energies of the states $|\varphi_1\rangle$ and $|\varphi_2\rangle$ would be identical and equal to E_0. However, there exists a nonzero amplitude for crossing this plane, and the Hamiltonian takes the form (5.2)

$$H = \begin{pmatrix} E_0 & -A \\ -A & E_0 \end{pmatrix} \tag{5.43}$$

with, of course, values of E_0 and A completely different from those in Section 5.1. The value of E_0 is irrelevant for our discussion. However, it is worth noting that the value

---

[12] The importance of this rotation for generating the two different configurations has been emphasized by Feynman, and it has often been neglected in later discussions by other authors. In fact, if this rotation were absent, it would be possible to pass continuously from one configuration to the other by a spatial rotation.

[13] The ammonia molecule possesses two rotational eigenfrequencies, one of which is degenerate. They correspond to the energies 0.8×10^{-3} eV and 1.2×10^{-3} eV (degenerate). There are four vibrational modes, two of which are degenerate; the energy of the lowest one is 0.12 eV. In addition, the complications arising from the hyperfine structure should be taken into account.

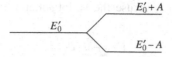

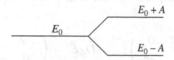

Fig. 5.11. Splitting of the two levels E_0 and E_0'.

of A in (5.43) differs from that in (5.2) by several orders of magnitude. We now have $2A \simeq 10^{-4}$ eV, whereas before $2A$ was of order 1 eV. This reflects the fact that it is easy for a π electron to jump from one atom to another, whereas it is very difficult for the nitrogen atom to cross the plane of the hydrogen atoms. This energy 10^{-4} eV corresponds to frequency 24 GHz or wavelength 1.25 cm. It is very low compared with the electron excitation energies (several eV), and also low compared with the vibrational (~ 0.1 eV) and rotational ($\sim 10^{-3}$ eV) energies (see Footnote 13). These numbers justify the approximation as a two-level system, because the difference between two adjacent rotational levels is of order $10A$ (Fig. 5.11). However, the molecule is not in its ground rotational state; since $k_B T \sim 0.025$ eV is large compared with $\sim 10^{-3}$ eV, the rotational levels are thermally excited. Following the discussion of Section 5.1.1, the energy levels of H are $E_0 \mp A$, corresponding to the stationary states (5.2) and (5.3):

$$E_0 - A : |\chi_+\rangle = \frac{1}{\sqrt{2}} \left(|\varphi_1\rangle + |\varphi_2\rangle \right) = \frac{1}{\sqrt{2}} \begin{pmatrix} 1 \\ 1 \end{pmatrix}, \tag{5.44}$$

$$E_0 + A : |\chi_-\rangle = \frac{1}{\sqrt{2}} \left(|\varphi_1\rangle - |\varphi_2\rangle \right) = \frac{1}{\sqrt{2}} \begin{pmatrix} 1 \\ -1 \end{pmatrix}. \tag{5.45}$$

The symmetric state $|\chi_+\rangle$ is the ground state of energy $(E_0 - A)$, and the antisymmetric state $|\chi_-\rangle$ is the excited state of energy $(E_0 + A)$.

5.3.2 The molecule in an electric field: the ammonia maser

The ammonia molecule possesses an electric dipole moment $\vec{d}$ which, by symmetry, is perpendicular to the plane of the hydrogen atoms. Since the hydrogen atoms tend to lose their electrons and the nitrogen atom tends to attract them, this dipole moment points from the nitrogen atom toward the plane of the hydrogen atoms (Fig. 5.10). Let us place the molecule in an electric field $\vec{\mathcal{E}}$ pointing in the Oz direction. The energy of a classical dipole $\vec{d}$ in an electric field $\vec{\mathcal{E}}$ (we use the script letter for the electric field to avoid confusion with the energy) is

$$E = -\vec{d} \cdot \vec{\mathcal{E}}. \tag{5.46}$$

In quantum mechanics the dipole moment is an operator $\vec{D}$ expressed as a function of the charges and the position operators of the various charged particles. We shall restrict $\vec{D}$ to our two-dimensional subspace, so that it is given by the following matrix in the $\{|\varphi_1\rangle, |\varphi_2\rangle\}$ basis:

$$-\vec{D} \rightarrow \begin{pmatrix} d & 0 \\ 0 & -d \end{pmatrix}, \quad -\vec{D} \cdot \vec{\mathcal{E}} \rightarrow \begin{pmatrix} d\mathcal{E} & 0 \\ 0 & -d\mathcal{E} \end{pmatrix}.$$

This corresponds to the diagram in Fig. 5.10. The energy of the state $|\varphi_1\rangle$ in this figure is $+d\mathcal{E}$ because the dipole moment is antiparallel to the field, and the energy of the state $|\varphi_2\rangle$ is $-d\mathcal{E}$ because the dipole moment is parallel to the field. The ultimate justification for the matrix form of this dipole moment lies in its agreement with experiment. The Hamiltonian then takes the form

$$H = \begin{pmatrix} E_0 + d\mathcal{E} & -A \\ -A & E_0 - d\mathcal{E} \end{pmatrix}. \tag{5.47}$$

Let us first study the case of a static electric field. The Hamiltonian is then independent of time. The eigenvalues can be calculated immediately:[14]

$$\det \begin{pmatrix} E_0 + d\mathcal{E} - E & -A \\ -A & E_0 - d\mathcal{E} - E \end{pmatrix} = (E - E_0)^2 - (d\mathcal{E})^2 - A^2 = 0,$$

giving

$$E_{\pm} = E_0 \mp \sqrt{A^2 + (d\mathcal{E})^2}. \tag{5.48}$$

These eigenvalues are shown in Fig. 5.12 as a function of $\mathcal{E}$. If $d\mathcal{E} \gg A$, the energies are $\simeq E_0 \pm d\mathcal{E}$ and the corresponding approximate eigenvectors are $|\varphi_1\rangle$ and $|\varphi_2\rangle$. In practice, the opposite case is the usual one: $d\mathcal{E} \ll A$. We can then expand the root in (5.48) as

$$E_{\pm} \simeq E_0 \mp A \mp \frac{1}{2} \frac{d^2 \mathcal{E}^2}{A}. \tag{5.49}$$

Up to terms of order $d\mathcal{E}/2A$ (cf. Exercise 5.5.4) the eigenvectors are $|\chi_+\rangle$ and $|\chi_-\rangle$. If the electric field is nonuniform, the molecule will be subject to a force

$$\vec{F}_{\pm} = -\vec{\nabla} E_{\pm} = \pm \frac{d^2}{2A} \vec{\nabla} \mathcal{E}^2. \tag{5.50}$$

As in the Stern–Gerlach experiment, it is possible to separate the eigenstates $|\chi_{\pm}\rangle$ of the Hamiltonian (5.47) experimentally, using a nonuniform electric field;[15] see Fig. 5.13.

---

[14] The results of Section 2.3.2 can also be used.

[15] In practice the field is chosen such that $|\chi_-\rangle$ is focused and the state $|\chi_+\rangle$ is defocused; cf. Basdevant and Dalibard [2002], Chapter 6.

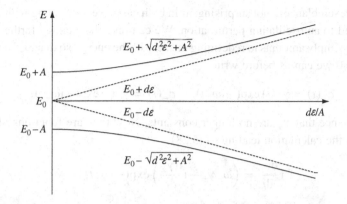

Fig. 5.12. Values of the energy as a function of the electric field $\mathcal{E}$.

Let us now assume that the electric field is an *oscillating* field:

$$\mathcal{E}(t) = \mathcal{E}_0 \cos \omega t = \frac{1}{2}\mathcal{E}_0 \left(e^{i\omega t} + e^{-i\omega t} \right), \quad \mathcal{E}_0 \text{ real} > 0. \tag{5.51}$$

The Hamiltonian depends explicitly on time. It will be convenient to take as the basis vectors the stationary states $|\chi_+\rangle$ and $|\chi_-\rangle$ ((5.44) and (5.45)) of the Hamiltonian (5.43), rather than $|\varphi_+\rangle$ and $|\varphi_-\rangle$. The Hamiltonian (5.47) in this new basis becomes

$$H(t) = \begin{pmatrix} E_0 - A & d\,\mathcal{E}(t) \\ d\,\mathcal{E}(t) & E_0 + A \end{pmatrix}. \tag{5.52}$$

Let us write down the general time-dependent state vector:

$$|\psi(t)\rangle = c_+(t)|\chi_+\rangle + c_-(t)|\chi_-\rangle. \tag{5.53}$$

The evolution equations (4.13) are

$$i\hbar\frac{dc_+}{dt} = (E_0 - A)c_+ + d\,\mathcal{E}(t)\,c_-,$$

$$i\hbar\frac{dc_-}{dt} = d\,\mathcal{E}(t)\,c_+ + (E_0 + A)c_-. \tag{5.54}$$

Thanks to our choice of basis vectors, when $\vec{\mathcal{E}} = 0$

$$c_+(t) = \gamma_+ \exp(-i\omega_+ t), \quad c_-(t) = \gamma_- \exp(-i\omega_- t),$$

where $\omega_+ = (E_0 - A)/\hbar$, $\omega_- = (E_0 + A)/\hbar$, and γ_+ and γ_- are constants. It will be convenient to set $\omega_0 = 2A/\hbar$, which physically represents the angular frequency, about 1.5×10^{12} rad s^{-1}, of the electromagnetic wave emitted when the molecule makes a transition from the excited level of energy $(E_0 + A)$ to the ground state of energy $(E_0 - A)$, so that $2A$ is the energy of the photon emitted in this transition. The frequency ω_0 is again called the resonance frequency, and the strong resemblance to the NMR equations should be noticed. This resemblance is not surprising, as in both cases we are dealing with a two-level system coupled to an oscillating perturbation. We can take the analogy farther by setting $E_0 = 0$, which simply amounts to redefining the zero of the energy so that $\omega_\pm = \mp\omega_0/2$.

When $\mathcal{E}_0 \neq 0$ we can as before write

$$c_+(t) = \gamma_+(t)\exp(-i\omega_+t), \quad c_-(t) = \gamma_-(t)\exp(-i\omega_-t),$$

with the difference that $\gamma_\pm$ are no longer constants. Now they are functions of time, and we can repeat the calculation leading to (5.28)

$$i\frac{dc_\pm}{dt} = \left(\omega_\pm\gamma_\pm + i\frac{d\gamma_\pm}{dt}\right)\exp(-i\omega_\pm t).$$

Substituting these into (5.54), we find

$$\begin{aligned}
i\frac{d\gamma_+}{dt} &= \frac{d\,\mathcal{E}(t)}{\hbar}\exp(-i\omega_0 t)\,\gamma_-(t),\\[4pt]
i\frac{d\gamma_-}{dt} &= \frac{d\,\mathcal{E}(t)}{\hbar}\exp(i\omega_0 t)\,\gamma_+(t).
\end{aligned} \tag{5.55}$$

We have obtained a system of coupled differential equations, which shows that the electric field induces transitions from the state $|\chi_+\rangle$ to the state $|\chi_-\rangle$ and back. Now let us substitute the electric field (5.51) into (5.55):

$$\begin{aligned}
i\frac{d\gamma_+}{dt} &= \frac{d\,\mathcal{E}_0}{2\hbar}\Big(\exp[i(\omega-\omega_0)t] + \exp[-i(\omega+\omega_0)t]\Big)\gamma_-(t),\\[4pt]
i\frac{d\gamma_-}{dt} &= \frac{d\,\mathcal{E}_0}{2\hbar}\Big(\exp[i(\omega+\omega_0)t] + \exp[-i(\omega-\omega_0)t]\Big)\gamma_+(t).
\end{aligned} \tag{5.56}$$

These equations are exact, but they cannot be solved analytically.[16] We shall obtain an approximate solution first assuming that the perturbation due to the electric field is weak: $d\,\mathcal{E}_0 \ll A$, or, equivalently, $d\,\mathcal{E}_0/\hbar \ll \omega_0$. The Rabi frequency is now $\omega_1 = d\,\mathcal{E}_0/\hbar$. The weak-field condition can therefore also be written as $\omega_1 \ll \omega_0$, which is (almost) always realized in practice. Under these conditions the functions $\gamma_\pm(t)$ vary slowly over a characteristic time ω_0^{-1}:

$$\left|\frac{d\gamma_\pm}{dt}\right| \sim \omega_1|\gamma_\mp| \ll \omega_0|\gamma_\mp|.$$

The second hypothesis needed for a simple approximate solution of (5.56) is that the frequency of the electric field be close to resonance, $\omega \simeq \omega_0$. This can be expressed as a function of the detuning $\delta = (\omega - \omega_0)$, so that we can state the preceding condition more precisely as $|\delta| \ll \omega_0$. Under these conditions the terms that behave as

$$\exp(\pm i(\omega + \omega_0)t) \sim \exp(\pm 2i\omega_0 t)$$

---

[16] Had we chosen a linearly polarized magnetic field in (5.22) instead of a circularly polarized field, we would also have needed to appeal to the rotating wave approximation: see Exercise 5.5.6.

in (5.56) vary very rapidly compared with the terms

$$\exp(\pm i(\omega - \omega_0)t) \sim \exp(\pm i\delta t),$$

and so their effect averaged over time is negligible. Omitting these terms, an approximation known as *the rotating-wave* or *quasi-resonant approximation*, we finally obtain the following system of coupled equations:

$$i\frac{d\gamma_\pm}{dt} = \frac{\omega_1}{2}\exp[i(\omega - \omega_0)t]\gamma_\mp(t). \tag{5.57}$$

This system of coupled differential equations, which is identical to that of (5.28) for NMR up to an unimportant overall sign, can now be solved analytically. Again we stress the fact that the two conditions $\omega_1 \ll \omega_0$ and $|\delta| \ll \omega_0$ are essential in going from (5.56) to (5.57).

Let us now take the frequency of the electric field equal to the transition frequency, so that we are sitting right on the resonance: $\omega = \omega_0$. We assume that at time $t = 0$ the molecule is in the state $|\chi_-\rangle$ of energy $(E_0 + A)$ ($a = 0$, $b = 1$).[17] To calculate the probability $p_\pm$ of finding the molecule in the state $|\chi_\pm\rangle$ at time t it is sufficient to copy (5.33):

$$p_-(t) = |\langle\chi_-|\psi(t)\rangle|^2 = |\gamma_-(t)|^2 = \cos^2\left(\frac{\omega_1 t}{2}\right),$$
$$p_+(t) = |\langle\chi_+|\psi(t)\rangle|^2 = |\gamma_+(t)|^2 = \sin^2\left(\frac{\omega_1 t}{2}\right). \tag{5.58}$$

The molecule goes from the state $|\chi_-\rangle$ to the state $|\chi_+\rangle$ with angular frequency $\omega_1/2 = d\mathcal{E}_0/2\hbar$.

Having put the molecule in the state $|\chi_-\rangle$ by means of the filter described above, the molecule is then allowed to pass through a cavity in which there is a field oscillating at the resonance frequency (Fig. 5.13). The molecule crosses the cavity in a time interval t. If this time is adjusted such that

$$\frac{d\mathcal{E}_0 t}{2\hbar} = \frac{\pi}{2},$$

that is, a π-pulse, at the exit from the cavity, all the molecules that have passed through will be in the state $|\chi_+\rangle$. By energy conservation the molecules *deliver* energy to the electromagnetic field. This process is called *stimulated (or induced) emission*. If the molecules are initially in the state $|\chi_+\rangle$, they will absorb energy from the electromagnetic field in going to the state $|\chi_-\rangle$, a process called *(stimulated) absorption*.

The process of stimulated emission can be used for amplifying an electromagnetic field provided that molecules can be produced in an excited state, that is, that a *population*

---

[17] In the case of NMR the spin is initially in the lowest energy state, while in the case of the maser we are interested in the opposite situation.

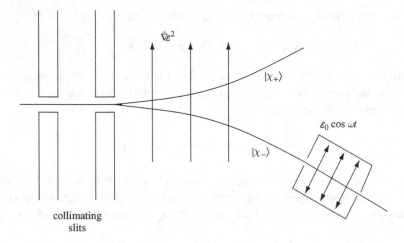

Fig. 5.13. The ammonia maser.

inversion can be generated.[18] The experimental apparatus shown schematically in Fig. 5.13 realizes such an amplification. The molecules selected in the state $|\chi_-\rangle$ cross a cavity of suitable length in which there is an electric field oscillating at the resonance frequency. This apparatus is a prototype of a maser.[19]

5.3.3 Off-resonance transitions

Now let us imagine the system is away from resonance, $\omega \simeq \omega_0$ but $\omega \neq \omega_0$, and start for example at time $t = 0$ from a molecule in the state $|\chi_+\rangle$. We wish to calculate the probability $\mathsf{p}(\omega; t)$ of finding the molecule in the state $|\chi_-\rangle$ at time t. Exact solution of Eqs. (5.57) gives the result (5.40) which can be written as

$$
\mathsf{p}(\omega; t) = \frac{\omega_1^2}{(\omega - \omega_0)^2 + \omega_1^2} \sin^2 \left(\frac{t}{2} \sqrt{(\omega - \omega_0)^2 + \omega_1^2} \right) . \tag{5.59}
$$

We recall that the Rabi frequency $\omega_1 = d\,\mathcal{E}_0/\hbar$. Although we can write down the exact solution, it is useful to find a simple approximate solution of (5.57) when the condition

$$
\frac{d\,\mathcal{E}_0 t}{\hbar} = \omega_1 t \ll 1 \quad \text{or} \quad t \ll \frac{\hbar}{d\mathcal{E}_0} = \frac{1}{\omega_1} = \tau_2 \tag{5.60}
$$

---

[18] As we have already seen in (5.42), if E_0 is the ground-state energy and E_1 the excited-state energy, the ratio $\mathsf{p}_1/\mathsf{p}_0$ of the probabilities of finding an atomic or molecular system in the state E_1 or E_0 is given by the Boltzmann law: $\mathsf{p}_1/\mathsf{p}_0 = \exp[(E_0 - E_1)/k_\mathrm{B}T] < 1$. It is therefore necessary to depart from thermal equilibrium to obtain such a population inversion.

[19] Maser is an acronym for "microwave amplification by stimulated emission of radiation," and laser for "light amplification by stimulated emission of radiation."

is satisfied: that is, for sufficiently short times. This approximate solution is interesting because it may be used in many problems that cannot be solved exactly and it sets the stage for Chapter 9. At $t = 0$ we have

$$\gamma_+ = 1, \quad \gamma_- = 0.$$

We are interested in a process in which the absorption of electromagnetic radiation makes it possible to go from the ground state to an excited state. In solving (5.57) for $\gamma_-(t)$ we can assume that $\gamma_+ \simeq 1$; in fact, owing to the condition (5.60) there is no time for γ_+ to vary appreciably. The approximate solution of the equation giving γ_- is then obvious:

$$\gamma_-(t) \simeq \frac{\omega_1}{2i} \int_0^t dt' \exp[-i(\omega - \omega_0)t'] = -\frac{\omega_1}{2} \left[\frac{1 - \exp[-i(\omega - \omega_0)t]}{\omega - \omega_0} \right]. \tag{5.61}$$

This gives the transition probability at frequency ω, $p(\omega; t)$:

$$p(\omega; t) = |\gamma_-(t)|^2 = \frac{1}{4} \omega_1^2 t^2 \frac{\sin^2[(\omega - \omega_0)t/2]}{[(\omega - \omega_0)t/2]^2}. \tag{5.62}$$

It thus appears that $p(\omega; t) \propto t^2$ for $|\delta| t \ll 1$, but this situation actually arises because we are considering a single frequency ω. In practice, the frequency spectrum is always continuous, and we are going to take this into account. The ratio of the above result and the result at resonance is

$$\frac{p(\omega; t)}{p(\omega_0; t)} = f(\omega - \omega_0; t) = \frac{\sin^2[(\omega - \omega_0)t/2]}{[(\omega - \omega_0)t/2]^2}.$$

The function $f(\omega - \omega_0; t)$ is plotted as a function of ω in Fig. 5.14. At $\omega = \omega_0$ it has a sharp peak of width $\sim 2\pi/t$. Using the fact that

$$\int_{-\infty}^{\infty} \frac{\sin^2 x}{x^2} dx = \pi,$$

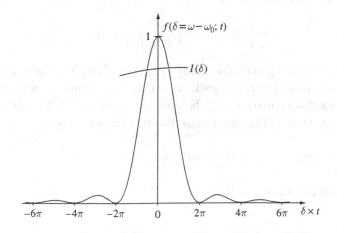

Fig. 5.14. The function $f(\omega - \omega_0; t)$.

the area under the peak is $2\pi/t$ and $f(\omega - \omega_0; t)$ is approximately a Dirac delta function:

$$f(\omega - \omega_0; t) = \frac{\sin^2[(\omega - \omega_0)t/2]}{[(\omega - \omega_0)t/2]^2} \simeq \frac{2\pi}{t} \delta(\omega - \omega_0). \tag{5.63}$$

These results allow us to calculate the rate of the transition from the state $|\chi_+\rangle$ to the state $|\chi_-\rangle$ due to absorption of electromagnetic radiation by the molecule in its ground state.[20] The incident energy flux $\mathcal{J}$ of an electromagnetic wave is given by the Poynting vector $\vec{S} = \varepsilon_0 c^2 \vec{E} \times \vec{B}$:

$$\mathcal{J} = \varepsilon_0 c^2 \langle \vec{E} \times \vec{B} \rangle = \frac{1}{2} \varepsilon_0 c \mathcal{E}_0^2, \tag{5.64}$$

where $\langle \bullet \rangle$ represents the time average and the electric field is of the form (5.51). Under these conditions

$$\mathsf{p}(\omega; t) = \left(\frac{d\mathcal{E}_0}{2\hbar}\right)^2 t^2 f(\omega - \omega_0; t) = 2\pi \left(\frac{d^2}{4\pi\varepsilon_0 \hbar^2 c}\right) \mathcal{J} t^2 f(\omega - \omega_0; t). \tag{5.65}$$

As we have already noted, the frequency of the electric field is not fixed exactly, but lies in a spectrum of frequencies whose typical variation scale is $\Delta\omega$. Let $\mathcal{J}(\omega)$ be the intensity per unit frequency and assume that $\Delta\omega \gg \pi/t$ (Fig. 5.14). The transition probability integrated over ω is then

$$\mathsf{p}(t) = 2\pi \left(\frac{d^2}{4\pi\varepsilon_0 \hbar^2 c}\right) t^2 \int_0^\infty d\omega \, \mathcal{J}(\omega) f(\omega - \omega_0; t)$$

$$\simeq 4\pi^2 \left(\frac{d^2}{4\pi\varepsilon_0 \hbar^2 c}\right) \mathcal{J}(\omega_0) t,$$

where we have used the approximation (5.63) for $f(\omega - \omega_0; t)$. The remarkable fact is that $\mathsf{p}(t)$ is proportional to t (and not to t^2!), and that $\mathsf{p}(t)/t$ can be interpreted as a *transition probability per unit time* Γ:

$$\Gamma = \frac{1}{t} \mathsf{p}(t) = 4\pi^2 \left(\frac{d^2}{4\pi\varepsilon_0 \hbar^2 c}\right) \mathcal{J}(\omega_0). \tag{5.66}$$

The fact that the transition probability is proportional to d^2 and $\mathcal{J}$ is characteristic of most processes of absorption of electromagnetic radiation by an atomic or molecular system. The conditions for this approximation to be valid are (i) $t \gg \tau_1 \sim 1/\Delta\omega$ and (ii) $\mathsf{p}(t) \ll 1$, that is, $t \ll \tau_2$ (see (5.60)). The time t must therefore lie in the range

$$\tau_1 \sim \frac{1}{\Delta\omega} \ll t \ll \tau_2 \sim \frac{1}{\omega_1}.$$

Of course this implies that $\omega_1 \ll \Delta\omega$.

---

[20] More precisely, these results apply to an ensemble of transitions from energy $(E_0 - A)$ to energy $(E_0 + A)$ (Fig. 5.11), where it is assumed that molecules in the state $(E_0 - A)$ are selected by the method described in Section 5.3.2.

5.4 The two-level atom

The calculation which we have just presented lays the foundations of a general theory of the absorption and emission of electromagnetic radiation by an atomic or molecular system, up to the following restrictions.

- The approximation by a two-level system must be valid. This will be the case if we are exclusively interested in transitions between two levels separated by an energy $\hbar\omega_0$ induced by an electromagnetic field of frequency $\omega \simeq \omega_0$, that is, if we are near resonance. We shall conventionally denote the state with the lowest energy as $|g\rangle$ (this will often be the ground state), and the second as $|e\rangle$ (the excited state; Fig. 5.15). In the case of an atom, this approximation is called the *two-level atom approximation*, and it provides a basic model for atomic physics and lasers.
- The transition must be an *electric dipole* transition, that is, controlled by the matrix element of the electric dipole moment operator $\vec{D}$ acting between the two levels, and the condition $\omega_1 \ll \omega_0$ must be satisfied.
- The electromagnetic field is treated as a classical field. The treatment which we have just presented is termed "semiclassical": the atom is treated as a quantum system, but the field remains classical. The "photon" behavior of the electromagnetic field is therefore ignored, and it is not possible in principle to take into account the spontaneous emission of radiation by an atom in an excited state (or at best it is possible to treat it heuristically).
- The results of Section 5.3.3 should be modified to take into account the finite lifetime of the excited state (Section 14.4).

When a two-level atom interacts with an electromagnetic field, in practice these days the field of a laser, the absorption probability is calculated following the scheme of Section 5.3.3, but the orders of magnitude are of course different from those in the case of the ammonia molecule. To take the example already mentioned in Section 1.5.3, the energy difference $\hbar\omega_0$ between the ground state and the first excited state of rubidium is about 1.6 eV, corresponding to a wavelength of 0.78 μm, at the limit of the infrared region. This order of magnitude is typical of atomic physics; the transitions generally used are in the visible region or in the near ultraviolet or near infrared.

We have already emphasized the fact that spontaneous emission cannot in principle be described by a semiclassical treatment, because it involves a transition from an initial state with zero photons to a final state with one photon – a photon is created at the instant the atom de-excites. Only a quantum theory of the electromagnetic field permits

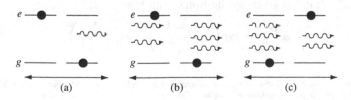

Fig. 5.15. (a) Spontaneous emission. (b) Stimulated emission. (c) Absorption.

the rigorous description of spontaneous emission. Although our classical treatment of the electromagnetic field does not admit an interpretation in terms of photons, we can nevertheless try to describe heuristically the process of Section 5.3.3 using this concept. For example, we can interpret the energy gain of the field as an increase of the number of photons in the cavity. The process

$$|\chi_-\rangle + n \text{ photons} \rightarrow |\chi_+\rangle + (n+1) \text{ photons} \qquad (5.67)$$

then represents stimulated emission. Stimulated absorption is the reverse process:

$$|\chi_+\rangle + n \text{ photons} \rightarrow |\chi_-\rangle + (n-1) \text{ photons.} \qquad (5.68)$$

Finally, the spontaneous emission of a photon occurs when the excited level $|\chi_-\rangle$ de-excites in the absence of an electromagnetic field:

$$|\chi_-\rangle + 0 \text{ photon} \rightarrow |\chi_+\rangle + 1 \text{ photon.} \qquad (5.69)$$

These processes are shown schematically in Fig. 5.15. It is important to distinguish between stimulated emission, which is coherent with the incident wave and proportional to the incident intensity, and spontaneous emission, which is random, as it has no phase relation to the applied field and is not influenced by external conditions.[21]

The necessity of spontaneous emission was first demonstrated by Einstein. Let us study a collection of atoms with two levels E_1 and E_2, $E_1 < E_2$, located in a cavity at temperature T. The cavity contains radiation obeying Planck's law (1.22). If N is the total number of atoms and $N_1(t)$ and $N_2(t)$ are the numbers of atoms in the states E_1 and E_2, then

$$N_1(t) + N_2(t) = N = \text{constant},$$

assuming that only the states E_1 and E_2 have significant populations.[22] The numbers $N_1(t)$ and $N_2(t)$ satisfy the kinetic equations

$$\frac{dN_1}{dt} = -\frac{dN_2}{dt} = (-AN_1 + BN_2)\epsilon(\omega), \qquad (5.70)$$

where $\hbar\omega = E_2 - E_1$, $A\,\epsilon(\omega)$ is the rate per unit time of $E_1 \rightarrow E_2$ transitions due to stimulated absorption in the state E_1, and $B\,\epsilon(\omega)$ is the rate per unit time of $E_2 \rightarrow E_1$ transitions due to stimulated emission. These rates are proportional to the energy density $\epsilon(\omega)$. At equilibrium

$$\frac{dN_1}{dt} = \frac{dN_2}{dt} = 0,$$

and the population ratio is given by the Boltzmann law (1.12):

$$\frac{A}{B} = \frac{N_1^{\text{eq}}}{N_2^{\text{eq}}} = \exp\left(-\frac{E_1 - E_2}{k_B T}\right) = \exp\left(\frac{\hbar\omega}{k_B T}\right). \qquad (5.71)$$

---

[21] Except in the following exceptional case: if the atom is trapped between highly reflective mirrors and held at a very low temperature, it is possible to modify spontaneous emission. This is called cavity electrodynamics; see, for example, Grynberg *et al.* [2005], Complement VI.1.

[22] This will be the case if, for example, the other states E_n are such that $E_n - E_1 \gg E_2 - E_1$ and $E_n - E_1 \gg k_B T$.

This result is not physically acceptable, because A and B can only depend on the characteristics of the interaction between the electromagnetic field and the atom, and not on temperature. Therefore, (5.70) must be corrected to include spontaneous emission *independent* of $\epsilon(\omega)$:

$$\frac{dN_1}{dt} = (-AN_1 + BN_2)\epsilon(\omega) + B'N_2. \tag{5.72}$$

The condition $dN_1/dt = 0$ combined with the Boltzmann equilibrium condition gives the following for $\epsilon(\omega)$:

$$\epsilon(\omega) = \frac{B'}{AN_1/N_2 - B} = \frac{B'}{A\exp\left(\dfrac{\hbar\omega}{kT}\right) - B}. \tag{5.73}$$

Comparison with (1.22) shows that $A = B$ and that

$$\frac{B'}{A} = \frac{\hbar\omega^3}{\pi^2 c^3}.$$

We note that we could just as well have based our arguments on the photon density $n(\omega) = \epsilon(\omega)/\hbar\omega$ or any quantity proportional to the energy density $\epsilon(\omega)$, at the price of a simple redefinition of A and B. Let us calculate B' explicitly. According to (1.16), $\epsilon(\omega)$ is an energy density per unit frequency, and the intensity $\mathcal{J}(\omega)$ in (5.66) is related to $\epsilon(\omega)$ as

$$\mathcal{J}(\omega) = c\,\epsilon(\omega),$$

which by comparison with (5.66) gives the probability of stimulated emission:

$$A = 4\pi^2 c\left(\frac{d^2}{4\pi\varepsilon_0\hbar^2 c}\right).$$

We can then derive the probability of spontaneous emission B':[23]

$$B' = \frac{\hbar\omega^3}{\pi^2 c^3} A = \frac{4\omega^3}{c^2}\left(\frac{d^2}{4\pi\varepsilon_0\hbar c}\right). \tag{5.74}$$

In the case of atomic physics, the order of magnitude of the dipole moment d is $d \sim q_e a$, where a is the radius of the electron orbit, and using the substitution $\omega \to \omega_0$ we obtain the estimate

$$B' \sim \alpha\frac{a^2\omega_0^3}{c^2} \sim \alpha^5\left(\frac{m_e c^2}{\hbar}\right), \tag{5.75}$$

where $\alpha = q_e^2/4\pi\varepsilon_0\hbar c$ is the fine-structure constant. This estimate agrees with (1.44), which was based on a classical calculation of the radiation. A complete calculation of B' will be given in Section 14.3.4, where we shall re-examine (5.75).

---

[23] Equation (5.74) is sometimes written with an additional overall factor $\frac{1}{3}$. This factor comes from an angular average. Alternatively one can replace d^2 by $\langle d^2\rangle$, where $\langle\,\rangle$ denotes an angular average; see (14.52).

Although NMR and two-level atoms display interesting analogies and analogous mathematical treatment, there are important differences of principle. Indeed the NMR measurement is not a projective measurement as defined in (4.7), but it uses a *collective* signal, built by collecting individual signals from a large number of molecules ($\sim 10^{20}$). The photon energy of the transition between the two Zeeman levels of the nuclear spin is much too small ($\sim 1\,\mu$eV) to be detected on a single molecule, and another consequence is that spontaneous emission is essentially negligible. The NMR detector is a coil of wire, wrapped around the sample (see Fig. 5.8). As the magnetization cuts across the wire, it induces an electromotive force which can be detected, and the detection method is best described classically.

5.5 Exercises

5.5.1 An orthonormal basis of eigenvectors

Show by explicit calculation that the vectors $|\chi_s\rangle$ (5.12) form an orthonormal basis: $\langle \chi_{s'}|\chi_s\rangle = \delta_{s's}$.

5.5.2 The electric dipole moment of formaldehyde

1. We wish to model the behavior of the two π electrons of the double bond in the formaldehyde molecule H_2–C=O. Using the fact that oxygen is more electronegative than carbon, show that the Hamiltonian of an electron takes the form

$$\begin{pmatrix} E_C & -A \\ -A & E_O \end{pmatrix}$$

with $E_O < E_C$, where E_C (E_O) is the energy of an electron localized at a carbon (oxygen) atom.

2. We define

$$B = \frac{1}{2}(E_C - E_O) > 0$$

and the angle θ by

$$B = \sqrt{A^2 + B^2}\cos\theta, \quad A = \sqrt{A^2 + B^2}\sin\theta.$$

Calculate as a function of θ the probability of finding a π electron localized at a carbon or oxygen atom.

3. We *assume* that the electric dipole moment d of formaldehyde is exclusively due to the charge distribution on the C=O axis. Express this dipole moment as a function of the distance l between the carbon and oxygen atoms, the proton charge q_p, and θ. The experimental values are $l = 0.121$ nm and $d = q_p \times 0.040$ nm.

5.5.3 Butadiene

The butadiene molecule C_4H_6 has a linear structure (Fig. 5.16). Its $(C_4H_6)^{4+}$ skeleton formed of σ electrons involves four carbon atoms numbered $n = 1$ to $n = 4$. The state

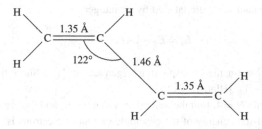

Fig. 5.16. The chemical formula of butadiene.

of a π electron localized near the nth carbon atom is designated $|\varphi_n\rangle$. It is convenient to generalize to a linear chain of N carbon atoms, numbering them $n = 1, \ldots, N$. The Hamiltonian of a π electron acts on the state $|\varphi_n\rangle$ as follows:

$$H|\varphi_n\rangle = E_0|\varphi_n\rangle - A(|\varphi_{n-1}\rangle + |\varphi_{n+1}\rangle) \quad \text{if} \quad n \neq 1, \ N,$$

$$H|\varphi_1\rangle = E_0|\varphi_1\rangle - A|\varphi_2\rangle,$$

$$H|\varphi_N\rangle = E_0|\varphi_N\rangle - A|\varphi_{N-1}\rangle,$$

where A is a positive constant. We note that the states $|\varphi_1\rangle$ and $|\varphi_N\rangle$ play a special role, because in contrast to benzene there is no cyclic symmetry in this molecule.

1. Write down the explicit matrix for H in the $|\varphi_n\rangle$ basis for $N = 4$.
2. The most general state for a π electron is

$$|\chi\rangle = \sum_{n=1}^{N} c_n|\varphi_n\rangle.$$

To adapt the method used in the case of cyclic symmetry to the present case, we introduce two fictitious states $|\varphi_0\rangle$ and $|\varphi_{N+1}\rangle$ and two components $c_0 = c_{N+1} = 0$, which allows us to rewrite $|\chi\rangle$ as

$$|\chi\rangle = \sum_{n=0}^{N+1} c_n|\varphi_n\rangle.$$

Show that the action of H on the state $|\chi\rangle$ is written as

$$H|\chi\rangle = E_0|\chi\rangle - A \sum_{n=1}^{N} (c_{n-1} + c_{n+1})|\varphi_n\rangle.$$

3. Inspired by the method used in the case of cyclic symmetry, we seek c_n in the form

$$c_n = \frac{c}{2i} \left(e^{in\delta} - e^{-in\delta} \right),$$

which ensures that $c_0 = 0$. Show that we must choose

$$\delta = \frac{\pi s}{N+1}, \quad s = 1, \ldots, N,$$

if we also wish to have $c_{N+1} = 0$.

4. Show that the eigenvalues of H are labeled by an integer s:

$$E_s = E_0 - 2A \cos \frac{\pi s}{N+1},$$

and give the expression for the corresponding eigenvectors $|\chi_s\rangle$. Show that the normalization constant c is $\sqrt{2/(N+1)}$. [Hint: cf. (5.15).]

5. In the case of butadiene $N = 4$, find the numerical values of E_s and the eigenvector components. Show that the ground-state energy of the ensemble of four π electrons is

$$E_0 \simeq 4(E_0 - A) - 0.48A.$$

Is the gain due to the delocalization of the π electrons belonging to the chain important as regards the chemical formula of Fig. 5.16? Qualitatively sketch the probability density for these electrons for $s = 1$ and $s = 2$.

6. What would the ground-state energy of a hypothetical cyclic (i.e., having the form of a square) molecule C_4H_4 be?

7. We define the order of a bond l between two carbon atoms n and $n+1$ as

$$l = 1 + \sum_s \langle \varphi_n | \chi_s \rangle \langle \chi_s | \varphi_{n+1} \rangle,$$

where the sum runs over the states $|\chi_s\rangle$ occupied by the π electrons. The factor 1 corresponds to the σ electrons. Show that the order of the bond is $l = 2$ for ethylene. Calculate the order of the bonds for benzene and of the various bonds of butadiene and comment on the results. Why is the central bond of butadiene shorter than a simple bond (1.46 Å instead of 1.54 Å)?

5.5.4 Eigenvectors of the Hamiltonian (5.47)

Show that in the case where the electric field is independent of time and when $d\mathcal{E}/A \ll 1$, the normalized eigenvector of H corresponding to the eigenvalue $E_0 - A$ is given to order $d\mathcal{E}/A$ by

$$|\chi'_+\rangle = \frac{1}{\sqrt{2}} \begin{pmatrix} 1 - d\mathcal{E}/(2A) \\ 1 + d\mathcal{E}/(2A) \end{pmatrix}.$$

What is the other eigenvector?

5.5.5 The hydrogen molecular ion H_2^+

The hydrogen molecular ion H_2^+ is formed of two protons and an electron. The two protons are located on an axis which we choose to be the x axis, at points $-r/2$ and $r/2$. They are assumed to be fixed, in agreement with the Born–Oppenheimer approximation.

1. Assuming that the electron is located on the x axis, express its potential energy $V(x)$ as a function of its position x and $e^2 = q_e^2/4\pi\varepsilon_0$, where q_e is the electron charge, and sketch it qualitatively.

2. If the two protons are very far apart, $r \gg l$, the electron is either localized near the proton on the right (the state $|\varphi_1\rangle$), or near the proton on the left (the state $|\varphi_2\rangle$). We assume that these states both correspond to the ground state of the hydrogen atom of energy

$$E_0 = -\frac{1}{2}\frac{m_e e^4}{\hbar^2} = -\frac{e^2}{2a_0},$$

where m_e is the electron mass and a_0 is the Bohr radius: $a_0 = \hbar^2/m_e e^2$. What is the relevant length scale l in the relation $r \gg l$?

3. We shall treat the ion H_2^+ as a two-level system with basis states $\{|\varphi_1\rangle, |\varphi_2\rangle\}$ and $\langle\varphi_i|\varphi_j\rangle = \delta_{ij}$. Justify the following form of the Hamiltonian with the choice $A > 0$:

$$H = \begin{pmatrix} E_0 & -A \\ -A & E_0 \end{pmatrix}.$$

What are the eigenstates $|\chi_+\rangle$ and $|\chi_-\rangle$ of H and the corresponding energies E_+ and E_-, $E_+ < E_-$? Qualitatively sketch the wave functions $\chi_\pm(x) = \langle x|\chi_\pm\rangle$ of the electron on the x axis.

4. The parameter A is a function of the distance r between the protons, $A(r)$. Justify the fact that A is an increasing function of r and $\lim_{r\to\infty} A(r) = 0$. The electron energy is then a function of r, $E_\pm(r)$.

5. Show that the total energy of the ion $E'_\pm(r)$ must contain an additional term $+e^2/r$. What is the physical origin of this term?

6. We parametrize $A(r)$ as

$$A(r) = c\,e^2\,\exp\left(-\frac{r}{b}\right),$$

where b is a length and c an inverse length. Give the expression for the two energy levels E'_+ and E'_- of the ion. Let

$$\Delta E(r) = E'_+(r) - E_0$$

be the energy difference between the ground state of the ion and that of the hydrogen atom. Show that $\Delta E(r)$ can pass through a minimum at a value $r = r_0$ and derive the expression

$$\Delta E(r_0) = \frac{e^2}{r_0}\left(1 - \frac{b}{r_0}\right).$$

What condition must hold for b and r_0 in order for the ion H_2^+ to be a bound state?

7. The experimental values are $r_0 \simeq 2a_0$ and $\Delta E(r_0) \simeq E_0/5 = -e^2/10a_0$. Compute b and c as functions of a_0.

5.5.6 The rotating-wave approximation in NMR

1. Instead of the rotating field of (5.22), we shall use a field $\vec{B}_1(t)$ parallel to Ox:

$$\vec{B}_1(t) = 2B_1\hat{x}\cos(\omega t - \phi).$$

We define the state vector $|\tilde{\varphi}(t)\rangle$ in the rotating frame with angular velocity ω as

$$|\tilde{\varphi}(t)\rangle = \exp\left[-\frac{i\omega\sigma_z t}{2}\right]|\varphi(t)\rangle \quad |\tilde{\varphi}(t=0)\rangle = |\varphi(t=0)\rangle.$$

Why can one call $|\tilde{\varphi}(t)\rangle$ the state vector in the rotating frame? Show that the time evolution of $|\tilde{\varphi}(t)\rangle$ is governed by a Hamiltonian $\tilde{H}(t)$

$$i\hbar \frac{d|\tilde{\varphi}\rangle}{dt} = \tilde{H}(t)|\tilde{\varphi}(t)\rangle$$

where

$$\tilde{H}(t) = \exp\left[-\frac{i\omega\sigma_z t}{2}\right] H(t) \exp\left[\frac{i\omega\sigma_z t}{2}\right].$$

More generally, for any operator $A(t)$, we have in the rotating frame

$$\tilde{A}(t) = \exp\left[-\frac{i\omega\sigma_z t}{2}\right] A(t) \exp\left[\frac{i\omega\sigma_z t}{2}\right].$$

2. Show that the preceding definition gives for the operators $\sigma_\pm = (\sigma_x \pm i\sigma_y)/2$

$$\tilde{\sigma}_\pm(t) = \exp\left[-\frac{i\omega\sigma_z t}{2}\right] \sigma_\pm \exp\left[-\frac{i\omega\sigma_z t}{2}\right] = e^{\mp i\omega t}\sigma_\pm.$$

Hint: establish the following differential equation from the definition of $\tilde{\sigma}_\pm(t)$

$$\frac{d\tilde{\sigma}_\pm(t)}{dt} = \mp i\omega\tilde{\sigma}_\pm(t).$$

Writing $\sigma_x = \sigma_+ + \sigma_-$, obtain the Hamiltonian in the rotating frame

$$\tilde{H}(t) = \frac{\hbar}{2}\delta\sigma_z - \frac{\hbar}{2}\omega_1(\sigma_x\cos\phi + \sigma_y\sin\phi) + \hbar\omega_1\left(\sigma_+ e^{-2i\omega t}e^{i\phi} + \sigma_- e^{2i\omega t}e^{-i\phi}\right)$$

where δ is the detuning, $\delta = \omega - \omega_0$. Use the rotating wave approximation to eliminate the terms between square brackets in the preceding equation. The Hamiltonian $\tilde{H}(t)$ is now time-independent!

3. Show that *at resonance*, the evolution operator $\tilde{U}(t)$ in the rotating frame given by

$$\tilde{U}(t) = \exp\left[\frac{-i\tilde{H}t}{\hbar}\right] = \exp\left[\frac{i\omega_1 t(\sigma_x\cos\phi + \sigma_y\sin\phi)}{2}\right]$$

is a rotation operator of angle $-\omega_1 t$ about an axis $\hat{n}$ of components

$$\hat{n}_x = \cos\phi \quad \hat{n}_y = \sin\phi \quad \hat{n}_z = 0.$$

Thus the angle ϕ allows one to choose the rotation axis. One may (rightly) be puzzled by the fact that ϕ could be eliminated by changing the origin of time. However, this angle is important in a *sequence* of pulses: then the *relative* phase between the pulses is physically relevant.

4. Let us now take for simplicity $\phi = 0$. In order to compute the matrix form of the evolution operator in the rotating frame, we write

$$\exp(-i\tilde{H}t/h) = \exp\left[-i\frac{\Omega t}{2}\left(\frac{\delta}{\Omega}\sigma_z - \frac{\omega_1}{\Omega}\sigma_x\right)\right]$$

with $\Omega = \sqrt{\delta^2 + \omega_1^2}$. The vector $\hat{n}$

$$\hat{n} = \left(\hat{n}_x = -\frac{\omega_1}{\Omega}, \hat{n}_y = 0, \hat{n}_z = \frac{\delta}{\Omega}\right)$$

is a unit vector. Using (3.67), obtain the following expression

$$\exp(-i\tilde{H}t) = \left(\cos\frac{\Omega t}{2} - i\frac{\delta}{\Omega} \sin\frac{\Omega t}{2} \right)|+\rangle\langle+| + i\frac{\omega_1}{\Omega} \sin\frac{\Omega t}{2} \left(|+\rangle\langle-| + |-\rangle\langle+| \right)$$

$$+ \left(\cos\frac{\Omega t}{2} + i\frac{\delta}{\Omega} \sin\frac{\Omega t}{2} \right)|-\rangle\langle-|.$$

5.6 Further reading

Discussions of elementary quantum chemistry can be found in Feynman *et al.* [1965], Vol. III, Chapter 15; F. Goodrich, *A Primer of Quantum Chemistry*, New York: Wiley (1972), Chapter 2; or C. Gatz, *Introduction to Quantum Chemistry*, Columbia: C. E. Merrill (1971), Chapters 10–12. Two-level systems with resonant and quasi-resonant interactions are discussed by Feynman *et al.* [1965], Vol. III, Chapters 8 and 9 and by Cohen-Tannoudji *et al.* [1977], Chapter IV. An excellent introduction to NMR can be found in, for example, J. W. Akitt, *NMR Chemistry: An Introduction to Modern NMR Spectroscopy*, New York: Chapman & Hall (1992) or Levitt [2001]. The interaction of a two-level atom with an electromagnetic field is studied at an advanced level by Grynberg *et al.* [2005], Chapter II. The reader will find additional details on the molecular ion H_2^+ in Cohen-Tannoudji *et al.* [1977], Complement G_{XI}.

6

Entangled states

Up to now we have limited ourselves to states of a single particle. In the present chapter we shall introduce the description of two-particle states. Once this case is understood, it will be easy to generalize to any number of particles. States of two (or more) particles lead to very rich configurations called entangled states. A remarkable feature is that two entangled quantum particles, even at arbitrarily large spatial separations, continue to form a single entity and no classical probabilistic model is able to reproduce the correlation between particles. In the first section we shall present the essential mathematical formalism, that of the tensor product. This will permit us in Section 6.2 to describe quantum mixtures using the state operator formalism. Section 6.3 is devoted to the study of important physical consequences like the Bell inequalities and interference experiments involving entangled states, which will lead us to a deeper understanding of quantum physics. Finally, in the last section we shall briefly review applications to measurement theory and quantum information theory. The latter is undergoing rapid development at present and has applications to quantum computing, cryptography, and teleportation.

6.1 The tensor product of two vector spaces

6.1.1 Definition and properties of the tensor product

We wish to construct the space of states of two physical systems which we assume initially to be completely independent. Let $\mathcal{H}_1^N$ and $\mathcal{H}_2^M$ be the spaces of states of the two systems, of dimension N and M, respectively. Since the two systems are independent, the global state is defined by specifying the state vector $|\varphi\rangle \in \mathcal{H}_1^N$ of the first system and the state vector $|\chi\rangle \in \mathcal{H}_2^M$ of the second. The pair $\{|\varphi\rangle, |\chi\rangle\}$ can be viewed as a vector belonging to a vector space of dimension NM, called the *tensor product* of the spaces $\mathcal{H}_1^N$ and $\mathcal{H}_2^M$ and denoted $\mathcal{H}_1^N \otimes \mathcal{H}_2^M$. It will be defined precisely below.

We choose an orthonormal basis $|n\rangle$ of $\mathcal{H}_1^N$ and an orthonormal basis $|m\rangle$ of $\mathcal{H}_2^M$ on which we decompose the arbitrary vectors $|\varphi\rangle \in \mathcal{H}_1^N$ and $|\chi\rangle \in \mathcal{H}_2^M$:

$$|\varphi\rangle = \sum_{n=1}^{N} c_n |n\rangle, \quad |\chi\rangle = \sum_{m=1}^{M} d_m |m\rangle. \tag{6.1}$$

The space $\mathcal{H}_1^N \otimes \mathcal{H}_2^M$ will be defined as a space of NM dimensions where the pairs $\{|n\rangle, |m\rangle\}$, denoted $|n \otimes m\rangle$ or $|n\rangle \otimes |m\rangle$, form an orthonormal basis

$$\langle n' \otimes m' | n \otimes m \rangle = \delta_{n'n} \delta_{m'm}, \tag{6.2}$$

and the tensor product of the vectors $|\varphi\rangle$ and $|\chi\rangle$, denoted $|\varphi \otimes \chi\rangle$ or $|\varphi\rangle \otimes |\chi\rangle$, is a vector with components $c_n d_m$ in this basis:

$$|\varphi \otimes \chi\rangle = \sum_{n,m} c_n d_m |n \otimes m\rangle. \tag{6.3}$$

The linearity of the tensor product can be verified immediately:

$$|\varphi \otimes (\chi_1 + \lambda \chi_2)\rangle = |\varphi \otimes \chi_1\rangle + \lambda |\varphi \otimes \chi_2\rangle,$$
$$|(\varphi_1 + \lambda \varphi_2) \otimes \chi\rangle = |\varphi_1 \otimes \chi\rangle + \lambda |\varphi_2 \otimes \chi\rangle. \tag{6.4}$$

We must also check that the definition of the tensor product is independent of the choice of basis. Let $|i\rangle$ and $|j\rangle$ be two orthonormal bases of $\mathcal{H}_1^N$ and $\mathcal{H}_2^M$ obtained from the bases $|n\rangle$ and $|m\rangle$ by the unitary transformations R $(R^{-1} = R^\dagger)$ and S $(S^{-1} = S^\dagger)$, respectively:

$$|i\rangle = \sum_n R_{in} |n\rangle, \quad |j\rangle = \sum_m S_{jm} |m\rangle.$$

According to (6.3), the tensor product $|i \otimes j\rangle$ is given by

$$|i \otimes j\rangle = \sum_{n,m} R_{in} S_{jm} |n \otimes m\rangle.$$

Moreover, the decomposition of $|\varphi\rangle$ and $|\chi\rangle$ in the bases $|i\rangle$ and $|j\rangle$, respectively, can be written as

$$|\varphi\rangle = \sum_{i=1}^N \overline{c}_i |i\rangle, \quad |\chi\rangle = \sum_{j=1}^M \overline{d}_j |j\rangle.$$

Direct calculation (Exercise 6.4.1) shows that

$$\sum_{i,j} \overline{c}_i \overline{d}_j |i \otimes j\rangle = |\varphi \otimes \chi\rangle,$$

where $|\varphi \otimes \chi\rangle$ is defined by (6.3). The result for $|\varphi \otimes \chi\rangle$ is then independent of the choice of basis. When the two systems are no longer independent, we must state a fifth postulate.

Postulate V

The space of states of two interacting quantum systems is $\mathcal{H}_1^N \otimes \mathcal{H}_2^M$.[1]

It is reasonable to assume that interactions cannot modify the space of states. The most general state vector will be of the form

$$|\Phi\rangle = \sum_{n,m} b_{nm} |n \otimes m\rangle. \tag{6.5}$$

---

[1] Nevertheless, we shall see in Chapter 13 that in the case of two identical particles (where $N = M$) only a part of $\mathcal{H}_1^N \otimes \mathcal{H}_2^N$ corresponds to physical states.

In general, the vector $|\Phi\rangle$ cannot be written as a tensor product $|\varphi \otimes \chi\rangle$. This would require that it be possible to factorize b_{nm} in the form $c_n d_m$, which is impossible except for independent systems. The state vectors which can be written as a tensor product form a subset (but not a subspace) of $\mathcal{H}_1^N \otimes \mathcal{H}_2^M$. A state vector which cannot be written in the form of a tensor product is termed *entangled state*.

The tensor product $C = A \otimes B$ of two linear operators A and B acting respectively in the spaces $\mathcal{H}_1^N$ and $\mathcal{H}_2^M$ is defined by its action on the tensor product vector $|\varphi \otimes \chi\rangle$:

$$(A \otimes B)|\varphi \otimes \chi\rangle = |A\varphi \otimes B\chi\rangle, \tag{6.6}$$

and its matrix elements in the basis $|n \otimes m\rangle$ of $\mathcal{H}_1^N \otimes \mathcal{H}_2^M$ are then

$$\langle n' \otimes m'|A \otimes B|n \otimes m\rangle = A_{n'n}B_{m'm}. \tag{6.7}$$

In general, an operator C acting on $\mathcal{H}_1^N \otimes \mathcal{H}_2^M$ will *not* be of the form $A \otimes B$. Its matrix elements will be

$$\langle n' \otimes m'|C|n \otimes m\rangle = C_{n'm';nm}, $$

and, except in special cases, it will not be possible to write $C_{n'm';nm}$ in the factorized form $A_{n'n}B_{m'm}$. Two interesting special cases of (6.6) are $A = I_1$ and $B = I_2$, where I_1 and I_2 are the identity operators of $\mathcal{H}_1^N$ and $\mathcal{H}_2^M$:

$$(A \otimes I_2)|\varphi \otimes \chi\rangle = |A\varphi \otimes \chi\rangle, \quad (I_1 \otimes B)|\varphi \otimes \chi\rangle = |\varphi \otimes B\chi\rangle. \tag{6.8}$$

In terms of the matrix elements, we have

$$\langle n' \otimes m'|A \otimes I_2|n \otimes m\rangle = A_{n'n}\delta_{m'm}, \quad \langle n' \otimes m'|I_1 \otimes B|n \otimes m\rangle = \delta_{n'n}B_{m'm}. \tag{6.9}$$

Finally, if $|\varphi\rangle$ is an eigenvector of A with eigenvalue a $(A|\varphi\rangle = a|\varphi\rangle)$, then $|\varphi \otimes \chi\rangle$ will be an eigenvector of $A \otimes I_2$ with eigenvalue a:

$$(A \otimes I_2)|\varphi \otimes \chi\rangle = a|\varphi \otimes \chi\rangle. \tag{6.10}$$

The identity operators I_1 and I_2 are often not written out explicitly, and one finds (6.10) written as

$$A|\varphi \otimes \chi\rangle = a|\varphi \otimes \chi\rangle \quad \text{or simply} \quad A|\varphi\chi\rangle = a|\varphi\chi\rangle, \tag{6.11}$$

with the symbol for the tensor product omitted. Since the notation $\otimes$ is rather cumbersome, it will often be omitted when there is no possibility of confusion.

6.1.2 A system of two spins 1/2

Let us illustrate the notion of the tensor product by constructing the space of states of a system of two spins 1/2. The spaces of states of the two spins are the two-dimensional spaces $\mathcal{H}_1$ and $\mathcal{H}_2$. The space of states of the system of two spins $\mathcal{H} = \mathcal{H}_1 \otimes \mathcal{H}_2$ is four-dimensional ($4 = 2 \times 2$). We choose the orthonormal bases of $\mathcal{H}_1$ and $\mathcal{H}_2$ to be the eigenstates $|\varepsilon_1\rangle$ and $|\varepsilon_2\rangle$, $\varepsilon_i = \pm 1$, of the operators S_{1z} and S_{2z} projecting the spin on the z axis, where

$$S_{1z}|\varepsilon_1\rangle = \frac{1}{2}\hbar\varepsilon_1|\varepsilon_1\rangle, \quad S_{2z}|\varepsilon_2\rangle = \frac{1}{2}\hbar\varepsilon_2|\varepsilon_2\rangle.$$

According to (6.5), the states of the two-spin system are decomposed on the orthonormal basis $|\varepsilon_1 \otimes \varepsilon_2\rangle$; furthermore, we have, for example,

$$(S_{1z}\otimes I_2)|\varepsilon_1 \otimes \varepsilon_2\rangle = \frac{1}{2}\hbar\varepsilon_1|\varepsilon_1 \otimes \varepsilon_2\rangle, \quad (S_{1z}\otimes S_{2z})|\varepsilon_1 \otimes \varepsilon_2\rangle = \frac{1}{4}\hbar^2\varepsilon_1\varepsilon_2|\varepsilon_1 \otimes \varepsilon_2\rangle.$$

Following (6.11), we shall often use the abbreviated notation $|\varepsilon_1\varepsilon_2\rangle$ instead of $|\varepsilon_1 \otimes \varepsilon_2\rangle$ and $S_{1z}S_{2z}$ instead of $S_{1z} \otimes S_{2z}$. In this notation the preceding equations become

$$S_{1z}|\varepsilon_1\varepsilon_2\rangle = \frac{1}{2}\hbar\varepsilon_1|\varepsilon_1\varepsilon_2\rangle, \quad S_{1z}S_{2z}|\varepsilon_1\varepsilon_2\rangle = \frac{1}{4}\hbar^2\varepsilon_1\varepsilon_2|\varepsilon_1\varepsilon_2\rangle. \tag{6.12}$$

Let $|\chi_1\rangle$ and $|\chi_2\rangle$ be two arbitrary (normalized) vectors of $\mathcal{H}_1$ and $\mathcal{H}_2$:

$$|\chi_1\rangle = \lambda_1|+_1\rangle + \mu_1|-_1\rangle, \quad |\lambda_1|^2 + |\mu_1|^2 = 1,$$

$$|\chi_2\rangle = \lambda_2|+_2\rangle + \mu_2|-_2\rangle, \quad |\lambda_2|^2 + |\mu_2|^2 = 1.$$

According to (6.3), the tensor product $|\chi_1 \otimes \chi_2\rangle$ is given by ($|+_1 \otimes +_2\rangle = |+ \otimes +\rangle$ etc.)

$$|\chi_1 \otimes \chi_2\rangle = \lambda_1\lambda_2|+\otimes+\rangle + \lambda_1\mu_2|+\otimes-\rangle + \lambda_2\mu_1|-\otimes+\rangle + \mu_1\mu_2|-\otimes-\rangle. \tag{6.13}$$

An arbitrary vector $|\Psi\rangle \in \mathcal{H}$ is

$$|\Psi\rangle = \alpha|+\otimes+\rangle + \beta|+\otimes-\rangle + \gamma|-\otimes+\rangle + \delta|-\otimes-\rangle. \tag{6.14}$$

This vector is not in general of the form (6.13); comparing (6.13) and (6.14), we see that a tensor product vector satisfies

$$\alpha\delta = \beta\gamma,$$

and a priori there is no reason for this condition (which is necessary and sufficient) to be valid. When $|\Psi\rangle$ is not of the form (6.13), we are thus dealing with an entangled state of two spins.

An important special case is the entangled state

$$|\Phi\rangle = \frac{1}{\sqrt{2}}\left(|+\otimes-\rangle - |-\otimes+\rangle\right),$$

or in abbreviated notation (6.12)

$$\boxed{|\Phi\rangle = \frac{1}{\sqrt{2}}\left(|+-\rangle - |-+\rangle\right)}. \tag{6.15}$$

This state is manifestly entangled because $\alpha = \delta = 0$ and $\beta = -\gamma = 1/\sqrt{2}$, and so $\alpha\delta \neq \beta\gamma$. A remarkable property of $|\Phi\rangle$ is its invariance under rotations, i.e., it is a scalar under rotations.[2] In fact, as we have seen in Section 3.2.4, the transform $|\chi\rangle_{\mathcal{R}}$ by a rotation $\mathcal{R}$ of a state $|\chi\rangle$ is obtained by applying the operator $D^{1/2}$ (3.58), which is an $SU(2)$ matrix, that is, a 2×2 unitary matrix of unit determinant (Exercise 3.3.6). The transforms of $|+\rangle$ and $|-\rangle$ are

$$|+\rangle_{\mathcal{R}} = a|+\rangle + b|-\rangle,$$
$$|-\rangle_{\mathcal{R}} = c|+\rangle + d|-\rangle, \qquad (6.16)$$

with $ad - bc = 1$.[3] We then obtain

$$|+-\rangle_{\mathcal{R}} = ac|++\rangle + ad|+-\rangle + bc|-+\rangle + bd|--\rangle, \qquad (6.17)$$

and, making the exchange $+ \leftrightarrow -$,

$$|-+\rangle_{\mathcal{R}} = ac|++\rangle + ad|-+\rangle + bc|+-\rangle + bd|--\rangle,$$

we see that $|\Phi\rangle$ transforms under rotations as

$$|\Phi\rangle_{\mathcal{R}} = \frac{1}{\sqrt{2}}\left(|+-\rangle_{\mathcal{R}} - |-+\rangle_{\mathcal{R}}\right) = (ad - bc)|\Phi\rangle = |\Phi\rangle. \qquad (6.18)$$

6.2 The state operator (or density operator)

6.2.1 Definition and properties

Let us consider a system of two particles described by a state vector $|\Psi\rangle \in \mathcal{H}_1 \otimes \mathcal{H}_2$. If $|\Psi\rangle$ is a tensor product $|\varphi_1 \otimes \varphi_2\rangle$, the state vector of particle 1 is $|\varphi_1\rangle$. But what happens if $|\Psi\rangle$ is not a tensor product, or, in other words, if $|\Psi\rangle$ is an entangled state? Can we still regard particle 1 as having a state vector? We shall see that the answer to this question is no: in general, a state vector cannot be associated with particle 1. This example shows that we must generalize our description of quantum systems, and this generalization will go well beyond the special case we have just mentioned. When a quantum system can be described by a vector in a Hilbert space of states, we say that we are dealing with a *pure state* or a *pure case*; this will be the situation if complete information about the system is available. When the information on the system is incomplete, we are dealing with a *mixture*, and a quantum system is then described mathematically by a *state operator*.[4] The introduction of the state operator will allow us to reformulate postulate **I** of Chapter 4 so as to describe physical situations more general than those imagined so far, such as cases in which only partial information is available on the system under consideration.

---

[2] In Section 10.6.1 we shall see that $|\Phi\rangle$ is a state of zero angular momentum and therefore a scalar under rotations.

[3] And also $c = -b^*$, $d = a^*$, but we shall not use these relations here.

[4] This is another instance where the common term "density operator" is inappropriate. This terminology was introduced in the case of wave mechanics (Chapter 9), where the diagonal elements of ρ in position space, $\langle x|\rho|x\rangle$, or in momentum space, $\langle p|\rho|p\rangle$, are indeed densities. However, "density operator" conceals the fact that the operator contains essential information on the phases. We prefer to use "state operator" by analogy with "state vector". "Statistical operator" would also be possible.

When we are dealing with a pure state, being given the state vector $|\varphi\rangle \in \mathcal{H}$ describing a quantum system is equivalent to being given the projector $\mathcal{P}_\varphi = |\varphi\rangle\langle\varphi|$ onto the state $|\varphi\rangle$. In some sense, $\mathcal{P}_\varphi$ is a better mathematical description because the arbitrary phase of $|\varphi\rangle$ disappears: $\mathcal{P}_\varphi$ is invariant when $|\varphi\rangle$ is multiplied by a phase factor

$$|\varphi\rangle \to e^{i\alpha}|\varphi\rangle,$$

and then there is a one-to-one correspondence between the physical state and $\mathcal{P}_\varphi$ rather than correspondence up to a phase. The expectation value of a physical property A is expressed simply as a function of $\mathcal{P}_\varphi$, which is, as we shall see, the simplest case of a state operator. Let us introduce an orthonormal basis $|n\rangle$ of $\mathcal{H}$ to compute this expectation value:

$$\langle A \rangle = \langle\varphi|A|\varphi\rangle = \sum_{n,m}\langle\varphi|n\rangle\langle n|A|m\rangle\langle m|\varphi\rangle$$

$$= \sum_{n,m}\langle m|\varphi\rangle\langle\varphi|n\rangle\langle n|A|m\rangle$$

$$= \sum_{m}\langle m|\mathcal{P}_\varphi A|m\rangle = \text{Tr}(\mathcal{P}_\varphi A). \tag{6.19}$$

Now we can generalize to a mixture. There we know only that the quantum system has probability p_α ($0 \le \mathsf{p}_\alpha \le 1$, $\sum_\alpha \mathsf{p}_\alpha = 1$) of being in the state $|\varphi_\alpha\rangle$. The states $|\varphi_\alpha\rangle$ are assumed to be normalized ($\langle\varphi_\alpha|\varphi_\alpha\rangle = 1$) but not necessarily orthogonal. By definition, the state operator ρ describing this quantum system is

$$\boxed{\rho = \sum_\alpha \mathsf{p}_\alpha|\varphi_\alpha\rangle\langle\varphi_\alpha| = \sum_\alpha \mathsf{p}_\alpha \mathcal{P}_{\varphi_\alpha}}. \tag{6.20}$$

The expectation value of a physical property A is obtained by immediate generalization of (6.19). In fact, $\langle A \rangle_\alpha$, the expectation value of A in the state $|\varphi_\alpha\rangle$, is

$$\langle A \rangle_\alpha = \langle\varphi_\alpha|A|\varphi_\alpha\rangle,$$

and it is associated with the weight p_α when calculating the global expectation value $\langle A \rangle$. The expectation value in the mixture is then

$$\boxed{\langle A \rangle = \sum_\alpha \mathsf{p}_\alpha\langle A \rangle_\alpha = \sum_\alpha \mathsf{p}_\alpha\langle\varphi_\alpha|A|\varphi_\alpha\rangle = \text{Tr}(\rho A)}. \tag{6.21}$$

The weights p_α are fixed by the physical problem under consideration. Let us give two important examples.

- The quantum system is a subsystem of a larger system in a pure state. The weights p_α are then determined by taking a partial trace according to the procedure defined in (6.30) below.
- The system is described by equilibrium statistical mechanics. The weights p_α are then obtained by maximizing the von Neumann entropy $S_{\text{vN}} = -\text{Tr}\,\rho\ln\rho$, which corresponds physically to maximizing the missing information.

The fundamental properties of ρ that follow immediately from the definition (6.20) are

- ρ is Hermitian: $\rho = \rho^\dagger$;
- ρ has unit trace: $\text{Tr}\,\rho = 1$;
- ρ is a positive operator:[5] $\langle\varphi|\rho|\varphi\rangle \geq 0$ for any $|\varphi\rangle$;
- a necessary and sufficient condition for ρ to describe a pure state is $\rho^2 = \rho$. In fact, since $\rho = \rho^\dagger$, the condition $\rho^2 = \rho$ implies that ρ is a projector. Since $\text{Tr}\,\rho = 1$, the dimension of the projection vector space is unity[6] and ρ has the form $|\varphi\rangle\langle\varphi|$.

Inversely, a Hermitian operator which is positive and has unit trace can be interpreted as a state operator. In fact, since ρ is Hermitian, we can write down its spectral decomposition (which is not unique if there are degenerate eigenvalues)

$$\rho = \sum_n p_n |n\rangle\langle n|,$$

and a possible way of preparing the quantum system is to construct a mixture of states $|n\rangle$ with probabilities p_n. However, whereas specifying p_α and $|\varphi_\alpha\rangle$ in (6.20) determines ρ uniquely, the reverse is not true: many different preparations can correspond to a single state operator, as we shall see explicitly for the example of spin 1/2. In other words, a state operator does not specify a unique microscopic configuration, but it is sufficient for calculating the expectation values of physical properties using (6.21).

6.2.2 The state operator for a two-level system

As an example, let us find the most general form of the state operator for a two-level quantum system, in which case the Hilbert space is two-dimensional. There are many applications of this: the description of the polarization of a massive spin-1/2 particle or of a photon, the state of a two-level atom, and so on. The standard two-level system is that of spin 1/2, and so we shall use this particular case to define the notation and terminology. Let us choose two basis vectors of the space of states, $|+\rangle$ and $|-\rangle$. These might be, for example, the eigenvectors of the z component of the spin. In this basis the state operator is represented by a 2×2 matrix, the *state matrix* (or *density matrix*) ρ. This matrix is Hermitian and has unit trace. The most general such matrix is

$$\rho = \begin{pmatrix} a & c \\ c^* & 1-a \end{pmatrix}, \tag{6.22}$$

where a is a real number and c is a complex number. Equation (6.22) does not yet define a state matrix, because in addition ρ must be positive. The eigenvalues λ_+ and λ_- of ρ satisfy

$$\lambda_+ + \lambda_- = 1, \quad \lambda_+\lambda_- = \det\rho = a(1-a) - |c|^2,$$

---

[5] A (strictly) positive operator is Hermitian and has (strictly) positive eigenvalues and vice versa; see Exercise 2.4.10.

[6] In general, if $\mathcal{P}$ is a projector, $\text{Tr}\,\mathcal{P}$ is equal to the dimension of the projection vector space. To see this it is sufficient to use a basis in which $\mathcal{P}$ is diagonal.

and we must have $\lambda_+ \geq 0$ and $\lambda_- \geq 0$. The condition $\det \rho \geq 0$ implies that λ_+ and λ_- have the same sign, and the condition $\lambda_+ + \lambda_- = 1$ implies that $\lambda_+ \lambda_-$ reaches its maximum for $\lambda_+ \lambda_- = 1/4$, so that finally

$$0 \leq a(1-a) - |c|^2 \leq \frac{1}{4}. \tag{6.23}$$

The necessary and sufficient condition for ρ to describe a pure state is

$$\det \rho = a(1-a) - |c|^2 = 0.$$

As an exercise, the reader should calculate a and c for the state matrix describing the normalized state vector $|\psi\rangle = \lambda|+\rangle + \mu|-\rangle$ with $|\lambda|^2 + |\mu|^2 = 1$, and show that the determinant of this matrix vanishes.

It is often convenient to decompose the state matrix (6.22) on the basis of Pauli matrices σ_i. In fact, any 2×2 matrix can be written as a linear combination of the unit matrix I and the σ_i (Exercise 3.3.5):

$$\rho = \frac{1}{2} \begin{pmatrix} 1 + b_z & b_x - i b_y \\ b_x + i b_y & 1 - b_z \end{pmatrix} = \frac{1}{2} \left(I + \vec{b} \cdot \vec{\sigma} \right). \tag{6.24}$$

The vector $\vec{b}$, called the *Bloch vector*, must satisfy $|\vec{b}|^2 \leq 1$ owing to (6.23). The pure state, which corrresponds to $|\vec{b}|^2 = 1$, is also termed completely polarized, the case $\vec{b} = 0$ unpolarized or of zero polarization, and the case $0 < |\vec{b}| < 1$ partially polarized. To obtain the physical interpretation of the vector $\vec{b}$, we calculate the expectation value of the spin $\vec{S} = \frac{1}{2} \hbar \vec{\sigma}$ using $\mathrm{Tr}\, \sigma_i \sigma_j = 2\delta_{ij}$. We find

$$\langle S_i \rangle = \mathrm{Tr}\, (\rho\, S_i) = \frac{1}{2} \hbar b_i, \tag{6.25}$$

so that $\hbar \vec{b}/2$ is the expectation value $\langle \vec{S} \rangle$ of the spin.

Let us show that several different preparations can lead to the same state matrix when $|\vec{b}| < 1$. We set $\vec{b} = \vec{OP}$, construct a sphere of center O and unit radius, and draw a chord of the sphere passing through the tip of $\vec{b}$. This chord cuts the sphere at two points P_1 and P_2, and we define the two unit vectors

$$\hat{n}_1 = \vec{OP}_1, \quad \hat{n}_2 = \vec{OP}_2.$$

The Bloch vector can be written as

$$\vec{b} = \hat{n}_1 + \lambda(\hat{n}_2 - \hat{n}_1) = (1 - \lambda)\hat{n}_1 + \lambda\hat{n}_2, \quad 0 < \lambda < 1.$$

The state matrix defined by the Bloch vector $\vec{b}$ then is

$$\rho = \frac{1}{2} \left(I + \vec{\sigma} \cdot \vec{b} \right) = \frac{1}{2} (1 - \lambda) \left(I + \vec{\sigma} \cdot \hat{n}_1 \right) + \frac{1}{2} \lambda \left(I + \vec{\sigma} \cdot \hat{n}_2 \right). \tag{6.26}$$

We can prepare the corresponding quantum state using a statistical mixture with probability $p_1 = (1 - \lambda)$ for the state $|+, \hat{n}_1\rangle$ and probability $p_2 = \lambda$ for the state $|+, \hat{n}_2\rangle$ (cf. (3.56)):

$$\rho = p_1 |+, \hat{n}_1\rangle\langle +, \hat{n}_1| + p_2 |+, \hat{n}_2\rangle\langle +, \hat{n}_2|.$$

Since there are an infinite number of chords passing through the tip of $\vec{b}$, there are an infinite number of ways of preparing the quantum state (6.26).

It is essential to clearly distinguish between a pure state and a mixture. Let us suppose, for example, that a spin 1/2 is in the pure state:

$$|\chi\rangle = \frac{1}{\sqrt{2}} (|+\rangle + |-\rangle). \tag{6.27}$$

Analysis using a Stern–Gerlach device in which the magnetic field $\vec{B}$ is parallel to Oz will give a 50% probability of upward deflection and 50% probability of downward deflection. However, the state (6.27) is an eigenstate of S_x, $|\chi\rangle = |+, \hat{x}\rangle$, and so if $\vec{B}$ is parallel to Ox, 100% of the spins must be deflected toward positive x; the Bloch vector is $\vec{b} = (1, 0, 0)$. When $\vec{b} = 0$, the unpolarized case with state matrix

$$\rho = \frac{1}{2} |+\rangle\langle+| + \frac{1}{2} |-\rangle\langle-|, \tag{6.28}$$

the probabilities of deflection toward positive and negative z will be of 50% as for (6.27). However, *for any orientation of the Stern–Gerlach apparatus*, there will always be 50% of the spins deflected in the $\vec{B}$ direction and 50% in the $-\vec{B}$ direction. The difference between the two cases is that in the pure state (6.27), where the state is completely polarized, there is a well-defined phase relation between the amplitudes for finding $|\chi\rangle$ in the states $|+\rangle$ and $|-\rangle$. The pure state $|\chi\rangle$ is a *coherent* superposition of the states $|+\rangle$ and $|-\rangle$, and the mixture (6.28) is an *incoherent* superposition of the same states. The phase information is lost, at least partially, in a mixture (because partially polarized states $0 < |\vec{b}| < 1$ can certainly exist), and it is completely lost in an unpolarized state. In a given basis, the phase information is contained in the off-diagonal elements of the matrix ρ. For this reason these elements are called *coherences of the state operator*.

The same remarks apply to the polarization of light, or the polarization of a photon. Unpolarized light is an incoherent superposition of light linearly polarized 50% in the Ox direction and 50% in the Oy direction, with no phase relation between the two. Light with right- or left-handed circular polarization, $|R\rangle$ or $|L\rangle$, is described by the vectors (3.24)

$$|R\rangle = -\frac{1}{\sqrt{2}} (|x\rangle + i|y\rangle), \quad |L\rangle = \frac{1}{\sqrt{2}} (|x\rangle - i|y\rangle).$$

Fifty percent of this light will be stopped by a linear polarizer oriented in the Ox direction, or, more generally, in any direction, just as for unpolarized light. However, the corresponding photons will be transmitted with either 100% or 0% probability by a circular polarizer, while if the photons are not polarized any (λ, μ) polarizer (see Section 3.1.1) will allow photons through with 50% probability.

In general, a characteristic of a pure state is that there exists a maximal test such that one of its outcomes occurs with 100% probability, whereas for a mixture there is no maximal test possessing this property (Exercise 6.4.3). In the case of spin 1/2, this means that for a mixture there is no orientation of $\vec{B}$ such that 100% of the spins will be deflected in the $\vec{B}$ direction, and in the case of the photon there is no (λ, μ) polarizer which allows all photons to pass through with unit probability.

6.2.3 The reduced state operator

As an application of the state operator formalism, let us consider a system of two particles described by a state operator ρ acting in the space $\mathcal{H}_1 \otimes \mathcal{H}_2$. What then is the state operator of particle 1? To answer this question, let us examine a physical property C which depends solely on this particle. Then C has the form $A \otimes I_2$, where A acts in $\mathcal{H}_1$. We want to find a state operator $\rho^{(1)}$ acting in $\mathcal{H}_1$ such that

$$\langle A \rangle = \mathrm{Tr}\,(\rho^{(1)} A). \tag{6.29}$$

In the space $\mathcal{H}_1 \otimes \mathcal{H}_2$ the expectation value of $A \otimes I_2$ is given by

$$\langle A \otimes I_2 \rangle = \mathrm{Tr}([A \otimes I_2]\rho) = \sum_{n_1 m_1; n_2 m_2} A_{n_1 m_1} \delta_{n_2 m_2} \rho_{m_1 m_2; n_1 n_2} = \sum_{n_1 m_1} A_{n_1 m_1} \sum_{n_2} \rho_{m_1 n_2; n_1 n_2}$$

$$= \sum_{n_1 m_1} A_{n_1 m_1} \rho^{(1)}_{m_1 n_1} = \mathrm{Tr}(A \rho^{(1)}).$$

The state operator of particle 1 is then given in the $|n_1\rangle$ basis of $\mathcal{H}_1$ by the matrix $\rho^{(1)}$ with elements

$$\boxed{\rho^{(1)}_{n_1 m_1} = \sum_{n_2} \rho_{n_1 n_2; m_1 n_2} \quad \text{or} \quad \rho^{(1)} = \mathrm{Tr}_2 \rho} . \tag{6.30}$$

The second expression is independent of the basis; Tr_2 represents the trace on the space $\mathcal{H}_2$, called the *partial trace* of the global state operator, while $\rho^{(1)}$ is *the reduced state operator*. It can be shown that the reduced state operator gives the unique solution of (6.29).[7] An important application of (6.29) is to calculate the probability of finding the eigenvalue a_n of a physical property A, which is given as a function of the projector $\mathcal{P}_n$ onto the subspace of the eigenvalue a_n by an expression which generalizes (4.4):

$$\boxed{\mathrm{p}(a_n) = \mathrm{Tr}_1\left(\mathcal{P}_n \rho^{(1)}\right) = \mathrm{Tr}_1 \mathrm{Tr}_2\left[(\mathcal{P}_n \otimes I_2)\rho\right]} . \tag{6.31}$$

It is important to understand that the prescription of taking the partial trace is a consequence of postulate **II**, because the expression giving the expectation values follows from this postulate.

As an example, let us give the reduced state operator starting from the most general pure state $|\Psi\rangle$ in the tensor product space $\mathcal{H}_1^{(N)} \otimes \mathcal{H}_2^{(M)}$:

$$|\Psi\rangle = \sum_{i=1}^{N} \sum_{j=1}^{M} c_{ij} |\varphi_i \otimes \chi_j\rangle, \quad \rho = |\Psi\rangle\langle\Psi|.$$

---

[7] See Nielsen and Chuang [2000], p. 107.

The reduced state operator can be calculated immediately if we observe that

$$\text{Tr} \, |a\rangle\langle b| = \sum_n \langle n|a\rangle\langle b|n\rangle = \sum_n \langle b|n\rangle\langle n|a\rangle = \langle b|a\rangle. \tag{6.32}$$

Writing out the explicit expression for $|\Psi\rangle\langle\Psi|$, we find that the reduced state operator $\rho^{(1)}$ in $\mathcal{H}_1^{(N)}$ is

$$\rho^{(1)} = \text{Tr}_2 |\Psi\rangle\langle\Psi| = \sum_{ijkl} c_{ij} c_{kl}^* |\varphi_i\rangle\langle\varphi_k| \langle\chi_l|\chi_j\rangle. \tag{6.33}$$

A commonly encountered special case is:

$$|\Psi\rangle = \sum_{i=1}^{N} c_i |\varphi_i \otimes \chi_i\rangle,$$

with $N = M$, but the dimension of $\mathcal{H}_2^{(M)}$ can be larger than N, $M \geq N$. Then (6.33) is simplified as

$$\boxed{\rho^{(1)} = \sum_i c_i c_j^* |\varphi_i\rangle\langle\varphi_j| \langle\chi_j|\chi_i\rangle}. \tag{6.34}$$

If the $|\chi_i\rangle$ are orthogonal, $\langle\chi_i|\chi_j\rangle = \delta_{ij}$, the coherences in $\rho^{(1)}$ vanish and we obtain an incoherent mixture:

$$\boxed{\rho^{(1)} = \sum_i |c_i|^2 |\varphi_i\rangle\langle\varphi_i| \quad \text{if} \quad \langle\chi_i|\chi_j\rangle = \delta_{ij}}. \tag{6.35}$$

Equations (6.34) and (6.35) will play an important role in the discussion of measurement in Appendix B1.

If two particles are in the tensor product state $|\Psi\rangle = |\varphi \otimes \chi\rangle$, then $\rho^{(1)} = |\varphi\rangle\langle\varphi|$ describes a pure state, as expected. However, (6.33) or (6.34) show that this is not the case when $|\Psi\rangle$ is not a tensor product: then it is not possible to attribute a well-defined state to either particle. Let us verify this explicitly in the case of two spin-1/2 particles in the state (6.15). The reduced state operator is readily obtained using (6.35)

$$\rho^{(1)} = \text{Tr}_2 \rho = \frac{1}{2} |+\rangle\langle+| + \frac{1}{2} |-\rangle\langle-| = \begin{pmatrix} 1/2 & 0 \\ 0 & 1/2 \end{pmatrix}, \tag{6.36}$$

which is nothing other than the unpolarized state (6.28). *Even if the two-spin system is in a pure state, the state of an individual spin is in general a mixture.* In fact, the state matrix (6.36) represents an extreme case of a mixture corresponding to maximal disorder and minimal information on the spin. It can be shown that a quantitative measure of the information contained in the state operator is given by the *von Neumann (or statistical) entropy* $S_{vN} = -\text{Tr} \, \rho \ln \rho$,[8] which is the larger the less the information. In the case of spin 1/2, it lies between 0 and $\ln 2$, 0 corresponding to the pure state and $\ln 2$ to the

---

[8] It should be noted that $\text{Tr} \, \rho \ln \rho \neq \sum_\alpha p_\alpha \ln p_\alpha$ except when the vectors $|\varphi_\alpha\rangle$ in (6.20) are orthogonal to each other. $-\sum_\alpha p_\alpha \ln p_\alpha$ is the Shannon entropy, S_{Sh}, and it can be shown that $S_{vN} \leq S_{Sh}$.

mixture (6.36), respectively; ln 2 is the maximum value of the von Neumann entropy for a spin 1/2, and the mixture (6.36) is that which contains the minimal information. If the Hilbert space of states of a quantum system has dimension N, the state operator corresponding to maximal disorder is $\rho = I/N$, and so the statistical entropy $S_{\text{vN}} = \ln N$. Further properties of entangled states and state operators will be examined in Chapter 15.

6.2.4 Time dependence of the state operator

It is not difficult to find the time dependence of the state operator for a closed quantum system.[9] If we first consider the state operator

$$\mathcal{P}_\varphi(t) = |\varphi(t)\rangle\langle\varphi(t)|$$

for a pure state, using (4.11) we have

$$i\hbar\frac{d}{dt}\mathcal{P}_{\varphi(t)} = i\hbar\frac{d}{dt}\Big(|\varphi(t)\rangle\langle\varphi(t)|\Big) = H(t)\mathcal{P}_{\varphi(t)} - \mathcal{P}_{\varphi(t)}H(t) = \big[H(t), \mathcal{P}_{\varphi(t)}\big].$$

Summing over the probabilities p_α as in (6.20), we obtain the evolution equation for $\rho(t)$:

$$\boxed{i\hbar\frac{d\rho(t)}{dt} = \big[H(t), \rho(t)\big]} \,. \qquad (6.37)$$

An equivalent law is obtained using the evolution operator $U(t, 0)$ in (4.14):

$$\rho(t) = U(t, 0)\,\rho(t = 0)\,U^{-1}(t, 0).$$

This type of time evolution of a state operator is called *Hamiltonian*, or *unitary evolution*. It is worth observing that a state of maximal disorder is a dynamical invariant because $[H, \rho] = 0$.

Let us discuss the important example of the evolution law of the state operator of a spin-1/2 particle in a constant magnetic field. With $\vec{B}$ parallel to Oz, the Hamiltonian (3.62) is written as

$$H = -\frac{1}{2}\gamma\sigma_z,$$

and the evolution equation (6.37) becomes, using the commutation relations (3.52),

$$\frac{d\rho}{dt} = \frac{1}{i\hbar}[H, \rho] = -\frac{1}{2}\gamma B(b_x\sigma_y - b_y\sigma_x),$$

which is equivalent to

$$\frac{db_x}{dt} = -\gamma Bb_y, \quad \frac{db_y}{dt} = \gamma Bb_x, \quad \frac{db_z}{dt} = 0,$$

[9] See the comments following (4.11).

or in vector form

$$\frac{d\vec{b}}{dt} = -\gamma \vec{B} \times \vec{b}. \tag{6.38}$$

This is exactly the classical differential equation (3.31) describing Larmor precession. The Bloch vector undergoes the same motion as a classical spin.

In our discussion of NMR in Section 5.2.2 we studied an isolated spin. In fact, the spins are located in an environment which fluctuates at temperature T, and in the absence of a radiofrequency field they are described by a state operator ρ corresponding to thermal equilibrium in a constant magnetic field $\vec{B}_0$:

$$\rho \simeq \frac{1}{2}\left(I + \frac{\hbar\omega_0}{2k_\mathrm{B}T}\sigma_z\right) = \frac{1}{2}\left(I + \frac{1}{2}\delta\mathsf{p}\,\sigma_z\right), \tag{6.39}$$

where $\delta\mathsf{p}$ is the difference of the populations $\delta\mathsf{p} = \mathsf{p}_+ - \mathsf{p}_-$ (5.42) in the levels $|+\rangle$ and $|-\rangle$. The Bloch vector has components $\vec{b} = (0, 0, \delta\mathsf{p}/2)$. The application of a resonant radiofrequency pulse during a time $t = \theta/\omega_1$ transforms ρ into ρ_θ:

$$\rho \to \rho_\theta = U[\mathcal{R}_x(-\theta)]\rho\,U^\dagger[\mathcal{R}_x(-\theta)]$$

owing to (5.32). It is easy to calculate the matrix product explicitly, but more elegant to use (2.54):

$$\mathrm{e}^{\mathrm{i}\theta\sigma_x/2}\sigma_z\mathrm{e}^{-\mathrm{i}\theta\sigma_x/2} = \sigma_z + \frac{\mathrm{i}\theta}{2}[\sigma_x, \sigma_z] + \frac{1}{2!}\left(\frac{\mathrm{i}\theta}{2}\right)^2[\sigma_x, [\sigma_x, \sigma_z]] + \cdots$$

$$= \sigma_z + \frac{\mathrm{i}\theta}{2}\sigma_y - \frac{1}{2!}\theta^2\sigma_z + \cdots = \cos\theta\,\sigma_z + \sin\theta\,\sigma_y,$$

which is just the transformation law for the y and z components of a vector rotated by an angle $-\theta$ about Ox. We then find

$$\rho_\theta = \frac{1}{2}\left[I + \frac{1}{2}\delta\mathsf{p}\big(\cos\theta\,\sigma_z + \sin\theta\,\sigma_y\big)\right]. \tag{6.40}$$

In the special case of a $\pi/2$ pulse ($\theta = \pi/2$) the result is

$$\rho_{\pi/2} = \frac{1}{2}\left[I + \frac{1}{2}\delta\mathsf{p}\,\sigma_y\right]. \tag{6.41}$$

Since the matrix σ_y is not diagonal, we have created coherences: the difference between the initial populations has been converted into coherences. Note first that a natural basis $\{|+\rangle, |-\rangle\}$ is defined by the $\vec{B}_0$ field, and second that the identity operator I is not affected by unitary evolution (6.37), and that it is permissible to start from σ_z in (6.39), although σ_z is not a state matrix! The return to equilibrium is controlled by the relaxation time T_2. In the case of a π-pulse we obtain an inversion of the populations of the levels $|+\rangle$ and $|-\rangle$, and the return to equilibrium is controlled by the relaxation time T_1.

6.2.5 *General form of the postulates*

The introduction of the state operator allows us to give a more general formulation of the postulates stated in Chapter 4.

- **Postulate Ia.** The state of a quantum system is represented mathematically by a state operator ρ acting in a Hilbert space of states $\mathcal{H}$; ρ is a positive operator with unit trace.
- **Postulate IIa.** The probability p_χ of finding the quantum system in the state $|\chi\rangle$ is given by

$$\mathsf{p}_\chi = \mathrm{Tr}\left(\rho|\chi\rangle\langle\chi|\right) = \mathrm{Tr}\left(\rho\,\mathcal{P}_\chi\right).$$

- **Postulate IVa.** The time evolution of the state operator is given by (6.37):

$$i\hbar\,\frac{\mathrm{d}\rho(t)}{\mathrm{d}t} = [H(t), \rho(t)].$$

Postulate **III** is unchanged, and the WFC (wave-function collapse) postulate (4.7) becomes

$$\rho \to \frac{\mathcal{P}_n\rho\mathcal{P}_n}{\mathrm{Tr}\,\rho\mathcal{P}_n}$$

when the result of a measurement of a physical property A is the eigenvalue a_n. We again stress the fact that (6.37) holds only for a closed system. The time evolution of the state operator of a system which is part of a larger quantum system is much more complicated and will be studied in Chapter 15. In statistical mechanics, the case of a system in contact with a heat bath represents a typical example of a system which is not closed. The evolution of the ensemble system + heat bath is unitary (if the ensemble itself is closed), but that of the system obtained by taking the trace over the variables of the heat bath is not.[10]

6.3 Examples

6.3.1 *The EPR argument*

Let us suppose that we are capable of making a state $|\Phi\rangle$ (6.15) of two identical spin-1/2 particles, with the two particles traveling with equal momenta in opposite directions. For example, they could originate in the decay of an unstable particle of zero spin and zero momentum, in which case momentum conservation implies that the particles move in opposite directions. An example which is simple theoretically (but not experimentally) is the decay of a π^0 meson into an electron and a positron:[11] $\pi^0 \to \mathrm{e}^+ + \mathrm{e}^-$. Two experimentalists, conventionally named Alice and Bob, measure the spin component of each particle on a certain axis (Fig. 6.1) when the particles are very far apart compared with the range of the force and have not interacted with each other for a long time. For clarity, in this figure the axes used for spin measurement are taken to be perpendicular to

---

[10] In Hamiltonian evolution, the von Neumann entropy $-\mathrm{Tr}\,\rho\ln\rho$ is conserved, but this is not the case for non-Hamiltonian evolution, where the von Neumann entropy of a system in contact with a heat bath is not constant.

[11] This decay mode is rare, but it is useful for our theoretical discussion.

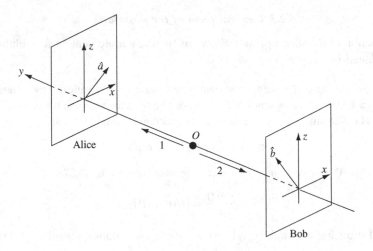

Fig. 6.1. Configuration of an EPR type of experiment.

the direction of propagation, though this is not essential.[12] Using a Stern–Gerlach device in which the magnetic field points in the direction $\hat{a}$, Alice measures the spin component on this axis for the particle traveling to the left, particle a, while Bob measures the component along the $\hat{b}$ axis of the particle traveling to the right, particle b. Let us first study the case where Alice and Bob both use the Oz axis, $\hat{a} = \hat{b} = \hat{z}$. We assume that the decays are well separated in time, and that each experimentalist can know if he or she is measuring the spins of particles emitted in the same decay. In other words, each *pair* (e^+, e^-) is perfectly well identified in the experiment.

   Using her Stern–Gerlach device, Alice measures the z component of the spin of particle a, $S_z^{(a)}$, with the result $+\hbar/2$ or $-\hbar/2$, and Bob measures $S_z^{(b)}$ of particle b. As we have seen in (6.36), neither of these particles is polarized; Alice and Bob observe a random series of results $+\hbar/2$ and $-\hbar/2$. After the series of measurements has been completed, Alice and Bob meet and compare their results. They conclude that the results for each pair exhibit a perfect (anti-)correlation. When Alice has measured $+\hbar/2$ for particle a, Bob has measured $-\hbar/2$ for particle b and vice versa. To explain this anticorrelation, let us calculate the result of a measurement in the state $|\Phi\rangle$ (6.15) of the physical property $[S_z^{(a)} \otimes S_z^{(b)}]$, a Hermitian operator acting in the tensor product space of the two spins. Taking into account (6.12), we immediately see that $|\Phi\rangle$ is an eigenvector of $[S_z^{(a)} \otimes S_z^{(b)}]$ with eigenvalue $-\hbar^2/4$:

$$[S_z^{(a)} \otimes S_z^{(b)}]\frac{1}{\sqrt{2}}\left(|+-\rangle - |-+\rangle\right) = -\frac{\hbar^2}{4}\frac{1}{\sqrt{2}}\left(|+-\rangle - |-+\rangle\right).$$

---

[12] See Footnote 15 of Chapter 3.

Measurement of $[S_z^{(a)} \otimes S_z^{(b)}]$ must then give the result $-\hbar^2/4$, which implies that Bob must measure the value $-\hbar/2$ if Alice has measured the value $+\hbar/2$ and vice versa.[13] Within the limit of accuracy of the experimental apparatus, it is impossible that Alice and Bob both measure the value $+\hbar/2$ or $-\hbar/2$.

Upon reflection, this result is not very surprising. It is a variation of the game of the two customs inspectors.[14] Two travelers a and b, each carrying a suitcase, depart in opposite directions from the origin and eventually are checked by two customs inspectors Alice and Bob. One of the suitcases contains a red ball and the other a green ball, but the travelers have picked up their closed suitcases at random and do not know what color the ball inside is. If Alice checks the suitcase of traveler a, she has a 50% chance of finding a green ball. But if in fact she finds a green ball, clearly Bob will find a red ball with 100% probability. Correlations between the two suitcases were introduced at the time of departure, and these correlations reappear as a correlation between the results of Alice and Bob.

However, as first noted by Einstein, Podolsky, and Rosen (EPR) in a celebrated paper[15] (which used a different example, ours being due to Bohm), the situation becomes much less commonplace if Alice and Bob decide to use the Ox axis instead of the Oz axis for another series of measurements.[16] Since $|\Phi\rangle$ is invariant under rotation, if Alice and Bob orient their Stern–Gerlach devices in the Ox direction, they will again find that their measurements are perfectly anticorrelated, because

$$[S_x^{(a)} \otimes S_x^{(b)}]|\Phi\rangle = -\frac{\hbar^2}{4}|\Phi\rangle.$$

The viewpoint underlying the EPR analysis of these results is that of "realism": EPR assume that microscopic systems possess intrinsic properties which must have a counterpart in the physical theory. More precisely, according to EPR, *if the value of a physical property can be predicted with certainty without disturbing the system in any way, there is an "element of reality" associated with this property.* For a particle of spin 1/2 in the state $|+\rangle$, S_z is a property of this type because it can be predicted with certainty that $S_z = \hbar/2$. However, the value of S_x in this same state cannot be predicted with certainty (it can be $+\hbar/2$ or $-\hbar/2$ with 50% probability of each); S_x and S_z cannot simultaneously have a physical reality. Since the operators S_x and S_z do not commute, in quantum physics it is impossible to attribute simultaneous values to them.

In performing their analysis, EPR used a second hypothesis, the *locality principle*, which stipulates that if Alice and Bob make their measurements in local regions of

---

[13] The following argument is sometimes encountered: if Alice obtains $+\hbar/2$ upon making the first measurement of $S_z^{(a)}$, the state $|\Phi\rangle$ is projected onto $|+-\rangle$ by wave-function collapse (the WFC postulate), and Bob then measures $S_z^{(b)} = -\hbar/2$. This reasoning is not satisfactory, because the statement "Alice makes the first measurement of the spin" is not Lorentz-invariant if Alice and Bob are separated by a distance L and if their measurements are separated by a time interval $\tau < L/c$. The temporal order of the measurements of Alice and Bob is irrelevant.

[14] Invented just for this occasion!

[15] A. Einstein, B. Podolsky, and N. Rosen, Can quantum-mechanical description of physical reality be considered complete? *Phys. Rev.* **77**, 777–780 (1935). The term "EPR paradox" is sometimes used, but there is nothing paradoxical in the EPR analysis.

[16] However, even in this case the result can be reproduced using a classical model; see Fig. 6.2.

spacetime which cannot be causally connected,[17] then it is not possible that an experimental parameter chosen by Alice, for example the orientation of her Stern–Gerlach device, can affect the properties of particle b.[18] According to the preceding discussion, this implies that without disturbing particle b in any way, a measurement of $S_z^{(a)}$ by Alice permits knowledge of $S_z^{(b)}$ with certainty, and a measurement of $S_x^{(a)}$ permits knowledge of $S_x^{(b)}$ with certainty. If the "local realism" of EPR is accepted, the result of Alice's measurement serves only to reveal a piece of information which was already stored in the local region of spacetime associated with particle b. A theory that is more complete than quantum mechanics should contain simultaneous information on the values of $S_x^{(b)}$ and $S_z^{(b)}$, and be capable of predicting with certainty all the results of measurements of these two physical properties in the local region of spacetime attached to particle b. The physical properties $S_x^{(b)}$ and $S_z^{(b)}$ then simultaneously have a physical reality, in contrast to the quantum description of the spin of a particle by a state vector. EPR do not dispute the fact that quantum mechanics gives predictions that are statistically correct, but quantum mechanics is not sufficient for describing the physical reality of an individual pair. Within the framework of local realism such as that defined above, the EPR argument is unassailable and the verdict incontestable: quantum mechanics is incomplete! Nevertheless, EPR do not suggest any way of "completing" it, and we shall see in what follows that local realism is in conflict with experiment.

6.3.2 Bell inequalities

According to local realism, even if an experiment does not permit the simultaneous measurement of $S_x^{(b)}$ and $S_z^{(b)}$, these two quantities still have a simultaneous physical reality in the local region of spacetime attached to particle b, and owing to symmetry the same is true for $S_x^{(a)}$ and $S_z^{(a)}$ of particle a. This ineluctable consequence of local realism makes it possible to prove the *Bell inequalities*, which fix the maximum possible correlations given this hypothesis. Let us return to the case of some given measurement axes $\hat{a}$ and $\hat{b}$, which as above we take to lie in the xOz plane perpendicular to the propagation direction Oy, in order to make the figures clear. We shall use $A(\hat{a})$ and $B(\hat{b})$ to denote the results of measuring $\vec{\sigma} \cdot \hat{a}$ and $\vec{\sigma} \cdot \hat{b}$; in order to eliminate the factor $\hbar/2$, it is convenient to use the Pauli matrices rather than the spin operators. In addition, we shall simplify the notation by omitting the indices (a) and (b) when the vectors $\hat{a}$ or $\hat{b}$ remove any ambiguity:

$$\sigma_{\hat{a}}^{(a)} = \vec{\sigma}^{(a)} \cdot \hat{a} \rightarrow \vec{\sigma} \cdot \hat{a}, \quad \sigma_{\hat{b}}^{(b)} = \vec{\sigma}^{(b)} \cdot \hat{b} \rightarrow \vec{\sigma} \cdot \hat{b}.$$

---

[17] For example, if Alice and Bob are separated by a distance L in a reference frame in which they are both at rest, and if the measurements take a time τ with $\tau \ll L/c$.

[18] This is not the same as saying that the results of Alice and Bob are not correlated. In the simple example of the two travelers, the opening of the suitcase of traveler a by Alice reveals the color of the ball in the suitcase of b. This opening does not disturb anything in the suitcase of b, but it determines the result of Bob, which means the results are correlated. The color of the ball in the suitcase of b existed before the suitcase of a was opened.

The possible results of the measurements are ± 1:

$$A(\hat{a}) = \varepsilon_a = \pm 1, \quad B(\hat{b}) = \varepsilon_b = \pm 1.$$

Let $\mathsf{p}_{\varepsilon_a \varepsilon_b}$ be the joint probability for Alice to find the result ε_a and Bob to find the result ε_b, and let $E(\hat{a}, \hat{b})$ be the expectation value $\langle \varepsilon_a \varepsilon_b \rangle$:

$$E(\hat{a}, \hat{b}) = \sum \varepsilon_a \varepsilon_b \, \mathsf{p}_{\varepsilon_a \varepsilon_b} = [\mathsf{p}_{++} + \mathsf{p}_{--}] - [\mathsf{p}_{+-} + \mathsf{p}_{-+}]. \qquad (6.42)$$

This quantity measures the correlation between the measurements of Alice and Bob when they use the axes $\hat{a}$ and $\hat{b}$. It is obtained experimentally by making a series of $N \gg 1$ measurements on N pairs. If $A_n(\hat{a})$ and $B_n(\hat{b})$ are the results of a measurement on the pair n for the orientations $(\hat{a}, \hat{b})$, then

$$E(\hat{a}, \hat{b}) = \lim_{N \to \infty} \frac{1}{N} \sum_{n=1}^{N} A_n(\hat{a}) B_n(\hat{b}).$$

This is an experimental result, independent of any a priori theoretical considerations. Let us now consider two possible orientations $\hat{a}$ and $\hat{a}'$ for the measurements of Alice, two possible ones $\hat{b}$ and $\hat{b}'$ for those of Bob, and use the abbreviated notation $A_n' = A_n(\hat{a}')$, $B_n' = B_n(\hat{b}')$ for the pair n. Let X_n be the combination

$$X_n = A_n B_n + A_n B_n' + A_n' B_n' - A_n' B_n = A_n(B_n + B_n') + A_n'(B_n' - B_n). \qquad (6.43)$$

In contrast to $E(\hat{a}, \hat{b})$, writing down X_n rests on an a priori theoretical idea, that of the EPR picture in which particles a and b "possess" the properties $A_n, \ldots, B_n'$. Only one of the four possible combinations $(A_n, B_n) \ldots (A_n', B_n')$ can be effectively measured in an experiment on the pair n, but the potential result for the three other experiments, although unknown, is well defined. This can be illustrated using the suitcase model, where each suitcase is composed of small angular sectors labeled $+$ or $-$, with opposite labels for Alice and Bob (Fig. 6.2). To measure $A_n(\hat{a})$ [$B_n(\hat{b})$], Alice [Bob] opens the angular sector marked by the direction $\hat{a}$ [$\hat{b}$], and if $\hat{a} = \hat{b}$, Alice and Bob find two opposite results, reproducing all the results of Section 6.3.1. If Alice opens the sector $\hat{a}$ and observes the result $(+)$ as in Fig. 6.2, the sector $\hat{a}'$ must contain the well-defined result $(-)$, which Alice would have observed had she opened that sector.

For each pair the combination X_n is ± 2. In fact, we have either $B_n = B_n'$, in which case $B_n' - B_n = 0$ and $B_n + B_n' = \pm 2$, or $B_n = -B_n'$, in which case $B_n + B_n' = 0$ and $B_n' - B_n = \pm 2$. Since the possible values of A_n and A_n' are ± 1, we necessarily have $X_n = \pm 2$. The average over a large number of experiments can only give an expectation value $\langle X \rangle$ whose absolute value is less than two:

$$\boxed{|\langle X \rangle| = \left| \lim_{N \to \infty} \frac{1}{N} \sum_{n=1}^{N} X_n \right| \le 2}. \qquad (6.44)$$

The result $|\langle X \rangle| \le 2$ is an example of a *Bell inequality*. We again stress the fact that this inequality depends crucially on local realism: particle a possesses the properties A_n and

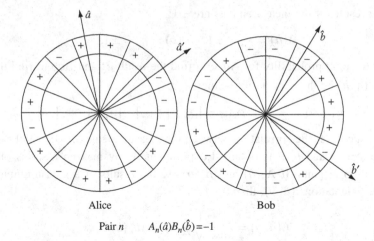

Alice   Bob

Pair n $A_n(\hat{a})B_n(\hat{b}) = -1$

Fig. 6.2. Classical model of EPR correlations. The suitcases of travelers A and B are circles divided into small angular sectors labeled by the orientations $\hat{a}, \ldots, \hat{b}'$ in the xOz plane and containing the result $(+)$, meaning spin in this direction, or the result $(-)$ meaning spin in the opposite direction. The figure corresponds to $A_n(\hat{a}) = (+)$, $A_n(\hat{a}') = (-)$, $B_n(\hat{b}) = (-)$, and $B_n(\hat{b}') = (-)$ for pair n.

A_n' *simultaneously*, particle b possesses the properties B_n and B_n', and the value of, for example, A_n cannot depend on the orientation $\hat{b}$ or $\hat{b}'$ of Bob's analyzer.

What are the predictions of quantum mechanics? To calculate $E(\hat{a}, \hat{b})$ defined in (6.42) we use the rotational invariance of $|\Phi\rangle$, which allows us to choose $\hat{a}$ in the Oz direction. The eigenstates of $S_{\hat{a}}$, or of $\vec{\sigma} \cdot \hat{a}$, are then the eigenstates $|+\rangle$ and $|-\rangle$ of $S_z^{(a)}$. Let θ be the angle between $\hat{b}$ and Oz. According to (3.56), in the basis $\{|+\rangle, |-\rangle\}$ we have

$$|+, \hat{b}\rangle = \cos\frac{\theta}{2}|+\rangle + \sin\frac{\theta}{2}|-\rangle.$$

The tensor product[19] $|+\otimes[+, \hat{b}]\rangle$ is then given by

$$|+\otimes[+, \hat{b}]\rangle = \cos\frac{\theta}{2}|+\otimes+\rangle + \sin\frac{\theta}{2}|+\otimes-\rangle, \qquad (6.45)$$

and the amplitude a_{++} in $\mathsf{p}_{++} = |a_{++}|^2$ will be

$$a_{++} = \langle+\otimes[+, \hat{b}]|\Phi\rangle = \langle+\otimes[+, \hat{b}]|+\otimes-\rangle = \frac{1}{\sqrt{2}}\sin\frac{\theta}{2}. \qquad (6.46)$$

By symmetry, under the exchange $+ \leftrightarrow -$ we have

$$\mathsf{p}_{++} = \mathsf{p}_{--} = \frac{1}{2}\sin^2\frac{\theta}{2}$$

and thus

$$\mathsf{p}_{+-} = \mathsf{p}_{-+} = \frac{1}{2}\cos^2\frac{\theta}{2},$$

[19] For clarity, we temporarily restore the notation for the tensor product.

as can be verified by explicit calculation (Exercise 6.4.7). We find that

$$E(\hat{a}, \hat{b}) = \sin^2 \frac{\theta}{2} - \cos^2 \frac{\theta}{2} = -\cos\theta = -\hat{a} \cdot \hat{b}. \qquad (6.47)$$

Another way of calculating $E(\hat{a}, \hat{b})$ is to note that $A(\hat{a}) = \varepsilon_a$ is the eigenvalue of $\vec{\sigma} \cdot \hat{a}$, $B(\hat{b}) = \varepsilon_b$ is that of $\vec{\sigma} \cdot \hat{b}$, and measurement of $(\vec{\sigma} \cdot \hat{a}) \otimes (\vec{\sigma} \cdot \hat{b})$ gives the result $\varepsilon_a \varepsilon_b$. Then $E(\hat{a}, \hat{b})$ is just the expectation value of $(\vec{\sigma} \cdot \hat{a}) \otimes (\vec{\sigma} \cdot \hat{b})$ in the state $|\Phi\rangle$:

$$E(\hat{a}, \hat{b}) = \langle (\vec{\sigma} \cdot \hat{a}) \otimes (\vec{\sigma} \cdot \hat{b}) \rangle_\Phi = \langle \Phi | (\vec{\sigma} \cdot \hat{a}) \otimes (\vec{\sigma} \cdot \hat{b}) | \Phi \rangle. \qquad (6.48)$$

Exercise 6.5.7 shows that we recover (6.47) starting from (6.48).

Let us now choose the axes on which the two spins are measured. We take $\hat{a}$ parallel to $\hat{z}$, $\hat{b}$ pointing along the second bisector of the axes $\hat{x}$ and $\hat{z}$ (Fig. 6.3), $\hat{a}'$ parallel to $\hat{x}$, and $\hat{b}'$ parallel to the first bisector and orthogonal to $\hat{b}$. The various expectation values are given by

$$E(\hat{a}, \hat{b}) = E(\hat{a}, \hat{b}') = E(\hat{a}', \hat{b}') = -\frac{1}{\sqrt{2}}, \quad E(\hat{a}', \hat{b}) = \frac{1}{\sqrt{2}}. \qquad (6.49)$$

The combination $\langle X \rangle$ of these expectation values will be $-2\sqrt{2}$ in quantum mechanics:

$$\langle X \rangle = E(\hat{a}, \hat{b}) + E(\hat{a}, \hat{b}') + E(\hat{a}', \hat{b}') - E(\hat{a}', \hat{b}) = -2\sqrt{2}. \qquad (6.50)$$

It can be shown that the choice of orientations in Fig. 6.2 gives the maximum value of $|\langle X \rangle|$, $|\langle X \rangle|_{\max} = 2\sqrt{2}$. This value violates the limit (6.44) $|\langle X \rangle| \leq 2$. Quantum mechanics is incompatible with the Bell inequalities, and therefore with the EPR hypothesis of local realism – the correlations of quantum mechanics are too strong. Theories with local hidden variables represent an example of a realistic local theory, and the predictions of quantum mechanics are therefore incompatible with any theory of this type. The contradiction between quantum mechanics and the EPR hypotheses arises because in quantum mechanics we cannot simultaneously attribute well-defined values to the four

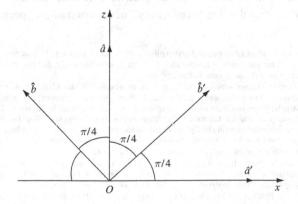

Fig. 6.3. Optimal configuration of the angles.

quantities A_n, B_n, A'_n, and B'_n of (6.43) for a single pair of spin-1/2 particles, because these quantities correspond to eigenvalues of operators that do not commute with each other. We can experimentally measure at most two of these quantities simultaneously, one per particle, and we cannot assume in any physical argument that these quantities exist although they are unknown. In contrast to the opening of suitcase a, measurement of the spin of particle a by Alice does not reveal a pre-existing property of particle b.[20] The quantity X_n in (6.43) is "counterfactual," that is, it cannot be measured in any realizable experiment.[21]

The first experiments comparing the predictions of local realism with those of quantum mechanics were performed using two photons originating in the successive de-excitation of two excited states of an atom (an atomic cascade), the polarizations of the two photons being entangled in a state[22]

$$|\Psi\rangle = \frac{1}{\sqrt{2}}(|RR\rangle + |LL\rangle) = \frac{1}{\sqrt{2}}(|xx\rangle + |yy\rangle). \qquad (6.51)$$

The experiments of Aspect *et al.*[23] in the early 1980s were the first to demonstrate convincingly the conflict with local realism. Nowadays much more precise experiments are carried out using parametric photon conversion. In an experiment performed in Innsbruck[24] an ultraviolet photon is converted in a nonlinear crystal into two photons in an entangled polarization state (Fig. 6.4). In this experiment the orientation of the analyzers can be changed randomly while the photons are traveling between their production point and the detectors. The two detectors are 400 m apart, a distance traveled by light in 1.3 µs, while the total time required to make the individual measurements and rotate the polarizers is less than 100 ns. It is impossible that the measurements of Alice and Bob are causally related, and any information on the orientation of the analyzers that could have been stored in advance is also erased. The only possible objection is that only 5% of the photon pairs are detected, and it must be assumed that this 5% constitutes a representative sample. A priori, there is no reason to dispute this.[25] It can very reasonably be stated that experiment has decided in favor of quantum mechanics and has eliminated Einstein's principle of local realism. One might be tempted to conclude that quantum physics is nonlocal, but in such a way that the "nonlocality" never contradicts special relativity and

---

[20] From this point of view, Fig. 2.18 of Lévy-Leblond and Balibar [1990] can be interpreted erroneously. It might be inferred that the quanton "possesses" the properties of a wave and of a particle simultaneously, and that observation revealing one or the other of these aspects only reveals a pre-existing reality.

[21] As stated by A. Peres [1993]: "Unperformed experiments have no results." It should not at all be concluded that it is necessarily forbidden to introduce quantities which are not directly observable into the theory. For example, the consequences of causality on a time-dependent dielectric constant are expressed most conveniently by taking its Fourier transform and showing that this transform is an analytic function of the frequency ω in the complex half-plane $\text{Im}\,\omega > 0$. However, a complex frequency is never observed experimentally! As Feynman has written (Feynman *et al.* [1965], Vol. III, Section 2.6), "it is not true that we can pursue science completely by using only those concepts directly subject to experiment."

[22] Great care must be taken with the orientation conventions; cf. Exercise 6.5.8.

[23] A. Aspect, P. Grangier, and G. Roger, Experimental realization of Einstein–Podolsky–Rosen gedanken experiment: a new violation of Bell's inequalities, *Phys. Rev. Lett.* **49**, 91–94 (1982); A. Aspect, J. Dalibard, and G. Roger, Experimental test of Bell's inequalities using time-varying analyzers, *Phys. Rev. Lett.* **49**, 1804–1807 (1982).

[24] G. Weihs *et al.*, Violation of Bell's inequality under strict locality conditions, *Phys. Rev. Lett.* **81**, 5039–5043 (1998).

[25] The result of an election for the President of the French Republic can be predicted with some degree of confidence from a sample of 1000 out of 30 million voters, that is, 0.003%.

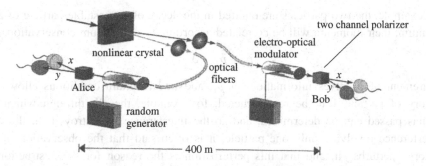

Fig. 6.4. Experiment involving entangled photons. A pair of entangled photons is produced in a nonlinear BBO crystal, and the two photons travel inside optical fibers which take them to polarization analyzers. After A. Zeilinger, Experiment and the foundations of quantum physics, *Rev. Mod. Phys.* **71**, S288–S297 (1999).

does not allow, for example, information transmission at speeds higher than the speed of light. Alice and Bob each observe a random sequence of $+1$ and -1, which does not contain any information, and it is only when their results transmitted by a classical path, that is, a speed lower than c, are compared that they can see they are correlated. Additional remarks on this point will be found in the comments following (6.69).

Rather than nonlocality, it is preferable to speak of *nonseparability of the state vector* $|\Phi\rangle$ (6.15), which does not contain any reference to spacetime. The experiments described above permit an inference of nonlocality only if "realism" is added: it is "local realism" which is refuted.

6.3.3 Interference and entangled states

In the discussion of interference experiments in Chapter 1, we emphasized the fact that interference is destroyed if it is possible, at least in principle, to know the particle trajectory and to determine which slit the particle has passed through. The qualification "at least in principle" is crucial: it doesn't matter whether or not the experimentalist actually makes the observation, or whether or not the observation can actually be made using the available technology. It is sufficient that the observation be possible in principle in the framework of the experimental setup. The use of entangled states will considerably enrich our possibilities, and allow us to better appreciate the astonishing strangeness of quantum mechanics relative to our prejudices gained from classical experience.

Let us imagine an experiment in which a particle 1 passes through a Young's slit apparatus, and let $|a\rangle$ ($|a'\rangle$) be the quantum state of this particle when it passes through slit a (a'), that is, the quantum state of the particle when slit a' (a) is closed. Let us suppose that the state of particle 1 is entangled with that of a particle 2, so that the global state $|\Psi\rangle$ is

$$|\Psi\rangle = \frac{1}{\sqrt{2}} (|a \otimes b\rangle + |a' \otimes b'\rangle). \tag{6.52}$$

If, for example, the two particles are emitted in the decay of an unstable particle of zero momentum, their momenta will be correlated according to momentum conservation:

$$\vec{p}_1 + \vec{p}_2 = 0.$$

Measurement of $\vec{p}_2$ gives information on $\vec{p}_1$, and under certain conditions allows the trajectory of particle 1 to be reconstructed; for example, the slit through which the latter has passed can be determined, and so the interference is destroyed. In the case of interference involving only one particle, it is often said that the observation of the trajectory "perturbs" it, and that this perturbation is the reason for the destruction of the interference. Our example of interference involving entangled particles confirms the discussion of Section 1.4.4 and shows that this "explanation" misses the essential point: in this new experiment, particle 1 is never observed, and it is the information on 1 provided by a measurement made (or not made) on 2 that leads to the conclusion that interference is destroyed. It is the possibility of labeling the different trajectories and not the perturbation due to observing them which is the origin of the destruction of the interference.

This labeling of trajectories has already been displayed in Exercise 3.3.9 for neutron diffraction by spin-1/2 nuclei. In fact, the possibility in theory of labeling the neutron trajectory owing to spin flip of a nucleus is sufficient to destroy the interference – instead of diffraction peaks, a continuous background is observed, as the spatial variables of the neutrons are not affected at all by spin flip. However, the experiment we are going to examine below is even more complete, because it provides the option of erasing this labeling and recovering the interference.

Before describing an experiment which has actually been performed, let us discuss its principle for a simplified geometry. Two photons 1 and 2 are emitted in the decay of an unstable particle assumed to be practically at rest; we shall return to this assumption later. The decay occurs in a plate of height d (Fig. 6.5). Photon 1 travels to the left and passes through a Young's slit device, while photon 2 travels to the right with opposite momentum, passes a convergent lens of focal length f, and then is detected by a detector array at screen E_2 located a distance $2f$ from the lens. The plane F of the Young's slits is also located a distance $2f$ from the lens. The position at which photon 2 arrives

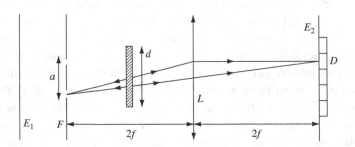

Fig. 6.5. The blurring of interference: the detection of photon 2 in the plane located a distance $2f$ from the lens makes it possible to trace back to the position of photon 1 in the plane of the Young's slits.

on the screen E_2 can be used to trace back to the position of photon 1 on the plane F, as the planes E_2 and F are conjugate to each other with respect to the lens. If photon 1 is detected on the screen E_1 after passing through the Young slits, photon 2 will be detected in coincidence with it on the screen E_2, which will give information on which slit it has passed through. Therefore, the photons 1 will not form an interference pattern. Even in the absence of the lens and the detector, there will be no interference pattern, because we can in principle install the lens and the detector array at E_2 and thus recover the information on the trajectory of photon 1. It is the existence of the accompanying photon that is crucial.

However, it is possible to erase this potential information by performing a *different* experiment, where a detector is placed in the focal plane of the lens (Fig. 6.6). The detection of photon 2 then determines the direction of the momentum of photon 2 before the lens, and as a consequence also that of photon 1. All the information on the position of photon 1 in the passage through the plane F of the slits is now erased – the detector functions like a "quantum eraser." The photons 1 detected *in coincidence* with photons 2 will again form an interference pattern on the screen E_1, with the position of the central fringe fixed by the position of the detector in the focal plane of the lens.

The following observation should be added. The characteristic angle in the experimental geometry is $\theta = a/D$, where a is the distance between the slits and D is the distance between the slits and the source. The spread Δp_z in the vertical component of the momentum of the photons produced in the plate of height d as a function of wavelength λ is

$$\Delta p_z \sim \frac{h}{d} \implies \frac{\Delta p_z}{p} \sim \frac{h}{dp} = \frac{\lambda}{d}.$$

In the discussion above it is assumed that this spread is negligible compared with θ:

$$\frac{\lambda}{d} \ll \theta. \tag{6.53}$$

On the other hand, for $\lambda/d \gg \theta$ we observe two sets of independent fringes if the two photons are allowed to pass through Young's slits (Exercise 6.5.9).

The experiment is performed in a slightly different geometry. The two photons are produced by parametric conversion in a nonlinear crystal from an ultraviolet photon of momentum $\vec{P}$, and the condition $\vec{p}_1 + \vec{p}_2 = 0$ is replaced by $\vec{p}_1 + \vec{p}_2 = \vec{P}$. The two photons

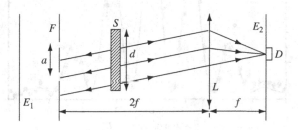

Fig. 6.6. Interference in coincidence. The detector of photon 2 is now located in a plane a distance f from the lens. The potential information on the trajectory of photon 1 is erased, and an interference pattern is observed if photon 1 is detected in coincidence with photon 2.

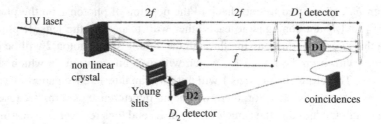

Fig. 6.7. Experiment of the Innsbruck group. The pair of entangled photons is produced in a nonlinear crystal. After A. Zeilinger, *Rev. Mod. Phys.* **71**, S288 (1999).

both travel to the right with a small variable angle between their trajectories (Fig. 6.7). In order to obtain the trajectory of photon 1, it is sufficient to reverse its direction of propagation when leaving the plate in Figs. 6.5 and 6.6. The experiment confirms the preceding discussion in all respects (Fig. 6.8).

6.3.4 Three-particle entangled states (GHZ states)

GHZ (Greenberger–Horne–Zeilinger) states are three-particle entangled states which exhibit nonclassical properties in an even more spectacular fashion than two-particle states. It is known how to create three-photon entangled states experimentally using parametric conversion. To simplify the discussion, we shall limit ourselves to the theory of entangled states of three spin-1/2 particles. We assume that an unstable particle decays

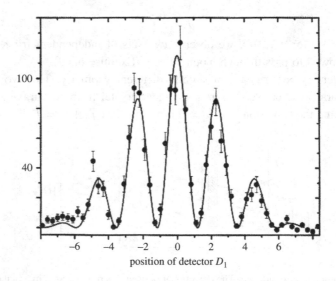

Fig. 6.8. Interference observed by the Innsbruck group. After A. Zeilinger, *Rev. Mod. Phys.* **71**, S288 (1999).

into three identical particles of spin 1/2 which are emitted in a plane in a configuration in which the three momenta lie at angles of $2\pi/3$ to each other, and the three particles are in the entangled spin state

$$|\Psi\rangle = \frac{1}{\sqrt{2}}\left(|+++\rangle - |---\rangle\right).$$
(6.54)

Three experimentalists, Alice (a), Bob (b), and Charlotte (c), can measure the spin component in the direction perpendicular to the direction of propagation of each particle (Fig. 6.9). The momenta lie in the horizontal plane, and the Oz axis is chosen to lie along the propagation direction (so that it depends on the particle), while Oy is vertical and $\hat{x} = \hat{y} \times \hat{z}$. Let us examine the three following operators:

$$\Sigma_a = \sigma_{ax}\sigma_{by}\sigma_{cy}, \quad \Sigma_b = \sigma_{ay}\sigma_{bx}\sigma_{cy}, \quad \Sigma_c = \sigma_{ay}\sigma_{by}\sigma_{cx}.$$
(6.55)

The matrices σ_i act in the space of spin states of particle i, $i = a, b, c$. The index i of Σ_i specifies the position of the matrix σ_x in the products (6.55). The three operators Σ_i commute with each other. To show this, we use the fact that σ matrices acting on different spaces commute, for example

$$\sigma_{ax}\sigma_{by} = \sigma_{by}\sigma_{ax}.$$

For matrices acting in the same space we use (3.48):

$$\sigma_x\sigma_y = -\sigma_y\sigma_x,$$

as well as

$$\sigma_x^2 = \sigma_y^2 = I.$$

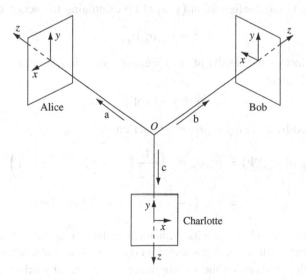

Fig. 6.9. Configuration of a GHZ type of experiment.

As an example, let us show that Σ_a and Σ_b commute owing to the fact that the two operators $\Sigma_a\Sigma_b$ and $\Sigma_b\Sigma_a$ differ by an even number of anticommutations:

$$\Sigma_a\Sigma_b = \sigma_{ax}\sigma_{by}\sigma_{cy}\sigma_{ay}\sigma_{bx}\sigma_{cy} = \sigma_{ax}\sigma_{by}\sigma_{ay}\sigma_{bx}$$

$$= -\sigma_{ay}\sigma_{ax}\sigma_{by}\sigma_{bx} = \sigma_{ay}\sigma_{bx}\sigma_{ax}\sigma_{by}$$

$$= \sigma_{ay}\sigma_{bx}\sigma_{cy}\sigma_{cy}\sigma_{ax}\sigma_{by} = \Sigma_b\Sigma_a.$$

The other commutation relations are demonstrated in a similar fashion. The squares of the operators Σ_i are unit operators ($\Sigma_i^2 = I$), their eigenvalues are ± 1, and, as they commute with each other, they can be simultaneously diagonalized. There then exists an eigenvector $|\Psi\rangle$ preserving the symmetry between the three particles constructed explicitly in (6.54) such that

$$\Sigma_a|\Psi\rangle = \Sigma_b|\Psi\rangle = \Sigma_c|\Psi\rangle = |\Psi\rangle. \tag{6.56}$$

Equation (6.56) can be shown directly by examining the action of Σ_i on $|\Psi\rangle$ using the following properties

$$\sigma_x|+\rangle = |-\rangle, \quad \sigma_x|-\rangle = |+\rangle,$$

$$\sigma_y|+\rangle = i|-\rangle, \quad \sigma_y|-\rangle = -i|+\rangle.$$

The spins are measured in the configurations (x, y, y), (y, x, y), and (y, y, x). For example, in the configuration (x, y, y), Alice orients her Stern–Gerlach apparatus in the direction Ox, and Bob and Charlotte orient theirs in the direction Oy. Measurements of σ_{ix} or of σ_{iy} always give the result ± 1, and if the particle triplet is in the state $|\Psi\rangle$, the product of the results of Alice, Bob, and Charlotte will be $+1$ for any configuration of measurement devices.

Let us now turn to the configuration (x, x, x) by examining the action of the operator

$$\Sigma = \sigma_{ax}\sigma_{bx}\sigma_{cx}$$

on $|\Psi\rangle$. The product of the results of spin measurements in the configuration (x, x, x) will always be -1 because

$$\Sigma|\Psi\rangle = -|\Psi\rangle, \tag{6.57}$$

as is easily checked by allowing $\sigma_{ax}\sigma_{bx}\sigma_{cx}$ to act on $|\Psi\rangle$:

$$\sigma_{ax}\sigma_{bx}\sigma_{cx}|\Psi\rangle = \sigma_{ax}\sigma_{bx}\sigma_{cx}\left(\frac{1}{\sqrt{2}}\left(|+++\rangle - |---\rangle\right)\right)$$

$$= \frac{1}{\sqrt{2}}\left(|---\rangle - |+++\rangle\right) = -|\Psi\rangle.$$

Let us now confront the above results with local realism. Once the three particles are sufficiently far apart, each of them possesses its own physical characteristics. We use A_x to denote the result of measuring the x component of the spin of particle a by Alice, $\ldots$, C_y the result of measuring the y component of the spin of particle c by Charlotte, and so on, with $A_x, \ldots, C_y = \pm 1$. When the x component is measured in conjunction with two measurements of the y component, we have seen (see (6.56)) that the product of the results is $+1$:

$$A_x B_y C_y = +1, \quad A_y B_x C_y = +1, \quad A_y B_y C_x = +1. \tag{6.58}$$

However, when the particles are in flight, two of the three experimentalists can decide to modify the direction of their analyzer axes, orienting them in the Ox direction. Then the product of the three spin components will be -1:

$$A_x B_x C_x = -1. \tag{6.59}$$

However, we note that

$$A_x B_x C_x = (A_x B_y C_y)(A_y B_x C_y)(A_y B_y C_x) = 1$$

because $A_y^2 = B_y^2 = C_y^2 = 1$. Equations (6.58) and (6.59) are incompatible. We do not have an inequality based on statistical correlations as in Section 6.3.2, but instead a perfect anticorrelation! Local realism would mean that the property σ_{ax} has a physical reality in the EPR sense, since it can be measured without disturbing particle a by measuring σ_{by} and σ_{cy}: $A_x = B_y C_y$. However, it is also possible to obtain A_x by measuring σ_{bx} and σ_{cx}: $A_x = -B_x C_x$. Local realism implies that it is *the same* A_x, but this is not the case in quantum mechanics. The value of A_x is contextual; it depends on physical properties incompatible with each other which are measured simultaneously with σ_{ax}, and A_x in (6.58) is not the same as A_x in (6.59). As in the case of the Bell inequalities, the problem arises because it is not possible to simultaneously measure the six quantities $A_x, \ldots, C_y$, which are the eigenvalues of operators which do not all commute with each other, and the simultaneous measurement of these six quantities is counterfactual: at most three can be measured in a given experiment. The operators Σ_a, Σ_b, Σ_c, and Σ all commute with each other, because Σ is a function of the commuting operators Σ_a, Σ_b, Σ_c

$$\Sigma = -\Sigma_a \Sigma_b \Sigma_c.$$

It is therefore possible to imagine an experiment where they are all four measured simultaneously. Such an experiment could not be performed by measuring the spins separately, and as in the case of teleportation (Section 6.4.2), it would be necessary to use an interaction between the spins. However, local realism also requires that measurement of the product $\Sigma_a \Sigma_b \Sigma_c$ gives a result identical to the product of the individual values of the spin operators, which is a statement incompatible with quantum physics.

6.4 Applications

6.4.1 Measurement and decoherence

In the Bohr or Copenhagen interpretation – or rather noninterpretation; see A. Leggett in Further Reading – of measurement in quantum mechanics, the measuring device operates according to macroscopic laws: the result of the measurement is read, for example, from the position of a needle on a meter. Furthermore, it is not meaningful to regard a quantum particle as possessing *any* intrinsic property, independent of the (classical) measuring apparatus used to observe it. This interpretation is remarkably useful, and is used unthinkingly by thousands of physicists. From the viewpoint of everyday practice, there is nothing left to be desired. However, if we think more deeply about this interpretation, the situation is not so clear. In fact, if we believe that the universal laws of physics are quantum laws, then classical physics is only an approximation,[26] under conditions which remain largely unknown today, except for models which are too crude to be realistic. It can be tentatively stated that macroscopic objects are classical, but this would not apply to macroscopic objects such as quantum fluids (for example, the ^{3}He and ^{4}He helium superfluids) or superconductors. The boundary between the quantum and classical worlds, which is an essential feature of Bohr's interpretation, is a fuzzy concept, which may even be dependent on the ability of experimentalists to manufacture quantum superpositions of "large" objects.

The measurement process certainly begins with a microscopic interaction which takes us into the quantum domain. Then, by some process whose details remain largely unknown to this day, the microscopic interaction is amplified and the measurement is translated into a classical effect like the position of a needle on a meter. von Neumann did not want to draw a boundary between the quantum and classical worlds, and he proposed, as above, that a measurement begins with an initial quantum interaction between the object being measured and the measurement device, which is also considered to be a quantum object. In the von Neumann theory it is easy to follow the first phase of the measurement process, that which is governed by the evolution equation (4.11) and which can be referred to as the *premeasurement phase* (Exercise 9.7.14). However, pursuing the process, one arrives at the so-called infinite-regress problem, so that the final stage of the measurement can be pushed as far as the brain of the experimentalist, a feature of von Neumann's theory which has been the subject of an abundant literature.

To obtain an actual measurement one must necessarily pass through a stage which is governed no longer by (4.11), but rather by an irreversible evolution. The interaction of the system being measured S with the measurement apparatus M creates an entangled state $S + M$. This does not present any problem as long as M remains microscopic, but it cannot persist until the end of the measurement process. To give a simple example, suppose that the initial state of the system is either $|\varphi_+\rangle$ or $|\varphi_-\rangle$, assumed to be orthogonal, and that of the apparatus is $|\Psi_0\rangle$. The interaction between the system and the apparatus leads to the following evolution

$$|\varphi_+ \otimes \Psi_0\rangle \rightarrow |\varphi_+ \otimes \Psi_+\rangle \quad |\varphi_- \otimes \Psi_0\rangle \rightarrow |\varphi_- \otimes \Psi_-\rangle \quad \langle\Psi_+|\Psi_-\rangle = 0.$$

---

[26] The "classical approximation" of a quantum system is fundamentally different from the classical approximation of relativistic mechanics by the Newtonian one. In the latter case, there is no conceptual difference in our description of the world, and it is a simple matter, at least in principle, to take the limit $v/c \rightarrow 0$. In the former case, we have two different conceptions of the world, and going from quantum to classical cannot be as simple as letting a small parameter go to zero.

Then observation of the apparatus, either in the state $|\Psi_+\rangle$ or in the state $|\Psi_-\rangle$, informs us of the initial state of the system. Now comes the difficulty: nothing prevents us from starting from an initial system state that is a linear superposition of $|\varphi_+\rangle$ and $|\varphi_-\rangle$, $\lambda|\varphi_+\rangle + \mu|\varphi_-\rangle$; then, from the linearity of quantum mechanics, the evolution leads to a final state

$$\lambda|\varphi_+ \otimes \Psi_+\rangle + \mu|\varphi_- \otimes \Psi_-\rangle$$

that is a linear superposition of macroscopic states if the measuring apparatus is macroscopic. This argument, first put forward by Schrödinger, is known as "the Schrödinger's cat paradox": in the original argument, the macroscopic states are the states $|\Psi_+\rangle$ and $|\Psi_-\rangle$ corresponding to a live and dead cat, so that the unfortunate cat is left in a superposition of alive and dead states. To take a less extreme example, we could have a measurement apparatus in a linear superposition, with, for example, a needle pointing to two positions on a meter at the same time. In such a situation which could lead (in principle) to interference effects, we could not say that the system was in just one state before it was observed. By contrast, in a classical mixture, each individual system is in either one state or the other, but we cannot tell which without observing it. Our experience with measurement devices (or cats) implies that they are described by a classical statistical ensemble and not a state vector, and it is widely believed that irreversible interactions of M with its environment, or *decoherence*, lead to this result. As we shall see in Chapter 15 on simple examples, decoherence selects a preferred basis which is linked to the particular form of interaction of the quantum system with its environment. Then, in this basis, the off-diagonal matrix elements of the state operator of the macroscopic quantum system which contain the information on the phases decay at a rate much faster than the "natural" decay rate, for example the characteristic decay rate of the energy. This process is irreversible for all practical purposes, and it leaves the system in a classical mixture, although information on the phases is, in principle, available in the system–environment quantum correlations. However, it should be emphasized that while decoherence is very likely an essential stage of the measurement process, it is not sufficient to account for the complete process. It explains how to pass from a quantum superposition to a statistical mixture, but has nothing to say about the origin of postulate **II** or about the fact that a particular experiment on a quantum system always gives a unique result (the problem of definite outcomes). It also appears that some degrees of freedom remain almost entirely decoupled from the environment and are thus not very sensitive to decoherence. This may be the case, for example, for the position of the center of mass of a heavy molecule. It cannot be excluded that superpositions of macroscopically distinguishable states be observed in the future, for example superpositions of macroscopic currents ($\sim 1 \ \mu A$) flowing in opposite directions in superconducting rings with Josephson junctions.

In order to make these ideas more concrete, let us discuss an experiment performed at the Ecole Normale Supérieure in 1996.[27] It is shown schematically in Fig. 6.10. Our

---

[27] M. Brune *et al.*, Observing the progressive decoherence of the "meter" in a quantum measurement, *Phys. Rev. Lett.* **77**, 4887–4890 (1996).

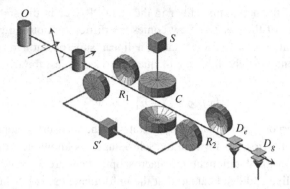

Fig. 6.10. An experiment on decoherence. Atoms leave an oven O and cross the first microwave cavity R_1. They then pass through a superconducting cavity C followed by a second microwave cavity R_2. The cavities R_1 and R_2 are fed by the same source S. Finally, the atoms are detected by two ionization detectors D_e and D_g, which are triggered by atoms in the states e and g, respectively. After M. Brune *et al.*, *Phys. Rev. Lett.* **77**, 4887 (1996).

discussion will be brief; details can be found in Appendix B and in the original article. In this experiment, the measurement is made by an electromagnetic field enclosed in a superconducting cavity C shown in Fig. 6.10. The quality factor of this cavity is very high, of order 5×10^7; the lifetime T_r of a photon in the cavity is several hundred microseconds and the resonance frequency ω_C is 3.21×10^{11} rad s^{-1} ($\nu_C = 51.1$ GHz). After the field is established in the cavity, all interaction with the field source S is cut off and one works with an average number of photons $\langle n \rangle$ between 3 and 10. The object that is measured is an atom which follows a trajectory from O to the detectors D in crossing the cavity. This atom can exist in two states, the ground state $|g\rangle$ and an excited state $|e\rangle$.[28] The passage of the atom through the cavity induces a phase shift $\pm\Phi$ of the electromagnetic field depending on the state of the atom.[29] We use $|G\rangle$ with phase shift $+\Phi$ ($|E\rangle$ with phase shift $-\Phi$) to denote the (quantum) state of the field after an atom in the state $|g\rangle$ ($|e\rangle$) has crossed the cavity. Depending on whether the atom is in the state $|e\rangle$ or $|g\rangle$, the atom + field state vector is

$$|eE\rangle \quad \text{or} \quad |gG\rangle.$$

Measurement of the state of the field makes it possible in principle – if not in practice – to measure the state of the atom.[30] If the field is found in the state $|E\rangle$, this would indicate that the atom is in the state $|e\rangle$. The state of the field is the needle which gives the measurement result: the needle position is either $+\Phi$ corresponding to $|g\rangle$, or $-\Phi$ corresponding to $|e\rangle$. However, we are still in the premeasurement stage: up to now the entire evolution has been governed by an equation of the type (4.11) for a closed

---

[28] These two states are the Rydberg states of a rubidium atom corresponding to a valence electron in a level $n \simeq 50$; see Exercise 14.5.4.

[29] The situation is off-resonance and the cavity photons are not absorbed by the atoms; see Section 5.3.3.

[30] The potential existence of such a measurement is confirmed by the disappearance of interference; see Appendix B.

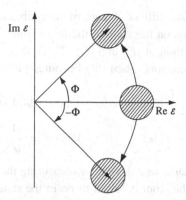

Fig. 6.11. Representation of the modulus and phase of the electric field in the cavity C. The shaded circles show the spread at the tip of the field vector.

atom + field system. The states $|G\rangle$ and $|E\rangle$ are "almost classical": if the number of photons were large, the modulus and phase of the field would be perfectly defined.[31] The modulus and phase of these states are shown in Fig. 6.11. In the complex plane of the electric field the field modulus is proportional to the square root $\langle n \rangle^{1/2}$ of the average number of photons. However, in contrast to the classical case, the tip of the electric field vector is not exactly fixed; it is affected by quantum fluctuations satisfying $\Delta n \Delta \Phi \sim 1$ (cf. Section 11.3.4).

Now in R_1 a microwave pulse of suitable duration $\omega_1 t = \pi/2$ (a $\pi/2$ pulse; see (5.35)), where ω_1 is the Rabi frequency (Section 5.3.2), is applied to the atom before it passes through C; see Fig. 6.10. This pulse has the following effect on the state vector of the atom:[32]

$$|e\rangle \rightarrow |a\rangle = \frac{1}{\sqrt{2}} \left(|e\rangle + |g\rangle \right),$$
$$|g\rangle \rightarrow |b\rangle = \frac{1}{\sqrt{2}} \left(-|e\rangle + |g\rangle \right). \tag{6.60}$$

If the atom is initially in the state $|e\rangle$, the microwave pulse sends it into the state $|a\rangle$, and the atom + field final state will be the entangled state

$$|\Psi\rangle = \frac{1}{\sqrt{2}} \left(|eE\rangle + |gG\rangle \right), \tag{6.61}$$

but the correspondence $E \rightarrow e$, $G \rightarrow g$ always holds. The difficulties will arise from the fact that we can perform linear transformations on the state of the atom *after* its passage through C at a time such that an actual measurement has not been completed and the atom + field system has remained closed. Nothing is yet final in the measurement when

---

[31] From a technical point of view these states are "coherent states"; see Section 11.2.
[32] Equations (6.60) are derived from (5.31) with $\omega_1 t/2 = \pi/4$. The factors $\pm i$ can be absorbed by redefining the basis vectors by a phase.

the atom exits from C; we are still in a stage of reversible evolution. It is possible to perform linear transformations on the state of the atom which have the effect of leaving the field in a linear superposition of $|E\rangle$ and $|G\rangle$. To do this, a second microwave pulse is applied at R_2 before the detectors. Then $|\Psi\rangle$ becomes $|\Psi'\rangle$:

$$|\Psi\rangle \rightarrow |\Psi'\rangle = \frac{1}{2}\Big[(|e\rangle + |g\rangle)|E\rangle + (-|e\rangle + |g\rangle)|G\rangle\Big]$$

$$= \frac{1}{\sqrt{2}}\Big[|e\rangle \frac{1}{\sqrt{2}}(|E\rangle - |G\rangle) + |g\rangle \frac{1}{\sqrt{2}}(|E\rangle + |G\rangle)\Big]. \qquad (6.62)$$

If we now decide to use the atom as a device for measuring the field, this equation shows that depending on whether the atom is found to be in the state $|e\rangle$ by D_e or in the state $|g\rangle$ by D_g, the field is in a linear superposition

$$\frac{1}{\sqrt{2}}(|E\rangle - |G\rangle) \quad \text{or} \quad \frac{1}{\sqrt{2}}(|E\rangle + |G\rangle). \qquad (6.63)$$

As in an experiment of the EPR type, the final state of the field is not fixed until *after* the interaction of the atom with the field, because this state is determined by manipulations (in the cavity R_2) after this interaction. This is an example of a "delayed choice" experiment. Equation (6.63) shows that the previous measurement device, the field, is projected in a state of linear superposition. In contrast to the states $|E\rangle$ and $|G\rangle$, the states (6.63) are not "almost classical" states, and they give an example of a Schrödinger's cat.[33] As we shall see in Section 15.4.5, linear superpositions of the kind in (6.63) are destroyed very rapidly by interactions with the environment, and this occurs the more quickly the larger the object. It is not yet possible to identify $|E\rangle$ and $|G\rangle$ as two positions of a needle, and this first measurement stage can in fact only be a premeasurement, because linear superpositions are not observed in a measurement which has been completed.

To learn more about the state of the field, a second atom is sent to probe the field inside the cavity (a mouse to test the cat). It is then possible to show experimentally that the linear superposition (6.63) is very fragile. The coherence between the states $|E\rangle$ and $|G\rangle$ vanishes in several tens of microseconds, a time much shorter than the field relaxation time, and the field returns to a statistical mixture of the states $|E\rangle$ and $|G\rangle$. This is the phenomenon of decoherence due to the dissipative coupling of the field with its environment. If we initially have the field in a pure state

$$|\Phi\rangle = \lambda|E\rangle + \mu|G\rangle, \quad |\lambda|^2 + |\mu|^2 = 1, \qquad (6.64)$$

the state operator in the basis $\{|E\rangle, |G\rangle\}$ will be

$$\rho_{\text{in}} = \begin{pmatrix} |\lambda|^2 & \lambda\mu^* \\ \lambda^*\mu & |\mu|^2 \end{pmatrix}. \qquad (6.65)$$

---

[33] Transposing the original discussion of Schrödinger, if the entangled state is (6.61), observation of the atom in the state $|e\rangle$ implies the death of the cat (the state $|E\rangle$), while observation of the atom in the state $|g\rangle$ means the cat is alive (the state $|G\rangle$). After the microwave pulse is applied and the state of the atom is observed, the cat is in a linear superposition alive + dead.

Decoherence transforms this state operator into

$$\rho_{\text{final}} = \begin{pmatrix} |\lambda|^2 & 0 \\ 0 & |\mu|^2 \end{pmatrix}. \tag{6.66}$$

In the present case, decoherence is principally due to the leakage of photons out of the cavity owing to imperfections of the mirrors, and the leakage of a single photon is enough to destroy the phase coherence. The off-diagonal elements of ρ in the preferred basis of coherent states, or coherences, contain information about the phase and tend to zero very rapidly. This evolution $\rho_{\text{in}} \to \rho_{\text{fin}}$ is nonunitary – it is not governed by a Hamiltonian. In fact, the interaction of the field with its environment leads to a field + environment entangled state, and the state operator of the field is obtained by taking a partial trace:

$$\rho_{\text{field}} = \text{Tr}_{\text{env}}[\rho_{\text{field+env}}].$$

This nonunitary evolution translates into a leakage of information to the environment degrees of freedom, corresponding to an increase of the von Neumann entropy of the field characteristic of a dissipative phenomenon:

$$S_{\text{vN}}(\rho_{\text{fin}}) \geq S_{\text{vN}}(\rho_{\text{in}}).$$

In summary, the measurement process begins with an interaction $S + M$ governed by (4.11), but this is not sufficient for performing the complete measurement. It is necessary to pass through a stage of irreversible evolution, with leakage of information to unobservable degrees of freedom. As long as the system $S + M$ remains closed, the measurement cannot be completed and we remain in the premeasurement stage. It is the interaction of M with the environment which is responsible for the irreversibility and decoherence. The Ecole Normale Supérieure experiment demonstrates this decoherence in a well-controlled experimental situation, even though there is still a considerable way to go from a cavity containing a few photons to a macroscopic measurement device. However, it seems clear that the interaction with the environment lies at the origin of the loss of the phase information and the absence of Schrödinger's cats. As we shall see in more detail in Section 15.4.5, most of the Hilbert space of states is extremely fragile owing to the environment, and after a very short time only a tiny fraction of this space survives, that which is selected by decoherence and defines the statistical mixtures of states possessing a classical limit, the states which are robust regarding dissipation in the environment.

6.4.2 Quantum information

Let us conclude this chapter with an examination of some applications of entangled states to the field of *quantum information*, that is, the theory of the processing and transmission of information using the features specific to quantum mechanics. As a preliminary result, let us demonstrate the *quantum no-cloning theorem*. The essential condition for the method of quantum encryption described in Section 3.1.3 to be perfectly secure is that the spy Eve should not be able to reproduce (clone) the state of the particle sent by Bob to Alice while leaving unchanged the result of Bob's measurement, so that interception of the message is undetectable. The impossibility of Eve reproducing the state is guaranteed by the quantum no-cloning theorem. To demonstrate this theorem, let us suppose that we wish to duplicate an unknown quantum state $|\chi_1\rangle$. The system on which we wish to print the copy is denoted $|\varphi\rangle$; it is the equivalent of a blank page. For example, if we wish to clone a spin-1/2 state $|\chi_1\rangle$, $|\varphi\rangle$ is also a spin-1/2 state. The evolution of the state vector in the cloning process must have the form

$$|\chi_1 \otimes \varphi\rangle \rightarrow |\chi_1 \otimes \chi_1\rangle. \tag{6.67}$$

This evolution is governed by a unitary operator U which we do not need to specify:

$$U|\chi_1 \otimes \varphi\rangle = |\chi_1 \otimes \chi_1\rangle. \tag{6.68}$$

U must be independent of $|\chi_1\rangle$, which is unknown by hypothesis. If we wish to clone a second original $|\chi_2\rangle$ we must have

$$U|\chi_2 \otimes \varphi\rangle = |\chi_2 \otimes \chi_2\rangle.$$

Let us now evaluate the scalar product

$$X = \langle \chi_1 \otimes \varphi | U^\dagger U | \chi_2 \otimes \varphi \rangle$$

in two different ways:

$$\begin{aligned}
&(1) \quad X = \langle \chi_1 \otimes \varphi | \chi_2 \otimes \varphi \rangle = \langle \chi_1 | \chi_2 \rangle, \\
&(2) \quad X = \langle \chi_1 \otimes \chi_1 | \chi_2 \otimes \chi_2 \rangle = (\langle \chi_1 | \chi_2 \rangle)^2.
\end{aligned} \tag{6.69}$$

It follows that either $|\chi_1\rangle \equiv |\chi_2\rangle$ or $\langle \chi_1 | \chi_2 \rangle = 0$, which prevents us from cloning any a priori given state. This proof of the no-cloning theorem explains why in quantum cryptography we cannot restrict ourselves to a basis of orthogonal polarization states $\{|x\rangle, |y\rangle\}$ for the photons. It is the use of linear superpositions of polarization states $|x\rangle$ and $|y\rangle$ that allows the presence of a spy to be detected. The no-cloning theorem also guarantees that Alice and Bob cannot communicate at speeds greater than the speed of light in the experiment of Fig. 6.1. If Bob were capable of cloning his spin 1/2, he would be able to measure its polarization and deduce the choice of axes used by Alice to measure her spin.

Let us now turn to the second subject in this subsection, quantum computing. In information theory the elementary unit is the bit, which can take two values, by convention 0 and 1. A bit is stored classically by a two-state system, for example, a capacitor which can be either uncharged (bit value 0) or charged (bit value 1). A bit of information typically implies 10^4 to 10^5 electrons in the RAM of an actual computer. An interesting question is then whether or not it is possible to store information using electrons (or other particles) which are isolated. As we have already seen, a two-state quantum system is capable of storing a bit of information. For example, in Section 3.1.3 we have used the two orthogonal polarization states of a photon to store a bit. To be specific, we are now going to use the two polarization states of a spin-1/2 particle. By convention, the up spin state $|+\rangle$ will correspond to the value 0 of the bit and the down spin state $|-\rangle$ to the value 1: $|+\rangle \equiv |0\rangle, |-\rangle \equiv |1\rangle$. However, in contrast to a classical system which can only exist in the state 0 or 1, the quantum system can exist in states $|\varphi\rangle$ that are linear superpositions of $|0\rangle$ and $|1\rangle$:

$$|\varphi\rangle = \lambda|0\rangle + \mu|1\rangle, \quad |\lambda|^2 + |\mu|^2 = 1. \tag{6.70}$$

Instead of an ordinary bit, the quantum system stores a *quantum bit* or a qubit whose value in the state (6.70) remains undetermined until the z component of the spin is measured. This measurement will give the result 1 with probability $|\mu|^2$ and the result 0 with probability $|\lambda|^2$, which itself is not a particularly useful property. The information stored by means of qubits is an example of quantum information. The no-cloning theorem implies that it is impossible to copy this information.

Suppose that we would like to store a number between 0 and 7 in a register. This would require three bits, as in a system of base 2 a number between 0 and 7 can be represented by a set of three numbers 0 or 1. A classical register would store one of the eight following configurations:

$$0 = \{000\} \quad 1 = \{001\} \quad 2 = \{010\} \quad 3 = \{011\}$$
$$4 = \{100\} \quad 5 = \{101\} \quad 6 = \{110\} \quad 7 = \{111\}.$$

A system of three spins 1/2 could also be used to store a number between 0 and 7, for example, by having these numbers correspond to the eight three-spin states

$$0 : |000\rangle \quad 1 : |001\rangle \quad 2 : |010\rangle \quad 3 : |011\rangle$$
$$4 : |100\rangle \quad 5 : |101\rangle \quad 6 : |110\rangle \quad 7 : |111\rangle. \tag{6.71}$$

We shall use $|x\rangle$, $x = 0, \ldots, 7$, to denote the eight states in (6.71), for example $|5\rangle = |101\rangle (=|-+-\rangle)$. These vectors form a basis in the Hilbert space of states of the three spins, which is called the *computational basis*. Since we can form a linear superposition of the states (6.71), we conclude that the state vector of a system of three spins will allow us to store $2^3 = 8$ numbers at a time, while a system of n spins will allow us to store 2^n numbers. However, a measurement of the components of the three spins on the Oz axis will necessarily give one of the eight states in (6.71). We possess some important virtual information, but when we seek to access it by making a measurement we do not do any better than with the classical system: the measurement gives one of eight numbers, not all eight at the same time.

The operations performed by a quantum computer are unitary transformations (4.14) acting in the Hilbert space of states $\mathcal{H}^{\otimes n}$ of the qubits. These operations are performed by *quantum logic gates*. It is possible to show that all unitary operations in $\mathcal{H}^{\otimes n}$ can be decomposed into

- unitary transformations on individual qubits;
- control-not (cNOT) gates acting on a pair of qubits, to be defined below.

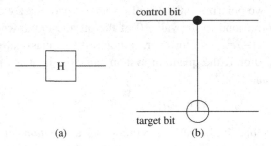

Fig. 6.12. Graphical representation of quantum logic gates. (a) Hadamard gate; (b) cNOT gate.

One frequently used unitary transformation on individual qubits is the *Hadamard gate* H (Fig. 6.12(a))

$$H = \frac{1}{\sqrt{2}} \begin{pmatrix} 1 & 1 \\ 1 & -1 \end{pmatrix},$$

so that

$$H|0\rangle = \frac{1}{\sqrt{2}}\left(|0\rangle + |1\rangle\right) \quad H|1\rangle = \frac{1}{\sqrt{2}}\left(|0\rangle - |1\rangle\right).$$

It is easy to see that by applying a gate H to each of the n qubits in the $|0\rangle$ state, we obtain the following linear combination $|\Phi\rangle$ of states in the computational basis

$$|\Phi\rangle = H^{\otimes n}|0\ldots 0\rangle = H^{\otimes n}|0^{\otimes n}\rangle = \frac{1}{2^{n/2}} \sum_{x=0}^{2^n - 1} |x\rangle. \tag{6.72}$$

The cNOT gate (Fig. 6.12(b)) has the following action on a two qubit state: if the first qubit, termed *control bit*, is in the $|0\rangle$ state, nothing happens to the second qubit, termed *target bit*. If the control qubit is in the $|1\rangle$ state, then the two basis states of the target qubit are exchanged: $|0\rangle \leftrightarrow |1\rangle$. The matrix representation of the cNOT gate is, in the basis $\{|00\rangle, |01\rangle, |10\rangle, |11\rangle\}$,

$$\text{cNOT} = \begin{pmatrix} 1 & 0 & 0 & 0 \\ 0 & 1 & 0 & 0 \\ 0 & 0 & 0 & 1 \\ 0 & 0 & 1 & 0 \end{pmatrix} = \begin{pmatrix} I & 0 \\ 0 & \sigma_x \end{pmatrix}. \tag{6.73}$$

What advantage can we expect from a quantum computer functioning with qubits? A quantum computer is capable of performing a large number of operations in parallel. The elementary operations on qubits and therefore on states of the type (6.72) are unitary evolutions governed by the evolution equation (4.11) or its integral version (4.14). In certain cases useful information can be extracted by these operations if parallel quantum computing can be used. Such computing is based on the following principle. An *input register* of n qubits is stored in a state $|\Phi\rangle$ (6.72). If we start from the state $|00\ldots 0\rangle = |0^{\otimes n}\rangle$, only n elementary operations are necessary for arriving at (6.72). Then we construct the tensor product $|\Psi\rangle$ of $|\Phi\rangle$ with the state $|0^{\otimes m}\rangle$ of an *output register* of m qubits

$$|\Psi\rangle = |\Phi \otimes 0^{\otimes m}\rangle = \frac{1}{2^{n/2}} \sum_x |x \otimes 0^{\otimes m}\rangle, \tag{6.74}$$

and a unitary operator U_f corresponding to a time evolution of the system transforms $|\Psi\rangle$ into $|\Psi'\rangle$:

$$|\Psi\rangle \to |\Psi'\rangle = U_f|\Psi\rangle = \frac{1}{2^{n/2}} \sum_x |x \otimes f(x)\rangle. \tag{6.75}$$

The ensemble of two registers *simultaneously* contains the 2^{n+m} values of the pair $(x, f(x))$. Of course, a measurement will give a unique pair, but it is possible to use the information stored in the state vector (6.75), for example to perform a Fourier transform of this superposition and then sample the power spectrum to find out the period of $f(x)$. A toy example of a quantum algorithm is given in Exercise 6.5.11.

An interesting example is the determination of the period of a function $f(x)$. Let us suppose that $f(x)$ is defined on Z_N, the additive group of integers modulo N. An algorithm executed by a classical computer must perform a number of operations of order $O(\exp[\ln N]^{1/3})$ to find the period, whereas if a quantum computer is used this number will be $O(\ln^2 N)$. This period determination forms the basis of the Shor algorithm for the decomposition of a number into primes, the function $f(x)$ in that case being $a^x \bmod N$, a integer.

Once the principle of algorithms which can be executed by quantum computers is mastered, there remains the question of the actual realization of such a computer. Opinions on this vary widely, from complete pessimism to measured optimism. A group at IBM has managed to obtain the period of $a^x \bmod 15$ using a quantum computer based on NMR,[34] but a computer that can give useful results is still far from realization. The main problem is decoherence. The calculations described above require that the evolution be unitary, which implies the absence of uncontrolled interactions with the environment. Of course, total isolation of this type is impossible. At best it is possible to minimize the perturbations due to the environment, and to develop algorithms for correcting the inevitable errors using redundant information. The field of quantum information is expanding rapidly, and the reader is referred to the articles and books cited in the References for further details. A promising technique, based on trapped ions, is described in Exercise 11.5.13.

Teleportation is an amusing application of entangled states which could serve as a method of transferring quantum information (Fig. 6.13).[35] Let us suppose that Alice wishes to transfer to Bob information about the spin state $|\varphi_A\rangle$ of particle A of spin 1/2

$$|\varphi_A\rangle = \lambda|0_A\rangle + \mu|1_A\rangle, \tag{6.76}$$

---

[34] L. Vandersypen *et al.*, Experimental realization of Shor's quantum factoring algorithm using nuclear magnetic resonance, *Nature* **414**, 883–887 (2001).

[35] Two recent experiments are described by M. Riebe *et al.*, Deterministic quantum teleportation with atoms, *Nature* **429**, 734–737 (2004) and M. Barret *et al.*, Deterministic quantum teleportation of atomic qubits, *Nature* **429**, 737–739 (2004).

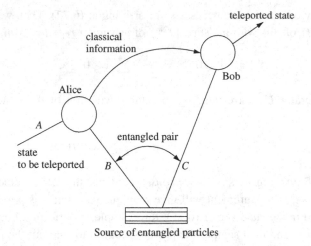

Fig. 6.13. Teleportation. Alice performs a Bell measurement on particles A and B and informs Bob of the result through a classical channel.

which is a priori unknown, without sending him this particle directly.[36] She cannot measure the spin, because she does not know the spin orientation of particle A, and any measurement would in general project $|\varphi_A\rangle$ onto another state. The principle of information transfer amounts to using a pair of entangled particles B and C of spin 1/2. Particle B is used by Alice and particle C is sent to Bob. Particles B and C are assumed to have been put in an entangled state, for example in the state $|\Psi_{BC}\rangle$

$$|\Psi_{BC}\rangle = \frac{1}{\sqrt{2}}\left(|0_B 0_C\rangle + |1_B 1_C\rangle\right). \tag{6.77}$$

The initial state of the three particles is thus $|\Phi_{ABC}\rangle$

$$
\begin{aligned}
|\Phi_{ABC}\rangle &= \left(\lambda|0_A\rangle + \mu|1_A\rangle\right)\frac{1}{\sqrt{2}}\left(|0_B 0_C\rangle + |1_B 1_C\rangle\right) \\
&= \frac{\lambda}{\sqrt{2}}|0_A\rangle\left(|0_B 0_C\rangle + |1_B 1_C\rangle\right) + \frac{\mu}{\sqrt{2}}|1_A\rangle\left(|0_B 0_C\rangle + |1_B 1_C\rangle\right).
\end{aligned}
\tag{6.78}
$$

Alice is now going to perform a measurement on the pair AB by applying first a cNOT gate (6.73), with the qubit A (B) as the control (target) qubit, followed by a Hadamard gate on qubit A (Fig. 6.14). The cNOT gate transforms the initial state (6.77) of the three qubits into $|\Phi'_{ABC}\rangle$

$$|\Phi'_{ABC}\rangle = \text{cNOT}|\Phi_{ABC}\rangle = \frac{\lambda}{\sqrt{2}}\left(|0_A\rangle\langle|0_B 0_C\rangle + |1_B 1_C\rangle\right) + \frac{\mu}{\sqrt{2}}\left(|1_A\rangle\langle|1_B 0_C\rangle + |0_B 1_C\rangle\right). \tag{6.79}$$

---

[36] For clarity, it is better to label the three particles A, B, and C, rather than 1, 2, and 3.

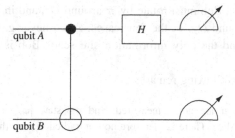

Fig. 6.14. Alice applies a cNOT gate on the pair AB, and then a Hadamard gate on qubit A.

Then the Hadamard gate has the following action

$$|\Phi''_{ABC}\rangle = H|\Phi'_{ABC}\rangle = \frac{1}{2}\Big[\lambda|0_A0_B0_C\rangle + \lambda|0_A1_B1_C\rangle + \lambda|1_A0_B0_C\rangle + \lambda|1_A1_B1_C\rangle$$

$$+\mu|0_A1_B0_C\rangle + \mu|0_A0_B1_C\rangle - \mu|1_A1_B0_C\rangle - \mu|1_A0_B1_C\rangle\Big]. \qquad (6.80)$$

This equation can be cast in the form

$$|\Phi''_{ABC}\rangle = \frac{1}{2}|0_A0_B\rangle\big(\lambda|0_C\rangle + \mu|1_C\rangle\big)$$

$$+\frac{1}{2}|0_A1_B\rangle\big(\mu|0_C\rangle + \lambda|1_C\rangle\big)$$

$$+\frac{1}{2}|1_A0_B\rangle\big(\lambda|0_C\rangle - \mu|1_C\rangle\big) \qquad (6.81)$$

$$+\frac{1}{2}|1_A1_B\rangle\big(-\mu|0_C\rangle + \lambda|1_C\rangle\big).$$

The last operation is a measurement by Alice of the two qubits in the $\{|0\rangle, |1\rangle\}$ basis. The whole measurement is termed *Bell measurement*. It projects the AB pair on one of the four states $|i_Aj_B\rangle$ $i, j = 0, 1$, and the state vector can be read on each of the lines of (6.81).

The simplest case is that where the result is $|0_A0_B\rangle$. The C qubit then arrives at Bob in the state

$$\lambda|0_C\rangle + \mu|1_C\rangle,$$

that is, exactly in the initial state of qubit A, with the *same* coefficients λ and μ. Alice informs Bob through a classical channel (telephone...) that he is going to receive qubit C in the same state as A. If, on the contrary, she measures $|0_A1_B\rangle$, qubit C is in the state

$$\mu|0_C\rangle + \lambda|1_C\rangle,$$

she informs Bob that he must rotate qubit C by π around Ox, or equivalently, apply the σ_x matrix

$$\exp\left(-i\frac{\pi\sigma_x}{2}\right) = -i\sigma_x.$$

In the third case ($|1_A 0_B\rangle$), Bob must rotate by π around Oz, and in the last case ($|1_A 1_B\rangle$) he must rotate by π around Oy. In the four cases, Alice never gains knowledge of the coefficients λ and μ, and the only information she sends Bob is the rotation he must perform.

It is useful to add the following remarks.

- The coefficients λ and μ are never measured, and the state $|\varphi_A\rangle$ is destroyed during the measurement made by Alice. There is therefore no contradiction with the no-cloning theorem.
- Bob does not "know" the state of particle C until he has received the result of Alice's measurement. This information must be transmitted by a classical channel, at a speed at most equal to the speed of light. Therefore, there is no instantaneous transmission of information at a distance.
- Teleportation never involves the transport of matter.

6.5 Exercises

6.5.1 Independence of the tensor product from the choice of basis

Verify that the definition (6.3) of the tensor product of two vectors is independent of the choice of basis in $\mathcal{H}_1$ and $\mathcal{H}_2$.

6.5.2 The tensor product of two 2×2 matrices

Write down explicitly the 4×4 matrix $A \otimes B$, the tensor product of the 2×2 matrices A and B:

$$A = \begin{pmatrix} a & b \\ c & d \end{pmatrix}, \quad B = \begin{pmatrix} \alpha & \beta \\ \gamma & \delta \end{pmatrix}.$$

6.5.3 Properties of state operators

1. The matrix elements ρ_{ii}, ρ_{ij}, ρ_{ji}, and ρ_{jj} of a state operator ρ can be used to construct the 2×2 matrix

$$A = \begin{pmatrix} \rho_{ii} & \rho_{ij} \\ \rho_{ji} & \rho_{jj} \end{pmatrix}.$$

Show that $\rho_{ii} \geq 0$, $\rho_{jj} \geq 0$, and $\det A \geq 0$, from which $|\rho_{ij}|^2 \leq \rho_{ii}\rho_{jj}$. Also deduce that if $\rho_{ii} = 0$, then $\rho_{ij} = \rho_{ji}^* = 0$.

2. Show that if there exists a maximal test giving 100% probability for the quantum state described by a state operator ρ, then this state is a pure state. Also show that if ρ describes a pure state, and if it can be written as

$$\rho = \lambda \rho' + (1 - \lambda)\rho'', \quad 0 \leq \lambda \leq 1,$$

then $\rho = \rho' = \rho''$. Hint: first demonstrate that if ρ' and ρ'' are generic state operators, then ρ is a state operator. The state operators form a convex subset of Hermitian operators.

6.5.4 Fine structure and the Zeeman effect in positronium

Positronium is an electron–positron bound state very similar to the electron–proton bound state of the hydrogen atom.

1. Calculate the energy of the ground state of positronium as a function of that of the hydrogen atom. We recall that the positron mass is equal to the electron mass.
2. In this exercise we are interested solely in the spin structure of the ground state of positronium. The space of states to be taken into account is then a four-dimensional space $\mathcal{H}$, the tensor product of the spaces of spin-1/2 states of the electron and the positron. Following the notation of Section 6.1.2, we use $|\varepsilon_1\varepsilon_2\rangle$ to denote a state in which the z component of the electron spin is $\hbar\varepsilon_1/2$ and that of the positron spin is $\hbar\varepsilon_2/2$, with $\varepsilon = \pm1$. Determine the action of the operators $\sigma_{1x}\sigma_{2x}$, $\sigma_{1y}\sigma_{2y}$, and $\sigma_{1z}\sigma_{2z}$ on the four basis states $|++\rangle$, $|+-\rangle$, $|-+\rangle$, and $|--\rangle$ of $\mathcal{H}$. Deduce the action of the operator

$$\vec{\sigma}_1\cdot\vec{\sigma}_2 = \sigma_{1x}\sigma_{2x} + \sigma_{1y}\sigma_{2y} + \sigma_{1z}\sigma_{2z}$$

on these states.
3. Show that the four vectors

$$|I\rangle = |++\rangle$$

$$|II\rangle = \frac{1}{\sqrt{2}}(|+-\rangle + |-+\rangle)$$

$$|III\rangle = |--\rangle$$

$$|IV\rangle = \frac{1}{\sqrt{2}}(|+-\rangle - |-+\rangle)$$

form an orthonormal basis of $\mathcal{H}$ and that these vectors are eigenvectors of $\vec{\sigma}_1\cdot\vec{\sigma}_2$ with eigenvalues 1 or -3.
4. Find the projectors $\mathcal{P}_1$ and $\mathcal{P}_{-3}$ onto the subspaces of the eigenvalues 1 and -3, writing these projectors in the form

$$\lambda I + \mu\vec{\sigma}_1\cdot\vec{\sigma}_2.$$

5. Show that the operator $\mathcal{P}_{12}$

$$\mathcal{P}_{12} = \frac{1}{2}(I + \vec{\sigma}_1\cdot\vec{\sigma}_2)$$

exchanges the values of ε_1 and ε_2:

$$\mathcal{P}_{12}|\varepsilon_1\varepsilon_2\rangle = |\varepsilon_2\varepsilon_1\rangle.$$

6. The Hamiltonian H_0 of the spin system in the absence of an external field is given by

$$H_0 = E_0 I + A\vec{\sigma}_1\cdot\vec{\sigma}_2, \quad A > 0,$$

where E_0 and A are constants. Find the eigenvectors and eigenvalues of H_0.
7. The positronium atom is placed in a uniform, constant magnetic field $\vec{B}$ parallel to Oz. Show that the Hamiltonian becomes

$$H = H_0 - \frac{q_e\hbar}{2m}B(\sigma_{1z} - \sigma_{2z}),$$

where m is the electron mass and q_e is its charge. Find the matrix representation of H in the basis $\{|I\rangle, |II\rangle, |III\rangle, |IV\rangle\}$. The parameter x is defined by

$$\frac{q_e \hbar}{2m} B = -Ax.$$

Find the eigenvalues of H and graph their behavior as a function of x.

6.5.5 Spin waves and magnons

NB: This exercise uses the notation and results of questions 2 to 5 in the preceding exercise.

A one-dimensional ferromagnet can be represented as a chain of N spins 1/2 numbered $n = 0, \ldots, N-1$, $N \gg 1$, fixed along a line with a spacing l between each. It is convenient to use periodic boundary conditions, where spin N is identified with spin 0: $N \equiv 0$. We suppose that each spin can interact only with its two nearest neighbors, and the Hamiltonian is written as a function of a constant A as

$$H = \frac{1}{2} NAI - \frac{1}{2} A \sum_{n=0}^{N-1} \vec{\sigma}_n \cdot \vec{\sigma}_{n+1}.$$

1. Show that all eigenvalues E of H satisfy $E \geq 0$ and that the minimum one E_0 corresponding to the ground state is obtained when all the spins point in the same direction. Throughout this exercise this is chosen to be the z direction. A possible choice for the ground state $|\Phi_0\rangle$ then is[37]

$$|\Phi_0\rangle = |+++\cdots+++\rangle.$$

2. Show that H can be written as

$$H = NAI - A \sum_{n=0}^{N-1} \mathcal{P}_{n,n+1} = A \sum_{n=0}^{N-1} (I - \mathcal{P}_{n,n+1}),$$

where

$$\mathcal{P}_{n,n+1} = \frac{1}{2} (I + \vec{\sigma}_n \cdot \vec{\sigma}_{n+1}).$$

Using the result of question 5 of the preceding exercise, show that the eigenvectors of H are linear combinations of vectors in which the number of up spins minus the number of down spins is a constant. Let $|\Psi_n\rangle$ be the state in which the spin n is down with all the other spins up. What is the action of H on $|\Psi_n\rangle$?

3. We seek eigenvectors $|k_s\rangle$ of H which are linear combinations of $|\Psi_n\rangle$. Taking into account the cyclic symmetry, we set

$$|k_s\rangle = \sum_{n=0}^{N-1} e^{ik_s nl} |\Psi_n\rangle$$

with

$$k_s = \frac{2\pi s}{Nl}, \quad s = 0, 1, \ldots, N-1.$$

---

[37] Any state obtained from $|\Phi_0\rangle$ by rotating the ensemble of spins by the same angle about Oz is also a possible ground state.

Show that $|k_s\rangle$ is an eigenvector of H and determine the corresponding energy E_k. Show that the energy is proportional to k_s^2 if $k_s \to 0$. An elementary excitation called a *magnon* is associated with the state $|k_s\rangle$ of (quasi-) wave vector k_s and energy E_k.

6.5.6 Spin echo and level splitting in NMR

1. For various purposes, it is important to be able to measure accurately the transverse relaxation time T_2 (Section 5.2.3) in NMR experiments. In the rotating frame of Exercise 5.5.6, the NMR signal $a(t)$ takes the form (δ is the detuning)

$$a(t) \propto e^{i\delta t/2} e^{-t/T_2}.$$

Compute the Fourier transform $\tilde{a}(\omega)$ of $a(t)$

$$\tilde{a}(\omega) = \int_0^\infty dt\, e^{i\omega t}\, a(t).$$

One could hope to deduce T_2 from the width $1/T_2$ of the peak of $\tilde{a}(\omega)$. However, the different molecules have different detunings, for example because the field $\vec{B}_0$ may be slightly inhomogeneous, leading to different Larmor frequencies, so that the signals from the different molecules interfere destructively and $a(t)$ decays with a characteristic time much smaller than T_2. In order to overcome this problem, one applies the following sequence of operations on the state matrix (6.41): free evolution during $t/2$, rotation by π about the y axis and free evolution during $t/2$. Show that in the absence of relaxation, the state matrix would evolve from $\rho(t=0)$ (6.41) as

$$\rho(t=0) \to \rho(t) = U(t)\,\rho(t=0)\,U^\dagger(t)$$

$$U(t) = \exp\left(\frac{-i\delta\sigma_z t}{4}\right)(-i\sigma_y)\exp\left(\frac{-i\delta\sigma_z t}{4}\right).$$

Show that $U(t) = -i\sigma_y$, and that, taking relaxation into account, $\rho(t)$ is

$$\rho(t) = \frac{1}{2}\left(I + \frac{1}{2}\,\delta p\sigma_y\, e^{-t/T_2}\right)$$

independently of the detuning δ. Show that measuring the time decay of the height of the peak in $\tilde{a}(\omega)$ allows a reliable determination of T_2, and explain why the sequence of operations described above is called a "spin echo experiment."

2. Let us consider two identical spin-1/2 nuclei (for example two protons) belonging to a single molecule which is being used in a NMR experiment. The two nuclear spins have an interaction Hamiltonian H_{12}, which, in the simplest case, has the following form

$$H_{12} = \hbar\omega_{12}\,\sigma_z^{(1)} \otimes \sigma_z^{(2)}.$$

Show that the corresponding evolution operator is given by

$$U_{12}(t) = \exp(-iH_{12}t/\hbar) = I_{12}\cos\omega_{12}t - i[\sigma_z^{(1)} \otimes \sigma_z^{(2)}]\sin\omega_{12}t.$$

Prove the following identity

$$U[\mathcal{R}_x^{(1)}(\pi)]\exp(-iH_{12}t/\hbar)U[\mathcal{R}_x^{(1)}(\pi)]\exp(-iH_{12}t/\hbar) = I_{12}$$

where $U[\mathcal{R}_x^{(1)}(\pi)]$ is a rotation by π of spin 1 around the x axis. From this equation, demonstrate that the sequence of operations

free evolution during $t \to \pi$ rotation about $Ox \to$ free evolution during $t \to \pi$ rotation about Ox

brings back the spins to their original orientation at time $t = 0$. The preceding sequence of operations is widely used in NMR quantum computing. It relies on the property that ω_{12}^{-1} is of the order of a hundred milliseconds, while a rotation takes only a few tens of microseconds.

3. In the rotating frame, show that the full Hamiltonian for the two spins is

$$H_{\text{tot}} = \frac{1}{2}\delta^{(1)}\sigma_z^{(1)} + \frac{1}{2}\delta^{(2)}\sigma_z^{(2)} - \frac{1}{2}\omega_1^{(1)}\sigma_x^{(1)} - \frac{1}{2}\omega_1^{(2)}\sigma_x^{(2)} + \hbar\omega_{12}\sigma_z^{(1)} \otimes \sigma_z^{(2)}$$

where $\delta^{(i)}$ is the detuning and $\omega_1^{(i)}$ the Rabi frequency for spin (i). The difference

$$\delta^{(1)} - \delta^{(2)} = \gamma(B_0^{(1)} - B_0^{(2)})$$

is the chemical shift (Section 5.2.3). What are the four energy levels in the absence of radio-frequency field ($\omega_1^{(1)} = \omega_1^{(2)} = 0$)? Let us introduce the operator[38]

$$\Sigma_z = \frac{1}{2}(\sigma_z^{(1)} + \sigma_z^{(1)}).$$

One can show that the only allowed transitions correspond to $\Delta\Sigma_z = \pm 1$, while $\Delta\Sigma_z = \pm 2$ and $\Delta\Sigma_z = 0$ are forbidden. Show that the four frequencies which appear in the NMR signal are

$$\delta^{(1)} + \omega_{12} \quad \delta^{(1)} - \omega_{12} \quad \delta^{(2)} + \omega_{12} \quad \delta^{(2)} - \omega_{12}.$$

Sketch the NMR spectrum and compare with Figure 5.9.

6.5.7 Calculation of $E(\hat{a}, \hat{b})$

1. Find the amplitudes a_{+-}, a_{-+}, and a_{--} (cf. (6.46)).
2. Show that (cf. (6.47))

$$E(\hat{a}, \hat{b}) = \langle (\vec{\sigma} \cdot \hat{a}) \otimes (\vec{\sigma} \cdot \hat{b}) \rangle_{\Phi} = \langle \Phi | (\vec{\sigma} \cdot \hat{a}) \otimes (\vec{\sigma} \cdot \hat{b}) | \Phi \rangle = -\hat{a} \cdot \hat{b},$$

where $|\Phi\rangle$ is the entangled state (6.15) of two spins 1/2:

$$|\Phi\rangle = \frac{1}{\sqrt{2}}(|+-\rangle - |-+\rangle).$$

Hint: using the rotational invariance of $|\Phi\rangle$, show that $\vec{\sigma}^{(a)}|\Phi\rangle = -\vec{\sigma}^{(b)}|\Phi\rangle$ and use (3.50).

---

[38] Σ_z is the z component of the total spin; see Chapter 10.

6.5.8 Bell inequalities involving photons

Let us consider two photons traveling in opposite directions, one (1) along Oz and the other (2) along $-Oz$, in an entangled polarization state:

$$|\Phi\rangle = \frac{1}{\sqrt{2}}\left(|x\rangle_1 \otimes |y\rangle_2 - |y\rangle_1 \otimes |x\rangle_2\right) = \frac{1}{\sqrt{2}}\left(|xy\rangle - |yx\rangle\right).$$

The states $|x\rangle$ and $|y\rangle$ are states of linear polarization in the Ox and Oy directions.

1. Let

$$|\theta\rangle = \cos\theta |x\rangle + \sin\theta |y\rangle$$

be the state of linear polarization in the direction $\hat{n}_\theta$ of the xOy plane (cf. (3.23)) and $|\theta_\perp\rangle$ be the orthogonal polarization state (3.24). Show that

$$|\Phi\rangle = \frac{1}{\sqrt{2}}\left(|\theta\,\theta_\perp\rangle - |\theta_\perp\,\theta\rangle\right).$$

The state $|\Phi\rangle$ is then invariant under rotation about Oz.

2. Write $|\Phi\rangle$ as a function of the circular polarization states $|R\rangle$ and $|L\rangle$ (3.11) paying attention to the orientation of the axes (Fig. 6.15). The sense of rotation depends on the propagation direction:

$$|\Phi\rangle = \frac{i}{\sqrt{2}}\left(|RR\rangle - |LL\rangle\right).$$

Use (3.27) to verify that the second form of $|\Phi\rangle$ is invariant under rotations about Oz.

3. Alice and Bob analyze the photon polarization using linear polarizers oriented in the direction $\hat{n}_\alpha$ for photon 1 and $\hat{n}_\beta$ for photon 2 in the xOy plane. We define

   - $p_{++}(\alpha, \beta)$, the probability for photon 1 to be polarized in the $\hat{n}_\alpha$ direction and photon 2 in the $\hat{n}_\beta$ direction;
   - $p_{+-}(\alpha, \beta)$, the probability for photon 1 to be polarized in the $\hat{n}_\alpha$ direction and photon 2 in the $\hat{n}_{\beta_\perp}$ direction.

   The probabilities $p_{-+}(\alpha, \beta)$ and $p_{--}(\alpha, \beta)$ are defined analogously. As for spin 1/2 (cf. (6.45)), we define

   $$E(\alpha, \beta) = [p_{++}(\alpha, \beta) + p_{--}(\alpha, \beta)] - [p_{+-}(\alpha, \beta) + p_{-+}(\alpha, \beta)].$$

   Show that

   $$E(\alpha, \beta) = -\cos[2(\alpha - \beta)].$$

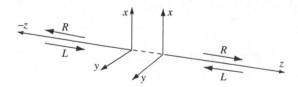

Fig. 6.15. Configuration of polarizations of entangled photons.

Use the rotational invariance of $|\Phi\rangle$ to simplify the calculation. What values of α, α', β, and β' should be used to obtain

$$X = E(\alpha, \beta) + E(\alpha, \beta') + E(\alpha', \beta') - E(\alpha', \beta) = -2\sqrt{2}$$

as in (6.50)?

4. Show that the state

$$|\Psi\rangle = \frac{1}{\sqrt{2}}(|xx\rangle + |yy\rangle)$$

is also invariant under rotations about Oz. Express it as a function of the circular polarization states.[39]

6.5.9 Two-photon interference

Let us consider the two-photon Young's slit interference experiment shown schematically in Fig. 6.16. The two photons are emitted in opposite directions with wave vectors of about $\pm\vec{k}$ by a source whose vertical position is defined with accuracy $\pm d/2$; we can assume, for example, that the two photons are created in the decay of a particle Ω of momentum close to 0 located on segment CD of height d. The distance between the slits is l and the distance between the slits and the source, as well as between the slits and the screens, is D, with $l \ll D$.

1. What is the spread Δk_x in the x component of the photon wave vector? It is always assumed that $\Delta k_x \ll k$.

2. The position of the source is specified by its x coordinate, and the impacts of photons 1 and 2 by their y and z coordinates. Show that for photon 1 the path difference $\delta(x, y)$ is

$$\delta(x, y) - \delta(0, 0) = \mp\frac{l}{2D}(x + y) = \mp\theta(x + y),$$

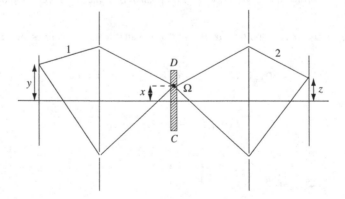

Fig. 6.16. Two-photon interference.

---

[39] The states $|\Phi\rangle$ and $|\Psi\rangle$ both have zero angular momentum. If the two photons originate in the decay of a spin-0 particle, the choice between the two states depends on the parity of the parent particle; see Exercise 13.4.4.

where the signs $\mp$ correspond to the passage of photon 1 through the upper $(-)$ or lower $(+)$ slit; $2\theta = l/D$ is the angle subtended by the space between the slits as seen from the source.

3. Show that the probability amplitude for detecting in coincidence photon 1 at y and photon 2 at z is proportional to

$$\overline{a}(x|y,z) = \cos[k\theta(y+x)]\cos[k\theta(x+z)]$$

when the source is located at point x.

4. Show that the total amplitude of detection in coincidence is proportional to

$$a(y,z) = \frac{1}{d}\int_{-d/2}^{d/2}\overline{a}(x|y,z)\,dx$$

and deduce that

$$a(y,z) = \frac{1}{2d}\left[\frac{1}{k\theta}\sin(k\theta d)\cos[k\theta(y+z)]+d\cos[k\theta(y-z)]\right].$$

Carefully justify the fact that the amplitudes must be added rather than the intensities, as would be the case for interference involving a single photon.

5. Show that for $d \gg 1/(k\theta) \sim \lambda/\theta$ the probability of detection in coincidence is

$$\mathsf{p}(y,z) \propto \cos^2[k\theta(y-z)].$$

How is this result interpreted in terms of conditional interference? What happens if only one screen is observed?

6. Show that when $d \ll 1/(k\theta)$ we have

$$\mathsf{p}(y,z) = \cos^2(k\theta y)\cos^2(k\theta z),$$

and two sets of independent fringes are obtained. What is the physical reason that the sets of individual fringes are restored?

7. What conditions on Δk_x do the limits $d \gg \lambda/\theta$ and $d \ll \lambda/\theta$ correspond to? How can the results of questions 5 and 6 be interpreted?

8. Instead of using Young's slits, photons can be made to interfere by means of two symmetric beam splitters S and S' (Fig. 6.17). The reflection and transmission probabilities are 50%. The phase shift between reflection and transmission by a beam splitter is $\pi/2$ (Exercise 2.4.12). We introduce the phase shifts α and β in the two arms of the interferometer and set $\phi = (\alpha - \beta)$. Let $\mathsf{p}(c,c')$ be the probability of detection in coincidence by the detectors c and c'. Show that

$$\mathsf{p}(c,c') = \frac{1}{2}\sin^2\frac{\phi}{2}$$

and that

$$E(\alpha,\beta) = [\mathsf{p}(c,c')+\mathsf{p}(d,d')]-[\mathsf{p}(c,d')+\mathsf{p}(c',d)] = -\cos\phi.$$

Construct a Bell inequality analogous to that obtained using spins 1/2 by allowing α and β to vary.

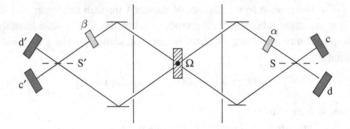

Fig. 6.17. Interference using beam splitters.

6.5.10 Interference of emission times

In an experiment performed by a Nice–Geneva collaboration,[40] a laser beam (from a pumped laser) of wavelength $\lambda = 655$ nm is incident on a nonlinear crystal (Fig. 6.18). A fraction of the incident photons is converted into pairs of photons of wavelength $2\lambda = 1310$ nm, each photon leaving via one of two optical fibers and then crossing a Mach–Zehnder (MZ) interferometer (cf. Exercise 1.6.6). These interferometers are chosen to have a short arm and a long arm, and the difference between the two is $\Delta l = 20$ cm. The optical path on the long arm of the right-hand interferometer can be varied by an amount δ by means of a plate. The coherence length $l_{\text{coh}} \simeq 40\,\mu$m of the converted photons is very small compared with Δl: $l_{\text{coh}} \ll \Delta l$ (whereas the coherence length of the pumped laser is around 100 m).

1. The phase δ on the long arm of the right-hand interferometer is allowed to vary. Show that the number of photons counted by the detector D_1 is independent of δ.
2. The two photons are detected in coincidence at D_1 and D_2 with a window of coincidence of order 0.1 ns. Since the pumped laser operates continuously, no other information about the creation time of the photon pair is available. Show that it is not possible to distinguish between the two paths, short–short and long–long, followed by the photons. Demonstrate that by varying δ it is possible to obtain a sinusoidal variation in the coincidence count, but that the numbers detected individually in D_1 and D_2 remain independent of δ. Hint: show that if the two beam splitters of the left-hand MZ interferometer are suppressed, it is possible to obtain one piece of

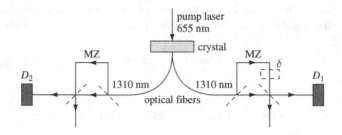

Fig. 6.18. Interference of emission times.

[40] S. Tanzilli *et al.*, PPLN waveguide for quantum communication, *Eur. Phys. J.* **D18**, 155–160 (2002).

information about the trajectory followed by the photon on the right. What happens if the entire apparatus on the left (MZ interferometer and detectors) is suppressed?

6.5.11 *The Deutsch algorithm*

This exercise gives the simplest example of a parallel quantum algorithm, the Deutsch algorithm. We are given a function $f(x)$, $x = 0$ or 1, which also takes two values, 0 or 1, so that we need one qubit for the input register and one qubit for the output register. We want to ask the following question: is $f(x)$ constant ($f(0) = f(1)$) or "balanced" ($f(0) \neq f(1)$)? With a classical computer, we need to compute the two values of $f(x)$ and compare. With a quantum computer, we can get the answer in only one operation. The quantum circuit is drawn on Fig. 6.19. The register (output) qubit is initially in state $|0\rangle$ ($|1\rangle$). Starting from

$$|\Psi\rangle = (H|0\rangle) \otimes (H|1\rangle)$$

show that (see (6.75))

$$U_f|\Psi\rangle = \frac{1}{2}\left(\sum_{x=0}^{1}(-1)^{f(x)}|x\rangle\right) \otimes (|0\rangle - |1\rangle).$$

What is the state $|\varphi\rangle$ of the input register in Fig. 6.19? Compute $H|\varphi\rangle$ and show that measuring the qubit of the input register allows us to decide whether $f(x)$ is constant or "balanced."

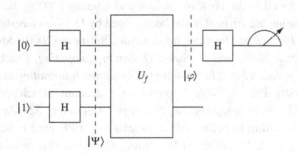

Fig. 6.19. Quantum circuit for implementing the Deutsch algorithm.

6.6 Further reading

The tensor product and the state operator are discussed by Messiah [1999], Chapters VII and VIII, and by Cohen-Tannoudji *et al.* [1977], Complements E_{III} and E_{IV}. Two more recent references are Isham [1995], Chapter 6, and Basdevant and Dalibard [2002], Appendix D. Applications of the state operator to statistical mechanics and the properties of the von Neumann entropy can be found in Balian [1991], Chapters 2 to 5, and Le Bellac *et al.* [2004], Chapter 2. Applications of the state operator to NMR are discussed, for example, by Levitt [2001], Chapter 10. There are many accounts of Bell inequalities, and we recommend those of Peres [1993], Chapters 6 and 7; Isham [1995], Chapters 8 and 9; N. Mermin, Hidden variables and the two theorems of John Bell, *Rev. Mod. Phys.* **65**, 803–815 (1993); and Laloë [2001]. These references also discuss the important theorems of Gleason and of Kochen-Specker. The original article corresponding to the experiment described in Section 6.4.1 is M. Brune *et al.*, Observing the progressive decoherence of the "meter" in a quantum measurement, *Phys. Rev. Lett.* **77**, 4887–4890 (1976). A popularized account is given by S. Haroche, Entanglement, decoherence and the quantum/classical boundary, *Phys. Today*, 36–42 (July 1998), and a pedagogical discussion by Omnès [1999], Chapter 22. Interference involving entangled states is described by D. Greenberger, M. Horne, and A. Zeilinger, Multiparticle interferometry and the superposition principle, *Phys. Today,* 22–29 (August 1993), and by A. Zeilinger, Experiment and the foundations of quantum physics, *Rev. Mod. Phys.* **71**, S288–S297 (1999). The 1989–90 Collège de France lecture course by C. Cohen-Tannoudji (in French, available on the website www.lkb.ens.fr) contains a very complete discussion of measurement theory and decoherence; see also W. Zurek, Decoherence and the transition from quantum to classical, *Phys. Today*, 36–44 (October 1991), p. 36 and Zurek [2003]. For a critical view of the "decoherence program," see A. Leggett, Testing the limits of quantum mechanics: motivation, state of play, prospects, *J. Phys. Cond. Mat.* **14**, R415–R451 (2002); and The quantum measurement problem, *Science* **307**, 871–872 (2005). See also M. Schlossauer, Decoherence, the measurement problem and interpretations of quantum mechanics, *Rev. Mod. Phys.* **76**, 1267 (2004). An excellent introduction to quantum computing can be found in the book of Nielsen and Chuang [2000]; the various aspects of quantum information are covered in the book edited by D. Bouwmeester, A. Ekert, and A. Zeilinger, *The Physics of Quantum Information*, Springer (2000). More recent (and shorter!) books are: J. Stolze and D. Suter, *Quantum Computing*, Chichester: J. Wiley (2004) and M. Le Bellac, *A Short Introduction to Quantum Information and Computation*, Cambridge University Press (2006). A popularized account of teleportation is given by A. Zeilinger, Quantum teleportation, *Scientific American*, 32 (April 2000). The "historical" articles (dating to before 1982, for example EPR etc.) have been collected in a book edited by J. A. Wheeler and W. Zurek, *Quantum Theory and Measurement*, Princeton: Princeton University Press (1983).

7

Mathematics of quantum mechanics II: infinite dimension

In Chapter 4 we saw that the canonical commutation relations force us to use a space of states of infinite dimension, in which rigor would require the use of advanced mathematical tools. Fortunately, physicists generally need only to carry the results for finite dimension over to infinite dimension with some simple modifications which we shall indicate here, without embarking on sophisticated mathematics. Nevertheless, it is useful to be aware of the lapses in rigor which are customarily made in physics in order to avoid possible unpleasant surprises.

The objective of this chapter is, on the one hand, to present some concrete examples illustrating the new features which arise in infinite dimension and, on the other, to give the rules for practical calculations, in particular to write down the spectral decomposition of Hermitian and unitary operators. The mathematics we use is a bit more detailed than commonly found in most quantum mechanics textbooks. The reader interested purely in the practical aspects can proceed directly to Section 7.3, where the results essential for later on are summarized.

7.1 Hilbert spaces

7.1.1 Definitions

The space of states of quantum mechanics is a Hilbert space $\mathcal{H}$, which in general is of infinite dimension. The axiomatic definition of a Hilbert space is the following.

1. It is a vector space which, for the needs of quantum mechanics, is defined on complex numbers. The vectors of this space are denoted $|\varphi\rangle$.
2. This space is endowed with a positive-definite scalar product; if $|\varphi\rangle$ and $|\chi\rangle$ are two vectors, the scalar product is denoted $\langle\chi|\varphi\rangle$ and satisfies

$$\langle\chi|\varphi\rangle = (\langle\varphi|\chi\rangle)^*, \tag{7.1}$$

$$\langle\chi|\varphi + \lambda\psi\rangle = \langle\chi|\varphi\rangle + \lambda\langle\chi|\psi\rangle, \tag{7.2}$$

$$\langle\varphi|\varphi\rangle = ||\varphi||^2 = 0 \iff |\varphi\rangle = 0, \tag{7.3}$$

where λ is an arbitrary complex number and $||\varphi||$ denotes the norm of $|\varphi\rangle$.

3. $\mathcal{H}$ is a *complete space*, that is, a space where every Cauchy series has a limit: if one series of vectors $|\varphi^{(l)}\rangle$ of $\mathcal{H}$ is such that $||\varphi^{(l)} - \varphi^{(m)}|| \to 0$ for $l, m \to \infty$, then there exists a vector $|\varphi\rangle$ of $\mathcal{H}$ such that $||\varphi^{(l)} - \varphi|| \to 0$ for $l \to \infty$.[1]

4. A Hilbert space is characterized by its dimension; all spaces of the same dimension are isomorphic. The dimension of a Hilbert space can be finite and equal to N, or it can be denumerably or nondenumerably infinite.

In Chapter 2 we studied Hilbert spaces of finite dimension in detail. If the dimension is N, it takes N orthogonal unit vectors $|n\rangle$, $n = 1, \ldots, N$, to form an orthonormal basis: $\{|1\rangle, |2\rangle, \ldots, |n\rangle, \ldots, |N\rangle\}$. In the denumerable case there exists a denumerable series of orthogonal unit vectors $|1\rangle, |2\rangle, \ldots, |n\rangle, \ldots$ forming a basis of $\mathcal{H}$, and any vector of $\mathcal{H}$ can be written as a linear combination of these basis vectors:

$$|\varphi\rangle = \sum_{n=1}^{\infty} c_n |n\rangle. \tag{7.4}$$

However, in contrast to the case of finite dimension, an arbitrary combination of the form (7.4) is not in general a vector of $\mathcal{H}$. In fact, the squared norm of $|\varphi\rangle$ is given by

$$||\varphi||^2 = \sum_{n=1}^{\infty} |c_n|^2, \tag{7.5}$$

and (7.4) defines a vector if and only if this norm is finite: the series in (7.5) must be convergent,

$$\sum_{n=1}^{\infty} |c_n|^2 < \infty.$$

Under these conditions, for any $\varepsilon > 0$ there exists an integer N such that the vector $|\varphi_N\rangle$ defined by the following *finite* combination of basis vectors

$$|\varphi_N\rangle = \sum_{n=1}^{N} c_n |n\rangle$$

satisfies

$$||\varphi - \varphi_N||^2 = \sum_{n=N+1}^{\infty} |c_n|^2 \leq \varepsilon. \tag{7.6}$$

In other words, it is possible to approximate $|\varphi\rangle$ by a vector $|\varphi_N\rangle$ whose norm differs by an arbitrarily small amount from that of $|\varphi\rangle$. We can now approximate the c_n by rational numbers, and we see that it is possible to construct in $\mathcal{H}$ a denumerable series of vectors which is dense in $\mathcal{H}$.[2] This property, which is common to spaces of finite and denumerably infinite dimension, is called the *separability* of the Hilbert space, not to be

---

[1] This axiom is in fact rather superfluous. It is automatically satisfied in the case of finite dimension, and for separable Hilbert spaces, we can always add the limit vectors of Cauchy series.

[2] A set of vectors $\{|\varphi^{(\alpha)}\rangle\}$ is dense in $\mathcal{H}$ if for any $\varepsilon > 0$ and for any vector $|\varphi\rangle$ of $\mathcal{H}$ it is possible to find a $|\varphi^{(\alpha)}\rangle$ such that $||\varphi - \varphi^{(\alpha)}|| < \varepsilon$.

confused with the separability of Section 6.3.2. The Hilbert spaces of quantum mechanics are separable.

The convergence defined by (7.6) is *convergence in the norm*, also called *strong convergence*. It is said that a series of vectors $|\varphi^{(l)}\rangle$ converges in the norm to $|\varphi\rangle$ for $l \to \infty$ if for any $\varepsilon > 0$ there exists an integer N such that for $l \geq N$

$$||\varphi - \varphi^{(l)}|| \leq \varepsilon \quad \forall \, l \geq N. \tag{7.7}$$

There exists another type of convergence, called *weak convergence*: a series of vectors $|\varphi^{(l)}\rangle$ converges weakly to $|\varphi\rangle$ if for any vector $|\chi\rangle$ of $\mathcal{H}$

$$\lim_{l \to \infty} \langle \varphi^{(l)} | \chi \rangle = \langle \varphi | \chi \rangle. \tag{7.8}$$

We shall not have occasion to use weak convergence,[3] but the existence of this convergence illustrates a difference from the case of finite dimension: the two types of convergence are identical for a space of finite dimension but not for a space of infinite dimension. Strong convergence implies weak convergence, but not the reverse (Exercise 7.4.1).

7.1.2 Realizations of separable spaces of infinite dimension

All separable Hilbert spaces of infinite dimension are isomorphic. However, their concrete realizations can *a priori* appear different and it is interesting to be able to identify them. We shall successively define the spaces $\ell^{(2)}$, $L^{(2)}[a, b]$, and $L^{(2)}(\mathbb{R})$, which are all separable and of infinite dimension.

1. *The space $\ell^{(2)}$.* A vector $|\varphi\rangle$ is defined by an infinite series of complex numbers $c_1, \dots, c_n \dots$ such that

$$||\varphi||^2 = \sum_{n=1}^{\infty} |c_n|^2 < \infty. \tag{7.9}$$

As in (7.4), the c_n are the coordinates of $|\varphi\rangle$. Let us verify that $|\varphi + \lambda\chi\rangle$ belongs to $\mathcal{H}$. If $|\chi\rangle$ has components d_n, as

$$|c_n + \lambda d_n|^2 \leq 2(|c_n|^2 + |\lambda|^2 |d_n|^2),$$

it follows that $||\varphi + \lambda\chi|| < \infty$. The scalar product of two vectors

$$\langle \chi | \varphi \rangle = \sum_{n=1}^{\infty} d_n^* c_n$$

is well defined because, according to the Schwartz inequality (2.10),

$$|\langle \chi | \varphi \rangle| = \left| \sum_{n=1}^{\infty} d_n^* c_n \right| \leq ||\chi|| \, ||\varphi||.$$

---

[3] It arises in, for example, certain problems of quantum field theory.

Let us now show that $\ell^{(2)}$ is complete. Let $|\varphi^{(l)}\rangle$ and $|\varphi^{(m)}\rangle$ be two vectors with components $c_n^{(l)}$ and $c_n^{(m)}$. If $||\varphi^{(m)} - \varphi^{(l)}|| < \varepsilon$ for $l, m > N$, this means that

$$\left(\sum_{n=1}^{\infty} \left| c_n^{(l)} - c_n^{(m)} \right|^2 \right)^{1/2} < \varepsilon.$$

The inequality is *a fortiori* true for each individual value of n and, for n fixed, the numbers $c_n^{(l)}$ form a Cauchy series which converges to c_n for $l \to \infty$. It is easy to show (Exercise 7.4.1) that the vector $\varphi^{(l)}$ converges to $|\varphi\rangle = \sum_n c_n |n\rangle$ for $l \to \infty$:

$$\lim_{l \to \infty} \sum_n \left| c_n - c_n^{(l)} \right|^2 = \lim_{l \to \infty} ||\varphi - \varphi^{(l)}||^2 = 0.$$

Finally, $\ell^{(2)}$ is of denumerable dimension by construction.

2. *The space $L^{(2)}[a, b]$.* Now we are going to introduce a class of vector spaces which will play a fundamental role, functional spaces. The simplest example is the space of functions which are square-integrable on the interval $[a, b]$. Let us consider complex functions $\varphi(x)$ satisfying[4]

$$\int_a^b dx |\varphi(x)|^2 < \infty, \tag{7.10}$$

or functions which are square-integrable on the interval $[a, b]$. These functions form a vector space denoted $L^{(2)}[a, b]$. In fact, (i) $\varphi(x) + \lambda \chi(x)$ is square-integrable if $\varphi(x)$ and $\chi(x)$ are, and (ii) the scalar product $\langle \chi | \varphi \rangle$,

$$\langle \chi | \varphi \rangle = \int_a^b dx \, \chi^*(x) \varphi(x), \tag{7.11}$$

is well defined owing to the Schwartz inequality:

$$\left| \int_a^b dx \, \chi^*(x) \varphi(x) \right|^2 \leq \int_a^b dx \, |\chi(x)|^2 \int_a^b dx \, |\varphi(x)|^2 = ||\chi||^2 \, ||\varphi||^2. \tag{7.12}$$

The fact that $L^{(2)}[a, b]$ is complete is a result of a theorem due to Riesz and Fischer, and the separability results from a standard theorem of Fourier analysis: any square-integrable function $\varphi(x)$ can be written, in the sense of convergence in the mean (or in the norm), as the sum of a Fourier series:

$$\varphi(x) = \sum_{n=-\infty}^{\infty} c_n \frac{1}{\sqrt{(b - a)}} \exp \left(\frac{2i\pi n x}{b - a} \right), \tag{7.13}$$

$$c_n = \frac{1}{\sqrt{(b - a)}} \int_a^b dx \, \varphi(x) \exp \left(-\frac{2i\pi n x}{b - a} \right). \tag{7.14}$$

The functions

$$\varphi_n(x) = \frac{1}{\sqrt{(b - a)}} \exp \left(\frac{2i\pi n x}{b - a} \right) \tag{7.15}$$

---

[4] Two functions $\varphi(x)$ and $\overline{\varphi}(x)$ such that

$$\int_a^b dx \, |\varphi(x) - \overline{\varphi}(x)|^2 = 0$$

represent the same vector of $\mathcal{H}$: $||\varphi - \overline{\varphi}|| = 0$.

form a denumerable orthonormal basis of $L^{(2)}[a, b]$, which is then a separable Hilbert space.

3. *The space* $L^{(2)}(\mathbb{R})$. When the interval $[a, b]$ is identified as the real line $\mathbb{R}$, $[a, b] \rightarrow [-\infty, +\infty]$, we obtain the Hilbert space $L^{(2)}(\mathbb{R})$ (or $L^{(2)}(-\infty, +\infty)$), the space of functions which are square-integrable on $[-\infty, +\infty]$. Although the proof is more delicate, it can be shown that $L^{(2)}(\mathbb{R})$ is still a separable space and is thus isomorphic to $\ell^{(2)}$.

7.2 Linear operators on $\mathcal{H}$

7.2.1 The domain and norm of an operator

Linear operators on $\mathcal{H}$ are defined as in the case of finite dimension. However, there are important differences. It can happen, and is very often the case in quantum mechanics, that an operator is not defined for any vector of $\mathcal{H}$, but only on a subset of vectors of $\mathcal{H}$. For example, let the operator A act in $\ell^{(2)}$ such that if $|\varphi\rangle$ has components $\{c_1, c_2, \ldots, c_n, \ldots\}$, then $A|\varphi\rangle$ has components $\{c_1, 2c_2, \ldots, nc_n, \ldots\}$. In $L^{(2)}[a, b]$ this operator corresponds to differentiation up to a multiplicative factor, as is seen immediately by examining the Fourier decomposition (7.13). It is clear that the squared norm of $A|\varphi\rangle$, given by

$$||A\varphi||^2 = \sum_n n^2 |c_n|^2,$$

can diverge, whereas $\sum_n |c_n|^2$ converges; it is sufficient, for example, to take $c_n = 1/n$. In other words, $A|\varphi\rangle$ is not a vector of $\mathcal{H}$. The *domain* of A, denoted $\mathcal{D}_A$, is defined as the set of vectors $|\varphi\rangle$ such that $A|\varphi\rangle$ is a vector of $\mathcal{H}$. In the example above, the domain of A is the set of vectors such that $\sum_n n^2 |c_n|^2 < \infty$, and it is easy to convince ourselves that this domain is dense in $\mathcal{H}$. In practice, an operator A is of interest only if its domain is dense in $\mathcal{H}$.

If $A|\varphi\rangle$ exists for any $|\varphi\rangle$, it is said that the operator A is *bounded*. We must then have $||A\varphi|| < \infty$ for any $|\varphi\rangle$. The maximum of $||A\varphi||/||\varphi||$ is called the *norm* of A and denoted $||A||$:

$$||A|| = \sup_{||\varphi||=1} ||A\varphi||. \tag{7.16}$$

If the norm of $||A||$ does not exist, then A is termed unbounded. Unbounded operators are much more delicate to handle than bounded operators. Unfortunately, they are omnipresent in quantum mechanics.

In $L^{(2)}[0, 1]$ the operator X which takes the function $\varphi(x)$ to $x\varphi(x)$,

$$\varphi(x) \rightarrow (X\varphi)(x) = x\varphi(x), \tag{7.17}$$

is a bounded operator of unit norm. On the other hand, the operator d/dx which takes $\varphi(x)$ to its derivative,

$$\varphi(x) \rightarrow \frac{d\varphi(x)}{dx}, \tag{7.18}$$

is not a bounded operator, as we have already seen. Another simple argument to show that d/dx is unbounded is to find a function such that the norm of $\varphi(x)$ is finite, but that of $\varphi'(x)$ is not. For example, we can choose

$$\varphi(x) = x^{-1/4}, \qquad \frac{d\varphi(x)}{dx} = -\frac{1}{4} x^{-5/4}$$

so that

$$\int_0^1 dx\, x^{-1/2} = 2, \quad \int_0^1 dx\, \frac{1}{16} x^{-5/2} \quad \text{diverges at } x = 0.$$

Domain problems can make the definition of the sum and product of two unbounded operators rather delicate. For example, it is not possible a priori to define the sum $A + B$ of two unbounded operators A and B except on the intersection $\mathcal{D}_A \cap \mathcal{D}_B$ of the two domains, which can become problematic if this intersection reduces to a null vector. When two operators A and B are equal on the same domain $\mathcal{D}_A$, but when the domain of B contains that of A, $\mathcal{D}_A \subseteq \mathcal{D}_B$, it is said that B is an extension of A, $A \subseteq B$. Let us give an example. The canonical commutation relation (4.33) between the position and momentum operators X and P written in one-dimensional space ($d = 1$),

$$[X, P] = i\hbar I, \tag{7.19}$$

implies that at least one of the two operators is unbounded (Exercise 7.4.3). The left-hand side $[X, P]$ of (7.19) is a priori defined only on a subset of $\mathcal{H}$, while the right-hand side $i\hbar I$ is defined for any vector of $\mathcal{H}$. The correct way to write the canonical commutation relation is then

$$[X, P] \subseteq i\hbar I.$$

Let us note another difference from the case of finite dimension. Whereas in a vector space of finite dimension the existence of a left inverse implies the existence of a right one and vice versa, this property no longer holds in infinite dimension.[5] For example, let the operators A and B be defined by their action on the components c_n of a vector $|\varphi\rangle$:

$$A(c_1, c_2, c_3 \ldots) = (c_2, c_3, c_4 \ldots), \quad B(c_1, c_2, c_3 \ldots) = (0, c_1, c_2, \ldots).$$

Then

$$BA(c_1, c_2, c_3 \ldots) = B(c_2, c_3, c_4 \ldots) = (0, c_2, c_3, \ldots),$$

$$AB(c_1, c_2, c_3 \ldots) = A(0, c_1, c_2, \ldots) = (c_1, c_2, c_3, \ldots),$$

and $AB = I$ but $BA \neq I$, although A and B are both bounded.

---

[5] An important example of such an operator in physics is the Møller operator of scattering theory in the presence of bound states.

7.2.2 *Hermitian conjugation*

In the case of a bounded operator there is no difficulty of principle in defining the Hermitian conjugate operator $A^\dagger$ of A by

$$\langle\chi|A\varphi\rangle = \langle A^\dagger\chi|\varphi\rangle. \tag{7.20}$$

As in the case of finite dimension, it is said that A is Hermitian if $A = A^\dagger$, and then

$$\langle\chi|A\varphi\rangle = \langle A\chi|\varphi\rangle. \tag{7.21}$$

The situation becomes more complicated if A is unbounded owing to domain problems. First, (7.20) can be used to define $A^\dagger$ only if $\mathcal{D}_A$ is dense in $\mathcal{H}$. Next, the domain in which $A^\dagger$ is defined is generally larger than that of A: $\mathcal{D}_A \subseteq \mathcal{D}_{A^\dagger}$. In an instant we shall give an example of this. In general, for an unbounded operator that satisfies (7.21) we will have not $A = A^\dagger$ but rather $A \subseteq A^\dagger$. Mathematicians reserve the term "Hermitian operators" for operators such that $A \subseteq A^\dagger$, and call operators satisfying $A = A^\dagger$ "self-adjoint."

Let us illustrate this by an example in $L^{(2)}[0, 1]$ which will familiarize us with the scalar product and Hermitian conjugation in this space. Let A_0 be the operator $-\mathrm{i}d/dx$ defined on the domain $\mathcal{D}_{A_0}$ of functions $\varphi(x)$ of $L^{(2)}[0, 1]$ which are differentiable and have square-integrable derivative and which also satisfy the boundary conditions $\varphi(0) = \varphi(1) = 0$, whence the subscript 0 of A_0. It is intuitively obvious and easily verified that this domain is dense in $L^{(2)}[0, 1]$. Let us first show that A_0 is Hermitian. Since $\chi(x)$ is a differentiable function of $L^{(2)}[0, 1]$ with derivative belonging to $L^{(2)}[0, 1]$,

$$\langle\chi|A_0\varphi\rangle = \int_0^1 dx\,\chi^*(x)\left(-\mathrm{i}\frac{d}{dx}\varphi(x)\right) = -\mathrm{i}\int_0^1 dx\,\chi^*(x)\varphi'(x),$$

$$\langle A_0\chi|\varphi\rangle = \int_0^1 dx\left(-\mathrm{i}\frac{d}{dx}\chi(x)\right)^*\varphi(x) = \mathrm{i}\int_0^1 dx\,(\chi'(x))^*\varphi(x).$$

Integration by parts shows that

$$\langle\chi|A_0\varphi\rangle - \langle A_0\chi|\varphi\rangle = -\mathrm{i}[\chi^*(x)\varphi(x)]_0^1 = 0. \tag{7.22}$$

We note that Hermiticity requires the presence of the factor i and the boundary conditions. We can define $A_0^\dagger$ on a domain larger than $\mathcal{D}_{A_0}$. In fact, for functions $\chi(x)$ that are not constrained by boundary conditions, that is, functions for which $\chi(0)$ and $\chi(1)$ are arbitrary,

$$\langle A_0^\dagger\chi|\varphi\rangle = \mathrm{i}\int_0^1 dx\,(\chi'(x))^*\varphi(x)$$

$$= \mathrm{i}[\chi^*(x)\varphi(x)]_0^1 - \mathrm{i}\int_0^1 dx\,\chi^*(x)\varphi'(x) = \langle\chi|A_0\varphi\rangle,$$

and consequently $A_0 \subseteq A_0^\dagger$. Finally, we define A_C as the operator $-\mathrm{i}\mathrm{d}/\mathrm{d}x$ acting in the domain $\mathcal{D}_{A_C}$ of functions $\varphi(x)$ of $L^{(2)}[0,1]$ that are differentiable with derivative belonging to $L^{(2)}[0,1]$ and satisfy the boundary conditions

$$\varphi(1) = C\varphi(0), \quad |C| = 1.$$

The operator A_C is self-adjoint. Indeed

$$\langle A_C \chi | \varphi \rangle - \langle \chi | A_C \varphi \rangle = -\mathrm{i}(C\chi^*(1) - \chi^*(0))\varphi(0).$$

The necessary and sufficient condition for the right-hand side to vanish[6] is that $\chi(1) = C\chi(0)$, which shows that the domain of the Hermitian conjugate operator is also $\mathcal{D}_{A_C}$: $A_C^\dagger = A_C$. The operators A_C represent different extensions of A_0 for each value of C. Even though the definition is superficially the same ($A = -\mathrm{i}\mathrm{d}/\mathrm{d}x$), owing to the difference of the domains A_C and $A_{C'}$ are different operators for $C \neq C'$. This can be confirmed by showing that the eigenvalues and eigenvectors of A_C and $A_{C'}$ are different for $C \neq C'$ (Exercise 7.4.3).

7.3 Spectral decomposition

7.3.1 Hermitian operators

The spectral decomposition theorem which generalizes (2.31) is rigorously valid only for self-adjoint operators.[7] Following physicists' practice, we shall no longer distinguish between Hermitian and self-adjoint, and speak only of Hermitian operators. If an operator A is Hermitian, the eigenvalue equation

$$A|\varphi\rangle = a|\varphi\rangle \tag{7.23}$$

does not always have a solution, even if A is a bounded operator. In $L^{(2)}(\mathbb{R})$ the operator $-\mathrm{i}\mathrm{d}/\mathrm{d}x$ is Hermitian, as seen by immediate generalization of (7.22). The equation

$$-\mathrm{i}\frac{\mathrm{d}}{\mathrm{d}x}\varphi(x) = a\varphi(x) \tag{7.24}$$

has plane-wave solutions

$$\varphi_a(x) = C\,\mathrm{e}^{iax} \tag{7.25}$$

where C is a constant, but $\varphi_a(x)$ does not belong to $L^{(2)}(\mathbb{R})$ because

$$\int_{-\infty}^{\infty} \mathrm{d}x\,|\varphi_a(x)|^2 = \int_{-\infty}^{\infty} \mathrm{d}x\,|C|^2$$

is a divergent integral. The operator $-\mathrm{i}\mathrm{d}/\mathrm{d}x$ is unbounded, however, even for a bounded operator such as x in $L^{(2)}[0,1]$, the equation

$$x\chi_a(x) = a\chi_a(x) \tag{7.26}$$

[6] Note that $C^* = 1/C$.
[7] More precisely, for operators that are "essentially self-adjoint," $(A^\dagger)^\dagger = A^\dagger$.

has no solution in $L^{(2)}[0, 1]$. In fact, the generalization of (7.23) to the case of infinite dimension is guaranteed only for a very special class of operators, compact operators.

In finite dimension, when $|\varphi\rangle$ is an eigenvector of A with eigenvalue a as in (7.23), it is said that a belongs to the *spectrum* of A. To generalize this idea to infinite dimension, we consider the operator $(zI - A)$, where z is a complex number and the equation

$$(zI - A)|\varphi\rangle = |\chi\rangle. \tag{7.27}$$

Let $\mathcal{D}$ be the domain of $(zI - A)$ and $\Delta(z)$ be its image. If $\Delta(z) = \mathcal{H}$, z is a regular value of A. The correspondence between $|\varphi\rangle$ and $|\chi\rangle$ is one-to-one and the resolvent (2.46) $R(z, A) = (zI - A)^{-1}$ exists. The spectrum of A is by definition the set of singular values of z. This definition coincides with that in finite dimension. If $|\varphi\rangle$ satisfies (7.23),

$$(zI - A)\Big|_{z=a} |\varphi\rangle = (aI - A)|\varphi\rangle = 0 \, ,$$

and the resolvent is not defined for $z = a$. If A is Hermitian, it is easy to show (Exercise 7.4.2) that $z = a + ib$ is a regular value when $b \neq 0$. The spectrum of A is then real, as for finite dimension. The values of a can either be labeled by a discrete index, $a_1, a_2, \dots, a_n, \dots$, or they can be continuous, for example all the values in an interval on the real line. These correspond to the cases of a *discrete spectrum* and a *continuous spectrum*. The values of a belonging to a discrete spectrum satisfy an eigenvalue equation (7.23), but those of a continuous spectrum do not. It may happen that the continuous spectrum and the discrete spectrum overlap. For example, if a takes all values between 0 and 1, it may happen that the spectrum of A contains some discrete eigenvalues $0 \leq a_n \leq 1$, although this case is exceptional in practice. In general, for most of the operators used in quantum physics the discrete and continuous spectra do not overlap.

Although the spectrum for infinite dimension presents some new properties compared to that for finite dimension, there exists a spectral decomposition theorem which generalizes (2.31):

$$A = \sum_n a_n \mathcal{P}_n.$$

The precise mathematical form of this theorem is complicated, and physicists resort to using "pseudoeigenvectors," that is, objects as in (7.25) that formally satisfy the eigenvalue equation but are not elements of $\mathcal{H}$. In the case of (7.26), the "solution" will be

$$\chi_a(x) = \delta(x - a), \quad \text{because} \quad x\delta(x - a) = a\delta(x - a), \tag{7.28}$$

where $\delta(x)$ is the Dirac delta function, which is not actually a function and is certainly not an element of $L^{(2)}[0, 1]$.

The examples we have just given hint at a general result. The "normalization" condition of the pseudoeigenvectors (7.25) of $-\mathrm{i}\mathrm{d}/\mathrm{d}x$ is, with the choice $C = 1/\sqrt{2\pi}$,

$$\langle \varphi_a | \varphi_b \rangle = \frac{1}{2\pi} \int_{-\infty}^{\infty} \mathrm{d}x \, \mathrm{e}^{-\mathrm{i}ax} \mathrm{e}^{\mathrm{i}bx} = \delta(a - b), \tag{7.29}$$

while for the eigenvalues (7.28) of x

$$\langle \chi_a | \chi_b \rangle = \int_{-\infty}^{\infty} dx \, \delta(x-a)\delta(x-b) = \delta(a-b). \tag{7.30}$$

The normalization of the "pseudoeigenvectors" is therefore given not by a Kronecker delta symbol, but by a Dirac delta function. The generalization of the spectral decomposition theorem is then stated (without proof) as follows.

- For the values a_n of the discrete spectrum labeled by a discrete index n, it is possible to write down an eigenvalue equation and normalization conditions analogous to those for finite dimension:

$$A|n,r\rangle = a_n|n,r\rangle, \tag{7.31}$$

$$\langle n,r|n',r'\rangle = \delta_{nn'}\,\delta_{rr'}, \tag{7.32}$$

where r is a discrete degeneracy index.
- For the values $a(\nu)$ of the continuous spectrum labeled by continuous index ν we have

$$A|\nu,s\rangle = a(\nu)|\nu,s\rangle, \tag{7.33}$$

$$\langle \nu,s|\nu',s'\rangle = \delta(\nu-\nu')\,\delta_{ss'}, \tag{7.34}$$

where $|\nu,s\rangle$ is not a vector of $\mathcal{H}$; s is a degeneracy index which can be either discrete or continuous, and here we have taken it to be discrete for the sake of clarity in the notation.
- Moreover, the eigenvectors of the discrete spectrum and of the continuous spectrum are orthogonal:

$$\langle n,r|\nu,s\rangle = 0. \tag{7.35}$$

The generalization of the decomposition of the identity, or the completeness relation (2.30), is written as

$$\boxed{I = \sum_{n,r} |n,r\rangle\langle n,r| + \sum_s \int d\nu \, |\nu,s\rangle\langle \nu,s|} \,, \tag{7.36}$$

while the spectral decomposition (2.31) of A becomes

$$\boxed{A = \sum_{n,r} |n,r\rangle a_n \langle n,r| + \sum_s \int d\nu \, |\nu,s\rangle a(\nu)\langle \nu,s|} \,. \tag{7.37}$$

We stress the fact that the existence of a discrete and/or continuous spectrum has no relation whatsoever to whether or not the operator A is bounded. There exist unbounded operators whose spectrum is entirely discrete, such as the Hamiltonian of the harmonic oscillator (Section 11.1.1) or the squared angular momentum $\vec{J}^2$ (Section 10.1), and there are bounded operators like multiplication by x on $L^{(2)}[0,1]$ ((7.26)) whose spectrum is entirely continuous.

7.3.2 Unitary operators

A unitary operator U is defined as

$$U^\dagger U = UU^\dagger = I \quad \text{or} \quad U^\dagger = U^{-1}. \tag{7.38}$$

It is immediately seen that unitary operators are necessarily bounded, as they have unit norm. As in the case of finite dimension, it is possible to construct unitary operators by exponentiating Hermitian operators. Using the spectral decomposition of A (7.37), we have

$$\boxed{U(\alpha) = \exp(i\alpha A) = \sum_{n,r} |n, r\rangle \exp(i\alpha a_n)\langle n, r| + \sum_s \int d\nu\, |\nu, s\rangle \exp[i\alpha a(\nu)]\langle \nu, s|}\,.$$

$$\tag{7.39}$$

This equation shows that the spectrum of $\exp(i\alpha A)$ is localized on the circle $|z| = 1$, and it is easy to verify that this property holds for any unitary operator. Moreover, (7.39) shows that $U(\alpha)$ satisfies the Abelian group property:

$$U(\alpha_1 + \alpha_2) = U(\alpha_1)U(\alpha_2), \quad U(0) = I. \tag{7.40}$$

The reciprocal of this property is the important Stone theorem.[8]

The Stone theorem. Given a set of unitary operators depending on a continuous parameter α and satisfying the Abelian group law (7.40), there exists a Hermitian operator T, called the infinitesimal generator of the transformation group $U(\alpha)$, such that $U(\alpha) = \exp(i\alpha T)$.

This theorem can be demonstrated heuristically by showing that $U(\alpha)$ satisfies a differential equation. If $\delta\alpha \to 0$, then

$$U(\alpha + \delta\alpha) = U(\delta\alpha)U(\alpha) \simeq \left(I + \delta\alpha \frac{dU}{d\alpha}\Big|_{\alpha=0}\right)U(\alpha). \tag{7.41}$$

If we take

$$T = -i\frac{dU}{d\alpha}\Big|_{\alpha=0}, \tag{7.42}$$

T must be Hermitian because

$$U(\delta\alpha)U^\dagger(\delta\alpha) \simeq (I + i\delta\alpha\, T)\left(I - i\delta\alpha\, T^\dagger\right)$$
$$\simeq I + i\delta\alpha(T - T^\dagger) = I,$$

from which we have $T = T^\dagger$. From (7.41) we deduce that

$$\frac{dU(\alpha)}{d\alpha} = iTU(\alpha), \tag{7.43}$$

which gives the Stone theorem by integrating and taking into account the boundary condition $U(0) = I$.

---

[8] Also known as the SNAG (Stone, Naimark, Ambrose, and Godement) theorem.

7.4 Exercises

7.4.1 Spaces of infinite dimension

1. Show that the space ℓ^2 is complete.
2. Show that strong convergence implies weak convergence, but not the reverse, except if the space is of finite dimension.

7.4.2 Spectrum of a Hermitian operator

Show that if $A = A^\dagger$ and $z = x + iy$, the vector

$$|\chi\rangle = (zI - A)|\varphi\rangle$$

cannot vanish if $y \neq 0$.

7.4.3 Canonical commutation relations

1. Let two Hermitian operators A and B satisfy the commutation relation $[B, A] = iI$. Show that at least one of these operators is unbounded. Without loss of generality (why?) it can be assumed that $||B|| = 1$. Hint: show that

$$[B, A^n] = inA^{n-1}$$

   and derive

$$||A^n|| \geq \frac{n}{2}||A^{n-1}||.$$

2. Assume that A possesses a normalizable eigenvector $|\varphi\rangle$

$$A|\varphi\rangle = a|\varphi\rangle, \quad a = a^*.$$

   On the one hand we have

$$\langle\varphi|(BA - AB)|\varphi\rangle = \langle\varphi|B|A\varphi\rangle - \langle A\varphi|B|\varphi\rangle$$
$$= a(\langle\varphi|B|\varphi\rangle - \langle\varphi|B|\varphi\rangle) = 0,$$

   while on the other

$$\langle\varphi|(BA - AB)|\varphi\rangle = \langle\varphi|[B, A]|\varphi\rangle = i||\varphi||^2 \neq 0.$$

   What is the solution of this pseudoparadox? Hint: examine the case where $B = x$ and $A = -\mathrm{id}/\mathrm{d}x$ on $L^2[0, 1]$ with the boundary conditions $\varphi(x = 0) = \varphi(x = 1)$.

3. Let us consider the operators A_C defined in Section 7.2.2. Find the eigenvalues and eigenvectors of A_C, and show that the spectrum of A_C varies depending on the values of C. The von Neumann theorem (Chapter 8) states that the canonical commutation relations are unique up to a unitary equivalence. However,

$$[X, A_C] = iI \quad \text{and} \quad [X, A_{C'}] = iI,$$

   and $A_C \neq A_{C'}$ if $C \neq C'$. What is the solution of this new pseudoparadox (which is not independent of the preceding one)?

7.4.4 Dilatation operators and the conformal transformation

1. Let A be the operator

$$A = -\mathrm{i}\, x\, \frac{\partial}{\partial x}.$$

Is A Hermitian? Show that

$$\left[\mathrm{e}^{-\mathrm{i}\alpha A}\,\Phi\right](x) = \Phi(\mathrm{e}^{-\alpha}\, x).$$

Method 1: use the variable $u = \ln x$.

2. Method 2: obtain the partial differential equation

$$\left(\frac{\partial}{\partial \alpha} + x\,\frac{\partial}{\partial x}\right)\left[\mathrm{e}^{-\mathrm{i}\alpha A}\,\Phi\right](x) = 0.$$

3. Let B be the operator

$$B = -\mathrm{i}\, x^2\, \frac{\partial}{\partial x}.$$

Show that

$$\left[\mathrm{e}^{-\mathrm{i}\alpha B}\,\Phi\right](x) = \Phi\left(\frac{x}{1+\alpha x}\right).$$

7.5 Further reading

Jauch [1968], Chapters 1–4, and Peres [1993], Chapter 4, contain a fairly detailed and mathematically rigorous exposition of useful notions about Hilbert spaces of infinite dimension and operators on these spaces. The reader interested in the mathematical aspects can plunge into the classic text of F. Riesz and B. Sz.-Nagy, *Functional Analysis*, New York: Ungar (1955).

8

Symmetries in quantum physics

The solution of problems in classical physics is simplified, sometimes considerably, by the presence of *symmetries*, that is, transformations that leave certain physical problems invariant. For example, in classical mechanics the problem of a particle in a time-independent central force field $\vec{F}(\vec{r}) = F(r)\hat{r}$ is invariant under time-translations and under rotations about any axis passing through the origin. Invariance under time-translations ensures the conservation of mechanical energy E, and invariance under rotations ensures the conservation of angular momentum $\vec{j}$. In the absence of symmetries, it is necessary to solve a system of three second-order differential equations (one for each component). When these symmetries are present the problem reduces to the solution of only a single first-order differential equation. Let us summarize the consequences of invariance principles in classical mechanics.

- Invariance of the potential energy under time-translations implies conservation of mechanical energy $E = K + V$, the sum of the kinetic energy K and the potential energy V.
- Invariance of the potential energy under spatial translations parallel to a vector $\hat{n}$ implies conservation of the momentum component $\vec{p} \cdot \hat{n} = \vec{p}_{\hat{n}}$.
- Invariance of the potential energy under rotations about an axis $\hat{n}$ implies conservation of the component $\vec{j} \cdot \hat{n} = \vec{j}_{\hat{n}}$ of the angular momentum.

Symmetry properties play an even more important role in quantum mechanics. They make it possible to obtain very general results which are independent of approximations made, for example, for the Hamiltonian (of course, as long as these approximations respect the symmetries of the problem). In this chapter we shall exploit the following invariance hypotheses, which we assume are valid for an isolated system.[1]

- The description of an isolated system should not depend on the origin of time; it must be invariant under translation of the time origin.
- Space is homogeneous, that is, the description of an isolated system should not depend on the origin of the axes; it must be invariant under space translations.

---

[1] These hypotheses are eminently plausible, but there may always exist subtle effects that violate one (or several) of the invariances. Before 1957, the vast majority of physicists believed that physics was invariant under the parity operation. Pauli himself vetoed plans for an experiment at CERN in Geneva designed to seek parity violation, as he found the idea of such violation so absurd. As a consequence, parity violation was discovered experimentally soon afterwards in the USA by C. S. Wu (cf. Section 8.3.3).

- Space is isotropic, that is, the description of an isolated system should not depend on the orientation chosen for the axes; it must be invariant under rotations.
- The form of the laws of physics should not change in going from one inertial reference frame to another.

This last hypothesis must be made more precise, because there exist two possible transformation laws between inertial reference frames, the Lorentz law and the Galilean law, the latter being valid when $v/c \to 0$. Naturally, the Lorentz transformation law is the more general one, but it would take us into quantum field theory. Since here we shall consider only particles with speeds much less than the speed of light, we can limit ourselves to Galilean transformations and work within the framework of what is conventionally, but improperly, called "nonrelativistic quantum mechanics."[2]

8.1 Transformation of a state in a symmetry operation

8.1.1 Invariance of probabilities in a symmetry operation

The viewpoint adopted implicitly in the introduction to this chapter is called *passive*: the physical system is not changed, but the set of axes is. It is in general equivalent to adopt the *active* point of view,[3] in which the set of axes is unchanged, but a symmetry operation is applied to the physical system. We have already used this equivalence in the discussion of Section 3.2.4: compare in Figure 3.11 the passive (a) and the active (b) points of view. In the rest of this chapter we shall adopt the active point of view, as it is perhaps more intuitive,[4] and it will be more convenient for certain discussions, for example that of Section 10.5.

We have seen in Chapter 4, postulate **I**, that the mathematical object in one-to-one correspondence with a physical state is a normalized ray in the space of states $\mathcal{H}$, that is, a normalized vector up to a phase. *In the present section only*, the distinction between vectors and rays will be crucial; afterwards, we shall forget it. It can be shown immediately that the relation between two vectors of $\mathcal{H}$

$$|\varphi'\rangle = e^{i\theta}|\varphi\rangle, \tag{8.1}$$

where θ is a real number, is an equivalence relation $|\varphi'\rangle \sim |\varphi\rangle$.[5] The equivalence class is a ray, which we denote $\tilde{\varphi}$. The scalar product of two rays $\tilde{\varphi}$ and $\tilde{\chi}$ is not defined, but the modulus of this scalar product, which we denote $|(\tilde{\chi}, \tilde{\varphi})|$, is well defined. We can choose two arbitrary representatives $|\varphi\rangle$ and $|\chi\rangle$ in the equivalence classes and write

$$|(\tilde{\chi}, \tilde{\varphi})| = |\langle \chi | \varphi \rangle|, \tag{8.2}$$

because the modulus does not depend on the phase factors. The result is independent of the choice of representatives in the equivalence classes.

---

[2] In fact, this theory is perfectly relativistic, as it satisfies Galilean relativity.

[3] For certain transformations like reflection in a plane it is simpler to use the passive point of view, which amounts to viewing the system in a mirror. One can also imagine constructing a setup symmetric to the original one with respect to a plane.

[4] At least it is for the author!

[5] In this subsection only, $\sim$ means "belongs to the same equivalence class", and not "of the order of."

Let us return to the spin 1/2 of Chapter 3. We have seen how to prepare a spin state oriented along Oz, represented by the ray $\tilde{\varphi}_+$, by using a Stern–Gerlach device with magnetic field pointing along Oz and selecting atoms which are deflected upwards (by choosing the appropriate sign of the field). Let us rotate the field by an angle α about the direction of propagation Oy to have it point in the direction $\hat{n}_\alpha$ making an angle α with Oz, $0 \leq \alpha < 2\pi$. In this way we prepare the physical spin state represented by the ray $\tilde{\varphi}_+(\hat{n}_\alpha)$, which by definition will be the state $\tilde{\varphi}_+$ transformed by rotation by α about Oy (Fig. 8.1). Using the notation of Chapter 3, the equivalence class of the vector $|+\rangle$ is the ray $\tilde{\varphi}_+$, and that of the vector $|+, \hat{n}_\alpha\rangle$ is the ray $\tilde{\varphi}_+(\hat{n}_\alpha)$. In general, the state $\tilde{\varphi}_{\mathcal{R}}$ obtained by a rotation $\mathcal{R}$ of the state $\tilde{\varphi}$ will be obtained by a rotation $\mathcal{R}$ of the apparatus that prepares the state $\tilde{\varphi}$.

Now let us suppose that after the first Stern–Gerlach apparatus (the polarizer), in which the field is parallel to Oz, we place a second device (the analyzer) with field parallel to the direction $\hat{n}_\beta$ obtained from Oz by rotation by an angle β about Oy (Fig. 8.2a). If along the trajectory there is no magnetic field that can rotate the spin, the probability

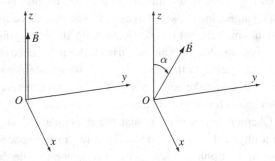

Fig. 8.1. Preparation of the physical states (rays) $\tilde{\varphi}_+$ and $\tilde{\varphi}_+(\hat{n}_\alpha)$.

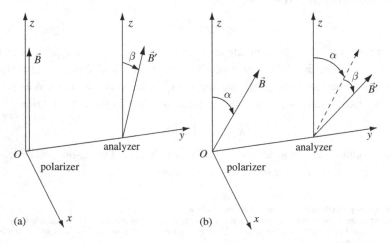

Fig. 8.2. Simultaneous rotations of the polarizer and the analyzer by an angle α.

for the spin to be deflected in the direction $\hat{n}_\beta$ is

$$|(\tilde{\varphi}_+(\hat{n}_\beta), \tilde{\varphi}_+)|^2.$$

Let us now perform the experiment after rotating the polarizer and the analyzer *at the same time* by an angle α (Fig. 8.2b). The probability of deflection in the direction $\hat{n}_{\alpha+\beta}$ is

$$|(\tilde{\varphi}_+(\hat{n}_{\alpha+\beta}), \tilde{\varphi}_+(\hat{n}_\alpha))|^2.$$

Since both the polarizer and the analyzer have undergone the same rotation, rotational invariance implies that the probabilities are unchanged:

$$|(\tilde{\varphi}_+(\hat{n}_{\alpha+\beta}), \tilde{\varphi}_+(\hat{n}_\alpha))|^2 = |(\tilde{\varphi}_+(\hat{n}_\beta), \tilde{\varphi}_+)|^2. \tag{8.3}$$

Let us generalize (8.3). If we make a transformation g on a state $\tilde{\varphi}$ by applying this transformation to the apparatus that prepares $\tilde{\varphi}$ to obtain the transformed state $\tilde{\varphi}_g$, $\tilde{\varphi} \to \tilde{\varphi}_g$, and if we perform the same operation on the measurement device for $\tilde{\chi}$, $\tilde{\chi} \to \tilde{\chi}_g$, then the probabilities must be unchanged if the physics is invariant under this operation:

$$|(\tilde{\chi}_g, \tilde{\varphi}_g)|^2 = |(\tilde{\chi}, \tilde{\varphi})|^2. \tag{8.4}$$

8.1.2 The Wigner theorem

The property (8.4) of *rays* is translated into a property of *vectors* owing to a very important theorem due to Wigner.

The Wigner theorem. If a transformation g on physical states is mathematically translated into a transformation law for the corresponding rays, $\tilde{\varphi} \to \tilde{\varphi}_g$, and if we assume that the probabilities are invariant under this transformation,

$$|(\tilde{\chi}_g, \tilde{\varphi}_g)|^2 = |(\tilde{\chi}, \tilde{\varphi})|^2 \quad \forall \tilde{\varphi}, \tilde{\chi},$$

it is then possible to choose a representative $|\varphi_g\rangle$ of $\tilde{\varphi}_g$ such that for any vector $|\varphi\rangle \in \mathcal{H}$

$$|\varphi_g\rangle = U(g)|\varphi\rangle, \tag{8.5}$$

where the operator $U(g)$ is unitary or antiunitary and is unique up to a phase.

The transformation law of rays thus becomes a transformation law of vectors by the application of an operator that depends only on the transformation g. If $U(g)$ is unitary, the Wigner theorem implies not only invariance of the norm of the scalar product, but also invariance of its phase, since

$$\langle U(g)\chi | U(g)\varphi \rangle = \langle \chi | \varphi \rangle.$$

Antiunitary operators transform the scalar product into its complex conjugate:

$$\langle U(g)\chi | U(g)\varphi \rangle = \langle \chi | \varphi \rangle^* = \langle \varphi | \chi \rangle. \tag{8.6}$$

The proof of the Wigner theorem involves only elementary concepts, but it is quite delicate, and we shall leave it to Appendix A. Antiunitary operators come in only when the transformation g includes time reversal; we shall say a bit more about this in Section 8.3.3, but we leave the detailed study to Appendix A. For the time being we limit ourselves to unitary transformations.

The Wigner theorem has particularly interesting consequences if the transformations g form a group $\mathcal{G}$. The product $g = g_2 g_1$ of two transformations, as well as the inverse transformation g^{-1}, is then a transformation of $\mathcal{G}$. The order of the transformations in $g_2 g_1$ is important because the group $\mathcal{G}$ is not in general Abelian: $g_2 g_1 \neq g_1 g_2$. If $g = g_2 g_1$, the rays $\tilde{\varphi}_g$ and $\tilde{\varphi}_{g_2 g_1}$ must be identical. For example, if $\mathcal{G}$ is the group of rotations about Oz, and if $\mathcal{R}_z(\theta)$ represents a rotation by angle θ about Oz, then we have

$$\mathcal{R}_z(\theta = \theta_2 + \theta_1) = \mathcal{R}_z(\theta_2)\mathcal{R}_z(\theta_1). \tag{8.7}$$

The physical state obtained by rotation by an angle $\theta = \theta_2 + \theta_1$ must be identical to that obtained by rotation by an angle θ_1 followed by rotation by an angle θ_2.

Let us now use the Wigner theorem to choose the phases of the vectors such that the correspondence between $|\varphi\rangle$ and $|\varphi_g\rangle$ will be given by (8.5). If $g = g_2 g_1$, on the one hand we have

$$|\varphi_g\rangle = U(g)|\varphi\rangle, \tag{8.8}$$

while on the other

$$|\varphi_{g_2 g_1}\rangle = U(g_2)|\varphi_{g_1}\rangle = U(g_2)U(g_1)|\varphi\rangle. \tag{8.9}$$

The vectors $|\varphi_g\rangle$ and $|\varphi_{g_2 g_1}\rangle$ represent identical physical states, and they must be equal up to a phase:

$$|\varphi_g\rangle = e^{i\alpha(g_2, g_1)}|\varphi_{g_2 g_1}\rangle. \tag{8.10}$$

The phase factor in (8.10) could a priori depend on $|\varphi\rangle$, but in fact it depends only on g_1 and g_2. This is easily seen by writing

$$|\varphi_g\rangle = e^{i\alpha}|\varphi_{g_2 g_1}\rangle, \quad |\chi_g\rangle = e^{i\beta}|\chi_{g_2 g_1}\rangle,$$

and by examining the scalar product $\langle\chi|\varphi\rangle$:

$$\langle\chi|\varphi\rangle = \langle\chi_g|\varphi_g\rangle = e^{i(\alpha-\beta)}\langle\chi_{g_2 g_1}|\varphi_{g_2 g_1}\rangle$$
$$= e^{i(\alpha-\beta)}\langle U(g_2)U(g_1)\chi|U(g_2)U(g_1)\varphi\rangle$$
$$= e^{i(\alpha-\beta)}\langle\chi|\varphi\rangle,$$

which implies that $\alpha = \beta$. Since the vector $|\varphi\rangle$ is arbitrary, (8.10) implies the corresponding relation for the operators $U(g)$:

$$U(g) = e^{i\alpha(g_2, g_1)}\, U(g_2)U(g_1). \tag{8.11}$$

This equation expresses a mathematical property: the operators $U(g)$ form a *projective representation* of the group $\mathcal{G}$. In the rest of this book we shall consider only two simple versions of (8.11). In one the phase factor is +1, and this corresponds to a *vector representation* of $\mathcal{G}$:

$$U(g) = U(g_2)U(g_1). \tag{8.12}$$

In the other the phase factor is ± 1:

$$U(g) = \pm U(g_2)U(g_1). \tag{8.13}$$

We shall see this factor $\pm$ arises in the case where $\mathcal{G}$ is the rotation group; the representations (8.13) of this group are called *spinor representations* of the rotation group.

8.2 Infinitesimal generators

8.2.1 Definitions

Two types of transformation group can be distinguished.

- Discrete groups, in which the number of elements is finite or denumerably infinite. Some simple examples are parity, the operation that changes the sign of the coordinates $\vec{r} \rightarrow -\vec{r}$ (cf. Section 8.3.3), and the crystallographic groups that play an important role in solid-state physics.
- Continuous groups, in which the elements are parametrized by one or more parameters that vary continuously.[6] For example, the rotation $\mathcal{R}_z(\theta)$ about Oz is parametrized by an angle θ which varies continuously between 0 and 2π.

The interesting continuous groups in physics are the Lie groups (Exercise 8.5.4), of which an example is the group of spatial rotations, or the $SO(3)$ group of orthogonal matrices $\mathcal{R}^T\mathcal{R} = \mathcal{R}\mathcal{R}^T = I$ of determinant +1 in three-dimensional space.[7] Here A^T stands for the transpose of the operator A. This group, which is a three-parameter group, will play a major role in the rest of the book. For example, a rotation can be parametrized by the two angles giving the direction $\hat{n}$ of the rotation axis in a reference frame $Oxyz$ plus the rotation angle, where all three angles can vary continuously. The rotation group possesses an infinite number of Abelian subgroups, rotations about a fixed axis. We shall show that it is sufficient to consider the three Abelian subgroups corresponding to rotations about Ox, Oy, and Oz; the number of these subgroups is equal to the number of independent parameters. Rotations belonging to these subgroups are parametrized by an angle θ, and according to (8.7) this parameter is additive: the product of two rotations by angles θ_1 and θ_2 is a rotation by an angle $\theta = \theta_1 + \theta_2$. In general, if a Lie group $\mathcal{G}$ is parametrized by n independent parameters, it is said that the *dimension of the group* is n, and we are

---

[6] It should be noted that in the case of a continuous group, the transformations $U(g)$ are necessarily unitary by continuity if any group element can be related in a continuous fashion to the neutral element e of the group, in other words, if the group is connected: $U(e) = I$ is unitary.

[7] The relation $\mathcal{R}^T\mathcal{R} = I$ implies that $\det \mathcal{R} = \pm 1$. When writing $SO(3)$ for the rotation group, S indicates that we must choose $\det \mathcal{R} = +1$, O means that the group is orthogonal, and the 3 denotes the spatial dimension. If inversion of the axes, or parity, is added to the rotations, we obtain the $O(3)$ group, which includes also matrices of determinant -1. The group $SO(3)$ is connected, but $O(3)$ is not: it is not possible to pass continuously from $\det \mathcal{R} = +1$ to $\det \mathcal{R} = -1$.

led to the study of n Abelian subgroups (Exercise 8.5.4). Let us take an Abelian subgroup of $\mathcal{G}$ whose elements h are parametrized by an additive parameter α:

$$h(\alpha_1 + \alpha_2) = h(\alpha_2)h(\alpha_1). \tag{8.14}$$

According to (8.12), the operators $U_h(\alpha)$ which transform the state vectors of $\mathcal{H}$ must satisfy

$$U_h(\alpha_1 + \alpha_2) = U_h(\alpha_2)U_h(\alpha_1). \tag{8.15}$$

The Stone theorem (Section 7.3.2) implies the existence of a Hermitian operator $T_h = T_h^\dagger$ such that

$$U_h(\alpha) = e^{-i\alpha T_h}. \tag{8.16}$$

The operator T_h is called the *infinitesimal generator* of the transformation in question. Since T_h is Hermitian, it is a good candidate for a physical property, and in fact all the transformations listed in the introduction to this chapter correspond to fundamental physical properties. The following correspondence can be established between the infinitesimal generators for these various transformations and physical properties, and we shall discuss all these in more detail later on in this chapter.

- Time translations by t: $U(t) = \exp(-itH/\hbar)$. The operator $T_h = H$ is the Hamiltonian; see Chapter 4.
- Space translations by $\vec{a} = a\hat{a}$: $U(\vec{a}) = \exp(-ia(\vec{P} \cdot \hat{a})/\hbar)$. The operator $T_h = \vec{P} \cdot \hat{a}$ is the component of the momentum $\vec{P}$ along $\hat{a}$.
- Rotations by θ about an axis $\hat{n}$: $U_{\hat{n}}(\theta) = \exp(-i\theta(\vec{J} \cdot \hat{n})/\hbar)$. The operator $T_h = \vec{J} \cdot \hat{n}$ is the component of the angular momentum $\vec{J}$ along $\hat{n}$.
- Galilean transformations of the velocity $\vec{v}$: $U(\vec{v}) = \exp(-i(\vec{v} \cdot \vec{G})/\hbar)$. The operator $\vec{G} = -m\vec{R}$, where $\vec{R}$ is the position and m is the mass.

In each case the presence of $\hbar$ in the exponential ensures that the exponent is dimensionless. If we choose precisely $\hbar$ and not $\hbar$ times a number, the preceding expressions *define* the operators representing the physical properties of energy, momentum, angular momentum, and position. In fact, these expressions give the most general definition of these operators.

8.2.2 Conservation laws

We are going to show that in quantum physics the conservation laws for the expectation values of physical properties correspond to the conservation laws of classical physics in the presence of a symmetry. Let us first generalize (4.26) to the case where the operator A depends explicitly on time. To the right-hand side of (4.26) we must add

$$\langle \varphi(t) \left| \frac{\partial A}{\partial t} \right| \varphi(t) \rangle = \left\langle \frac{\partial A}{\partial t} \right\rangle_\varphi,$$

and this equation gives the general form of the *Ehrenfest theorem*:

$$\frac{\mathrm{d}}{\mathrm{d}t}\langle A\rangle_\varphi(t) = \frac{\mathrm{i}}{\hbar}\langle[H,A]\rangle_\varphi + \left\langle\frac{\partial A}{\partial t}\right\rangle_\varphi . \tag{8.17}$$

When the operator A is time-independent, $(\partial A/\partial t) = 0$, we recover (4.26):

$$\frac{\mathrm{d}}{\mathrm{d}t}\langle A\rangle_\varphi(t) = \frac{\mathrm{i}}{\hbar}\langle[H,A]\rangle_\varphi . \tag{8.18}$$

Since this equation is valid for any $|\varphi\rangle$, we obtain the following theorem (assuming that H is independent of time).

Theorem of conservation of the expectation value. When a physical property A is independent of time, the condition $\mathrm{d}\langle A\rangle/\mathrm{d}t = 0$ implies that $[H,A] = 0$ and the reverse:

$$\text{If } \frac{\partial A}{\partial t} = 0, \quad \frac{\mathrm{d}}{\mathrm{d}t}\langle A\rangle_\varphi = 0 \Longleftrightarrow [H,A] = 0. \tag{8.19}$$

As an application, let us assume that the properties of a physical system are invariant under spatial translations. This will be the case, for example, for an isolated system of two particles whose potential energy depends only on the difference of their positions $(\vec{r}_1 - \vec{r}_2)$. The expectation value of the Hamiltonian must be the same for the state $|\varphi\rangle$ and the state $|\varphi_{\vec{a}}\rangle = \exp[-\mathrm{i}(\vec{P}\cdot\vec{a})/\hbar]|\varphi\rangle$ obtained by translation by $\vec{a}$, where $\vec{a}$ is an arbitrary vector:

$$\langle\varphi_{\vec{a}}|H|\varphi_{\vec{a}}\rangle = \langle\varphi|\exp\left(\mathrm{i}\frac{\vec{P}\cdot\vec{a}}{\hbar}\right) H \exp\left(-\mathrm{i}\frac{\vec{P}\cdot\vec{a}}{\hbar}\right)|\varphi\rangle = \langle\varphi|H|\varphi\rangle .$$

Allowing $\vec{a}$ to tend to zero, we deduce that

$$\boxed{\text{Invariance under spatial translation} \Longleftrightarrow [H,\vec{P}] = 0} . \tag{8.20}$$

The notation $[H,\vec{P}] = 0$ indicates that the three components of the momentum commute with H. According to (8.18), this equation implies that the expectation value $\langle\vec{P}\rangle$ of $\vec{P}$ is independent of time: invariance under translation implies conservation of momentum (more precisely, its expectation value). The same reasoning shows that

$$\boxed{\text{Invariance under rotation} \Longleftrightarrow [H,\vec{J}] = 0} . \tag{8.21}$$

The expectation value $\langle\vec{J}\rangle$ of $\vec{J}$ is independent of time: invariance under rotation implies the conservation of angular momentum (more precisely, its expectation value).

It is also useful to note the following.

- If $[H,A] = 0$, A and H can be diagonalized simultaneously and, in particular, it is possible to find the stationary states among the eigenvectors of A.
- The condition $[H,A] = 0$ implies that A commutes with the evolution operator $U(t - t_0)$ (4.20). If $|\varphi(t_0)\rangle$ is an eigenvector of A at time t_0,

$$A|\varphi(t_0)\rangle = a|\varphi(t_0)\rangle ,$$

then $|\varphi(t)\rangle$ is an eigenvector of A with the same eigenvalue:

$$A|\varphi(t)\rangle = AU(t-t_0)|\varphi(t_0)\rangle = U(t-t_0)A|\varphi(t_0)\rangle = a|\varphi(t)\rangle.$$

The eigenvalue a is conserved; it is a constant of the motion. We could have obtained this result directly from (8.19), because in this case $\langle A \rangle = a$.

8.2.3 *Commutation relations of infinitesimal generators*

Most of the properties of a Lie group can be determined by examining the neighborhood of the identity; more precisely, by studying the commutation relations of the infinitesimal generators. The set of these commutation relations constitutes the *Lie algebra* of the group (Exercise 8.5.4). However, two Lie groups that are isomorphic in the neighborhood of the identity may differ in their global topological properties; we shall soon give an example of this. Let us examine in more detail the case of the rotation group.[8] The operator $\mathcal{R}_{\hat{n}}(\theta)$ which rotates by an angle θ about the axis $\hat{n}$ is an orthogonal operator of three-dimensional space: $\mathcal{R}^T \mathcal{R} = \mathcal{R} \mathcal{R}^T = I$. The rotations $\mathcal{R}_{\hat{n}}(\theta)$ form an Abelian subgroup of the rotation group, and according to the Stone theorem we can always write

$$\mathcal{R}_{\hat{n}}(\theta) = \exp\left(-i\theta(\vec{T}\cdot\hat{n})\right), \tag{8.22}$$

where $\vec{T}\cdot\hat{n}$ is a Hermitian operator. Since $\mathcal{R}$ is orthogonal and real, it is also unitary. In this notation a vector $\vec{V}$ is transformed into $\vec{V}'$ (Fig. 8.3):

$$\vec{V}' = (1-\cos\theta)(\hat{n}\cdot\vec{V})\hat{n} + \cos\theta\,\vec{V} + \sin\theta(\hat{n}\times\vec{V}). \tag{8.23}$$

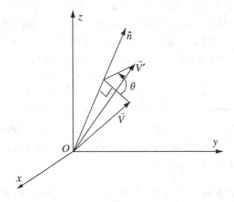

Fig. 8.3. Rotation of a vector $\vec{V}$ by an angle θ about the axis $\hat{n}$.

---

[8] Unless explicitly stated otherwise, we are always dealing with the $SO(3)$ group of rotations in three-dimensional Euclidean space.

This transformation law can be written in matrix form as

$$V_i' = \sum_{j=1}^{3} [\mathcal{R}_{\hat{n}}(\theta)]_{ij} V_j. \tag{8.24}$$

The explicit determination of the matrix $[\mathcal{R}_{\hat{n}}(\theta)]_{ij}$ is proposed in Exercise 8.5.1. We shall not need it, because we are going to take the limit $\theta \to 0$, that is, the limit of infinitesimal rotations:

$$\vec{V}' = \vec{V} + \theta(\hat{n} \times \vec{V}) + O(\theta)^2. \tag{8.25}$$

Expansion of the exponential in (8.22) and comparison with (8.25) gives

$$(\vec{T} \cdot \hat{n})\vec{V} = i \begin{pmatrix} 0 & -n_z & n_y \\ n_z & 0 & -n_x \\ -n_y & n_x & 0 \end{pmatrix} \begin{pmatrix} V_x \\ V_y \\ V_z \end{pmatrix},$$

and by identification the Hermitian operators T_x, T_y, and T_z:

$$T_x = \begin{pmatrix} 0 & 0 & 0 \\ 0 & 0 & -i \\ 0 & i & 0 \end{pmatrix}, \quad T_y = \begin{pmatrix} 0 & 0 & i \\ 0 & 0 & 0 \\ -i & 0 & 0 \end{pmatrix}, \quad T_z = \begin{pmatrix} 0 & -i & 0 \\ i & 0 & 0 \\ 0 & 0 & 0 \end{pmatrix}. \tag{8.26}$$

When θ is finite, the exponential in (8.22) can easily be calculated by noting that $(\vec{T} \cdot \hat{n})^3 = \vec{T} \cdot \hat{n}$ (Exercise 8.5.1) and we recover (8.23). Direct calculation (Exercise 8.5.1) gives the following commutation relations,[9] which form the Lie algebra of $SO(3)$:

$$[T_x, T_y] = i T_z, \quad [T_y, T_z] = i T_x, \quad [T_z, T_x] = i T_y, \tag{8.27}$$

or, using the notation of (3.52),

$$\boxed{[T_i, T_j] = i \sum_k \varepsilon_{ijk} T_k}. \tag{8.28}$$

Now let us give a quicker and more instructive demonstration of (8.27) using the following expression for a rotation by an angle θ about an axis $\hat{n}(\phi)$ in the yOz plane, obtained starting from the Oy axis by rotating by an angle ϕ about Ox (Fig. 8.4):

$$\mathcal{R}_{\hat{n}(\phi)}(\theta) = \mathcal{R}_x(\phi)\mathcal{R}_y(\theta)\mathcal{R}_x(-\phi). \tag{8.29}$$

The rotation $\mathcal{R}_x(-\phi)$ first takes the axis $\hat{n}(\phi)$ onto Oy. We then rotate by an angle θ about Oy and finally return to the initial position of the axis by the rotation $\mathcal{R}_x(\phi)$. Let us express $\mathcal{R}_{\hat{n}(\phi)}(\theta)$ and $\mathcal{R}_y(\theta)$ in exponential form (8.22) and expand to first order in θ:

$$\vec{T} \cdot \hat{n}(\phi) = \cos\phi\, T_y + \sin\phi\, T_z = e^{-i\phi T_x} T_y e^{i\phi T_x}.$$

Then expanding to first order in ϕ we find

$$[T_x, T_y] = i T_z,$$

---

[9] In fact, it is sufficient to prove only the first, and the other two follow by circular permutation.

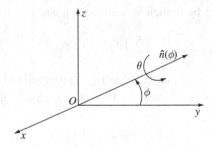

Fig. 8.4. The rotation $\mathcal{R}_{\hat{n}(\phi)}(\theta)$.

and the two other commutation relations (8.27) follow by circular permutation.

Now let us consider operators that perform rotations on physical states in $\mathcal{H}$. We have seen that the operator which performs a rotation by an angle θ about an axis $\hat{n}$ is

$$U_{\hat{n}}(\theta) = \exp\left(-i\theta\,\frac{\vec{J}\cdot\hat{n}}{\hbar}\right). \tag{8.30}$$

Since these operators form a representation of the rotation group, from (8.12) and (8.29) we deduce that

$$U\left[\mathcal{R}_{\hat{n}(\phi)}(\theta)\right] = U\left[\mathcal{R}_x(\phi)\right]U\left[\mathcal{R}_y(\theta)\right]U\left[\mathcal{R}_x(-\phi)\right].$$

Again expanding the exponentials to first order in θ and then in ϕ, we obtain the angular momentum commutation relations:

$$[J_x, J_y] = i\hbar J_z, \quad [J_y, J_z] = i\hbar J_x, \quad [J_z, J_x] = i\hbar J_y, \tag{8.31}$$

where

$$[J_i, J_j] = i\hbar \sum_k \varepsilon_{ijk} J_k. \tag{8.32}$$

The commutation relations of the J_i are, up to a factor of $\hbar$, identical to those of the T_i. The infinitesimal generators of rotations in $\mathcal{H}$ have the same commutation relations as the infinitesimal generators of the rotation group in ordinary space. Our demonstration of the relations (8.31) or (8.32) emphasizes their geometrical origin.

The commutation relations of scalar and vector operators with $\vec{J}$ are of great practical importance. A *scalar operator* $\mathcal{S}$ is an operator whose expectation value is invariant under rotation. If $U(\mathcal{R})$ is the operator performing a rotation $\mathcal{R}$ in the space of states

$$|\varphi_{\mathcal{R}}\rangle = U(\mathcal{R})|\varphi\rangle,$$

we must have

$$\langle\varphi_{\mathcal{R}}|\mathcal{S}|\varphi_{\mathcal{R}}\rangle = \langle\varphi|U^{\dagger}(\mathcal{R})\mathcal{S}U(\mathcal{R})|\varphi\rangle = \langle\varphi|\mathcal{S}|\varphi\rangle,$$

and therefore for a rotation $R_{\hat{n}}(\theta)$,

$$\exp\left(i\theta\frac{\vec{J}\cdot\hat{n}}{\hbar}\right)\mathcal{S}\exp\left(-i\theta\frac{\vec{J}\cdot\hat{n}}{\hbar}\right)=\mathcal{S}.$$

Taking θ to be infinitesimal, we can state that $\mathcal{S}$ commutes with $\vec{J}$:

$$\boxed{[\vec{J},\mathcal{S}]=0}\ . \tag{8.33}$$

A scalar operator commutes with the angular momentum.

By similar reasoning we can determine the commutation relations for $\vec{J}$ with $\vec{R}$ or $\vec{P}$, and more generally with any *vector operator*. By definition, a vector operator $\vec{V}$ is an operator whose expectation value transforms under rotation according to the law (8.24). We must then have

$$\langle\varphi_{\mathcal{R}}|V_i|\varphi_{\mathcal{R}}\rangle=\langle\varphi|U^{\dagger}(\mathcal{R})V_iU(\mathcal{R})|\varphi\rangle=\sum_{j=1}^{3}\mathcal{R}_{ij}\langle\varphi|V_j|\varphi\rangle,$$

and consequently for a rotation $\mathcal{R}_{\hat{n}}(\theta)$,

$$\exp\left(i\theta\frac{\vec{J}\cdot\hat{n}}{\hbar}\right)V_i\exp\left(-i\theta\frac{\vec{J}\cdot\hat{n}}{\hbar}\right)=\sum_{j=1}^{3}[\mathcal{R}_{\hat{n}}(\theta)]_{ij}V_j. \tag{8.34}$$

Let us take $\hat{n}=\hat{x}$ and θ to be infinitesimal. According to (8.25), $\vec{V}'$ has the components

$$(V_x,\ V_y-\theta V_z,\ V_z+\theta V_y),$$

and then we have, for example, for the component $i=y$ of (8.34),

$$\left(I+\frac{i}{\hbar}\theta J_x\right)V_y\left(I-\frac{i}{\hbar}\theta J_x\right)=V_y-\theta V_z$$

whence $i[J_x,V_y]=-\hbar V_z$. Examining the other components, we find

$$[J_x,V_x]=0,\quad[J_x,V_y]=i\hbar V_z,\quad[J_x,V_z]=-i\hbar V_y,$$

or in the general form

$$\boxed{[J_i,V_j]=i\hbar\sum_k\varepsilon_{ijk}V_k}\ . \tag{8.35}$$

These relations are valid, in particular, for the position operator $\vec{R}$ and the momentum operator $\vec{P}$, which are vector operators.

The attentive reader will have noticed that the commutation relations (3.53) for spin 1/2, $\vec{S}=\frac{1}{2}\hbar\vec{\sigma}$, are identical to (8.31), and spin 1/2 is therefore an angular momentum. Let us give some other evidence for this identification without entering into the mathematical details which would take us too far afield. The Lie algebra (3.52) of the Pauli matrices is that of the $SU(2)$ group of 2×2 unitary matrices of determinant $+1$ (Exercise 8.5.2). The Lie algebras of $SU(2)$ and $SO(3)$ are identical; the two groups coincide in the neighborhood of the identity. However, the two groups are not *globally* identical. This can be seen by considering a rotation of 2π about an axis $\hat{n}$. Using (3.67)

$$\exp\left(-i\frac{\theta}{2}\vec{\sigma}\cdot\hat{n}\right) = \cos\frac{\theta}{2}I - i\vec{\sigma}\cdot\hat{n}\sin\frac{\theta}{2}$$

we see that

$$\exp\left(-i\frac{\theta}{2}\vec{\sigma}\cdot\hat{n}\right) = -I \quad \text{for} \quad \theta = 2\pi.$$

The identity is recovered only for $\theta = 4\pi$! The identity rotation of $SO(3)$ therefore corresponds to *two* elements of $SU(2)$, $+I$ and $-I$. The correspondence between $SU(2)$ and $SO(3)$ is a homomorphism such that two elements of $SU(2)$ correspond to one element of $SO(3)$, and so for spin 1/2 we have a projective representation (8.13) of the rotation group. This property results from the fact that the $SO(3)$ group is connected, but not simply connected.[10] A continuous curve drawn in the parameter space of the group cannot always be continuously deformed to a point. This property is seen in rotations in ordinary space[11] and is not peculiar to quantum mechanics, as there is sometimes a tendency to suggest.[12] The real identity rotation of an object *in relation to its environment* is not a rotation by 2π but a rotation by 4π.

8.3 Canonical commutation relations

8.3.1 Dimension $d = 1$

Let us first place ourselves in one dimension, on the x axis, and let X be the position operator. We consider a particle in a state $|\varphi\rangle$ such that the particle is localized in the neighborhood of an average position x_0 with dispersion Δx:

$$\langle\varphi|X|\varphi\rangle = \langle X\rangle = x_0, \quad \langle\varphi|(X - x_0)^2|\varphi\rangle = (\Delta x)^2. \tag{8.36}$$

The particle is localized, for example, in the interval $[x_0 - \Delta x, x_0 + \Delta x]$ (Fig. 8.5). If we

---

[10] A disk in a plane is simply connected. If a hole is made in the disk, the resulting region of the plane is no longer simply connected, because a curve encircling the hole can no longer be shrunk to a point.

[11] Cf. Lévy-Leblond and Balibar [1990], Chapter 3.D; the argument is due to Dirac.

[12] A word about the conditions under which projective representations are necessary. Two cases can arise. (i) As for the correspondence between $SU(2)$ and $SO(3)$, a projective representation may become necessary owing to global topological properties. The phase factor in (8.11) then takes discrete values, as in (8.13). (ii) If

$$[T_i, T_j] = i\sum_k C_{ijk}T_k$$

is the algebra of the Lie group (of which (8.28) for $SO(3)$ is an example; see also Exercise 8.5.4), it can happen that it is possible to construct another Lie algebra with right-hand side differing by a multiple of the identity:

$$[T_i', T_j'] = i\sum_k C_{ijk}T_k' + iD_{ij}I, \quad D_{ij} = -D_{ji}.$$

This extra term is called a central extension of the initial Lie algebra. If the term $D_{ij}I$ can be eliminated by a redefinition of the infinitesimal generators T_i', then only vector representations exist (with perhaps discrete phase factors due to the global topological properties, as in (i)). In the contrary case, for example, that of the Galilean group (Exercise 8.5.7), there exist projective representations in which the phase factor varies continuously; see, for example, Weinberg [1995], Chapter 2.

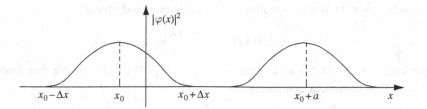

Fig. 8.5. A particle localized in the neighborhood of $x = x_0$ and translated by a.

apply to this state a translation a,

$$|\varphi\rangle \rightarrow |\varphi_a\rangle = \exp\left(-i\frac{Pa}{\hbar}\right)|\varphi\rangle = U(a)|\varphi\rangle,$$

where P is the momentum operator and $U(a)$ is the translation operator,

$$U(a) = \exp\left(-i\frac{Pa}{\hbar}\right), \quad U^{-1}(a) = U^\dagger(a) = \exp\left(i\frac{Pa}{\hbar}\right), \tag{8.37}$$

then after translation the particle will be localized in the interval $[x_0 + a - \Delta x, x_0 + a + \Delta x]$:

$$\langle X\rangle_a = \langle\varphi_a|X|\varphi_a\rangle = \langle\varphi|U^{-1}(a)XU(a)|\varphi\rangle = x_0 + a = \langle X\rangle + a.$$

Since the state $|\varphi\rangle$ is arbitrary, equality of the expectation values implies that of the operators:

$$U^{-1}(a)XU(a) = X + aI, \tag{8.38}$$

and if we allow a to tend to zero we obtain the *canonical commutation relation* between X and P:

$$\boxed{[X, P] = i\hbar I}. \tag{8.39}$$

As an application, let us calculate the commutator of P and some function $f(X)$. We expand $f(X)$ in a Taylor series:

$$f(X) = c_0 + c_1 X^2 + \cdots + c_n X^n + \cdots.$$

According to (8.38),

$$U^{-1}(a)X^2 U(a) = U^{-1}(a)XU(a)U^{-1}(a)XU(a) = (X + aI)^2,$$

and this generalizes immediately to X^n:

$$U^{-1}(a)X^n U(a) = (X + aI)^n.$$

We then obtain

$$\boxed{U^{-1}(a)f(X)U(a) = f(X + aI)}. \tag{8.40}$$

Using a well-proven technique we allow a to tend to zero and obtain

$$[P, f(X)] = -i\hbar \frac{\partial f(X)}{\partial X}. \tag{8.41}$$

As a particular case of (8.40) we can choose $f(X) = \exp(i\beta X)$, β real. We then find the Weyl form of the canonical commutation relations:

$$\exp\left(i\frac{Pa}{\hbar}\right)\exp(i\beta X)\exp\left(-i\frac{Pa}{\hbar}\right) = \exp(i\beta X)\exp(i\beta a). \tag{8.42}$$

The Weyl form is more interesting mathematically than (8.39), because the unitary operators involved in (8.42) are bounded (Section 7.2.1), in contrast to the operators X and P.

From (8.39) we immediately derive the Heisenberg inequality relating the dispersions in position and momentum. According to (4.10),

$$\boxed{\Delta x \, \Delta p = \sqrt{\langle (X-x)^2 \rangle}\sqrt{\langle (P-p)^2 \rangle} \geq \frac{1}{2}\hbar}. \tag{8.43}$$

8.3.2 Explicit realization and von Neumann's theorem

An explicit realization or *representation* of the canonical commutation relations (8.39) can be given in the space $L^{(2)}(\mathbb{R})$ of differentiable functions $\varphi(x)$ which are square-integrable on the real line in the range $[-\infty, +\infty]$. This representation is

$$\boxed{(X\varphi)(x) = x\varphi(x), \quad (P\varphi)(x) = -i\hbar\frac{\partial \varphi}{\partial x}}. \tag{8.44}$$

In these equations $(X\varphi)$ and $(P\varphi)$ stand for functions, for example, $(X\varphi)(x) = g(x)$ and $(P\varphi)(x) = h(x)$. Let us verify (8.44):

$$([XP - PX]\varphi)(x) = -i\hbar x\frac{\partial \varphi}{\partial x} + i\hbar\frac{\partial}{\partial x}(x\varphi)(x) = i\hbar\varphi(x)$$

or

$$([X, P]\varphi)(x) = i\hbar\varphi(x).$$

It is legitimate to ask whether or not the representation (8.44) for the canonical commutation relations is unique: is (8.44) a unique solution of (8.39)? Obviously, two representations should not be considered distinct if they are related by a unitary transformation, which is just a simple change of orthonormal basis in $\mathcal{H}$. Let U be a unitary operator. The operators P' and X' obtained by a unitary transformation

$$P' = U^\dagger PU, \quad X' = U^\dagger XU$$

also obey the canonical commutation relations

$$[X', P'] = U^\dagger XUU^\dagger PU - U^\dagger PUU^\dagger XU = U^\dagger[X, P]U = i\hbar I.$$

The representation (X', P') of the canonical commutation relations is said to be unitarily equivalent to the representation (X, P). The importance of (8.44) comes from the following theorem, which we state without proof.

The unitary equivalence theorem of von Neumann. All representations of the canonical commutation relations of the Weyl form[13] (8.42) are unitarily equivalent to the representation (8.44) on $L^{(2)}(\mathbb{R})$. Moreover, this representation is irreducible, that is, any operator on $\mathcal{H}$ can be written as a function of X and P. Any operator that commutes with X (P) is a function of X (P). Any operator that commutes with X and P is a multiple of the identity I.

This theorem implies that we do not have to worry about the choice of representation in (8.39), because any two choices are related to each other by a unitary transformation.

In three dimensions the position and momentum operators $\vec{R}$ and $\vec{P}$ are vector operators with components X, Y, Z and P_x, P_y, P_z, which we denote collectively as X_i and P_i, $i = x, y, z$. The different components of $\vec{R}$ and $\vec{P}$ commute, and only identical components have nonzero commutation relations:

$$\boxed{[X_i, P_j] = i\hbar \, \delta_{ij} I} \, . \tag{8.45}$$

8.3.3 The parity operator

The parity operator reverses the sign of the coordinates: $\vec{x} \to -\vec{x}$. It can also be viewed as a combination of reflection with respect to a plane followed by a rotation by π about an axis perpendicular to this plane. Let us take for example the xOy plane and call M the reflection with respect to this plane and $\mathcal{R}_z(\pi)$ the rotation about Oz:

$$\begin{pmatrix} x \\ y \\ z \end{pmatrix} \xrightarrow{M} \begin{pmatrix} x \\ y \\ -z \end{pmatrix} \xrightarrow{\mathcal{R}_z(\pi)} \begin{pmatrix} -x \\ -y \\ -z \end{pmatrix} . \tag{8.46}$$

Since rotational invariance is valid in general, parity invariance can be imagined as follows: the mirror image of a physics experiment must appear as being physically possible. The action of the parity operator on true vectors, or *polar vectors*, such as the position $\vec{r}$, momentum $\vec{p}$, or electric field $\vec{E}$,

$$\vec{r} \to -\vec{r}, \quad \vec{p} \to -\vec{p}, \quad \vec{E} \to -\vec{E}, \tag{8.47}$$

is different from that on *pseudovectors*, or *axial vectors*, such as the angular momentum $\vec{j}$ or the magnetic field $\vec{B}$, which are associated with a rotational sense rather than a direction:

$$\vec{j} \to \vec{j}, \quad \vec{B} \to \vec{B}. \tag{8.48}$$

---

[13] This precision is important, because otherwise the operators A_C of Section 7.2.2 would permit the construction of a counterexample to the theorem.

We recall that the vector product of two polar vectors is an axial vector;[14] for example, $\vec{j} = \vec{r} \times \vec{p}$ is an axial vector.

Weak interactions (see Section 1.1.4) are not parity-invariant; this was first shown by C. S. Wu using the β-decay (1.4) of polarized cobalt (^{60}Co) nuclei to an excited state of ^{60}Ni:

$$^{60}\text{Co} \rightarrow {}^{60}\text{Ni}^* + e^- + \bar{\nu}.$$

In the Wu experiment, the expectation value of the ^{60}Co angular momentum $\langle \vec{J} \rangle$ has a fixed orientation (Fig. 8.6). The decay electrons are emitted preferentially in the direction opposite to that of the angular momentum: if $\vec{P}$ is the electron momentum, $\langle \vec{J} \cdot \vec{P} \rangle < 0$. However, $\langle \vec{J} \cdot \vec{P} \rangle$, the expectation value of the scalar product of a polar vector and an axial vector, is a pseudoscalar which changes sign under the parity operation. The mirror image of the experiment (Fig. 8.6) does not appear to be physically possible: in the mirror image the rotations are reversed in sense, and the electrons are emitted preferentially in the direction of $\vec{J}$.

The group $\mathcal{G}$ corresponding to the parity operation is the multiplicative group of two elements $\{+1, -1\}$, the group Z_2. Since -1 cannot be continuously connected to the identity, it is necessary to find an argument for deciding if the operator Π representing the parity operation in the space of states is unitary or antiunitary. Let χ and φ be two

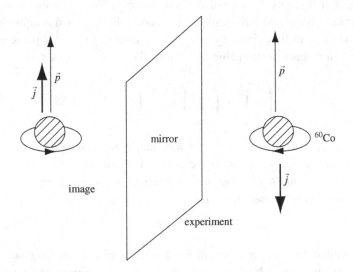

Fig. 8.6. Experiment on the decay of polarized cobalt.

---

[14] The existence of axial vectors is a peculiarity of three-dimensional space $d = 3$. An axial vector is in fact an antisymmetric tensor of rank 2 with $d(d-1)/2$ components in general. For $d = 3$ the number of components is three, so that it can correspond to a (pseudo)vector. In four dimensions it is not possible to make this identification, because an antisymmetric tensor of rank 2 like the electromagnetic field has six components.

arbitrary vectors and (χ, φ) be their scalar product (we switch to the mathematicians' notation until the end of this section). If parity is a symmetry, then

$$|(\Pi\chi, \Pi\varphi)| = |(\chi, \varphi)|.$$

Since in the parity operation the position and momentum operators must both transform as vectors:

$$\vec{R} \to \Pi^{-1}\vec{R}\Pi = -\vec{R}, \quad \vec{P} \to \Pi^{-1}\vec{P}\Pi = -\vec{P}, \tag{8.49}$$

their commutator is unchanged:

$$\Pi[X_i, P_j]\Pi^{-1} = i\hbar\delta_{ij}I.$$

Let us examine the matrix element

$$(\Pi\chi, \Pi[X_i, P_j]\varphi) = (\Pi\chi, \Pi[X_i, P_j]\Pi^{-1}\Pi\varphi)$$

$$= (\Pi\chi, i\hbar\delta_{ij}\Pi\varphi) = i\hbar\delta_{ij}(\Pi\chi, \Pi\varphi). \tag{8.50}$$

On the other hand, we also have

$$(\Pi\chi, \Pi[X_i, P_j]\varphi) = (\Pi\chi, \Pi i\hbar\delta_{ij}\varphi)$$

$$= i\hbar\delta_{ij}(\Pi\chi, \Pi\varphi) \tag{8.51}$$

if we assume that Π is unitary. In fact, for a unitary operator

$$(U\chi, Ui\varphi) = (\chi, i\varphi) = i(\chi, \varphi),$$

while for an antiunitary operator

$$(U\chi, Ui\varphi) = (i\varphi, \chi) = -i(\varphi, \chi).$$

The equations (8.50) and (8.51) are compatible only if Π is unitary. On the other hand, if instead of parity Π we consider time reversal Θ, $\vec{R} \to \vec{R}$ and $\vec{P} \to -\vec{P}$ (See Appendix A.2), then

$$\Theta[X_i, P_j]\Theta^{-1} = -[X_i, P_j] = -i\hbar\delta_{ij}$$

and the change of sign implies that Θ is antiunitary.

If parity is a symmetry, which as far as we know is the case in strong and electromagnetic interactions, Π must commute with the Hamiltonian: $[\Pi, H] = 0$. Since $\Pi^2 = I$, two successive parity operations take the system of axes back to its initial position, and the eigenvalues of Π are ± 1. Since Π and H commute, it is possible to find a set of eigenvectors $|\varphi_\pm\rangle$ common to H and Π:

$$H|\varphi_\pm\rangle = E_\pm|\varphi_\pm\rangle, \quad \Pi|\varphi_\pm\rangle = \pm|\varphi_\pm\rangle. \tag{8.52}$$

The states $|\varphi_+\rangle$ are said to have *positive parity* and the states $|\varphi_-\rangle$ to have *negative parity*.

8.4 Galilean invariance

8.4.1 The Hamiltonian in dimension $d = 1$

We are now going to examine the consequences of the one invariance that so far we have not used, invariance under a change of inertial reference frame. First we limit ourselves to one dimension, taking the case of a particle on the x axis. The equations of nonrelativistic physics must preserve their form under a Galilean transformation

$$x' = x - vt, \tag{8.53}$$

which takes one reference frame into another moving at speed v relative to the first. The transformation law (8.53) corresponds to the passive point of view of changing the axes. In order to be consistent with the preceding sections, we shall choose the active point of view, in which the speeds of all the particles are modified by v; it is said that the particles are "boosted"[15] by an amount v. If the initial position, speed, momentum p, and kinetic energy K of a classical particle of mass m are

$$x, \quad \dot{x}, \quad p = m\dot{x}, \quad K = \frac{1}{2}m\dot{x}^2,$$

these variables when boosted by v become

$$x' = x + vt, \quad \dot{x}' = \dot{x} + v, \quad p' = m\dot{x}', \quad K' = \frac{1}{2}m(\dot{x}')^2. \tag{8.54}$$

In contrast to the case of translations and rotations, the energy is not invariant under a Galilean transformation. The only requirement that can be imposed is that the form of the equations of physics remains invariant.

Let us now turn to the quantum case and place ourselves at time $t = 0$, which corresponds to an instantaneous Galilean transformation. The transformation law for the state vectors under a Galilean transformation will be a unitary transformation $U(v)$

$$U(v) = \exp\left(-i\frac{vG}{\hbar}\right), \tag{8.55}$$

where $G = G^\dagger$ is the infinitesimal generator of Galilean transformations. Galilean transformations in one dimension form an additive group, because the composition of two transformations with velocities v and v' is a transformation with velocity $v'' = v + v'$. Once again, the Stone theorem guarantees the existence of a Hermitian infinitesimal generator G. If $\langle A \rangle$ is the expectation value of a physical property in the state $|\varphi\rangle$, the expectation value $\langle A \rangle_v$ in the transformed state $|\varphi_v\rangle = U(v)|\varphi\rangle$ will be

$$\langle \varphi_v | A | \varphi_v \rangle = \langle A \rangle_v = \langle \varphi | U^{-1}(v) A U(v) | \varphi \rangle. \tag{8.56}$$

---

[15] This term originates from the idea of a rocket booster.

From (8.54) (for $t = 0$) we expect that the expectation values of the position X, momentum P, and velocity operators $\dot{X}$ will transform as

$$\langle X \rangle \rightarrow \langle X \rangle_v = \langle \varphi | U^{-1}(v) X U(v) | \varphi \rangle = \langle X \rangle, \tag{8.57}$$

$$\langle \dot{X} \rangle \rightarrow \langle \dot{X} \rangle_v = \langle \varphi | U^{-1}(v) \dot{X} U(v) | \varphi \rangle = \langle \dot{X} \rangle + v, \tag{8.58}$$

$$\langle P \rangle \rightarrow \langle P \rangle_v = \langle \varphi | U^{-1}(v) P U(v) | \varphi \rangle = \langle P \rangle + mv. \tag{8.59}$$

The strong hypothesis,[16] even though it seems natural, is in fact (8.58), because in quantum mechanics $\dot{X}$ is *defined* as $(i/\hbar)[H, X]$, and (8.58) leads to constraints on the possible Hamiltonians. Since (8.58) is valid for any $| \varphi \rangle$, we obtain

$$\exp\left(i \frac{vG}{\hbar} \right) P \exp\left(-i \frac{vG}{\hbar} \right) = P + mvI, \tag{8.60}$$

and by making v tend to zero,

$$[G, P] = -i\hbar m I.$$

It is therefore possible to choose $G = -mX$. According to the von Neumann theorem, any other choice will be unitarily equivalent.

Let us now consider the operator $\dot{X}$ describing the speed, which according to (8.18) for $A = X$ is *defined* by

$$\dot{X} = \frac{i}{\hbar} [H, X]. \tag{8.61}$$

From (8.58) we have

$$\exp\left(i \frac{vG}{\hbar} \right) \dot{X} \exp\left(-i \frac{vG}{\hbar} \right) = \dot{X} + vI, \tag{8.62}$$

and subtracting (8.60) (divided by m) from (8.62) we find

$$\exp\left(i \frac{vG}{\hbar} \right) \left[\dot{X} - \frac{1}{m} P \right] \exp\left(-i \frac{vG}{\hbar} \right) = \dot{X} - \frac{1}{m} P, \tag{8.63}$$

which implies that the operator $[\dot{X} - P/m]$ commutes with G and therefore with X. Again using the von Neumann theorem, $[\dot{X} - P/m]$ must be a function of X:

$$\dot{X} - \frac{P}{m} = \frac{1}{m} f(X). \tag{8.64}$$

In the one-dimensional case, and in general only in this case, the function f can be eliminated by a unitary transformation. Let $F(x)$ be a primitive of $f(x)$, $F'(x) = f(x)$,

---

[16] See H. Brown and P. Holland, The Galilean covariance of quantum mechanics in the case of external fields, *Am. J. Phys.* **67**, 204 (1999) for a critical evaluation of this hypothesis.

and consider a unitary transformation, which is in fact a local gauge transformation (cf. Section 11.4.1):

$$S = \exp\left(\frac{i}{\hbar} F(X)\right). \tag{8.65}$$

In the unitary transformation $X' = S^{-1}XS$ the quantity X is obviously not changed, $X' = X$. Let us calculate P'. Using (8.41), we find

$$[P, S] = -i\hbar \frac{\partial S}{\partial X} = (-i\hbar)\left(\frac{i}{\hbar}\right) f(X)S = f(X)S,$$

from which we deduce

$$S^{-1}PS - P = S^{-1}(PS - SP) = S^{-1}[P, S] = S^{-1}f(X)S = f(X).$$

This gives $P' = S^{-1}PS = P + f(X)$ and, according to (8.64),

$$\dot{X} = \frac{1}{m} P'.$$

We can therefore always choose the momentum operator to be $P = m\dot{X}$. This choice is unitarily equivalent to any other. We shall use these results to determine the most general form of the Hamiltonian compatible with the Galilean transformation laws. We define the operator K, which will of course be the quantum version of the kinetic energy, as

$$K = \frac{1}{2} m\dot{X}^2 = \frac{P^2}{2m} \tag{8.66}$$

and calculate its commutator with X:

$$[K, X] = \frac{1}{2m}[P^2, X] = -\frac{i\hbar}{2m}\frac{\partial P^2}{\partial P} = -i\hbar\frac{P}{m}. \tag{8.67}$$

Interchanging the roles of P and X, equation (8.41) implies that

$$[X, f(P)] = i\hbar \frac{\partial f(P)}{\partial P}.$$

However,

$$\frac{1}{m} P = \dot{X} = \frac{i}{\hbar}[H, X],$$

and subtracting this equation from (8.67) gives

$$[H - K, X] = 0.$$

The operator $(H - K)$ is a function only of X, which we denote as $V(X)$. This then gives the most general form of the Hamiltonian compatible with Galilean invariance:

$$H = K + V(X) = \frac{P^2}{2m} + V(X). \tag{8.68}$$

This is what we would have obtained using the correspondence principle and starting from the classical analog of the energy, equal to the sum of the kinetic and potential energies:

$$E = \frac{p^2}{2m} + V(x).$$

Galilean invariance is ensured by the fact that the Hamiltonian preserves its *form* after transformation. If the initial Hamiltonian is a function of X and P, the transformed Hamiltonian is *the same* function of $X_v = X$ and $P_v = P + mv$:

- The initial state:

$$H = \frac{P^2}{2m} + V(X).$$

- The transformed state:

$$H_v = \frac{P_v^2}{2m} + V(X_v) = H + Pv + \frac{1}{2}mv^2 + V(X). \tag{8.69}$$

8.4.2 The Hamiltonian in dimension $d = 3$

Repeating the argument of the preceding subsection for the case of three space dimensions, we easily arrive at the generalization of (8.64):

$$\frac{d\vec{R}}{dt} = \frac{1}{m}\vec{P} - \frac{1}{m}\vec{f}(\vec{R}), \tag{8.70}$$

but we cannot in general eliminate $\vec{f}(\vec{R})$. It would be necessary to find a unitary transformation

$$S = \exp\left(\frac{i}{\hbar}F(\vec{R})\right)$$

such that

$$\vec{f}(\vec{R}) = \vec{\nabla}F(\vec{R}),$$

which is only possible if $\vec{\nabla} \times \vec{f} = 0$.[17] The equation (8.70) implies the commutation relation

$$[\dot{X}_i, X_j] = -\frac{i\hbar}{m}\delta_{ij}. \tag{8.71}$$

The kinetic energy K is defined by

$$K = \frac{1}{2}m\left(\frac{d\vec{R}}{dt}\right)^2 = \frac{1}{2m}(\vec{P} - \vec{f}(\vec{R}))^2. \tag{8.72}$$

---

[17] This condition is necessary but not sufficient in a domain that is not simply connected.

It is easy to calculate the commutator of K and X_i. We find

$$[K, X_i] = \frac{1}{2} m \sum_j [\dot{X}_j^2, X_i] = \frac{1}{2} m \sum_j (\dot{X}_j[\dot{X}_j, X_i] + [\dot{X}_j, X_i]\dot{X}_j) = -i\hbar\dot{X}_i.$$

Comparing the commutators

$$[K, X_i] = -i\hbar\dot{X}_i \text{ and } [H, X_i] = -i\hbar\dot{X}_i,$$

we obtain

$$[H - K, X_i] = 0,$$

and so $(H - K)$ is a function only of $\vec{R}$: $H = K + V(\vec{R})$. The most general Hamiltonian compatible with Galilean invariance is then of the form

$$H = \frac{1}{2m} \left(\vec{P} - \vec{f}(\vec{R}) \right)^2 + V(\vec{R}). \tag{8.73}$$

It is important to emphasize the difference between $\vec{P}/m$ and $d\vec{R}/dt$: it is the latter that gives the kinetic energy K,

$$K = \frac{1}{2m} \left(\frac{d\vec{R}}{dt} \right)^2 \neq \frac{\vec{P}^2}{2m}.$$

We can now make the connection with classical physics. In classical mechanics the Hamiltonian of a particle of charge q in a magnetic field $\vec{B}(\vec{r}) = \vec{\nabla} \times \vec{A}(\vec{r})$ and an electric field $\vec{E}(\vec{r}) = -\vec{\nabla}\varphi(\vec{r})$ which may be time-dependent is[18]

$$H_{\text{cl}} = \frac{1}{2m} \left(\vec{p} - q\vec{A} \right)^2 + q\varphi(\vec{r}). \tag{8.74}$$

We then find (8.73) using the correspondence principle and making the identification $q\vec{A} = \vec{f}$ and $q\varphi = V$. The significance of this Hamiltonian will be examined more deeply in Section 11.4.1, when we discuss local gauge invariance; the transformation (8.65) and its generalization to three dimensions are local gauge transformations. If $f(\vec{R})$ can be eliminated by such a transformation, this would imply that $\vec{B} = 0$. However, one should not conclude that $\vec{f}$ and V are necessarily identified with electromagnetic potentials, because $\vec{f}$ and V are arbitrary functions which need not obey Maxwell's equations, and the particle need not be charged. All we have shown is that the classical Hamiltonian (8.74) can be quantized with a result *consistent* with Galilean invariance.

Let us summarize what has been achieved in this chapter. By assuming that expectation values of physical properties (Hermitian operators) transform in the same manner as the corresponding classical quantities, we have been able to derive the canonical commutation relations and the form of the Hamiltonian. We never made use of the correspondence principle, but we checked the consistency of this principle with our results.

---

[18] Cf. Jackson [1999], Chapter 12.

8.5 Exercises

8.5.1 Rotations

1. Let $\mathcal{R}_{\hat{n}}(\theta)$ be the 3×3 matrix representing a rotation by an angle θ about $\hat{n}$. Show that $\operatorname{Tr} \mathcal{R}_{\hat{n}}(\theta) = 1 + 2\cos\theta$. Hint: use (8.29).

2. Starting from (8.23), write out the matrix $\mathcal{R}_{\hat{n}}(\theta)$ explicitly as a function of the components of $\hat{n}$,

$$\hat{n} = (\alpha, \beta, \gamma), \quad \alpha^2 + \beta^2 + \gamma^2 = 1.$$

3. Explicitly verify the commutation relation $[T_x, T_y] = iT_z$ using the matrix forms (8.26).

4. Show that

$$(\vec{T} \cdot \hat{n})^3 = \vec{T} \cdot \hat{n}$$

and that

$$e^{-i\theta(\vec{T} \cdot \hat{n})} = I - i\sin\theta(\vec{T} \cdot \hat{n}) - (1 - \cos\theta)(\vec{T} \cdot \hat{n})^2.$$

Compare with (8.23).

8.5.2 Rotations and SU(2)

The $SU(2)$ group is the group of 2×2 unitary matrices of unit determinant.

1. Show that if $U \in SU(2)$, then U has the form

$$U = \begin{pmatrix} a & b \\ -b^* & a^* \end{pmatrix}, \quad |a|^2 + |b|^2 = 1.$$

2. Show that in the neighborhood of the identity we can write

$$U = I - i\tau \quad \text{with} \quad \tau = \tau^\dagger,$$

and that τ is expressed as a function of the Pauli matrices as

$$\tau = \frac{1}{2} \sum_{i=1}^{3} \theta_i \sigma_i, \quad \theta_i \to 0.$$

3. We take $\theta = (\sum_i \theta_i^2)^{1/2}$ and $\theta_i = \theta \hat{n}_i$, where $\hat{n}$ is a unit vector. Assuming that the θ_i are finite, we define $U_{\hat{n}}(\theta)$ as

$$U_{\hat{n}}(\theta) = \lim_{N \to \infty} \left[U_{\hat{n}}\left(\frac{\theta}{N}\right) \right]^N.$$

Show that

$$U_{\hat{n}}(\theta) = e^{-i\theta\vec{\sigma} \cdot \hat{n}/2}.$$

Conversely, any $SU(2)$ matrix is of this form (Exercise 3.3.6).

4. Let $\vec{V}$ be a vector of $\mathbb{R}^3$ and $\mathcal{V}$ be a Hermitian matrix of zero trace:

$$\mathcal{V} = \begin{pmatrix} V_z & V_x - iV_y \\ V_x + iV_y & -V_z \end{pmatrix} = \vec{\sigma} \cdot \vec{V}.$$

What is the determinant of $\mathcal{V}$? Let $\mathcal{W}$ be the matrix $[U \in SU(2)]$

$$\mathcal{W} = U\mathcal{V}U^{-1}.$$

Show that $\mathcal{W}$ has the form $\vec{\sigma} \cdot \vec{W}$ and that $\vec{W}$ is derived from $\vec{V}$ by a rotation. Has this property been completely proved at this stage?

5. We define $\vec{V}(\theta)$ as

$$\vec{\sigma} \cdot \vec{V}(\theta) = U_{\hat{n}}(\theta) [\vec{\sigma} \cdot \vec{V}] U_{\hat{n}}^{-1}(\theta), \quad \vec{V}(\theta = 0) = \vec{V}.$$

Show that

$$\frac{d\vec{V}(\theta)}{d\theta} = \hat{n} \times \vec{V}(\theta).$$

Show that $\vec{V}(\theta)$ is obtained from $\vec{V}$ by rotation by an angle θ about $\hat{n}$. This result establishes a correspondence between the matrices $\mathcal{R}_{\hat{n}}(\theta)$ of $SO(3)$ and the matrices $U_{\hat{n}}(\theta)$ of $SU(2)$. Is this a one-to-one or a two-to-one correspondence?

8.5.3 Commutation relations between momentum and angular momentum

This exercise gives another demonstration of the commutation relations (8.35) between momentum and angular momentum if we choose the vector operator $\vec{V} = \vec{P}$. Let $\mathcal{T}_y(a)$ be a translation by a parallel to Oy:

$$\mathcal{T}_y(a)\vec{r} = \vec{r} + a\hat{y}.$$

If $\mathcal{R}_x(\theta)$ is a rotation by an angle θ about Ox, show that

$$\mathcal{R}_x(\theta) \mathcal{T}_y(a) \mathcal{R}_x(-\theta)$$

is a translation along an axis to be determined. From the result, derive the commutation relation

$$[J_x, P_y] = i\hbar P_z.$$

8.5.4 The Lie algebra of a continuous group

Let us consider a group $\mathcal{G}$ whose elements g are parametrized by N coordinates θ_a, $a = 1, \ldots, N$, where $g(\theta_a = 0)$ is the neutral element of the group. The variables θ_a are collectively denoted θ: $\theta = \{\theta_a\}$. If $\mathcal{G}$ is a Lie group, the composition law is given by an infinitely differentiable function f:

$$g(\bar{\theta})g(\theta) = g(f(\bar{\theta}, \theta)).$$

Again, f is collective notation for the set of N functions f: $f(\bar{\theta}, \theta) = \{f_a(\bar{\theta}_b, \theta_c)\}$. Given a set of unitary matrices $U(\theta_a)$ with the multiplication law

$$U(\bar{\theta})U(\theta) = U(f(\bar{\theta}, \theta)),$$

the matrices $U(\theta)$ then form a representation of the group $\mathcal{G}$; see (8.12).

1. Show that $f_a(\bar{\theta}, \theta = 0) = \bar{\theta}_a$ and that $f_a(\bar{\theta} = 0, \theta) = \theta_a$. Show that for $\bar{\theta}, \theta \to 0$, the function $f_a(\bar{\theta}, \theta)$ has the form

$$f_a(\bar{\theta}, \theta) = \theta_a + \bar{\theta}_a + f_{abc}\bar{\theta}_b\theta_c + O(\theta^3, \theta^2\bar{\theta}, \theta\bar{\theta}^2, \bar{\theta}^3),$$

where we have used the convention of summation over repeated indices:

$$f_{abc}\bar{\theta}_b\theta_c = \sum_{b,c} f_{abc}\bar{\theta}_b\theta_c.$$

2. In the neighborhood of $U(\theta) = I$ we expand $U(\theta)$ for $\theta \to 0$:

$$U(\theta) = I - i\theta_a T_a - \frac{1}{2}\theta_b\theta_c T_{bc} + O(\theta)^3.$$

Compute the product $U(\bar{\theta})U(\theta)$ to order $(\bar{\theta}^2, \theta^2)$ and show that the equation

$$U(\bar{\theta})U(\theta) = U(f(\bar{\theta}, \theta))$$

for the terms in $\bar{\theta}_a\theta_b$ implies that

$$T_{bc} = T_c T_b - i f_{abc} T_a.$$

Using the symmetry of T_{bc}, obtain

$$[T_b, T_c] = i C_{abc} T_a$$

with $C_{abc} = -C_{acb}$. Express C_{abc} as a function of f_{abc}. The preceding commutation relations constitute the Lie algebra of the group defined by the composition law $f(\bar{\theta}, \theta)$.

8.5.5 The Thomas–Reiche–Kuhn sum rule

Let us take a particle of mass m in a potential $V(\vec{r})$. The Hamiltonian is

$$H = \frac{\vec{P}^2}{2m} + V(\vec{R}).$$

Let $|\varphi_n\rangle$ be a complete set of eigenvectors of H:

$$H|\varphi_n\rangle = E_n|\varphi_n\rangle, \quad \sum_n |\varphi_n\rangle\langle\varphi_n| = I,$$

and $|\varphi_0\rangle$ be a bound, and therefore normalizable, state of energy E_0. We set

$$\langle\varphi_n|X|\varphi_0\rangle = X_{n0}.$$

1. Demonstrate the commutation relation

$$[[H, X], X] = -\frac{\hbar^2}{m}.$$

2. Show that

$$\sum_n \frac{2m|X_{n0}|^2}{\hbar^2}(E_n - E_0) = 1.$$

8.5.6 The center of mass and the reduced mass

Let us take two particles of masses m_1 and m_2 moving on a line. We use X_1 and X_2 to denote their position operators and P_1 and P_2 to denote their momentum operators. The position and momentum operators of two different particles commute. We define the operators X and P as

$$X = \frac{m_1 X_1 + m_2 X_2}{m_1 + m_2}, \quad P = P_1 + P_2$$

and $\tilde{X}$ and $\tilde{P}$ as

$$\tilde{X} = X_1 - X_2, \quad \tilde{P} = \frac{m_2 P_1 - m_1 P_2}{m_1 + m_2}.$$

1. Calculate the commutators $[X, P]$ and $[\tilde{X}, \tilde{P}]$ and show that

$$[X, \tilde{P}] = [\tilde{X}, P] = 0.$$

2. Write the Hamiltonian

$$H = \frac{P_1^2}{2m_1} + \frac{P_2^2}{2m_2} + V(X_1 - X_2)$$

as a function of the operators $X, P, \tilde{X}, \tilde{P}$. Show that, as in classical mechanics, it is possible to separate the motion of the center of mass and the motion of a particle of reduced mass $\mu = m_1 m_2/(m_1 + m_2)$ about the center of mass. Generalize this to three dimensions.

3. The following example of an entangled state was used in the original article of Einstein, Podolsky, and Rosen (Section 6.2.1). The wave function of two particles is written as

$$\psi(x_1, x_2; p_1, p_2) = \delta(x_1 - x_2 - L)\,\delta(p_1 + p_2),$$

where L is a constant length. Why is it possible to write such a wave function? What is its physical interpretation? Measurement of x_1 determines x_2, and measurement of p_1 determines p_2. Develop the analogy with the example of Section 6.3.1.

8.5.7 The Galilean transformation

1. Let $W(a, v)$ be the product of a Galilean transformation of velocity v and of a one-dimensional translation by a, both along Ox:

$$W(a, v) = \exp\left(-i\frac{Pa}{\hbar}\right)\exp\left(i\frac{mvX}{\hbar}\right).$$

Show that

$$W(a_1, v_1)W(a_2, v_2) = \exp\left(-i\frac{mv_1a_2}{\hbar}\right)W(a_1 + a_2, v_1 + v_2).$$

2. Calculate

$$W(a, v)W(-a, -v)$$

and show that it is necessary to use projective representations for the Galilean group.

8.6 Further reading

Useful complementary information on symmetries in quantum physics can be found in Jauch [1968], Chapters 9 and 10; Ballentine [1998], Chapter 3; and Merzbacher [1970], Chapter 16. Chapter 2 of Weinberg [1995] also contains an excellent summary of all the basic concepts. The canonical commutation relations and Galilean invariance are discussed by Jauch [1968], Chapters 12 and 13. There are many books devoted to the use of group theory in quantum mechanics, one of which is M. Tinkham, *Group Theory and Quantum Mechanics*, New York: McGraw Hill (1964).

9

Wave mechanics

In this chapter we shall study a particular realization of quantum mechanics of great practical importance, namely *wave mechanics*, used to describe the motion of one[1] quantum particle in three-dimensional space $\mathbb{R}^3$. It is this realization which serves as the introduction to the fundamentals of quantum mechanics in most textbooks. It amounts to taking the "eigenvectors"[2] $|\vec{r}\rangle$ of the position operator $\vec{R}$ as the basis in $\mathcal{H}$, or, in other words, choosing a basis in which the position operator is diagonal. In wave mechanics a state vector can be identified with an element $\varphi(\vec{r})$ of the Hilbert space $L^2_{\vec{r}}(\mathbb{R}^3)$ of functions which are square-integrable in three-dimensional space $\mathbb{R}^3$. This state vector is called the *wave function*, and we shall see that it is identified with the probability amplitude $\langle \vec{r}|\varphi\rangle$ for finding the particle in the state $|\varphi\rangle$ localized at position $\vec{r}$. The wave function is normalized by the integrability condition (7.10)

$$\int_{-\infty}^{\infty} \mathrm{d}^3 r\, |\varphi(\vec{r})|^2 = 1. \tag{9.1}$$

Owing to the symmetric roles played by the position and momentum operators, it is also possible to use eigenvectors of $\vec{P}$ and "momentum-space wave functions" $\tilde{\varphi}(\vec{p}) = \langle \vec{p}|\varphi\rangle$, which we shall see are the Fourier transforms of the $\varphi(\vec{r})$. After examining the principal properties of the wave functions, we shall study some applications: bound states, scattering, tunneling, and the periodic potential. These applications will first be treated in the simplest case of one dimension. The generalization to three dimensions will permit us to discuss the important notion of the density of states and its use in Fermi's Golden Rule.

9.1 Diagonalization of X and P and wave functions

9.1.1 Diagonalization of X

We wish to study the motion of a quantum particle, and for the time being we restrict this motion to the real line $\mathbb{R}$, on which the particle moves between $-\infty$ and $+\infty$. The relevant

---

[1] Or more; see the generalization in Section 9.9.3 and Chapter 13.
[2] As we have seen in Section 7.3.1, these objects are not vectors of the Hilbert space, which we have stressed by using quotation marks. However, since we shall make intensive use of these "vectors" in what follows, we shall drop the quotation marks in order to simplify the notation.

physical properties are a priori the position and momentum of the particle, represented mathematically by the operators X and P whose properties we have established in Section 8.3. We shall study the eigenvectors of X starting from the canonical commutation relation between X and P in the form (8.40):

$$\exp\left(i\frac{Pa}{\hbar}\right) X \exp\left(-i\frac{Pa}{\hbar}\right) = X + aI. \tag{9.2}$$

Let us first of all show that the spectrum of X is continuous. We take $|x\rangle$ to be an eigenvector of X

$$X|x\rangle = x|x\rangle, \tag{9.3}$$

and examine the action of X on the vector $\exp(-iPa/\hbar)|x\rangle$:

$$X\left[\exp\left(-i\frac{Pa}{\hbar}\right)|x\rangle\right] = \exp\left(-i\frac{Pa}{\hbar}\right)(X + aI)|x\rangle$$

$$= (x + a)\left[\exp\left(-i\frac{Pa}{\hbar}\right)|x\rangle\right]. \tag{9.4}$$

We have used the commutation relation (9.2) and the definition (9.3) of the eigenvector $|x\rangle$. The vector $\exp(-iPa/\hbar)|x\rangle$ is an eigenvector of X with eigenvalue $(x + a)$, and since a is arbitrary, this shows that all real values of x between $-\infty$ and $+\infty$ are eigenvalues of X. This also proves that the spectrum of x is continuous, and consequently the normalization must be written as in (7.34) using Dirac delta functions:

$$\langle x'|x\rangle = \delta(x - x'). \tag{9.5}$$

In view of the arguments of Section 8.3.1, the result (9.4), which can be written as

$$\exp\left(-i\frac{Pa}{\hbar}\right)|x\rangle = |x + a\rangle,$$

is not surprising, since $\exp(-iPa/\hbar)$ is the operator for translation by a which transforms the state $|x\rangle$ exactly localized at x into the state $|x + a\rangle$ exactly localized at $(x + a)$: P is the infinitesimal generator of translations. The vector $|x + a\rangle$ satisfies a normalization condition analogous to (9.5) because the operator $\exp(-iPa/\hbar)$ is unitary. If we wish, we can fix the phase of the basis vectors $|x\rangle$ by the condition

$$|x\rangle = \exp\left(-i\frac{Px}{\hbar}\right)|x = 0\rangle. \tag{9.6}$$

Let us return to the physical interpretation. What exactly does the vector $|x\rangle$ represent? According to the postulates of Chapter 4, $|x\rangle$ represents a state in which the position of the particle is known with absolute precision: the particle is localized exactly at the point x on the real line. However, in quantum mechanics it is impossible to realize such a state physically. As we shall soon see, such a state has all possible momenta between $p = -\infty$ and $p = +\infty$ with equal probabilities. The mathematical property "$|x\rangle$ is not an element of the Hilbert space" corresponds to the physical property "$|x\rangle$ is not a realizable physical state." Physically realizable states are always represented by "true" vectors of $\mathcal{H}$, that is, normalizable vectors.

We have implicitly assumed that the eigenvalues x of X are nondegenerate. Of course, this is not necessarily the case; for example, the particle can have spin 1/2, in which case it is necessary to specify whether the particle is in a state with spin up $|+\rangle$ or one with spin down $|-\rangle$, and every eigenvalue of X will be doubly degenerate. Under these conditions, the Hilbert space of states will be the tensor product $L_x^{(2)}(\mathbb{R}) \otimes \mathcal{H}_2$ of the space of position states $L_x^{(2)}(\mathbb{R})$ and the two-dimensional space of spin states $\mathcal{H}_2$. A basis in this space might, for example, be constructed from the states $|x \otimes +\rangle$ and $|x \otimes -\rangle$ with

$$(X \otimes \sigma_z)|x \otimes \pm\rangle = \pm x|x \otimes \pm\rangle.$$

Even though the use of eigenvectors that are not true elements of $\mathcal{H}$ is mathematically questionable, it is extremely convenient and we shall do it often in what follows without any particular precautions. We shall also generalize the notion of a matrix element. Since the operator X is diagonal in the basis $|x\rangle$, we can write down the "matrix elements" of X:

$$\langle x'|X|x\rangle = x\langle x'|x\rangle = x\,\delta(x - x'), \tag{9.7}$$

and more generally those of a function $F(X)$:

$$\langle x'|F(X)|x\rangle = F(x)\langle x'|x\rangle = F(x)\,\delta(x - x'). \tag{9.8}$$

The completeness relation (7.37) is written as

$$\int_{-\infty}^{\infty} |x\rangle\,dx\,\langle x| = I. \tag{9.9}$$

The projector $\mathcal{P}[a, b]$ onto the subspace of eigenvalues of X in the interval $[a, b]$ is obtained by restricting the integration over x to this interval:

$$\mathcal{P}[a, b] = \int_a^b |x\rangle\,dx\,\langle x|. \tag{9.10}$$

This expression generalizes that for a finite-dimensional space. If Δ is the subspace of a set of eigenvalues of a Hermitian operator A, the projector $\mathcal{P}(\Delta)$ onto this subspace is

$$\mathcal{P}(\Delta) = \sum_{n \in \Delta} |n\rangle\langle n|.$$

9.1.2 Realization in $L_x^{(2)}(\mathbb{R})$

Now let us make the connection between the Dirac formalism which we have just made explicit in the basis in which X is diagonal and the realization given in Section 8.3.2 of the operators X and P as operators acting in the space $L^{(2)}(\mathbb{R})$ of square-integrable functions on $\mathbb{R}$. Let $|\varphi\rangle$ be a normalized vector of $\mathcal{H}$ representing a physical state. Using the completeness relation (9.9), we can decompose $|\varphi\rangle$ in the basis $|x\rangle$,

$$|\varphi\rangle = \int_{-\infty}^{\infty} |x\rangle\,dx\,\langle x|\varphi\rangle, \tag{9.11}$$

where $\langle x|\varphi\rangle$ is thus a component of $|\varphi\rangle$ in the basis $|x\rangle$, or, in physical terms, the probability amplitude of finding the particle localized at point x. Let us examine the matrix elements of the operators X and $\exp(-iPa/\hbar)$:

$$\langle x|[X|\varphi\rangle] = \langle Xx|\varphi\rangle = x\langle x|\varphi\rangle = x\,\varphi(x), \tag{9.12}$$

$$\langle x|\left[\exp\left(-i\frac{Pa}{\hbar}\right)|\varphi\rangle\right] = \langle x-a|\varphi\rangle = \varphi(x-a). \tag{9.13}$$

These equations show that $\langle x|\varphi\rangle$ can be identified with a function $\varphi(x)$ of $L_x^{(2)}(\mathbb{R})$ such that the action of the operators X and P will be given by (8.44). The equation (9.12) then is

$$\boxed{[X\varphi](x) = x\varphi(x)}, \tag{9.14}$$

and (9.13) is written as

$$\left[\exp\left(-i\frac{Pa}{\hbar}\right)\varphi\right](x) = \varphi(x-a). \tag{9.15}$$

Expanding to first order in a, we have

$$\boxed{[P\varphi](x) = -i\hbar\frac{\partial\varphi}{\partial x}}. \tag{9.16}$$

We recover the action of the operators X and P as defined in Section 8.3.2. Let us check that the scalar product is correctly given by (7.11) using the completeness relation (9.9):

$$\langle \chi|\varphi\rangle = \int_{-\infty}^{\infty} dx\,\langle\chi|x\rangle\langle x|\varphi\rangle = \int_{-\infty}^{\infty} dx\,\chi^*(x)\varphi(x). \tag{9.17}$$

The function $\varphi(x-a)$ in (9.15) is just the function $\varphi(x)$ translated by $+a$, and not by $-a$. If, for example, $\varphi(x)$ has a maximum at $x = x_0$, then $\varphi(x-a)$ has a maximum at $x-a = x_0$, that is, at $x = x_0 + a$ (Fig. 9.1). We emphasize the fact that the choice $\varphi_a(x) = \varphi(x-a)$ for the translated wave function is the simplest one, but it is not unique. The function

$$\varphi'_a(x) = e^{i\theta(x)}\varphi(x-a)$$

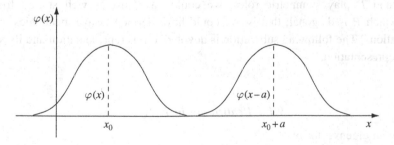

Fig. 9.1. Translation by a of a particle localized in the neighborhood of x_0.

is derived from $\varphi(x-a)$ by a local gauge transformation (8.65). The choice $\varphi(x-a)$ is related to that of the infinitesimal translation generator, and the phase transformation $\varphi_a(x) \to \varphi'_a(x)$ will correspond to using an infinitesimal translation generator derived from (9.16) by the local gauge transformation

$$P' = e^{-i\theta(x)} \left(i\hbar \frac{\partial}{\partial x} \right) e^{i\theta(x)}.$$

In summary, the physical state of a particle moving on the x axis is described by a normalized wave function $\varphi(x)$ belonging to $L_x^{(2)}(\mathbb{R})$:

$$\int_{-\infty}^{\infty} dx\, |\varphi(x)|^2 = 1, \tag{9.18}$$

which is interpreted physically as the probability amplitude $\langle x|\varphi \rangle$ of finding the particle localized at the point x. The action of the position and momentum operators X and P on $\varphi(x)$ is given by (9.14) and (9.16). The squared modulus

$$|\varphi(x)|^2 = |\langle x|\varphi \rangle|^2$$

is called the *probability for the particle to be found* at a point x; it is actually a *probability density*, in this case a probability per unit length. According to (9.10), the probability $\mathsf{p}([a,b])$ of finding the particle localized in the interval $[a, b]$ is

$$\mathsf{p}([a, b]) = \langle \varphi|\mathcal{P}[a, b]|\varphi \rangle = \int_a^b dx\, |\varphi(x)|^2. \tag{9.19}$$

This probability is normalized to unity by construction since $\langle \varphi|\varphi \rangle = 1$, which is the same as (9.18). If we take the interval $[x, x+dx]$ to be infinitesimal, $|\varphi(x)|^2 dx$ is the probability of finding the particle in this interval.

When the particle possesses extra degrees of freedom, for example, a spin 1/2, its quantum state can be described using the wave functions $\varphi_\pm(x)$:

$$\varphi_+(x) = \langle x \otimes +|\varphi \rangle, \quad \varphi_-(x) = \langle x \otimes -|\varphi \rangle.$$

We have just defined what is customarily called "wave mechanics in the x representation," as we have chosen to start from the basis $|x\rangle$ in which the position operator is diagonal. Since X and P play symmetric roles, we could have just as well started from the basis in which P is diagonal; that is, we could have defined "wave mechanics in the p representation." The following subsection is devoted to this representation and its relation to the x representation.

9.1.3 Realization in $L_p^{(2)}(\mathbb{R})$

Let $|p\rangle$ be an eigenvector of P:

$$P|p\rangle = p|p\rangle. \tag{9.20}$$

First we shall determine the corresponding wave functions

$$\chi_p(x) = \langle x|p \rangle \tag{9.21}$$

in the x representation:

$$\langle x|[P|p\rangle] = p\langle x|p \rangle = p\chi_p(x)$$

$$= -i\hbar \frac{\partial}{\partial x} \chi_p(x).$$

We have used (9.16) to obtain the second line of the preceding equation. For any p in the interval $[-\infty, +\infty]$, the differential equation

$$-i\hbar \frac{\partial}{\partial x} \chi_p(x) = p\chi_p(x)$$

has the solution

$$\chi_p(x) = \frac{1}{\sqrt{2\pi\hbar}} e^{ipx/\hbar}, \tag{9.22}$$

which shows that the spectrum of P is continuous, like that of x. The normalization factor $(2\pi\hbar)^{-1/2}$ in (9.22) was chosen such that $\chi_p(x)$ is normalized to a Dirac delta function:

$$\int_{-\infty}^{\infty} dx\, \chi_{p'}^*(x)\chi_p(x) = \frac{1}{2\pi\hbar} \int_{-\infty}^{\infty} dx\, \exp\left[i \frac{(p-p')x}{\hbar}\right] = \delta(p-p'), \tag{9.23}$$

and the completeness relation is written as

$$\int_{-\infty}^{\infty} dp\, \chi_p(x)\chi_p^*(x') = \frac{1}{2\pi\hbar} \int_{-\infty}^{\infty} dp\, \exp\left[i \frac{p(x-x')}{\hbar}\right] = \delta(x-x'). \tag{9.24}$$

We could equally well have started from the completeness relation in the form

$$\int_{-\infty}^{\infty} |p\rangle\, dp\, \langle p| = I \tag{9.25}$$

and written

$$\int_{-\infty}^{\infty} \langle x'|p\rangle\, dp\, \langle p|x\rangle = \langle x'|I|x\rangle = \delta(x-x'),$$

which also leads to (9.24).

If $|\varphi\rangle$ is the state vector of a particle, the "wave function in the p representation" will be $\tilde{\varphi}(p) = \langle p|\varphi\rangle$. This wave function in the p representation is just the Fourier transform of the wave function $\varphi(x) = \langle x|\varphi\rangle$ in the x representation. Since $|\langle x|p\rangle|^2$ is a constant, the $|x\rangle$ and $|p\rangle$ bases are complementary according to a slight generalization of the definition in Section 3.1.2. Using the completeness relation (9.9) as well as (9.21) and (9.22), we find

$$\boxed{\tilde{\varphi}(p) = \langle p|\varphi\rangle = \int_{-\infty}^{\infty} \langle p|x\rangle\, dx\, \langle x|\varphi\rangle = \frac{1}{\sqrt{2\pi\hbar}} \int_{-\infty}^{\infty} dx\, e^{-ipx/\hbar}\, \varphi(x)}\,, \tag{9.26}$$

and conversely

$$\boxed{\varphi(x) = \frac{1}{\sqrt{2\pi\hbar}} \int_{-\infty}^{\infty} \mathrm{d}p \, e^{ipx/\hbar} \, \tilde{\varphi}(p)} \, . \tag{9.27}$$

The action of the operators X and P in the p representation is easily obtained:

$$[X\tilde{\varphi}](p) = i\hbar \frac{\partial}{\partial p} \tilde{\varphi}(p), \tag{9.28}$$

$$[P\tilde{\varphi}](p) = p \, \tilde{\varphi}(p). \tag{9.29}$$

An expression analogous to (9.19) holds in momentum space: the probability $\mathsf{p}([k, q])$ for the particle to have momentum in the interval $[k, q]$ is

$$\mathsf{p}([k, q]) = \int_{k}^{q} \mathrm{d}p \, |\tilde{\varphi}(p)|^{2}, \tag{9.30}$$

where $|\tilde{\varphi}(p)|^{2}$ is a probability density in momentum space.

9.1.4 Evolution of a free wave packet

Let us start from the Fourier representation (9.27) of the wave function $\varphi(x)$ of a physical state. The Fourier transform $\tilde{\varphi}(p)$, like $\varphi(x)$, satisfies the normalization condition

$$\int_{-\infty}^{\infty} \mathrm{d}p \, |\tilde{\varphi}(p)|^{2} = 1. \tag{9.31}$$

Such a physical state is often called a *wave packet*, because according to (9.27) it is a superposition of plane waves. The expectation values of position $\langle X \rangle$ and momentum $\langle P \rangle$ are calculated by inserting the completeness relations (9.9) and (9.25) twice:[3]

$$\langle X \rangle = \langle \varphi | X | \varphi \rangle = \int \mathrm{d}x \, \mathrm{d}x' \langle \varphi | x \rangle \langle x | X | x' \rangle \langle x' | \varphi \rangle = \int_{-\infty}^{\infty} \mathrm{d}x \, x |\varphi(x)|^{2}, \tag{9.32}$$

$$\langle P \rangle = \langle \varphi | P | \varphi \rangle = \int \mathrm{d}p \, \mathrm{d}p' \langle \varphi | p \rangle \langle p | P | p' \rangle \langle p' | \varphi \rangle = \int_{-\infty}^{\infty} \mathrm{d}p \, p |\tilde{\varphi}(p)|^{2}. \tag{9.33}$$

We have also used (9.7) and an analogous equation in momentum space. The dispersions ΔX and ΔP are given by a similar calculation:

$$(\Delta X)^{2} = \langle \varphi | (X - \langle X \rangle)^{2} | \varphi \rangle = \int_{-\infty}^{\infty} \mathrm{d}x \, (x - \langle X \rangle)^{2} |\varphi(x)|^{2}, \tag{9.34}$$

$$(\Delta P)^{2} = \langle \varphi | (P - \langle P \rangle)^{2} | \varphi \rangle = \int_{-\infty}^{\infty} \mathrm{d}p \, (p - \langle P \rangle)^{2} |\tilde{\varphi}(p)|^{2}. \tag{9.35}$$

---

[3] The explicit notation would be $\langle X \rangle_{\varphi}$ and $\langle P \rangle_{\varphi}$; we have suppressed the index φ to simplify the notation.

According to the general argument of Section 4.1.3, these dispersions satisfy the Heisenberg inequality:

$$\Delta x\, \Delta p \geq \frac{1}{2}\hbar \quad , \tag{9.36}$$

where we have used the usual notation $\Delta x\, \Delta p$ instead of $\Delta X\, \Delta P$. A direct demonstration of (9.36) is proposed in Exercise 9.7.1.

Let us introduce a time dependence in the state vector: the state vector is $|\varphi(0)\rangle \equiv |\varphi\rangle$ at time $t = 0$ and $|\varphi(t)\rangle$ at time t. The wave function $\varphi(x, t)$ at time t then is $\varphi(x, t) = \langle x|\varphi(t)\rangle$. To obtain $|\varphi(t)\rangle$ as a function of $|\varphi(0)\rangle$, we need the evolution equation (4.11) and also the Hamiltonian H. Until the end of this section, we shall restrict ourselves to the case where the potential energy is zero and the Hamiltonian reduces to the kinetic energy term K (8.66):

$$H = K = \frac{P^2}{2m}. \tag{9.37}$$

Since K and P commute, the eigenstates of H can be chosen among those of P:

$$P|p\rangle = p|p\rangle \quad H|p\rangle = \frac{P^2}{2m}|p\rangle = \frac{p^2}{2m}|p\rangle = E(p)|p\rangle, \tag{9.38}$$

and consequently

$$\exp\left[-i\frac{Ht}{\hbar}\right]|p\rangle = \exp\left[-i\frac{E(p)t}{\hbar}\right]|p\rangle. \tag{9.39}$$

Then it is natural to express $\langle x|\varphi(t)\rangle$ as a function of the components of $|\varphi(t)\rangle$ in the basis $|p\rangle$:

$$\langle x|\varphi(t)\rangle = \langle x|\exp\left(-i\frac{Ht}{\hbar}\right)|\varphi(0)\rangle = \int dp\, \langle x|p\rangle\langle p|\exp\left(-i\frac{Ht}{\hbar}\right)|\varphi\rangle$$

$$= \frac{1}{\sqrt{2\pi\hbar}}\int_{-\infty}^{\infty} dp\, \exp\left(i\frac{px}{\hbar} - i\frac{E(p)t}{\hbar}\right)\tilde{\varphi}(p). \tag{9.40}$$

In order to eliminate the factors of $\hbar$, we introduce the wave vector $k = p/\hbar$ and the frequency $\omega(k)$:

$$k = \frac{p}{\hbar}, \quad \omega(k) = \frac{E(\hbar k)}{\hbar} = \frac{\hbar k^2}{2m}, \quad A(k) = \sqrt{\hbar}\,\tilde{\varphi}(\hbar k)$$

so that $\varphi(x, t)$ can be written as

$$\varphi(x, t) = \frac{1}{\sqrt{2\pi}}\int_{-\infty}^{\infty} dk\, A(k)\exp\left(ikx - i\omega(k)t\right). \tag{9.41}$$

The qualitative behavior of $|A(k)|^2$ and $|\varphi(x, 0)|^2$ is shown in Fig. 9.2. The function $|A(k)|^2$ is centered at $k \simeq \overline{k}$ and has width Δk. The Heisenberg inequality (9.36) becomes

$$\Delta x\, \Delta k \geq \frac{1}{2}. \tag{9.42}$$

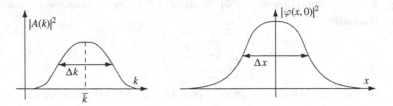

Fig. 9.2. Spread of a wave packet in k and in x.

The limiting cases are

- A particle of sharply defined wave vector (or momentum), which is a plane wave:

$$A(k) = \delta(k - \overline{k}), \quad \varphi(x, 0) = \frac{1}{\sqrt{2\pi}} \, e^{i\overline{k}x}. \tag{9.43}$$

- A particle localized exactly at $x = x_0$:

$$A(k) = \frac{1}{\sqrt{2\pi}} \, e^{-ikx_0}, \quad \varphi(x, 0) = \delta(x - x_0). \tag{9.44}$$

We recall that neither a plane wave (9.43) nor a perfectly localized state (9.44) corresponds to a physically realizable state. In the case (9.44) of a localized particle, the probability $|A(k)|^2$ of observing momentum $\hbar k$ is independent of k, and so the probability distribution cannot be normalized. Similarly, for the case (9.43) of fixed momentum we have $|\varphi(x)|^2 =$ const. and the probability density is uniform on the x axis, so that again the probability distribution cannot be normalized. According to (9.31), for a state to be physically realizable we must have

$$\int_{-\infty}^{\infty} \mathrm{d}k \, |A(k)|^2 < \infty.$$

Let us now study the time evolution of a wave packet. We shall use the stationary phase approximation to evaluate (9.41). Defining $A(k) = |A(k)| \exp[i\phi(k)]$, the phase $\theta(k)$ of the exponential in (9.41) becomes

$$\theta(k) = kx - \omega(k)t + \phi(k).$$

We obtain the leading contribution to the integral (9.41) if the phase $\theta(k)$ is stationary in the region $k \simeq \overline{k}$ where $|A(k)|$ has a maximum; if $\theta(k)$ is not stationary, the exponential oscillates rapidly and the contribution to the integral (9.41) averages to zero. We then must have

$$\left.\frac{\mathrm{d}\theta}{\mathrm{d}k}\right|_{k=\overline{k}} = x - t \left.\frac{\mathrm{d}\omega}{\mathrm{d}k}\right|_{k=\overline{k}} + \left.\frac{\mathrm{d}\phi}{\mathrm{d}k}\right|_{k=\overline{k}} = 0.$$

The center of the wave packet will move according to the law

$$x = v_{\mathrm{g}}(t - \tau), \tag{9.45}$$

where v_g is the group velocity, which is just the average velocity $\bar{v}$ of the particle:

$$v_g = \bar{v} = \left.\frac{d\omega}{dk}\right|_{k=\bar{k}} = \left.\frac{d}{dk}\frac{\hbar k^2}{2m}\right|_{k=\bar{k}} = \frac{\hbar\bar{k}}{m} = \frac{\bar{p}}{m}. \tag{9.46}$$

The time τ determining the $t = 0$ position $x_0 = -v_g\tau$ of the center of the wave packet is

$$\tau = \left.\frac{1}{v_g}\frac{d\phi}{dk}\right|_{k=\bar{k}} = \left.\hbar\frac{d\phi}{dE}\right|_{k=\bar{k}}. \tag{9.47}$$

In order to obtain a more precise result, we can rewrite the phase by expanding $\omega(k)$ in the neighborhood of $k = \bar{k}$:

$$\theta(k) = kx - \omega(\bar{k})t - (k-\bar{k})v_g t - \frac{1}{2}(k-\bar{k})^2\frac{\hbar}{m}t + \phi(k)$$

$$= \omega(\bar{k})t + k(x - v_g t) - \frac{1}{2}(k-\bar{k})^2\frac{\hbar}{m}t + \phi(k).$$

We obtain a very simple form for $\varphi(x, t)$ if it is possible to neglect the quadratic term in $(k-\bar{k})^2$:

$$\varphi(x, t) = \frac{1}{\sqrt{2\pi}}\exp[i\omega(\bar{k})t]\int dk A(k)\exp[ik(x - v_g t)]$$

$$= \exp[i\omega(\bar{k})t]\varphi(x - v_g t, 0). \tag{9.48}$$

This equation shows that aside from the phase factor $\exp[i\omega(\bar{k})t]$, the wave function at time t is obtained from that at time $t = 0$ by the substitution $x \to x - v_g t$, that is, if $v_g > 0$ the wave packet propagates without deformation in the direction of positive x with velocity v_g. However, this result is only approximate since we have neglected the quadratic term in $(k-\bar{k})^2$. This term gives a contribution to the phase

$$-\frac{1}{2}(k-\bar{k})^2\frac{\hbar}{m}t$$

which must remain $\ll 1$ in the domain where $|A(k)|$ is sizable if we want to remain within the linear approximation. The contribution of this term can be neglected if

$$\frac{1}{2}(k-\bar{k})^2\frac{\hbar t}{m} \ll 1$$

in a region of extent Δk about $\bar{k}$. For the deformation of the wave packet to be small, we must have

$$t \ll \frac{2m}{\hbar(\Delta k)^2} = \frac{2m\hbar}{(\Delta p)^2}. \tag{9.49}$$

If this condition is not satisfied, the wave packet is deformed and broadens, with its center continuing to move at speed v_g. This phenomenon is called *wave-packet spreading*.

Let us conclude this section by showing how the Heisenberg inequality (9.36) can be used as a heuristic tool to estimate the energy of the ground state of the hydrogen atom (see Section 1.5.2). If the electron describes a circular orbit of radius r with momentum $p = mv$, its classical energy will be

$$E = \frac{p^2}{2m} - \frac{e^2}{r}. \tag{9.50}$$

In classical physics, the orbital radius of the electron tends to zero (it is said that the "electron falls into the nucleus") with the emission of electromagnetic radiation. In fact, in classical physics the energy of a circular orbit $E = -e^2/(2r)$ is not bounded below and nothing prevents the orbit radius from becoming arbitrarily small. The decrease in the energy of the orbit is compensated for by the emission of energy in the form of electromagnetic radiation, which ensures energy conservation. However, in an orbit of radius r the spread Δx of the position on the x axis is of order r, which makes the momentum spread at least $\sim \hbar/\Delta x = \hbar/r$. We find $rp \sim \hbar$, and the expression for the energy (9.50) becomes

$$E \sim \frac{\hbar^2}{2mr^2} - \frac{e^2}{r}.$$

Let us seek the minimum of E:

$$\frac{dE}{dr} \sim -\frac{\hbar^2}{mr^3} + \frac{e^2}{r^2} = 0,$$

so that a minimum occurs at

$$r = a_0 = \frac{\hbar^2}{me^2}, \tag{9.51}$$

which is just the Bohr radius (1.34) of the hydrogen atom. Naturally, the fact that we obtain exactly a_0 in this order-of-magnitude calculation is a happy coincidence. It leads to the ground-state energy (1.35):

$$E_0 = -\frac{e^2}{2a_0} = -\frac{me^4}{2\hbar^2}. \tag{9.52}$$

While this calculation can give only the order of magnitude, the accompanying physics explains the deep reason for the stability of the atom: owing to the Heisenberg inequalities, the electron cannot exist in an orbit of very small radius without acquiring a large momentum, which makes its kinetic energy high. The energy of the ground state is obtained by finding the best possible compromise between the kinetic and potential energy so as to obtain the minimum total energy.

9.2 The Schrödinger equation

9.2.1 The Hamiltonian of the Schrödinger equation

We have seen in Section 8.4.1 that the most general time-independent Hamiltonian compatible with Galilean invariance in dimension $d = 1$ is given by (8.68):

$$H = \frac{P^2}{2m} + V(X), \tag{9.53}$$

where $K = P^2/2m$ is the kinetic energy operator and $V(X)$ is the potential energy operator, or briefly the *potential*. We also recall the evolution equation (4.11):

$$i\hbar \frac{d|\varphi(t)\rangle}{dt} = H|\varphi(t)\rangle. \tag{9.54}$$

We multiply both sides of this equation on the left by the bra $\langle x|$ taking (9.53) as the Hamiltonian:

$$i\hbar \frac{d}{dt} \langle x|\varphi(t)\rangle = i\hbar \frac{\partial}{\partial t} \varphi(x, t),$$

$$\langle x|P^2|\varphi(t)\rangle = (P^2\varphi)(x, t) = \left(-i\hbar \frac{\partial}{\partial x}\right)^2 \varphi(x, t) = -\hbar^2 \frac{\partial^2 \varphi(x, t)}{\partial x^2},$$

$$\langle x|V(X)|\varphi(t)\rangle = V(x)\varphi(x, t),$$

where we have used (9.8) and (9.16). We thus obtain the time-dependent Schrödinger equation:

$$\boxed{i\hbar \frac{\partial \varphi(x, t)}{\partial t} = -\frac{\hbar^2}{2m} \frac{\partial^2 \varphi(x, t)}{\partial x^2} + V(x)\varphi(x, t)}, \tag{9.55}$$

which is a wave equation for the wave function $\varphi(x, t)$.

Since the potential $V(X)$ is independent of time, we know that there exist stationary solutions of (9.54):

$$|\varphi(t)\rangle = \exp\left(-i\frac{Et}{\hbar}\right)|\varphi(0)\rangle, \quad H|\varphi(0)\rangle = E|\varphi(0)\rangle. \tag{9.56}$$

Multiplying on the left by the bra $\langle x|$, the equation $H|\varphi\rangle = E|\varphi\rangle$ becomes the time-independent Schrödinger equation:

$$\boxed{\left[-\frac{\hbar^2}{2m} \frac{\partial^2}{\partial x^2} + V(x)\right] \varphi(x) = E\varphi(x)}. \tag{9.57}$$

Equation (9.55) can be generalized in two ways. While remaining compatible with Galilean invariance, it is possible to add a time dependence to the potential: $V(x) \rightarrow V(x, t)$. It is also possible to use velocity-dependent potentials, for example to approximate relativistic effects. In this case the Galilean invariance is lost, and moreover ambiguities may be introduced when it is necessary to choose the ordering of a product of position and momentum operators.

9.2.2 The probability density and the probability current density

With the probability density $|\varphi(x, t)|^2$ we can associate a current density $j(x, t)$ by analogy with hydrodynamics or electrodynamics. Let us recall the example of hydrodynamics to see how this works. Let $\rho(\vec{r}, t)$ be the mass density of a compressible fluid of total mass M

flowing with local velocity $\vec{v}(\vec{r}, t)$.[4] The *current density* (or simply current) $\vec{j}(\vec{r}, t)$ is defined as

$$\vec{j}(\vec{r}, t) = \rho(\vec{r}, t)\,\vec{v}(\vec{r}, t).\tag{9.58}$$

We consider a surface $\mathcal{S}$ surrounding the volume $\mathcal{V}$, which contains a mass $M(\mathcal{V})$ of fluid (Fig. 9.3). The mass $\mathrm{d}M(\mathcal{V})/\mathrm{d}t$ of fluid leaving $\mathcal{V}$ per unit time is equal to the flux of current through $\mathcal{S}$:

$$\frac{\mathrm{d}M(\mathcal{V})}{\mathrm{d}t} = \int_{\mathcal{S}} \vec{j}\cdot\mathrm{d}\vec{\mathcal{S}} = \int_{\mathcal{V}} (\vec{\nabla}\cdot\vec{j})\,\mathrm{d}^3 r,$$

where we have used Green's theorem. This fluid mass is also equal to minus the time derivative of the integral of the density over $\mathcal{V}$:

$$\frac{\mathrm{d}M(\mathcal{V})}{\mathrm{d}t} = -\frac{\mathrm{d}}{\mathrm{d}t}\int_{\mathcal{V}}\mathrm{d}^3 r\,\rho(\vec{r}, t) = -\int_{\mathcal{V}}\mathrm{d}^3 r\,\frac{\partial\rho(\vec{r}, t)}{\partial t}.$$

The two expressions for $\mathrm{d}M(\mathcal{V})/\mathrm{d}t$ must be equal for any volume $\mathcal{V}$, which implies that the integrands must be equal. This leads to the *continuity equation*:

$$\boxed{\frac{\partial\rho}{\partial t} + \vec{\nabla}\cdot\vec{j} = 0}.\tag{9.59}$$

In electrodynamics ρ is the charge density and $\vec{j}$ is the current density, which also satisfy a continuity equation of the type (9.59) expressing the local conservation of electric charge. Returning to dimension $d = 1$,

$$\frac{\partial\rho}{\partial t} + \frac{\partial j}{\partial x} = 0.\tag{9.60}$$

In quantum mechanics we expect to find a continuity equation of the type (9.59), or (9.60) in one dimension. If

$$\int_a^b \mathrm{d}x\,|\varphi(x, t)|^2$$

is the probability of finding the particle at time t in the interval $[a, b]$, this probability will in general depend on the time. If, for example, this probability decreases, this indicates

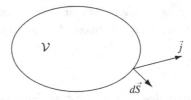

Fig. 9.3. Current and flux leaving a volume $\mathcal{V}$.

---

[4] We temporarily revert to the dimension $d = 3$.

that the probability of finding the particle in the union of the two intervals $[-\infty, a]$ and $[b, +\infty]$ must increase, because for any t the integral

$$\int_{-\infty}^{\infty} dx \, |\varphi(x, t)|^2$$

is constant and equal to unity. Similarly, the integral of the fluid density over all space remains constant and equal to the total mass M, whereas in electrodynamics the integral of the charge density over all space remains constant and equal to the total charge Q. The analog of the density in quantum mechanics is $\rho(x, t) = |\varphi(x, t)|^2$; however, this is a probability density and not an actual density. We shall seek a current $j(x, t)$ satisfying (9.60); this also will be a probability current and not an actual current. The form of this current is suggested by the following argument. In hydrodynamics, the average velocity $\langle v(t) \rangle$ of a fluid (or the velocity of the center of mass) is given by

$$\langle v(t) \rangle = \frac{1}{M} \int \rho(x, t) v(x, t) dx = \frac{1}{M} \int j(x, t) dx. \tag{9.61}$$

In quantum mechanics, the velocity operator according to (8.61) is

$$\dot{X} = \frac{i}{\hbar}[H, X] = \frac{P}{m},$$

and its expectation value is

$$\langle \dot{X} \rangle(t) = \langle \varphi(t) | \frac{P}{m} | \varphi(t) \rangle = \int dx \, \varphi^*(x, t) \frac{\hbar}{im} \frac{\partial \varphi(x, t)}{\partial x},$$

where we have used (9.9) and (9.16). The integrand in this equation is in general complex and is not suitable for the current density. Integration by parts allows us to construct a current which is a real function of x:

$$\langle \dot{X} \rangle(t) = \frac{\hbar}{2im} \int dx \left(\varphi^*(x, t) \frac{\partial \varphi(x, t)}{\partial x} - \varphi(x, t) \frac{\partial \varphi^*(x, t)}{\partial x} \right). \tag{9.62}$$

Comparison of (9.61) for $M = 1$ with (9.62) suggests the following form for the current $j(x, t)$:

$$\boxed{j = \frac{\hbar}{2im} \left(\varphi^*(x, t) \frac{\partial \varphi(x, t)}{\partial x} - \varphi(x, t) \frac{\partial \varphi^*(x, t)}{\partial x} \right) = \text{Re}\left(\frac{\hbar}{im} \varphi^*(x, t) \frac{\partial \varphi(x, t)}{\partial x} \right)}.$$

$$\tag{9.63}$$

In order to familiarize ourselves with this rather unintuitive expression, let us examine the case of a plane wave:

$$\varphi(x) = A \, e^{ipx/\hbar}.$$

The density is $\rho(x) = |A|^2$. The current becomes

$$j(x) = \text{Re}\left(\frac{\hbar}{im} A^* e^{-ipx/\hbar} \left[\frac{ip}{\hbar} \right] A e^{ipx/\hbar} \right) = |A|^2 \frac{p}{m} \tag{9.64}$$

and is interpreted as current = density × velocity. The current points to the right if $p > 0$ and to the left if $p < 0$. When the wave function is independent of time, as in the case of a plane wave, the current is necessarily independent of x since $\partial\rho/\partial t = 0 \Rightarrow \partial j/\partial x = 0$. We still need to check that the current (9.63) is actually the current that satisfies the continuity equation (9.60). On the one hand

$$\frac{\partial j}{\partial x} = \frac{\hbar}{2im}\left[\varphi^*\frac{\partial^2\varphi}{\partial x^2} - \varphi\frac{\partial^2\varphi^*}{\partial x^2}\right] = \frac{i}{\hbar}[\varphi^*(H\varphi) - \varphi(H\varphi)^*],$$

where we have used

$$\frac{\hbar}{2im}\frac{\partial^2\varphi}{\partial x^2} = \frac{i}{\hbar}[(H-V)\varphi]$$

and the fact that V is a real function of x and t. On the other hand

$$\frac{\partial}{\partial t}|\varphi(x,t)|^2 = \varphi^*\frac{\partial\varphi}{\partial t} + \varphi\frac{\partial\varphi^*}{\partial t} = \frac{1}{i\hbar}[\varphi^*(H\varphi) - (H\varphi)\varphi^*],$$

which shows that

$$\frac{\partial}{\partial t}|\varphi(x,t)|^2 + \frac{\partial}{\partial x}j(x,t) = 0. \tag{9.65}$$

9.3 Solution of the time-independent Schrödinger equation

9.3.1 Generalities

The sections 9.3 to 9.5 will be devoted to finding the solutions of the time-independent Schrödinger equation (9.57), that is, the eigenvalues E and the corresponding eigenfunctions $\varphi(x)$. We start with the simplest case where the potential $V(x) = 0$. The equation (9.57) becomes

$$\left(\frac{\partial^2}{\partial x^2} + \frac{2mE}{\hbar^2}\right)\varphi(x) = 0. \tag{9.66}$$

The general solution of this equation is a combination of plane waves with $p = \sqrt{2mE} > 0$,

$$\varphi(x) = A\,e^{ipx/\hbar} + B\,e^{-ipx/\hbar} \tag{9.67}$$

propagating toward the positive x direction with amplitude A and the negative x direction with amplitude B. Since the solution (9.67) is independent of time, it generates a stationary current,[5] which according to (9.64) consists of a term $|A|^2 p/m$ pointing to positive x and a term $-|B|^2 p/m$ pointing to negative x. To the time-independent solutions $\exp(\pm ipx/\hbar)$ there correspond time-dependent solutions of (9.55), namely, $\exp[i(\pm px - E(p)t)/\hbar]$, which are traveling waves propagating in the positive or negative x direction. The traveling waves $\exp[i(+px - E(p)t)/\hbar]$ can be combined to form wave packets propagating in the positive x direction, and we say that these wave packets originate from a *source* of particles at $x = -\infty$. From the traveling waves $\exp[i(-px - E(p)t)/\hbar]$ we can construct

---

[5] An example of a stationary current is the d.c. electric current.

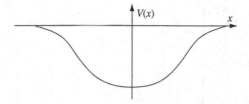

Fig. 9.4. A potential well.

wave packets propagating in the negative x direction, corresponding to a source of particles at $x = +\infty$.

Let us consider the case $V(x) \neq 0$ and, to be specific, assume that $V(x)$ has the form in Fig. 9.4, that of a "potential well" with $V(x) \to 0$ if $x \to \pm\infty$. In classical mechanics, from the discussion of Section 1.5.1, this potential has bound states if $E < 0$ and scattering states if $E > 0$. For $E < 0$ the classical particle remains confined in a finite range of the x axis, and for $E > 0$ it travels to infinity. The range of the x axis allowed for the classical particle is that for which $E > V(x)$ and the momentum $p(x)$ is real:

$$p(x) = \pm\sqrt{2m(E - V(x))}, \tag{9.68}$$

while the region $E < V(x)$ where the momentum is imaginary,

$$p(x) = \pm i\sqrt{2m(V(x) - E)}, \tag{9.69}$$

is forbidden. We shall see that this classical behavior is reflected in the quantum behavior: the form of the solutions of (9.57) will differ depending on whether $p(x)$ is real or imaginary. For $\varphi(x)$ to be an acceptable solution, it is not sufficient that it formally satisfies (9.57); $\varphi(x)$ must also be normalizable:

$$\int_{-\infty}^{\infty} dx \, |\varphi(x)|^2 < \infty.$$

It is this condition which we shall use to obtain the bound states. However, it is too strong for the scattering states. We have seen that for $V(x) = 0$ the solutions of (9.57) are non-normalizable plane waves. For $x \to \pm\infty$ we expect the solutions of (9.57) to have plane-wave behavior because the potential vanishes at infinity. For the scattering states $E > 0$ of the potential in Fig. 9.4 we shall demand only plane-wave behavior at infinity: one should not require more from the solution in the presence of the potential than in its absence!

9.3.2 Reflection and transmission by a potential step

In the rest of this section we shall be interested in the case where the potential is piecewise-constant, that is, $V(x)$ is constant in some range and then jumps abruptly to another constant value at certain points (Fig. 9.5). This type of potential represents a good approximation of an actual potential in certain cases and can be used to approximate

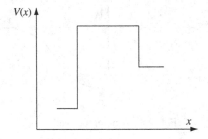

Fig. 9.5. A piecewise-constant potential.

a potential which varies continuously in other cases (Fig. 9.6). Since the potential has discontinuities, it is necessary to examine the behavior of the wave function in the neighborhood of one. We shall show that the wave function $\varphi(x)$ and its derivative $\varphi'(x)$ are continuous if the potential has a *finite* discontinuity V_0 at the point $x = x_0$ (Fig 9.7). Since $|\varphi(x)|^2$ must be integrable at x_0, $|\varphi(x)|$ must be also. It will be convenient to rewrite the time-independent Schrödinger equation (9.57) as

$$\left(\frac{\partial^2}{\partial x^2} + \frac{2m(E - V(x))}{\hbar^2} \right) \varphi(x) = 0. \tag{9.70}$$

We can find the behavior of $\varphi'(x)$ in the neighborhood of the discontinuity using

$$\varphi'(x_0 + \varepsilon) - \varphi'(x_0 - \varepsilon) = \int_{x_0 - \varepsilon}^{x_0 + \varepsilon} \frac{\partial^2 \varphi(x)}{\partial x^2} \, dx = \int_{x_0 - \varepsilon}^{x_0 + \varepsilon} \left[\frac{2m(V(x) - E)}{\hbar^2} \right] \varphi(x).$$

The second integral is well defined because $\varphi(x)$ is integrable. This integral must tend to zero with ε, which shows that $\varphi'(x)$ and *a fortiori* $\varphi(x)$ are continuous as long as the discontinuity V_0 is finite.

Instead of writing down the continuity equations for $\varphi(x)$ and $\varphi'(x)$, it is often convenient to write them down for $\varphi(x)$ and its logarithmic derivative $\varphi'(x)/\varphi(x)$. An immediate consequence of these conditions is that the current $j(x)$ is equal to the same

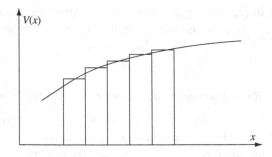

Fig. 9.6. Approximation of a potential by a sequence of steps.

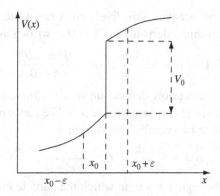

Fig. 9.7. A discontinuity in the potential.

constant on both sides of x_0. As an application of these continuity conditions, we take the case of a "step potential" (Fig. 9.8):

$$\text{region I}: V(x) = 0 \quad \text{for} \quad x < 0,$$
$$\text{region II}: V(x) = V_0 \quad \text{for} \quad x > 0.$$

To be specific we first choose $0 < E < V_0$. If we define k and κ as

$$k = \sqrt{\frac{2mE}{\hbar^2}}, \quad \kappa = \sqrt{\frac{2m(V_0 - E)}{\hbar^2}}, \tag{9.71}$$

the solutions of (9.70) are written in regions I and II as

$$\text{I}: \varphi(x) = A\,e^{ikx} + B\,e^{-ikx}, \tag{9.72}$$

$$\text{II}: \varphi(x) = C\,e^{-\kappa x} + D\,e^{\kappa x}. \tag{9.73}$$

If $V(x)$ remains equal to V_0 for all $x > 0$, the behavior (9.73) of the wave function remains unchanged for any $x > 0$. It is then necessary that $D = 0$, because otherwise the function $|\varphi(x)|^2$ behaves as $\exp(2\kappa x)$ for $x \to \infty$. Behavior of constant modulus like that of a

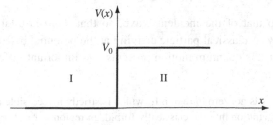

Fig. 9.8. A step potential.

plane wave is acceptable, but behavior this divergent is not. Under these conditions, the continuity of φ and its logarithmic derivative at $x = 0$ is written as

$$\varphi : C = A + B, \quad \frac{\varphi'}{\varphi} : -\kappa = \frac{ik(A - B)}{A + B}.$$

The coefficients A and B are a priori defined up to a multiplicative constant since we have not made any hypotheses about the region $x > 0$. We can arbitrarily set $A = 1$, and then the solution for the other two coefficients becomes

$$B = -\frac{\kappa + ik}{\kappa - ik}, \quad C = -\frac{2ik}{\kappa - ik}. \tag{9.74}$$

Since $C \neq 0$, we see that the region $x > 0$, in which the particle momentum is imaginary (see (9.69)), is not strictly forbidden to the quantum particle. From these expressions we can derive the limiting case of $V_0 \to \infty$, which corresponds to a barrier insurmountable by a classical particle no matter what its energy – that is an infinite potential barrier. Equation (9.71) then shows that $\kappa \to \infty$ and (9.74) that $B \to -1$ and $C \to 0$. The wave function vanishes in region II and remains continuous at the point $x = 0$. However, its derivative $\varphi'(x)$ is discontinuous at this point.

Let us now discuss the physical interpretation of these results. We assume that at $x = -\infty$ we have a source of particles of unit amplitude: $A = 1$. The corresponding incident wave will be partly reflected and partly transmitted by the potential step. If we take as above the case $0 < E < V_0$, we expect that the quantum particle will be reflected with 100% probability, since the corresponding classical particle cannot cross the potential step. On the other hand, in the case $E > V_0$ we can show that the solution of the quantum problem corresponds to partial reflection and partial transmission, whereas a classical particle is 100% transmitted. Let us compare these two cases.

The potential step: total reflection

We have as above $E < V_0$. The wave functions in regions I and II are

$$\text{I} : \varphi(x) = e^{ikx} + B e^{-ikx},$$

$$\text{II} : \varphi(x) = C e^{-\kappa x}.$$

The values of B and C are given by (9.74). We note that $|B| = 1$, and so B is a phase factor, $B = \exp(-i\phi)$. This shows that the reflected wave

$$B e^{-ikx} = e^{-ikx - i\phi}$$

has intensity equal to that of the incident wave, so that there is total reflection at the potential discontinuity. A classical particle arriving at the potential discontinuity will also be reflected. However, the quantum motion presents two important differences compared to the classical motion.

- The probability density is nonzero in region II, which is strictly inaccessible to the classical particle: the depth of penetration into the classically forbidden region is $\ell = 1/\kappa$. This phenomenon parallels that of an evanescent wave in optics.

- If we construct an incident wave packet, the particle will be reflected with a delay τ given by (9.47):

$$\tau = -\hbar \frac{d\phi}{dE},$$

whereas the reflection of the classical particle is instantaneous.

The potential step: reflection and transmission

Now we turn to the case $E > V_0$, assuming as before that the particles are incident from the left and arrive at the potential step, so that in region II the particles can travel only to the right:[6] there is no source of particles at $x = +\infty$, only at $x = -\infty$. We define

$$k' = \sqrt{\frac{2m(E - V_0)}{\hbar^2}}.$$

The wave functions in regions I and II are now

$$I : \varphi(x) = e^{ikx} + B e^{-ikx},$$

$$II : \varphi(x) = C e^{ik'x}.$$

The continuity conditions are

$$1 + B = C, \quad ik' = \frac{ik(1 - B)}{1 + B},$$

so that

$$B = \frac{k - k'}{k + k'}, \quad C = \frac{2k}{k + k'}. \tag{9.75}$$

A classical particle will always cross the potential step (and in the process lose kinetic energy), but in quantum mechanics there exists a reflection probability $|B|^2 \neq 0$, so that $|B|^2 = R$ is the reflection coefficient and $T = 1 - R$ is the transmission coefficient:

$$R = \left(\frac{k - k'}{k + k'} \right)^2, \quad T = 1 - R = \frac{4kk'}{(k + k')^2}. \tag{9.76}$$

It is important to note that $T \neq |C|^2$. In fact, it is not the probability density which must be conserved, but the particle current (or flux). In Fig. 9.9 the particle flux entering the hatched area must be equal to the flux leaving it, or

$$\frac{\hbar k}{m} = \frac{\hbar k}{m} |B|^2 + \frac{\hbar k'}{m} |C|^2, \tag{9.77}$$

which is satisfied for the values (9.75) of B and C. The transmission coefficient is not $|C|^2$, but

$$T = \frac{k'}{k} |C|^2.$$

---

[6] As we have already emphasized, to be completely rigorous it is necessary to construct wave packets from superpositions of plane waves in order to have a truly time-dependent problem describing the motion of a quantum particle.

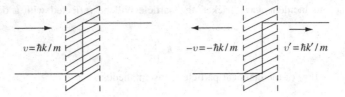

Fig. 9.9. Conservation of the current in crossing a potential step.

It takes into account the change of velocity in crossing the potential step: $v'/v = k'/k$. The loss of kinetic energy is of course the same as in classical mechanics.

9.3.3 The bound states of the square well

As the first example of bound states, let us study those of the infinite square well (Fig. 9.10):

$$V(x) = 0, \qquad 0 \le x \le a,$$

$$V(x) = +\infty, \quad x < 0 \ \text{ or } \ x > a.$$

The potential barriers at $x = 0$ and $x = a$ are infinite: a classical particle is confined to the region $0 \le x \le a$ for any energy. According to the preceding discussion, the wave function of a quantum particle vanishes outside the range $[0, a]$ and so the quantum particle is also strictly confined to the interval $[0, a]$; its probability density is zero outside the range $[0, a]$. Since the wave function vanishes at $x = 0$, the solutions of (9.70) have the form

$$\varphi(x) = A \sin(kx), \quad k = \sqrt{\frac{2mE}{\hbar^2}},$$

and they must also vanish at $x = a$. The values of k then are

$$k = k_n = \frac{\pi(n+1)}{a}, \quad n = 0, 1, 2, 3, \dots \tag{9.78}$$

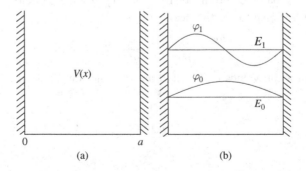

Fig. 9.10. The infinite square well and the wave functions of its first two levels.

We see that the energy takes discrete values labeled by a positive integer n:[7]

$$E_n = \frac{\hbar^2 k_n^2}{2m} = \frac{\hbar^2}{2m} \left(\frac{\pi}{a}\right)^2 (n+1)^2, \quad n = 0, 1, 2, 3, \ldots \, . \qquad (9.79)$$

In other words, we have just shown that the energy levels of the infinite well are *quantized*, and this is the first example in which we have explicitly demonstrated this quantization. The correctly normalized wave function corresponding to the level E_n is

$$\varphi_n(x) = \sqrt{\frac{2}{a}} \sin \frac{\pi(n+1)x}{a}. \qquad (9.80)$$

It is easy to check that two wave functions $\varphi_n(x)$ and $\varphi_m(x)$ are orthogonal for $n \neq m$. The values k_n and $-k_n$ correspond to the same physical state, because the substitution $k_n \to -k_n$ leads to a simple change of sign of the wave function, and a minus sign is a phase factor. This is why we have not included negative values of n in (9.78). We also note that the wave function $\varphi_n(x)$ vanishes n times in the interval $[0, a]$: it is said that the wave function has n *nodes* in this interval. The number of nodes gives a classification of the levels according to increasing energy: the higher the energy, the more nodes there are in the wave function. This is a general result when the potential $V(x)$ is sufficiently regular, which we always assume is the case: if E_n is the energy of the nth level, the corresponding wave function will have n nodes. The ground state wave function E_0 does not vanish. Another remark is that the Heisenberg inequality can be used to find the order of magnitude of the ground-state energy. It gives $p \sim \hbar/x \sim \hbar/a$, from which we find

$$E = \frac{p^2}{2m} \sim \frac{\hbar^2}{2ma^2},$$

in agreement with (9.79) for $n = 0$ up to a factor of π^2. In contrast to the case of the hydrogen atom, the heuristic result differs from the exact result by a factor of ~ 10. This originates in the strong variation of the potential at $x = 0$ and $x = a$ which makes the wave function vanish abruptly, resulting in a large kinetic energy. The expectation value of the kinetic energy in the state φ is

$$\langle K \rangle_\varphi = \langle \varphi | K | \varphi \rangle = -\frac{\hbar^2}{2m} \int dx \, \varphi^*(x) \frac{d^2 \varphi(x)}{dx^2},$$

and it is larger the larger the second derivative of $\varphi(x)$.

Let us now find the energy levels of the finite square well (Fig. 9.11):

$$V(x) = 0, \qquad |x| > a/2,$$

$$V(x) = -V_0, \qquad |x| < a/2.$$

---

[7] Our convention is that $n = 0$ corresponds to the ground state, so as to conform with the usual convention: in general, the ground-state energy is denoted E_0.

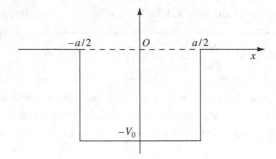

Fig. 9.11. The finite square well.

We seek the bound states, and so we must choose the energy to lie in the range $[-V_0, 0]$. We define k and κ as

$$\kappa = \sqrt{-\frac{2mE}{\hbar^2}}, \quad k = \sqrt{\frac{2m(V_0+E)}{\hbar^2}}, \quad 0 \le \kappa^2 \le \frac{2mV_0}{\hbar^2}. \tag{9.81}$$

The potential $V(x)$ is invariant under the parity operation $\Pi: x \to -x$, as $V(x)$ is an even function of x, $V(-x) = V(x)$, and so the Hamiltonian is also parity-invariant: $H(-x) = H(x)$. Following the discussion of Section 8.3.3, we can seek the eigenvectors $|\varphi_\pm\rangle$ of H which are even or odd under the parity operation:

$$\Pi|\varphi_\pm\rangle = \pm|\varphi_\pm\rangle.$$

In terms of the wave function, if $\langle x|\varphi_\pm\rangle = \varphi_\pm(x)$, then

$$\varphi_+(-x) = \varphi(x), \quad \varphi_-(-x) = -\varphi_-(x)$$

where we have used $\Pi|x\rangle = |-x\rangle$:

$$\langle x|\Pi|\varphi_\pm\rangle = \langle -x|\varphi_\pm\rangle = \varphi_\pm(-x)$$
$$= \pm\langle x|\varphi_\pm\rangle = \pm\varphi_\pm(x).$$

The solutions of the Schrödinger equation (9.57) split up into even and odd ones. In the following display we give these solutions for region I where $x < -a/2$, region II where $|x| < a/2$, and region III where $x > a/2$. The middle column gives the wave functions of the even solutions, and the right-hand column gives the wave functions of the odd ones:

$$\text{I}: A\,\mathrm{e}^{-\kappa|x|} \qquad -A'\mathrm{e}^{-\kappa|x|}$$

$$\text{II}: B\cos(kx) \quad B'\sin(kx)$$

$$\text{III}: A\,\mathrm{e}^{-\kappa x} \qquad A'\mathrm{e}^{-\kappa x}.$$

The continuity conditions on φ'/φ at the point $x = a/2$ give

$$\kappa = k\tan(ka/2) \qquad \text{for even solutions,} \tag{9.82}$$

$$\kappa = -k\cot(ka/2) \quad \text{for odd solutions.} \tag{9.83}$$

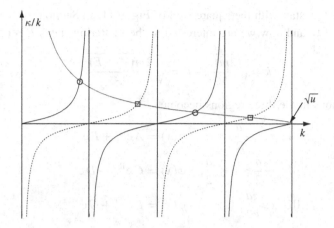

Fig. 9.12. Graphical solution for the bound states of the finite square well, located at points where the curves $\tan ka/2$ (solid line) and $-\cot ka/2$ (dotted line) intersect the curve $\sqrt{U-k^2}/k$, with $U = 2mV_0/\hbar^2$.

The graphical solution of these equations is shown in Fig. 9.12. We see that the number of bound states is finite, and there always exists at least one.

9.4 Potential scattering

9.4.1 The transmission matrix

Now that we have studied bound states, let us turn to scattering states. We shall study the behavior of a particle when it passes over a square well (Fig. 9.11) or a square barrier (Fig. 9.13) using explicit expressions based on the continuity of the wave function and its derivative at a discontinuity of the potential. In the course of our discussion, we shall also be able to derive results which are general as they are independent of the shape of

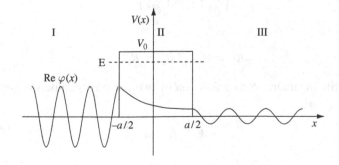

Fig. 9.13. Behavior of the real part of the wave function in the presence of the tunnel effect.

the potential. Let us start with the square well of Fig. 9.11. In Section 9.3.3 we found its bound states $E < 0$, and now we are interested in the scattering states $E > 0$. Defining

$$k = \sqrt{\frac{2mE}{\hbar^2}}, \quad k' = \sqrt{\frac{2m(V_0 + E)}{\hbar^2}}, \tag{9.84}$$

the wave functions in the three regions become

$$\text{I}: x < -\frac{a}{2}, \qquad \varphi(x) = A\,e^{ikx} + B\,e^{-ikx}, \tag{9.85}$$

$$\text{II}: -\frac{a}{2} \leq x \leq \frac{a}{2}, \quad \varphi(x) = C\,e^{ik'x} + D\,e^{-ik'x}, \tag{9.86}$$

$$\text{III}: x > \frac{a}{2}, \qquad \varphi(x) = F\,e^{ikx} + G\,e^{-ikx}. \tag{9.87}$$

Let us first study the passage from region I to region II, that is, the point $x = -a/2$. Since the Schrödinger equation is linear, A and B are linearly related to C and D, which we can write in matrix form:[8]

$$\begin{pmatrix} A \\ B \end{pmatrix} = R \begin{pmatrix} C \\ D \end{pmatrix}, \tag{9.88}$$

where R is a 2×2 matrix. The properties of R can be determined without explicitly writing down the continuity conditions. A first observation is that if $\varphi(x)$ is a time-independent solution of the Schrödinger equation (9.70), then the complex conjugate $\varphi^*(x)$ is also a solution of this equation because the potential $V(x)$ is real. This property is related to the invariance under time reversal; see Section 9.4.3 and Appendix A. The function $\varphi^*(x)$ in regions I and II is

$$\text{I}: \varphi^*(x) = A^*e^{-ikx} + B^*e^{ikx}, \tag{9.89}$$

$$\text{II}: \varphi^*(x) = C^*e^{-ik'x} + D^*e^{ik'x}. \tag{9.90}$$

Comparing the coefficients of $\exp(\pm ikx)$ and $\exp(\pm ik'x)$ with those of (9.85) and (9.86), from (9.88) we find that

$$\begin{pmatrix} B^* \\ A^* \end{pmatrix} = R \begin{pmatrix} D^* \\ C^* \end{pmatrix},$$

or

$$R_{11}^* = R_{22}, \quad R_{12}^* = R_{21}.$$

We can then write the matrix R as a function of two complex numbers α and β:

$$R = \sqrt{\frac{k'}{k}} \begin{pmatrix} \alpha & \beta \\ \beta^* & \alpha^* \end{pmatrix}. \tag{9.91}$$

---

[8] One can also observe that the continuity conditions linearly relate (A, B) to (C, D).

The reason for the introduction of the a priori arbitrary factor $\sqrt{k'/k}$ will become apparent shortly. The current conservation in regions I and II is expressed as (cf. (9.77))

$$k(|A|^2 - |B|^2) = k'(|C|^2 - |D|^2).$$

Let us calculate the current in region I, writing A and B as functions of C and D:

$$k(|A|^2 - |B|^2) = k \frac{k'}{k} \left(|\alpha C + \beta D|^2 - |\beta^* C + \alpha^* D|^2 \right)$$

$$= k' \left(|\alpha|^2 - |\beta|^2 \right) \left(|C|^2 - |D|^2 \right),$$

which implies that $|\alpha|^2 - |\beta|^2 = 1$: the matrix $\sqrt{k/k'}\, R$ has unit determinant. We see why the coefficient $\sqrt{k'/k}$ in (9.91) is of interest: owing to the variation of the velocity between regions I and II, it is the matrix $\sqrt{k/k'}\, R$ which possesses the simplest properties.

Let us now return to the explicit calculation of the continuity conditions in order to find the parameters α and β of the matrix R. It is convenient to choose $C = 1$ and $D = 0$, which corresponds to the situation where there is no source of particles at $x = +\infty$ (see Footnote 6). The continuity conditions then become

$$e^{-ik'a/2} = A e^{-ika/2} + B e^{ika/2},$$

$$k' e^{-ik'a/2} = kA e^{-ika/2} - kB e^{ika/2}.$$

Multiplying the first equation by k' and then adding and subtracting the two equations, we immediately obtain A and B:

$$\alpha = \sqrt{\frac{k}{k'}} A = \frac{k + k'}{2\sqrt{kk'}} \, e^{i(k-k')a/2}, \tag{9.92}$$

$$\beta = \sqrt{\frac{k}{k'}} B^* = \frac{k - k'}{2\sqrt{kk'}} \, e^{i(k+k')a/2}. \tag{9.93}$$

These values of α and β satisfy $|\alpha|^2 - |\beta|^2 = 1$. The continuity equations for $x = a/2$ are obtained by the substitutions $a \to -a$ and $k \leftrightarrow k'$. The matrix $\tilde{R}$ satisfying

$$\begin{pmatrix} C \\ D \end{pmatrix} = \tilde{R} \begin{pmatrix} F \\ G \end{pmatrix}$$

is written as

$$\tilde{R} = \sqrt{\frac{k}{k'}} \begin{pmatrix} \tilde{\alpha} & \tilde{\beta} \\ \tilde{\beta}^* & \tilde{\alpha}^* \end{pmatrix}$$

with

$$\tilde{\alpha} = \frac{k + k'}{2\sqrt{kk'}} \, e^{i(k-k')a/2} = \alpha,$$

$$\tilde{\beta} = -\frac{k - k'}{2\sqrt{kk'}} \, e^{-i(k+k')a/2} = -\beta^*.$$

The transmission matrix M for regions I and III relates the coefficients A and B to the coefficients F and G:

$$\begin{pmatrix} A \\ B \end{pmatrix} = R \begin{pmatrix} C \\ D \end{pmatrix} = R\tilde{R} \begin{pmatrix} F \\ G \end{pmatrix} = M \begin{pmatrix} F \\ G \end{pmatrix}, \tag{9.94}$$

and so we have $M = R\tilde{R}$. The arguments used above immediately give two properties of M.

(i) Since $\varphi^*(x)$ is a solution of (9.57) (invariance under time reversal), we find relations identical to those for R:

$$M_{11} = M_{22}^*, \quad M_{12} = M_{21}^*.$$

(ii) Current conservation implies that $\det M = 1$. There is no factor $\sqrt{k'/k}$ because the velocity is the same in regions I and III.

The general form of M is therefore

$$M = \begin{pmatrix} \gamma & \delta \\ \delta^* & \gamma^* \end{pmatrix}, \quad |\gamma|^2 - |\delta|^2 = 1. \tag{9.95}$$

This expression for M is independent of the form of the potential provided that the latter vanishes sufficiently rapidly for $x \to \pm\infty$; for example, it is valid for the potential of Fig. 9.4. Let us explicitly calculate M for the potential well of Fig. 9.11 using the results obtained for the matrices R and $\tilde{R}$:

$$M_{11} = \gamma = \alpha^2 - \beta^2 = \frac{e^{ika}}{4kk'} \left[(k+k')^2 e^{-ik'a} - (k-k')^2 e^{ik'a} \right]$$

$$= e^{ika} \left[\cos k'a - i \frac{k^2 + k'^2}{2kk'} \sin k'a \right], \tag{9.96}$$

$$M_{12} = \delta = -\alpha\beta^* + \alpha^*\beta = i \frac{k'^2 - k^2}{2kk'} \sin k'a. \tag{9.97}$$

It is instructive to check, using (9.95), that the expressions (9.96) and (9.97) satisfy $|\gamma|^2 - |\delta|^2 = 1$.

There is a general property of M which we have not yet used. When the potential is parity-invariant, $V(x) = V(-x)$, the parity operation $x \to -x$ exchanges regions I and III. If $\varphi(x)$ is the initial solution and $\chi(x) = \varphi(-x)$, we have

$$\text{I}: \chi(x) = F e^{-ikx} + G e^{ikx},$$

$$\text{III}: \chi(x) = A e^{-ikx} + B e^{ikx},$$

and the relation between the various coefficients is now

$$\begin{pmatrix} G \\ F \end{pmatrix} = M \begin{pmatrix} B \\ A \end{pmatrix}$$

or

$$\begin{pmatrix} B \\ A \end{pmatrix} = M^{-1} \begin{pmatrix} G \\ F \end{pmatrix} = \begin{pmatrix} M_{22} & -M_{12} \\ -M_{21} & M_{11} \end{pmatrix} \begin{pmatrix} G \\ F \end{pmatrix}.$$

We have used $\det M = 1$. Comparing with (9.94), we find that M is an antisymmetric matrix, $M_{12} = -M_{21}$, which together with $M_{12}^* = M_{21}$ implies that δ is purely imaginary, $\delta = i\eta$, with η real. This property is satisfied by (9.97). The general form of M for an even potential $[V(x) = V(-x)]$ then is

$$M = \begin{pmatrix} \gamma & i\eta \\ -i\eta & \gamma^* \end{pmatrix} \quad |\gamma|^2 - \eta^2 = 1, \tag{9.98}$$

with γ complex and η real.

All of these results can be used to calculate the reflection and transmission coefficients for the potential well of Fig. 9.11 and to understand their behavior. We shall return to this subject in Exercise 9.7.8. Now we go directly to the case of a potential barrier, which will lead to discussion of the tunnel effect.

9.4.2 The tunnel effect

Let us consider the potential barrier of Fig. 9.13:

$$V(x) = V_0, \quad |x| \le \frac{a}{2},$$

$$V(x) = 0, \quad |x| > \frac{a}{2}, \tag{9.99}$$

for energy $E < V_0$ (the case $E > V_0$ is solved immediately using the results of the preceding subsection). The quantity k' then is purely imaginary:

$$k' = i\kappa, \quad \kappa = \sqrt{\frac{2m(V_0 - E)}{\hbar^2}}, \tag{9.100}$$

and the wave function in region II, $|x| \le a/2$, is

$$\varphi(x) = C e^{-\kappa x} + D e^{\kappa x}. \tag{9.101}$$

The element M_{11} of the transmission matrix is obtained without calculation by replacing k' by $i\kappa$ in (9.96); this gives, for example,

$$\sin k'a = \frac{1}{2i} \left(e^{ik'a} - e^{-ik'a} \right) \rightarrow \frac{1}{2i} \left(e^{-\kappa a} - e^{\kappa a} \right) = i \sinh \kappa a$$

and similarly $\cos k'a \rightarrow \cosh \kappa a$. The result for M_{11} then is

$$M_{11} = e^{ika} \left[\cosh \kappa a + i \frac{\kappa^2 - k^2}{2\kappa k} \sinh \kappa a \right]. \tag{9.102}$$

We assume that the particle source is located at $x = -\infty$ and we adopt the normalization $A = 1$. Since there is no particle source at $x = +\infty$, we must have $G = 0$, which gives

$$\begin{pmatrix} 1 \\ B \end{pmatrix} = M \begin{pmatrix} F \\ 0 \end{pmatrix} = \begin{pmatrix} M_{11}F \\ M_{21}F \end{pmatrix}$$

or $F = 1/M_{11}$:

$$F = \frac{e^{-ika}}{\cosh \kappa a + i \dfrac{\kappa^2 - k^2}{2\kappa k} \sinh \kappa a}. \tag{9.103}$$

This leads to an important physical result, namely, the transmission coefficient $T = |F|^2$:

$$\boxed{T = |F|^2 = \frac{1}{1 + \dfrac{q^4}{4k^2\kappa^2} \sinh^2 \kappa a}}, \tag{9.104}$$

where we have defined $q^2 = k^2 + \kappa^2 = 2mV_0/\hbar^2$. The essential point is that $T \neq 0$. Whereas region III is inaccessible to a classical particle incident from $x = -\infty$ with an energy $E < V_0$, a quantum particle has a nonzero probability of passing through the potential barrier. This is called the *tunnel effect*. The origin of this effect is easy to understand: the wave function does not vanish in the region $|x| \leq a/2$ and it can be matched to a plane wave in the region $x > a/2$ (Fig. 9.13).

An approximate expression for T can be obtained in the commonly encountered case $\kappa a \gg 1$:

$$T \simeq \frac{16k^2\kappa^2}{q^4} e^{-2\kappa a}. \tag{9.105}$$

The dominant factor in this equation is the exponential $\exp(-2\kappa a)$. It is possible to derive heuristically a widely used approximation for a potential barrier of any shape when $E < \mathrm{Max}\, V(x)$. Approximating the barrier as a sequence of steps of length Δx as in Fig. 9.6, we can calculate the transmission factor in the range $[x_i, x_i + \Delta x]$:

$$T(x_i) \simeq e^{-2\kappa(x_i)\Delta x}, \quad \kappa(x_i) = \sqrt{\frac{2m(V(x_i) - E)}{\hbar^2}},$$

and for the total transmission factor we find

$$T \simeq \prod_i e^{-2\kappa(x_i)\Delta x} = \exp\left(-2\Delta x \sum_i \kappa(x_i) \right).$$

We recognize this as a Riemann sum, and in the limit $\Delta x \to 0$

$$\boxed{T \simeq \exp\left(-2 \int_{x_1}^{x_2} \sqrt{\frac{2m(V(x) - E)}{\hbar^2}}\, dx \right)}. \tag{9.106}$$

The points x_1 and x_2 are defined by $V(x_1) = V(x_2) = E$. The demonstration we have just given is not rigorous, because the treatment of the turning points x_1 and x_2 is actually rather delicate. An important observation is that the exponential dependence in (9.106) makes the transmission coefficient T extremely sensitive to the height of the barrier and the value of the energy.

The tunnel effect has numerous applications in quantum physics. Here we shall consider only two, α-radioactivity and tunneling microscopy. Alpha-radioactivity is the decay of a heavy nucleus with the emission of an α-particle, that is, a ^{4}He nucleus. Using Z and N to denote the numbers of protons and neutrons in the initial nucleus ($A = Z + N$) (in general, $Z \gtrsim 80$), the nuclear α-decay reaction can be written as

$$(Z, N) \to (Z - 2, N - 2) + {}^4\text{He}. \tag{9.107}$$

An example is the decay of polonium into lead:

$$^{214}_{84}\text{Po} \to {}^{210}_{82}\text{Pb} + {}^4_2\text{He} + 7.8 \text{ MeV}. \tag{9.108}$$

In an approximate theory of α radioactivity, it is assumed that the α-particle pre-exists inside the initial nucleus and for simplicity the problem is assumed to be one-dimensional. If $R \simeq 1.2 \times A^{1/3} \simeq 7 \, \text{fm}$ is the nuclear radius, the α-particle will be subjected to the nuclear potential and the repulsive Coulomb potential between the ^{4}He nucleus of charge 2 (in units of the proton charge) and the final nucleus of charge $(Z - 2)$ assuming that the charge distribution is spherically symmetric. If r is the distance between the helium nucleus and the final nucleus, for $r > R$ we will have

$$V_{\text{Coul}}(r) = \frac{2(Z - 2)e^2}{r^2}.$$

When $r < R$ the attractive nuclear forces dominate the Coulomb forces and the latter can be neglected. The result is the potential shown schematically in Fig. 9.14. It has a potential barrier which would prevent the α-particle from leaving the nucleus if its motion were governed by classical physics. It is the tunnel effect that allows the α-particle to

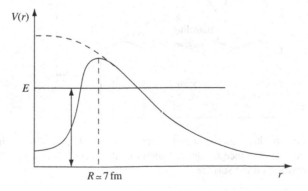

Fig. 9.14. Potential barrier of α-radioactivity.

leave the nucleus. This argument can be used to obtain a theoretical estimate of the lifetime of the initial nucleus, but the approximations we have made are crude and the tunnel effect is very sensitive to the details. While the underlying physics is undoubtedly correct, we cannot expect to obtain results in quantitative agreement with experiment. The reverse of radioactive decay is the fusion reaction; an example is the reaction mentioned in Section 1.1.2:

$$^2H + {}^3H \rightarrow {}^4He + n + 17.6 \text{ MeV},$$

which also involves the tunnel effect and is studied in Exercise 12.5.1.

A very important application of the tunnel effect is scanning tunneling microscopy (STM). In such a microscope a very fine tip is moved over the surface of the conducting sample very close to it (Fig. 9.15). Owing to the tunnel effect, electrons can pass from the tip to the sample, thus producing a macroscopic current that depends very sensitively on the distance between the tip and the sample (the dependence (9.105) is exponential). This allows a very precise mapping of the surface of the sample with a resolution of about 0.01 nm. An extension of this technique can be used to manipulate atoms and molecules deposited on a substrate (Fig. 9.16).

9.4.3 The S matrix

In Chapter 12 we shall study the theory of scattering in three-dimensional space. We shall see that an important tool in this theory is the S matrix, which we introduce here in the simplest case of one dimension. We assume a potential of arbitrary shape which vanishes in the region $|x| > L.$[9] Particle sources at $x = -\infty$ and $x = +\infty$ generate plane waves

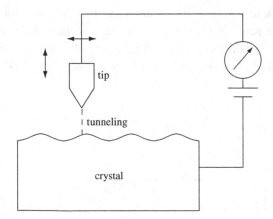

Fig. 9.15. The principle of the scanning tunneling microscope (STM). A fine tip is moved near the surface of a crystal and the distance is adjusted such that the current is constant. This gives a map of the electron distribution on the surface.

---

[9] We can generalize to the case of a potential which vanishes sufficiently rapidly for $x \rightarrow \pm\infty$.

Fig. 9.16. Deposition of atoms by scanning tunneling microscopy. Iron atoms (peaks) are deposited in a circle on a copper substrate and form resonant electron states (waves) on the copper surface. Copyright: IBM.

$\exp(ikx)$ and $\exp(-ikx)$ in the regions $x < -L$ and $x > L$, respectively; we call these the *incoming waves*. These incoming waves can be reflected or transmitted, resulting in outgoing waves $\exp(-ikx)$ in the region $x < -L$ and $\exp(ikx)$ in the region $x > L$. By definition, the S matrix relates the coefficients B and F of the outgoing waves to the coefficients A and G of the incoming waves (cf. (9.85) and (9.87)):

$$\begin{pmatrix} B \\ F \end{pmatrix} = S \begin{pmatrix} A \\ G \end{pmatrix} = \begin{pmatrix} S_{11} & S_{12} \\ S_{21} & S_{22} \end{pmatrix} \begin{pmatrix} A \\ G \end{pmatrix}. \tag{9.109}$$

The S matrix can be expressed as a function of M. However, before deriving the expressions for going from M to S, it is instructive to repeat the arguments that led us to the general properties of M.

(i) Current conservation:

$$|A|^2 - |B|^2 = |F|^2 - |G|^2 \implies |A|^2 + |G|^2 = |B|^2 + |F|^2.$$

This equation shows that the norm of S is conserved and so S is unitary.[10]

(ii) $\varphi^*(x)$ is a solution of the Schrödinger equation, so that

$$\begin{pmatrix} A^* \\ G^* \end{pmatrix} = S \begin{pmatrix} B^* \\ F^* \end{pmatrix} \implies \begin{pmatrix} B \\ F \end{pmatrix} = (S^*)^{-1} \begin{pmatrix} A \\ G \end{pmatrix},$$

---

[10] This argument is valid only for finite dimension: we have proved only that S is an isometry, which is sufficient to make it a unitary operator in finite dimension. It turns out that S is unitary also in infinite dimension, but the proof of this requires additional arguments.

from which we find

$$S = (S^*)^{-1} = (S^{-1})^* = (S^\dagger)^* = S^T.$$

The S matrix is symmetric: $S_{12} = S_{21}$. The operation of complex conjugation exchanges the incoming and outgoing waves, which corresponds to time reversal. The symmetry property $S_{12} = S_{21}$ is therefore related to invariance under time reversal.

Now let us relate S and M in the form (9.95) by calculating the coefficient B:

$$B = S_{11}A + S_{12}G = S_{11}(\gamma F + \delta G) + S_{12}G$$

$$= S_{11}\gamma F + (S_{11}\delta + S_{12})G.$$

We identify

(a) $S_{11}\gamma = \delta^*, \quad S_{11} = \dfrac{\delta^*}{\gamma};$

(b) $S_{12} + S_{11}\delta = \gamma^*, \quad S_{12} = \gamma^* - S_{11}\delta = \dfrac{1}{\gamma},$

or

$$S = \frac{1}{\gamma} \begin{pmatrix} \delta^* & 1 \\ 1 & -\delta \end{pmatrix}. \tag{9.110}$$

If the potential is even $V(x) = V(-x)$, $\delta = i\eta$ with η real and S becomes

$$S = \frac{1}{\gamma} \begin{pmatrix} -i\eta & 1 \\ 1 & -i\eta \end{pmatrix}. \tag{9.111}$$

To write S in the most transparent form possible, we set

$$\gamma = |\gamma|e^{-i\phi}, \quad \frac{\eta}{|\gamma|} = \cos\theta, \quad \frac{1}{|\gamma|} = \sin\theta.$$

The S matrix becomes

$$S = -ie^{i\phi} \begin{pmatrix} \cos\theta & i\sin\theta \\ i\sin\theta & \cos\theta \end{pmatrix}. \tag{9.112}$$

However, we cannot have $\theta = 0$, as this would correspond to $|\gamma| \to \infty$. On the other hand, it is possible to have $\theta = \pm\pi/2$ if $\eta = 0$.

An interesting aspect of the S matrix is that it can be used to relate scattering to bound states and, more generally, to resonances (Exercise 12.5.4). Taking a potential well of arbitrary shape (but such that $V(x) = 0$ outside some finite range in order to simplify the discussion), we choose $E < 0$ with $\kappa = -ik$ given by (9.81). The wave functions in regions I and III are

$$\text{I} : \varphi(x) = A e^{-\kappa x} + B e^{\kappa x},$$

$$\text{III} : \varphi(x) = F e^{-\kappa x} + G e^{\kappa x}.$$

We must have $A = G = 0$ in order for $\varphi(x)$ to be normalizable. Using the relation (9.109), if we want to have $(B, F) \neq 0$, S must have a pole[11] at $k = i\kappa$. This property is general and can be verified for the square well of Fig. 9.11. According to (9.96),

$$\gamma(i\kappa) = e^{-\kappa a} \left[\cos k'a - \frac{k'^2 - \kappa^2}{2\kappa k'} \sin k'a \right].$$

Since S contains an overall factor of $1/\gamma$ (cf. (9.111)), γ must vanish for a bound state. Setting $v = \tan(k'a/2)$, the equation $\gamma = 0$ is equivalent to

$$\kappa k'v^2 + v(k'^2 - \kappa^2) - \kappa k' = 0,$$

whose solutions are $v = \kappa/k'$ and $v = -k'/\kappa$, that is, precisely the relations (9.82) and (9.83) found directly for the finite square well.

9.5 The periodic potential

9.5.1 The Bloch theorem

As a final example of the one-dimensional Schrödinger equation, let us take the case of a periodic potential of spatial period l:

$$V(x) = V(x+l). \tag{9.113}$$

The results that we shall obtain are of great importance in solid-state physics, as an electron in a crystal is subjected to a periodic potential due to its interactions with the ions of the crystal lattice. That case is, of course, three-dimensional, but the results obtained for one dimension generalize to three. The periodicity of the potential leads to the existence of energy bands which, in combination with the Pauli principle, form the basis of our understanding of electrical conductivity. If the potential has the form (9.113), the problem is invariant under any translation $x \rightarrow x + l$, and according to the Wigner theorem there exists a unitary operator T_l acting in the Hilbert space of states, here the space of wave functions $L_x^{(2)}(\mathbb{R})$, such that

$$(T_l\varphi)(x) = \varphi(x-l), \quad T_l^\dagger = T_l^{-1}. \tag{9.114}$$

We recall that the function obtained from $\varphi(x)$ by translation by l is $\varphi(x - l)$. Since the operator T_l is unitary, its eigenvalues t_l have unit modulus and can be written as a function of a parameter q as

$$t_l(q) = e^{-iql}. \tag{9.115}$$

The parameter q is defined up to an integer multiple of $2\pi/l$; if

$$q \rightarrow q' = q + \frac{2\pi p}{l}, \quad p = 0, \pm 1, \pm 2, \ldots, \tag{9.116}$$

---

[11] Or, more generally, a singularity, but it can be shown that bound states and resonances correspond to poles.

the value of t_l is unchanged. Since T_l commutes with the Hamiltonian owing to the periodicity (9.113) of the potential, T_l and H can be diagonalized simultaneously. Let $\varphi_q(x)$ be the common eigenfunctions of T_l and H:

$$T_l\varphi_q(x) = t_l(q)\varphi_q(x) = e^{-iql}\,\varphi_q(x),$$

$$H\varphi_q(x) = E_q\varphi_q(x). \tag{9.117}$$

The first of these equations shows that

$$\varphi_q(x-l) = e^{-iql}\,\varphi_q(x),$$

and we derive the *Bloch theorem*,[12] which states that the stationary states in a periodic potential (9.113) have the form

$$\varphi_q(x) = e^{iqx}\,u_{sq}(x), \quad u_{sq}(x) = u_{sq}(x+l), \tag{9.118}$$

where $u_{sq}(x)$ is a periodic function with period l. The index s is needed because several possible solutions correspond to each value of q; we shall see below that s labels the energy bands. It is easy to write down the differential equation satisfied by $u_{sq}(x)$. Since $P = -i\hbar d/dx$, we have

$$Pe^{iqx} = \hbar q\,e^{iqx},$$

$$P\varphi_q(x) = e^{iqx}\,(P+\hbar q)u_{sq}(x),$$

$$P^2\varphi_q(x) = e^{iqx}\,(P+\hbar q)^2 u_{sq}(x),$$

from which

$$H\varphi_q(x) = e^{iqx}\left[-\frac{\hbar^2}{2m}\frac{d^2}{dx^2} - i\frac{\hbar^2 q}{m}\frac{d}{dx} + \frac{\hbar^2 q^2}{2m} + V(x)\right]u_{sq}(x) = E_{sq}e^{iqx}\,u_{sq}(x),$$

or, dividing by $\exp(iqx)$,

$$\left[-\frac{\hbar^2}{2m}\frac{d^2}{dx^2} - i\frac{\hbar^2 q}{m}\frac{d}{dx} + \frac{\hbar^2 q^2}{2m} + V(x)\right]u_{sq}(x) = E_{sq}u_{sq}(x). \tag{9.119}$$

The wave function in a periodic potential is obtained by solving (9.119) in, for example, the range $[0, l]$ with the boundary condition $u_{sq}(0) = u_{sq}(l)$. The quantity $\hbar q$ has the dimensions of momentum and is in some ways analogous to a momentum. However, it is not a true momentum, because according to (9.116) q is not unique; $\hbar q$ is therefore called a *quasi-momentum*. Finally, we note that if the potential is even, $V(x) = V(-x)$, then (9.119) is unchanged under the simultaneous transformations $x \to -x$, $q \to -q$; $u_{s,-q}(x)$ is therefore a solution of (9.119) with the same value of the energy, $E_{sq} = E_{s,-q}$, and all levels are doubly degenerate.

---

[12] This theorem is also known as the Floquet theorem in the case of periodicity in time.

9.5.2 Energy bands

Let us now examine the properties of the solutions of the Schrödinger equation (9.119) for the periodic potential of Fig. 9.17. Here $V(x)$ is a series of potential barriers and $V(x)$ is nonzero in intervals centered on $x = pl$, $p = \ldots, -2, -1, 0, 1, 2 \ldots$ and vanishes in the intervals[13]

$$\left(p - \frac{1}{2}\right)l - \Delta x \leq x \leq \left(p - \frac{1}{2}\right)l + \Delta x. \tag{9.120}$$

In the intervals where $V(x)$ vanishes a solution $\varphi(x)$ of the Schrödinger equation is a superposition of plane waves with wave vector $\pm k$, $k = (2mE/\hbar^2)^{1/2}$. To the left of the nth barrier and in the interval (9.120) for $p = n$, $\varphi(x)$ is written as

$$\varphi(x) = A_n e^{ikx} + B_n e^{-ikx},$$

and to the right of this barrier, in the interval (9.120) with $p = n + 1$,

$$\varphi(x) = A_{n+1} e^{ikx} + B_{n+1} e^{-ikx}.$$

The coefficients (A_n, B_n) are related to the coefficients (A_{n+1}, B_{n+1}) as in (9.94) by the transmission matrix M (9.95) corresponding to a barrier $V(x)$:

$$\begin{pmatrix} A_n \\ B_n \end{pmatrix} = \begin{pmatrix} \gamma & \delta \\ \delta^* & \gamma^* \end{pmatrix} \begin{pmatrix} A_{n+1} \\ B_{n+1} \end{pmatrix}. \tag{9.121}$$

However, using the Bloch theorem (9.118) we find

$$\varphi(x + l) = e^{iql} \varphi(x),$$

so that

$$A_{n+1} e^{ikl} e^{ikx} + B_{n+1} e^{-ikl} e^{-ikx} = e^{iql} \left(A_n e^{ikx} + B_n e^{-ikx} \right)$$

or

$$e^{iql} \begin{pmatrix} A_n \\ B_n \end{pmatrix} = \begin{pmatrix} e^{ikl} A_{n+1} \\ e^{-ikl} B_{n+1} \end{pmatrix} = D \begin{pmatrix} A_{n+1} \\ B_{n+1} \end{pmatrix} = DM^{-1} \begin{pmatrix} A_n \\ B_n \end{pmatrix}. \tag{9.122}$$

Here D is a diagonal matrix with elements $D_{11} = \exp(ikl)$, $D_{22} = \exp(-ikl)$ and

$$DM^{-1} = \begin{pmatrix} \gamma^* e^{ikl} & -\delta e^{ikl} \\ -\delta^* e^{-ikl} & \gamma e^{-ikl} \end{pmatrix}. \tag{9.123}$$

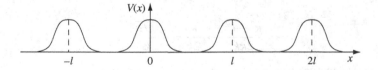

Fig. 9.17. A periodic potential of period l in one dimension.

---

[13] In fact, it is not necessary to assume this vanishing to obtain the following results, but it simplifies the discussion.

Equation (9.122) implies that (A_n, B_n) is an eigenvector of the matrix $\tilde{M} = DM^{-1}$ with eigenvalue $\exp(iql)$, which has unit modulus. The eigenvalues λ of the matrix $\tilde{M}$ are given by ($\det \tilde{M} = 1$)

$$\lambda^2 - 2\lambda \mathrm{Re}\,(\gamma^* e^{ikl}) + 1 = 0,$$

and setting $x = \mathrm{Re}\,[\gamma^* \exp(ikl)]$ the eigenvalues $\lambda_\pm$ become

$$\lambda_\pm = x \pm \sqrt{x^2 - 1}, \quad |x| > 1,$$
$$\lambda_\pm = x \pm i\sqrt{1 - x^2}, \quad |x| \leq 1.$$

The case $|x| > 1$ is excluded because the roots cannot have unit modulus as their product is equal to unity and they are real. However, the two complex roots have unit modulus for $|x| \leq 1$; they are nondegenerate if $|x| < 1$ and degenerate if $|x| = 1$.

To study the energy eigenvalues we could use the example of the rectangular barrier $V(x)$ (9.99) of Fig. 9.13. In order to simplify the calculations as much as possible, we shall study a limiting case of (9.99) where the barrier becomes a delta function. Our results can be qualitatively generalized to any periodic potential. The periodic potential (9.113) then is

$$V(x) = \sum_{p=-\infty}^{\infty} \frac{\hbar^2 g}{2m} \delta(x - lp). \tag{9.124}$$

The delta-function potential is obtained by taking the limit $a \to 0$ of the barrier (9.99) while keeping the product $V_0 a$ constant:

$$V_0 a = \frac{\hbar^2 g}{2m}.$$

The arbitrary factor $\hbar^2/2m$ is chosen so as to simplify the expressions which follow. Taking $V_0 \gg E$, we find that κ (9.100) has the limit

$$\kappa \to \sqrt{\frac{2mV_0}{\hbar^2}} = \sqrt{\frac{g}{a}},$$

which gives

$$\frac{\kappa^2 - k^2}{2\kappa k} \to \frac{\kappa}{2k} = \frac{\sqrt{g/a}}{2k},$$

while $\gamma = M_{11}$ in (9.102) becomes (see also Exercise 9.7.7)

$$\gamma \to 1 + i\frac{\sqrt{g/a}}{2k}\sqrt{ga} = 1 + i\frac{g}{2k}. \tag{9.125}$$

We then find

$$x = \text{Re}\,(\gamma^* e^{ikl}) = \cos kl + \frac{g}{2k}\sin kl,$$

and the eigenvalue equation is written as

$$x = \cos ql = \cos kl + \frac{g}{2k}\sin kl. \tag{9.126}$$

It should be noted that q is not fixed uniquely by (9.126), as $q' = q + 2\pi p/l$ with integer p also satisfies (9.126). This equation shows that certain ranges of k, and therefore certain energy ranges owing to $E = \hbar^2 k^2/2m$, are excluded because the right-hand side of (9.126) can have modulus greater than unity. These ranges are called *forbidden bands*. Let us demonstrate this explicitly in the region $k \simeq 0$. We set $y = kl$ and

$$f(y) = \cos y + \frac{gl}{2y}\sin y.$$

Since $f(0) = 1 + gl/2$, we see that the range $0 \leq y < y_0$ or $0 \leq k < k_0$ is forbidden. Assuming that $gl \ll 1$ in order to make an analytic estimate, we find

$$y_0 \simeq \sqrt{gl} \quad \text{or} \quad k_0 \simeq \sqrt{g/l}.$$

Other forbidden bands exist; in fact, if

$$y = n\pi + \varepsilon, \quad |\varepsilon| \ll 1,$$

then

$$|f(y)| \simeq 1 + \frac{gl}{2y}\varepsilon,$$

and we see that there is a forbidden region where $|f(y)| > 1$ for $0 < \varepsilon \ll 1$. These remarks allow us to qualitatively sketch the curve $f(y)$ in Fig. 9.18. We adopt the convention where E is a function of q (recalling that $\hbar q$ is the quasi-momentum), which gives Fig. 9.19, in which the allowed bands labeled by s are displayed. Using (9.116), q can be restricted to the range $[0, 2\pi/l]$, or, equivalently, the range $[-\pi/l, \pi/l]$, which is called the *first Brillouin zone*. In certain regions E can be expressed simply as a function of q. For example, let us examine the region $k \simeq k_0$. Since $\cos ql = 1$ for $k = k_0$, (9.126) becomes, taking $f(k_0 l) = 1$,

$$-\frac{1}{2}q^2 l^2 \simeq (k - k_0)lf'(k_0 l).$$

This allows us to estimate $(E - E_0)$:

$$E - E_0 = \frac{\hbar^2}{2m}(k^2 - k_0^2) \simeq \frac{\hbar^2 k_0(k - k_0)}{m},$$

or

$$E - E_0 = \frac{\hbar^2 l k_0}{2m|f'(k_0 l)|}q^2 = \frac{\hbar^2}{2m^*}q^2. \tag{9.127}$$

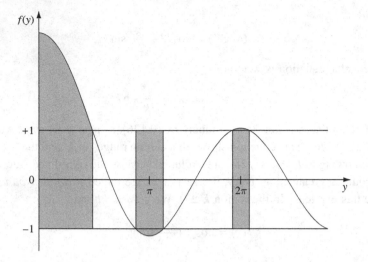

Fig. 9.18. Solutions of (9.126).

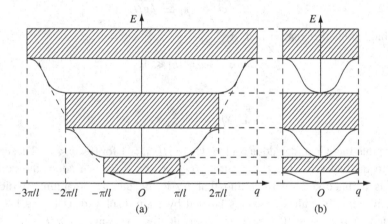

Fig. 9.19. Energy bands. (a) q varies without restrictions; (b) q is limited to the first Brillouin zone. The hatched regions correspond to forbidden bands.

In the neighborhood of $k = k_0$ the behavior of the energy is that of a particle of effective mass m^*:

$$m^* = \frac{m|f'(k_0 l)|}{l k_0}. \tag{9.128}$$

This effective mass plays an important role in the theory of electrical conductivity. To a first approximation the effect of the crystal lattice amounts to a simple change of the mass.

9.6 Wave mechanics in dimension $d = 3$

9.6.1 Generalities

Let $\vec{R}$ and $\vec{P}$ be the position and momentum operators in three-dimensional space with components X_j and P_j, $j = x, y, z$.[14] We recall the canonical commutation relations (8.45):

$$[X_j, P_k] = i\hbar \delta_{jk} I. \tag{9.129}$$

The components of $\vec{R}$ and $\vec{P}$ commute if $j \neq k$. We can then construct the space of states as the tensor product of the spaces $L_x^{(2)}(\mathbb{R})$, $L_y^{(2)}(\mathbb{R})$, and $L_z^{(2)}(\mathbb{R})$:

$$L_{\vec{r}}^{(2)}(\mathbb{R}^3) = L_x^{(2)}(\mathbb{R}) \otimes L_y^{(2)}(\mathbb{R}) \otimes L_z^{(2)}(\mathbb{R}). \tag{9.130}$$

In this space the X component of $\vec{R}$ will be the operator

$$X \otimes I_y \otimes I_z.$$

If $\varphi_n(x)$ is an orthonormal basis of $L_x^{(2)}(\mathbb{R})$, we can construct a basis $\varphi_{nlm}(x, y, z)$ of $L_{\vec{r}}^{(2)}(\mathbb{R}^3)$ by taking the products[15]

$$\varphi_{nlm}(x, y, z) = \varphi_n(x)\varphi_m(y)\varphi_l(z). \tag{9.131}$$

The construction of the space of states and the orthonormal basis is strictly parallel to that of the space of states of two spins 1/2. In Section 6.2.3 we observed that the most general state vector of the space of states of two spins 1/2 is not in general a tensor product $|\varphi_1 \otimes \varphi_2\rangle$ of two state vectors of the individual spins. Similarly, a function $\psi(x, y, z)$ of $L_{\vec{r}}^{(2)}(\mathbb{R}^3)$ is not in general a product $\varphi(x)\chi(y)\eta(z)$, but $\psi(x, y, z)$ can be decomposed on the basis (9.131):

$$\psi(x, y, z) = \sum_{n,m,l} c_{nml} \varphi_n(x)\varphi_m(y)\varphi_l(z), \tag{9.132}$$

$$c_{nlm} = \int d^3r \, \varphi_n^*(x)\varphi_m^*(y)\varphi_l^*(z)\psi(x, y, z). \tag{9.133}$$

We can immediately write down the three-dimensional generalization of the equations in Section 9.1. We shall just give a few examples, leaving it to the reader to derive the other expressions.

• The eigenstates $|\vec{r}\rangle$ of $\vec{R}$ (cf. (9.3)):

$$\vec{R}|\vec{r}\rangle = \vec{r}|\vec{r}\rangle. \tag{9.134}$$

• The completeness relation (cf. (9.9)):

$$\int d^3r \, |\vec{r}\rangle\langle\vec{r}| = I. \tag{9.135}$$

---

[14] The components of $\vec{R}$ will also be denoted as (X, Y, Z) and those of $\vec{r}$ will be denoted as (x, y, z).
[15] To simplify the notation, we have taken the same basis functions in the (x, y, z) spaces, but we could of course have chosen three different bases.

- The probability amplitude $\varphi(\vec{r})$ for finding a particle in the state $|\varphi\rangle$ at the point $\vec{r}$, that is, the wave function of the particle:

$$\varphi(\vec{r}) = \langle \vec{r}|\varphi\rangle. \tag{9.136}$$

- The probability density: $|\varphi(\vec{r})|^2 d^3 r$ is the probability of finding the particle in the volume $d^3 r$ about the point $\vec{r}$.
- The action of the operators $\vec{R}$ and $\vec{P}$ on $\varphi(\vec{r})$ [cf. (9.14) and (9.16)]:

$$\boxed{\left(\vec{R}\varphi\right)(\vec{r}) = \vec{r}\varphi(\vec{r}), \quad \left(\vec{P}\varphi\right)(\vec{r}) = -i\hbar\vec{\nabla}\varphi(\vec{r})} \;. \tag{9.137}$$

- The Fourier transform (cf. (9.26)):

$$\boxed{\tilde{\varphi}(\vec{p}) = \frac{1}{(2\pi\hbar)^{3/2}} \int d^3 r \, \varphi(\vec{r}) \, e^{-i\vec{p}\cdot\vec{r}/\hbar}} \;. \tag{9.138}$$

The factor $(2\pi\hbar)^{-1/2}$ for each space dimension should be noted.

In Section 8.4.2 we determined the general form of the Hamiltonian in dimension $d = 3$. In the rest of this section we assume that $\vec{A}$ is a gradient: $\vec{A} = \vec{\nabla}\Lambda(\vec{r})$. Physically, this means that there is no magnetic field; the case of nonzero magnetic field will be studied in Section 11.3. The Hamiltonian (8.74) is simply

$$H = \frac{\vec{P}^2}{2m} + V(\vec{R}). \tag{9.139}$$

The time-independent Schrödinger equation[16] generalizing (9.57) to three dimensions is

$$\boxed{\left(-\frac{\hbar^2}{2m}\vec{\nabla}^2 + V(\vec{r})\right)\varphi(\vec{r}) = E\varphi(\vec{r})} \;. \tag{9.140}$$

The generalization of the probability current (9.63) is

$$\boxed{\vec{j}(\vec{r}, t) = \mathrm{Re}\left[\frac{\hbar}{im}\,\varphi^*(\vec{r}, t)\vec{\nabla}\varphi(\vec{r}, t)\right]}\,, \tag{9.141}$$

which satisfies the continuity equation (Exercise 9.7.10)

$$\frac{\partial|\varphi(\vec{r}, t)|^2}{\partial t} + \vec{\nabla}\cdot\vec{j}(\vec{r}, t) = 0. \tag{9.142}$$

---

[16] We leave to the reader the task of writing down the time-dependent Schrödinger equation that generalizes (9.55) to three dimensions.

9.6.2 The phase space and level density

In many problems it is necessary to know how to count the number of energy levels in a certain region of space $(\vec{r}, \vec{p})$; this space is called the *phase space*. Let us return to the infinite well of Section 9.3.3 and use L_x to denote the width of the well. The energy levels are labeled by a positive integer n, and we shall consider the case where $n \gg 1$ and L_x is large. Then the energy levels (9.79) are very closely spaced and the sums over n can be replaced by integrals. Let us take a wave vector (9.78) with $k_n = \pi(n+1)/L_x$. We shall calculate the number of energy levels in a range of k: $[k_n, k_n + \Delta k]$. According to (9.78) for $a \to L_x$, the number of levels Δn $(1 \ll \Delta n \ll n)$ in the range $[k, k+\Delta k]$ is

$$\Delta n = \frac{L_x}{\pi} \Delta k. \tag{9.143}$$

Instead of vanishing boundary conditions for the wave function at the points $x = 0$ and $x = L_x$, it is often more convenient to choose *periodic boundary conditions*, $\varphi(0) = \varphi(L_x)$, leading to the wave functions[17]

$$\varphi_n(x) = \frac{1}{\sqrt{L_x}} e^{ik_n x}, \quad k_n = \frac{2\pi n}{L_x}, \quad n = \ldots, -2, -1, 0, 1, 2, \ldots, \tag{9.144}$$

and therefore

$$\Delta n = \frac{L_x}{2\pi} \Delta k. \tag{9.145}$$

At first sight (9.145) differs from (9.143) by a factor of $1/2$.[18] However, we have already observed that for the wave functions (9.78) the values k_n and $-k_n$ correspond to the same physical state because the substitution $k_n \to -k_n$ leads to a simple change of sign of the wave function. By contrast, the substitution $k_n \to -k_n$ in (9.144) leads to a different physical state; thus the division by two in (9.145) is compensated for by doubling the number of possible values of k_n. Periodic and vanishing boundary conditions are equivalent for counting the energy levels (see also Footnote 19).

Let us now turn to the infinite square well in dimension $d = 3$. The wave functions vanish outside the ranges where $V(\vec{x}) = 0$, i.e., outside

$$0 \le x \le L_x, \quad 0 \le y \le L_y, \quad 0 \le z \le L_z. \tag{9.146}$$

The wave functions inside the well take the form

$$\varphi_{[n_x, n_y, n_z]}(x, y, z) = \sqrt{\frac{8}{L_x L_y L_z}} \sin\left(\frac{\pi(n_x+1)x}{L_x}\right) \sin\left(\frac{\pi(n_y+1)y}{L_y}\right) \sin\left(\frac{\pi(n_z+1)z}{L_z}\right)$$

$$\tag{9.147}$$

---

[17] This choice of wave function is sometimes called "quantization in a box." It makes it possible to avoid working with plane waves of the continuum, since the "plane waves" of (9.144) are normalizable. However, the Fourier integrals of the continuum case then are replaced by Fourier sums, making the calculations more cumbersome.
[18] Since $n \gg 1$, no distinction is made between n and $(n+1)$.

with $(n_x, n_y, n_z) = 0, 1, 2, \ldots$. The corresponding energies are

$$E(n_x, n_y, n_z) = \frac{\hbar^2 \pi^2}{2m} \left(\frac{(n_x + 1)^2}{L_x^2} + \frac{(n_y + 1)^2}{L_y^2} + \frac{(n_z + 1)^2}{L_z^2} \right). \tag{9.148}$$

When $L_x = L_y = L_z = L$, these eigenvalues are in general degenerate (Exercise 9.7.9). Let us count the levels in three dimensions. It will be convenient to use periodic boundary conditions:

$$\varphi(x, y, z) = \varphi(x + L_x, y + L_y, z + L_z). \tag{9.149}$$

Let $\Delta \mathcal{K}$ be the volume element $\Delta k_x \Delta k_y \Delta k_z$ of $\vec{k}$ space such that the tip of the wave vector $\vec{k}$ lies in $\Delta \mathcal{K}$. The (x, y, z) components of this vector lie in the ranges

$$[k_x, k_x + \Delta k_x], \ [k_y, k_y + \Delta k_y], \ [k_z, k_z + \Delta k_z].$$

The number of energy levels in $\Delta \mathcal{K}$ is found by generalizing (9.145):

$$\Delta n = \left(\frac{L_x}{2\pi} \right) \Delta k_x \left(\frac{L_y}{2\pi} \right) \Delta k_y \left(\frac{L_z}{2\pi} \right) \Delta k_z = \frac{L_x L_y L_z}{(2\pi)^3} \Delta \mathcal{K}. \tag{9.150}$$

Taking $\Delta \mathcal{K}$ to be infinitesimal, $\Delta \mathcal{K} = d^3 k$, we define the *level density* (or *density of states*) $\mathcal{D}(\vec{k})$ in $\vec{k}$ space as follows: $\mathcal{D}(\vec{k}) d^3 k$ is the number of levels in the volume $d^3 k$ centered on $\vec{k}$. According to (9.150),

$$\boxed{\mathcal{D}(\vec{k}) d^3 k = \frac{\mathcal{V}}{(2\pi)^3} d^3 k}, \tag{9.151}$$

where $\mathcal{V} = L_x L_y L_z$ is the volume of the box with sides (L_x, L_y, L_z).[19] Using $\vec{p} = \hbar \vec{k}$, for the level density[20] in $\vec{p}$ space we find

$$\boxed{\mathcal{D}(\vec{p}) = \frac{\mathcal{V}}{(2\pi\hbar)^3} = \frac{\mathcal{V}}{h^3}}. \tag{9.152}$$

This is a very often used result. Now let us find the level density per unit energy.[21] Since $\mathcal{D}(\vec{p})$ depends only on $p = |\vec{p}|$, we have

$$\mathcal{D}(p) = \frac{4\pi \mathcal{V}}{(2\pi\hbar)^3} p^2 = \frac{\mathcal{V}}{2\pi^2 \hbar^3} p^2. \tag{9.153}$$

---

[19] This result is also valid for a box which is not a parallelepiped. The correction terms are powers of $(kL)^{-1}$, where L is the typical scale of the box. The first correction represents a surface term. The difference between periodic and vanishing boundary conditions, which is a surface effect, is also included by this type of correction. Such corrections are negligible in a sufficiently large box.

[20] To be rigorous we should use different notation for the various level densities; however, we use the same letter $\mathcal{D}$ everywhere so as to reduce the amount of notation.

[21] When vanishing boundary conditions on the wave function are used, a factor of $1/8$ is introduced in (9.151) to take into account the fact that the components of $\vec{k}$ are positive. The final result will in any case be the same, because of the factor of $1/2$ difference between (9.143) and (9.145): $(1/2)^3 = 1/8$.

The level density per unit energy $\mathcal{D}(E)$ is

$$\mathcal{D}(E) = \frac{\mathcal{V}}{2\pi^2\hbar^3} p^2 \frac{\mathrm{d}p}{\mathrm{d}E} = \frac{\mathcal{V}}{2\pi^2\hbar^3} mp$$

or

$$\mathcal{D}(E) = \frac{\mathcal{V}m}{2\pi^2\hbar^3}(2mE)^{1/2} \,. \tag{9.154}$$

The number of levels in $[E, E+\mathrm{d}E]$ is $\mathcal{D}(E)\mathrm{d}E$. It is also possible to calculate $\mathcal{D}(E)$ starting from $\Phi(E)$, which is the number of energy levels below E: $\mathcal{D}(E) = \Phi'(E)$ (Exercise 9.7.11). The quantity $\mathcal{D}/\mathcal{V}$ is the level density per unit volume and is independent of the volume.

Noting that $\mathcal{V} = \int_\mathcal{V} \mathrm{d}^3 r$, from (9.152) we find that the number of levels in $\mathrm{d}^3 r\,\mathrm{d}^3 p$ is

$$\mathrm{d}N = \frac{\mathrm{d}^3 r\,\mathrm{d}^3 p}{(2\pi\hbar)^3} = \frac{\mathrm{d}^3 r\,\mathrm{d}^3 p}{h^3} \,, \tag{9.155}$$

where $\mathrm{d}^3 r\,\mathrm{d}^3 p$ is an infinitesimal volume in phase space $(\vec{r}, \vec{p})$. Equation (9.155) can be interpreted as follows: h^3 is the volume of an elementary cell in phase space, and one can assign one energy level to each elementary cell. The Heisenberg inequality explains this: if a particle is confined within a range Δx, its momentum satisfies $p \sim h/\Delta x$, and then (9.155) can be expressed more pictorially as follows. Whereas a classical particle whose state is defined by its position $\vec{r}$ and its momentum $\vec{p}$ occupies a point $(\vec{r}, \vec{p})$ in phase space, a quantum particle must occupy at least a volume $\sim h^3$.

The results (9.153) or (9.154) are very important in quantum statistical mechanics: the probability that a system in thermal equilibrium has energy E (see (1.12) and Footnote 16 of Chapter 1) is

$$\mathsf{p}(E) = \mathcal{N}\,\mathcal{D}(E)\,\mathrm{e}^{-\beta E},$$

where $\mathcal{N}$ is a normalization constant fixed by

$$\int \mathrm{d}E\,\mathsf{p}(E) = 1.$$

9.6.3 The Fermi Golden Rule

The concept of level density will be used in the proof of one of the most important formulas of quantum physics, the *Fermi Golden Rule*, which allows us to calculate the probabilities of transition to scattering states. These are also called continuum states because they belong to the continuous spectrum of the Hamiltonian, which in the present case is $H^{(0)}$ (9.156). Let us consider a physical system governed by a time-dependent Hamiltonian $H(t)$:

$$H(t) = H^{(0)} + W(t), \tag{9.156}$$

where $H^{(0)}$ is time-independent and has known spectrum with eigenvalues E_n and eigen-vectors $|n\rangle$:

$$H^{(0)}|n\rangle = E_n|n\rangle. \tag{9.157}$$

We wish to solve the following problem. At time $t = 0$ the system is in the initial state $|\psi(0)\rangle = |i\rangle$, an eigenstate of $H^{(0)}$ with energy E_i, and we want to calculate the probability $\mathsf{p}_{i \to f}(t)$ of finding it at time t in the eigenstate $|f\rangle$ of $H^{(0)}$ with energy E_f. For this we must find the state vector $|\psi(t)\rangle$ of the system at time t, because

$$\mathsf{p}_{i \to f}(t) = |\langle f|\psi(t)\rangle|^2 \quad \text{with} \quad |\psi(t=0)\rangle = |i\rangle. \tag{9.158}$$

We have already encountered this problem in a simple case. In Chapter 5 we calculated the probability of transition from one level to another for an ammonia molecule in an oscillating electromagnetic field. The Hamiltonian (9.156) generalizes (5.52), with $H^{(0)}$ being the analog of (5.43). We follow the method of Section 5.3.2 adapted to any number of levels. Generalizing (5.53), we decompose the state vector $|\psi(t)\rangle$ on the basis $|l\rangle$ of eigenstates of $H^{(0)}$:

$$|\psi(t)\rangle = \sum_l c_l(t)|l\rangle. \tag{9.159}$$

Multiplying (9.159) on the left by the bra $\langle n|H^{(0)}$, we obtain

$$\langle n|H^{(0)}|\psi(t)\rangle = \sum_l \langle n|H^{(0)}|l\rangle \langle l|\psi(t)\rangle = \sum_l H_{nl}^{(0)} c_l(t)$$

$$= E_n \langle n|\psi(t)\rangle = c_n(t)E_n. \tag{9.160}$$

The system of differential equations obeyed by the coefficients $c_n(t)$ is, according to (4.13),

$$i\hbar\dot{c}_n(t) = \sum_l \left(H_{nl}^{(0)} + W_{nl}(t) \right) c_l(t). \tag{9.161}$$

Still following the method of Section 5.3.2, we eliminate the trivial dependence on t, the factor $\exp(-iE_n t/\hbar)$ in $c_n(t)$ arising from the time evolution due to $H^{(0)}$, by setting

$$c_n(t) = e^{-iE_n t/\hbar} \gamma_n(t), \tag{9.162}$$

which transforms (9.161) into

$$i\hbar\dot{\gamma}_n(t)e^{-iE_n t/\hbar} + E_n c_n(t) = \sum_l H_{nl}^{(0)} c_l(t) + \sum_l W_{nl}(t)\,\gamma_l(t)e^{-iE_l t/\hbar}.$$

Using (9.160), this equation simplifies to become

$$i\hbar\dot{\gamma}_n(t) = \sum_l W_{nl}e^{i\omega_{nl}t}\,\gamma_l(t), \quad \omega_{nl} = \frac{E_n - E_l}{\hbar}. \tag{9.163}$$

The system of differential equations (9.163) generalizes (5.55). The equations are exact, but they are not solvable analytically, except in special cases, and approximations must be made. We shall use the method called *time-dependent perturbation theory*. It is convenient to introduce a real parameter λ, $0 \leq \lambda \leq 1$, multiplying the perturbation W. Then $W \to \lambda W$, which allows the strength of the perturbation to be varied by hand.[22] Perturbation theory amounts to obtaining an approximate solution of the Schrödinger equation in the form of a series in powers of λ and taking $\lambda = 1$ at the end of the calculation. In what follows we shall limit ourselves to first order in λ.[23] At time $t = 0$ the system is assumed to be in the state $|i\rangle$:

$$\gamma_n(0) = \delta_{ni},$$

and we write

$$\gamma_n(t) = \delta_{ni} + \gamma_n^{(1)}(t).$$

When t is sufficiently small, $|\gamma_n^{(1)}(t)| \ll 1$ because the system does not have time to evolve appreciably. Upon introduction of the parameter λ, (9.163) becomes

$$i\hbar \frac{d}{dt} \left(\delta_{ni} + \gamma_n^{(1)}(t) \right) = \sum_l \lambda W_{nl}(t) \left[\delta_{li} + \gamma_l^{(1)}(t) \right] e^{i\omega_{nl}t}.$$

We observe that $\gamma_l^{(1)}(t)$ is of order λ, and that the term $\sum_l \lambda W_{nl}(t) \gamma_l^{(1)}(t)$ will therefore be of order λ^2. This term is negligible to first order in λ, and taking $\lambda = 1$ we find

$$i\hbar \dot{\gamma}_n^{(1)}(t) \simeq W_{ni}(t) \, e^{i\omega_{ni}t}. \tag{9.164}$$

An important special case is that of an oscillating potential:

$$W(t) = A \, e^{-i\omega t} + A^\dagger e^{i\omega t}, \tag{9.165}$$

where A is an operator. It is this type of potential that describes, for example, the interaction of an atom with an oscillating electromagnetic field:

$$\mathcal{E}(t) = \mathcal{E}_0 \, e^{-i\omega t} + \mathcal{E}_0^* \, e^{i\omega t}.$$

If as in Chapter 5 we are interested in a transition $i \to f$ to a well-defined final level $|f\rangle$, the probability amplitude $\langle f|\psi(t)\rangle$ is given up to a phase by $\gamma_f(t) \simeq \gamma_f^{(1)}(t)$, which is the solution of the differential equation (9.164),

$$i\hbar \dot{\gamma}_f^{(1)}(t) = A_{fi} \, e^{-i(\omega - \omega_0)t} + A_{if}^* \, e^{i(\omega + \omega_0)t}, \tag{9.166}$$

with $\omega_0 = \omega_{fi} = (E_f - E_i)/\hbar$. This differential equation can be integrated immediately because the coefficients $A_{fi} = \langle f|A|i\rangle$ are independent of time:

$$\gamma_f^{(1)}(t) = \frac{1}{\hbar} \left[A_{fi} \frac{e^{-i(\omega - \omega_0)t} - 1}{\omega - \omega_0} - A_{if}^* \frac{e^{i(\omega + \omega_0)t} - 1}{\omega + \omega_0} \right]. \tag{9.167}$$

This probability amplitude will be important if $\omega \simeq \pm\omega_0$, that is, as in Chapter 5, at resonance. For $\omega \simeq \omega_0$ we have

$$E_f \simeq E_i + \hbar\omega,$$

---

[22] If the perturbation is due to an interaction with an external field, it can be varied by varying the field.
[23] The complexity of the expressions grows rapidly with increasing powers of λ.

and the system absorbs an energy $\hbar\omega$. If we consider the situation of interaction with an electromagnetic wave, the system absorbs a photon of energy $\hbar\omega$. In the case $\omega \simeq -\omega_0$

$$E_f \simeq E_i - \hbar\omega,$$

and the system gives up an energy $\hbar\omega$, for example, by emitting a photon of energy $\hbar\omega$. To clarify these ideas let us study the first case. The transition probability $\mathsf{p}_{i\to f}(t)$ will be

$$\mathsf{p}_{i\to f}(t) = |\gamma_f^{(1)}(t)|^2 = \frac{1}{\hbar^2}|A_{fi}|^2 t^2 f(\omega - \omega_0; t), \qquad (9.168)$$

where the function f was defined in (5.63):

$$f(\omega - \omega_0; t) = \frac{\sin^2[(\omega - \omega_0)t/2]}{[(\omega - \omega_0)t/2]^2} \simeq \frac{2\pi}{t}\delta(\omega - \omega_0). \qquad (9.169)$$

We recover the results of Section 5.3.3 in a more general case. Within our approximations, a necessary condition for (9.168) to be valid is that $\mathsf{p}_{i\to f}(t) \ll 1$.

However, it is in general impossible to isolate a transition to any particular final state f, and so we are usually interested in a transition to a set of final states close in energy:

$$\Gamma = \sum_f \Gamma_{i\to f}.$$

The summation over f is equivalent to integration over energy if we include the level density $\mathcal{D}(E)$:

$$\sum_f \to \int dE\, \mathcal{D}(E).$$

For example, if the final state corresponds to that of a free particle and if $|A_{fi}|^2$ is isotropic, the level density will be given by (9.154). If $|A_{fi}|^2$ is not isotropic but depends, for example, on the direction of the momentum $\vec{p}$ of the final particle, we will use

$$\mathcal{D}(E) = \frac{\mathcal{V}m}{2\pi^2\hbar^3}(2mE)^{1/2}\frac{d\Omega}{4\pi},$$

where $\Omega = (\theta, \phi)$ defines the direction of $\vec{p}$. Using (9.168) and (9.169), we obtain a *transition probability per unit time* Γ

$$\Gamma = \frac{1}{\hbar^2}\int dE\, |A_{fi}|^2\, \mathcal{D}(E)\, t\, \frac{\sin^2[(\omega - \omega_0)t/2]}{[(\omega - \omega_0)t/2]^2}$$

$$\simeq \frac{1}{\hbar}\int dE\, |A_{fi}|^2\, \mathcal{D}(E)\, 2\pi\, \delta[E - (E_i + \hbar\omega)].$$

Performing the integration, we obtain the Fermi Golden Rule with energy absorption:

$$\Gamma = \frac{2\pi}{\hbar} |A_{fi}|^2 \mathcal{D}(E_f), \quad E_f = E_i + \hbar\omega \quad . \tag{9.170}$$

This equation holds also in the case of energy emission if we take $E_f = E_i - \hbar\omega$, and for a constant potential $V(t)$ if $E_f = E_i$ (Exercise 9.7.12). The calculation is valid under the following conditions.

- The probability of finding the system in the initial state (i) must be close to unity, or

$$\sum_{f \neq i} \mathsf{p}_{i \to f}(t) \ll 1 \quad \text{or, in terms of } \Gamma_{i \to f}, \quad \left(\sum_{f \neq i} \Gamma_{i \to f} \right) t \ll 1,$$

which implies that t must be sufficiently short: $t \ll \tau_2$.
- In the integral over energy E the quantity $f(\omega - (E - E_i)/\hbar; t)$ may be replaced by a delta function:

$$\int \mathrm{d}E\, g(E) f\left(\omega - \frac{E - E_i}{\hbar}; t \right) \to \int \mathrm{d}\omega\, g(E) \frac{2\pi}{t} \delta(E - \hbar\omega_0) = \frac{2\pi}{t} g(E_f).$$

If ΔE_1 is the characteristic range of variation of $g(E) = |A_{fi}|^2 \mathcal{D}(E)$, $\tau_1 = \hbar/\Delta E_1$ must be small compared to t: $t \gg \tau_1$.

In summary, t must lie in the range $\tau_1 \ll t \ll \tau_2$. When the condition $t \ll \tau_2$ is not satisfied, it is sometimes possible to use the resonance approximation to reduce the problem to one of two levels, for which an exact solution exists (Exercise 9.7.12).

An important application of the Fermi Golden Rule is to the decay of an unstable state i (an excited state of an atom or a nucleus, an unstable particle, and so on) to a continuum of states f. The perturbation is then time-independent and $E_f \simeq E_i$ in (9.170). For sufficiently short times the probability of finding the system in the initial unstable state i (survival probability) is

$$\mathsf{p}_{ii}(t) = 1 - \Gamma t \simeq \mathrm{e}^{-\Gamma t}, \quad t \ll \tau_2, \tag{9.171}$$

and it is tempting to identify Γ as the inverse of the lifetime τ: $\Gamma = \hbar/\tau$. The calculation we have just done does not permit us to make this identification, because it is not a priori valid for any t. However, the exponential decay law (9.171) can be generalized to long times using a method due to Wigner and Weisskopf described in Appendix C. This method shows that the spread ΔE of the energy E_f of the final states is $\Delta E = \hbar/\tau = \hbar\Gamma/2$.

9.7 Exercises

9.7.1 The Heisenberg inequalities

1. Let $\varphi(x)$ be a square-integrable function normalized to unity and $I(\alpha)$ the non-negative quantity:

$$I(\alpha) = \int_{-\infty}^{\infty} \mathrm{d}x \left| x\varphi(x) + \alpha \frac{\mathrm{d}\varphi}{\mathrm{d}x} \right|^2 \geq 0,$$

with α a real number. Integrating by parts, show that

$$I(\alpha) = \langle X^2 \rangle - \alpha + \alpha^2 \langle K^2 \rangle,$$

where $K = -\mathrm{i}\,\mathrm{d}/\mathrm{d}x$ and

$$\langle X^2 \rangle = \int_{-\infty}^{\infty} \mathrm{d}x\, x^2 |\varphi(x)|^2, \quad \langle K^2 \rangle = -\int_{-\infty}^{\infty} \mathrm{d}x\, \varphi^*(x) \frac{\mathrm{d}^2\varphi}{\mathrm{d}x^2}.$$

Derive the expression

$$\langle X^2 \rangle \langle K^2 \rangle \geq \frac{1}{4}.$$

2. How should the argument of the preceding question be modified to obtain the Heisenberg inequality

$$\Delta x\, \Delta k \geq \frac{1}{2} \; ?$$

Show that $\Delta x\, \Delta k = 1/2$ implies that $\varphi(x)$ is a Gaussian:

$$\varphi(x) \propto \exp\left(-\frac{1}{2}\sigma^2 x^2\right).$$

9.7.2 Wave-packet spreading

1. Show that $[P^2, X] = -2\mathrm{i}\hbar\, P$.
2. Let $\langle X^2 \rangle(t)$ be the mean square position in the state $|\varphi(t)\rangle$:

$$\langle X^2 \rangle(t) = \langle \varphi(t) | X^2 | \varphi(t) \rangle.$$

Show that

$$\frac{\mathrm{d}}{\mathrm{d}t} \langle X^2 \rangle(t) = \frac{1}{m} \langle PX + XP \rangle$$

$$= \frac{\mathrm{i}\hbar}{m} \int_{-\infty}^{\infty} \mathrm{d}x\, x \left[\varphi \frac{\partial \varphi^*}{\partial x} - \varphi^* \frac{\partial \varphi}{\partial x} \right].$$

Are these results valid if the potential $V(x) \neq 0$?

3. Show that if the particle is free ($V(x) = 0$), then

$$\frac{\mathrm{d}^2}{\mathrm{d}t^2} \langle X^2 \rangle(t) = \frac{2}{m^2} \langle P^2 \rangle = 2v_1^2 = \text{const.}$$

4. Use these results to derive

$$\langle X^2 \rangle(t) = \langle X^2 \rangle(t = 0) + \xi_0 t + v_1^2 t^2, \quad \xi_0 = \left. \frac{\mathrm{d}\langle X^2 \rangle}{\mathrm{d}t} \right|_{t=0},$$

as well as the expression for $(\Delta x(t))^2$:

$$(\Delta x(t))^2 = (\Delta x(t = 0))^2 + [\xi_0 - 2v_0 \langle X \rangle(t = 0)]t + (v_1^2 - v_0^2)t^2$$

with $v_0 = \langle P/m \rangle = \text{const.}$

9.7.3 A Gaussian wave packet

1. We assume that the function $A(k)$ in (9.41) is a Gaussian:

$$A(k) = \frac{1}{(\pi\sigma^2)^{1/4}} \exp\left[-\frac{(k-\bar{k})^2}{2\sigma^2}\right].$$

Show that

$$\int |A(k)|^2 dk = 1, \quad \Delta k = \frac{1}{\sqrt{2}}\sigma,$$

and that the wave function $\varphi(x, t = 0)$ is

$$\varphi(x, t = 0) = \frac{\sigma^{1/2}}{\pi^{1/4}} \exp\left[i\bar{k}x - \frac{1}{2}\sigma^2 x^2\right].$$

Sketch the curve of $|\varphi(x, t = 0)|^2$. What is the width of this curve? Identify the dispersion Δx and show that $\Delta x \Delta k = 1/2$.

2. Calculate $\varphi(x, t)$. Show that if $\hbar\sigma^2 t/m \ll 1$ we have

$$\varphi(x, t) = \exp\left(\frac{i\hbar\bar{k}^2}{2m}t\right)\varphi(x - v_g t, 0), \quad v_g = \frac{\hbar\bar{k}}{m}.$$

3. Calculate $\varphi(x, t)$ exactly:

$$\varphi(x, t) = \left(\frac{1}{\pi\sigma^2}\right)^{1/4}\sigma' \exp\left[i\bar{k}x - i\omega(\bar{k})t - \frac{1}{2}\sigma'^2(x - v_g t)^2\right]$$

with

$$\frac{1}{\sigma'^2} = \frac{1}{\sigma^2} + \frac{i\hbar t}{m}$$

and find $|\varphi(x, t)|^2$. Show that

$$\Delta x^2(t) = \frac{1}{2\sigma^2}\left(1 + \frac{\hbar^2\sigma^4 t^2}{m^2}\right).$$

Interpret this result physically.

4. A neutron leaves a nuclear reactor with a wavelength of 0.1 nm. We assume that the wave function at $t = 0$ is a Gaussian wave packet of width $\Delta x = 1$ nm. How long does it take for the width to double? What distance does the neutron travel during this time?

9.7.4 Heuristic estimates using the Heisenberg inequality

1. If the electron emitted in neutron β decay

$$n \rightarrow p + e^- + \bar{\nu}_e$$

were initially confined inside the neutron with radius of about 0.8 fm, what would its kinetic energy be? What conclusion can be drawn?

2. A quantum particle of mass m moves on the x axis in the harmonic potential

$$V(x) = \frac{1}{2}m\omega^2 x^2.$$

Use the Heisenberg inequality to estimate the energy of its ground state.

9.7.5 The Lennard–Jones potential for helium

1. The potential energy of two atoms separated by a distance r is often well represented by the Lennard–Jones potential:

$$V(r) = \varepsilon\left[\left(\frac{\sigma}{r}\right)^{12} - 2\left(\frac{\sigma}{r}\right)^6\right],$$

where ε and σ are parameters with the dimensions of energy and length, respectively. Calculate the position r_0 of the potential minimum and sketch $V(r)$ qualitatively. Show that near $r = r_0$

$$V(r) \simeq -\varepsilon\left[1 - 36\left(\frac{r - r_0}{r_0}\right)^2\right] = \frac{1}{2}m\omega^2(r - r_0)^2 + V_0.$$

2. In the case of helium, $\varepsilon \simeq 10^{-3}\,\text{eV}$ and $r_0 \simeq 0.3\,\text{nm}$. Calculate the vibration frequency ω and the energy $\hbar\omega/2$ of the ground state. Why does helium remain a liquid even if the temperature $T \to 0$? Does the reasoning hold for the two isotopes ^3He and ^4He?
3. For hydrogen, $\varepsilon \simeq 4\,\text{eV}$. Why does hydrogen become a solid at low temperature? What about the rare gases (argon, neon, etc.)?

9.7.6 Reflection delay

1. The equation (9.74) gives the coefficient B of the reflected wave when an incident wave $\exp(ikx)$ of energy $E = \hbar^2 k^2/2m < V_0$ arrives at a potential step, where V_0 is the step height. Show that $|B| = 1$ and B can be written as $B = \exp(-i\phi)$. Find ϕ and $d\phi/dE$.
2. We assume that the incident wave is a wave packet of the type (9.41),

$$\varphi(x, t) = \int \frac{dk}{\sqrt{2\pi}} A(k)\exp[ikx - i\omega(k)t].$$

What will the reflected wave packet be? Show that the reflection occurs with a delay

$$\tau = -\hbar\frac{d\phi}{dE} > 0.$$

9.7.7 A delta-function potential

We consider a one-dimensional potential of the form

$$V(x) = \frac{\hbar^2 g}{2m}\delta(x),$$

where m is the mass of the particle subject to the potential. This potential sometimes can be used as a convenient approximation. For example, it can represent a potential barrier of width a and height V_0 in the limit $a \to 0$ and $V_0 \to \infty$ with $V_0 a$ constant and equal to $\hbar^2 g/2m$. In the case of a barrier (a repulsive potential) $g > 0$, but we can also model a well (an attractive potential), in which case $g < 0$.

1. Show that g has the dimensions of an inverse length.
2. The function $\varphi(x)$ obeys the Schrödinger equation

$$\left[-\frac{d^2}{dx^2} + g\,\delta(x) \right] \varphi(x) = \frac{2mE}{\hbar^2}\,\varphi(x).$$

Show that the derivative of $\varphi(x)$ satisfies the following equation near $x = 0$:

$$\varphi'(0^+) - \varphi'(0^-) = g\,\varphi(0), \quad \varphi(0^\pm) = \lim_{\varepsilon \to 0^\pm} \varphi(\varepsilon).$$

Assuming $g < 0$, show that there exists one and only one bound state. Determine its energy and the corresponding wave function. Show that we recover these results by taking the limit of a square well with $V_0 a \to \hbar^2 |g|/2m$ and $a \to 0$.

3. *Model of a diatomic molecule.* Assuming always that $g < 0$, we can very crudely model the potential felt by an electron of a diatomic molecule as

$$V(x) = \frac{\hbar^2 g}{2m}\left[\delta(x+l) + \delta(x-l) \right].$$

The nuclear axis is taken as the x axis, and the two nuclei are located at $x = -l$ and $x = +l$. Show that the solutions of the Schrödinger equation can be classified as even and odd. If the wave function is even, show that there exists a single bound state given by

$$\kappa = \frac{|g|}{2}\left(1 + e^{-2\kappa l}\right), \quad \kappa = \sqrt{\frac{2m|E|}{\hbar^2}}.$$

Draw a qualitative sketch of its wave function.
   If the wave function is odd, find the equation giving the energy of the bound state:

$$\kappa = \frac{|g|}{2}\left(1 - e^{-2\kappa l}\right).$$

Is there always a bound state? If not, what condition must be obeyed for there to be one? Qualitatively sketch the wave function when there is a bound state.

4. *The double well and the tunnel effect.* Let us consider the preceding question assuming that $\kappa l \gg 1$. Show that the two bound states form a two-level system whose Hamiltonian is

$$H = \begin{pmatrix} E_0 & -A \\ -A & E_0 \end{pmatrix},$$

and relate A to $\sqrt{T}$, where T is the transmission coefficient due to tunneling between the two wells.

5. *The potential barrier.* Now we are interested in the case $g > 0$, which models a potential barrier. Directly calculate the transmission matrix and show that it is the limit of that in the case of a square barrier if $V_0 a \to g$ and $a \to 0$. Give the expression for the transmission coefficient.

6. *A periodic potential.* An electron moves in a one-dimensional crystal in a periodic potential of period l modeled as

$$V(x) = \sum_{n=-\infty}^{\infty} \frac{\hbar^2 g}{2m} \delta(x - nl).$$

For convenience we take $g > 0$. Show that the periodicity of the potential implies that the wave function, labeled by q, has the form

$$\varphi_q(x - l) = e^{-iql} \varphi_q(x).$$

Hint: examine the action of the operator T_l which translates by l. It is therefore possible to limit ourselves to study of the range $[-l/2, l/2]$. Outside the point $x = 0$ the wave functions are complex exponentials:

$$-\frac{l}{2} \leq x < 0: \quad \varphi_q(x) = A e^{ikx} + B e^{-ikx},$$

$$0 < x \leq \frac{l}{2}: \quad \varphi_q(x) = F e^{ikx} + G e^{-ikx}.$$

Use the conditions on $\varphi'(x)$ to obtain

$$\cos ql = \cos kl + \frac{g}{2k} \sin kl.$$

Show that there exist forbidden regions of energy. Qualitatively sketch the energy E_q as a function of q.

9.7.8 Transmission by a well

1. Show that the transmission coefficient T for the square well of Fig. 9.11 is

$$T = \frac{1}{1 + \left(\dfrac{q^2}{4kk'}\right)^2 \sin^2 k'a}, \quad q^2 = \frac{2mV_0}{\hbar^2}.$$

Show that T passes through a maximum if the de Broglie wavelength in the well $\lambda' = 2\pi/k'$ is of the form $2a/n$, n integer.

2. Qualitatively sketch the curves giving T and the reflection coefficient $1 - T$. This behavior explains, among other things, the Ramsauer–Townsend effect.[24]

9.7.9 Energy levels of an infinite cubic well in dimension $d = 3$

Find the energies of the first six energy levels of the infinite cubic well as a function of the length L of a side of the cube along with their degeneracies.

---

[24] Cf. Lévy-Leblond and Balibar [1990], page 314.

9.7.10 The probability current in three dimensions

Show that the continuity equation

$$\frac{\partial \rho}{\partial t} + \vec{\nabla} \cdot \vec{j} = 0, \quad \rho = |\varphi(\vec{r}, t)|^2$$

holds for the current (9.141).

9.7.11 The level density

1. Calculate the energy level density $\mathcal{D}(E)$ in dimension $d = 2$. Show that it is independent of E.
2. Calculate directly the number of levels $\Phi(E)$ of energy lower than E by counting the number of possible levels in a sphere of radius $|p| = \sqrt{2mE}$ in momentum space and taking into account the boundary conditions. Recover the expression (9.154) for $\mathcal{D}(E)$:

$$\mathcal{D}(E) = \frac{d\Phi(E)}{dE}.$$

3. Calculate the energy level density $\mathcal{D}(E)$ for an ultrarelativistic particle of energy $E = cp$. Generalize to the case $E = (p^2 c^2 + m^2 c^4)^{1/2}$. Show that $d^3 p / E$ is a Lorentz invariant. Owing to this invariance, this expression is often taken as the level density.

9.7.12 The Fermi Golden Rule

1. *Comparison with the Rabi formula.* In a two-level system, the Rabi formula (5.40) gives the exact transition probability between the two levels in the presence of a harmonic perturbation, for example,

$$p_{+\to-}(t) = \frac{\omega_1^2}{\Omega^2} \sin^2 \frac{\Omega t}{2}, \quad \Omega^2 = \left[(\omega - \omega_0)^2 + \omega_1^2\right]^{1/2}.$$

Show that the approximate expression (9.168) is obtained as the limit of the Rabi formula if

- $|\omega - \omega_0| \gg \omega_1$, that is, far from resonance, or
- $\omega_1 t \ll 1$, that is, for sufficiently short times.

2. *A constant potential.* Give the expression for the amplitude (9.167) $\gamma^{(1)}(t)$ and the transition probability per unit time Γ when the potential $W(t)$ of (9.165) is time-independent.

9.7.13 Study of the Stern–Gerlach experiment

1. *Classical study.* We use the notation of Section 3.2.2. The trajectory of the silver atoms (Fig. 3.8) is assumed to lie in the symmetry plane yOz and along the y axis. Show that $\partial B_z / \partial x|_{x=0} = 0$ and $\partial B_z / \partial y = 0$ if edge effects are neglected. Show that an approximate form of the magnetic field satisfying the Maxwell equations between the magnet poles near $x = 0$ and $z = 0$ is

$$\vec{B} = B_0 \hat{z} + b(z\hat{z} - x\hat{x}),$$

where $b = \partial B_z / \partial z|_{z=0}$. The classical expression for the force is $\vec{F} = -\vec{\nabla}(\vec{\mu} \cdot \vec{B})$. Find the components F_x, F_y, and F_z. Show that under the influence of B_0 the magnetic moment $\vec{\mu}$ precesses about the z axis with frequency $\omega = |\gamma B_0|$, where γ is the gyromagnetic ratio, and that if $1/\omega$ is very small compared with the time for the atom to travel between the magnet poles, then the component μ_x gives a vanishing average force. Therefore, it is as though the magnetic moment were subject to an effective force $\vec{F} = b\mu_z \hat{z}$.

2. *Numerical data.* Silver atoms of mass $m = 1.8 \times 10^{-27}$ kg leave an oven with a speed $v \simeq 500\,\mathrm{m\,s^{-1}}$ and a velocity spread $\Delta v \sim 10\,\mathrm{m\,s^{-1}}$. The collimating slits have height $\Delta z = 10^{-4}$ m, the length of the gap is $L = 5 \times 10^{-2}$ m, the magnetic field is $B_0 = 1$ T, and $b = 10^4\,\mathrm{T\,m^{-1}}$. Show that at the exit from the magnet poles the spacing δ between the two trajectories corresponding to $S_z = \hbar/2$ and $S_z = -\hbar/2$ is

$$\delta = \frac{\mu b}{m}\left(\frac{L}{v}\right)^2.$$

Evaluate δ numerically. Calculate the product $\Delta z \Delta p_z$ and show that $\Delta z \Delta p_z \gg \hbar$. The atomic trajectories can therefore be treated classically.

3. *The quantum description.* Let $\varphi_{\pm}(\vec{r}, t)$ be the wave function of an atom with spin in the state $|\pm\rangle$. Show that $\varphi_{\pm}$ satisfies the Schrödinger equation

$$i\hbar \frac{\partial \varphi_{\pm}}{\partial t} = \left(-\frac{\hbar^2}{2m}\nabla^2 \mp \mu B\right)\varphi_{\pm}.$$

We define the average position $\langle \vec{r}_{\pm}\rangle(t)$ and the average momentum $\langle \vec{p}_{\pm}\rangle(t)$ of the wave packets $\varphi_{\pm}(\vec{r}, t)$ as

$$\langle \vec{r}_{\pm}\rangle(t) = \int d^3r\, \vec{r}\, |\varphi_{\pm}(\vec{r}, t)|^2,$$

$$\langle \vec{p}_{\pm}\rangle(t) = \int d^3r\, \varphi_{\pm}^*(\vec{r}, t)\left[-i\hbar \vec{\nabla}\varphi_{\pm}(\vec{r}, t)\right].$$

Write down the evolution equations for these average values by calculating $d\langle \vec{r}_{\pm}\rangle(t)/dt$ and $d\langle \vec{p}_{\pm}\rangle(t)/dt$ using the Ehrenfest theorem (4.26). Show that the spacing δ between the centers of the two wave packets is the same as that calculated in question 2 for classical trajectories.

4. *Parity invariance.* In an experimental configuration for analyzing a spin pointing in the z direction using a Stern–Gerlach apparatus such that $\vec{B} \parallel Ox$, we assume that the spin is deflected preferentially in the direction $x > 0$, for example, $\langle S_x \rangle > 0$. By examining the image of the experiment in a mirror located in the xOy plane, show that such a preferred deflection is excluded if the relevant interactions in the experiment are invariant under parity (which is indeed the case).

9.7.14 The von Neumann model of measurement

1. In the model of quantum measurement imagined by von Neumann, a physical property A of a quantum system S is measured by allowing the system to interact with a (quantum) particle Π whose momentum operator is P. For simplicity we consider the case of one spatial dimension. The interaction Hamiltonian is assumed to be of the form

$$H = g(t)AP,$$

where $g(t)$ is a positive function with a sharp peak of width τ at $t = 0$ and

$$g = \int_{-\infty}^{\infty} g(t)dt \simeq \int_{-\tau/2}^{\tau/2} g(t)dt.$$

We assume that the evolution of S and Π can be neglected during the very short time τ of the interaction between S and Π, which occurs between times t_i and t_f: $t_i \simeq -\tau/2$ and $t_f \simeq \tau/2$. Find the evolution operator (4.14):

$$U(t_f, t_i) \simeq e^{-igAP/\hbar}.$$

2. We assume that the $S + \Pi$ initial state is

$$|\psi(t_i)\rangle = |n \otimes \varphi\rangle,$$

where $|n\rangle$ is an eigenvector of A with, for simplicity, nondegenerate spectrum, $A|n\rangle = a_n|n\rangle$, and $|\varphi\rangle$ is a state of the particle localized near the point $x = x_0$ with dispersion Δx. Show that the final state is

$$|\psi(t_f)\rangle = |n \otimes \varphi_n\rangle \quad \text{with} \quad |\varphi_n\rangle = e^{-igAP/\hbar}|\varphi\rangle.$$

Let $\varphi_n(x) = \langle x|\varphi_n\rangle$ be the final wave function of the particle. Show that

$$\varphi_n(x) = \varphi(x - ga_n).$$

The function $\varphi_n(x)$ then is localized near the point $x_0 - ga_n$, and if $g|a_n - a_m| \gg \Delta x$ for any $n \neq m$, the position of the particle allows one to deduce the value a_n of A so that a measurement of A is obtained. The final state of the particle is perfectly correlated with the value of A and the final state of S because the states $|\varphi_n\rangle$ and $|\varphi_m\rangle$ are orthogonal for $n \neq m$: $\langle \varphi_n|\varphi_m\rangle = \delta_{nm}$.

3. What is the final state of Π if the initial state of S is the linear superposition

$$|\chi\rangle = \sum_n c_n|n\rangle?$$

Show that the probability of observing S in the final state $|n\rangle$ is $|c_n|^2$. The measurement is ideal because it does not modify the probabilities $|c_n|^2$.

9.7.15 The Galilean transformation

Let us consider a classical plane wave, for example a sound wave, propagating along the x axis:

$$f(x, t) = A\cos(kx - \omega t)$$

and a Galilean transformation of velocity v:

$$x' = x + vt, \quad t' = t.$$

1. Show that for a classical wave the transformed amplitude $f'(x', t')$ satisfies

$$f'(x', t') = f(x, t),$$

from which we extract the transformation law of the wave vectors and frequencies:

$$k' = k, \quad \omega' = \omega + vk.$$

What is the physical interpretation of the frequency transformation law? Now let us assume that we are dealing with the de Broglie wave of a particle of mass m. Are the preceding relations compatible with the momentum and energy transformation laws

$$p' = p + mv, \quad E' = E + pv + \frac{1}{2} mv^2?$$

2. Show that for a de Broglie wave we should not require

$$\varphi'(x', t') = \varphi(x, t)$$

but rather

$$\varphi'(x', t') = \exp\left[\frac{\mathrm{i}f(x, t)}{\hbar}\right] \varphi(x, t).$$

Using the relations (prove them)

$$\frac{\partial}{\partial t'} = \frac{\partial}{\partial t} - v\frac{\partial}{\partial x},$$

$$\frac{\partial}{\partial x'} = \frac{\partial}{\partial x},$$

determine the form of the function $f(x, t)$ by requiring that if $\varphi(x, t)$ obeys the Schrödinger equation, $\varphi'(x', t')$ must also.

9.8 Further reading

The results of this chapter are classic and can be found in similar form in most texts on quantum mechanics. One of the clearest expositions is that of Merzbacher [1970], Chapter 6. Lévy-Leblond and Balibar [1990], Chapter 6, also give a very complete discussion with many illustrative examples. See also Messiah [1999], Chapter III; Cohen-Tannoudji et al. [1977], Chapter I; or Basdevant and Dalibard [2002], Chapter 2; this last reference comes with a CD made by M. Joffre which allows the motion of wave packets to be visualized. For the Fermi Golden Rule the reader can consult Messiah [1999], Chapter XVII, or Cohen-Tannoudji et al. [1977], Chapter XIII.

10

Angular momentum

In this chapter we shall study the properties of angular momentum, which we have introduced already in Chapter 8. The fundamental property of angular momentum is that it is the infinitesimal generator of rotations. All the results that we shall obtain in this chapter will be more or less direct consequences of this property. In Section 10.1 we explicitly construct a basis of eigenvectors common to $\vec{J}^2$ and J_z, which are compatible Hermitian operators. The rotation of a physical state, which we have already introduced in Chapter 3 for the photon polarization and for spin 1/2, will be studied in the general case in Section 10.2. Section 10.3 is devoted to orbital angular momentum, which originates in the spatial motion of particles. In Section 10.4 we extend the classical results on motion in a central force field to quantum mechanics, and in Section 10.5 we discuss applications to particle decay and excited states. Finally, in Section 10.6 we study the addition of angular momenta.

NB Throughout this chapter we use a system of units in which $\hbar = 1$.

10.1 Diagonalization of $\vec{J}^2$ and J_z

In Chapter 8 we established the commutation relations (8.31) and (8.32) between the various components of angular momentum. Here we give them again in a system of units in which $\hbar = 1$ (we recall that angular momentum has the same dimensions as $\hbar$, which is why the notation is simpler in this system of units):

$$\boxed{[J_x, J_y] = iJ_z, \qquad [J_y, J_z] = iJ_x, \qquad [J_z, J_x] = iJ_y} \,, \qquad (10.1)$$

or

$$\boxed{[J_k, J_l] = i\sum_m \varepsilon_{klm} J_m} \,. \qquad (10.2)$$

Knowledge of only these commutation relations will permit us to diagonalize the angular momentum, that is, to find the eigenvectors and eigenvalues of suitable combinations of

J_x, J_y, and J_z. Since these three operators do not commute with each other, they cannot be diagonalized simultaneously: the three components of $\vec{J}$ are mutually incompatible physical properties. To choose our combinations of J_x, J_y, and J_z, we observe that $\vec{J}^2$ is a scalar operator (cf. (8.33)) and, according to the result of Section 8.2.3, must commute with the three components of $\vec{J}$:

$$[\vec{J}^2, J_k] = 0,\tag{10.3}$$

as can be verified by explicit calculation (Exercise 10.7.1). The usual choice is to simultaneously diagonalize $\vec{J}^2$ and J_z, and this is often referred to as *quantization of the angular momentum in the z direction*. It is also said that Oz is *chosen as the angular momentum quantization axis*. It is convenient to define the operators $J_\pm = J_\mp^\dagger$ and J_0 as

$$J_\pm = J_x \pm iJ_y, \quad J_0 = J_z.\tag{10.4}$$

We can immediately verify the commutation relations and the following identities:

$$[J_0, J_\pm] = \pm J_\pm,\tag{10.5}$$

$$[J_+, J_-] = 2J_0,\tag{10.6}$$

$$\vec{J}^2 = \frac{1}{2}(J_- J_+ + J_+ J_-) + J_0^2,\tag{10.7}$$

$$J_+ J_- = \vec{J}^2 - J_0(J_0 - 1),\tag{10.8}$$

$$J_- J_+ = \vec{J}^2 - J_0(J_0 + 1).\tag{10.9}$$

These relations will be useful for the diagonalization. Let $|jm\rangle$ be an eigenvector of $\vec{J}^2$ and J_z, where j labels the eigenvalue of $\vec{J}^2$ and m labels those of J_z. Since $\vec{J}^2$ is a positive operator, its eigenvalues are ≥ 0. We write them in the form $j(j+1)$ with $j \geq 0$; this notation for the eigenvalues of $\vec{J}^2$ will be justified below. The number m is called the *magnetic quantum number*. In summary:

$$\vec{J}^2|jm\rangle = j(j+1)|jm\rangle,\tag{10.10}$$

$$J_0|jm\rangle = m|jm\rangle.\tag{10.11}$$

According to (10.5), the vectors $J_\pm|jm\rangle$ are eigenvectors of J_0 with eigenvalue $m \pm 1$:

$$J_0[J_\pm|jm\rangle] = (J_\pm J_0 \pm J_\pm)|jm\rangle = J_\pm(m \pm 1)|jm\rangle$$

$$= (m \pm 1)[J_\pm|jm\rangle].$$

Similarly, since $[\vec{J}^2, J_\pm] = 0$,

$$\vec{J}^2[J_\pm|jm\rangle] = j(j+1)[J_\pm|jm\rangle].$$

We have just shown that the vectors $J_\pm|jm\rangle$ are eigenvectors of $\vec{J}^2$ with eigenvalue $j(j+1)$ and of J_0 with eigenvalue $m \pm 1$. Moreover, assuming that $|jm\rangle$ is normalized, $\langle jm|jm\rangle = 1$, we can calculate the norm of $J_+|jm\rangle$ using (10.9):

$$||J_+|jm\rangle||^2 = \langle jm|J_-J_+|jm\rangle = \langle jm|\vec{J}^2 - J_0(J_0 + 1)|jm\rangle$$
$$= j(j+1) - m(m+1) = (j-m)(j+m+1) \geq 0, \qquad (10.12)$$

and that of $J_-|jm\rangle$ using (10.8):

$$||J_-|jm\rangle||^2 = \langle jm|J_+J_-|jm\rangle = \langle jm|\vec{J}^2 - J_0(J_0 - 1)|jm\rangle$$
$$= j(j+1) - m(m-1) = (j+m)(j-m+1) \geq 0. \qquad (10.13)$$

The simultaneous positivity of the two norms is guaranteed only if $-j \leq m \leq j$. Starting from $|jm\rangle$, by repeated application of J_+ we obtain a series of eigenvectors common to $\vec{J}^2$ and J_0, labeled by $(j, m+1)$, $(j, m+2)$, etc. These eigenvectors have positive norm as long as $m \leq j$, but the norm becomes negative for $m > j$. The series must therefore terminate, which is possible only if one of the vectors $(J_+)^n|jm\rangle$ vanishes for an integer value of $n = n_1 + 1$ such that $m + n_1 = j$:

$$J_+[(J_+)^{n_1}|jm\rangle] = 0.$$

The same argument for J_- shows that there must exist an integer n_2 such that

$$J_-[(J_-)^{n_2}|jm\rangle] = 0.$$

From the relations

$$j = m + n_1, \qquad -j = m - n_2$$

we find that $2j$, and therefore $(2j+1)$, must be an integer, which leads to the diagonalization theorem for $\vec{J}^2$ and J_z.

Theorem. The possible values of j are integers or half-integers: $j = 0, 1/2, 1, 3/2, \ldots$. If $|jm\rangle$ is an eigenvector common to $\vec{J}^2$ and J_0, m necessarily takes one of $(2j+1)$ values:

$$m = -j, -j+1, -j+2, \ldots, j-2, j-1, j.$$

When j takes the values $0, 1, 2, \ldots$ we have so-called integer angular momentum, and when $j = 1/2, 3/2, \ldots$ we have half-integer angular momentum.[1] Let us study the normalization and phase of the vectors $|jm\rangle$. Starting from a vector $|jm\rangle$, by repeated application of J_+ and J_- we construct a series of $(2j+1)$ orthogonal vectors which span a vector

---

[1] Although half of an even integer is also a half-integer...

subspace of $(2j+1)$ dimensions $\mathcal{E}(j)$ of $\mathcal{H}$. These vectors do not have unit norm, but if we define $|j, m-1\rangle$ by

$$|j, m-1\rangle = [j(j+1) - m(m-1)]^{-1/2} J_-|jm\rangle, \tag{10.14}$$

then $|j, m-1\rangle$ has unit norm according to (10.13). Moreover, using (10.8),

$$J_+ J_-|jm\rangle = [j(j+1) - m(m-1)]^{1/2} J_+|j, m-1\rangle$$
$$= [j(j+1) - m(m-1)]|jm\rangle$$

or

$$J_+|j, m-1\rangle = [j(j+1) - m(m-1)]^{1/2}|jm\rangle,$$

and with the replacement $m \to m+1$ we have

$$J_+|jm\rangle = [j(j+1) - m(m+1)]^{1/2}|j, m+1\rangle. \tag{10.15}$$

The relations (10.14) or (10.15) completely fix the relative phase of the vectors $|j, j\rangle, |j, j-1\rangle, \ldots, |j, -j\rangle$. A basis of $\mathcal{E}(j)$ formed from vectors $|jm\rangle$ satisfying (10.14) or (10.15) is called the *standard basis* $|jm\rangle$.

It can happen that knowing (j, m) is not sufficient for uniquely specifying a vector of $\mathcal{H}$: $\vec{J}^2$ and J_z do not form a complete set of compatible physical properties. We shall see an example of this in Section 10.4.2 where we discuss the hydrogen atom. There the values of the (orbital) angular momentum, denoted l, are not sufficient for specifying a bound state; an additional quantum number $n = l+1, l+2, \ldots$, called the principal quantum number, must also be given. In general, it is necessary to use a quantum number or a set of supplementary quantum numbers τ to label the eigenvectors $|j, m = j\rangle$ of $\vec{J}^2$ and J_z, and these are normalized by the condition

$$\langle \tau, j, j | \tau', j, j \rangle = \delta_{\tau, \tau'}.$$

By repeated application of J_- we form the standard basis of $\mathcal{E}(\tau, j)$:

$$|\tau, j, j\rangle, \ |\tau, j, j-1\rangle, \ldots, |\tau, j, -j+1\rangle, \ |\tau, j, -j\rangle.$$

Let us summarize the essential properties of a standard basis $|\tau, jm\rangle$:

$$\vec{J}^2|\tau, jm\rangle = j(j+1)|\tau, jm\rangle, \quad J_z|\tau, jm\rangle = m|\tau, jm\rangle, \tag{10.16}$$

$$J_+|\tau, jm\rangle = [j(j+1) - m(m+1)]^{1/2}|\tau, j, m+1\rangle, \tag{10.17}$$

$$J_-|\tau, jm\rangle = [j(j+1) - m(m-1)]^{1/2}|\tau, j, m-1\rangle, \tag{10.18}$$

$$J_+|\tau, j, j\rangle = 0, \quad J_-|\tau, j, -j\rangle = 0, \tag{10.19}$$

$$\langle \tau', j'm' | \tau, jm \rangle = \delta_{\tau'\tau}\delta_{j'j}\delta_{m'm}. \tag{10.20}$$

In what follows we shall suppress the index τ, as it plays no role in this chapter (except in Section 10.4). The matrix elements of $\vec{J}^2$, J_0, and J_- in a standard basis are

$$\langle j'm'|\vec{J}^2|jm\rangle = j(j+1)\delta_{j'j}\delta_{m'm}, \tag{10.21}$$

$$\langle j'm'|J_0|jm\rangle = m\,\delta_{j'j}\delta_{m'm}, \tag{10.22}$$

$$\langle j'm'|J_\pm|jm\rangle = [j(j+1) - mm']^{1/2}\delta_{j'j}\delta_{m',m\pm1}. \tag{10.23}$$

In the subspace $\mathcal{E}(j)$ in which $\vec{J}^2$ has fixed eigenvalue $j(j+1)$, the operators J_0 and $J_\pm$ are represented by $(2j+1)\times(2j+1)$ matrices, and the matrix representing J_0 is diagonal. It is instructive (Exercise 10.7.4) to write out these matrices explicitly in the case $j = 1/2$ and recover the 2×2 matrices of spin 1/2 (3.47) as well as those of the case $j = 1$. In the latter case we recover the infinitesimal generators of rotations in three-dimensional space: the transformation law of a vector in $\mathbb{R}^3$ is that of angular momentum $j = 1$. Equation (10.23) gives the following for the infinitesimal generators (Exercise 10.7.4):

$$J_x = \frac{1}{\sqrt{2}}\begin{pmatrix} 0 & 1 & 0 \\ 1 & 0 & 1 \\ 0 & 1 & 0 \end{pmatrix}, \quad J_y = \frac{1}{\sqrt{2}}\begin{pmatrix} 0 & -i & 0 \\ i & 0 & -i \\ 0 & i & 0 \end{pmatrix}, \quad J_z = \begin{pmatrix} 1 & 0 & 0 \\ 0 & 0 & 0 \\ 0 & 0 & -1 \end{pmatrix}.$$

$$\tag{10.24}$$

These infinitesimal generators superficially differ in form from the generators T_i found in (8.26). In fact, the two sets are related by the unitary transformation (10.64) which transforms the Cartesian components of $\hat{r}$ into spherical components; see Exercise 10.7.4.

10.2 Rotation matrices

In Chapter 3 we saw how to rotate a spin 1/2. Starting from a state $|+\rangle$ obtained by means of a Stern–Gerlach apparatus in which the magnetic field is parallel to Oz, we know from (3.57) how to construct the state $|+, \hat{n}\rangle$ obtained using a Stern–Gerlach apparatus with magnetic field parallel to $\hat{n}$. We apply to the state $|+\rangle$ a rotation operator $U[\mathcal{R}]$ which transforms $|+\rangle$ into $|+, \hat{n}\rangle$:

$$|+, \hat{n}\rangle = U[\mathcal{R}]|+\rangle = |+\rangle_\mathcal{R}.$$

The rotation $\mathcal{R}$ aligns Oz in the direction $\hat{n}$. This rotation is not unique, and we shall see that this nonuniqueness corresponds to an arbitrary phase in the definition of $|+, \hat{n}\rangle$. Another example of the rotation of a physical state was given in Chapter 3 in the case of photon polarization. Starting from a linear polarization state $|x\rangle$, we obtain a linear polarization state $|\theta\rangle$ by applying to the former a rotation operator $U[\mathcal{R}_z(\theta)]$ corresponding to rotation by an angle θ about the photon's direction of propagation Oz (3.29):

$$|\theta\rangle = \exp(-i\theta\Sigma_z)|x\rangle = U[\mathcal{R}_z(\theta)]|x\rangle.$$

In the general case, the state $|\varphi\rangle_{\mathcal{R}}$ transformed by a rotation $\mathcal{R}$ from a state $|\varphi\rangle$ is

$$|\varphi\rangle_{\mathcal{R}} = U[\mathcal{R}]|\varphi\rangle.$$

We now give the explicit matrix form of the rotation operator $U[\mathcal{R}]$ in the basis $|jm\rangle$. The rotation operator $U[\mathcal{R}]$ is expressed as a function of the infinitesimal generators J_x, J_y, and J_z; cf. (8.30). Since the components of $\vec{J}$ commute with $\vec{J}^2$, the commutator $[U(\mathcal{R}), \vec{J}^2] = 0$ and the matrix elements of U are zero if $j \neq j'$:

$$\langle j'm'|U[\mathcal{R}]|jm\rangle \propto \delta_{j'j}.$$

In the subspace $\mathcal{E}(j)$, the operator $U(\mathcal{R})$ will be represented by a $(2j+1) \times (2j+1)$ matrix denoted $D^{(j)}[\mathcal{R}]$. Its elements are

$$D^{(j)}_{m'm}[\mathcal{R}] = \langle jm'|U[\mathcal{R}]|jm\rangle. \tag{10.25}$$

The matrices $D^{(j)}$ are called *rotation matrices*, or *Wigner matrices*. Let us examine the rotational transformation of a state $|jm\rangle$ giving the vector $|jm\rangle_{\mathcal{R}}$:

$$|jm\rangle_{\mathcal{R}} = U[\mathcal{R}]|jm\rangle = \sum_{m'} |jm'\rangle \langle jm'|U[\mathcal{R}]|jm\rangle,$$

where we have used the fact that in the completeness relation

$$\sum_{j',m'} |j'm'\rangle \langle j'm'| = I$$

only the terms with $j = j'$ contribute. We can then write

$$\boxed{\; |jm\rangle_{\mathcal{R}} = \sum_{m'} D^{(j)}_{m'm}[\mathcal{R}]|jm'\rangle \;} . \tag{10.26}$$

Let us recall the group properties of the operators $U[\mathcal{R}]$. In the case of a vector representation (8.12)

$$U[\mathcal{R}_2]U[\mathcal{R}_1] = U[\mathcal{R}_2\mathcal{R}_1], \tag{10.27}$$

while for a spinor representation (8.13)

$$U[\mathcal{R}_2]U[\mathcal{R}_1] = \pm U[\mathcal{R}_2\mathcal{R}_1]. \tag{10.28}$$

At the end of this section we shall show that (10.27) corresponds to the case of integer angular momentum and (10.28) to the half-integer case. The multiplication law for rotation matrices is determined by the group property for the operators U:

$$D^{(j)}_{m'm}[\mathcal{R}_2\mathcal{R}_1] = \pm \sum_{m''} D^{(j)}_{m'm''}[\mathcal{R}_2]D^{(j)}_{m''m}[\mathcal{R}_1].$$

Let us return to the study of the rotation which takes Oz to the direction $\hat{n}$ described by the polar and azimuthal angles (θ, ϕ):

$$\hat{n}_x = \sin\theta\cos\phi, \quad \hat{n}_y = \sin\theta\sin\phi, \quad \hat{n}_z = \cos\theta. \tag{10.29}$$

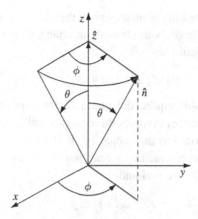

Fig. 10.1. The rotation $\mathcal{R}(\theta, \phi)$ aligns the axis Oz with $\hat{n}$.

We shall adopt the following *convention* for the rotation: $\mathcal{R}$, denoted $\mathcal{R}(\theta, \phi)$, will be the product of a rotation by an angle θ about Oy followed by one by an angle ϕ about Oz (Fig. 10.1):

$$\boxed{\mathcal{R}(\theta, \phi) = \mathcal{R}_z(\phi)\mathcal{R}_y(\theta)}.$$ (10.30)

Using (10.30) and the group law, the rotation operator $U[\mathcal{R}(\theta, \phi)]$ is given as a function of the infinitesimal generators J_y and J_z by

$$U[\mathcal{R}(\theta, \phi)] = e^{-i\phi J_z}e^{-i\theta J_y},$$ (10.31)

and its matrix elements in the basis $|jm\rangle$ are

$$\boxed{D^{(j)}_{m'm}[\mathcal{R}(\theta, \phi)] = \langle jm'|e^{-i\phi J_z}e^{-i\theta J_y}|jm\rangle}.$$ (10.32)

This equation can be simplified:

$$D^{(j)}_{m'm}[\mathcal{R}(\theta, \phi)] \equiv D^{(j)}_{m'm}(\theta, \phi) = e^{-im'\phi}\langle jm'|e^{-i\theta J_y}|jm\rangle \qquad (10.33)$$

$$= e^{-im'\phi}d^{(j)}_{m'm}(\theta). \qquad (10.34)$$

We have defined the matrix $d^{(j)}(\theta)$ as

$$\boxed{d^{(j)}_{m'm}(\theta) = \langle jm'|e^{-i\theta J_y}|jm\rangle}.$$ (10.35)

The matrices $d^{(j)}$ satisfy a group property derived from that of the matrices $D^{(j)}$:

$$d^{(j)}_{m'm}(\theta_2 + \theta_1) = \sum_{m''} d^{(j)}_{m'm''}(\theta_2)d^{(j)}_{m''m}(\theta_1).$$

There is no sign $\pm$ in this equation because the rotation angle can be greater than 2π.

We have already mentioned the arbitrariness in the choice of rotation (θ, ϕ); we could have first rotated by an angle ψ about Oz without changing the final axis $\hat{n}$. In that case the new rotation operator would be

$$U[\mathcal{R}'] = U[\mathcal{R}(\theta, \phi)]e^{-i\psi J_z},$$

and the result (10.26) would acquire the phase factor $\exp(-im\psi)$. The most general definition of the rotation matrices involves three angles, called the Euler angles (ϕ, θ, ψ), and our convention corresponds to the choice $(\phi, \theta, 0)$.[2]

In the basis $|jm\rangle$, iJ_y is represented by a real matrix, because according to (10.23) the matrix elements of J_+ and J_- are real and

$$J_y = -\frac{i}{2}(J_+ - J_-).$$

The matrix $\exp(-i\theta J_y)$ is also a real matrix and the group property

$$U^\dagger[\mathcal{R}] = U^{-1}[\mathcal{R}] = U[\mathcal{R}^{-1}]$$

becomes

$$\left[d^{(j)}(\theta)\right]^\dagger = \left[d^{(j)}(-\theta)\right],$$

which gives the following for the matrix elements:

$$\boxed{d^{(j)}_{m'm}(\theta) = d^{(j)}_{mm'}(-\theta)}. \tag{10.36}$$

There exists another symmetry property (Exercise 10.4.12):

$$\boxed{d^{(j)}_{m'm}(\theta) = (-1)^{m-m'} d^{(j)}_{-m',-m}(\theta)}. \tag{10.37}$$

Finally, it can be shown that the matrices $D^{(j)}$ form a so-called *irreducible* representation of the rotation group, that is, any vector of $\mathcal{E}(j)$ can be obtained from an arbitrary vector of this space by application of a rotation matrix $D^{(j)}$, and any matrix that commutes with all the matrices $D^{(j)}$ is a multiple of the identity matrix.

Whether or not the factor $\pm$ occurs in (10.28) can be checked by studying rotations by 2π, as this factor arises when a rotation by 2π is represented by the operator $-I$ in the space $\mathcal{E}(j)$. Let us consider a rotation by 2π about the z axis:

$$\langle jm'|U[\mathcal{R}_z(2\pi)]|jm\rangle = e^{-2i\pi m}\delta_{m'm} = \delta_{m'm}, \qquad \text{integer } j$$

$$= e^{-2i\pi m}\delta_{m'm} = -\delta_{m'm}, \qquad \text{half-integer } j.$$

---

[2] The usual notation for the rotation matrices is

$$D^{(j)}(\theta, \phi) \rightarrow D^{(j)}(\phi, \theta, \psi = 0).$$

Since the choice of axis Oz is arbitrary, the operator rotating by 2π will be I for integer j and $-I$ for half-integer j. However, operators that rotate by 4π are all equal to I for any value of j. Let us examine two successive rotations by angles θ_1 and θ_2 about an axis $\hat{n}$, with

$$\theta_1 + \theta_2 = \theta + 2\pi n, \quad 0 \le \theta < 2\pi, \quad \text{integer } n \ge 0.$$

From the equations

$$e^{-i(\theta_1+\theta_2)\vec{J}\cdot\hat{n}} = e^{-i\theta(\vec{J}\cdot\hat{n})}e^{-2i\pi n(\vec{J}\cdot\hat{n})} = e^{-i\theta(\vec{J}\cdot\hat{n})}, \quad \text{integer } j$$

$$= (-1)^n e^{-i\theta(\vec{J}\cdot\hat{n})}, \quad \text{half-integer } j,$$

we find that (10.27) is valid for integer j and (10.28) for half-integer j. In other words, to any rotation $\mathcal{R}$ there correspond two rotation operators of opposite sign for half-integer j and only one for integer j.

Let us check that in the case of spin 1/2 we recover the matrix $D^{(1/2)}(\theta, \phi)$ already calculated in Chapter 3. The matrix $d^{(1/2)}(\theta)$ according to Exercise 3.3.6 is

$$d^{(1/2)}(\theta) = \exp(-i\theta\sigma_y/2) = \cos\frac{\theta}{2}I - i\sigma_y \sin\frac{\theta}{2},$$

or in explicit form

$$d^{(1/2)}(\theta) = \begin{pmatrix} \cos\theta/2 & -\sin\theta/2 \\ \sin\theta/2 & \cos\theta/2 \end{pmatrix}, \tag{10.38}$$

where the rows and columns are arranged in the order $m = 1/2, -1/2$. Then (10.33) gives the following for the matrix $D^{(1/2)}(\theta, \phi)$:

$$D^{(1/2)}(\theta, \phi) = \begin{pmatrix} e^{-i\phi/2}\cos\theta/2 & -e^{-i\phi/2}\sin\theta/2 \\ e^{i\phi/2}\sin\theta/2 & e^{i\phi/2}\cos\theta/2 \end{pmatrix},$$

in agreement with (3.58).

The rotation matrix $d^{(1)}(\theta)$ for angular momentum $j = 1$ is obtained from the infinitesimal generators (10.24), with the rows and columns arranged in the order $m = 1, 0, -1$ (Exercise 10.7.4):

$$d^{(1)}(\theta) = \begin{pmatrix} \frac{1}{2}(1+\cos\theta) & -\frac{1}{\sqrt{2}}\sin\theta & \frac{1}{2}(1-\cos\theta) \\ \frac{1}{\sqrt{2}}\sin\theta & \cos\theta & -\frac{1}{\sqrt{2}}\sin\theta \\ \frac{1}{2}(1-\cos\theta) & \frac{1}{\sqrt{2}}\sin\theta & \frac{1}{2}(1+\cos\theta) \end{pmatrix}. \tag{10.39}$$

The reader should verify that the matrices $d^{(1/2)}$ and $d^{(1)}$ possess the symmetry properties (10.36) and (10.37).

10.3 Orbital angular momentum

10.3.1 The orbital angular momentum operator

Let us consider a classical scalar field $\psi(\vec{r})$ and subject it to a rotation $\mathcal{R}_z(\phi)$ by an angle ϕ about Oz, with $\vec{r}\,' = \mathcal{R}\vec{r}$ being the vector transformed from $\vec{r}$ by this rotation:

$$x' = x\cos\phi - y\sin\phi,$$
$$y' = x\sin\phi + y\cos\phi,$$
$$z' = z.$$

The value of the transformed scalar field $\psi'(\vec{r})$ at the point $\vec{r}\,'$ must be identical to that of the initial field at the point $\vec{r}$:

$$\psi'(\vec{r}\,') = \psi(\vec{r}),$$

or

$$\boxed{\psi'(\vec{r}) = \psi(\mathcal{R}^{-1}\vec{r})}.\tag{10.40}$$

This transformation law is correct for a (scalar) classical field, but if $\psi(\vec{r})$ is the wave function of a particle $\psi(\mathcal{R}^{-1}\vec{r})$ and $\psi'(\vec{r})$ can a priori differ by a phase:

$$\psi'(\vec{r}) = e^{i\theta(\vec{r})}\psi(\mathcal{R}^{-1}\vec{r})$$

(see the discussion following (9.17)). We know only that $|\psi'(\vec{r}\,')| = |\psi(\vec{r})|$, and our goal is to show that the phase factor that might arise is actually absent. The vector $U(\mathcal{R})|\vec{r}\rangle$ physically represents an eigenstate of the position operator $\vec{R}$, obtained from the eigenstate $|\vec{r}\rangle$ of $\vec{R}$ by a rotation $U(\mathcal{R})$. Let us show this explicitly using the fact that $\vec{R}$ is a vector operator whose components X_k transform as the components of V in (8.34):

$$X_k[U(\mathcal{R})|\vec{r}\rangle] = U(\mathcal{R})U^{-1}(\mathcal{R})X_k U(\mathcal{R})|\vec{r}\rangle$$
$$= U(\mathcal{R})\Big(\sum_l \mathcal{R}_{kl}X_l\Big)|\vec{r}\rangle = U(\mathcal{R})\Big(\sum_l \mathcal{R}_{kl}x_l\Big)|\vec{r}\rangle$$
$$= (\mathcal{R}\vec{r})_k[U(\mathcal{R})|\vec{r}\rangle],$$

which shows that the state vector $|\mathcal{R}\vec{r}\rangle$ can be defined, that is, its phase can be fixed, as

$$|\mathcal{R}\vec{r}\rangle \equiv U(\mathcal{R})|\vec{r}\rangle.\tag{10.41}$$

If $|\psi'\rangle$ is the transform of $|\psi\rangle$ by $U(\mathcal{R})$, $|\psi'\rangle = U(\mathcal{R})|\psi\rangle$, then

$$\psi'(\vec{r}) = \langle\vec{r}|\psi'\rangle = \langle\vec{r}|U(\mathcal{R})|\psi\rangle = \langle U^\dagger(\mathcal{R})\vec{r}|\psi\rangle$$
$$= \langle U^{-1}(\mathcal{R})\vec{r}|\psi\rangle = \langle\mathcal{R}^{-1}\vec{r}|\psi\rangle = \psi(\mathcal{R}^{-1}\vec{r}),$$

which demonstrates (10.40). At first sight the argument $\mathcal{R}^{-1}$ in (10.40), which can also be written as

$$[U(\mathcal{R})\psi](\vec{r}) = \psi(\mathcal{R}^{-1}\vec{r}),$$

may seem surprising, but we have already encountered a similar situation in the case of translations in (9.15), which in three dimensions with $\hbar = 1$ is written as

$$\left[e^{-i\vec{P}\cdot\vec{a}}\psi\right](\vec{r}) = \psi(\vec{r}-\vec{a}),$$

even though[3]

$$e^{-i\vec{P}\cdot\vec{a}}|\vec{r}\rangle = |\vec{r}+\vec{a}\rangle.$$

The function $\psi(\vec{r})$ transformed by a translation $\vec{a}$ is $\psi(\vec{r}-\vec{a})$ and not $\psi(\vec{r}+\vec{a})$! If the rotation angle ϕ becomes infinitesimal for a rotation about Oz, then

$$U[\mathcal{R}_z(\phi)] \simeq I - i\phi J_z,$$

and according to (10.40)

$$[(I-i\phi J_z)\psi](\vec{r}) \simeq \psi(x+y\phi, -x\phi+y, z)$$

$$\simeq \psi(\vec{r}) + \phi\left(y\frac{\partial\psi}{\partial x} - x\frac{\partial\psi}{\partial y}\right)$$

$$= \psi(\vec{r}) - i\phi(XP_y - YP_x)\psi,$$

from which we find

$$[J_z\psi](\vec{r}) = [(XP_y - YP_x)\psi](\vec{r}) = [(\vec{R}\times\vec{P})_z\psi](\vec{r}) = (L_z\psi)(\vec{r}). \tag{10.42}$$

The angular momentum operator of the particle described by a wave function $\psi(\vec{r})$ is called the *orbital angular momentum* (because it is associated with the motion of the particle in a spatial orbit), and is in general denoted $\vec{L}$:

$$\vec{L} = \vec{R}\times\vec{P}. \tag{10.43}$$

The operator $\vec{L}$ has been constructed as the infinitesimal generator of rotations and necessarily satisfies the angular momentum commutation relations (10.1) or (10.2):

$$[L_j, L_k] = i\sum_l \varepsilon_{jkl}L_l. \tag{10.44}$$

These relations can be verified by explicit calculation using the canonical commutation relations (8.45); see Exercise 10.7.5. We use $|lm\rangle$ to denote the eigenvectors of $\vec{L}^2$ and L_z:

$$\vec{L}^2|lm\rangle = l(l+1)|lm\rangle, \tag{10.45}$$

$$L_z|lm\rangle = m|lm\rangle. \tag{10.46}$$

These equations can be transformed into differential equations by writing the operators L_j as differential operators acting in $L^{(2)}(\mathbb{R}^3)$. The calculation is lengthy if we make the change of variables $(x, y, z) \to (r, \theta, \phi)$, but it is simplified if we use the fact that the L_i

---

[3] We note that this equation fixes the phase of the vector $|\vec{r}+\vec{a}\rangle$ relative to that of $|\vec{r}\rangle$, in the same way as (10.41) fixes the phase of $|\mathcal{R}\vec{r}\rangle$ relative to that of $|\vec{r}\rangle$.

are infinitesimal generators of rotations. The case of L_z is particularly simple. Considering ψ as a function of (r, θ, ϕ), we have

$$\left(e^{-i\alpha L_z}\psi\right)(r, \theta, \phi) = \psi(r, \theta, \phi - \alpha),$$

and taking α to be infinitesimal,

$$([I - i\alpha L_z]\psi)(r, \theta, \phi) = \psi(r, \theta, \phi) - \alpha\frac{\partial\psi}{\partial\phi}$$

or $L_z\psi = -i(\partial\psi/\partial\phi)$. The calculation of L_x and L_y takes a few more lines, because both θ and ϕ vary in a rotation about Ox or Oy. The result is (Exercise 10.7.5)

$$L_z = -i\frac{\partial}{\partial\phi}, \tag{10.47}$$

$$L_\pm = ie^{\pm i\phi}\left(\cot\theta\frac{\partial}{\partial\phi} \mp i\frac{\partial}{\partial\theta}\right), \tag{10.48}$$

$$\vec{L}^2 = -\left(\frac{1}{\sin\theta}\frac{\partial}{\partial\theta}\left[\sin\theta\frac{\partial}{\partial\theta}\right] + \frac{1}{\sin^2\theta}\frac{\partial^2}{\partial\phi^2}\right). \tag{10.49}$$

The operators L_j depend only on angles and not on r, hence the name *angular momentum*. The eigenfunctions of $\vec{L}^2$ and L_z depend only on the angles θ and ϕ or, equivalently, on $\hat{r}$. These eigenfunctions are called the *spherical harmonics*:

$$Y_l^m(\theta, \phi) = Y_l^m(\hat{r}) = \langle\hat{r}|lm\rangle. \tag{10.50}$$

Equations (10.45) and (10.46) become

$$[\vec{L}^2 Y_l^m](\hat{r}) = \langle\hat{r}|\vec{L}^2|lm\rangle = l(l+1)Y_l^m(\hat{r}), \tag{10.51}$$

$$[L_z Y_l^m](\hat{r}) = \langle\hat{r}|L_z|lm\rangle = mY_l^m(\hat{r}), \tag{10.52}$$

while (10.15) is written as

$$[L_\pm Y_l^m](\hat{r}) = \langle\hat{r}|L_\pm|lm\rangle = [l(l+1) - m(m+1)]^{1/2}Y_l^{m\pm1}(\hat{r}).$$

Equation (10.52) becomes, using (10.47),

$$[L_z Y_l^m](\theta, \phi) = -i\frac{\partial}{\partial\phi}Y_l^m(\theta, \phi) = mY_l^m(\theta, \phi),$$

which implies that

$$Y_l^m(\theta, \phi) = e^{im\phi}f_l^m(\theta). \tag{10.53}$$

The tranformation law (10.40) shows that in a rotation by 2π the wave function is unchanged, and so no minus sign is introduced. This implies that *orbital angular momenta are always integers*.

   A simple and important application is the spherical rotator. We consider a diatomic molecule rotating about its center of mass, taken to be the coordinate origin (Fig. 10.2 and Exercise 1.6.1). Its moment of inertia is $I = \mu r_0^2$, where μ is the reduced mass and

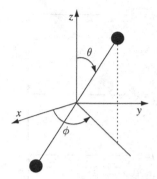

Fig. 10.2. The spherical rotator.

r_0 is the distance between the nuclei (the electron contribution is negligible). If ω is the angular velocity of the rotation, the classical Hamiltonian H_{cl} is

$$H_{\text{cl}} = \frac{1}{2}I\omega^2 = \frac{1}{2}\frac{(I\omega)^2}{I} = \frac{l^2}{2I},$$

where $l = I\omega$ is the angular momentum. The quantum version of the Hamiltonian is

$$H = \frac{\vec{L}^2}{2I},$$

and the energies are

$$E_l = \frac{l(l+1)}{2I}. \tag{10.54}$$

The eigenfunctions are the $Y_l^m(\theta, \phi)$, where the angles θ and ϕ specify the orientation of the line joining the two nuclei; $Y_l^m(\theta, \phi)$ is the amplitude for finding this line oriented in the direction (θ, ϕ). The spectrum of rotational levels is given in Fig. 10.3, and well reproduces the experimental results for the spectra of diatomic molecules.

10.3.2 Properties of the spherical harmonics

Let us now summarize, in some cases without proof, the properties of the spherical harmonics that are most frequently used.

1. Basis on the unit sphere

The spherical harmonics form an orthonormal basis for square-integrable functions on the unit sphere $\vec{r}^2 = 1$:

$$\int \sin\theta\, d\theta\, d\phi [Y_{l'}^{m'}(\theta, \phi)]^* Y_l^m(\theta, \phi) = \int d\Omega [Y_{l'}^{m'}(\theta, \phi)]^* Y_l^m(\theta, \phi) = \delta_{l'l}\delta_{m'm}. \tag{10.55}$$

We shall frequently use the notation $\Omega = (\theta, \phi)$ and

$$d\Omega = \sin\theta\, d\theta\, d\phi = d^2\hat{r}. \tag{10.56}$$

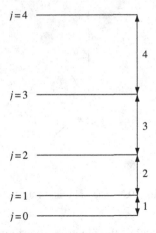

Fig. 10.3. Spectrum of the spherical rotator. The jth level is separated from the $(j-1)$th level by an amount j/I, or $\hbar^2 j/I$ if $\hbar$ is restored.

If a function $f(\theta, \phi)$ is square-integrable on the unit sphere, we can write down an expansion analogous to a Fourier series:

$$f(\theta, \phi) = \sum_{l,m} c_{lm} Y_l^m(\theta, \phi),$$

$$c_{lm} = \int d\Omega [Y_l^m(\theta, \phi)]^* f(\theta, \phi). \tag{10.57}$$

2. Relation to the Legendre polynomials

One definition of the Legendre polynomials $P_l(u)$ is

$$P_l(u) = \frac{1}{2^l l!} \frac{d^l}{du^l} (u^2 - 1)^l, \tag{10.58}$$

where $P_l(u)$ is a polynomial of degree l and parity $(-1)^l$:

$$P_l(-u) = (-1)^l P_l(u).$$

The Legendre polynomials form a complete set of orthogonal polynomials in the interval $[-1, +1]$. The first few Legendre polynomials are

$$P_0(u) = 1, \quad P_1(u) = u, \quad P_2(u) = \frac{1}{2}(3u^2 - 1). \tag{10.59}$$

The associated Legendre functions $P_l^m(u)$ are defined as

$$P_l^m(u) = (1 - u^2)^{m/2} \frac{d^m}{du^m} P_l(u), \quad P_l^0(u) = P_l(u), \tag{10.60}$$

and it can be shown that the spherical harmonics are related to the P_l^m as

$$Y_l^m(\theta, \phi) = (-1)^m \left[\frac{(2l+1)}{4\pi} \frac{(l-m)!}{(l+m)!} \right]^{1/2} P_l^m(\cos\theta)\, e^{im\phi}, \quad m > 0,$$

$$Y_l^m(\theta, \phi) = (-1)^m [Y_l^{-m}(\theta, \phi)]^*, \qquad\qquad\qquad m < 0.$$

(10.61)

According to (10.53), Y_l^0 is independent of ϕ and proportional to $P_l(\cos\theta)$:

$$Y_l^0(\theta, \phi) = \sqrt{\frac{2l+1}{4\pi}} P_l(\cos\theta).$$

(10.62)

As a special case, we write down the Y_l^m for $l = 0$ and $l = 1$:

$$l = 0: \quad Y_0^0 = \sqrt{\frac{1}{4\pi}},$$

$$l = 1: \quad Y_1^0 = \sqrt{\frac{3}{4\pi}} \cos\theta = \sqrt{\frac{3}{4\pi}}\, \hat{r}_0,$$

(10.63)

$$Y_1^{\pm} = \mp \sqrt{\frac{3}{8\pi}}\, e^{\pm i\phi} \sin\theta = \sqrt{\frac{3}{4\pi}}\, \hat{r}_{\pm 1}.$$

Up to the normalization factor $\sqrt{3/4\pi}$ the Y_1^m are just the *spherical components* of the unit vector $\hat{r}$:

$$\hat{r} = (\sin\theta\cos\phi, \sin\theta\sin\phi, \cos\theta),$$

$$Y_1^0 = \sqrt{\frac{3}{4\pi}}\, \hat{r}_0, \quad Y_1^{\pm} = \mp\sqrt{\frac{3}{4\pi}} \frac{\hat{r}_x \pm i\hat{r}_y}{\sqrt{2}} = \sqrt{\frac{3}{4\pi}}\, \hat{r}_{\pm 1}.$$

(10.64)

These expressions justify the phase conventions used for right- and left-handed polarization in (3.11).

3. Transformation under rotation

Multiplying (10.26) for $j = l$ on the left by the bra $\langle\hat{r}|$, we find

$$Y_l^m(\mathcal{R}^{-1}\hat{r}) = \sum_{m'} D_{m'm}^{(l)}(\mathcal{R}) Y_l^{m'}(\hat{r}).$$

(10.65)

We can also obtain (Exercise 10.7.6) a relation between the spherical harmonics and the rotation matrices:

$$D_{m0}^{(l)}(\theta, \phi) = \sqrt{\frac{4\pi}{2l+1}} [Y_l^m(\theta, \phi)]^*.$$

(10.66)

From these two equations we can derive the addition theorem for the spherical harmonics. Taking $\hat{r}$ in the direction given by the polar angles (α, β), let $\mathcal{R}$ be the rotation by angles (θ, ϕ) aligning $\hat{z}$ with $\hat{n}$ and Θ be the angle between $\hat{r}$ and the direction defined by the angles (θ, ϕ) (Fig. 10.4):

$$\cos\Theta = \cos\alpha\cos\theta + \sin\alpha\sin\theta\cos(\beta - \phi).$$

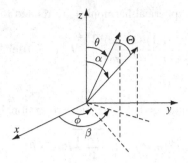

Fig. 10.4. Angular configuration in (10.67).

The angle Θ between $\mathcal{R}^{-1}\hat{r}$ and the z axis is the same as the angle between $\hat{n}$ and $\hat{r}$. It is then sufficient to take $m = 0$ in (10.65) to obtain

$$\boxed{P_l(\cos\Theta) = \frac{4\pi}{2l+1} \sum_{m=-l}^{l} [Y_l^m(\theta,\phi)]^* Y_l^m(\alpha,\beta)} . \tag{10.67}$$

4. Parity of the spherical harmonics

The parity operator Π defined in Section 8.3.3 acts on a wave function $\psi(\vec{r})$ as

$$[\Pi\psi](\vec{r}) = \psi(-\vec{r}). \tag{10.68}$$

Π commutes with the orbital angular momentum $\vec{L}$ and, more generally, with $\vec{J}$. In fact, the representation matrix of the parity operator in three-dimensional space $\mathbb{R}^3$ is the matrix $-I$, which commutes with any rotation matrix $\mathcal{R}$, from which we infer

$$[U(\mathcal{R}), \Pi] = 0 \Rightarrow [\vec{J}, \Pi] = 0 \quad \text{and} \quad [\vec{L}, \Pi] = 0. \tag{10.69}$$

This implies the equations

$$\vec{L}^2 \Pi Y_l^m = \Pi \vec{L}^2 Y_l^m = l(l+1)\Pi Y_l^m,$$

$$L_z \Pi Y_l^m = \Pi L_z Y_l^m = m\Pi Y_l^m,$$

which show that ΠY_l^m is proportional to Y_l^m:

$$\Pi Y_l^m = \alpha(l,m) Y_l^m.$$

Y_l^m is therefore an eigenfunction of Π, and since $\Pi^2 = I$, $\alpha(l,m) = \pm 1$. Let us show that $\alpha(l,m)$ is in fact independent of m using the fact that L_+ commutes with Π:

$$L_+\Pi Y_l^m = \alpha(l,m)L_+ Y_l^m = \alpha(l,m)[l(l+1) - m(m+1)]^{1/2} Y_l^{m+1}$$

$$= \Pi L_+ Y_l^m = [l(l+1) - m(m+1)]^{1/2} \Pi Y_l^{m+1}$$

$$= [l(l+1) - m(m+1)]^{1/2} \alpha(l,m+1) Y_l^{m+1},$$

which implies that $\alpha(l, m+1) = \alpha(l, m)$. Therefore, $\alpha(l, m)$ is independent of m and

$$[\Pi Y_l^m](\hat{r}) = \alpha(l) Y_l^m(\hat{r}) = Y_l^m(-\hat{r}).$$

The transformation $\hat{r} \to -\hat{r}$ corresponds to

$$\theta \to \pi - \theta, \quad \phi \to \phi + \pi. \tag{10.70}$$

If $m = 0$, then $Y_l^0 \propto P_l(\cos\theta)$; using (10.62) and $P_l(-u) = (-1)^l P_l(u)$, we find $\alpha(l) = (-1)^l$ and

$$\boxed{Y_l^m(\theta, \phi) = (-1)^l Y_l^m(\pi - \theta, \phi + \pi) \quad \text{or} \quad Y_l^m(\hat{r}) = (-1)^l Y_l^m(-\hat{r})} \tag{10.71}$$

10.4 Particle in a central potential

10.4.1 The radial wave equation

We shall use the preceding results to show that the three-dimensional Schrödinger equation, which is a partial differential equation, can be reduced to an ordinary differential equation when the potential is central, that is, invariant under rotation:

$$V(\vec{r}) = V(|\vec{r}|) = V(r).$$

In this case, since the kinetic energy is a scalar operator, the full Hamiltonian for a particle of mass M

$$H = \frac{\vec{P}^2}{2M} + V(r) \tag{10.72}$$

is invariant under rotation: $[H, \vec{J}] = 0$. Our problem involves only the orbital angular momentum, since the only operators at our disposal are $\vec{P}$ and $\vec{R}$:

$$[H, \vec{L}] = 0 \quad \text{or} \quad [H, L_x] = [H, L_y] = [H, L_z] = 0. \tag{10.73}$$

In the space $L_{\hat{r}}^{(2)}(\mathbb{R}^3)$ the kinetic energy operator is proportional to the Laplacian ∇^2:

$$-\vec{P}^2 = -(-i\nabla)^2 = \nabla^2 = \frac{1}{r}\frac{\partial^2}{\partial r^2}r + \frac{1}{r^2}\left[\frac{1}{\sin\theta}\frac{\partial}{\partial\theta}\left(\sin\theta\frac{\partial}{\partial\theta}\right) + \frac{1}{\sin^2\theta}\frac{\partial^2}{\partial\phi^2}\right], \tag{10.74}$$

where we have written the Laplacian in polar coordinates. Comparing with (10.49), we recognize in the operator $\vec{L}^2$ the angular part of the Laplacian:

$$\nabla^2 = \frac{1}{r}\frac{\partial^2}{\partial r^2}r - \frac{1}{r^2}\vec{L}^2. \tag{10.75}$$

This equation confirms the commutation relation $[H, \vec{L}] = 0$, since $[\vec{L}^2, \vec{L}] = 0$ and the radial part of the Laplacian, which does not depend on angles, obviously commutes with $\vec{L}$. We can therefore write the Hamiltonian (10.72) as

$$H = -\frac{1}{2M}\frac{1}{r}\frac{\partial^2}{\partial r^2}r + \frac{1}{2Mr^2}\vec{L}^2 + V(r). \tag{10.76}$$

Owing to these commutation relations, we know that it is possible to simultaneously diagonalize H, $\vec{L}^2$, and L_z. Let $\psi_{lm}(\vec{r})$ be an eigenfunction common to these three operators. Since there is only one spherical harmonic (l, m), if

$$\vec{L}^2\psi_{lm} = l(l+1)\psi_{lm} \quad \text{and} \quad L_z\psi_{lm} = m\psi_{lm},$$

then ψ_{lm} must be proportional to Y_l^m:[4]

$$\psi_{lm}(r, \theta, \phi) = f_l(r)Y_l^m(\theta, \phi) = \frac{u_l(r)}{r}Y_l^m(\theta, \phi). \tag{10.77}$$

It is convenient to factorize $1/r$; $u_l(r)$ is the *radial wave function*. Let us examine the action of H on ψ_{lm}:

$$H\psi_{lm}(r, \theta, \phi) = \left[-\frac{1}{2M}\frac{1}{r}\frac{\partial^2}{\partial r^2}u_l(r) + \left(\frac{l(l+1)}{2Mr^2} + V(r)\right)\frac{u_l(r)}{r}\right]Y_l^m(\theta, \phi).$$

The eigenvalue equation

$$H\psi_{lm} = E_l\psi_{lm}$$

becomes the radial equation

$$\boxed{\left[-\frac{1}{2M}\frac{d^2}{dr^2} + \frac{l(l+1)}{2Mr^2} + V(r)\right]u_l(r) = E_l u_l(r)}. \tag{10.78}$$

The radial wave function and the energy are labeled by only the index l and not m, because according to (10.78) they are independent of m. Each value of the energy will therefore be at least $(2l+1)$-fold degenerate. This could have been foreseen from the commutation relation $[H, L_\pm] = 0$. If

$$H\psi_{lm} = E_{lm}\psi_{lm},$$

by reasoning similar to that which enabled us to show that $\alpha(l, m)$ is independent of m, we deduce that E_{lm} is also independent of m (Exercise 10.7.7). For each value of the angular momentum l, or for each *partial wave l*, we have reduced the Schrödinger equation to an ordinary differential equation in a single variable r. Following historical tradition, the partial waves are labeled $s, p, d, f, g, h, \ldots$:

$$l = 0: s \text{ wave}, \quad l = 1: p \text{ wave}, \quad l = 2: d \text{ wave}, \quad l = 3: f \text{ wave},$$

and so on in alphabetical order: $l = 4$: g wave, etc. In each partial wave, (10.78) shows that the potential $V(r)$ must be replaced by an *effective potential $V_l(r)$* (Fig. 10.5):

$$V_l(r) = V(r) + \frac{l(l+1)}{2Mr^2}. \tag{10.79}$$

[4] We anticipate the fact, proved a few lines later, that f_l is independent of m.

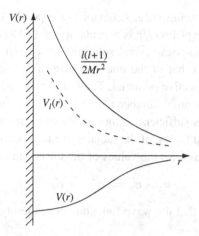

Fig. 10.5. An effective potential. The solid lines represent the potential $V(r)$ and the centrifugal barrier $l(l+1)/2mr^2$, and the dashed lines represent their sum, the effective potential $V_l(r)$ in the partial wave of angular momentum l.

The term $l(l+1)/2Mr^2$ is called the *centrifugal barrier* term. It is also present in classical mechanics, where the energy can be written as

$$E = \frac{1}{2}M\vec{v}^2 + V(r) = \frac{1}{2}M(v_r^2 + \omega^2 r^2) + V(r),$$

where v_r is the radial velocity and ω the angular velocity. Since[5] $l = M\omega r^2$ and $\vec{l}$ is constant in the case of a central force, we have

$$E = \frac{1}{2}Mv_r^2 + \frac{l^2}{2Mr^2} + V(r) = \frac{1}{2}Mv_r^2 + V_l(r).$$

The term $l^2/2Mr^2$ corresponds to the centrifugal force:

$$-\frac{d}{dr}\left(\frac{l^2}{2Mr^2}\right) = \frac{l^2}{Mr^3} = M\omega^2 r.$$

This term tends to push the particle away from the force center in the rotating frame and corresponds to a repulsive potential. In quantum mechanics we replace the operator $\vec{L}^2$ by its eigenvalue $l(l+1)$ for each value of l, and to the potential $V(r)$ we add the repulsive potential $l(l+1)/2Mr^2$.

Not all functions $\psi_{lm}(\vec{r})$ of the type (10.77) with $u_l(r)$ a solution of (10.78) are physically acceptable. If the function $\psi_{lm}(\vec{r})$ represents a bound state, it must satisfy the normalization condition

$$\int d^3 r\, |\psi_{lm}(\vec{r})|^2 = 1. \qquad (10.80)$$

---

[5] Following our usual convention, lower-case letters denote classical quantities (numbers) or quantum numbers.

If $\psi_{lm}(\vec{r})$ represents a scattering state, behavior corresponding to a plane wave plus a spherical wave at infinity $\exp(\pm ikr)/r$ is acceptable [cf. (10.81)]. In the case of a bound state, (10.78) in general possesses several solutions for l fixed. In fact, since $0 \le r < \infty$ this equation is identical to that of the one-dimensional problem in the range $[0, +\infty]$ with $V_l(r)$ (10.79) as the effective potential. The radial wave function and the energy are labeled by an additional quantum number n', $n' = 0, 1, 2, \ldots$, and denoted as $u_{n'l}(r)$ and $E_{n'l}$. If the potential $V(r)$ is sufficiently smooth, it can be shown that n' is equal to the number of zeros, also called nodes, of the radial wave function $u_{n'l}(r)$ (cf. Section 9.3.3). The quantum number n' classifies the values of the energy in increasing order:

$$n'_1 > n'_2 \ \Rightarrow\ E_{n'_1 l} > E_{n'_2 l}.$$

In Chapter 12 we shall see that the wave functions of scattering states are labeled by the wave vector $\vec{k}$:

$$r \to \infty : \psi_{\vec{k}}(\vec{r}) \simeq e^{i\vec{k}\cdot\vec{r}} + f(\theta, \phi)\frac{e^{\pm ikr}}{r}. \tag{10.81}$$

It is possible to analyze the behavior of the wave functions $u_{n'l}(r)$ for $r \to 0$. In all cases of physical interest the centrifugal barrier term is the most singular term when $r \to 0$ and it controls the behavior of $u_{nl}(r)$ in this limit. If we assume a power-law behavior[6]

$$r \to 0 : u_l(r) \propto r^\alpha$$

and substitute it into (10.78), for the two most singular terms in $r^{\alpha-2}$ we obtain

$$-\frac{1}{2M}\alpha(\alpha - 1)r^{\alpha-2} + \frac{l(l+1)}{2M}r^{\alpha-2} = 0,$$

which implies that

$$\alpha(\alpha - 1) = l(l+1),$$

i.e., $\alpha = l+1$ or $\alpha = -l$. The second value is excluded because the integral (10.80) diverges at the origin unless $l = 0$. However, for $l = 0$ a solution $u_0(r) \propto$ const., or $\psi_l(\vec{r}) \propto 1/r$, although normalizable, is not acceptable because it cannot be a solution of the Schrödinger equation owing to

$$\nabla^2 \frac{1}{r} = -4\pi\delta(\vec{r}).$$

In summary, the behavior of the radial wave functions for $r \to 0$ is

$$\boxed{r \to 0 : u_l(r) \propto r^{l+1}}\ . \tag{10.82}$$

The radial wave function vanishes at the origin. This can be seen intuitively: since $0 \le r < \infty$, it is as though there were an infinite potential barrier at $r = 0$, and we know that in this case (see Section 9.3.2) the wave function must vanish. Nevertheless, the

---

[6] The power law giving the behavior at the origin is independent of the quantum numbers n' and k, and so we suppress them.

solutions involving r^{-l} may be useful in solving the Schrödinger equation in a region where r is strictly positive.

The example of the hydrogen atom, which is studied in the following subsection, leads to a redefinition of the radial quantum number, which becomes the principal quantum number:

$$n' \rightarrow n = n' + l + 1. \tag{10.83}$$

10.4.2 The hydrogen atom

The results of the preceding subsection can be used to calculate the energy levels and wave functions of the hydrogen atom, which is one of the few physical problems for which an analytic solution is available. The mass M in (10.78) is the electron mass m_e, or, more precisely, the reduced mass μ (Exercise 8.5.6):

$$\mu = \frac{m_e m_p}{m_e + m_p} \simeq m_e, \tag{10.84}$$

where m_p is the proton mass. However, we shall use m_e rather than μ in the equations in order to emphasize the order of magnitude of the masses which are relevant to this problem. The potential $V(r)$ is the attractive Coulomb potential between the electron and the proton:

$$V(r) = -\frac{q_e^2}{4\pi\varepsilon_0 r} = -\frac{e^2}{r}, \tag{10.85}$$

and (10.78) becomes

$$\left[-\frac{1}{2m_e} \frac{d^2}{dr^2} + \frac{l(l+1)}{2m_e r^2} - \frac{e^2}{r} \right] u_{nl}(r) = E_{nl} u_{nl}(r). \tag{10.86}$$

In physics it is always advisable to make equations dimensionless by an appropriate change of variable. In the present problem the natural unit of length is the Bohr radius (1.34) $a_0 = 1/m_e e^2$, and the natural unit of energy is the Rydberg (1.35) $R_\infty = e^2/2a_0 = m_e e^4/2$.[7] This suggests that we define the dimensionless quantities x and ε_{nl}:

$$x = \frac{r}{a_0} = me^2 r, \qquad \varepsilon_{nl} = -\frac{E_{nl}}{R_\infty} = -\frac{2a_0 E_{nl}}{e^2}. \tag{10.87}$$

In what follows we limit ourselves to bound states for which $E_{nl} < 0$ and therefore $\varepsilon_{nl} > 0$, whence the choice of the minus sign. Also defining

$$v_{nl}(x) = u_{nl}(r) = u_{nl}(a_0 x),$$

after simplification by $(2m_e a_0^2)^{-1}$ we obtain

$$\left[-\frac{d^2}{dx^2} + \frac{l(l+1)}{x^2} - \frac{2}{x} \right] v_{nl}(x) = -\varepsilon_{nl} v_{nl}(x). \tag{10.88}$$

---

[7] We recall that we have chosen a system of units in which $\hbar = 1$. If $\hbar$ is restored, then $a_0 = \hbar^2/m_e e^2$ and $R_\infty = m_e e^4/2\hbar^2$.

We shall limit ourselves to finding the solution in the case $l = 0$, that is, in the s wave, and leave the general case to Exercise 10.7.9. To simplify the notation, we set

$$v_{n0}(x) = v(x), \quad \varepsilon_{n0} = \varepsilon,$$

and (10.88) becomes

$$\frac{d^2 v(x)}{dx^2} = \left(\varepsilon - \frac{2}{x} \right) v(x).$$

We know from the preceding subsection that $v(x) \propto x$ for $x \to 0$. Let us now find the dominant behavior for $x \to \infty$ neglecting the term involving $2/x$. We then have[8]

$$v(x) \sim \exp(\pm\sqrt{\varepsilon}\, x).$$

The $\exp(\sqrt{\varepsilon}\, x)$ behavior is unacceptable because the wave function will not be normalizable owing to the exponential divergence. The only possible behavior is $\exp(-\sqrt{\varepsilon}\, x)$. In order to include the information contained in the behavior at infinity, we define a new function $f(x)$ as

$$v(x) = e^{-\alpha x} f(x), \quad \alpha^2 = \varepsilon.$$

This change of function transforms the differential equation for $v(x)$ into

$$\frac{d^2 f}{dx^2} - 2\alpha \frac{df}{dx} + \frac{2}{x} f = 0. \tag{10.89}$$

Let us seek $f(x)$ in the form of a series in powers of x. Since we know that $f(x) \propto x$ for $x \to 0$,

$$f(x) = \sum_{k=1}^{\infty} a_k x^k. \tag{10.90}$$

Equation (10.89) determines a recursion relation for the coefficients a_k:

$$\sum_{k=1}^{\infty} k(k-1)a_k x^{k-2} - 2\alpha \sum_{k=1}^{\infty} k a_k x^{k-1} + 2 \sum_{k=1}^{\infty} a_k x^{k-1} = 0.$$

Noting that for $k = 1$ the first term in the preceding equation vanishes and relabeling k, we have

$$\sum_{k=1}^{\infty} \left[k(k+1)a_{k+1} - 2(\alpha k - 1)a_k \right] x^{k-1} = 0. \tag{10.91}$$

The cancellation of the coefficient of x^{k-1} gives a relation between a_{k+1} and a_k:

$$a_{k+1} = \frac{2(\alpha k - 1)}{k(k+1)} a_k.$$

---

[8] In fact, this behavior is determined only up to a multiplicative polynomial.

If we arbitrarily fix a_1, all the a_k can be derived from a_1. For $k \gg 1$ the recursion relation is approximately

$$a_{k+1} \simeq \frac{2\alpha}{k} a_k \Rightarrow a_k \simeq \frac{(2\alpha)^k}{k!} a_1$$

and

$$\sum_{k=1}^{\infty} a_k x^k \sim \sum_{k=1}^{\infty} \frac{(2\alpha)^k}{k!} a_1 x^k \sim a_1 e^{2\alpha x}.$$

This implies that for $x \to \infty$

$$v(x) \sim e^{2\alpha x} e^{-\alpha x} \sim a_1 e^{\alpha x},$$

which makes the wave function non-normalizable. The only way to avoid the exponential divergence is to have the series (10.90) terminate at some integer $k = n$, which can happen only if $\alpha n = 1$. The possible values of ε then are labeled by an integer n:

$$\varepsilon_n = \alpha^2 = \frac{1}{n^2},$$

as are those of the energy:

$$E_n = E_{n0} = -\frac{me^4}{2} \frac{1}{n^2} = -\frac{R_\infty}{n^2}. \tag{10.92}$$

Exercise 10.7.9 shows that the possible energies for $l \neq 0$ have the form

$$E_{nl} = -\frac{R_\infty}{n^2}, \quad n = l+1, l+2, \ldots \tag{10.93}$$

The first two ($n = 1, 2$) radial wave functions $v_{n0}(x)$ of the bound states of the hydrogen atom in the s wave, normalized to unity, are

$$v_{10}(x) = 2x e^{-x}, \tag{10.94}$$

$$v_{20}(x) = \frac{1}{\sqrt{2}} x \left(1 - \frac{x}{2}\right) e^{-x/2}. \tag{10.95}$$

The radial wave function in the state $n = 2$, $l = 1$ (the p wave) is

$$v_{21}(x) = \frac{1}{2\sqrt{6}} x^2 e^{-x/2}. \tag{10.96}$$

The spectrum of the hydrogen atom that we have found is shown in Fig. 10.6. The notation for the levels is $ns, np, \ldots$: $1s$ denotes the ground state, $2s$ and $2p$ the first excited (degenerate) levels etc. All the levels are degenerate, except in the case $n = 1$. For a given value of n, all values of l lying between $l = 0$ and $l = n - 1$ are possible, and the degeneracy is

$$G(n) = \sum_{l=0}^{n-1} (2l+1) = n^2.$$

This degeneracy is peculiar to the Coulomb potential. The spectrum of the outer electron of an alkali atom (Fig. 10.7) qualitatively resembles that of the hydrogen atom, except that

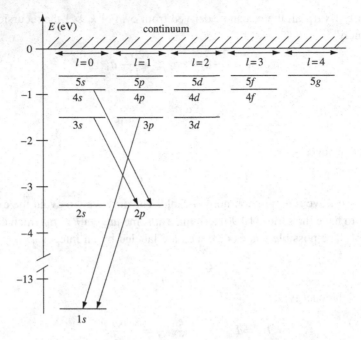

Fig. 10.6. Spectrum of the hydrogen atom.

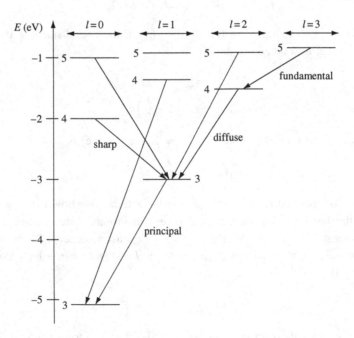

Fig. 10.7. Spectrum of the sodium atom.

there is no degeneracy. The Coulomb potential also presents a remarkable peculiarity in classical mechanics: it is the only potential, along with the harmonic potential $V(r) \propto r^2$, for which the trajectories close on themselves.[9] This feature of the classical motion as well as the degeneracies associated with the quantum problem are due to the presence of an extra symmetry. This symmetry leads to an additional conservation law, that of the Lenz vector in the Coulomb case.

10.5 Angular distributions in decays

10.5.1 Rotations by π, parity, and reflection with respect to a plane

In this section we shall study decays of a particle C into two particles A and B:

$$C \to A + B. \tag{10.97}$$

We shall choose a reference frame in which particle C is at rest; particles A and B then have equal and opposite momenta $\vec{p}$ and $-\vec{p}$, respectively. The process (10.97) includes *radiative decays* (or *transitions*) with the emission of a photon, in which an excited level A^* of an atom, a molecule, or a nucleus emits a photon γ as the system undergoes a transition to a lower energy level A, which may or may not be the ground state:

$$A^* \to A + \gamma. \tag{10.98}$$

The states A^* and A may also correspond to different particles, as, for example, in the decay

$$\Sigma^0 \to \Lambda^0 + \gamma, \tag{10.99}$$

where the particles Σ^0 and Λ^0 are neutral particles formed from an up quark, a down quark, and a strange quark (Exercise 10.7.17).

The invariance under rotation of the Hamiltonian responsible for the decay implies conservation of angular momentum, which leads to constraints on the decay amplitudes and to important consequences for the angular distribution of the final particles. If the Hamiltonian governing the decay is invariant under parity, which is the case for the electromagnetic and strong interactions but not for weak interactions, we obtain additional constraints. It is convenient to introduce the operator $\mathcal{Y}$, the product of a rotation by π about the y axis and the parity operator Π (Section 8.3.3):

$$Y = e^{-i\pi J_y}, \quad \mathcal{Y} = Y\,\Pi = e^{-i\pi J_y}\Pi = \Pi e^{-i\pi J_y}. \tag{10.100}$$

This operator is just reflection with respect to the plane xOz; $\mathcal{Y}$ is the reflection operator with respect to this plane. Let us first study the action of Y. This operator transforms J_x into $-J_x$ and J_z into $-J_z$ while leaving J_y unchanged:

$$Y^{-1}J_z Y = -J_z, \quad Y^{-1}J_\pm Y = -J_\mp. \tag{10.101}$$

---

[9] The two cases are related; cf. Basdevant and Dalibard [2002], Chapter 11, Exercise 3. The extra symmetry can be used to find the energy levels and the wave functions, see e.g. E. Abers, *Quantum Mechanics*, New Jersey: Pearson Education (2004), Chapter 3.

Let us examine the action of Y on the state $|jm\rangle$:

$$J_z(Y|jm\rangle) = -YJ_z|jm\rangle = -m(Y|jm\rangle).$$

The state $Y|jm\rangle$ is then equal to $|j, -m\rangle$ up to a phase:

$$Y|jm\rangle = e^{i\alpha(j,m)}|j, -m\rangle,$$

because Y is unitary and preserves the norm. This result is not surprising, because the action of Y is equivalent to reversing the direction of the angular momentum quantization axis. Following the procedure used above in the case of parity, we apply J_+ to relate $\alpha(j, m)$ to $\alpha(j, m+1)$:

$$\begin{aligned}
J_+Y|jm\rangle &= e^{i\alpha(j,m)}J_+|j, -m\rangle = \sqrt{j(j+1) - m(m-1)}\; e^{i\alpha(j,m)}\;|j, -m+1\rangle \\
&= -YJ_-|jm\rangle = -\sqrt{j(j+1) - m(m-1)}\; Y|j, m-1\rangle \\
&= -\sqrt{j(j+1) - m(m-1)}\; e^{i\alpha(j,m-1)}|j, -m+1\rangle,
\end{aligned}$$

or

$$e^{i\alpha(j,m-1)} = -e^{i\alpha(j,m)}.$$

Since Y is a rotation by π, Y^2 is a rotation by 2π, $Y^2 = (-1)^{2j}$, and

$$Y^2|jm\rangle = e^{i\alpha(j,m)}e^{i\alpha(j,-m)}|jm\rangle = e^{2i\alpha(j,m)}(-1)^{2m}|jm\rangle = (-1)^{2j}|jm\rangle,$$

from which we find the two possible solutions

$$e^{i\alpha(j,m)} = (-1)^{j-m} \quad \text{or} \quad e^{i\alpha(j,m)} = (-1)^{j+m}.$$

These two solutions are identical for integer j, while for $j = 1/2$ we can check using (10.38) that the first solution is the good one. It can be shown that this is also the case for all half-integer j. In the end, we have

$$\boxed{Y|jm\rangle = (-1)^{j-m}|j, -m\rangle, \quad Y^{-1}|jm\rangle = (-1)^{j+m}|j, -m\rangle} \ . \qquad (10.102)$$

10.5.2 Dipole transitions

Now let us study radiative transitions of the type (10.98). First we return to the description of the photon polarization studied in Chapter 3, placing it within the general context of angular momentum. We have determined the infinitesimal generator of rotations of the polarization when the rotation is made about the propagation direction, taken to be the z axis. In the basis of linear polarization states $|x\rangle$ and $|y\rangle$ this infinitesimal generator is given by (3.26):

$$\Sigma_z = \begin{pmatrix} 0 & -i \\ i & 0 \end{pmatrix}.$$

We have already seen in (3.29) that $\exp(-i\theta\Sigma_z)$ performs a rotation of the polarization in the xOy plane by an angle θ, and we can identify Σ_z as the z component of the photon angular momentum: $\Sigma_z = J_z$. Then according to (3.27) the action of the operator $\exp(-i\theta\Sigma_z)$ on the right- and left-handed polarization states $|R\rangle$ and $|L\rangle$ (3.11) is

$$\exp(-i\theta\Sigma_z)|R\rangle = e^{-i\theta}|R\rangle, \quad \exp(-i\theta\Sigma_z)|L\rangle = e^{i\theta}|L\rangle,$$

which proves that the states $|R\rangle$ and $|L\rangle$ have the magnetic quantum numbers $m = 1$ and $m = -1$, respectively.[10] Furthermore, the description of the electromagnetic field by a vector potential shows that the photon has a vector nature and therefore spin 1, which permits $|R\rangle$ and $|L\rangle$ to be identified as the states $|jm\rangle$ (Fig. 10.8):

$$|R\rangle = |j = 1, m = 1\rangle = |11\rangle, \quad |L\rangle = |j = 1, m = -1\rangle = |1, -1\rangle, \qquad (10.103)$$

where the angular momentum quantization axis Oz is taken to lie along the photon propagation direction. The value of m is called the photon *helicity*: $m = +1$ corresponds to positive helicity and $m = -1$ to negative helicity. Since angular momentum 1 corresponds to three possible values of the magnetic quantum number, $m = +1, 0, -1$, we might wonder what has happened to the value $m = 0$ for the photon. A general analysis due to Wigner shows that for a particle of zero mass and spin j, the only allowed eigenvalues of J_z are $m = j$ and $m = -j$, where the axis Oz is taken to lie along the particle propagation direction. When parity is not a symmetry of the Hamiltonian, the two possible values are independent. If the spin-1/2 neutrino had zero mass,[11] it would always have $m = -1/2$, while the antineutrino, which is a *different* particle, would always have $m = +1/2$. The photon interactions conserve parity as they are electromagnetic interactions, and so the same particle can have both $m = 1$ or $m = -1$.

We still need to check that the definition (10.103) corresponds to a standard angular momentum basis. We shall use the operator $Y = \exp(-i\pi J_y)$ which changes the direction

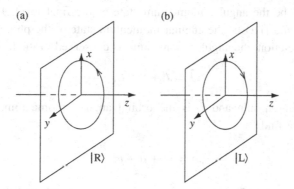

Fig. 10.8. (a) Right-handed circular polarization; (b) left-handed circular polarization.

---

[10] An equivalent argument is to note that $\Sigma_z|R\rangle = |R\rangle$ and $\Sigma_z|L\rangle = -|L\rangle$.
[11] Which for a long time seemed possible, but apparently is not the case; see Exercise 4.3.6 and Footnote 4 of Chapter 1.

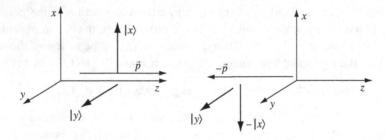

Fig. 10.9. Action of Y on linear polarization states.

of the photon propagation while leaving Oz unchanged. Its action on linear polarization states is (Fig. 10.9)

$$Y|x\rangle = -|x\rangle, \quad Y|y\rangle = |y\rangle.$$

We can derive its action on the circular polarization states $|R\rangle$ and $|L\rangle$ (3.11):

$$Y|R\rangle = Y\left[\frac{-1}{\sqrt{2}}\left(|x\rangle + i|y\rangle\right)\right] = \frac{1}{\sqrt{2}}\left(|x\rangle - i|y\rangle\right) = |L\rangle. \tag{10.104}$$

The relative phase of the states $|R\rangle$ and $|L\rangle$ corresponds to that of a standard basis since, according to (10.102),

$$Y|R\rangle = Y|1,1\rangle = (-1)^{1-1}|1,-1\rangle = |L\rangle.$$

The choice (3.11) is also confirmed by the fact that $|R\rangle$ and $|L\rangle$ are given by the same combinations as the spherical components $\hat{r}_1$, $\hat{r}_{-1}$, and $\hat{r}_0$ (10.64) of $\hat{r}$.

Let us use $\vec{p}$ to denote the photon momentum, which we choose to lie along the z axis, and let $|jm\rangle$ be the angular momentum state of A^* (it is often said that the excited state has spin j), $|j'm'\rangle$ be the angular momentum state of the final level A (or the spin of the final level A), and $|1\mu\rangle$ be the angular momentum state of the photon. Owing to the invariance under rotation, the angular momentum is conserved in the transition:

$$\vec{J} = \vec{J}' + \vec{S} + \vec{L},$$

where $\vec{S}$ is the photon spin and $\vec{L}$ is the orbital angular momentum. Projecting this equation on Oz, we find

$$m = m' + \mu + m_l.$$

It is easy to convince ourselves that the magnetic quantum number of the orbital angular momentum is zero: $m_l = 0$. In fact, the spatial wave function of the photon is a plane wave

$$e^{i\vec{p}\cdot\vec{r}} = e^{ipz} = e^{ipr\cos\theta},$$

which is invariant under rotation about Oz. The z component of the orbital angular momentum must be zero. Another justification follows from (10.47):

$$L_z e^{ipr\cos\theta} = -i\frac{\partial}{\partial\phi}e^{ipr\cos\theta} = 0.$$

The conservation of the angular momentum in the z direction gives

$$\text{right-handed final photon: } m = m' + 1,$$
$$\text{left-handed final photon: } m = m' - 1. \qquad (10.105)$$

If A and A^* have zero spin ($j = j' = 0$), then $m = m' = 0$ and the equations (10.105) have no solution: there is no single-photon radiative transition $j = 0 \to j' = 0$, often called a $0 \to 0$ transition. Radiative $0 \to 0$ transitions are possible only with the emission of at least two photons, and the probability of such a transition is suppressed by a power of the fine-structure constant $\alpha \simeq 1/137$.

A more interesting case which is often encountered in practice is that of $j = 1$ and $j' = 0$. If the photon is emitted in the z direction with helicity $\mu = \pm 1$, there are two possible cases taking into account $j' = m' = 0$:

$$\text{right-handed final photon: } m = 1, \quad \mu = 1, \qquad (10.106)$$
$$\text{left-handed final photon: } m = -1, \quad \mu = -1. \qquad (10.107)$$

Let a be the probability amplitude of (10.106) and b that of (10.107). It should be clearly understood that we are dealing with the amplitude of a *transition* probability, analogous to that calculated in (9.167), and not with probability amplitudes like those defined in postulate **II** of Chapter 4. The squared modulus of a transition amplitude gives the transition probability per unit time. The amplitudes a and b can be viewed as matrix elements of an operator T called the *transition matrix*, which can be calculated, at least formally, as a function of the Hamiltonian and which has the same symmetries as the Hamiltonian. We define the angle θ between the photon emission direction, taken to lie in the xOz plane, and the z axis, and we write the transition amplitudes a and b as (in (10.105) $m' = 0$ because $j' = 0$) for $\theta = 0$

$$a = \langle R, \theta = 0 | T | j = 1, m = 1 \rangle = \langle R, \theta = 0 | T | 11 \rangle,$$
$$b = \langle L, \theta = 0 | T | j = 1, m = -1 \rangle = \langle L, \theta = 0 | T | 1, -1 \rangle. \qquad (10.108)$$

If parity is a symmetry of the Hamiltonian responsible for the transition, then T commutes with $\mathcal{Y}$ (10.100). Since the two amplitudes a and b correspond to transitions which are deduced from each other by reflection with respect to the plane xOz (Fig. 10.10(a) and (b)), we must have $|a| = |b|$. To determine the phase in this relation we use

$$a = \langle R, \theta = 0 | \mathcal{Y}^{-1}T\mathcal{Y} | 1, 1 \rangle = \eta_\gamma \eta_A \eta_{A^*} \langle L, \theta = 0 | T | 1, -1 \rangle$$
$$= \eta_\gamma \eta_A \eta_{A^*} b, \qquad (10.109)$$

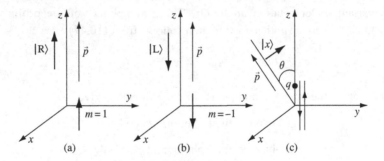

Fig. 10.10. Emission of photons with $\vec{p} \parallel Oz$. The amplitudes in (a) and (b) are deduced from each other by reflection with respect to the plane xOz. (c) Linear polarization of the final photon. The charge q undergoes oscillations along Oz.

where $\eta_X = \pm 1$ is the parity of the particle X. If X has momentum $\vec{p}$ and we write its state vector as $|X, \vec{p}\rangle$, then

$$\Pi |X, \vec{p}\rangle = \eta_X |X, -\vec{p}\rangle. \tag{10.110}$$

The description of the electromagnetic field by a vector potential, which is a polar vector, shows that the photon parity is $\eta_\gamma = -1$. Let $\eta = \eta_A \eta_{A^*}$. Then there are two possible cases:

1. $\eta = -1$, $a = b$;
2. $\eta = +1$, $a = -b$.

We are going to show that the first case is that of an electric dipole transition and the second is that of a magnetic dipole transition.[12] We do this by comparing with the simplest classical case, that of the radiation of a charge undergoing harmonic motion along the z axis. The classical angular momentum of this charge relative to the origin, and in particular its component in the z direction, is always zero, and the quantum case most similar to this situation is that where the excited state A^* possesses zero angular momentum in the z direction, that is, it is in the state $|j = 1, m = 0\rangle$. In order to compare the photon angular distribution with that of the classical radiation, we must imagine the case where the photon emission angle $\theta \neq 0$, the initial state of the atom being $|10\rangle$. We obtain the state $|R, \theta\rangle$ ($|L, \theta\rangle$) of the photon by rotation by an angle θ about Oy starting from $|R, \theta = 0\rangle$ ($|L, \theta = 0\rangle$):

$$|R, \theta\rangle = U[\mathcal{R}_{\hat{y}}(\theta)]|R, \theta = 0\rangle,$$

$$|L, \theta\rangle = U[\mathcal{R}_{\hat{y}}(\theta)]|L, \theta = 0\rangle.$$

---

[12] This result depends on the sign conventions used for the states $|R\rangle$ and $|L\rangle$; we find the sign opposite to that of Feynman *et al.* [1965], Vol III, Section 18.1 owing to the different sign convention in the definition of $|R\rangle$.

The emission amplitude in the θ direction, for example, for a right-handed photon and initial state $|j = 1, m = 0\rangle$, is

$$a_{\mathrm{R}}^{m=0}(\theta) = \langle \mathrm{R}, \theta | T | 10 \rangle = \langle \mathrm{R}, \theta = 0 | U^{\dagger}[\mathcal{R}_{\hat{y}}(\theta)] T | 10 \rangle$$

$$= \langle \mathrm{R}, \theta = 0 | T U^{\dagger}[\mathcal{R}_{\hat{y}}(\theta)] | 10 \rangle$$

$$= \langle \mathrm{R}, \theta = 0 | T | 11 \rangle \langle 11 | U^{\dagger}[\mathcal{R}_{\hat{y}}(\theta)] | 10 \rangle$$

$$= a d_{01}^{(1)}(\theta) = \frac{a}{\sqrt{2}} \sin \theta. \tag{10.111}$$

We have used the rotational invariance of T, introduced a set of intermediate states $\sum_m |1m\rangle \langle 1m|$ in the $j = 1$ subspace, and obtained the rotation matrix element using (10.39). A similar calculation gives the following for the emission of a left-handed photon:

$$a_{\mathrm{L}}^{m=0}(\theta) = b d_{0-1}^{(1)}(\theta) = -\frac{b}{\sqrt{2}} \sin \theta. \tag{10.112}$$

If the final polarization is linear, we can decompose it on the states $|x\rangle$ polarized in the plane xOz and $|y\rangle$ polarized along Oy (Fig. 10.10 (c)):[13]

$$|x\rangle = \frac{1}{\sqrt{2}} \left(-|\mathrm{R}\rangle + |\mathrm{L}\rangle \right), \quad |y\rangle = \frac{i}{\sqrt{2}} \left(|\mathrm{R}\rangle + |\mathrm{L}\rangle \right), \tag{10.113}$$

and we find

$$a_x^{m=0}(\theta) = \langle x, \theta | T | 10 \rangle = -\frac{1}{2}(a+b) \sin \theta,$$

$$a_y^{m=0}(\theta) = \langle y, \theta | T | 10 \rangle = \frac{i}{2}(a-b) \sin \theta. \tag{10.114}$$

In the electric dipole case $a = b$ the photons are polarized along Ox, while in the magnetic dipole case they are polarized along Oy. This corresponds to the classical case. If, for example, we take a charge undergoing harmonic oscillations along Oz with zero z component of angular momentum, the radiation is polarized in the plane xOz. On the other hand, a magnetic dipole will produce radiation polarized along Oy. An electric dipole transition corresponds to $\eta = -1$, and therefore *to initial and final states with opposite parities*, while a magnetic dipole transition corresponds to initial state and final state with the same parity. In both cases the angular distribution is $\sin^2 \theta$.

10.5.3 Two-body decays: the general case

Let us return to the general case of two-body decay (10.97), using j_A, j_B, and j_C to denote the spins of the particles A, B, and C. We define the transition amplitude for the initial

---

[13] The states $|x\rangle$ and $|y\rangle$ are defined with respect to the propagation direction $\vec{p}$; see Fig. 10.10 (c).

state $|j_C m_C\rangle$ of particle C to the final states $|j_A m_A\rangle$ and $|j_B m_B\rangle$ of particles A and B, assuming that particle A is emitted with momentum $\vec{p}$ in the direction (θ, ϕ):

$$a^{mc}_{m_A m_B}(\theta, \phi) = \langle m_A m_B; (\theta, \phi)|T|m_C\rangle. \tag{10.115}$$

If particle A is emitted in the direction $\hat{p} = (\theta, \phi)$, the state

$$|m_A m_B; (\theta, \phi)\rangle = U(\mathcal{R})|m_A m_B; (\theta = 0, \phi = 0)\rangle$$

is the state $|m_A m_B; (\theta = 0, \phi = 0)\rangle$ transformed by the rotation (θ, ϕ) aligning the z axis in the direction of $\hat{p}$. It should be emphasized that in this state we have chosen the angular momentum quantization axis to lie along $\hat{p}$, and m_A and m_B are the eigenvalues of $(\vec{J} \cdot \hat{p})$ and not J_z (Fig. 10.11). When particle A is emitted in the z direction, $\theta = \phi = 0$, conservation of the z component of angular momentum implies, as in the preceding subsection, that $m_C = m_A + m_B$. The only nonzero transition amplitudes are

$$b_{m_A, m_B} = \langle m_A m_B; (\theta = 0, \phi = 0)|T|m_C = m_A + m_B\rangle. \tag{10.116}$$

Using the same arguments as for (10.111), we find

$$\begin{aligned}
a^{mc}_{m_A, m_B}(\theta, \phi) &= \langle m_A m_B; (\theta, \phi)|T|m_C\rangle \\
&= \langle m_A, m_B; (\theta = 0, \phi = 0)|U^\dagger(\mathcal{R})T|m_C\rangle \\
&= \langle m_A, m_B; (\theta = 0, \phi = 0)|T|m'_C = m_A + m_B\rangle\langle m'_C = m_A + m_B|U^\dagger(\mathcal{R})|m_C\rangle \\
&= b_{m_A, m_B}\left(D^{(j_C)}_{m_C; m_A + m_B}(\theta, \phi)\right)^* \tag{10.117} \\
&= b_{m_A, m_B} d^{(j_C)}_{m_C; m_A + m_B}(\theta)e^{im_C\phi}. \tag{10.118}
\end{aligned}$$

If parity is conserved in the decay, then

$$\begin{aligned}
b_{m_A, m_B} &= \langle m_A, m_B; (\theta = 0, \phi = 0)|\mathcal{Y}^\dagger T \mathcal{Y}|m_C = m_A + m_B\rangle \\
&= \eta(-1)^{j_C - j_A - j_B} b_{-m_A, -m_B}, \tag{10.119}
\end{aligned}$$

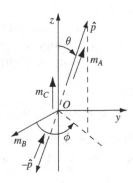

Fig. 10.11. The decay $C \rightarrow A + B$.

where $\eta = \eta_A \eta_B \eta_C$ is the product of the parities of the three particles. Parity conservation halves the number of independent amplitudes. The amplitudes defined in (10.118) are called *helicity amplitudes*. However, it should be noted that the angular momentum quantization axis of particle B is often taken to be aligned with its momentum $-\hat{p}$, which causes $m_B \to -m_B$. The magnetic quantum numbers m_A and $-m_B$ (with our definition) are the helicities of particles A and B.

10.6 Addition of two angular momenta

10.6.1 Addition of two spins 1/2

In Section 6.1.2 we constructed a four-dimensional space $\mathcal{H}_1 \otimes \mathcal{H}_2$ by taking the tensor product of the two-dimensional spaces of two spins 1/2, $\vec{S}_1$ and $\vec{S}_2$. A possible basis in this space is formed from the eigenvectors $|\varepsilon_1 \varepsilon_2\rangle$, $\varepsilon = \pm$, of S_{1z} and S_{2z}:

$$|++\rangle, \quad |+-\rangle, \quad |-+\rangle, \quad \text{and} \quad |--\rangle. \tag{10.120}$$

The physical properties that are diagonal in this basis are $\vec{S}_1^2$, $\vec{S}_2^2$, S_{1z}, and S_{2z}:

$$\vec{S}_1^2 |\varepsilon_1 \varepsilon_2\rangle = \frac{3}{4}|\varepsilon_1 \varepsilon_2\rangle, \quad S_{1z}|\varepsilon_1 \varepsilon_2\rangle = \varepsilon_1 |\varepsilon_1 \varepsilon_2\rangle, \tag{10.121}$$

$$\vec{S}_2^2 |\varepsilon_1 \varepsilon_2\rangle = \frac{3}{4}|\varepsilon_1 \varepsilon_2\rangle, \quad S_{2z}|\varepsilon_1 \varepsilon_2\rangle = \varepsilon_2 |\varepsilon_1 \varepsilon_2\rangle. \tag{10.122}$$

This basis corresponds to the following choice of complete set of compatible operators: $\{\vec{S}_1^2, \vec{S}_2^2, S_{1z}, S_{2z}\}$. It is possible to construct another interesting basis using the total angular momentum $\vec{S}$ obtained by adding $\vec{S}_1$ and $\vec{S}_2$:

$$\vec{S} = \vec{S}_1 + \vec{S}_2. \tag{10.123}$$

Here $\vec{S}$ is actually the total angular momentum, because it can be used to construct the infinitesimal generator in the tensor product space $\mathcal{H}_1 \otimes \mathcal{H}_2$ of a rotation $\mathcal{R}_{\hat{n}}(\theta)$ by an angle θ about the $\hat{n}$ axis:

$$U[\mathcal{R}_{\hat{n}}(\theta)] = e^{-i\theta(\vec{S}_1 \cdot \hat{n})} e^{-i\theta(\vec{S}_2 \cdot \hat{n})} = e^{-i\theta(\vec{S} \cdot \hat{n})}, \tag{10.124}$$

where we have used $[\vec{S}_1, \vec{S}_2] = 0$. Since $\vec{S}_1^2$ and $\vec{S}_2^2$ are scalar operators, they commute with $\vec{S}$, and another set of compatible operators is $\{\vec{S}_1^2, \vec{S}_2^2, \vec{S}^2, S_z\}$. We shall show below that this set is also complete. Let us find the basis vectors of this new set. Setting $|1, 1\rangle = |++\rangle$, we can show that

$$S_z |1, 1\rangle = |1, 1\rangle,$$

$$S_+ |1, 1\rangle = (S_{1+} + S_{2+})|++\rangle = 0,$$

$$S_- |1, 1\rangle = (S_{1-} + S_{2-})|++\rangle = |+-\rangle + |-+\rangle = \sqrt{2}\,|1, 0\rangle.$$

This last equation defines the normalized state vector $|1, 0\rangle$, which satisfies

$$S_z|1, 0\rangle = (S_{1z} + S_{2z})\frac{1}{\sqrt{2}}\big(|+-\rangle + |-+\rangle\big) = 0.$$

Finally,

$$S_-|1, 0\rangle = (S_{1-} + S_{2-})\frac{1}{\sqrt{2}}(|+-\rangle + |-+\rangle) = \sqrt{2}\,|--\rangle = \sqrt{2}\,|1, -1\rangle,$$

$$S_z|1, -1\rangle = -|1, -1\rangle, \qquad S_-|1, -1\rangle = 0.$$

These equations show that the three state vectors

$$\{|1, 1\rangle, \ |1, 0\rangle, \ |1, -1\rangle\}$$

form a standard basis for angular momentum 1. It is sufficient to check the properties of the standard basis for S_z and S_-, because $S_+ = S_-^\dagger$ and $\vec{S}^2 = \frac{1}{2}(S_+S_- + S_-S_+) + S_z^2$. The above calculation shows that we have indeed constructed a standard basis; for example,

$$S_-|1, 1\rangle = \sqrt{j(j+1) - m(m-1)}\,|1, 0\rangle = \sqrt{2}\,|1, 0\rangle.$$

Finally, to obtain a basis of $\mathcal{H}_1 \otimes \mathcal{H}_2$, we need to construct a fourth vector orthogonal to the other three:

$$|0, 0\rangle = \frac{1}{\sqrt{2}}(|+-\rangle - |-+\rangle).$$

This vector is just the vector $|\Phi\rangle$ (6.15). As it is invariant under rotation, it corresponds to angular momentum zero, and it can be verified explicitly that

$$S_z|0, 0\rangle = 0, \quad S_\pm|0, 0\rangle = 0.$$

In summary, when two angular momenta 1/2 are added, we obtain the angular momenta $s = 1$ and $s = 0$. A standard basis of $\vec{S}^2$ and S_z is formed from the vectors corresponding to $s = 1$:

$$s = 1 \quad \begin{cases} |1, 1\rangle = |++\rangle, \\ |1, 0\rangle = \frac{1}{\sqrt{2}}(|+-\rangle + |-+\rangle), \\ |1, -1\rangle = |--\rangle, \end{cases} \qquad (10.125)$$

and $s = 0$:

$$s = 0 \quad |0, 0\rangle = \frac{1}{\sqrt{2}}(|+-\rangle - |-+\rangle). \qquad (10.126)$$

Since we have found four orthogonal vectors, they form a basis of $\mathcal{H}_1 \otimes \mathcal{H}_2$, and the set of compatible operators $\{\vec{S}_1^2, \vec{S}_2^2, \vec{S}^2, S_z\}$, or simply $\{\vec{S}^2, S_z\}$, is complete. The $s = 1$ states are called *triplet states* and the $s = 0$ state is called the *singlet state*.

As an application, let us rederive the results of Exercise 6.5.4, where we diagonalized the operator $(\vec{\sigma}_1 \cdot \vec{\sigma}_2)$. This operator is diagonal in the basis $\{\vec{S}^2, S_z\}$. We have

$$\vec{S}^2 = \frac{1}{4}(\vec{\sigma}_1 + \vec{\sigma}_2)^2 = \frac{3}{2} + \frac{1}{2}\vec{\sigma}_1 \cdot \vec{\sigma}_2, \qquad (10.127)$$

whence

$$\vec{\sigma}_1 \cdot \vec{\sigma}_2 = 2\vec{S}^2 - 3I = [2s(s+1) - 3]I.$$

The operator $\vec{\sigma}_1 \cdot \vec{\sigma}_2$ is equal to I in the triplet state and $-3I$ in the singlet state. We can find the projectors $\mathcal{P}_1$ and $\mathcal{P}_0$ on the triplet and singlet states:

$$\mathcal{P}_0 + \mathcal{P}_1 = I, \quad \vec{\sigma}_1 \cdot \vec{\sigma}_2 = -3\mathcal{P}_0 + \mathcal{P}_1,$$

from which

$$\boxed{\mathcal{P}_0 = \frac{1}{4}(I - \vec{\sigma}_1 \cdot \vec{\sigma}_2), \quad \mathcal{P}_1 = \frac{1}{4}(3 + \vec{\sigma}_1 \cdot \vec{\sigma}_2)}. \tag{10.128}$$

The operator $\vec{\sigma}_1 \cdot \vec{\sigma}_2$ is a scalar operator which commutes with $\vec{S}$, but not with the individual spin operators $\vec{S}_1$ and $\vec{S}_2$. It should also be noted that the triplet states are symmetric (that is, they do not change sign) under the interchange of spins 1 and 2, while the singlet state is antisymmetric under this interchange.

10.6.2 The general case: addition of two angular momenta $\vec{J}_1$ and $\vec{J}_2$

Now let us generalize the preceding discussion to the addition of two angular momenta $\vec{J}_1$ and $\vec{J}_2$. The reasoning used in (10.124) can be repeated to show that $\vec{J} = \vec{J}_1 + \vec{J}_2$ is the total angular momentum. As in the preceding subsection, we construct the $(2j_1 + 1) \times (2j_2 + 1)$-dimensional tensor product space:

$$\mathcal{E} = \mathcal{E}(j_1) \otimes \mathcal{E}(j_2).$$

A possible basis of this space is constructed from the eigenvectors

$$|j_1 j_2 m_1 m_2\rangle = |j_1 m_1\rangle \otimes |j_2 m_2\rangle \tag{10.129}$$

common to $\vec{J}_1^2$, $\vec{J}_2^2$, J_{1z}, and J_{2z}:

$$\vec{J}_1^2 |j_1 j_2 m_1 m_2\rangle = j_1(j_1 + 1)|j_1 j_2 m_1 m_2\rangle,$$

$$\vec{J}_2^2 |j_1 j_2 m_1 m_2\rangle = j_2(j_2 + 1)|j_1 j_2 m_1 m_2\rangle,$$

$$J_{1z} |j_1 j_2 m_1 m_2\rangle = m_1 |j_1 j_2 m_1 m_2\rangle,$$

$$J_{2z} |j_1 j_2 m_1 m_2\rangle = m_2 |j_1 j_2 m_1 m_2\rangle.$$

This basis corresponds to the complete set of commuting operators $\{\vec{J}_1^2, \vec{J}_2^2, J_{1z}, J_{2z}\}$. We shall construct another basis of $\mathcal{E}$ in which the operators $\{\vec{J}_1^2, \vec{J}_2^2, \vec{J}^2, J_z\}$ are diagonal. We start with the two following observations.

- Any vector $|j_1 j_2 m_1 m_2\rangle$ is an eigenvector of J_z with eigenvalue $m = m_1 + m_2$.
- If a value of j is allowed, by applying J_+ and J_- we generate a series of $(2j+1)$ vectors $|jm\rangle$. A priori, we could have several series of vectors of this type, and we use $N(j)$ to denote the number of such series for a given value of j.

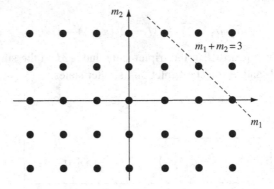

Fig. 10.12. Addition of two angular momenta.

Let $n(m)$ be the degeneracy of the eigenvalue m of J_z. Since m occurs if and only if $j \geq |m|$, we have (Fig. 10.12)

$$n(m) = \sum_{j \geq |m|} N(j),$$

and consequently

$$N(j) = n(j) - n(j+1).$$

However, $n(m)$ is equal to the number of pairs (m_1, m_2) such that $m = m_1 + m_2$. Assuming, for example, that $j_1 \geq j_2$,

$$n(m) = \begin{cases} 0 & \text{if} & |m| > j_1 + j_2, \\ j_1 + j_2 + 1 - |m| & \text{if} & j_1 - j_2 \leq m \leq j_1 + j_2, \\ 2j_2 + 1 & \text{if} & 0 \leq |m| \leq j_1 - j_2. \end{cases}$$

We then conclude that

$$N(j) = 1 \quad \text{for} \quad j_1 - j_2 \leq j \leq j_1 + j_2$$

and $N(j) = 0$ otherwise. To deal with the case $j_2 > j_1$ it is sufficient to replace $(j_1 - j_2)$ by $|j_1 - j_2|$. We can then state the following theorem.

The angular momentum addition theorem

In the tensor product space $\mathcal{E} = \mathcal{E}(j_1) \otimes \mathcal{E}(j_2)$

1. The possible values of j are

$$|j_1 - j_2|, \ |j_1 - j_2| + 1, \ldots, j_1 + j_2 - 1, \ j_1 + j_2; \qquad (10.130)$$

2. To each value of j there corresponds only one series of eigenvectors $|jm\rangle$:

$$\vec{J}^2 |jm\rangle = j(j+1)|jm\rangle, \quad J_z |jm\rangle = m|jm\rangle. \qquad (10.131)$$

■ It is instructive to verify that the dimension of $\mathcal{E}$ is indeed correct ($j_1 \geq j_2$):

$$\dim\mathcal{E} = \sum_{j_1-j_2 \leq j \leq j_1+j_2} (2j+1)$$

$$= (j_1+j_2)(j_1+j_2+1) - (j_1-j_2-1)(j_1-j_2) + (j_1+j_2) - (j_1-j_2-1)$$

$$= (2j_1+1)(2j_2+1).$$

Let us now go from the orthonormal basis $|j_1 j_2 m_1 m_2\rangle$ to the orthonormal basis $|jm\rangle$ by means of a unitary transformation. The elements of the unitary matrix that performs this transformation are called the *Clebsch–Gordan (CG) coefficients* $C_{m_1 m_2; jm}^{j_1 j_2}$:

$$|jm\rangle = \sum_{m_1+m_2=m} C_{m_1 m_2; jm}^{j_1 j_2} |j_1 j_2 m_1 m_2\rangle. \tag{10.132}$$

They can be nonzero only if $m = m_1 + m_2$ and $|j_1 - j_2| \leq j \leq j_1 + j_2$. We choose the following phase convention:

$$C_{m_1 m_2; j, m=j}^{j_1 j_2} \quad \text{real} \geq 0,$$

and then by application of J_- it can be shown that all the CG coefficients are real. The Clebsch–Gordan coefficients are the elements of a unitary real matrix with the matrix indices $(m_1 m_2)$ and (jm). They therefore satisfy the orthogonality conditions

$$\sum_{m_1=-j_1}^{j_1} \sum_{m_2=-j_2}^{j_2} C_{m_1 m_2; jm}^{j_1 j_2} C_{m_1 m_2; j'm'}^{j_1 j_2} = \delta_{jj'} \delta_{mm'}, \tag{10.133}$$

and inversely

$$\sum_{j=|j_1-j_2|}^{j_1+j_2} \sum_{m=-j}^{j} C_{m_1 m_2; jm}^{j_1 j_2} C_{m_1' m_2'; jm}^{j_1 j_2} = \delta_{m_1 m_1'} \delta_{m_2 m_2'}. \tag{10.134}$$

Equations (10.125) and (10.126) give examples of CG coefficients:

$$C_{\frac{1}{2}\frac{1}{2};11}^{\frac{1}{2}\frac{1}{2}} = 1, \quad C_{\frac{1}{2}-\frac{1}{2};10}^{\frac{1}{2}\frac{1}{2}} = \frac{1}{\sqrt{2}}.$$

As an application of angular momentum addition, let us study *spin–orbit coupling*. Owing to relativistic effects, the orbital angular momentum and the spin of an atomic electron, for example the electron of the hydrogen atom or the valence electron of an alkali atom, are not independent, as we shall see in Section 14.2.2. The total angular momentum of the electron is the sum of its orbital angular momentum $\vec{L}$ and its spin $\vec{S}$:

$$\vec{J} = \vec{L} + \vec{S}. \tag{10.135}$$

The possible values of j then are $j = l + 1/2$ and $j = l - 1/2$ (except if $l = 0$, in which case $j = s = 1/2$). The orbital angular momentum and the spin are coupled by a spin–orbit potential:

$$V_{\text{so}}(r) = V(r)\vec{L} \cdot \vec{S}. \tag{10.136}$$

This potential takes different values depending on whether $j = l + 1/2$ or $j = l - 1/2$. We can write

$$(\vec{L} + \vec{S})^2 = \vec{J}^2 = \vec{L}^2 + \vec{S}^2 + 2\vec{L} \cdot \vec{S}$$

and so

$$\vec{L} \cdot \vec{S} = \frac{1}{2}\big[j(j+1) - l(l+1) - s(s+1) \big], \tag{10.137}$$

which gives for the spin–orbit potential

$$V_{so}(r) = \begin{cases} \frac{1}{2}V(r)l & \text{for } j = l + 1/2, \\ -\frac{1}{2}V(r)(l+1) & \text{for } j = l - 1/2. \end{cases} \tag{10.138}$$

10.6.3 Composition of rotation matrices

The rule for the addition of angular momentum is reflected in a composition law for rotation matrices. Let us consider the matrix elements of the rotation operator $U(\mathcal{R})$ taken between states $|jm\rangle$ and $|jm'\rangle$ of the type (10.132):

$$\langle jm|U(\mathcal{R})|jm'\rangle = D^{(j)}_{mm'}(\mathcal{R})$$

$$= \sum_{m_1 m_2} \sum_{m'_1 m'_2} C^{j_1 j_2}_{m_1 m_2; jm} C^{j_1 j_2}_{m'_1 m'_2; jm'} \langle j_1 j_2 m_1 m_2 | U(\mathcal{R}) | j_1 j_2 m'_1 m'_2 \rangle,$$

from which

$$D^{(j)}_{mm'}(\mathcal{R}) = \sum_{m_1 m_2} \sum_{m'_1 m'_2} C^{j_1 j_2}_{m_1 m_2; jm} C^{j_1 j_2}_{m'_1 m'_2; jm'} D^{(j_1)}_{m_1 m'_1}(\mathcal{R}) D^{(j_2)}_{m_2 m'_2}(\mathcal{R}). \tag{10.139}$$

Using the orthogonality relations (10.133) and (10.134) of the CG coefficients, we can invert (10.139):

$$D^{(j_1)}_{m_1 m'_1}(\mathcal{R}) D^{(j_2)}_{m_2 m'_2}(\mathcal{R}) = \sum_{|j_1 - j_2|}^{j_1 + j_2} C^{j_1 j_2}_{m_1 m_2; jm} C^{j_1 j_2}_{m'_1 m'_2; jm'} D^{(j)}_{mm'}(\mathcal{R}). \tag{10.140}$$

These equations can be interpreted in the following manner. In the space $\mathcal{E}(j_1) \otimes \mathcal{E}(j_2)$ we construct the matrix $\Delta(\mathcal{R})$, the tensor product of $D^{(j_1)}(\mathcal{R})$ and $D^{(j_2)}(\mathcal{R})$:

$$\Delta_{m_1 m_2; m'_1 m'_2}(\mathcal{R}) = D^{(j_1)}_{m_1 m'_1}(\mathcal{R}) \otimes D^{(j_2)}_{m_2 m'_2}(\mathcal{R}).$$

By a change of basis made using a unitary matrix whose elements are the CG coefficients $C^{j_1 j_2}_{m_1 m_2; jm}$, the matrix

$$\Delta'(\mathcal{R}) = C\Delta(\mathcal{R})C^{-1}$$

becomes a block-diagonal matrix:

$$C\Delta(\mathcal{R})C^{-1} = \begin{pmatrix} D^{(j_1+j_2)} & 0 & \cdots & 0 \\ 0 & D^{(j_1+j_2-1)} & \cdots & \vdots \\ 0 & 0 & \cdots & \vdots \\ 0 & \cdots & 0 & D^{(|j_1-j_2|)} \end{pmatrix}.$$

In mathematical terms, this is referred to as reducing the product of two representations $D^{(j_1)}$ and $D^{(j_2)}$ of the rotation group to irreducible components:

$$D^{(j_1)} \otimes D^{(j_2)} = D^{(j_1+j_2)} \oplus D^{(j_1-j_2-1)} \oplus \cdots \oplus D^{(|j_1-j_2|)}. \tag{10.141}$$

10.6.4 The Wigner–Eckart theorem (scalar and vector operators)

In Section 8.2.3 we defined a scalar operator $\mathcal{S}$ as an operator which commutes with $\vec{J}$: $[\mathcal{S}, \vec{J}] = 0$. Let us examine the matrix elements $\langle j'm'|\mathcal{S}|jm\rangle$ of $\mathcal{S}$ in a standard angular momentum basis:

$$[\mathcal{S}, \vec{J}^2] = 0 \Rightarrow j' = j, \quad [\mathcal{S}, J_z] = 0 \Rightarrow m' = m.$$

In addition,

$$[\mathcal{S}, J_\pm] = 0 \Rightarrow \langle jm|\mathcal{S}|jm\rangle = \langle j||\mathcal{S}||j\rangle \text{ is independent of } m. \tag{10.142}$$

The quantity $\langle j||\mathcal{S}||j\rangle$ is called the *reduced matrix element* of $\mathcal{S}$.

Now let us turn to vector operators $\vec{V}$, which we have defined in Section 8.2.3. The Cartesian components V_k of a vector operator transform under rotation as

$$U^\dagger(\mathcal{R})V_k U(\mathcal{R}) = \sum_l \mathcal{R}_{kl} V_l. \tag{10.143}$$

By considering infinitesimal rotations, in Section 8.2.3 we derived the commutation relations involving the components of angular momentum:

$$[J_k, V_l] = i \sum_p \varepsilon_{klp} V_p. \tag{10.144}$$

Equations (10.143) and (10.144) are strictly equivalent and either can be used to define a vector operator. It is convenient to use spherical components V_q of $\vec{V}$:

$$V_1 = -\frac{1}{\sqrt{2}}(V_x + iV_y), \quad V_0 = V_z, \quad V_{-1} = \frac{1}{\sqrt{2}}(V_x - iV_y). \tag{10.145}$$

These components are also called the standard components of $\vec{V}$, because when $\vec{V}$ is the position operator, $\vec{V} = \vec{R}$, the components $\hat{r}_1$, $\hat{r}_0$, and $\hat{r}_{-1}$ of the vector $\hat{r}$ are just the spherical harmonics $Y_1^\pm$ and Y_1^0 up to a factor of $\sqrt{3/4\pi}$ (cf. (10.64)). According to (10.65), this implies the transformation law

$$(\mathcal{R}\hat{r})_m = \sum_{m'} D^{(1)}_{m'm}(\mathcal{R}^{-1})\hat{r}_{m'}. \tag{10.146}$$

The transformation law of the spherical components of $\vec{V}$ then is[14]

$$U(\mathcal{R})V_q U^\dagger(\mathcal{R}) = \sum_{q'} D^{(1)}_{q'q}(\mathcal{R})V_{q'}. \tag{10.147}$$

This can easily be checked using the explicit expressions for $D^{(1)}$ and the definition of the spherical components.

Our goal is to relate the matrix elements of the various components of a vector operator to the states $|jm\rangle$. To do this, let us study the properties of the vector $|1jqm\rangle = V_q|jm\rangle$ under rotation:

$$U(\mathcal{R})|1jqm\rangle = U(\mathcal{R})V_q U^\dagger(\mathcal{R})U(\mathcal{R})|jm\rangle$$

$$= \sum_{q'm'} D^{(1)}_{q'q}(\mathcal{R})D^{(j)}_{m'm}(\mathcal{R})|1jq'm'\rangle.$$

The vectors $|1jqm\rangle$ transform under rotation in exactly the same way as the vectors $|j_1 j_2 m_1 m_2\rangle$ with $j_1 = 1, j_2 = j, m_1 = q, m_2 = m$. We can then construct the vectors

$$|\tilde{j}\tilde{m}\rangle = \sum_{m+q=\tilde{m}} C^{1j}_{qm;\tilde{j}\tilde{m}}|1jqm\rangle, \tag{10.148}$$

which transform under rotation as

$$U(\mathcal{R})|\tilde{j}\tilde{m}\rangle = \sum_{\tilde{m}'} D^{(\tilde{j})}_{\tilde{m}'\tilde{m}}|\tilde{j}\tilde{m}'\rangle.$$

This equation shows that the vectors $|\tilde{j}\tilde{m}\rangle$ form a standard basis of the space $\mathcal{E}(\tilde{j})$ up to a global multiplicative factor. These vectors will not in general be normalized, but they will have the same norm for any $\tilde{m}$:

$$\langle \tilde{j}\tilde{m}|\tilde{j}'\tilde{m}'\rangle = \delta_{\tilde{j}\tilde{j}'}\delta_{\tilde{m}\tilde{m}'}\alpha(\tilde{j}).$$

Inverting (10.148),

$$V_q|jm\rangle = |1qjm\rangle = \sum_{\tilde{j}=j-1}^{j+1} C^{1j}_{qm;\tilde{j}\tilde{m}}|\tilde{j}\tilde{m}\rangle,$$

from which

$$\langle j'm'|V_q|jm\rangle = \sum_{\tilde{j}} C^{1j}_{qm;\tilde{j}\tilde{m}}\langle j'm'|1j\tilde{j}\tilde{m}\rangle$$

$$= \sum_{\tilde{j}} C^{1j}_{qm;\tilde{j}\tilde{m}}\delta_{j'\tilde{j}}\delta_{m'\tilde{m}}\beta(j',j) = C^{1j}_{qm;j',m'}\beta(j',j).$$

Defining the reduced matrix element $\langle j'||V_q||j\rangle$ as

$$\langle j'||V||j\rangle = \beta(j',j),$$

---

[14] We note that the ordering of U and $U^\dagger$, as well as that of the indices, is different from that in (10.143).

we obtain the *Wigner–Eckart theorem* for vector operators:

$$\boxed{\langle j'm'|V_q|jm\rangle = C^{1j}_{qm;\,j'm'}\,\langle j'||V||j\rangle}\,.$$
(10.149)

All the dependence on the magnetic quantum numbers m, m', and q is contained in the Clebsch–Gordan coefficient $C^{1j}_{qm;\,j'm'}$, which can be looked up in tables. For fixed j, the only possible values of j' are $j' = j-1,\; j,\; j+1$. This theorem can be generalized to *irreducible tensor operators*; see Exercise 10.7.18.

As an application, let us calculate the matrix elements of a vector operator when $j = j'$, using the fact that $\vec{J}$ is a vector operator with matrix elements satisfying (10.149):

$$\langle jm'|J_q|jm\rangle = C^{1j}_{qm;\,jm'}\,\langle j||J||j\rangle\,.$$

This leads to a proportionality relation for the Cartesian components V_k:

$$\langle jm'|V_k|jm\rangle = K\langle jm'|J_k|jm\rangle\,.$$

To evaluate the constant K, we calculate the scalar product $\vec{J}\cdot\vec{V}$, which is a scalar operator:

$$\langle jm|(\vec{J}\cdot\vec{V})|jm\rangle = \sum_{k,m''}\langle jm|J_k|jm''\rangle\langle jm''|V_k|jm\rangle$$

$$= K\sum_{k,m''}\langle jm|J_k|jm''\rangle\langle jm''|J_k|jm\rangle$$

$$= K\langle jm|\vec{J}^2|jm\rangle = Kj(j+1)\,.$$

Combining these equations, we obtain for the matrix elements of V_k

$$\boxed{\langle jm'|V_k|jm\rangle = \frac{1}{j(j+1)}\langle j||(\vec{J}\cdot\vec{V})||j\rangle\langle jm'|J_k|jm\rangle}\,.$$
(10.150)

Since $(\vec{J}\cdot\vec{V})$ is a scalar operator, $\langle jm|(\vec{J}\cdot\vec{V})|jm\rangle$ is independent of m and equal to the reduced matrix element $\langle j||(\vec{J}\cdot\vec{V})||j\rangle$.

10.7 Exercises

10.7.1 Properties of $\vec{J}$

Show by explicit calculation that $[\vec{J}^2, J_z] = 0$. Also verify the identities (10.5) to (10.9).

10.7.2 Rotation of angular momentum

Let $\mathcal{R}$ be a rotation (10.30) by angles (θ, ϕ). Show that the vector

$$U(\mathcal{R})|jm\rangle = e^{-i\phi J_z}e^{-i\theta J_y}|jm\rangle$$

is an eigenvector of the operator

$$J_x \sin\theta\cos\phi + J_y \sin\theta\sin\phi + J_z \cos\theta = \vec{J}\cdot\hat{n}$$

with eigenvalue m. Here $\hat{n}$ is the unit vector in the direction (θ,ϕ). Hint: adapt (8.29).

10.7.3 Rotations (θ,ϕ)

Show that the rotation (10.30) $\mathcal{R}(\theta,\phi)$ can be written as

$$\mathcal{R}(\theta,\phi) = \mathcal{R}_{y'}(\theta)\mathcal{R}_z(\phi),$$

where Oy' is the axis obtained from Oy by a rotation by ϕ about Oz. Hint: show that

$$\mathcal{R}_{y'}(\theta) = \mathcal{R}_z(\phi)\mathcal{R}_y(\theta)\mathcal{R}_z(-\phi).$$

10.7.4 The angular momenta $j = \frac{1}{2}$ and $j = 1$

1. Use (10.23) to find the operators S_x, S_y, and S_z for spin 1/2.
2. Again using (10.23), calculate the 3×3 matrix representations of J_x, J_y, and J_z for angular momentum $j = 1$.
3. Show that for $j = 1$, J_x, J_y, and J_z are related to the infinitesimal generators (8.26) T_x, T_y, and T_z by a unitary transformation which takes the Cartesian components of $\hat{r}$ to the spherical components (10.64): $J_i = U^\dagger T_i U$ with

$$U = \frac{1}{\sqrt{2}}\begin{pmatrix} -1 & 0 & 1 \\ -i & 0 & -i \\ 0 & \sqrt{2} & 0 \end{pmatrix}.$$

4. Calculate the rotation matrix $d^{(1)}(\theta)$:

$$d^{(1)}(\theta) = \exp(-i\theta J_y),$$

and verify (10.39). Hint: show that $J_y^3 = J_y$.

10.7.5 Orbital angular momentum

1. Use the canonical commutation relations

$$[X_i, P_j] = i\delta_{ij}I$$

and the expression $\vec{L} = \vec{R}\times\vec{P}$ to show that

$$[L_x, L_y] = iL_z.$$

2. Prove Equations (10.47) to (10.49). Hint: show that for an infinitesimal rotation by an angle $d\alpha$ about Ox, the angles θ and ϕ vary by

$$d\theta = -\sin\phi\, d\alpha, \quad d\phi = -\frac{\cos\phi}{\tan\theta} d\alpha.$$

Find L_x and $L_y = i[L_x, L_z]$.

3. Since $L_z = -i\partial/\partial\phi$, the following Heisenberg inequality should be valid:

$$\Delta\phi\Delta L_z \geq \frac{1}{2}.$$

In an eigenstate of L_z where m is fixed $\Delta L_z = 0$, whereas $\Delta\phi \leq 2\pi$ since $0 \leq \phi \leq 2\pi$. The Heisenberg inequality is therefore violated in this state. Where is the flaw in this argument? Hint: see Exercise 7.4.3, question 2. Why does the argument of Exercise 9.7.1 break down?

10.7.6 Relation between the rotation matrices and the spherical harmonics

1. Let $\varphi(\vec{r}) = \varphi(x, y, z)$ be the wave function of a particle. Show that

$$\left(e^{-i\alpha L_z}\varphi\right)(0, 0, z) = \varphi(0, 0, z)$$

and that if a particle is localized on the z axis, the z component of its orbital angular momentum is zero. Interpret this result qualitatively.

2. We assume that the orbital angular momentum of the particle is l and write the wave function as the product of a spherical harmonic and a radial wave function $g_l(r)$ depending only on $r = |\vec{r}|$:

$$\psi_{lm}(\vec{r}) = Y_l^m(\theta, \phi)g_l(r) = \langle\theta, \phi|lm\rangle g_l(r).$$

We are interested uniquely in the angular part. Using

$$|\theta, \phi\rangle = U(\mathcal{R})|\theta = 0, \phi = 0\rangle,$$

where $\mathcal{R}$ is a rotation by the angles (θ, ϕ), show that

$$Y_l^m(\theta, \phi) \propto \left[D_{m0}^{(l)}(\theta, \phi)\right]^*.$$

It can be shown that the proportionality coefficient is $\sqrt{(2l+1)/(4\pi)}$:

$$Y_l^m(\theta, \phi) = \sqrt{\frac{2l+1}{4\pi}}\left[D_{m0}^{(l)}(\theta, \phi)\right]^*.$$

10.7.7 Independence of the energy from m

Assuming that the potential $V(r)$ is invariant under rotation, let ψ_{lm} be a solution of the time-independent Schrödinger equation:

$$H\psi_{lm} = E_{lm}\psi_{lm}.$$

Use the commutation relation $[L_+, H] = 0$ to show that the energy E_{lm} is in fact independent of m.

10.7.8 The spherical well

1. We are given a potential $V(\vec{r})$ which is spherically symmetric (see Fig. 12.4):

$$V(\vec{r}) = -V_0, \quad 0 \leq r \leq R,$$

$$= 0, \quad r > R,$$

called a spherical well. Find the equation giving the s-wave ($l = 0$) bound states. Is there always a bound state? Compare with the case of a one-dimensional well.

2. The neutron–proton potential can be modeled by a spherical well of radius $R \simeq 2\,\text{fm}$. There is a single neutron–proton bound state in the s-wave, namely, the deuteron,[15] with binding energy $B \simeq 2.2\,\text{MeV}$. Calculate the depth V_0 of the well needed for there to be just a single bound state. Compare V_0 with the binding energy and show that $V_0 \gg B$.

3. Find the s-wave energy levels of a particle in the potential

$$V(r) = \frac{A}{r^2} - \frac{B}{r}, \quad A, B > 0.$$

10.7.9 The hydrogen atom for $l \neq 0$

1. Write down the equation that generalizes (10.89) when the orbital angular momentum $l \neq 0$. Show that it is necessary to add to (10.91) the term

$$-l(l+1) \left[\frac{a_1}{x} + \sum_{k=1}^{\infty} a_{k+1} x^{k-1} \right].$$

2. Prove the recursion relation

$$a_{k+1} = \frac{2(\alpha k - 1)}{k(k+1) - l(l+1)} a_k$$

and derive

- $\alpha = \dfrac{1}{n}$,
- $k \geq l+1$,

so that $l + 1 \leq k \leq n$. Show that the spectrum of the hydrogen atom is given by (10.93).

10.7.10 Matrix elements of a potential

The external electron of an atom is assumed to be in a p state ($l = 1$). Its wave function is

$$\psi_{1m}(\vec{r}) = Y_1^m(\theta, \phi) \frac{u_1(r)}{r}.$$

It is placed in an external potential of the form

$$V(\vec{r}) = Ax^2 + By^2 - (A+B)z^2,$$

where A and B are constants.

---

[15] In fact, the deuteron also has a small d-wave component.

1. Show *without calculation* that the matrix representing V in the basis $|lm\rangle$ has the form

$$V_{m'm} = \begin{pmatrix} \alpha & 0 & \beta \\ 0 & \gamma & 0 \\ \beta & 0 & \alpha \end{pmatrix},$$

where the rows and columns are arranged in the order $m', m = 1, 0, -1$.

2. Determine the eigenvalues and eigenvectors of V. Show that $\langle L_z \rangle = 0$ in an eigenstate of V.

3. Use (10.63) to calculate α, β, and γ explicitly as functions of A, B, and

$$I = \int_0^\infty |u_1(r)|^2 r^2 dr.$$

10.7.11 The radial equation in dimension $d = 2$

We wish to write the equivalent of (10.78) in two-dimensional space when the potential is rotationally invariant. The time-independent Schrödinger equation is

$$\left[-\frac{1}{2M} \nabla^2 + V(r) \right] \psi(\vec{r}) = E\psi(\vec{r}).$$

We use polar coordinates in the plane xOy:

$$x = r\cos\theta, \quad y = r\sin\theta.$$

We recall the expression for the Laplacian in polar coordinates:

$$\nabla^2 = \frac{1}{r} \frac{\partial}{\partial r} r \frac{\partial}{\partial r} + \frac{1}{r^2} \frac{\partial^2}{\partial \theta^2}$$

and the expression for the angular momentum

$$L_z = XP_y - YP_x = -i\frac{\partial}{\partial \theta}.$$

1. Show that the eigenfunctions of L_z have the form $\exp(im\theta)$.

2. We seek solutions of the Schrödinger equation of the form

$$\psi_{nm}(\vec{r}) = \frac{1}{\sqrt{r}} e^{im\theta} u_{nm}(r).$$

Show that $u_{nm}(r)$ and E_{nm} satisfy the radial equation

$$\left[-\frac{1}{2M} \frac{d^2}{dr^2} + V(r) + \frac{m^2 + 1/4}{2Mr^2} \right] u_{nm}(r) = Eu_{nm}(r).$$

What is the interpretation of n? What is the behavior of $u_{nm}(r)$ when $r \to 0$?

10.7.12 Symmetry property of the matrices $d^{(j)}$

Using the operator Y (10.100), demonstrate the symmetry property of the rotation matrices $d^{(j)}(\beta)$:

$$d_{m'm}^{(j)}(\beta) = (-1)^{m-m'} d_{-m',-m}^{(j)}(-\beta).$$

10.7.13 Light scattering

1. Let us resume the study of the radiative transition $A^* \to A + \gamma$ with $j = j(A^*) = 1$ and $j' = j(A) = 0$. Show in the electric dipole case that the transition amplitudes are the following for an initial state $m = 1$ when circularly polarized photons are emitted in the plane xOz with momentum $\vec{p}$ making an angle θ with the z axis:

$$a_R^{m=1}(\theta) = \frac{1}{2}a(1 + \cos\theta),$$

$$a_L^{m=1}(\theta) = \frac{1}{2}a(1 - \cos\theta).$$

Generalize to the case where the photon is emitted in the direction (θ, ϕ).

2. We assume that photons of momentum $\vec{p} \parallel Oz$ arrive on the atom in its ground state A. The atom absorbs a photon and makes a transition to its excited state A^*. It then returns to the ground state by emitting a photon in the plane xOz at an angle θ with respect to Oz. We use b to denote the absorption amplitude of a photon of right-handed circular polarization R:

$$b = \langle j = 1, m = 1 | T' | R \rangle.$$

Show that if the transitions are of the electric dipole kind, we also have

$$b = \langle j = 1, m = -1 | T' | L \rangle.$$

Let $c_{P \to P'}(\theta)$ be the transition amplitude for the scattering of the initial photon of circular polarization P (P = R or L) at an angle θ with final polarization P'. Show that

$$c_{P \to P'}(\theta) = \frac{ab}{2}(1 \pm \cos\theta),$$

where the $+$ sign corresponds to P = P' and the $-$ sign to P $\neq$ P'. Derive the transition amplitudes for linear polarization $|x\rangle$ of the initial photon and linear polarization $|x'\rangle$ or $|y\rangle$ of the scattered photon, defined with respect to the photon propagation direction:

$$c_{x \to x'}(\theta) = ab\cos\theta,$$

$$c_{x \to y}(\theta) = 0.$$

Give a classical analogy which also leads to a $\cos^2\theta$ angular distribution with radiation polarized in the plane xOz. Generalize to the case where the photon is emitted in the direction (θ, ϕ).

10.7.14 Measurement of the Λ^0 magnetic moment

The Λ^0 is a particle of zero charge, mass $M \simeq 1115$ MeV c^{-2}, spin 1/2, and lifetime $\tau \simeq 2.5 \times 10^{-10}$ s. One of its principal decay modes (66% of cases) is

$$\Lambda^0 \to \text{proton} + \pi^- \text{meson},$$

where the proton has spin 1/2 and the π^- meson has spin 0.

1. In the reference frame where the Λ^0 is at rest, we assume that the proton is emitted with momentum $\vec{p}$ in the direction Oz, chosen to be the angular momentum quantization axis. Let m be the projection of the Λ^0 spin on the z direction and m' be that of the proton. Why must we have $m = m'$? Let a and b be the probability amplitudes of the transitions

$$a: \ \Lambda^0\left(m = \frac{1}{2}\right) \to \text{proton} \left(m' = \frac{1}{2};\ \vec{p} \parallel Oz\right),$$

$$b: \ \Lambda^0\left(m = -\frac{1}{2}\right) \to \text{proton} \left(m' = -\frac{1}{2};\ \vec{p} \parallel Oz\right).$$

Show that $|a| = |b|$ if parity is conserved in the decay. Hint: examine the action of a reflection with respect to the plane xOz.

2. The proton is now emitted with momentum $\vec{p}$ in the plane xOz parallel to the direction $\hat{n}$ making an angle θ with Oz. Let m' be the projection of the proton spin on the direction $\hat{n}$ and $a_{m'm}(\theta)$ be the amplitude:

$$a_{m'm}(\theta): \Lambda^0\left(m = \frac{1}{2}\right) \to \text{proton}\ (m';\ \vec{p} \parallel \hat{n}).$$

Express

$$a_{\frac{1}{2},\frac{1}{2}}(\theta) = a_{++}(\theta) \ \text{ and } \ a_{-\frac{1}{2},\frac{1}{2}}(\theta) = a_{-+}(\theta)$$

as functions of a, b, and θ.

3. We assume that the Λ^0 is produced in the spin state $m = 1/2$. Show that the proton angular distribution is of the form

$$w(\theta) = w_0(1 + \alpha \cos\theta).$$

Calculate α as a function of a and b. Experiment shows that

$$\alpha \simeq -0.645 \pm 0.016.$$

What can be concluded about parity conservation in the decay?

4. The Λ^0 is produced by bombarding a target of protons at rest by a π^--meson beam in the reaction (Fig. 10.13)

$$\pi^- \text{ meson} + \text{proton} \to \Lambda^0 + K^0 \text{ meson}.$$

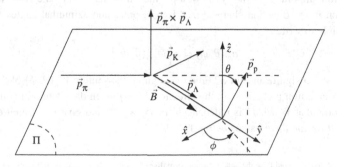

Fig. 10.13. Kinematics of Λ^0 production.

By momentum conservation, $\vec{p}_{\pi^-}$, $\vec{p}_{\Lambda^0}$, and $\vec{p}_{K^0}$ are located in the same plane. We choose the axis Oz to be perpendicular to this plane:

$$\hat{z} = \frac{\vec{p}_{\pi^-} \times \vec{p}_{\Lambda^0}}{|\vec{p}_{\pi^-} \times \vec{p}_{\Lambda^0}|},$$

and the axis Oy to be the direction $\vec{p}_{\Lambda^0}$ of the Λ^0 momentum. Given that parity is conserved in the production reaction and that the target protons are not polarized, show that if $\vec{S}$ is the Λ^0 spin operator, then the average values of the components S_x and S_y are zero: $\langle S_x \rangle = \langle S_y \rangle = 0$.

5. To simplify the situation, we assume that[16] $\langle S_z \rangle = 1/2$ and that all the Λ^0 have the same lifetime τ and decay at the same point. The system is located in a uniform, constant magnetic field $\vec{B}$ parallel to Oy. The Λ^0 possesses a magnetic moment $\vec{\mu}$ related to its spin $\vec{S}$ by the gyromagnetic ratio γ: $\vec{\mu} = \gamma \vec{S}$. Qualitatively describe the motion of the Λ^0 spin. Determine its orientation at the instant the decay occurs as a function of τ, B, and γ. Show that the angular distribution of the proton emitted in the decay is

$$w(\theta, \phi) = w_0(1 + \alpha \cos \Theta)$$

with

$$\cos \Theta = \cos \lambda \cos \theta + \sin \lambda \sin \theta \cos \phi,$$

where the angles θ and ϕ are the polar and azimuthal angles of the proton momentum. What is the value of the angle λ? Show that determination of $w(\theta, \phi)$ allows measurement of the gyromagnetic ratio γ. Neglect the curvature of the proton trajectory due to the magnetic field as well as the transformations of angles due to the motion of the Λ^0.

10.7.15 Production and decay of the ρ^+ meson

1. The ρ^+ meson is a particle of spin 1 which decays into two π mesons, particles of spin 0:

$$\rho^+ \rightarrow \pi^+ + \pi^0.$$

We choose a reference frame in which the ρ^+ meson is at rest, and assume that its spin is quantized on the z axis and that it is initially in the spin state $|1m\rangle$, $m = -1, 0, 1$. Let

$$a_m(\theta, \phi) = \langle \theta, \phi | T | 1m \rangle$$

be the transition amplitude for the decay of a ρ^+ meson in the initial state $|1m\rangle$ with emission of a π^+ meson in the direction characterized by the polar and azimuthal angles (θ, ϕ). Show that it is possible to write

$$a_m(\theta, \phi) = a \left[D^{(1)}_{m0}(\theta, \phi) \right]^*.$$

What is the physical significance of a? Find the angular distribution $W_m(\theta, \phi)$ of the π^+ meson, that is, the π^+ emission probability in the direction (θ, ϕ) when the ρ^+ meson is initially in the state $|1m\rangle$. Show that $W_m(\theta, \phi)$ is independent of ϕ (why?) and give its explicit expression as a function of θ for the three values $m = -1, 0, 1$.

---

[16] In fact, $|\langle S_z \rangle| < 1/2$ and we should use the state operator formalism for spin 1/2; see Section 6.2.2, where the Bloch vector $\vec{b}$ is identified with $2\langle \vec{S} \rangle$.

2. If the initial state of the ρ^+ meson is a linear combination of the states $|1m\rangle$,

$$|\lambda\rangle = \sum_{m=-1,0,1} c_m |1m\rangle, \qquad \sum_{m=-1,0,1} |c_m|^2 = 1,$$

what will the angular distribution $W_\lambda(\theta, \phi)$ be?

3. In general, the ρ^+ is not produced in a pure state, but in a mixture described by a state operator ρ:

$$\rho = \sum_\lambda p_\lambda |\lambda\rangle\langle\lambda|, \quad p_\lambda \geq 0, \quad \sum_\lambda p_\lambda = 1.$$

Show that the angular distribution is then

$$W(\theta, \phi) = \rho_{00} \cos^2\theta + \frac{1}{2}\sin^2\theta \left(\rho_{11} + \rho_{-1,-1}\right)$$

$$+ \frac{1}{\sqrt{2}} \sin 2\theta \operatorname{Re}\left(\rho_{-10} e^{-i\phi} - \rho_{10} e^{i\phi}\right) - \sin^2\theta \operatorname{Re}\left(\rho_{1,-1} e^{2i\phi}\right).$$

4. The ρ^+ meson is produced in the reaction π^+ meson $(\vec{p}_1)$ + proton $(\vec{p} = 0) \rightarrow \rho^+$ meson $(\vec{p}_2)$ + proton $(\vec{p}_3)$, where $\vec{p}_i$ denotes the particle momentum. We choose the normal $\hat{n}$ to the reaction plane as the z axis:

$$\hat{n} = \frac{\vec{p}_1 \times \vec{p}_2}{|\vec{p}_1 \times \vec{p}_2|}.$$

The parity Π is conserved in this reaction and we assume that the target protons are not polarized. Show that the expectation value $\langle\vec{J}\rangle = \operatorname{Tr}(\rho\vec{J})$ of the ρ^+ spin points in the direction $\hat{n}$: $\langle\vec{J}\rangle = c\hat{n}$. Show that

$$\operatorname{Tr}(\rho J_x) = \operatorname{Tr}(\rho J_y) = 0.$$

Use the fact that the kinematics of the production reaction is invariant under the operation

$$\mathcal{Z} = \Pi e^{-i\pi J_z}, \quad [\rho, \mathcal{Z}] = 0,$$

to show

$$\rho_{mm'} = (-1)^{m-m'} \rho_{mm'},$$

so that ρ in fact depends only on four *real* parameters and has a checkerboard pattern

$$\begin{pmatrix} \rho_{11} & 0 & \rho_{1,-1} \\ 0 & \rho_{00} & 0 \\ \rho_{1,-1}^* & 0 & \rho_{-1,-1} \end{pmatrix}.$$

10.7.16 Interaction of two dipoles

The interaction Hamiltonian of two magnetic dipoles carried by particles of spin 1/2 is written as

$$H = \frac{K}{r^3}\left[3(\vec{\sigma}_1 \cdot \hat{r})(\vec{\sigma}_2 \cdot \hat{r}) - \vec{\sigma}_1 \cdot \vec{\sigma}_2\right] = \frac{K}{r^3} S_{12},$$

where $\vec{r}$ is the vector joining the two dipoles and $\vec{\sigma}_1$ and $\vec{\sigma}_2$ are the Pauli matrices of these particles. Let

$$\vec{\Sigma} = \frac{1}{2}(\vec{\sigma}_1 + \vec{\sigma}_2)$$

be the total spin. Show that

$$S_{12} = 2\left(3Q^2 - \vec{\Sigma}^2\right), \quad Q^2 = (\vec{\Sigma} \cdot \hat{r})^2,$$

and that $Q^4 = Q^2$, i.e., Q^2 is a projector. Show that $S_{12}^2 = 4\vec{\Sigma}^2 - 2S_{12}$ and that the eigenvalues of S_{12} are 0, 2, and -4.

10.7.17 Σ^0 decay

The Σ^0 particle is composed of an up quark, a down quark, and a strange quark and has mass $1192 \text{ MeV } c^{-2}$ and spin 1/2. It decays via a radiative transition to a Λ^0 particle, also composed of an up quark, a down quark, and a strange quark and having mass $1115 \text{ MeV } c^{-2}$ and spin 1/2:

$$\Sigma^0 \to \Lambda^0 + \gamma.$$

The Σ^0 is assumed to be at rest, its spin is quantized along the z axis, and the spin projection on this axis is m. The photon momentum $\vec{p}$ lies in the plane xOz and makes an angle θ with the z axis.

1. First we assume that the photon is emitted in the z direction ($\theta = 0$). If m' is the projection of the Λ^0 spin on Oz, show that the nonzero amplitudes are (T is the transition operator)

$$a = \langle R, m' = -\frac{1}{2}; \theta = 0|T|m = \frac{1}{2}\rangle,$$

$$b = \langle L, m' = \frac{1}{2}; \theta = 0|T|m = -\frac{1}{2}\rangle,$$

while

$$c = \langle R, m' = \frac{1}{2}; \theta = 0|T|m = \frac{1}{2}\rangle = 0,$$

$$d = \langle L, m' = -\frac{1}{2}; \theta = 0|T|m = -\frac{1}{2}\rangle = 0,$$

in other words, $m' = m$ is forbidden and the allowed transitions correspond to $m' = -m$ when $\theta = 0$. The notation (R, L) specifies the right- or left-handed circular polarization state of the photon.

2. The transition operator T is invariant under the parity operation. Show that $|a| = |b|$. If η is the product of the Σ^0 and Λ^0 parities, also called the relative parity of the two particles

$$\eta = \eta_{\Sigma^0}\eta_{\Lambda^0},$$

show that $a = \eta b$. Experiment gives $\eta = 1$ and so $a = b$.

3. We assume that the initial value of the projection of the Σ^0 spin is $m = 1/2$. Let $a_R^{m'}(\theta)$ and $a_L^{m'}(\theta)$ be the transition amplitudes, where m' is the projection of the Λ^0 spin on the direction of $\vec{p}$, and therefore the eigenvalue of $\vec{S} \cdot \hat{p}$. Calculate $a_R^{m'}$ and $a_L^{m'}$ as functions of a and θ. What are the allowed values of m'?

10.7.18 Irreducible tensor operators

An irreducible tensor operator of order k, $T^{(k)}$, possesses $(2k+1)$ components $T_q^{(k)}$:

$$q = -k, -k+1, \ldots, k-1, k$$

and transforms under a rotation $\mathcal{R}$ as

$$U(\mathcal{R})T_q^{(k)}U^\dagger(\mathcal{R}) = \sum_{q'} D_{q'q}^{(k)}(\mathcal{R})T_{q'}^{(k)}.$$

Show that the vector

$$|kjqm\rangle = T_q^{(k)}|jm\rangle$$

transforms under rotation as the vector $|j_1 j_2 m_1 m_2\rangle$ with $j_1 = k$, $j_2 = j$, $m_1 = q$, and $m_2 = m$. Using the vectors

$$|kj\tilde{j}\tilde{m}\rangle = \sum_{q+m=\tilde{m}} C_{qm;\tilde{j}\tilde{m}}^{kj}|kjqm\rangle$$

as intermediaries, prove the general form of the Wigner–Eckart theorem:

$$\langle j'm'|T_q^{(k)}|jm\rangle = C_{qm;j'm'}^{kj}\langle j'||T^{(k)}||jm\rangle$$

and show that

$$|j-k| \le j' \le j+k.$$

10.8 Further reading

The presentation in this chapter, inspired by that of Feynman *et al.* [1965], Vol. III, Chapters 17 and 18, places particular emphasis on the properties and use of the rotation matrices. For a more classical presentation the reader can consult Messiah [1999], Chapter XIII, Cohen-Tannoudji *et al.* [1977], Chapter VII, or Basdevant and Dalibard [2002], Chapter 10. Numerous applications to elementary particle physics can be found in the book by S. Gasiorowicz, *Elementary Particle Physics*, New York: Wiley (1966). In addition, Chapter 4 of that book describes the Wigner analysis based on invariance under the Poincaré group, which shows in particular that a particle of zero mass has only two helicity states, whatever its spin. On this last subject see also Weinberg [1995], Chapter 2.

11

The harmonic oscillator

The harmonic oscillator describes small oscillations about a stable equilibrium position, and is a very important system in classical mechanics. It is just as important in quantum mechanics. To be specific, let us consider a simple example of motion in one dimension, the vibration of a diatomic molecule whose two nuclei have masses m_1 and m_2. We choose the line connecting the two nuclei as the x axis and use $x = x_1 - x_2$ to denote the relative particle coordinate (Exercise 8.5.6). At equilibrium the two nuclei are separated by a distance $x = x_0$. In classical physics the Hamiltonian of the relative particle is written as

$$H_{cl} = \frac{p^2}{2m} + V(x), \tag{11.1}$$

where $m = m_1 m_2 / (m_1 + m_2)$ is the mass of the relative particle. We expand $V(x)$ in a series about $x = x_0$:

$$V = V(x_0) + (x - x_0)V'(x_0) + \frac{1}{2}(x - x_0)^2 V''(x_0) + \cdots$$

The constant $V(x_0)$ is in general uninteresting and we can set it equal to zero by redefining the zero of the energy. Since x_0 is an equilibrium position $V'(x_0) = 0$, and if this equilibrium position is stable $V''(x_0) > 0$. Setting

$$q = x - x_0, \quad C = V''(x_0), \quad \omega = \sqrt{\frac{C}{m}},$$

the classical Hamiltonian (11.1) becomes

$$\boxed{H_{cl} = \frac{p^2}{2m} + \frac{1}{2}m\omega^2 q^2}, \tag{11.2}$$

where ω is the frequency of oscillations about the equilibrium position.

We shall start with the simplest example, that of an isolated oscillator. In Section 11.1 we study the quantum version of this case using a particular basis, that of the energy eigenstates. Another "basis," that of the coherent states, will be studied in the following section. It has many applications in quantum optics. A slightly more complicated case is that of coupled oscillators, which also has important applications. An example will be given in Section 11.3, where we study a simple model of vibrations in a solid which will allow us to introduce the concept of phonon. The generalization to photons will also be discussed for a simple situation.

It might be surprising to find, in the last section of this chapter, a study of the motion of a charged particle in a magnetic field. We shall see that in the case of constant magnetic field the equations of motion become those of two independent harmonic oscillators. We will define local gauge invariance, which fixes the form of the interaction of a charged particle with an electromagnetic field, and then study the energy levels in a magnetic field, called the Landau levels.

11.1 The simple harmonic oscillator

11.1.1 Creation and annihilation operators

Our starting point will be the Hamiltonian (11.2). It can be carried over to quantum mechanics if p and q are interpreted as operators: $p \to P$, $q \to Q$, and the canonical commutation relations are imposed:

$$[Q, P] = i\hbar I. \tag{11.3}$$

As is often the case in physics, it is useful to define dimensionless quantities, and so we introduce the dimensionless operators $\hat{P}$ and $\hat{Q}$:

$$Q = \left(\frac{\hbar}{m\omega}\right)^{1/2} \hat{Q}, \quad P = (m\hbar\omega)^{1/2}\hat{P}, \tag{11.4}$$

which obey the commutation relation

$$[\hat{Q}, \hat{P}] = iI. \tag{11.5}$$

We shall construct the eigenvectors of H by an algebraic method similar in spirit to that used for angular momentum. It is based on the principle of introducing the operators a and $a^\dagger$, respectively called the *annihilation* (or *destruction*) *operator* and the *creation operator* of the harmonic oscillator, which take us from one eigenvalue of H to another, reminiscent of how J_- and J_+ take us from one eigenvalue of J_z to another. We therefore define the operators[1]

$$a = \frac{1}{\sqrt{2}}\left(\hat{Q} + i\hat{P}\right), \tag{11.6}$$

$$a^\dagger = \frac{1}{\sqrt{2}}\left(\hat{Q} - i\hat{P}\right). \tag{11.7}$$

The commutation relations of a and $a^\dagger$ can be obtained by direct calculation:

$$\boxed{[a, a^\dagger] = I}, \tag{11.8}$$

---

[1] In order to conform to the standard notation, we depart from our rule of denoting operators by upper-case letters.

as can three useful expressions for H:

$$H = \frac{1}{2}\hbar\omega\left(\hat{P}^2 + \hat{Q}^2\right) = \hbar\omega\left(a^\dagger a + \frac{1}{2}\right) = \hbar\omega\left(N + \frac{1}{2}\right).$$ (11.9)

We have introduced the operator N, called the *number operator*:[2]

$$N = a^\dagger a \;,$$ (11.10)

which satisfies the following commutation relations with a and $a^\dagger$:

$$[N, a] = -a, \quad [N, a^\dagger] = a^\dagger.$$ (11.11)

Using (11.9), we see that diagonalizing N is equivalent to diagonalizing H.

11.1.2 Diagonalization of the Hamiltonian

Let us assume that we have found an eigenvector $|\nu\rangle$ of N which is normalizable but not necessarily of unit norm and has eigenvalue ν:

$$N|\nu\rangle = \nu|\nu\rangle.$$

We must have $\nu \geq 0$; actually,

$$0 \leq ||a|\nu\rangle||^2 = \langle\nu|a^\dagger a|\nu\rangle = \langle\nu|N|\nu\rangle = \nu\langle\nu|\nu\rangle,$$

which implies that if $\nu = 0$, then $a|\nu\rangle = 0$. In the contrary case, $a|\nu\rangle$ is a vector of squared norm $\nu\langle\nu|\nu\rangle$, and it is an eigenvector of N with eigenvalue $(\nu - 1)$ because it can be shown using (11.11) that

$$Na[|\nu\rangle] = a(N-1)|\nu\rangle = (\nu - 1)[a|\nu\rangle].$$

Finally, $a^\dagger|\nu\rangle$ is certainly a non-null vector; it has squared norm $(\nu + 1)\langle\nu|\nu\rangle$ and is an eigenvector of N with eigenvalue $(\nu + 1)$. On the one hand

$$0 \leq ||a^\dagger|\nu\rangle||^2 = \langle\nu|aa^\dagger|\nu\rangle = \langle\nu|(N+1)|\nu\rangle = (\nu + 1)\langle\nu|\nu\rangle,$$

while on the other

$$N[a^\dagger|\nu\rangle] = a^\dagger(N+1)|\nu\rangle = (\nu + 1)[a^\dagger|\nu\rangle].$$

If $\nu > 0$, we have seen that $a|\nu\rangle$ is an eigenvector of N with eigenvalue $(\nu - 1)$. If $(\nu - 1) = 0$, then $a^2|\nu\rangle = 0$. If $(\nu - 1) > 0$, we can construct a non-null vector $a^2|\nu\rangle$ of eigenvalue $(\nu - 2)$ and continue the process if $(\nu - 2) > 0$. The set of vectors

$$a^0|\nu\rangle, \; a^1|\nu\rangle, \; a^2|\nu\rangle, \ldots, a^p|\nu\rangle \ldots$$

---

[2] This terminology will be justified in Section 11.3.1.

is a set of eigenvectors of N corresponding to the eigenvalues

$$\nu, \ \nu - 1, \ldots, (\nu - p)\ldots$$

This shows that ν is necessarily an integer. If it were not, $(\nu - p)$ would become negative for p sufficiently large and the vector $a^p|\nu\rangle$ would have negative norm. The series must therefore terminate at an *integer* $\nu = p$ such that the vector $a^{p+1}|\nu\rangle = 0$.

The set of vectors

$$(a^\dagger)^0|\nu\rangle, \ (a^\dagger)^1|\nu\rangle, \ (a^\dagger)^2|\nu\rangle, \ldots, (a^\dagger)^p|\nu\rangle \ldots$$

forms a set of eigenvectors of N corresponding to the eigenvalues

$$\nu, \ \nu + 1, \ldots, (\nu + p)\ldots$$

In summary, *the eigenvalues of N are integers*:

$$n = 0, 1, 2, \ldots, n, \ldots$$

We use $|n\rangle$ to denote an eigenvector of N corresponding to the eigenvalue n:

$$N|n\rangle = a^\dagger a|n\rangle = n|n\rangle, \tag{11.12}$$

or, equivalently for H,

$$H|n\rangle = \hbar\omega\left(n + \frac{1}{2}\right)|n\rangle. \tag{11.13}$$

The energy eigenvalues E_n labeled by the integer n have the form

$$\boxed{E_n = \hbar\omega\left(n + \frac{1}{2}\right)}. \tag{11.14}$$

In contrast to the case of the classical oscillator, the ground-state level E_0 is nonzero rather than zero, as would be expected for a particle at rest at the equilibrium position. The value $E_0 = \hbar\omega/2$ is called the *zero-point energy* of the harmonic oscillator. This can be explained qualitatively using the Heisenberg inequalities (Exercise 9.7.4). We warn that the ground-state eigenvector $|0\rangle$ should not be confused with the null vector of the Hilbert space $\mathcal{H}$, $|\varphi\rangle = 0$! We also note that the energy levels are equidistant from each other, and this is what is found experimentally in a first approximation for the vibrational levels of a molecule.

The vectors $|n\rangle$ are of course orthogonal if $n \neq n'$, and from now on we assume that they have unit norm. We still need to show that they are nondegenerate, that they form a basis in the Hilbert space $\mathcal{H}$, and above all that N has at least one eigenvector, which is not guaranteed for an operator, even a Hermitian one, in a space of infinite dimension. In the following section we shall explicitly construct the vector $|0\rangle$ and show that it is unique. This will be sufficient for showing that the series of vectors

$$|0\rangle, (a^\dagger)^1|0\rangle, (a^\dagger)^2|0\rangle, \ldots, (a^\dagger)^n|0\rangle \ldots \tag{11.15}$$

is unique. Actually, we can argue recursively, assuming that the vector $|n\rangle$ is nondegenerate. Let $|n+1\rangle$ be an eigenvector of N corresponding to the eigenvalue $(n+1)$: $N|n+1\rangle = (n+1)|n+1\rangle$. Then, with c a nonzero complex number,

$$a|n+1\rangle = c|n\rangle \Rightarrow a^\dagger a|n+1\rangle = ca^\dagger|n\rangle \Rightarrow |n+1\rangle = \frac{ca^\dagger}{n+1}|n\rangle,$$

which shows that $|n+1\rangle \propto a^\dagger|n\rangle$. Therefore, if $|0\rangle$ is unique, which we shall prove to be the case in Section 11.1.3, the vector $|n\rangle$ is also unique up to a phase.

As in the case of the standard angular momentum basis $|jm\rangle$, it is convenient to fix the relative phase of the eigenvectors of H once and for all. If $|n\rangle$ has unit norm, the vector $a^\dagger|n\rangle$ has norm $\sqrt{n+1}$ and consequently

$$a^\dagger|n\rangle = e^{i\alpha}\sqrt{n+1}\,|n+1\rangle.$$

The simplest choice of phase is $\alpha = 0$ and we then have

$$a^\dagger|n\rangle = \sqrt{n+1}\,|n+1\rangle, \tag{11.16}$$

$$a|n\rangle = \sqrt{n}\,|n-1\rangle. \tag{11.17}$$

Equations (11.16) and (11.17) display the creation and destruction role of the operators $a^\dagger$ and a: the operator $a^\dagger$ increases n by unity, while a decreases n by unity. The vectors $|n\rangle$ are derived from $|0\rangle$ by

$$\boxed{|n\rangle = \frac{1}{\sqrt{n!}}\,(a^\dagger)^n\,|0\rangle}. \tag{11.18}$$

We still need to show that the vectors $|n\rangle$ form a basis of $\mathcal{H}$. This important issue is the subject of Exercise 11.5.1.

11.1.3 Wave functions of the harmonic oscillator

In wave mechanics, the Hamiltonian of the harmonic oscillator is written as

$$H = -\frac{\hbar^2}{2m}\frac{d^2}{dq^2} + \frac{1}{2}m\omega^2 q^2. \tag{11.19}$$

The wave mechanics representation of $\hat{Q}$ in (11.4) is the dimensionless variable u,

$$q = \left(\frac{\hbar}{m\omega}\right)^{1/2} u, \quad -i\hbar\frac{d}{dq} = -i(\hbar m\omega)^{1/2}\frac{d}{du}, \tag{11.20}$$

and the Hamiltonian (11.19) becomes

$$H = \frac{1}{2}\hbar\omega \left[-\frac{d^2}{du^2} + u^2 \right]. \tag{11.21}$$

We could have obtained this form of H directly starting from the first of Equations (11.9) and using the fact that u and $-id/du$ are just the realizations of the operators $\hat{Q}$ and $\hat{P}$ in the space $L_u^{(2)}(\mathbb{R})$. We could directly seek solutions of

$$H\varphi_n(u) = \frac{1}{2}\hbar\omega \left[-\frac{d^2}{du^2} + u^2 \right] \varphi_n(u) = E_n\varphi_n(u) \tag{11.22}$$

with $\varphi_n(u) = \langle u|n \rangle$, but instead we shall limit ourselves to showing that the vector $|0\rangle$ is unique, a feature which we need to check. Since $\langle u|0 \rangle = \varphi_0(u)$, the equation $a|0\rangle = 0$ becomes

$$\langle u|a|0 \rangle = \frac{1}{\sqrt{2}} \left[u + \frac{d}{du} \right] \varphi_0(u) = 0,$$

which can be integrated immediately to give

$$\varphi_0(u) = \frac{1}{\pi^{1/4}} e^{-u^2/2}. \tag{11.23}$$

The factor $\pi^{-1/4}$ ensures that φ_0 is normalized to unity. This solution is unique, which proves that the eigenvectors given by the series (11.15) are nondegenerate. It can be verified immediately that $\varphi_0(u)$ obeys (11.22) with eigenvalue $\hbar\omega/2$. The function $\varphi_0(u)$ possesses the property characteristic of a ground-state wave function: it does not vanish or, equivalently, it has no nodes.

Finally, let us determine the explicit form of the wave functions $\varphi_n(u) = \langle u|n \rangle$. We multiply (11.18) written as

$$|n\rangle = \frac{1}{\sqrt{2^n \, n!}} \left(\hat{Q} - i\hat{P} \right)^n |0\rangle$$

on the left by the bra $\langle u|$:

$$\varphi_n(u) = \langle u|n \rangle = \frac{1}{\pi^{1/4}} \frac{1}{\sqrt{2^n \, n!}} \left(u - \frac{d}{du} \right)^n e^{-u^2/2}. \tag{11.24}$$

The functions $\varphi_n(u)$ are orthogonal for $n \neq n'$ and normalized to unity because $\langle n|n' \rangle = \delta_{nn'}$. The functions defined in (11.24) are related to the *Hermite polynomials* $H_n(u)$:

$$e^{-u^2/2} H_n(u) = \left(u - \frac{d}{du} \right)^n e^{-u^2/2} \tag{11.25}$$

as

$$\varphi_n(u) = \frac{1}{\pi^{1/4}} \frac{1}{\sqrt{2^n \, n!}} e^{-u^2/2} H_n(u). \tag{11.26}$$

The first few Hermite polynomials are

$$H_0(u) = 1, \quad H_1(u) = 2u, \quad H_2(u) = 4u^2 - 2.$$

In summary, we can compile a "dictionary" which allows us to go from the "N representation" of Section 11.1.2 to the representation of Section 11.1.3 using as eigenstates of H the wave functions $\varphi_n(u)$. In the following summary the first equation is written in the basis $|n\rangle$, and the second is the equivalent equation in wave mechanics.

- The eigenvalue equation:

$$\frac{1}{2}\left(\hat{P}^2 + \hat{Q}^2\right)|n\rangle = \left(n + \frac{1}{2}\right)|n\rangle \iff \frac{1}{2}\left(-\frac{\mathrm{d}^2}{\mathrm{d}u^2} + u^2\right)\varphi_n(u) = \left(n + \frac{1}{2}\right)\varphi_n(u).$$

- The orthonormalization relations:

$$\langle n|m\rangle = \delta_{nm} \iff \int_{-\infty}^{\infty} \mathrm{d}u\, \varphi_n^*(u)\varphi_m(u) = \delta_{nm}.$$

- The completeness relation:

$$\sum_n |n\rangle\langle n| = I \iff \sum_n \varphi_n(u)\varphi_n^*(v) = \delta(u - v).$$

Complex conjugation is in fact superfluous because the functions $\varphi_n(u)$ are real.

11.2 Coherent states

Coherent states, or semi-classical states, are remarkable quantum states of the harmonic oscillator. In these states the expectation values of the position and momentum operators have properties identical to the classical values of position $q(t)$ and momentum $p(t)$. Exercise 11.5.3 shows that the expression for coherent states follows from the requirement that the dynamics of the quantum expectation values of Q, P, and H be identical to that of the classical variables. Below we shall give an a priori definition of these states. Let $z(t)$ be a complex number, a combination of $q(t)$ and $p(t)$:

$$z(t) = \sqrt{\frac{m\omega}{2\hbar}}\, q(t) + \frac{\mathrm{i}}{\sqrt{2m\hbar\omega}}\, p(t). \tag{11.27}$$

Starting from the classical equations of motion

$$\frac{\mathrm{d}q(t)}{\mathrm{d}t} = \frac{1}{m}\, p(t), \quad \frac{\mathrm{d}p(t)}{\mathrm{d}t} = -m\omega^2 q(t), \tag{11.28}$$

we show that $z(t)$ satisfies the differential equation

$$\frac{\mathrm{d}z}{\mathrm{d}t} = -\mathrm{i}\omega z(t), \tag{11.29}$$

which has the solution

$$z(t) = z_0 e^{-\mathrm{i}\omega t}.$$

The complex number $z(t)$ traces out a circular trajectory in the complex z plane with uniform speed. From $z(t)$ we can derive the position $q(t)$, the momentum $p(t)$, and the energy of the oscillator:

$$q(t) = \sqrt{\frac{2\hbar}{m\omega}} \, \mathrm{Re}\, z(t),$$

$$p(t) = \sqrt{2m\hbar\omega} \, \mathrm{Im}\, z(t), \tag{11.30}$$

$$E = \hbar\omega |z_0|^2.$$

It is easy to show that the expectation value $\langle a \rangle(t)$ of the annihilation operator a satisfies the same differential equation as $z(t)$ (Exercise 11.5.3). This suggests that we seek the eigenvectors of the operator a, which we shall show do exist,[3] because the corresponding eigenvalues will then obey (11.29). These eigenvectors will in fact be the coherent states. A *coherent state* $|z\rangle$ is defined as

$$\boxed{|z\rangle = e^{-|z^2|/2} \sum_{n=0}^{\infty} \frac{z^n}{\sqrt{n!}} \, |n\rangle = e^{-|z|^2/2} \, e^{a^\dagger z} |0\rangle} \,. \tag{11.31}$$

Let us list some properties of coherent states, after verifying that $|z\rangle$ is an eigenvector of a.

- The coherent state $|z\rangle$ is an eigenvector of the (non-Hermitian) annihilation operator a with eigenvalue z:

$$\boxed{a|z\rangle = z|z\rangle} \,. \tag{11.32}$$

This can be proved using (11.31) directly, but it is also possible to use the identity (2.54) of Exercise 2.4.11, which here we write as

$$e^{a^\dagger z} a \, e^{-a^\dagger z} = a + z[a^\dagger, a] = a - z,$$

$$e^{a^\dagger z} a = (a - z) e^{a^\dagger z}.$$

It is sufficient to apply both sides of the last equation to the vector $|0\rangle$ to obtain (11.32).

- The vector $|z\rangle$ has unit norm: $\langle z|z \rangle = 1$ and the squared modulus of the scalar product $\langle z|z' \rangle$,

$$|\langle z|z' \rangle|^2 = \exp\left(-|z - z'|^2\right), \tag{11.33}$$

is a measure of the "distance" between two coherent states.

- The probability distribution of n is given by a Poisson distribution:

$$\mathsf{p}(n) = |\langle n|z \rangle|^2 = \frac{|z|^{2n}}{n!} e^{-|z|^2}, \tag{11.34}$$

which gives the expectation value $\langle n \rangle = |z|^2$ and the dispersion $\Delta n = |z|$.

---

[3] It is not evident a priori that a, which is not a Hermitian operator, has eigenvalues, and even less that these eigenvectors form a basis of $\mathcal{H}$.

- The action of $\exp(\lambda N)$ on a coherent state, where λ is purely imaginary ($|\exp \lambda| = 1$), again gives a coherent state:

$$e^{\lambda N}|z\rangle = e^{\lambda N}e^{-|z|^2/2}\sum_{n=0}^{\infty}\frac{z^n}{\sqrt{n!}}|n\rangle = e^{-|z|^2/2}\sum_{n=0}^{\infty}\frac{z^n}{\sqrt{n!}}e^{\lambda n}|n\rangle$$

$$= e^{-|z|^2/2}\sum_{n=0}^{\infty}\frac{(e^{\lambda}z)^n}{\sqrt{n!}}|n\rangle = |e^{\lambda}z\rangle. \tag{11.35}$$

The relation $|\exp \lambda| = 1$ has been used only to obtain the last equality.

- The coherent states form an "overcomplete" basis:

$$\int \frac{d\mathrm{Re}\,z\,d\mathrm{Im}\,z}{\pi}|z\rangle\langle z| = I. \tag{11.36}$$

To prove this identity, we sandwich it between the bra $\langle n|$ and the ket $|m\rangle$. Setting $z = \rho \exp(i\theta)$, we have

$$\int \frac{d\mathrm{Re}\,z\,d\mathrm{Im}\,z}{\pi}\langle n|z\rangle\langle z|m\rangle = \int_0^{\infty}\rho\,d\rho\int_0^{2\pi}\frac{d\theta}{\pi}\frac{z^n z^{*m}}{\sqrt{n!m!}}e^{-\rho^2}$$

$$= \int_0^{\infty}\rho\,d\rho\int_0^{2\pi}\frac{d\theta}{\pi}\frac{\rho^{n+m}}{\sqrt{n!m!}}e^{i(n-m)\theta}e^{-\rho^2} = \delta_{nm},$$

where we have used the change of variable $\rho^2 = u$ and

$$\int_0^{\infty}du\,u^n e^{-u} = n!.$$

A direct consequence of (11.36) is that the "diagonal matrix elements" $\langle z|A|z\rangle$ are sufficient to completely define an operator A (Exercise 11.5.3).

These properties allow us easily to calculate the expectation values:

$$\langle z|Q|z\rangle = \sqrt{\frac{\hbar}{2m\omega}}\langle z|\left(a+a^{\dagger}\right)|z\rangle = \sqrt{\frac{2\hbar}{m\omega}}\,\mathrm{Re}\,z,$$

$$\langle z|P|z\rangle = \sqrt{2\hbar m\omega}\,\mathrm{Im}\,z, \tag{11.37}$$

$$\langle z|H|z\rangle = \hbar\omega\left(|z|^2+\frac{1}{2}\right).$$

This is the classical result (11.30) if we ignore the zero-point energy $\hbar\omega/2$ in the expression for $\langle H\rangle$. Moreover, if the state of the harmonic oscillator is a coherent state at time $t = 0$, this property is conserved by the time evolution. Let us assume that the oscillator at time $t = 0$ is in the coherent state $|\varphi(t=0)\rangle = |z_0\rangle$ and calculate $|\varphi(t)\rangle$:

$$|\varphi(t)\rangle = e^{-iHt/\hbar}|z_0\rangle = e^{-i\omega t/2}e^{-i\omega Nt}|z_0\rangle = e^{-i\omega t/2}|z_0 e^{-i\omega t}\rangle, \tag{11.38}$$

where we have used (11.35). We obtain the classical evolution $z(t) = z_0\exp(-i\omega t)$ up to a phase $\exp(-i\omega t/2)$ multiplying the state vector. If we start from a coherent state at time $t = 0$, the evolution of the expectation values $\langle Q\rangle$, $\langle P\rangle$, and $\langle H\rangle$ follows very exactly the classical evolution of $q(t)$, $p(t)$, and E. We have therefore shown that the expectation values in a coherent state obey the classical laws.

It is also instructive to calculate the dispersions. Let us evaluate, for example, $\langle Q^2 \rangle$ in the coherent state $|z\rangle$:

$$\langle Q^2 \rangle_z = \frac{\hbar}{2m\omega} \langle z|a^2 + (a^\dagger)^2 + aa^\dagger + a^\dagger a|z\rangle = \frac{\hbar}{2m\omega} \langle z|a^2 + (a^\dagger)^2 + 2a^\dagger a + 1|z\rangle$$

$$= \frac{\hbar}{2m\omega} \left[1 + (z + z^*)^2\right] = \frac{\hbar}{2m\omega} \left[1 + 4(\text{Re}\,z)^2\right].$$

A similar calculation (Exercise 11.5.3) gives $\langle P^2 \rangle$ and $\langle H^2 \rangle$, from which we derive the dispersions[4] in the coherent state $|z\rangle$:

$$\boxed{\quad \Delta_z Q = \sqrt{\frac{\hbar}{2m\omega}}, \quad \Delta_z P = \sqrt{\frac{m\hbar\omega}{2}}, \quad \Delta_z H = \hbar\omega|z| \quad}. \tag{11.39}$$

The dispersion $\Delta_z H$ can be obtained from (11.34) using $\Delta H = \hbar\omega\Delta_z N$ and $\Delta_z N = \Delta n = |z|$, but it is also possible to calculate $\langle z|N^2|z\rangle$ directly. We note that the Heisenberg inequality is saturated in a coherent state: $\Delta_z Q \Delta_z P = \hbar/2$, and for $|z| \gg 1$

$$\frac{\Delta_z H}{\langle H \rangle} \simeq \frac{1}{|z|} \to 0 \quad \text{if} \quad |z| \to \infty.$$

In summary, for $|z| \gg 1$ the dispersions about the expectation values are the smallest possible.

11.3 Introduction to quantized fields

11.3.1 Sound waves and phonons

When the vibration amplitudes are small, a system of coupled oscillators can be decomposed into normal modes and treated as a set of independent harmonic oscillators. An interesting case is that of vibrations in a solid, and we shall use it to introduce quantized fields. The first quantum model of vibrations in a crystalline solid was constructed by Einstein, who assumed that each atom can vibrate independently of the others about its equilibrium position with a frequency ω. In quantum physics each atom is therefore associated with a quantized harmonic oscillator of frequency ω. This model was the first to qualitatively explain the behavior of the specific heat of solids at low temperature: whereas the Dulong–Petit law predicts a specific heat independent of temperature, experiment shows that in fact this law is valid only at a sufficiently high temperature, and the specific heat actually decreases with temperature. However, the Einstein model does not give quantitatively correct results. This is not surprising, because the hypothesis of independent atomic vibrations is not realistic. If it were the case, vibrations would not be able to propagate in a solid and there would be no such thing as sound waves.

---

[4] We shall use either notation $(\Delta P, \Delta Q)$ or $(\Delta p, \Delta q)$ for the dispersions, as there is no possible ambiguity.

Let us study the simplest possible model of a chain of coupled oscillators, limiting ourselves to the case of one dimension. At equilibrium N atoms are located at regular intervals l along a line. The N equilibrium positions have abscissas $x_n = nl$, $n = 0, 1, \ldots, N-1$. It will be convenient to use periodic boundary conditions $x_{n+N} \equiv x_n$, but it is also possible to take vanishing ones: $x_0 = x_{N+1} = 0$. As before, we shall use q_n to denote the displacement from equilibrium of the nth atom. The coupling between the nth and $(n+1)$th atoms is described by the term $(K/2)(q_n - q_{n+1})^2$, where K is a constant, and the classical Hamiltonian of the ensemble is

$$H_{\mathrm{cl}} = \sum_{n=0}^{N-1} \frac{p_n^2}{2m} + \frac{1}{2} K \sum_{n=0}^{N-1} (q_{n+1} - q_n)^2 .$$ (11.40)

This is in fact the Hamiltonian of N identical masses m connected by identical springs with spring constant K (Fig. 11.1). In (11.40) $p_n = m\dot{q}_n$ is the momentum of the atoms. The first term in H_{cl} is the kinetic energy and the second is the potential energy. The equations of motion corresponding to the Hamiltonian (11.40) are written as

$$m\ddot{q}_n = -K \left[(q_n - q_{n-1}) + (q_n - q_{n+1}) \right].$$ (11.41)

Let us begin with the classical problem. To decouple the modes q_n, we seek the normal modes by taking the discrete (or lattice) Fourier transform of q_n and p_n:

$$q_k = \frac{1}{\sqrt{N}} \sum_{n=0}^{N-1} e^{ikx_n} q_n = \sum_n U_{kn} q_n, \quad k = j \times \frac{2\pi}{Nl}, \quad j = 0, \ldots, N-1.$$ (11.42)

To reduce the amount of notation we have not used $\tilde{q}_k$ to designate the Fourier transform, as the subscript k or n allows the Fourier components q_k and positions q_n on the lattice to be unambiguously distinguished. The matrix U_{kn} performs a discrete Fourier transform, and it is a unitary matrix:

$$\sum_n U_{kn} U_{nk'}^\dagger = \sum_n U_{kn} U_{k'n}^* = \frac{1}{N} \sum_n e^{ikx_n} e^{-ik'x_n} = \frac{1}{N} \sum_n \exp\left[\frac{2i\pi}{Nl} (j - j') x_n \right]$$

$$= \frac{1}{N} \frac{1 - \exp(2i\pi(j - j'))}{1 - \exp(2i\pi(j - j')/N)} = \delta_{jj'},$$

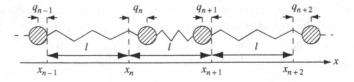

Fig. 11.1. Model for vibrations of a solid: a chain of springs.

that is, noting that $U_{nk}^{\dagger} = U_{kn}^{*} = U_{-kn}$,

$$\sum_n U_{kn} U_{nk'}^{\dagger} = \sum_n U_{kn} U_{-k'n} = \delta_{kk'}. \tag{11.43}$$

The range of variation of k is

$$0 \le k \le \frac{2\pi(N-1)}{Nl},$$

but, making use of the periodicity, we can replace this by the interval

$$-\frac{\pi}{l} \le k \le \frac{\pi}{l},$$

which is the first Brillouin zone already encountered in Section 9.5.2. Since we assume that $N \gg 1$, we neglect edge effects. The unitarity of the U_{kn} allows us to write down the inverse Fourier transform of (11.42):

$$q_n = \frac{1}{\sqrt{N}} \sum_{k=-\pi/l}^{\pi/l} e^{-ikx_n} q_k = \sum_k U_{nk}^{\dagger} q_k = \sum_k U_{-kn} q_k. \tag{11.44}$$

The Fourier transform (11.42) and its inverse (11.44) also apply to the momentum; we need only make the substitutions $q_n \to p_n$, $q_k \to p_k$. We obtain the desired expression for the Hamiltonian by expressing p_n and q_n as functions of p_k and q_k. The kinetic energy term is the simplest to evaluate:

$$\sum_n p_n^2 = \sum_n \sum_{k,k'} U_{-kn} U_{-k'n} p_k p_{k'} = \sum_{k,k'} \delta_{k,-k'} p_k p_{k'} = \sum_k p_k p_{-k}.$$

This is just the Parseval relation. Next we study the potential energy term:

$$\sum_n (q_{n+1} - q_n)^2 = \sum_n \sum_{k,k'} \left(e^{-ikl} - 1 \right) \left(e^{-ik'l} - 1 \right) U_{-kn} U_{-k'n} q_k q_{k'}$$

$$= \sum_k \left(e^{-ikl} - 1 \right) \left(e^{ikl} - 1 \right) q_k q_{-k} = 4 \sum_k \sin^2 \left(\frac{kl}{2} \right) q_k q_{-k}.$$

Combining these two equations, we arrive at an expression for H_{cl} in which the modes are nearly decoupled:

$$H_{cl} = \sum_k \frac{p_k p_{-k}}{2m} + \frac{1}{2} K \sum_k 4 \sin^2 \left(\frac{kl}{2} \right) q_k q_{-k} = \sum_k \frac{p_k p_{-k}}{2m} + \frac{1}{2} m \sum_k \omega_k^2 q_k q_{-k}. \tag{11.45}$$

We have defined the frequency ω_k of the kth mode as

$$\omega_k = 2\sqrt{\frac{K}{m}} \, \sin \frac{|k|l}{2}. \tag{11.46}$$

The law (11.46) giving the frequency ω_k as a function of k is the *dispersion law* for the normal modes (Fig. 11.2). The expression (11.45) for H_{cl} as a function of the normal modes was obtained within the framework of classical physics. It can be generalized

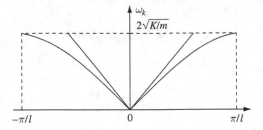

Fig. 11.2. Dispersion law of the normal modes.

immediately to the quantum version by replacing the numbers p_n and q_n in (11.40) by the operators P_n and Q_n obeying the commutation relations

$$[Q_n, P_{n'}] = i\hbar \delta_{nn'} I, \tag{11.47}$$

because the operators corresponding to different atoms n and n' commute. The Fourier transforms can be carried over without modification to the quantum version of the problem, and we obtain

$$H = \sum_k \frac{P_k P_{-k}}{2m} + \frac{1}{2} K \sum_k 4 \sin^2\left(\frac{kl}{2}\right) Q_k Q_{-k} = \sum_k \frac{P_k P_{-k}}{2m} + \frac{1}{2} m \sum_k \omega_k^2 Q_k Q_{-k}.$$

The commutation relations of the Q_k and P_k are

$$[Q_k, P_{k'}] = \sum_{nn'} U_{kn} U_{k'n'} [Q_n, P_{n'}] = i\hbar I \sum_n U_{kn} U_{k'n} = i\hbar \, \delta_{k,-k'} I. \tag{11.48}$$

We still need to decouple the modes k and $-k$. To do this we introduce the annihilation and creation operators of the normal modes by analogy with (11.4) and (11.6)–(11.7):

$$Q_k = \sqrt{\frac{\hbar}{2m\omega_k}} \left(a_k + a_{-k}^\dagger\right), \quad P_k = \frac{1}{i} \sqrt{\frac{\hbar m \omega_k}{2}} \left(a_k - a_{-k}^\dagger\right). \tag{11.49}$$

It can immediately be verified that the commutation relations (11.48) are satisfied when[5]

$$\boxed{[a_k, a_{k'}^\dagger] = \delta_{kk'} I}. \tag{11.50}$$

The factors $\delta_{k,-k'}$ in (11.48) and $\delta_{kk'}$ in (11.50) should be noted. They originate in the periodic boundary conditions, which imply plane waves with $k > 0$ and $k < 0$. If vanishing boundary conditions are used, we have only $k > 0$ and we find the factor $\delta_{kk'}$;

---

[5] Equivalently, a_k and $a_k^\dagger$ can be expressed as functions of Q_k and P_k and then the commutation relations (11.50) derived.

see Exercise 11.5.9. Substituting the relations (11.49) into the expression for H and using the commutation relations (11.50), we arrive at the final form of H:

$$H = \sum_{k=-\pi/l}^{\pi/l} \hbar\omega_k \left(a_k^\dagger a_k + \frac{1}{2} \right).$$

(11.51)

The Hamiltonian is a sum of independent harmonic oscillators of frequency ω_k. Let $|r\rangle$ be an eigenstate of H, $H|r\rangle = E_r|r\rangle$. Using the commutation relations (11.11), we have

$$Ha_k|r\rangle = (a_k H + [H, a_k]) |r\rangle = (E_r - \hbar\omega_k)a_k|r\rangle,$$

$$Ha_k^\dagger|r\rangle = (a_k^\dagger H + [H, a_k^\dagger]) |r\rangle = (E_r + \hbar\omega_k)a_k^\dagger|r\rangle.$$

The creation operator $a_k^\dagger$ *increases* the energy by $\hbar\omega_k$, and the annihilation operator a_k *decreases* it by $\hbar\omega_k$. This energy is associated with an *elementary excitation* or a *quasi-particle*, called a *phonon*. The operator $N_k = a_k^\dagger a_k$, which commutes with H, counts the number of phonons in the mode k. Let $|0_k\rangle$ be the ground state of the kth mode: $a_k|0_k\rangle = 0$. This state corresponds to zero phonons in the kth mode. Let us construct the state $|n_k\rangle$ containing n_k phonons in the kth mode using (11.18):

$$|n_k\rangle = \frac{1}{\sqrt{n_k!}} (a_k^\dagger)^{n_k}|0_k\rangle,$$

(11.52)

and the eigenstates of H by forming the tensor product of the states $|n_k\rangle$:

$$|r\rangle = \bigotimes_{k=-\pi/l}^{k=\pi/l}|n_k\rangle,$$

(11.53)

$$H|r\rangle = \sum_{k=-\pi/l}^{\pi/l} \left(n_k + \frac{1}{2} \right) \hbar\omega_k |r\rangle.$$

(11.54)

The Hilbert space thus constructed is called the *Fock space*. The state $|r\rangle$ is specified by its *occupation numbers* n_k, or the number of phonons in the kth mode. The formalism that we have developed allows us to describe situations in which the number of particles is variable; in fact, we have just constructed a quantized field using the simplest possible nontrivial example.

11.3.2 Quantization of a scalar field in one dimension

Now that we have quantized elasticity, our objective is to do the same with the electromagnetic field. We shall pass through an intermediate stage where we quantize a simple model, that of the *scalar field* in one dimension, which we define below. This model is relevant to the physical case of vibrations of an elastic rod considered as a continuous medium. When $|k|l \ll 1$, the dispersion law (11.46) becomes linear in $|k|$:

$$|k|l \ll 1: \omega_k \simeq \sqrt{\frac{K}{m}} \, |k|l = c_s|k|, \tag{11.55}$$

where $c_s = l\sqrt{K/m}$ is the speed of sound at low frequencies. It will prove useful to rewrite this equation as a relation between the speed of sound, Young's modulus $Y = Kl$,[6] and the mass per unit length $\mu = m/l$:

$$c_s = \sqrt{\frac{Y}{\mu}}. \tag{11.56}$$

Our scalar field will be the long-wavelength limit $\lambda \gg l$ (or $|k|l \ll 1$) of the lattice model of the preceding subsection, and the linear dispersion law (11.55) $\omega_k = c_s|k|$ will be assumed valid for all k. In fact, our ultimate goal is to take the limit $l \to 0$, also called the *continuum limit* of the lattice model. We introduce two functions $\varphi(x, t)$ and $\pi(x, t)$ such that

$$q_n(t) = \varphi(x_n, t), \quad p_n(t) = l\pi(x_n, t). \tag{11.57}$$

In the long-wavelength limit, the displacements $q_n(t)$ and momenta $p_n(t)$ vary only slightly from one site to another, and so we can use the following approximation for the derivative of $\varphi(x, t)$ with respect to x:

$$\left.\frac{\partial \varphi}{\partial x}\right|_{x=x_n} \simeq \frac{1}{l}\left[\varphi(x_{n+1}, t) - \varphi(x_n, t)\right] = \frac{1}{l}\left[q_{n+1}(t) - q_n(t)\right]. \tag{11.58}$$

The equation of motion (11.41) becomes

$$\left.\mu\frac{\partial^2 \varphi}{\partial t^2}\right|_{x=x_n} = \frac{Y}{l^2}\left\{[\varphi(x_{n+1}) - \varphi(x_n)] + [\varphi(x_{n-1}) - \varphi(x_n)]\right\}.$$

A Taylor series expansion through order l^2 gives

$$\varphi(x+l) + \varphi(x-l) - 2\varphi(x) \simeq l^2\frac{\partial^2 \varphi}{\partial x^2},$$

and we obtain a wave equation describing the propagation of vibrations at speed c_s:

$$\frac{\partial^2 \varphi}{\partial t^2} - c_s^2\frac{\partial^2 \varphi}{\partial x^2} = 0. \tag{11.59}$$

The classical Hamiltonian is written as a function of φ_n and π_n as

$$H_{cl} = l\sum_n \left\{\frac{\pi^2(x_n)}{2\mu} + \frac{1}{2}Kl\left[\frac{\varphi(x_{n+1}) - \varphi(x_n)}{l}\right]^2\right\},$$

---

[6] In one dimension, the change of length ΔL of a rod of length L acted on by a force $F = K\Delta x$ satisfies

$$\frac{\Delta L}{L} = \frac{F}{Y} = \frac{\Delta x}{l} = \frac{F}{Kl},$$

which gives $Y = Kl$. In three dimensions $\Delta L/L = F/YS$, where S is the cross-sectional area of the rod and $Y = K/l$, $c_s = \sqrt{Y/\mu}$ with $\mu = m/l^3$.

which is an approximation to the integral

$$H_{\text{cl}} = \int_0^L dx \left[\frac{1}{2\mu} \pi^2(x) + \frac{1}{2} \mu c_s^2 \left(\frac{\partial \varphi}{\partial x} \right)^2 \right], \tag{11.60}$$

where $L = Nl$ is the length of the rod: H_{cl} in (11.60) is the continuum version of (11.40).[7] We have suppressed the time dependence: $\varphi(x) = \varphi(x, t = 0)$ and $\pi(x) = \pi(x, t = 0)$ because H_{cl} is independent of time.

As in the preceding subsection, we shall decompose $\varphi(x)$ and $\pi(x)$ into normal modes by means of a Fourier transform. We define φ_k as

$$\varphi_k = \varphi_{-k}^* = \frac{1}{\sqrt{L}} \int_0^L dx \, e^{ikx} \varphi(x) \simeq \frac{l}{\sqrt{Nl}} \sum_n e^{ikx_n} \varphi(x_n) = \sqrt{l} \, q_k \tag{11.61}$$

by comparison with (11.42). The inverse of φ_k is given by

$$\varphi(x) = \frac{1}{\sqrt{L}} \sum_k e^{-ikx} \varphi_k. \tag{11.62}$$

The relation for p_k corresponding to (11.83) is $\pi_k = l^{-1/2} p_k$. Now let us go to the quantum version, replacing the numbers φ_k and π_k by the *operators* Φ_k and Π_k obeying commutation relations derived from (11.48):[8]

$$[\Phi_k, \Pi_{k'}] = i\hbar \delta_{k,-k'} I. \tag{11.63}$$

As a consequence, if the numbers φ_k and π_k in (11.62) and in the corresponding equation for $\pi(x)$ are replaced by the operators Φ_k and Π_k, the functions $\varphi(x)$ and $\pi(x)$ become operators $\Phi(x)$ and $\Pi(x)$. Here $\Phi(x)$ is called a *field operator* or a *quantized field*.[9] We note that $\Phi(x, t)$ and $\Pi(x, t)$ are labeled by a continuous variable x, whereas their Fourier transforms Φ_k and Π_k are labeled by a discrete index k. This property follows from the use of boundary conditions in a box: $0 \leq x \leq L$. The variable x is *not* a dynamical variable which is transformed into an operator in the quantum version of the problem, but rather the *label* of a point on the rod, and the fundamental operators are Φ and Π.

---

[7] The reader familiar with analytical mechanics will note that the Hamilton equations are

$$\frac{\delta H}{\delta \pi(x)} = \frac{1}{\mu} \pi = \dot{\varphi}, \qquad \frac{\delta H}{\delta \varphi(x)} = -Y \frac{\partial^2 \varphi}{\partial x^2} = -\mu \ddot{\varphi},$$

which give the wave equation (11.59).

[8] The usual procedure is to derive these relations from the equal-time canonical commutation relations postulated between the field $\Phi(x, t)$ and its "conjugate momentum" $\Pi(x, t)$:

$$[\Phi(x, t), \Pi(x', t)] = i\hbar \delta(x - x') I,$$

which will be demonstrated below in (11.69) starting from (11.63). This procedure is – mistakenly – considered by some authors to be more "rigorous"; in fact, it is just as heuristic as the one we follow here.

[9] The procedure we have followed is sometimes called "second quantization." This expression is completely misleading. Clearly, there is only a single quantization, and so "second quantization" should definitively be banished.

Now we can express the quantum Hamiltonian H as a function of the Fourier components of Π and Φ. We write, for example, the potential energy term as a function of the Φ_k as

$$\int_0^L dx \left(\frac{\partial \Phi}{\partial x} \right)^2 = \frac{1}{L} \int dx \sum_{k,k'} \Phi_k \Phi_{k'} (-ik)(-ik') e^{-ikx} e^{-ik'x}$$

$$= - \sum_k \Phi_k \Phi_{k'} kk' \delta_{k,-k'} = \sum_k k^2 \Phi_k \Phi_{-k}.$$

This leads to the following expression for the quantum Hamiltonian H:

$$H = \sum_k \left(\frac{1}{2\mu} \Pi_k \Pi_{-k} + \frac{1}{2} \mu c_s^2 k^2 \Phi_k \Phi_{-k} \right). \tag{11.64}$$

Finally, as in (11.49), we introduce the operators a_k and $a_k^\dagger$ satisfying the commutation relations (11.50):

$$\Phi_k = \sqrt{\frac{\hbar}{2\mu\omega_k}} \left(a_k + a_{-k}^\dagger \right), \quad \Pi_k = \frac{1}{i} \sqrt{\frac{\hbar\mu\omega_k}{2}} \left(a_k - a_{-k}^\dagger \right), \tag{11.65}$$

and H again takes the form of a sum of independent harmonic oscillators:

$$H = \sum_k \hbar\omega_k \left(a_k^\dagger a_k + \frac{1}{2} \right). \tag{11.66}$$

The result is superficially identical to (11.51), but there is an essential difference. The earlier wave vectors k were bounded as $|k| \leq \pi/l$. Now in the continuum limit there is no longer a bound on k and the zero-point energy

$$E_0 = \sum_k \frac{1}{2} \hbar\omega_k$$

is infinite. However, this infinite result is artificial in this particular case (Exercise 11.5.6). Actually, when the wave vector k becomes large or, equivalently, when the wavelength $\lambda = 2\pi/k$ becomes small, of the order of the lattice spacing l, the continuum theory is no longer valid. It is only when the wavelength of a vibration satisfies $\lambda \gg l$ that the wave does not "see" the underlying crystal lattice. We shall encounter this problem of infinite energy again in the case of the electromagnetic field, where k will be genuinely unbounded.

Let us conclude this subsection by giving the Fourier expansion of the quantized field $\Phi_H(x, t)$ in the Heisenberg picture (4.31), with $\Phi_H(x, t = 0) = \Phi_S(x) = \Phi(x)$. The time dependence is found using the equations

$$a_k(t) = e^{iHt/\hbar} a_k e^{-iHt/\hbar} = a_k e^{-i\omega_k t},$$

$$a_k^\dagger(t) = e^{iHt/\hbar} a_k^\dagger e^{-iHt/\hbar} = a_k^\dagger e^{-i\omega_k t}, \tag{11.67}$$

which follow from

$$\frac{da_k}{dt} = -i[a_k(t), H] = -i\omega_k a_k(t),$$

and we obtain from (11.62) and (11.65)

$$\Phi_{\mathrm{H}}(x, t) = \sqrt{\frac{\hbar}{2\mu L}} \sum_k \frac{1}{\sqrt{\omega_k}} \left[a_k e^{i(kx - \omega_k t)} + a_k^\dagger e^{-i(kx - \omega_k t)} \right]. \qquad (11.68)$$

We check from this expression that the field operator $\Phi_{\mathrm{H}}(x, t)$ (which has the dimensions of a length) is Hermitian as it should be. The commutation relations of $\Phi_{\mathrm{H}}(x, t)$ and $\Pi_{\mathrm{H}}(x', t)$ can be calculated immediately. First we take $t = 0$, $\Phi(x) = \Phi_{\mathrm{H}}(x, t = 0)$, $\Pi(x') = \Pi_{\mathrm{H}}(x', t = 0)$:

$$[\Phi(x), \Pi(x')] = -\frac{i\hbar}{2L} \sum_{k,k'} \sqrt{\frac{\omega_{k'}}{\omega_k}} \left[a_k e^{ikx} + a_k^\dagger e^{-ikx}, a_{k'} e^{ik'x} - a_{k'}^\dagger e^{-ik'x} \right]$$

$$= \frac{i\hbar}{L} \sum_k e^{ik(x - x')} I = i\hbar\delta(x - x')I, \qquad (11.69)$$

where we have used (9.145) to obtain the last expression. Since this commutator is a multiple of the identity, we trivially obtain the same result for the equal-time commutator $[\Phi_{\mathrm{H}}(x, t), \Pi_{\mathrm{H}}(x', t)]$.

11.3.3 Quantization of the electromagnetic field

The quantization of the electromagnetic field follows that of the scalar field in the preceding subsection with three modifications: we must work in three dimensions, we must take into account the vector nature of the electromagnetic field, and we must replace the speed of sound c_s by the speed of light c. Let us recall the Maxwell equations (1.8)–(1.9) for electric field $\vec{E}$ and magnetic field $\vec{B}$:

$$\vec{\nabla} \cdot \vec{B} = 0, \qquad \vec{\nabla} \times \vec{E} = -\frac{\partial \vec{B}}{\partial t}, \qquad (11.70)$$

$$\vec{\nabla} \cdot \vec{E} = \frac{\rho_{\mathrm{em}}}{\varepsilon_0}, \qquad c^2 \vec{\nabla} \times \vec{B} = \frac{\partial \vec{E}}{\partial t} + \frac{1}{\varepsilon_0} \vec{j}_{\mathrm{em}}. \qquad (11.71)$$

The two equations (11.70) are *constraints* on the fields $\vec{E}$ and $\vec{B}$, and the two equations (11.71) depend on the *sources* of the electromagnetic field, that is, the charge density ρ_{em} and the current density $\vec{j}_{\mathrm{em}}$. From the Maxwell equations we can derive the continuity equation:

$$\frac{\partial \rho_{\mathrm{em}}}{\partial t} + \vec{\nabla} \cdot \vec{j}_{\mathrm{em}} = 0. \qquad (11.72)$$

One could dream of quantizing the fields $\vec{E}$ and $\vec{B}$ directly. However, there are two technical difficulties with this. First, $\vec{E}$ and $\vec{B}$ are related by the constraints (11.70), which means that their six components are not independent and, moreover, as shown by the

Bohm–Aharonov effect,[10] the interaction of the electromagnetic field with the charges is not local. It is preferable to use the intermediary of the scalar and vector potentials[11] $\overline{V}$ and $\vec{A}$ and obtain the fields by partial differentiation:

$$\vec{E} = -\vec{\nabla}\overline{V} - \frac{\partial \vec{A}}{\partial t}, \quad \vec{B} = \vec{\nabla} \times \vec{A}. \tag{11.73}$$

The use of potentials instead of fields should not be surprising; in quantum mechanics we have never used forces, which are related directly to the fields $\vec{E}$ and $\vec{B}$ by the Lorentz law (1.11); instead, we used the potential energy. In quantum mechanics it is the energy and momentum that play the fundamental role, because they directly influence the phase of the wave function. In the presence of an electric field $\vec{E}$, it is the potential $\overline{V}$ that shows up in the Schrödinger equation via the potential energy $V = q\overline{V}$. It is therefore not surprising that in the presence of a magnetic field $\vec{B}$, it is the vector potential $\vec{A}$ that is involved directly in the Schrödinger equation rather than the field $\vec{B}$.

The potentials are not unique. Under a *gauge transformation*

$$\boxed{\vec{A} \to \vec{A}' = \vec{A} - \vec{\nabla}\Lambda, \quad \overline{V} \to \overline{V}' = \overline{V} + \frac{\partial \Lambda}{\partial t}}, \tag{11.74}$$

where $\Lambda(\vec{r}, t)$ is a scalar function of space and time, the fields $\vec{E}$ and $\vec{B}$ are unchanged. To eliminate this arbitrariness in the potentials $(\vec{A}, \overline{V})$, it is usual to choose a gauge by imposing a condition on $(\vec{A}, \overline{V})$. A common choice (but not the only one possible!) which we shall use here is the *Coulomb gauge*, or the *radiation gauge*:

$$\boxed{\vec{\nabla} \cdot \vec{A} = 0}. \tag{11.75}$$

With this choice, the vector potential becomes transverse: in Fourier space, the condition (11.75) becomes $\vec{k} \cdot \vec{A}(\vec{k}) = 0$ (see also Exercise 11.5.7). According to the first equation in (11.71) and (11.73),

$$\vec{\nabla} \cdot \left(\vec{\nabla}\overline{V} + \frac{\partial \vec{A}}{\partial t}\right) = \nabla^2\overline{V} + \frac{\partial}{\partial t}(\vec{\nabla} \cdot \vec{A}) = \nabla^2\overline{V} = -\frac{\rho_{em}}{\varepsilon_0},$$

from which we derive the scalar potential $\overline{V}$:

$$\overline{V}(\vec{r}, t) = \frac{1}{4\pi\varepsilon_0} \int \frac{\rho_{em}(\vec{r}', t)}{|\vec{r} - \vec{r}'|} d^3r'. \tag{11.76}$$

This expression for the scalar potential is called the *instantaneous Coulomb potential*, because the retardation effects are not explicit: the time t in $\overline{V}$ is the same as that of the source ρ_{em}. This might seem to be incompatible with relativity, but it should be born in mind that a potential is not directly observable, and so the contradiction is only apparent.[12]

---

[10] See, for example, Feynman *et al.* [1965], Vol. II, Chapter 15.

[11] We use the notation $\overline{V}$ for the electric potential so as not to create confusion with the potential energy V. A particle of charge q in a potential $\overline{V}$ has potential energy $V = q\overline{V}$.

[12] Cf. Weinberg [1995], Chapter 8.

In the absence of sources, $\rho_{em} = \vec{\jmath}_{em} = 0$, the second of Equations (11.71) is written as

$$c^2 \vec{\nabla} \times (\vec{\nabla} \times \vec{A}) = c^2 \vec{\nabla} \cdot (\vec{\nabla} \cdot \vec{A}) - c^2 \nabla^2 \vec{A} = -\frac{\partial(\vec{\nabla}V)}{\partial t} - \frac{\partial^2 \vec{A}}{\partial t^2},$$

or, using (11.75) and the fact that $\overline{V} = 0$,

$$\frac{\partial^2 \vec{A}}{\partial t^2} - c^2 \nabla^2 \vec{A} = 0. \tag{11.77}$$

This wave equation is analogous to (11.59) with the three following differences: (i) the spatial dimension is three rather than one; (ii) it involves the speed of light c rather than the speed of sound c_s; (iii) the field $\vec{A}$ is a vector field and not a scalar one. Using the classical expression for the energy density of the electromagnetic field, the expression for the classical Hamiltonian becomes

$$H_{cl} = \frac{1}{2} \varepsilon_0 \int d^3 r \left(\vec{E}^2 + c^2 \vec{B}^2 \right). \tag{11.78}$$

If $\vec{A}$ is the analog of φ, then $\vec{E} = -\partial \vec{A}/\partial t$ will be the analog[13] of π and the term $c^2 \vec{B}^2$, which depends on spatial derivatives of $\vec{A}$, will be the analog of $c_s^2(\partial\varphi/\partial x)^2$. We can immediately write down a Fourier expansion for the quantized electromagnetic field $\vec{A}_H(\vec{r}, t)$ by analogy with (11.68),[14] making the replacements $L \to L^3$ and $\mu \to \varepsilon_0$. The last substitution is determined by comparing the terms $\varepsilon_0 c^2 (\vec{\nabla} \times \vec{A})^2$ in (11.78) and $\mu c_s^2 (\partial\varphi/\partial x)^2$ in (11.60). The final difference from (11.68) is that $\vec{A}$ is a vector. A priori, a Fourier component of $\vec{A}$ should be decomposed on an orthonormal basis of three unit vectors $\hat{k}$, $\vec{e}_1(\hat{k})$, and $\vec{e}_2(\hat{k})$ with $\hat{k} \cdot \vec{e}_i(\hat{k}) = 0$. This is effectively the case for sound vibrations in three dimensions in an isotropic medium,[15] where the vibrations can be either compression waves, which are longitudinal waves parallel to $\hat{k}$, or shear waves, which are transverse and perpendicular to $\hat{k}$. In the case of an electromagnetic field, the gauge condition (11.75) becomes $\hat{k} \cdot \vec{A}(\vec{k}) = 0$ in Fourier space and there is no longitudinal component. Taking into account all these considerations, we can generalize (11.68) and write the quantized electromagnetic field[16] in the Heisenberg picture (we continue to use periodic boundary conditions in a box of volume $\mathcal{V} = L^3$, or *quantization in a box*):

$$\vec{A}_H(\vec{r}, t) = \sqrt{\frac{\hbar}{2\varepsilon_0 L^3}} \sum_{\vec{k}} \sum_{s=1}^{2} \frac{1}{\sqrt{\omega_k}} \left[a_{\vec{k}s} \vec{e}_s(\hat{k}) e^{i(\vec{k}\cdot\vec{r} - \omega_k t)} + a_{\vec{k}s}^\dagger \vec{e}_s^*(\hat{k}) e^{-i(\vec{k}\cdot\vec{r} - \omega_k t)} \right]. \tag{11.79}$$

---

[13] In fact, in a formulation of electromagnetism like that used in analytical mechanics (cf. Footnote 7), it is $-\varepsilon_0\vec{E}$ that plays the role of the momentum conjugate to $\vec{A}$, as seen from (11.85).

[14] In order to distinguish quantized fields from classical ones, we shall designate the former by sans serif letters: $\vec{A}$, $\vec{E}$, $\vec{B}$.

[15] Our discussion is actually oversimplified, because the speed of compression waves is different from that of shear waves.

[16] We have glossed over several delicate problems; see, for example, Weinberg [1995], Chapter 8, for a full discussion.

The unit vectors $\vec{e}_s(\hat{k})$ orthogonal to $\vec{k}$ describe the polarization. It is possible to choose a complex polarization basis, for example, a basis of circular polarization states: $s = $ R, L, which makes it necessary to perform the complex conjugation in the second term of (11.79), thus ensuring that $\vec{A}$ is Hermitian. The expression for the projector onto the subspace orthogonal to $\vec{k}$ is often useful:

$$\sum_s e_{si}(\hat{k})\, e_{sj}^*(\hat{k}) = \delta_{ij} - \hat{k}_i \hat{k}_j. \tag{11.80}$$

The operators $a_{\vec{k}s}$ ($a_{\vec{k}s}^\dagger$) destroy (create) photons of wave vector $\vec{k}$ and polarization s. They satisfy the commutation relations

$$\boxed{[a_{\vec{k}s}, a_{\vec{k}'s'}^\dagger] = \delta_{\vec{k},\vec{k}'}\delta_{ss'} I}\,. \tag{11.81}$$

From (11.79) we derive the expression for the quantized electric field $\vec{E}_H = -\partial\vec{A}_H/\partial t$:

$$\vec{E}_H(\vec{r}, t) = i\sqrt{\frac{\hbar}{2\varepsilon_0 L^3}} \sum_{\vec{k}} \sum_{s=1}^2 \sqrt{\omega_k} \left[a_{\vec{k}s}\vec{e}_s(\hat{k})e^{i(\vec{k}\cdot\vec{r}-\omega_k t)} - a_{\vec{k}s}^\dagger \vec{e}_s^*(\hat{k})e^{-i(\vec{k}\cdot\vec{r}-\omega_k t)} \right] \tag{11.82}$$

and, using the expression

$$\vec{\nabla} \times \left(\vec{e}_s(\hat{k})e^{i\vec{k}\cdot\vec{r}} \right) = i\vec{k} \times \vec{e}_s(\hat{k})\, e^{i\vec{k}\cdot\vec{r}}, \tag{11.83}$$

that for the magnetic field:

$$\vec{B}_H(\vec{r}, t) = \sqrt{\frac{\hbar}{2\varepsilon_0 L^3}} \sum_{\vec{k}} \sum_{s=1}^2 \frac{i}{c} \sqrt{\omega_k}\, \hat{k} \times \left[\vec{e}_s(\hat{k})a_{\vec{k}s}e^{i(\vec{k}\cdot\vec{r}-\omega_k t)} - \vec{e}_s^*(\hat{k})a_{\vec{k}s}^\dagger e^{-i(\vec{k}\cdot\vec{r}-\omega_k t)} \right]. \tag{11.84}$$

Just like for a classical plane wave, $\vec{B} = (\hat{k}/c) \times \vec{E}$. It is easy, as in the case of a scalar field, to calculate the commutators of the various components of the field at $t = 0$. We then find the following commutation relations between the field component A_i and the component $-\varepsilon_0 E_j$ of the conjugate momentum (Exercise 11.5.8):

$$[A_i(\vec{r}), -\varepsilon_0 E_j(\vec{r}')] = i\hbar \int \frac{d^3 k}{(2\pi)^3}\, e^{i\vec{k}\cdot(\vec{r}-\vec{r}')} \left(\delta_{ij} - \hat{k}_i \hat{k}_j \right) I, \tag{11.85}$$

where we have used (9.151). We then deduce that E_x commutes with B_x, but not with B_y or B_z, which shows that it is not possible to measure simultaneously the x component of the electric field and the y component of the magnetic field at the same point.

The expression for the Hamiltonian (Exercise 11.5.8) is a trivial generalization of (11.66):

$$\boxed{H = \sum_{\vec{k},s} \hbar\omega_k \left(a_{\vec{k},s}^\dagger a_{\vec{k},s} + \frac{1}{2} \right)}\,. \tag{11.86}$$

We then find the (infinite) zero-point energy:

$$E_0 = \frac{1}{2} \sum_{\vec{k},s} \hbar \omega_k \rightarrow \frac{L^3}{(2\pi)^3} \int d^3k \, \hbar ck = \frac{\hbar c L^3}{2\pi^2} \int_0^\infty k^3 dk, \tag{11.87}$$

where we have used (9.151). In the case of black-body radiation, it was shown that the thermal fluctuations leading to infinite energy in classical statistical mechanics can be controlled by quantum mechanics. However, we eliminated that infinity by introducing another one, an infinity associated with *quantum fluctuations*. These quantum fluctuations have observable effects: for example, they lead to the Casimir effect (Exercise 11.5.12). The zero-point energy is also called the *vacuum energy*; it may play an important role in cosmology, where it might be related to the so-called dark energy, whose properties are still far from being understood.

It is possible to couple the quantized field to a classical source $\vec{j}_{em}(\vec{r}, t)$ by writing

$$W(t) = -\int d^3r \, \vec{j}_{em}(\vec{r}, t) \cdot \vec{A}(\vec{r}). \tag{11.88}$$

This coupling generalizes that of (11.124) for the forced harmonic oscillator of Exercise 11.5.4, with the force $f(t)$ replaced by the source $\vec{j}_{em}$ and the position operator Q replaced by the quantized field $\vec{A}$. It can then be shown[17] that if we start from a state with zero photons and if the source acts for a finite time, we obtain a coherent state of the electromagnetic field in which the number of photons in a mode $\vec{k}$ obeys a Poisson law with average given by $|\vec{j}_{em}(\vec{k}, \omega_k)|^2$, where $\vec{j}_{em}(\vec{k}, \omega_k)$ is the four-dimensional Fourier transform of $\vec{j}_{em}(\vec{r}, t)$.

The quantized field $\vec{A}$ was written down in the Coulomb gauge. This is the gauge most convenient for elementary problems, but it is not convenient for a general study of quantum electrodynamics. The condition $\vec{\nabla} \cdot \vec{A} = 0$ distinguishes a particular reference frame, and so the Lorentz invariance of the theory is not manifest. Naturally, this is not a fundamental defect, because it is possible to show that the physical results are consistent with Lorentz invariance. The real fault of the Coulomb gauge is that it leads to inextricable calculations because the renormalization procedure (elimination of infinities) requires that Lorentz invariance be maintained explicitly in order for the calculations to be manageable.[18] A gauge in which Lorentz invariance is manifest is the Lorentz gauge:[19]

$$\frac{\partial \overline{V}}{\partial t} + \vec{\nabla} \cdot \vec{A} = 0.$$

However, the Lorentz gauge introduces unphysical states, which must be correctly interpreted and eliminated from the physical results. These unphysical states do not appear in the Coulomb gauge, which is an example of a "physical gauge." Unfortunately, it is not possible to use a physical gauge and preserve formal Lorentz invariance at the same time.

---

[17] See Exercise 11.5.4. A detailed discussion can be found, for example, in Le Bellac [1991], Chapter 9, or C. Itzykson and J.-B. Zuber, *Quantum Field Theory*, New York: McGraw-Hill (1980), Chapter 4.

[18] From a technical point of view, the counter-terms that eliminate the infinities are constrained by the Lorentz invariance if the gauge choice respects this formal invariance.

[19] This formal Lorentz invariance is manifest in four-dimensional notation: $\partial_\mu A^\mu = 0$, $A^\mu = (\overline{V}, \vec{A})$.

11.3.4 Quantum fluctuations of the electromagnetic field

In the formalism of the preceding subsection, the electromagnetic field is an operator and quantum fluctuations should be present. In the zero-photon state, or vacuum state $|0\rangle$, the expectation values of the electric field (11.82) and the magnetic field (11.84) vanish:

$$\langle 0|\vec{E}_H(\vec{r}, t)|0\rangle = \langle 0|\vec{B}_H(\vec{r}, t)|0\rangle = 0,$$

because $\langle 0|a_{\vec{k}s}|0\rangle = \langle 0|a_{\vec{k}s}^\dagger|0\rangle = 0$. However, the vanishing of an expectation value does not imply that there are no fluctuations. These fluctuations have important physical consequences, and we shall study them for several types of state of the electromagnetic field: the vacuum, states with a fixed number of photons, coherent states, and squeezed states. In order to simplify the discussion, we shall concentrate on a single mode with wave vector $\vec{k}$ and fixed polarization s, and so $a_{\vec{k}s} \to a$, $\omega_k \to \omega$. In addition, we take $\vec{r} = 0$. This restriction to a single mode is often a good approximation, for example in the case of a single-mode laser when transverse effects due to diffraction are neglected, or for a mode in a superconducting cavity of the type studied in Appendix B. The electric field in a cavity reduced to a single mode is written as

$$E(t) = i\sqrt{\frac{\hbar\omega}{2\varepsilon_0 \mathcal{V}}} \left(a\,e^{-i\omega t} - a^\dagger e^{i\omega t} \right), \tag{11.89}$$

where $\mathcal{V}$ is the cavity volume; the expression (11.89) can be derived immediately from (11.82). Here we have suppressed the label H and the vector notation in order to simplify the notations. The operators a and $a^\dagger$ satisfy the commutation relation $[a, a^\dagger] = I$. First let us calculate the fluctuations of E in the vacuum state using

$$\left(a\,e^{-i\omega t} - a^\dagger e^{i\omega t} \right)^2 = a^2 e^{-2i\omega t} + (a^\dagger)^2 e^{2i\omega t} - 2a^\dagger a - I. \tag{11.90}$$

Only the last term gives a nonzero result when the vacuum expectation value is taken, and we find

$$\langle 0|E^2(t)|0\rangle = \frac{\hbar\omega}{2\varepsilon_0 \mathcal{V}},$$

which gives the dispersion

$$\Delta_0 E = \left[\langle 0|E^2(t)|0\rangle - \langle 0|E(t)|0\rangle^2 \right]^{1/2} = \sqrt{\frac{\hbar\omega}{2\varepsilon_0 \mathcal{V}}}. \tag{11.91}$$

The quantum fluctuations of the electromagnetic field have important physical consequences. In addition to the Casimir effect (Exercise 11.5.12), they also lead to a splitting between the $2s_{1/2}$ and $2p_{1/2}$ levels of the hydrogen atom, which are degenerate in the approximation of the relativistic Dirac theory (cf. Section 14.2.2). This is called the *Lamb shift*. This shift of about 4.38×10^{-6} eV is roughly 10^{-7} of the difference between the energies of the 1s and 2s levels, and amounts to 1058 MHz in frequency units.[20] These quantum fluctuations are also responsible for the anomalous magnetic moment of the electron. Whereas the Dirac theory predicts an electron gyromagnetic ratio of $\gamma_e = q_e/m_e$, the actual one is

$$\gamma_e = \frac{q_e}{m_e}\left(1 + \frac{\alpha}{2\pi} + O(\alpha^2)\right),$$

where $\alpha \simeq 1/137$ is the fine-structure constant.

In a state with a fixed number of photons n (in the mode under consideration), the expectation value of $E(t)$ is zero because $\langle n|a|n\rangle = \langle n|a^\dagger|n\rangle = 0$, while that of $E^2(t)$ is, according to (11.90) and (11.12),

$$\langle n|E^2(t)|n\rangle = \frac{\hbar\omega(2n+1)}{2\varepsilon_0 V}.$$

This leads to the dispersion $\Delta_n E$ in the state $|n\rangle$:

$$\Delta_n E = \left[\langle n|E^2|n\rangle - \langle n|E|n\rangle^2\right]^{1/2} = \sqrt{\frac{\hbar\omega(2n+1)}{2\varepsilon_0 V}}. \tag{11.92}$$

This dispersion grows as the square root of the number of photons when $n \gg 1$.

States which are more interesting in practice than those with a fixed number of photons are coherent states $|z\rangle$. Most ordinary light sources emit states of the electromagnetic field that are very close to a coherent state (lasers), or to a statistical mixture of coherent states (classical sources). Let us calculate the expectation value of $E(t)$ in a coherent state setting $z = |z|\exp(i\phi)$:

$$\langle z|E(t)|z\rangle = i\sqrt{\frac{\hbar\omega}{2\varepsilon_0 V}}\left(ze^{-i\omega t} - z^*e^{i\omega t}\right) = \sqrt{\frac{2\hbar\omega}{\varepsilon_0 V}}\,|z|\sin(\omega t - \phi), \tag{11.93}$$

and

$$\langle z|E^2(t)|z\rangle = -\frac{\hbar\omega}{2\varepsilon_0 V}\left[\left(ze^{-i\omega t} - z^*e^{i\omega t}\right)^2 - 1\right].$$

The dispersion $\Delta_z E$ in a coherent state is identical to that in vacuum:

$$\Delta_z E = \left[\langle z|E^2(t)|z\rangle - \langle z|E(t)|z\rangle^2\right]^{1/2} = \sqrt{\frac{\hbar\omega}{2\varepsilon_0 V}} = \Delta_0 E. \tag{11.94}$$

The average number of photons is $\langle N\rangle_z = \langle z|N|z\rangle = |z|^2$ and the dispersion $\Delta_z N = |z|$. These two results follow from the Poisson distribution (11.34) for the number of photons, which makes it possible to predict the statistics of results of photon-counting experiments.

In the present section only, we define the Hermitian operators Q and P as

$$Q = \frac{1}{2}\left(a + a^\dagger\right), \quad P = \frac{1}{2i}\left(a - a^\dagger\right). \tag{11.95}$$

---

[20] A small part of this shift ($-27\,\mathrm{MHz} \simeq 3\%$) arises not from fluctuations of the electromagnetic field, but from fluctuations of the electron–positron field. The creation of (virtual) electron–positron pairs has the effect of screening the Coulomb field and acts as a vacuum dielectric constant. This effect is much more important in muonic atoms; cf. Exercise 14.5.3 and Footnote 36 of Chapter 1.

They satisfy the commutation relation $[Q, P] = i/2$, which leads to the Heisenberg inequality

$$\Delta P\, \Delta Q \geq \frac{1}{4}. \tag{11.96}$$

Direct calculation shows that

$$\mathsf{E}(t) = \sqrt{\frac{2\hbar\omega}{\varepsilon_0 V}} \left[Q \sin \omega t - P \cos \omega t\right], \tag{11.97}$$

whereas, according to (11.37) and (11.39),

$$\langle Q \rangle_z = \operatorname{Re} z, \quad \langle P \rangle_z = \operatorname{Im} z, \quad \Delta_z P = \Delta_z Q = \frac{1}{2}.$$

The Heisenberg inequality (11.96) is therefore saturated when the field is in a coherent state, in agreement with the results of Section 11.2. The expectation value $\langle \mathsf{E}(t) \rangle_z$ of the field is given by (11.93). To interpret the fluctuations about this expectation value it is convenient to use a Fresnel representation, in which the field is the projection on a fixed axis of a rotating vector. The Fresnel vector of the expectation value is a vector of length

$$|z|\sqrt{\frac{2\hbar\omega}{\varepsilon_0 V}} = \lambda |z|$$

which rotates in a plane with angular velocity ω. To be specific, let us take $\phi = 0$ in (11.93). At time $t = 0$, $\langle \mathsf{E}(t) \rangle_z = \lambda |z|$ and, according to (11.94), the dispersion about this expectation value is $\Delta_z \mathsf{E} = \lambda/2$. At time $t = \pi/2\omega$ we have $\langle \mathsf{E}(t) \rangle_z = 0$ and, as always, $\Delta_z \mathsf{E} = \lambda/2$. In general, we see that fluctuations may be visualized by imagining that the tip of the Fresnel vector is not actually a point, but rather a fuzzy area: the tip is centered at the end of a vector of length $\lambda |z|$, but fluctuates within a circle of radius

$$R = \frac{\lambda}{2} = \sqrt{\frac{\hbar\omega}{2\varepsilon_0 V}}.$$

These fluctuations of the tip of the Fresnel vector are interpreted as the dispersion in the phase $\Delta_z \phi$, and, as shown by Fig. 11.3,

$$\Delta_z \phi \simeq \frac{\Delta_z \mathsf{E}}{\lambda |z|} = \frac{1}{2|z|}. \tag{11.98}$$

According to (11.39), the fluctuation of the number of photons is precisely $\Delta_z N = |z|$. For a coherent state we then obtain a relation between the dispersion $\Delta_z \phi$ of the phase and the dispersion $\Delta_z N$ of the number of photons:

$$\Delta_z \phi\, \Delta_z N \simeq \frac{1}{2}.$$

These fluctuations are very weak for a single-mode laser where $|z| \gg 1$, but they are important for the superconducting cavity studied in Appendix B, where $|z| \lesssim 3$.

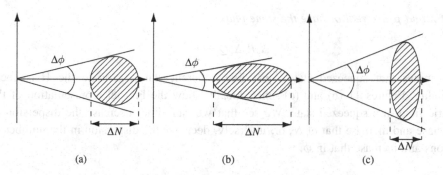

Fig. 11.3. Fresnel representation of the electric field. The shaded region represents the dispersion at the tip of the field. (a) A coherent state; (b) and (c) squeezed states.

We would like to obtain a Heisenberg inequality for the product $\Delta\phi\,\Delta N$, but a derivation similar to that of Section 4.1.3 is impossible because we do not know how to define a phase operator. Nevertheless, we can try to simulate quantum fluctuations by taking as a model a classical field whose amplitude and phase are random functions. Then it is possible to prove the inequality

$$\boxed{\Delta\phi\,\Delta N \geq \frac{1}{2}}. \tag{11.99}$$

Coherent states saturate this inequality.

There is another type of interesting state, a *squeezed state*. Such states are obtained by a Bogolyubov transformation of the operators a and $a^\dagger$.[21] Let b and $b^\dagger$ be the operators

$$b = \lambda a + \mu a^\dagger, \quad b^\dagger = \lambda^* a^\dagger + \mu^* a, \tag{11.100}$$

where the complex numbers λ and μ satisfy

$$|\lambda|^2 - |\mu|^2 = 1.$$

It is straightforward to show that the operators b and $b^\dagger$ satisfy $[b, b^\dagger] = I$. It is said that the Bogolyubov transformation is a *canonical transformation*, as it preserves the commutation relations. Since the operators b and $b^\dagger$ satisfy the same algebra as a and $a^\dagger$, there exist states $|\tilde{z}\rangle$ such that $b|\tilde{z}\rangle = \tilde{z}|\tilde{z}\rangle$. The transformation inverse to (11.100) is

$$a = \lambda^* b - \mu b^\dagger, \quad a^\dagger = \lambda b^\dagger - \mu^* b.$$

A simple but cumbersome calculation (Exercise 11.5.5) shows that the dispersions in the state $|\tilde{z}\rangle$ are

$$\Delta_{\tilde{z}} P = \frac{1}{2}|\lambda - \mu|, \quad \Delta_{\tilde{z}} Q = \frac{1}{2}|\lambda + \mu|,$$

---

[21] This transformation was first used by Bogolyubov in the early 1950s in the theory of superfluidity.

or, *if λ and μ are real or have the same phase,*

$$\Delta_{\tilde{z}} P \, \Delta_{\tilde{z}} Q = \frac{1}{4}.$$

This shows that squeezed states, just like coherent states, saturate the Heisenberg inequality. Figures 11 (b) and (c) schematically show the Fresnel representation of the electric field in a squeezed state. We see that we can either decrease the dispersion of the phase and increase that of N, or, inversely, decrease the dispersion in the number of photons and increase that in ϕ.

11.4 Motion in a magnetic field

11.4.1 Local gauge invariance

Now let us return to the *classical* electromagnetic field $(\vec{E}, \vec{B})$ with the objective of determining the form of the interaction between this field and a quantum particle of charge q. In classical electrodynamics the electric charge density $\rho_{em}(\vec{r}, t)$ and the current density

$$\vec{J}_{em}(\vec{r}, t) = \rho_{em}(\vec{r}, t)\vec{v}(\vec{r}, t) \tag{11.101}$$

satisfy the continuity equation (11.72). We want to generalize the expression for the current to quantum physics. In Chapter 9 we found the expression for the particle current (9.141):

$$\vec{j}(\vec{r}, t) = \text{Re} \left\{ \varphi^*(\vec{r}, t) \left[\frac{-i\hbar}{m} \vec{\nabla} \right] \varphi(\vec{r}, t) \right\}$$

$$= \varphi^*(\vec{r}, t) \left[\frac{-i\hbar}{2m} \vec{\nabla} \right] \varphi(\vec{r}, t) - \varphi(\vec{r}, t) \left[\frac{-i\hbar}{2m} \vec{\nabla} \right] \varphi^*(\vec{r}, t). \tag{11.102}$$

The electromagnetic current created by the motion of a quantum particle of charge q should a priori be $\vec{j}_{em} = q\vec{j}$, the charge density ρ_{em} being $q|\varphi|^2$. The particle current in this form obeys the continuity equation (11.72) when the wave function $\varphi(\vec{r}, t)$ satisfies the Schrödinger equation:

$$i\hbar \frac{\partial \varphi}{\partial t} = \left(-\frac{\hbar^2}{2m} \nabla^2 + V \right) \varphi,$$

and similarly for the associated electromagnetic current

$$\rho_{em} = q|\varphi|^2, \quad \vec{J}_{em} = q\vec{j},$$

which satisfies (11.72). However, we shall see that the expression for the current (11.102) must be modified when a vector potential is present. The current (11.102) is invariant under a *global gauge transformation*, which consists of multiplying φ by a phase factor

$$\boxed{\varphi(\vec{r}, t) \rightarrow \varphi'(\vec{r}, t) = \exp\left(-i\frac{q}{\hbar} \Lambda \right) \varphi(\vec{r}, t) = \Omega\varphi(\vec{r}, t)} \,, \tag{11.103}$$

where Λ is a real number. When Λ is a function of $\vec{r}$ and t, we have the case of a *local gauge transformation*; the connection to (11.74) will soon become clear. We are going to deduce the form of the current from a principle of local gauge invariance. This might a priori seem arbitrary, but in fact this principle is very general, and it is now believed that all the fundamental interactions of elementary particle physics can be derived from it (Exercise 11.5.11). A local gauge transformation is obtained by replacing the constant Λ in (11.103) by a *function* of $\vec{r}$ and t:

$$\varphi(\vec{r}, t) \to \varphi'(\vec{r}, t) = \exp\left(-i\frac{q}{\hbar}\Lambda(\vec{r}, t)\right)\varphi(\vec{r}, t) = \Omega(\vec{r}, t)\varphi(\vec{r}, t)\,.$$

(11.104)

This transformation is manifestly unitary. We can immediately verify that the current (11.102) is not invariant under a local gauge transformation, because the gradient acts on $\exp(iq\Lambda/\hbar)$. We shall modify the expression for the current by replacing the gradient $\vec{\nabla}$ by the *covariant derivative* $\vec{D}$:

$$-i\hbar\vec{D} = -i\hbar\vec{\nabla} - q\vec{A}\,.$$

(11.105)

In contrast to the ordinary derivative, the covariant derivative has a simple behavior under a local gauge transformation (11.104):

$$-i\hbar\vec{D}\varphi = -i\hbar\vec{D}(\Omega^{-1}\varphi') = (-i\hbar\vec{\nabla} - q\vec{A})\exp\left(i\frac{q}{\hbar}\Lambda(\vec{r}, t)\right)\varphi'(\vec{r}, t)$$

$$= \Omega^{-1}(-i\hbar\vec{\nabla} - q\vec{A} + q\vec{\nabla}\Lambda)\varphi'$$

$$= \Omega^{-1}(-i\hbar\vec{\nabla} - q\vec{A}')\varphi' = \Omega^{-1}(-i\hbar\vec{D}'\varphi'), \quad (11.106)$$

where $\vec{D}'$ is the covariant derivative calculated using the transformed vector potential (11.74). The covariant derivatives $\vec{D}$ and $\vec{D}'$ are physically equivalent because $\vec{A}$ and $\vec{A}'$ are. The expression for the current becomes invariant under a local gauge transformation if the ordinary derivative in (11.102) is replaced by the covariant derivative:

$$\vec{j}(\vec{r}, t) = \mathrm{Re}\left\{\varphi^*(\vec{r}, t)\left[\frac{-i\hbar}{m}\vec{\nabla} - \frac{q}{m}\vec{A}\right]\varphi(\vec{r}, t)\right\} = \mathrm{Re}\left\{\varphi^*(\vec{r}, t)\left[\frac{-i\hbar}{m}\vec{D}\varphi\right]\right\}\,.$$

(11.107)

Indeed, if φ is expressed as a function of φ' using (11.104) and (11.106), then the current is invariant:

$$\vec{j}(\vec{r}, t) = \mathrm{Re}\left\{\varphi'^*(\vec{r}, t)\Omega\,\Omega^{-1}\left[\frac{-i\hbar}{m}\vec{D}'\varphi'\right]\right\} = \mathrm{Re}\left\{\varphi'^*(\vec{r}, t)\left[\frac{-i\hbar}{m}\vec{D}'\varphi'\right]\right\} = \vec{j}'(\vec{r}, t).$$

This suggests that the velocity operator $d\vec{R}/dt$ is not simply $d\vec{R}/dt = \vec{P}/m = -(i\hbar/m)\vec{\nabla}$ but rather

$$\frac{d\vec{R}}{dt} = -\frac{i\hbar}{m}\vec{D} = -\frac{i\hbar}{m}\vec{\nabla} - \frac{q}{m}\vec{A}. \qquad (11.108)$$

Knowing that the velocity operator is given by the commutator of $\vec{R}$ and the Hamiltonian, let us study its x component. According to (8.61) and the expression (11.108) for $d\vec{R}/dt$,

$$\dot{X} = \frac{i}{\hbar}[H, X] = \frac{1}{m}(P_x - qA_x),$$

which, according to the reasoning of Section 8.4, gives the most general form of H:

$$H = \frac{1}{2m}\left(\vec{P} - q\vec{A}\right)^2 + q\overline{V} = \frac{1}{2m}\left(-i\hbar\vec{\nabla} - q\vec{A}\right)^2 + q\overline{V} = \frac{1}{2m}(-i\hbar\vec{D})^2 + q\overline{V}, \quad (11.109)$$

where $V = q\overline{V}$ is an arbitrary function of $\vec{R}$ and t. Requiring local gauge invariance of the current allows us to recover the generic form (8.73) of the Hamiltonian compatible with Galilean invariance. The substitution $-i\hbar\vec{\nabla} \rightarrow -i\hbar\vec{D}$ in the Schrödinger equation in the absence of an electromagnetic field gives this equation in the presence of an electromagnetic field; this is called *minimal coupling*.[22] The minimal-coupling prescription extends to non-Abelian gauge theories (Exercise 11.5.11) and can be used to write down all the interactions of the Standard Model of elementary particle physics between the spin-1/2 particles ("matter particles") and spin-1 particles (gauge bosons) listed in Section 1.1.3.

In analytical mechanics, it can be shown that the Hamiltonian leading to the Lorentz force (1.11) is

$$H_{\text{cl}} = \frac{1}{2m}\left(\vec{p} - q\vec{A}\right)^2 + q\overline{V}.$$

Another method of obtaining (11.109) is to start from this classical form and use the correspondence principle to replace $\vec{p}$ and $\vec{r}$ by operators: $\vec{p} \rightarrow \vec{P} = -i\hbar\vec{\nabla}$, $\vec{r} \rightarrow \vec{R}$.

If φ is a solution of the Schrödinger equation with the potential $(\vec{A}, \overline{V})$, then φ' will be a solution of it with the gauge-transformed potential $(\vec{A}', \overline{V}')$ (11.74). The Schrödinger equation for φ can be written as

$$i\hbar\frac{\partial\varphi}{\partial t} = \frac{1}{2m}(-i\hbar\vec{D})^2\varphi + q\overline{V}\varphi.$$

However, on the one hand

$$\frac{\partial\varphi}{\partial t} = \frac{\partial}{\partial t}\left(\exp\left(\frac{iq\Lambda}{\hbar}\right)\varphi'\right) = \Omega^{-1}\left(\frac{iq}{\hbar}\frac{\partial\Lambda}{\partial t}\varphi' + \frac{\partial\varphi'}{\partial t}\right),$$

---

[22] The interaction $W = -\gamma\vec{S}\cdot\vec{B}$ between a spin magnetic moment and a magnetic field does not appear to be derived from minimal coupling. In fact, this interaction is derived from the relativistic Dirac equation and the use of the minimal-coupling prescription in that equation, which leads to the gyromagnetic ratio $\gamma = q_e/m_e$. The corrections of the anomalous magnetic moment type are derived from minimal coupling applied to quantum electrodynamics.

while on the other

$$\frac{1}{2m}(-i\hbar\vec{D})^2\varphi = \frac{1}{2m}(-i\hbar\vec{D})^2\Omega^{-1}\varphi' = \Omega^{-1}\frac{1}{2m}(-i\hbar\vec{D}')^2\varphi'.$$

Dropping the factor Ω^{-1} from the two sides of the Schrödinger equation for φ', we find

$$i\hbar\frac{\partial\varphi'}{\partial t} = \frac{1}{2m}(-i\hbar\vec{D}')^2\varphi' + q\overline{V}\varphi'.$$

It can also be verified (Exercise 11.5.10) that $\vec{j}$ obeys the continuity equation:

$$\frac{\partial|\varphi|^2}{\partial t} + \vec{\nabla}\cdot\vec{j} = 0. \tag{11.110}$$

11.4.2 A uniform magnetic field: Landau levels

As an application, let us study the motion of a charged particle in a uniform constant magnetic field. We shall ignore spin effects, as the interaction of a magnetic moment related to the spin has already been studied in Section 3.2.5. We assume that $\vec{B}$ points along Oz, and to simplify the discussion we also assume that the motion is confined to the plane xOy. This case is in fact of great practical interest, because two-dimensional structures having important applications like the quantum Hall effect can be manufactured in the laboratory.[23] A classical particle under the action of a force

$$\vec{F} = q\vec{v}\times\vec{B}$$

moves in a circle of radius $\rho = mv/|q|B$ with frequency $\omega = |q|B/m$,[24] the Larmor frequency (cf. (3.61)). If, to be specific, we assume that $q < 0$, the circle is traced in the counterclockwise direction. The motion is then

$$x(t) = x_0 + \rho\cos\omega t,$$
$$y(t) = y_0 + \rho\sin\omega t, \tag{11.111}$$

where x_0 and y_0 are the coordinates of the center of the circle. The projection of this uniform circular motion on the axes Ox and Oy gives two independent harmonic oscillators, which we shall recover in quantum mechanics. A possible choice for the vector potential is

$$\vec{A} = \frac{1}{2}\vec{B}\times\vec{r}, \tag{11.112}$$

---

[23] Cf. Ph. Taylor and O. Heinonen, *Condensed Matter Physics*, Cambridge: Cambridge University Press (2002), Chapter 10.
[24] If the motion occurs in three dimensions, the trajectory is a helix whose projection on the plane xOy is a circle of radius ρ traced out with frequency ω.

or $A_x = -yB/2$, $A_y = xB/2$, $A_z = 0$. This choice is obviously not unique, and another common choice is $A_x = A_z = 0$, $A_y = xB$.[25] Let us calculate the commutator of the velocity components:

$$[\dot{X}, \dot{Y}] = \frac{1}{m^2}\left[P_x + \frac{q}{2} YB, P_y - \frac{q}{2} XB\right]$$

$$= \frac{1}{m^2}\frac{qB}{2}\left(-[P_x, X] + [Y, P_y]\right) = -\frac{i\hbar\omega}{m} I. \tag{11.113}$$

Since the Hamiltonian H can be written as

$$H = \frac{1}{2} m \left(\dot{X}^2 + \dot{Y}^2\right), \tag{11.114}$$

we can recover the form (11.9) by defining

$$\hat{Q} = \sqrt{\frac{m}{\hbar\omega}}\,\dot{Y}, \quad \hat{P} = \sqrt{\frac{m}{\hbar\omega}}\,\dot{X},$$

so that

$$H = \frac{1}{2}\hbar\omega\left(\hat{P}^2 + \hat{Q}^2\right). \tag{11.115}$$

The energy levels are labeled by an integer n:

$$E_n = \hbar\omega\left(n + \frac{1}{2}\right), \quad n = 0, 1, 2, \ldots \tag{11.116}$$

These levels are called *Landau levels*. Guided by the analogy with the classical case, we define an operator R^2 which is the analog of the squared radius ρ^2 of the circular trajectory:

$$R^2 = \frac{1}{\omega^2}\left(\dot{X}^2 + \dot{Y}^2\right) = \frac{2H}{m\omega^2}. \tag{11.117}$$

The expectation value of R^2 in the state $|n\rangle$ is

$$\langle R^2 \rangle_n = \frac{2}{m\omega^2}\langle n|H|n\rangle = \frac{2\hbar}{m\omega}\left(n + \frac{1}{2}\right).$$

If the particle is in an eigenstate of H, the dispersion of R^2 is zero. The flux Φ of the magnetic field through an orbit is quantized in units of $h/|q|$. We can write

$$\Phi = \pi\langle R^2\rangle_n B = \frac{h}{|q|}\left(n + \frac{1}{2}\right).$$

The second characteristic of the motion is the position of the center of the circle. Following (11.111), we define the operators X_0 and Y_0 as

$$X_0 = X - \frac{1}{\omega}\dot{Y}, \quad Y_0 = Y + \frac{1}{\omega}\dot{X}. \tag{11.118}$$

---

[25] This gauge is used by, for example, Landau and Lifschitz [1958], Section 111.

Using $[X, \dot{X}] = [Y, \dot{Y}] = i\hbar I/m$ and (11.113), the commutator $[X_0, Y_0]$ becomes

$$[X_0, Y_0] = \frac{i\hbar}{m\omega} I.$$

It can immediately be verified that

$$[X_0, \dot{X}] = [X_0, \dot{Y}] = [Y_0, \dot{X}] = [Y_0, \dot{Y}] = 0,$$

and so $[H, X_0] = [H, Y_0] = 0$. The operator R_0^2,

$$R_0^2 = X_0^2 + Y_0^2, \tag{11.119}$$

commutes with R^2; R^2 and R_0^2 are Hermitian and can be diagonalized simultaneously. Setting

$$\hat{Q}_0 = \sqrt{\frac{m\omega}{\hbar}} X_0, \quad \hat{P}_0 = \sqrt{\frac{m\omega}{\hbar}} Y_0,$$

we find

$$R_0^2 = \frac{\hbar}{m\omega} \left(\hat{Q}_0^2 + \hat{P}_0^2 \right),$$

and the eigenvalues r_0^2 of R_0^2 are

$$r_0^2 = \frac{2\hbar}{m\omega} \left(p + \frac{1}{2} \right), \quad p = 0, 1, 2, \ldots \tag{11.120}$$

We have again found two harmonic oscillators. The first gives the value n of the Landau level, that is, the radius of the orbit, and the second gives the position of the center of the orbit. Let us assume that the particle is located in the plane inside a circle of radius r_0 and that $\rho^2 \ll r_0^2$. The values of p will then be limited to

$$p \le \frac{m\omega}{2\hbar} r_0^2 = \frac{m\omega}{2\pi\hbar} \mathcal{S},$$

where $\mathcal{S} = \pi r_0^2$ is the area of the circle. The degeneracy g of a Landau level n is given by the number of possible values of p:

$$g = \frac{m\omega}{2\pi\hbar} \mathcal{S} = \frac{|q|B}{2\pi\hbar} \mathcal{S}. \tag{11.121}$$

This result must be multiplied by a factor of 2 if we wish to take spin into account. To be rigorous, it is necessary to check that there is no extra degeneracy by showing that any operator commuting with H (or R^2) and R_0^2 is a function of H and R_0^2, so that it is not possible to find additional physical properties which are compatible and independent. The demonstration is similar to that for the simple harmonic oscillator (Exercise 11.5.2).

It is not difficult to generalize to the case of three-dimensional motion. Actually, since $A_z = 0$ it is sufficient to add to the Hamiltonian a term $P_z^2/2m$ whose eigenvalues are $p_z^2/2m$. The total energy is a function of n and p_z:

$$E_{n,p_z} = \hbar\omega \left(n + \frac{1}{2} \right) + \frac{p_z^2}{2m}. \tag{11.122}$$

If the vertical motion of the particle is limited to the range $0 \leq z \leq L_z$, the number of Landau levels in the range $[p_z, p_z + \Delta p_z]$ is

$$g = \frac{L_z}{2\pi\hbar} \frac{|q|B}{2\pi\hbar} S \Delta p_z. \qquad (11.123)$$

11.5 Exercises

11.5.1 Matrix elements of Q and P

1. Calculate the matrix elements $\langle n|Q|m\rangle$ and $\langle n|P|m\rangle$ of the operators Q and P in the basis $|n\rangle$.
2. Calculate the expectation value $\langle n|Q^4|n\rangle$ of Q^4 in the state $|n\rangle$. Hint: calculate

$$|\varphi_n\rangle = \left(a + a^\dagger\right)^2 |n\rangle$$

and $||\varphi_n||^2$.

11.5.2 Mathematical properties

1. Prove the commutation relations

$$[N, a^p] = -pa^p \quad \text{and} \quad [N, a^{\dagger p}] = pa^{\dagger p}.$$

Show that the only functions of a and $a^\dagger$ that commute with N are functions of N, and that the eigenvalues of N are nondegenerate.

2. Let $\mathcal{H}'$ be the subspace of $\mathcal{H}$ spanned by the vectors $|n\rangle$ and let $\mathcal{H}'_\perp$ be the orthogonal space: $\mathcal{H} = \mathcal{H}' \oplus \mathcal{H}'_\perp$. We use $\mathcal{P}'$ to denote the projector onto $\mathcal{H}'$. Show that $\mathcal{P}'$ commutes with a and $a^\dagger$ and prove, using the von Neumann theorem of Section 8.3.2, that either $\mathcal{P}' = 0$ or $\mathcal{P}' = I$. Since the first possibility is excluded, $\mathcal{P}' = I$ and the vectors $|n\rangle$ form a basis of $\mathcal{H}$.

11.5.3 Coherent states

1. Calculate $\langle z|P^2|z\rangle$ and $\langle z|H^2|z\rangle$ and derive the dispersions (11.39).
2. Let us study states $|\varphi(t)\rangle$ such that the expectation values of a and H have properties identical to the classical properties. First, if $\langle a\rangle(t) = \langle\varphi(t)|a|\varphi(t)\rangle$, show that

$$i\frac{d}{dt}\langle a\rangle(t) = \omega\langle a\rangle(t),$$

so that $\langle a\rangle(t)$ must satisfy the same differential equation (11.29) as $z(t)$. We define the complex number z_0 as

$$z_0 = \langle a\rangle(t = 0) = \langle\varphi(0)|a|\varphi(0)\rangle,$$

and so we then have the following solution of the differential equation for $\langle a\rangle(t)$:

$$\langle a\rangle(t) = z_0 e^{-i\omega t}.$$

3. The second condition concerns the expectation value of the Hamiltonian. Using (11.30) and adding the zero-point energy, we require that

$$\langle\varphi(0)|H|\varphi(0)\rangle = \hbar\omega\left(|z_0|^2 + \frac{1}{2}\right),$$

or, equivalently,

$$\langle a^\dagger a\rangle = \langle\varphi(0)|a^\dagger a|\varphi(0)\rangle = |z_0|^2.$$

Let the operator $b(z_0) = a - z_0$. Show that

$$\langle\varphi(0)|b^\dagger(z_0)b(z_0)|\varphi(0)\rangle = 0$$

and that

$$a|\varphi(0)\rangle = z_0|\varphi(0)\rangle.$$

The state $|\varphi(0)\rangle$ then is the coherent state $|z_0\rangle$.

4. Let $D(z)$ be a unitary operator (prove this!):

$$D(z) = \exp(-z^*a + za^\dagger).$$

Using (2.55), show that

$$D(z) = \exp\left(-\frac{1}{2}|z|^2\right)\exp(za^\dagger)\exp(-z^*a), \quad D(z)|0\rangle = |z\rangle.$$

5. *The wave function of a coherent state.* Express $D(z)$ as a function of the operators P and Q and calculate the wave function $\psi_z(q) = \langle q|z\rangle$. Hint: write $D(z)$ in the form

$$D(z) = f(z, z^*)\exp[c(z - z^*)Q]\,\exp[ic'(z + z^*)P],$$

find the constants c and c', and use the fact that P is the infinitesimal generator of translations (cf. Section 9.1.1):

$$\exp\left(-i\frac{Pl}{\hbar}\right)|q\rangle = |q + l\rangle.$$

Express $\psi_z(q)$ as a function of the wave function $\varphi_0(q)$ (11.23) of the ground state.

6. Show that an operator A is fully determined by its "diagonal elements" $\langle z|A|z\rangle$. Hint: use

$$\langle z|A|z\rangle = e^{-|z|^2}\sum_{n,m}\frac{A_{nm}z^n z^{*m}}{\sqrt{n!m!}}.$$

11.5.4 Coupling to a classical force

Coherent states can be used for a simple treatment of the quantum version of the forced harmonic oscillator. In elementary classical mechanics, the action of an external force $F(t)$ on a harmonic oscillator

$$m\ddot{q}(t) = -m\omega^2 q + F(t)$$

is carried over into the Hamiltonian by a coupling $-qF(t)$ between the displacement q and the force $F(t)$. In the quantum version a coupling between the displacement Q and the external force is added to the Hamiltonian of the simple harmonic oscillator (11.9):

$$W(t) = -Q\sqrt{\frac{2m\omega}{\hbar}}\,f(t), \tag{11.124}$$

where the multiplicative factor $f(t)$ is chosen so as to simplify the later expressions. Here Q is an operator, but $f(t)$ is a number which, with our definition (11.124), has the dimensions of energy. It is conventionally referred to as the *classical force* or the *classical source*. We shall use H_0 to denote the Hamiltonian (11.9) of the simple harmonic oscillator and $H(t)$ the total Hamiltonian:

$$H(t) = H_0 + W(t). \tag{11.125}$$

1. The problem greatly resembles that encountered in Section 9.6.3 (cf. (9.156)), and we can attempt to solve it using perturbation theory. However, it turns out that it is possible to calculate the time evolution defined by (11.125) exactly. Show that

$$H(t) = H_0 - \left(a + a^\dagger\right)f(t).$$

We rewrite the evolution operator $U(t) = U(t, t_0 = 0)$ (4.14) in the form

$$U(t) = U_0(t)U_I(t),$$

where $U_0(t) = \exp[-\mathrm{i}H_0 t/\hbar]$. In order to simplify the notation, we have chosen the reference time $t_0 = 0$ and we write $U(t)$ instead of $U(t, 0)$. Show that $U_I(t)$ satisfies the differential equation

$$\mathrm{i}\hbar\frac{\mathrm{d}U_I}{\mathrm{d}t} = U_0^{-1}W(t)U_0 U_I = W_I(t)U_I. \tag{11.126}$$

The operator $W_I(t)$,

$$W_I(t) = U_0^{-1}W(t)U_0 = \mathrm{e}^{\mathrm{i}H_0 t/\hbar}\,W(t)\mathrm{e}^{-\mathrm{i}H_0 t/\hbar}, \tag{11.127}$$

is the perturbation in the *Dirac picture* or the *interaction picture*, hence the subscript I. This picture is intermediate between those of Schrödinger and Heisenberg (cf. Section 4.2.5). The results (11.126) and (11.127) are quite general and do not depend on the specific form of H_0 or $W(t)$. In fact, we have reformulated the method of Section 9.6.3 in operator language.

2. Show that the operator a in the interaction picture is given by

$$a_I(t) = \mathrm{e}^{\mathrm{i}H_0 t/\hbar}\,a\,\mathrm{e}^{-\mathrm{i}H_0 t/\hbar} = a\mathrm{e}^{-\mathrm{i}\omega t}$$

given that $f(t)$ is a number and not an operator. Hint: cf. (11.67). Derive the differential equation for $U_I(t)$:

$$\mathrm{i}\hbar\frac{\mathrm{d}U_I}{\mathrm{d}t} = -\left(a\mathrm{e}^{-\mathrm{i}\omega t} + a^\dagger\mathrm{e}^{\mathrm{i}\omega t}\right)f(t)U_I(t) = W_I(t)U_I(t), \quad U_I(0) = I. \tag{11.128}$$

In (4.19) we already noted that (11.126) cannot be simply integrated as

$$U_I(t) = \exp\left(-\frac{\mathrm{i}}{\hbar}\int_0^t W_I(t')\mathrm{d}t'\right),$$

because in general the commutator $[W_I(t'), W_I(t'')] \neq 0$. In the present case this commutator is not zero but rather a multiple of the identity, which allows (11.128) to be integrated. From the identity (2.55) of Exercise 2.4.11, valid if $[A_i, A_j] = c_{ij}I$, derive

$$e^{A_n}e^{A_{n-1}}\cdots e^{A_1} = e^{A_n+\cdots+A_1}\, e^{\frac{1}{2}\sum_{j>i}[A_j, A_i]}.$$

3. Divide the interval $[0, t]$ into n infinitesimal intervals Δt and, starting from

$$U_I(t) \simeq \prod_{j=1}^{n}\left[\exp\left(-\frac{i}{\hbar}W_I(t_j)\Delta t\right)\right],$$

show that

$$U_I(t) \simeq \exp\left(-\frac{i}{\hbar}\Delta t\sum_{j=1}^{n}W_I(t_j)\right)\exp\left(-\frac{(\Delta t)^2}{2\hbar^2}\sum_{t_j>t_i}[W_I(t_j), W_I(t_i)]\right).$$

What is the commutator $[W_I(t'), W_I(t'')]$? Show that we obtain $U_I(t)$ by taking the limit $\Delta t \to 0$:

$$\Delta t\sum_{j=1}^{n}W_I(t_j) \to \int_0^t dt'\, W_I(t') = -\int_0^t dt'\,\left(a\,e^{-i\omega t'} + a^\dagger e^{i\omega t'}\right)f(t')$$

$$= -\hbar a z^*(t) - \hbar a^\dagger z(t),$$

where the complex number $z(t)$ is defined as

$$z(t) = \frac{1}{\hbar}\int_0^t dt'\, e^{i\omega t'}\, f(t').$$

4. Obtain the $\Delta t \to 0$ limit of

$$(\Delta t)^2\sum_{t_j>t_i}[W_I(t_j), W_I(t_i)]$$

and show that

$$U_I(t) = \exp\left[i\left(az^*(t) + a^\dagger z(t)\right)\right]\exp\left[-\frac{X}{2\hbar^2}\right],$$

$$X = \int_0^t dt'\int_0^t dt''\, e^{-i\omega(t'-t'')}\, f(t')f(t'')\varepsilon(t'-t''),$$

where $\varepsilon(t)$ is the sign function: $\varepsilon(t) = 1$ if $t > 0$, $\varepsilon(t) = -1$ if $t < 0$.

5. This result can be written in a more convenient form. Show that

$$\exp\left[i\left(az^*(t) + a^\dagger z(t)\right)\right] = \exp\left[ia^\dagger z(t)\right]\exp\left[iaz^*(t)\right]\exp\left[-\frac{1}{2}z(t)z^*(t)\right]$$

and, noting that $2\theta(t) - \varepsilon(t) = 1$, where $\theta(t)$ is the Heaviside function, show that

$$U_I(t) = e^{ia^\dagger z(t)}\, e^{iaz^*(t)}e^{-Y/\hbar^2}, \tag{11.129}$$

$$Y = \int_0^t dt'\int_0^t dt''\, e^{-i\omega(t'-t'')}\, f(t')f(t'')\theta(t'-t''). \tag{11.130}$$

Verify by explicit calculation that (11.129)–(11.130) obey the original differential equation (11.128).

6. Let us study the case where the initial state at time $t = 0$ is an eigenstate $|n\rangle$ of H_0 assuming that the force acts only during a finite time interval $[t_1, t_2]$ and that we choose to observe the oscillator at a time $t > t_2$, where $0 < t_1 < t_2 < t$. Defining the Fourier transform $\tilde{f}(\omega)$ of $f(t)/\hbar$,

$$\tilde{f}(\omega) = \frac{1}{\hbar} \int_{-\infty}^{\infty} dt' \, e^{i\omega t'} f(t') = \frac{1}{\hbar} \int_{t_1}^{t_2} dt' \, e^{i\omega t'} f(t'),$$

and using the Fourier representation of the θ function,

$$\theta(t) = \lim_{\eta \to 0^+} \int_{-\infty}^{+\infty} \frac{dE}{2i\pi} \frac{e^{itE}}{E - i\eta} \quad \text{and} \quad \frac{1}{E - i\eta} = \mathbb{P}\frac{1}{E} + i\pi\delta(E), \tag{11.131}$$

where $\mathbb{P}$ designates the principal part, show that Y is given by

$$\frac{1}{\hbar^2} Y = \mathbb{P} \int \frac{dE}{2i\pi E} |\tilde{f}(E - \omega)|^2 + \frac{1}{2} |\tilde{f}(\omega)|^2$$

$$= i\phi + \frac{1}{2} |\tilde{f}(\omega)|^2.$$

7. Show that the final result for $U_{\mathrm{I}}(t)$ is independent of t for $t > t_2$:

$$U_{\mathrm{I}}(t) = \exp\left(ia^\dagger \tilde{f}(\omega)\right) \exp\left(ia\tilde{f}^*(\omega)\right) \exp(-i\phi) \exp\left(-\frac{1}{2}|\tilde{f}(\omega)|^2\right). \tag{11.132}$$

Show that if the oscillator is in its ground state at time $t = 0$, the final state vector is a coherent state:

$$U_{\mathrm{I}}(t)|0\rangle = e^{-i\phi} |i\tilde{f}(\omega)\rangle. \tag{11.133}$$

Show that the probability of observing a final state $|m\rangle$ is given by a Poisson law (11.34):

$$p(m) = \frac{\left(|\tilde{f}(\omega)|^2\right)^m \exp\left(-|\tilde{f}(\omega)|^2\right)}{m!}. \tag{11.134}$$

8. Generalize the above results to the coupling (11.88) of a quantized electromagnetic field to a classical source $\vec{J}_{\mathrm{em}}(\vec{r}, t)$ by writing the perturbation in the form (see Footnote 17)

$$W(t) = -\int \frac{d^3k}{(2\pi)^3} \vec{A}_{\vec{k}} \cdot \vec{J}_{\mathrm{em}}(\vec{k}, t).$$

11.5.5 Squeezed states

Replacing a and $a^\dagger$ by their expression (11.100) as functions of b and $b^\dagger$, calculate

$$\langle\tilde{z}| \left(a + a^\dagger\right) |\tilde{z}\rangle = \tilde{z}(\lambda^* - \mu^*) + \tilde{z}^*(\lambda - \mu)$$

and

$$\langle\tilde{z}|(a + a^\dagger)^2|\tilde{z}\rangle = \langle\tilde{z}| \left[a^2 + (a^\dagger)^2 + 2a^\dagger a + I\right] |\tilde{z}\rangle.$$

Show that

$$(\Delta_{\tilde{z}}Q)^2 = \frac{1}{4}(1 + 2|\mu|^2 - \lambda^*\mu - \lambda\mu^*) = \frac{1}{4}|\lambda - \mu|^2.$$

Also calculate $\Delta_{\bar{z}} P$. Writing

$$\lambda = \cosh\theta, \quad \mu = \sinh\theta\, e^{i\phi},$$

show that

$$|\lambda - \mu|^2\, |\lambda + \mu|^2 = \cosh^4\theta - 2\cosh^2\theta \sinh^2\theta \cos 2\phi + \sinh^4\theta$$

and derive

$$\Delta_{\bar{z}} Q\, \Delta_{\bar{z}} P = \frac{1}{4}$$

if $\phi = 0$ or $\phi = \pi$.

11.5.6 Zero-point energy of the Debye model

1. In the Debye model it is assumed that the dispersion law $\omega(k) = c_s|k|$ is valid for all $k \leq k_D$. Using

$$\frac{L}{2\pi}\, dk = \frac{L}{2\pi c_s}\, d\omega,$$

show that $0 \leq \omega \leq \omega_D$ with $\omega_D = c_s k_D = 2\pi c_s / l$. The quantity ω_D is called the *Debye frequency*. Derive the zero-point energy

$$E_0 = \frac{1}{4} N\hbar\omega_D.$$

2. Generalize to three dimensions and show that in this case

$$E_0 = \frac{9}{8} N\hbar\omega_D.$$

11.5.7 The scalar and vector potentials in Coulomb gauge

We can write the expression (11.76) giving the instantaneous Coulomb potential formally as

$$\overline{V} = -\frac{1}{\varepsilon_0}\, (\nabla^2)^{-1} \rho_{em},$$

which is the inverse of $\nabla^2 \overline{V} = -\rho_{em}/\varepsilon_0$. Use the second Maxwell equation (11.71) in the form

$$c^2 \vec{\nabla} \times (\vec{\nabla} \times \vec{A}) = \frac{\partial}{\partial t}\left(-\frac{\partial \vec{A}}{\partial t} - \vec{\nabla}\overline{V}\right) + \frac{1}{\varepsilon_0}\vec{j}_{em}$$

to show that $\vec{A}$ satisfies

$$\frac{1}{c^2}\frac{\partial^2 \vec{A}}{\partial t^2} - \nabla^2\vec{A} = \mu_0 \vec{j}_{em}^{\,T},$$

where the "transverse electromagnetic current" $\vec{j}_{em}^{\,T}$ is

$$\vec{j}_{em}^{\,T} = \vec{j}_{em} - \vec{\nabla}\cdot[(\nabla^2)^{-1}(\vec{\nabla}\cdot\vec{j}_{em})].$$

11.5.8 Commutation relations and Hamiltonian of the electromagnetic field

1. Taking $t = 0$, evaluate the commutator (11.85):

$$[A_i(\vec{r}), -\varepsilon_0 E_j(\vec{r}')] = i\hbar \int \frac{d^3k}{(2\pi)^3} \left(\delta_{ij} - \hat{k}_i \hat{k}_j \right) e^{i\vec{k}\cdot(\vec{r}-\vec{r}')}.$$

Show that these relations are also valid for the *equal-time* commutator:

$$[A_{Hi}(\vec{r}, t), -\varepsilon_0 E_{Hj}(\vec{r}', t)].$$

2. Derive the commutation relations between E_{Hi} and B_{Hj}.

3. Express the Hamiltonian (11.78) with $\vec{E} \to \vec{E}$ and $\vec{B} \to \vec{B}$ as a function of the operators $a_{\vec{k}s}$ and $a^\dagger_{\vec{k}s}$ at $t = 0$. Hint: for a polarization s write

$$\vec{E}_s = i\sqrt{\frac{\hbar}{2\varepsilon_0 V}} \sum_{\vec{k}} \sqrt{\omega_k} \left(a_{\vec{k}s}\vec{e}_s - a^\dagger_{-\vec{k}s}\vec{e}_s^{\,*} \right) e^{i\vec{k}\cdot\vec{r}}$$

$$= \sum_{\vec{k}} \vec{E}_{\vec{k}s}\, e^{i\vec{k}\cdot\vec{r}}$$

and use the Parseval relation

$$\int d^3r\, \vec{E}_s^2 = V \sum_{\vec{k}} \vec{E}_{\vec{k}s} \cdot \vec{E}_{-\vec{k}s}.$$

Proceed in the same way for $\vec{B}$ noting that

$$(\vec{e}_s \times \hat{k}) \cdot (\vec{e}_s \times \hat{k}) = 1.$$

11.5.9 Quantization in a cavity

1. We consider the classical scalar field $\varphi(\vec{r}, t)$ of Section 11.3.2 in the three-dimensional case, assuming that this field is enclosed in a cavity. Let ω_j be an eigenfrequency of the cavity and

$$\varphi_j(\vec{r}, t) = \varphi_j(\vec{r}) \cos(\omega t - \phi)$$

be the corresponding field, which obeys the wave equation (11.59) with appropriate boundary conditions, for example, vanishing on the cavity walls: $\varphi_j(\vec{r}) = 0$ at the walls. The eigenfunctions $\varphi_j(\vec{r})$ are assumed to be real and form a complete orthogonal set:

$$\int d^3r\, \varphi_j(\vec{r})\varphi_k(\vec{r}) = \delta_{jk}, \quad \sum_j \varphi_j(\vec{r})\varphi_j(\vec{r}') = \delta(\vec{r} - \vec{r}').$$

Show that the quantized field in the Heisenberg picture

$$\Phi_H(\vec{r}, t) = \sqrt{\frac{\hbar}{2\mu}} \sum_j \sqrt{\frac{1}{\omega_j}} \left(a_j e^{-i\omega_j t} + a_j^\dagger e^{i\omega_j t} \right) \varphi_j(\vec{r}) \tag{11.135}$$

satisfies the equal-time commutation relations

$$[\Phi_H(\vec{r}, t), \Pi_H(\vec{r}', t)] = [\Phi_H(\vec{r}, t), \mu\,\dot{\Phi}_H(\vec{r}', t)] = i\hbar\delta(\vec{r} - \vec{r}')I$$

if the operators a_j and $a_j^\dagger$ satisfy the commutation relations $[a_j, a_k^\dagger] = \delta_{jk}I$.

2. *Application to dimension* $d = 1$. The field is contained in the interval $[0, L]$ with vanishing boundary conditions at the ends $\varphi(x = 0) = \varphi(x = L) = 0$. Show that in this case the eigenmodes are labeled by a wave vector k:

$$\varphi_k(x) = \sqrt{\frac{2}{L}} \sin kx, \quad k = \frac{\pi j}{L}, \quad j = 1, 2, \ldots$$

Verify the orthogonality and completeness relations:

$$\frac{2}{L} \int_0^L dx \sin kx \sin k'x = \delta_{kk'}, \quad \frac{2}{L} \sum_k \sin kx \sin kx' = \delta(x - x').$$

Derive the expression for $\Phi_H(x, t)$.

3. *The electromagnetic field.* We take the case of three dimensions assuming that the field is enclosed in a cavity which is a parallelepiped of sides L_x, L_y, L_z and volume $V = L_x L_y L_z$. Show that instead of (11.82) we have

$$\vec{E}_H(\vec{r}, t) = i \sqrt{\frac{4\hbar}{\varepsilon_0 V}} \sum_{\vec{k}} \sum_{s=1}^{2} \sqrt{\omega_k} \left[a_{\vec{k}s} \vec{e}_s(\hat{k}) e^{-i\omega_k t} - a_{\vec{k}s}^\dagger \vec{e}_s^*(\hat{k}) e^{i\omega_k t} \right] \sin(xk_x) \sin(yk_y) \sin(zk_z)$$

(11.136)

with

$$\vec{k} = \left(\frac{\pi}{L_x} n_x, \frac{\pi}{L_y} n_y, \frac{\pi}{L_z} n_z \right), \quad n_x, n_y, n_z = 1, 2, \ldots$$

11.5.10 Current conservation in the presence of a magnetic field

Using the Schrödinger equation in a magnetic field, show that the current $\vec{j}$ (11.107) obeys the continuity equation

$$\frac{\partial \rho}{\partial t} + \vec{\nabla} \cdot \vec{j} = 0.$$

11.5.11 Non-Abelian gauge transformations

The fundamental interactions of elementary particle physics are all based on non-Abelian gauge theories, which we shall define in an elementary case by generalizing the gauge transformation (11.104). Omitting the time dependence in order to simplify the discussion, we shall assume that the wave function $\varphi(\vec{r})$ is a two-component vector $\Phi(\vec{r}) = [\varphi_1(\vec{r}), \varphi_2(\vec{r})]$ in a two-dimensional complex Hilbert space and that in this space there exists a symmetry operation called an *internal symmetry* that leaves the physics invariant:

$$\Phi(\vec{r}) \to \Phi'(\vec{r}) = \Omega\Phi \quad \text{or} \quad \varphi_\alpha' = \sum_{\beta=1}^{2} \Omega_{\alpha\beta}\varphi_\beta,$$

generalizing (11.103). Ω is a 2×2 unitary matrix with unit determinant, i.e., an $SU(2)$ matrix. The symmetry is called *gauge symmetry* and the $SU(2)$ group is the *gauge group*. In general, the gauge group is a compact Lie group. The gauge group of electromagnetism is the group of phase transformations (11.103), denoted $U(1)$, which is Abelian:

electromagnetism is an Abelian gauge theory. When the gauge group is non-Abelian, the gauge theory will be termed non-Abelian. The gauge groups of the Standard Model of elementary particle physics are the groups $SU(2) \times U(1)$ for the electroweak interactions and $SU(3)$ for quantum chromodynamics. These are all non-Abelian groups.

According to the results of Exercise 3.3.6, the matrix Ω can be written as a function of the Pauli matrices as

$$\Omega = \exp\left(-i\,q \sum_{a=1}^{3} \Lambda_a \frac{1}{2} \sigma_a\right).$$

When the functions Λ_a are independent of $\vec{r}$, we are dealing with a *global* gauge symmetry, and if the Λ_a are functions of $\vec{r}$, we have a *local* gauge symmetry. In order to simplify the notation, we use a system of units in which $\hbar = m = 1$.

1. The analog of the vector potential of electromagnetism is a vector field with components $\vec{A}_a$ in the internal symmetry space. The matrix $\vec{\mathbf{A}}$ is defined as

$$\vec{\mathbf{A}} = \sum_{a=1}^{3} \vec{A}_a \frac{1}{2}\, \sigma_a,$$

and it simultaneously has the ordinary components $i = (x, y, z)$ and components a in the internal symmetry space: $\vec{\mathbf{A}} = \{A_{ia}\}$. The expression for the current $\vec{\jmath}$ generalizes (11.113):

$$\vec{\jmath} = \mathrm{Re}\left[\Phi^\dagger(-i\vec{\nabla} - q\vec{\mathbf{A}})\Phi\right] = \mathrm{Re}\left[\Phi^\dagger(-i\vec{\mathbf{D}}\Phi)\right],$$

where

$$\vec{\mathbf{D}} = -i\vec{\nabla} - q\vec{\mathbf{A}}$$

is the covariant derivative. Show that the gauge transformation $\Phi \to \Phi'$ leaves $\vec{\jmath}$ invariant if this gauge transformation is global with the condition that $\vec{\mathbf{A}}$ is also transformed into $\vec{\mathbf{A}}'$:

$$\vec{\mathbf{A}}' = \Omega\vec{\mathbf{A}}\Omega^{-1}.$$

If the gauge transformation is local, show that invariance of the current

$$\vec{\jmath} = \vec{\jmath}' = \mathrm{Re}\left[\Phi'^\dagger(-i\vec{\nabla} - q\vec{\mathbf{A}}')\Phi'\right]$$

implies the transformation law $\vec{\mathbf{A}} \to \vec{\mathbf{A}}'$:

$$\vec{\mathbf{A}}' = \Omega\vec{\mathbf{A}}\Omega^{-1} - \frac{i}{q}(\vec{\nabla}\Omega)\Omega^{-1}.$$

Recover the transformation law (11.74) in the Abelian case.

2. We choose an infinitesimal gauge transformation: $|q\Lambda_a(\vec{r})| \ll 1$. Derive the transformation law for $\vec{A}_a$:

$$\delta\vec{A}_a = \vec{A}_a' - \vec{A}_a = -\vec{\nabla}\Lambda_a + q\sum_{b,c} \varepsilon_{abc}\Lambda_b\vec{A}_c.$$

The (crucial) difference from the Abelian case is that the *gauge field* $\vec{A}_a$ depends nontrivially on the internal symmetry index a of the gauge group.[26] In electromagnetism the photons do not carry charge, but the gauge bosons of a non-Abelian theory do: they are "charged" because they carry the quantum numbers of the internal symmetry.

3. Show that if Φ obeys the time-independent Schrödinger equation

$$\frac{1}{2}\left(-i\nabla - q\vec{A}\right)^2 \Phi = \frac{1}{2}(-i\hbar\vec{D})^2\Phi = E\Phi,$$

we have the same for Φ' if the field $\vec{A}'$ is used.

11.5.12 The Casimir effect

Owing to quantum fluctuations of the electromagnetic field, there is an attractive force between two parallel conducting plates separated by a distance L, even if the two plates are located in a vacuum and are electrically neutral. This is known as the *Casimir effect*. We assume that the dimensions of the plates are very large compared to their separation L.

1. Using a dimensional argument, show that the force P on a plate per unit surface area is of the form

$$P = A\frac{\hbar c}{L^4},$$

where A is a numerical coefficient. The surprise is that $A \neq 0$!

2. The two plates are rectangles parallel to the plane xOy and separated by a distance L, the lengths of their sides are L_x and L_y with $L_x, L_y \gg L$, and their area is $S = L_x L_y$. We choose periodic conditions along the axes Ox and Oy and define the wave vector $\vec{k}$ of xOy as

$$\vec{k} = \left(\frac{2\pi n_x}{L_x}, \frac{2\pi n_y}{L_y}\right),$$

where n_x and n_y are relative integers, $n_x, n_y \in \mathbb{Z}$. Show that if the plates are perfect conductors, then the possible frequencies of standing waves have the form

$$\omega_n(\vec{k}) = c\sqrt{\frac{\pi^2 n^2}{L^2} + \vec{k}^2}, \quad n = 0, 1, 2, \ldots$$

We recall that for a perfect conductor the transverse component of the electric field vanishes at the surface of the metal.[27] Explain why for $n = 0$ there is only one possible polarization mode.

3. Show that the zero-point energy (11.87) is

$$E_0(L) = \frac{\hbar}{2}\left(2\sum_{n,\vec{k}}' \omega_n(\vec{k})\right),$$

where

$$\sum_{n,\vec{k}}' = \frac{1}{2}\sum_{n=0,\vec{k}} + \sum_{n\geq 1,\vec{k}}.$$

---

[26] Since the field $\vec{A}$ is a vector field, the associated particles have spin 1, like the photon, and are called *gauge bosons*. The photon, Z^0 and $W^{\pm}$ are the gauge bosons of the electroweak interactions, and the gluons are those of chromodynamics.
[27] See, for example, Jackson [1999], Section 8.1.

4. It is necessary to take into account the fact that there is no such thing as a perfect conductor. The approximation that the conductor is perfect is excellent at low frequencies, but at high frequencies any real conductor becomes transparent. It is therefore necessary to modify the zero-point energy to include a cutoff $\chi(\omega/\omega_c)$, where $\chi(0) = 1$ and $\lim_{u\to\infty} \chi(u) = 0$; $\chi(u)$ is a regular function which decreases from unity at $u = 0$ to zero for $u \to \infty$. Show that

$$E_0(L) = \frac{\hbar S}{(2\pi)^2} \sum_{n=0}^{\infty}{}' \int d^2k\, \omega_n(\vec{k}) \chi\left(\frac{\omega_n(\vec{k})}{\omega_c}\right)$$

$$= \frac{\hbar S}{2\pi c^2} \sum_{n=0}^{\infty}{}' \int_{\omega_n}^{\infty} d\omega\, \omega^2 \chi\left(\frac{\omega}{\omega_c}\right), \qquad \omega_n = \frac{\pi c n}{L}.$$

Owing to the cutoff, this energy is finite.

5. Calculate the pressure on the right-hand plate

$$P_{\text{int}} = -\frac{1}{S}\frac{dE_0}{dL} = -\frac{\pi^2 \hbar c}{2L^4} \sum_n{}' g(n),$$

where

$$g(n) = n^3 \chi\left(\frac{\omega_n}{\omega_c}\right).$$

To obtain the total pressure on this plate it is necessary to subtract the pressure in the opposite direction due to the vacuum outside the space between the two plates. Calculate the corresponding energy and find the pressure

$$P_{\text{ext}} = -\frac{\pi^2 \hbar c}{2L^4} \int_0^{\infty} dn\, g(n).$$

The total pressure on the plate is $P_{\text{tot}} = P_{\text{int}} - P_{\text{ext}}$. Use the Euler–Maclaurin formula

$$\sum_{n=0}^{\infty}{}' g(n) - \int_0^{\infty} g(n) = -\frac{1}{12} g'(0) + \frac{1}{6!} g'''(0) + \cdots$$

to show that the result in the limit where the cutoff factor becomes unity is

$$P_{\text{tot}} = -\frac{\pi^2}{240}\frac{\hbar c}{L^4}.$$

This pressure is attractive, and moreover it is finite. By carefully taking into account all the physical effects, we have derived a quantity which is *finite and measurable* from a quantity which is a priori infinite, the zero-point energy.[28]

11.5.13 *Quantum computing with trapped ions*

1. Trapped ions may turn out to be a promising technique for building a quantum computer. In an experiment performed by a group in Innsbruck, $^{40}\text{Ca}^+$ ions are confined in an approximately one-dimensional harmonic trap.[29] The ground state $S_{1/2} = |g\rangle$ is identified with the state $|0\rangle$

---

[28] A recent reference is U. Mohiden and A. Roy, Precision measurement of the Casimir force from 0.1 to 0.9 μm, *Phys. Rev. Lett.* **81**, 4549 (1998). The accuracy with which the Casimir effect has been measured is of order 1%, and the measurements confirm the validity of the theoretical expression.

[29] F. Schmid-Kaler *et al.*, Realization of the Cirac-Zoller controlled-NOT gate, *Nature* **422**, 408 (2003).

of quantum computation (Section 6.4.2), and the excited state $D_{5/2} = |e\rangle$ with $|1\rangle$. The excited state is long-lived (~ 1 s) because the transition $D_{5/2} \to S_{1/2}$ is an electric quadrupole transition. Let us first consider a single ion in the trap. Its Hamiltonian is approximately

$$H_{\text{trap}} = \frac{1}{2M} p_z^2 + \frac{1}{2} M\omega_z^2 z^2,$$

where M is the ion mass and ω_z the frequency of the trap. In the absence of applied external field, one may write the total Hamiltonian as

$$H_0 = -\frac{1}{2} \hbar\omega_0\sigma_z + \hbar\omega_z a^\dagger a,$$

where ω_0 is the frequency of the transition $|0\rangle \leftrightarrow |1\rangle$. One applies to the ions the electric field of a laser wave

$$\vec{E} = E_1\hat{x}\cos(\omega t - kz - \phi),$$

and the Rabi frequency is denoted ω_1. The coupling between the field and the ion is

$$H_{\text{int}} = -\hbar\omega_1\sigma_x\cos(\omega t - kz - \phi)$$

and the state vector in the interaction picture (see Exercise 5.5.6 or 11.5.4) is

$$|\tilde{\varphi}(t)\rangle = e^{iH_0 t/\hbar}|\varphi(t)\rangle \quad |\tilde{\varphi}(t=0)\rangle = |\varphi(t=0)\rangle.$$

Show that the Hamiltonian $\tilde{H}_{\text{int}}$ in the interaction picture is

$$\tilde{H}_{\text{int}}(t) = e^{iH_0 t/\hbar} H_{\text{int}} e^{-iH_0 t/\hbar},$$

and that in the rotating wave approximation, with $\sigma_\pm = (\sigma_x \pm i\sigma_y)/2$

$$\tilde{H}_{\text{int}} \simeq -\frac{\hbar}{2}\omega_1\left[\sigma_+ e^{i(\delta t - \phi)} e^{-ik\bar{z}} + \sigma_- e^{-i(\delta t - \phi)} e^{ik\bar{z}}\right]$$

where $\delta = \omega - \omega_0$ is as usual the detuning. Since

$$z = \sqrt{\frac{\hbar}{2M\omega_z}}\left(a + a^\dagger\right)$$

$\exp(\pm ik\bar{z})$ couples the internal levels $|0\rangle$ and $1\rangle$ to the vibrational levels in the trap. The internal levels will be labeled n, $n = 0, 1$, the vibrational levels m, $m = 0, 1, 2, \ldots$ and the product state $|n, m\rangle$

2. Let us define the dimensionless Lamb–Dicke parameter η by

$$\eta = k\sqrt{\frac{\hbar}{2M\omega_z}}.$$

Give the physical interpretation of η. Consider two vibrations levels m and $m + m'$ and show that the Rabi frequency $\omega_1^{m \to m+m'}$ is given by

$$\omega_1^{m \to m+m'} = \omega_1|\langle m + m'|e^{i\eta(a+a^\dagger)}|m\rangle|.$$

3. We limit ourselves to the case $m' = \pm 1$. Transitions corresponding to frequencies $\omega = \omega_0 + \omega_z$ ($\omega = \omega_0 - \omega_z$) are called blue sideband (red sideband) transitions, while transitions with $\omega = \omega_0$ are called carrier transitions. We also assume that $\eta \ll 1$ and work to first order in η. Write the expression of $\tilde{H}_{int}$ on the two sidebands and show that for the blue one

$$\tilde{H}_{int}^+ = \frac{i}{2}\,\eta\,\omega_1\sqrt{m+1}\Big[\sigma_+ a_b\,e^{-i\phi} - \sigma_- a_b^\dagger e^{i\phi}\Big],$$

while for the red one

$$\tilde{H}_{int}^- = \frac{i}{2}\,\eta\,\omega_1\sqrt{m}\Big[\sigma_+ a_r^\dagger e^{-i\phi} - \sigma_- a_r\,e^{i\phi}\Big].$$

The operators $a_b \ldots a_r^\dagger$ are defined so as to preserve the norm of the state vectors

$$a_b = \frac{a}{\sqrt{m+1}} \qquad a_b^\dagger = \frac{a^\dagger}{\sqrt{m+1}} \qquad a_r = \frac{a}{\sqrt{m}} \qquad a_r^\dagger = \frac{a^\dagger}{\sqrt{m}}.$$

4. The levels used in the following discussion are $|0,0\rangle$, $|0,1\rangle$, $|1,0\rangle$, $|1,1\rangle$ and $|1,2\rangle$. Draw the level scheme and identify the blue sideband and the red sideband transitions. Show that the operator

$$R_{\alpha\beta}^+ = R^+(\alpha, \pi/2)\,R^+(\beta, 0)\,R^+(\alpha, \pi/2)\,R^+(\beta, 0),$$

is equal to $-I$ for $\alpha = \pi$ whatever β, or $\beta = \pi$ whatever α. $R^\pm(\theta, \phi)$ is a rotation by θ about an axis in the xOy plane which makes an angle ϕ with the x axis and which uses the blue (+) or red (−) sideband. Use the fact that the Rabi frequency for the transition $|0,1\rangle \leftrightarrow |1,2\rangle$ is $\sqrt{2}$ times that for the $|0,0\rangle \leftrightarrow |1,1\rangle$ transition to determine α and β in such a way that $R_{\alpha\beta}^+ = -I$ for both transitions. Show that a cZ gate (up to a sign) has been built in the preceding operation (a cZ gate is obtained from (6.73) by the substitution $\sigma_x \to \sigma_z$)

$$|0,0\rangle \leftrightarrow -|0,0\rangle \qquad |0,1\rangle \leftrightarrow -|0,1\rangle \qquad |1,0\rangle \leftrightarrow +|1,0\rangle \qquad |1,1\rangle \leftrightarrow -|1,1\rangle.$$

5. It is now necessary to "transfer" the cZ gate to the computational basis of product states $|n, n'\rangle$, $n, n' = 0, 1$ being ground and excited states of two different ions. Show that the desired result is obtained by sandwiching the rotation operator $R_{\alpha\beta}^{+(1)}$ on ion number one using the blue sideband between two rotations by π on ion number two using the red sideband

$$\big[R^{-(2)}(\pi, \pi/2)\big]R_{\alpha\beta}^{+(1)}\big[R^{-(2)}(-\pi, \pi/2)\big],$$

A slightly more complicated operation allows one to build a cNOT gate.

11.6 Further reading

The diagonalization of the Hamiltonian of the one-dimensional harmonic oscillator by the algebraic method is classic and can be found in any quantum mechanics textbook. The theory of coherent states is discussed by Cohen-Tannoudji *et al.* [1977], Complement G_V. Applications of phonons in thermodynamics are given by Le Bellac *et al.* [2004], Chapter 4. Additional material on the quantization of the scalar field and the electromagnetic field can be found in C. Itzykson and J.-B. Zuber, *Quantum Field Theory*, New York:

McGraw-Hill (1980), Chapter 3; Le Bellac [1991], Chapter 9; Grynberg *et al.* [2005], Chapter V; or Weinberg [1995], Chapter 8. Fluctuations of the electromagnetic field and squeezed states are treated by Ballentine [1998], Chapter 19; by Grynberg *et al.* [2005], Chapter V and Complement V.1; and by Mandel and Wolf [1995], Chapters 10–12. Feynman *et al.* [1965], Vol. III, Chapter 21 gives a physical discussion of the difference between the velocity and $\vec{p}/m$ in the presence of an electromagnetic field. The Landau levels are discussed by Cohen-Tannoudji *et al.* [1977], Complement E_{VI}, and applications to solid-state physics can be found in K. Huang, *Statistical Mechanics*, New York: Wiley (1963), Chapter 11.

12

Elementary scattering theory

Up to now we have mainly studied bound states, except for the brief mention of one-dimensional scattering in Section 9.4. However, essential information on interactions between particles, atoms, molecules, etc., as well as on the structure of composite objects, can be obtained from scattering experiments. Bound states – when they exist, which is not always the case – give only partial information on such interactions, whereas it is nearly always possible to perform scattering experiments. In this chapter we shall limit ourselves to potential scattering, which can be used to describe elastic collisions of two particles of masses m_1 and m_2. Indeed, in the center-of-mass frame the problem is reduced to that of a particle of mass $m = (m_1 m_2)/(m_1 + m_2)$ in a potential (Exercise 8.5.6).[1]

In Sections 12.1 and 12.2 we develop the elementary formalism of elastic scattering theory with emphasis on the low-energy limit, which plays an extremely important role in practice. In Section 12.3 we generalize the formalism to the inelastic case; more precisely, we examine the effect of inelastic channels on elastic scattering. Finally, Section 12.4 is devoted to some more formal aspects of scattering theory.

12.1 The cross section and scattering amplitude

12.1.1 The differential and total cross sections

A scattering experiment is shown schematically in Fig. 12.1. A beam of particles of mass m_1 and well-defined momentum moving along the z axis collides with a target composed of particles of mass m_2. To simplify the discussion, we assume that $m_1 \ll m_2$ and we neglect the recoil of the target in the collision. In general, it is necessary to go from the laboratory frame to the center-of-mass frame via a simple kinematic transformation (Exercise 8.5.6). A fraction of the incident particles is deflected in the collision with the target, and these particles are recorded by detectors placed at polar angles (θ, ϕ), called the *scattering angles* and collectively denoted by Ω. Let ΔS be the surface area of a detector located a distance r from the target. This detector is seen from the target as subtending a solid angle $\Delta \Omega \simeq \Delta S / r^2$. We assume that the density n_t of target particles is

---

[1] In ring accelerators such as LEP (the Large Electron–Positron collider), the $e^+ - e^-$ accelerator operating at CERN between 1990 and 2000, the center-of-mass frame is the same as the laboratory frame.

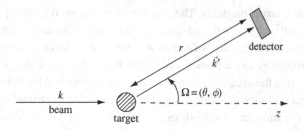

Fig. 12.1. Schematic view of a scattering experiment.

low enough that multiple collisions can be neglected. Under these conditions, the number of particles $\Delta\mathcal{N}(\Omega)$ per unit time and unit target volume that have undergone a collision and are recorded by the detector is proportional to

- the flux $\mathcal{F}$ of incident particles, that is, the number of particles crossing a unit surface perpendicular to Oz per unit time: $\mathcal{F} = n_i v$, where n_i is the incident particle density and v is the particle speed;
- the density n_t of target particles;
- the solid angle $\Delta\Omega$ the detector subtends as seen from the target (Fig. 12.1). In what follows we shall assume that this solid angle is infinitesimal: $\Delta\Omega \to d\Omega$.

We then have

$$d\mathcal{N}(\Omega) = \mathcal{F} \, n_t \frac{d\sigma}{d\Omega} \, d\Omega. \tag{12.1}$$

The proportionality factor $d\sigma/d\Omega$ is called the *differential cross section* of the scattering. Dimensional analysis shows that $d\sigma/d\Omega$ has the dimensions of a surface and is measured in m^2 per steradian. By integrating over Ω we obtain the *total cross section* σ_{tot}:

$$\boxed{\sigma_{tot} = \int d\Omega \, \frac{d\sigma}{d\Omega}}. \tag{12.2}$$

The product $\mathcal{F} n_t \sigma_{tot}$ is equal to the number of collisions recorded per second for a target of unit volume. The total cross section is a priori a function of the speed v of the incident particle, or, equivalently, its energy. The differential cross section is a function of the energy and the angles θ and ϕ. When the physical problem is invariant under rotation about the z axis,[2] the differential cross section depends only on θ.

Let us give an intuitive illustration of the idea of cross section by studying a collision between two billiard balls of radii R_1 and R_2 in classical mechanics. First we assume that the incident particles (here, the billiard balls) have radius R and the target particles are point particles. During one second an incident particle sweeps out a volume $\pi R^2 v$, and so

---

[2] Such invariance does not occur if, for example, the potential is not rotationally invariant or the target particles have spin polarized along an axis perpendicular to Oz and the scattering is spin-dependent.

it encounters $n_t \pi R^2 v$ target particles. The number of collisions recorded per second in the experiment is $n_i n_t \pi R^2 v = \mathcal{F} n_t \pi R^2$, which gives the total cross section $\sigma_{tot} = \pi R^2$. Geometrically, this is the area of a disk of radius R. This is also the cross section for the scattering of point particles by target particles of radius R, in which case the geometrical origin of πR^2 is obvious: it is the area of the target as seen by an incident particle. The total cross section for incident particles of radius R_1 and target particles of radius R_2 can be derived from this result: the number of collisions is the same as if the incident particles were point particles and the target particles had radius $(R_1 + R_2)$. The total cross section then is

$$\sigma_{tot} = \pi (R_1 + R_2)^2. \tag{12.3}$$

The differential cross section is easily obtained in the case of incident point particles (Fig. 12.2) colliding with target particles of radius R. The *impact parameter b* of the collision is the smallest distance between the incident trajectory in the absence of a collision and the center of the target. Figure 12.2 shows that the impact parameter and the scattering angle θ are related as

$$b = R \cos \frac{\theta}{2},$$

while

$$d\sigma = 2\pi b \, db = \pi R^2 \sin \frac{\theta}{2} \cos \frac{\theta}{2} \, d\theta = \frac{1}{2} \pi R^2 d(\cos \theta),$$

from which we find the differential cross section

$$\frac{d\sigma}{d\Omega} = \frac{1}{2\pi} \frac{d\sigma}{d\cos\theta} = \frac{1}{4} R^2, \tag{12.4}$$

because the integration over ϕ gives a factor of 2π. This cross section, which is called the *cross section for hard-sphere scattering*, is therefore independent of the scattering angle, i.e., it is isotropic. It can be checked that integration over Ω again gives πR^2.

12.1.2 The scattering amplitude

Now let us turn to the quantum description of scattering by a potential V which we assume to be spherically symmetric, $V = V(r)$. We shall return to the general potential $V(\vec{r})$ in

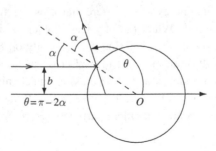

Fig. 12.2. Classical collision between a point particle and a sphere of radius R.

Section 12.3.2. We ignore possible spin degrees of freedom, except in Section 12.2.4. Scattering is a time-dependent process: an incident particle described by a wave packet $\varphi(\vec{r}, t)$ leaves from $z = -\infty$, travels along the z axis, and encounters the potential at time $t \sim 0$. This wave packet has a certain probability of being scattered in a direction θ, and a detector located at this angle has a certain probability of recording the particle. The rigorous quantum description can be obtained only by using wave packets. Nevertheless, this description is rather cumbersome, and at first we shall simplify the discussion by considering a stationary process. Later on in Section 12.4.2 we will return to wave packets. We start with an incident plane wave of wave vector $\vec{k} = (0, 0, k)$ parallel to Oz:

$$\varphi(\vec{r}) = A\, e^{ikz}, \quad k^2 = \frac{2m}{\hbar^2} E, \tag{12.5}$$

where m is the mass of the incident particles, E is their energy, and $|A|^2 = n_i$ is their density. The current $\vec{j}$ associated with a plane wave (12.5) is given by (9.141):

$$\vec{j} = \frac{\hbar}{2mi} \left[\varphi^* \vec{\nabla} \varphi - (\vec{\nabla} \varphi)^* \varphi \right] = |A|^2 \frac{\hbar \vec{k}}{m} = |A|^2 \vec{v}. \tag{12.6}$$

The flux of incident particles is $\mathcal{F} = |\vec{j}| = |A|^2 v$. The plane wave $\varphi(\vec{r})$ is a solution of the time-independent Schrödinger equation in the absence of a potential $[V(r) = 0]$:

$$-\frac{\hbar^2}{2m} \nabla^2 \varphi(\vec{r}) = \frac{\hbar^2 k^2}{2m} \varphi(\vec{r}) = E\varphi(\vec{r}). \tag{12.7}$$

In Section 12.4.1 we shall show that when $V(r) \neq 0$, for the *same value of the energy E* there exist solutions of the Schrödinger equation $\psi_{\vec{k}}^{(+)}(\vec{r})$ labeled by the wave vector $\vec{k}$,

$$\left[-\frac{\hbar^2}{2m} \nabla^2 + V(r) \right] \psi_{\vec{k}}^{(+)}(\vec{r}) = E\psi_{\vec{k}}^{(+)}(\vec{r}), \tag{12.8}$$

which for $r \to \infty$ behave as

$$\boxed{\psi_{\vec{k}}^{(+)}(\vec{r}) = A \left[e^{i\vec{k}\cdot\vec{r}} + f(\Omega) \frac{e^{ikr}}{r} \right],} \tag{12.9}$$

where f is a complex function of Ω (in our case only of θ, owing to the invariance under rotation about Oz) called the *scattering amplitude*. The first term in (12.9) is the incident plane wave $\exp(i\vec{k} \cdot \vec{r}) = \exp(ikz)$, and the second corresponds to an *outgoing spherical wave*, as we shall show shortly. It is essential to note that it is the absolute values of k and r that are involved in the second term. The expression (12.9) is valid provided that the potential $V(r)$ falls off sufficiently rapidly for $r \to \infty$. It is not valid for the Coulomb potential, whose $1/r$ falloff is too slow. There also exist solutions of the Schrödinger equation with an incoming spherical-wave term:

$$\psi_{\vec{k}}^{(-)}(\vec{r}) = A \left[e^{i\vec{k}\cdot\vec{r}} + f(\Omega) \frac{e^{-ikr}}{r} \right]. \tag{12.10}$$

Such solutions are useful in some cases, but we shall not need them here.

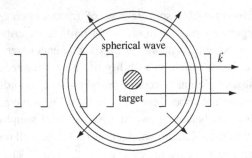

Fig. 12.3. Large-distance behavior of an incident plane wave.

Let us calculate the total current for the asymptotic wave function (12.9). This current is composed of the plane-wave current, the spherical-wave current, and an interference term. Here we must appeal to a physical argument, relying on the observation that the transverse extent of the incident wave is actually limited and not infinite, as in a plane wave (Fig. 12.3), and the interference term should be neglected except in the region where the incident wave packet and the spherical wave overlap.[3] For a direction $\theta \neq 0$, that is, away from the direction of the incident wave $\theta = 0$, it is always possible to place the detector far enough from the target that the interference term is negligible, and then it is sufficient to calculate the current of the spherical wave. Using $\vec{\nabla} g(r) = \hat{r}\, g'(r)$, we obtain

$$\vec{\nabla}\left[\frac{e^{ikr}}{r} f(\Omega)\right] = ik\hat{r}\, \frac{e^{ikr}}{r} f(\Omega) + O\left(\frac{1}{r^2}\right),$$

because

$$\left|\vec{\nabla}\frac{1}{r}\right| \propto \frac{1}{r^2} \quad \text{and} \quad |\vec{\nabla} f(\Omega)| \propto \frac{1}{r},$$

so that the final expression for $\vec{j}$ is

$$\vec{j} = \frac{|A|^2 \hbar k}{m}\, |f(\Omega)|^2\, \frac{\hat{r}}{r^2} = |A|^2 v\, |f(\Omega)|^2\, \frac{\hat{r}}{r^2}. \tag{12.11}$$

If we draw a very large sphere of radius r about the target, the current associated with the second term in (12.9) at the surface of this sphere points along $\hat{r}$ away from the center of the sphere and represents an outgoing wave. The current associated with the term $\exp(-ikr)/r$ in (12.10) will point toward the inside of the sphere and corresponds to an incoming spherical wave. The number of particles $\Delta \mathcal{N}(\Omega)$ recorded by the detector per unit time is equal to the integral of the current over the surface of the detector $\Delta \mathcal{S} \simeq r^2 \Delta\Omega$:

$$\Delta \mathcal{N}(\Omega) = \int_{\Delta \mathcal{S}} \vec{j} \cdot d\vec{\mathcal{S}} = r^2 \int_{\Delta\Omega} \vec{j} \cdot \hat{r}\, d\Omega,$$

where the detector is located at a distance r from the target. For infinitesimal $\Delta\Omega$ this gives

$$d\mathcal{N}(\Omega) = |A|^2 v\, |f(\Omega)|^2\, d\Omega = \mathcal{F}\, |f(\Omega)|^2\, d\Omega.$$

---

[3] This interference term is essential for understanding the optical theorem (12.54); cf. Lévy-Leblond and Balibar [1990].

It is in fact the $1/r$ behavior of the outgoing spherical wave term that ensures that the flux in a solid angle $\Delta\Omega$ is independent of r. The definition (12.1) of the differential cross section permits the following identification for $n_t = 1$:

$$\boxed{\frac{d\sigma}{d\Omega} = |f(\Omega)|^2}. \tag{12.12}$$

12.2 Partial waves and phase shifts

12.2.1 The partial-wave expansion

In Section 10.4.1 we presented a method for solving the Schrödinger equation when the potential $V(r)$ is spherically symmetric. The method consists of expanding the wave function in spherical harmonics as in (10.77):

$$\psi(r, \theta, \phi) = \sum_{l,m_l} \frac{u_l(r)}{r} Y_l^{m_l}(\theta, \phi).$$

The cylindrical symmetry about Oz in the present problem allows us to limit ourselves to terms independent of ϕ, $m_l = 0$, and take into account the proportionality (10.62) of the spherical harmonics with $m_l = 0$ to the Legendre polynomials. We can then write[4]

$$\psi(r, \theta) = \sum_{l=0}^{\infty} \frac{u_l(r)}{r} P_l(\cos \theta), \tag{12.13}$$

where $u_l(r)$ is the solution of the radial equation (10.78):

$$\left[-\frac{\hbar^2}{2m} \frac{d^2}{dr^2} + \frac{l(l+1)}{2mr^2} + V(r) \right] u_l(r) = E_l u_l(r), \tag{12.14}$$

with the boundary condition $u_l(0) = 0$, or, more precisely using (10.82),

$$r \to 0 : u_l(r) \propto r^{l+1}. \tag{12.15}$$

Since the Legendre polynomials form a basis for functions defined on the interval $[-1, +1]$, we can write the following series expansion for $f(\theta)$:

$$f(\theta) = \sum_{l=0}^{\infty} f_l P_l(\cos \theta), \quad f_l = \frac{2l+1}{2} \int_{-1}^{+1} f(\theta) P_l(\cos \theta) \, d\cos \theta. \tag{12.16}$$

The series (12.16) is called *the partial-wave expansion of the scattering amplitude*.

---

[4] We have modified the normalization of $u_l(r)$ by the unimportant factor $\sqrt{4\pi/(2l+1)}$ in going from one equation to the other.

If $V(r)$ tends to zero sufficiently rapidly for $r \to \infty$,[5] we can neglect $V(r)$ and the centrifugal barrier term in (12.14). The asymptotic behavior of $u_l(r)$ will then be

$$r \to \infty : u_l(r) \propto \sin(kr + \hat{\delta}_l).$$

Let us compare this behavior to that of a plane wave. A plane wave $\exp(ikz) = \exp(ikr\cos\theta)$ is a cylindrically symmetric solution of the Schrödinger equation when $V(r) = 0$. We can then expand $\exp(ikz)$ in a series of Legendre polynomials of the type (12.13). The coefficients of this series are calculated using (12.16) and are called the *spherical Bessel functions* $j_l(kr)$:

$$e^{ikz} = \sum_{l=0}^{\infty} (2l+1)i^l j_l(kr) P_l(\cos\theta). \tag{12.17}$$

The spherical Bessel functions can be expressed in terms of sines and cosines and are given by the recursion relation

$$j_l(x) = (-1)^l x^l \left(\frac{1}{x}\frac{d}{dx}\right)^l \frac{\sin x}{x} = (-1)^l x^l \left(\frac{1}{x}\frac{d}{dx}\right)^l j_0(x). \tag{12.18}$$

When $r \to 0$ we have $krj_l(kr) \propto (kr)^{l+1}$, which is a special case of the behavior (12.15) since $rj_l(kr)$ is a solution of the radial Schrödinger equation with $V(r) = 0$. When $r \to \infty$ it can be shown that[6]

$$r \to \infty : j_l(kr) \simeq \frac{1}{kr}\sin\left(kr - \frac{1}{2}l\pi\right). \tag{12.19}$$

Comparison with the behavior of $u_l(r)$ leads to the definition

$$\delta_l = \hat{\delta}_l - \frac{1}{2}l\pi,$$

which allows us to write down the asymptotic behavior of $u_l(r)$:

$$r \to \infty : u_l(r) \simeq a_l \sin\left(kr - \frac{1}{2}l\pi + \delta_l\right). \tag{12.20}$$

The number δ_l is the *phase shift* in the lth partial wave, and is a function of k: $\delta_l(k)$. To express $f(\theta)$ as a function of the phase shifts, it is sufficient to compare the asymptotic expansions of (12.9) and (12.13) at $r \to \infty$, choosing $A = 1$. Taking into account (12.17), the series (12.9) can be written as

$$e^{ikz} + f(\theta)\frac{e^{ikr}}{r} = \sum_{l=0}^{\infty} X_l(r) P_l(\cos\theta),$$

$$X_l(r) = (2l+1)i^l j_l(kr) + f_l \frac{e^{ikr}}{r}.$$

---

[5] This restriction on the potential should be made more precise. All the results of the present chapter are valid if $V(r)$ has finite range [$V(r) = 0$ if $r > R$] or decreases at infinity faster than any power. If $V(r)$ falls off at infinity as $r^{-\alpha}$, certain results will be valid only if $\alpha \geq \alpha_0$. The discussion of this problem is rather technical, and we refer the reader to the references cited in Further Reading.

[6] See, for example, Cohen-Tannoudji *et al.* [1977], Complement A_{VIII}.

The asymptotic form (12.19) of the j_l gives

$$i^l j_l(kr) \simeq \frac{1}{2ikr}\left[(-1)^{l+1}e^{-ikr} + e^{ikr}\right],$$

and we obtain

$$X_l = \frac{2l+1}{2ikr}\left[(-1)^{l+1}e^{-ikr} + \left(1 + \frac{2ik}{2l+1}f_l\right)e^{ikr}\right]. \tag{12.21}$$

The function $X_l(r)$ must asymptotically be equal to $u_l(r)/r$, and so according to (12.20)

$$\frac{u_l(r)}{r} \simeq \frac{a_l}{2ir}\left[(-1)^{l+1}e^{-ikr} + e^{2i\delta_l}e^{ikr}\right]. \tag{12.22}$$

The expressions (12.21) and (12.22) can be equal only if

$$e^{2i\delta_l} = 1 + \frac{2ik}{2l+1}f_l$$

or

$$f_l = \frac{2l+1}{2ik}\left(e^{2i\delta_l} - 1\right) = \frac{2l+1}{k}e^{i\delta_l}\sin\delta_l. \tag{12.23}$$

This equation gives the *partial wave expansion* for $f(\theta)$) as a function of the phase shifts:

$$\boxed{f(\theta) = \frac{1}{k}\sum_{l=0}^{\infty}(2l+1)e^{i\delta_l}\sin\delta_l P_l(\cos\theta)}. \tag{12.24}$$

We can obtain the differential cross section from (12.12) and then the total cross section by integrating over angles using the orthogonality relation of the Legendre polynomials derived from (10.62) and the orthogonality (10.55) of the spherical harmonics:

$$\int d\Omega\, P_l(\cos\theta)\, P_{l'}(\cos\theta) = \frac{4\pi}{2l+1}\delta_{ll'}.$$

The result for σ_{tot} can be written as

$$\boxed{\sigma_{\text{tot}} = \frac{4\pi}{k^2}\sum_{l=0}^{\infty}(2l+1)\sin^2\delta_l}. \tag{12.25}$$

The function

$$\boxed{S_l(k) = e^{2i\delta_l(k)}}, \tag{12.26}$$

where we have noted explicitly the dependence on k, is called the *S-matrix element in the lth partial wave*. It plays an important role in scattering, which can be understood by comparing the behavior (12.21) of a free spherical wave $j_l(kr)$ with that of the wave function in the presence of a potential (12.22):

$$j_l(kr) \ \propto \ \left[(-1)^{l+1}e^{-ikr} + e^{ikr}\right],$$

$$u_l(r) \ \propto \ \left[(-1)^{l+1}e^{-ikr} + e^{2i\delta_l}e^{ikr}\right].$$

The effect of the potential is to multiply the outgoing spherical wave by the phase factor $S_l = \exp(2i\delta_l)$ while not affecting the incoming wave. This is a result of the boundary conditions that have been imposed, since the incident plane wave is composed of an incoming spherical wave and an outgoing spherical wave. The outgoing part is modified by the scattering, because the particles are scattered by the target and diverge from it. However, the incoming wave is not modified by the interaction with the target. In Section 12.3.1 we shall show that the condition $|S_l| = 1$ takes into account the fact that the number of particles entering a sphere of large radius drawn about the target is equal to the number of particles leaving the sphere when the scattering is elastic.

   Each term of (12.25) corresponds to the scattering cross section in the lth partial wave. It is obviously impossible to identify the contribution of each partial wave except in the total cross section, because the various partial waves interfere in the differential cross section. We note that the contribution to the total cross section from each partial wave is bounded:

$$\sigma_l = \frac{4\pi}{k^2}\,(2l+1)\sin^2\delta_l \le \sigma_l^{\max} = \frac{4\pi}{k^2}\,(2l+1). \tag{12.27}$$

Let us give a semi-classical interpretation of this result. Classically, the angular momentum $\hbar l$ and the impact parameter are related as $l = kb$, and so

$$\frac{l}{k} \le b \le \frac{l+1}{k}.$$

The maximum classical cross section is the area between the circles of radii l and $l+1$:

$$\sigma_l \le \frac{\pi}{k^2}\left[(l+1)^2 - l^2\right] = \frac{\pi}{k^2}\,(2l+1) = \frac{1}{4}\,\sigma_l^{\max}.$$

The *classical* cross section is at most a quarter of the maximum quantum cross section. If the potential has finite range, $V(r) = 0$ for $r > R$, then, from the classical point of view, an incident particle can interact only if its impact parameter is less than R, $b < R$, and only partial waves with $l \lesssim kR$ will contribute. We see that the phase-shift method will work well if the energy is low, because in this case only a limited number of partial waves will contribute. In particular, only the s-wave ($l = 0$) will contribute appreciably when $k \to 0$. In quantum mechanical terms, the probability density $\propto r^2 j_l^2(kr)$ of a free spherical wave is negligible for $kr \lesssim [l(l+1)]^{1/2}$, and this wave does not penetrate into regions where the potential is important for small k unless $l = 0$, when $r^2 j_0^2(kr) \propto \text{const}$

if $r \to 0$. It can be rigorously shown[7] that for a potential of finite range the phase shift δ_l behaves as

$$\delta_l(k) \propto (kR)^{2l+1} \tag{12.28}$$

when $k \to 0$ or $l \to \infty$.

12.2.2 Low-energy scattering

When the potential has finite range, the s-wave will be the only one to contribute significantly to the low-energy cross section, and so the latter will be isotropic. In the rest of this section we shall take into account only the $l = 0$ wave and use the notation $\delta_{l=0}(k) = \delta(k)$, $S_{l=0}(k) = S(k)$, $f_{l=0}(k) = f(k)$, $u_{l=0}(r) = u(r)$. Using the behavior (12.28) for $l = 0$, $\delta(k) \propto k$, we can define the *scattering length a* as

$$a = -\lim_{k \to 0} \frac{\delta(k)}{k}. \tag{12.29}$$

The minus sign is chosen by convention and will be justified below.

As an example of a calculation of the phase shift and scattering length, let us consider the spherical well (Fig. 12.4):

$$V(r) = -V_0, \quad 0 \le r \le R,$$

$$V(r) = 0, \quad r > R.$$

Such a spherical well gives an approximate description of neutron–proton scattering with the following parameters (Exercises 10.7.8 and 12.5.3):

$$R \simeq 2 \text{ fm}, \quad V_0 \simeq 26 \text{ MeV}.$$

The radial Schrödinger equation is written as

$$\left(-\frac{d^2}{dr^2} + \frac{2m}{\hbar^2} V(r) \right) u(r) = \frac{2m}{\hbar^2} E u(r), \tag{12.30}$$

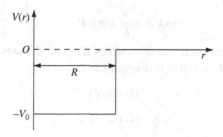

Fig. 12.4. The spherical well.

---

[7] See, for example, Messiah [1999], Chapter X.

which gives

$$r > R : \left(\frac{d^2}{dr^2} + k^2 \right) u(r) = 0,$$

$$r < R : \left(\frac{d^2}{dr^2} + k'^2 \right) u(r) = 0,$$

with $k^2 = 2mE/\hbar^2$ and $k'^2 = 2m(E + V_0)/\hbar^2$, from which, taking into account the condition $u(r = 0) = 0$, we find

$$r > R : u(r) = C \sin(kr + \delta),$$

$$r < R : u(r) = D \sin k'r.$$

The continuity of the logarithmic derivative of $u(r)$ at $r = R$ imposes the condition

$$k' \cot k'R = k \cot(kR + \delta). \tag{12.31}$$

The equation

$$\cot x = i \frac{e^{2ix} + 1}{e^{2ix} - 1}$$

can be used to determine the S-matrix element $S(k)$. An easy calculation gives

$$S(k) = e^{2i\delta(k)} = e^{-2ikR} \frac{\cos k'R + i\dfrac{k}{k'} \sin k'R}{\cos k'R - i\dfrac{k}{k'} \sin k'R}. \tag{12.32}$$

As expected, the expression for $S(k)$ has unit modulus. The phase shift is determined only up to a factor of π, and to learn the "true" value of the phase shift it is necessary to allow the potential to increase from 0 to V_0 while following the evolution of δ between zero and its final value.

As in the one-dimensional case (cf. Section 9.4.3), there exists a remarkable relation between the S-matrix and bound states. Let us set $k = i\kappa$ (in an instant we shall see that we must choose $k = i\kappa$, $\kappa > 0$ and not $k = -i\kappa$). The function $S(k)$ has poles for

$$\cos k'R + \frac{\kappa}{k'} \sin k'R = 0, \tag{12.33}$$

but this is also just the equation that determines the bound states. The wave function of a bound state of energy $E = -B < 0$ is given by

$$r > R : u(r) = Ce^{-\kappa r},$$

$$r < R : u(r) = D \sin k'r,$$

with $\kappa = (2mB/\hbar^2)^{1/2}$ and $k' = [2m(V_0 - B)]^{1/2}/\hbar$, and the continuity of the logarithmic derivative at $r = R$ is written as

$$-\kappa = k' \cot k'R, \tag{12.34}$$

which is exactly the equation for the poles of $S(k)$. The result is general for potentials that fall off sufficiently rapidly at infinity and is valid for any partial wave: the poles of $S_l(k)$ for $k = i\kappa$ give the position of the bound states in the lth partial wave.

It is easy to derive the scattering length from (12.31). This equation can also be written as

$$\tan(kR + \delta) = \frac{k}{k'} \tan k'R.$$

In the limit $k \to 0$ and $kR \to 0$, $\delta \to 0$ and $k' \to k_0 = (2mV_0/\hbar^2)^{1/2}$, from which we have

$$kR + \delta(k) \simeq \frac{k}{k_0} \tan k_0 R,$$

or

$$\delta(k) \simeq -k \left(R - \frac{\tan k_0 R}{k_0} \right),$$

which according to the definition (12.29) gives

$$a = R \left(1 - \frac{\tan k_0 R}{k_0 R} \right). \tag{12.35}$$

Another case of particular interest is that of hard-sphere scattering: $V(r) = 0$ if $r > R$ and $V(r) = +\infty$ if $r < R$. The radial wave function $u(r)$ must vanish at $r = R$:

$$r > R : u(r) = C \sin(kR + \delta),$$
$$r < R : u(r) = 0,$$

so that $kR + \delta = n\pi$ and for k sufficiently small,

$$\delta = -kR, \quad a = R. \tag{12.36}$$

The minus sign in the definition (12.29) has been chosen such that the scattering length of a hard sphere is $+R$ rather than $-R$. From the qualitative behavior of $u(r)$ in Fig. 12.5 we see that $a > 0$ for any repulsive potential. The situation is more complicated for an attractive potential. When there is no bound state an attractive potential gives a negative scattering length. The appearance of a bound state changes the sign of a, which becomes positive. The sign changes again with the appearance of a second bound state, and so on. This is confirmed by (12.35): the condition for the appearance of a first bound state is $k_0 R = \pi/2$ and the scattering length is negative for $k_0 R < \pi/2$. It becomes infinite when $k_0 R = \pi/2$, positive when $k_0 R > \pi/2$, and remains positive for $\pi/2 < k_0 R < 3\pi/2$. The appearance of a second bound state corresponds to $k_0 R = 3\pi/2$, and the scattering length is negative beyond this value after having again become infinite. A large positive scattering length indicates the presence of a low-energy bound state, and a scattering length that is large and negative indicates that a bound state is about to appear. It is sometimes said that there is an antibound or virtual state.

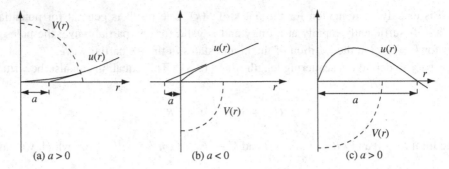

(a) $a > 0$ (b) $a < 0$ (c) $a > 0$

Fig. 12.5. Behavior of the wave function and the scattering length for various potentials: (a) a repulsive potential; (b) an attractive potential without a bound state; (c) an attractive potential with a single bound state.

According to (12.12) the low-energy cross section is isotropic, and the total cross section is

$$\sigma_{\text{tot}} = 4\pi a^2. \tag{12.37}$$

It is interesting to note that the quantum cross section of a hard sphere $(a = R)$ is four times the classical cross section πR^2, in agreement with the inequality mentioned previously. Measurement of the total cross section gives only the absolute value of a. However, the sign of the scattering length is an important quantity. For example, the effective potential which we shall define in the following paragraph is attractive for $a < 0$ and repulsive for $a > 0$, which has direct consequences, for example, for the possibility of forming Bose–Einstein condensates of atomic gases. Another important case is neutron–proton scattering (Section 12.2.4).

The low-energy form $\delta(k) \simeq -ka$ is actually the first term of an expansion of the phase shift in powers of k^2. Exercise 12.5.3 shows that the function $k \cot \delta(k)$ is an analytic function[8] of k^2 for which we can write down a Taylor series for $k^2 \to 0$:

$$k \cot \delta(k) = -\frac{1}{a} + \frac{1}{2} r_0 k^2 + O(k^4). \tag{12.38}$$

The distance r_0 is called the *effective range*. We often use the low-energy form of the scattering amplitude:

$$f(k) = \frac{e^{2i\delta(k)} - 1}{2ik} = \frac{1}{k[\cot \delta(k) - i]},$$

or, expressing $\cot \delta(k)$ as a function of a if $r_0 k \ll 1$,

$$\boxed{f(k) = \frac{-a}{1 + ika}}. \tag{12.39}$$

---

[8] If $V(r)$ falls off at least as fast as $\exp(-\mu r)$. Equation (12.38) is valid provided that $V(r)$ falls off at least as r^{-5}.

This form can be made more precise by using the effective-range approximation (12.38):

$$f(k) = \frac{-a}{1 + ika - \frac{1}{2} r_0 a k^2}. \qquad (12.40)$$

12.2.3 The effective potential

The scattering length makes it possible to introduce the very useful concept of effective potential, not to be confused with the effective potential $V_l(r)$ of (10.79). When studying a system of low-energy particles, it is convenient to be able to replace the actual potential $V(r)$ by a simpler potential $V_{\text{eff}}(r)$, called the *effective potential*, which gives the same results for low-energy scattering. An effective potential is used, for example, for the theoretical study of low-energy neutron scattering or Bose–Einstein condensates of atomic gases. We shall show that low-energy scattering is described by choosing an effective potential proportional to a δ function:

$$V_{\text{eff}}(r)\psi(r) = g\delta(\vec{r}) \frac{\mathrm{d}}{\mathrm{d}r} [r\psi(r)], \qquad (12.41)$$

where g is a constant to be determined. To justify this potential and find g, let us examine the Schrödinger equation for a wave function $\psi(r) = u(r)/r$. The expression for the Laplacian applied to a function of r

$$\nabla^2 f(r) = \frac{1}{r} \frac{\mathrm{d}^2}{\mathrm{d}r^2} [r f(r)] \qquad (12.42)$$

is valid only for a function $f(r)$ that is regular at $r = 0$, and for $f(r) \propto 1/r$ the familiar equation from electrostatics is used:

$$\nabla^2 \frac{1}{r} = -4\pi\delta(\vec{r}). \qquad (12.43)$$

Let us study the Schrödinger equation taking (12.41) as the potential:

$$-\frac{\hbar^2}{2m} \nabla^2 \frac{u(r)}{r} + V_{\text{eff}}(r) \frac{u(r)}{r} = \frac{\hbar^2 k^2}{2m} \frac{u(r)}{r},$$

and write down the kinetic energy term

$$\nabla^2 \frac{u(r)}{r} = \nabla^2 \left[\frac{u(r) - u(0)}{r} \right] + u(0)\nabla^2 \frac{1}{r}$$

$$= \frac{1}{r} \frac{\mathrm{d}^2}{\mathrm{d}r^2} r \left[\frac{u(r) - u(0)}{r} \right] - 4\pi u(0)\delta(\vec{r}) = \frac{1}{r} \frac{\mathrm{d}^2 u(r)}{\mathrm{d}r^2} - 4\pi u(0)\delta(\vec{r}),$$

where we have noted that $[u(r) - u(0)]/r$ is a regular function at $r = 0$. Moreover, if we write

$$u(r) = a + br + cr^2 + \cdots,$$

then

$$\frac{1}{r} \frac{\mathrm{d}^2 u}{\mathrm{d}r^2} = \frac{2c}{r} + \cdots$$

and the integral of this term in a sphere of radius R about the origin tends to zero with R. We then have

$$-\frac{\hbar^2}{2mr}\frac{d^2u(r)}{dr^2} - \frac{\hbar^2k^2}{2m}\frac{u(r)}{r} = \left[-\frac{4\pi\hbar^2}{2m}u(0) - gu'(0)\right]\delta(\vec{r}).$$

The two sides of this equation must vanish separately, which for the left-hand side implies

$$u(r) = C\sin(kr + \delta(k)), \quad r > 0,$$

and so $u'(0)/u(0) = k\cot\delta(k)$. The vanishing of the coefficient of $\delta(\vec{r})$ imposes the condition

$$-\frac{2\pi\hbar^2}{m} = gk\cot\delta(k),$$

and the $k \to 0$ limit of this equation makes it possible to relate g and a:[9]

$$\boxed{g = \frac{2\pi\hbar^2}{m}a, \quad V_{\text{eff}}(\vec{r}) = \frac{2\pi\hbar^2 a}{m}\delta(\vec{r})\frac{d}{dr}r}. \tag{12.44}$$

The effective potential depends on a single parameter, the scattering length a; we take it to be that of a more realistic potential or simply use the experimental value. Let us also study the bound states of the effective potential. The radial wave function of a bound state must have the form

$$u(r) = Ce^{-\kappa r},$$

and so $u'(0)/u(0) = -\kappa$. We can derive a relation between the binding energy B and the scattering length:

$$\kappa = \sqrt{\frac{2mB}{\hbar^2}} = \frac{2\pi\hbar^2 g}{m} = \frac{1}{a}. \tag{12.45}$$

The bound state of the effective potential is unique, and we again find that $a > 0$ for a single bound state. In summary, an effective potential for which $a > 0$ may correspond either to a hard sphere or to an attractive potential with a single bound state. These two potentials lead to the same behavior for an ensemble of low-energy particles, but the behavior will be different if $a < 0$: it is the sign of the scattering length that is crucial. The function $k\cot\delta(k)$ is a constant:

$$k\cot\delta(k) = -\frac{2\pi\hbar^2}{mg} = -\frac{1}{a},$$

and the scattering amplitude of the effective potential is given *exactly* by (12.39):

$$f_{\text{eff}}(k) = \frac{-a}{1 + ika}.$$

---

[9] It should be born in mind that if we consider the scattering of identical particles of mass M, the reduced mass is $m = M/2$ and $g = (4\pi\hbar^2/M)a$.

12.2.4 Low-energy neutron–proton scattering

Low-energy neutron–proton scattering provides a very important practical example of the formalism we have just developed. The proton and the neutron are spin-1/2 particles and the scattering is spin-dependent, and so we shall generalize the above results to take this into account. In low-energy scattering the total spin $\vec{S}_{tot}$ is conserved. The orbital angular momentum is zero, because the scattering occurs in the s-wave, and the conservation of total angular momentum is equivalent to the conservation of total spin. The scattering amplitude can be written as an operator $\hat{f}$ acting in the four-dimensional space $\mathcal{H}$, the tensor product of the two spaces of spin-1/2 states, as a function of the projectors $\mathcal{P}_s = \mathcal{P}_0$ and $\mathcal{P}_t = \mathcal{P}_1$ on the singlet (total spin zero) and triplet (total spin one) states given in (10.128):

$$\hat{f}(k) = f_s(k)\mathcal{P}_s + f_t(k)\mathcal{P}_t.$$

This form of $\hat{f}$ ensures that the total spin remains unchanged in the scattering: a singlet state remains a singlet and a triplet state remains a triplet. We shall limit ourselves to the case $ka \ll 1$. According to (12.39),

$$f_s(k) = -a_s, \quad f_t(k) = -a_t,$$

where a_s and a_t are the scattering lengths in the singlet and triplet states. When the condition $ka \ll 1$ is not satisfied, it is possible to use expressions analogous to (12.39), or even (12.40), for $f_s(k)$ and $f_t(k)$, thus introducing the effective ranges r_{0s} and r_{0t}. In summary, in the approximation where $ka \ll 1$

$$\hat{f} = -a_s\mathcal{P}_s - a_t\mathcal{P}_t, \tag{12.46}$$

or, introducing the Pauli matrices $\vec{\sigma}_p$ and $\vec{\sigma}_n$ acting in the space of the proton and neutron spin states,

$$-\hat{f} = \hat{a} = \frac{1}{4}(a_s + 3a_t)I + \frac{1}{4}(a_t - a_s)\vec{\sigma}_p \cdot \vec{\sigma}_n. \tag{12.47}$$

The differential cross section is isotropic and the total cross section for a state of initial spin $|i\rangle$ and final spin $|f\rangle$ is

$$\sigma_{fi} = 4\pi|\langle f|\hat{a}|i\rangle|^2. \tag{12.48}$$

If the final spins are not measured and the initial state is a mixture for which we know only the probability p_i of finding the initial spins in the state $|i\rangle$, it is necessary to sum over the states $|f\rangle$ and the probabilities p_i:

$$\sigma = 4\pi\sum_i p_i \sum_f \langle i|\hat{a}|f\rangle\langle f|\hat{a}|i\rangle$$

$$= 4\pi\sum_i p_i\langle i|\hat{a}^2|i\rangle = 4\pi\,\mathrm{Tr}\,(\rho_{init}\,\hat{a}^2),$$

where we have used the completeness relation in $\mathcal{H}$, $\sum_f |f\rangle\langle f| = I$, and the definition of the state operator of the initial state:

$$\rho_{\text{init}} = \sum_i \mathsf{p}_i |i\rangle\langle i|.$$

The most frequently encountered case is that of unpolarized initial state, so that the states $|++\rangle$, $|+-\rangle$, $|-+\rangle$, and $|--\rangle$ have the same probability. In this case $\rho_{\text{init}} = I/4$ and

$$\sigma_{\text{unpol}} = \pi\,\text{Tr}\,\hat{a}^2 = \pi\,\text{Tr}\left(a_s^2 \mathcal{P}_s + a_t^2 \mathcal{P}_t\right)$$

$$= 4\pi\left(\frac{1}{4}a_s^2 + \frac{3}{4}a_t^2\right) = \frac{1}{4}\sigma_s + \frac{3}{4}\sigma_t. \tag{12.49}$$

The physical interpretation is straightforward: if the initial state is unpolarized, the probability of having a singlet state is 1/4 and that of having a triplet state is 3/4, which gives the weights 1/4 and 3/4 of the singlet and triplet cross sections in (12.49).

The unpolarized cross section gives only the combination $a_s^2 + 3a_t^2$ of the scattering lengths. Additional information can be obtained from the existence of a bound state in the triplet state, the deuteron, which allows the approximate determination of a_t. A precise relation between the deuteron parameters and the low-energy scattering parameters in the triplet state is obtained in Exercise 12.5.3 using the effective-range approximation. An approximate expression is obtained by noting that the deuteron wave function extends far beyond the range of the potential, $\kappa^{-1} \gg R$, which makes it possible to use the effective potential and the relation (12.45). Using the fact that $B \simeq 2.22\,\text{MeV}$, we obtain $\kappa^{-1} \simeq 4.2\,\text{fm}$, while the exact value of a_t is 5.4 fm. However, this argument is sufficient for determining the sign of a_t: $a_t > 0$.

Knowledge of a_t from the deuteron parameters and measurement of the unpolarized cross section make it possible to determine the modulus of the scattering length in the singlet state $|a_s|$, but not its sign. A possible method for finding the sign of a_s is to use neutron scattering on a hydrogen molecule; this is studied in Exercise 12.5.2. It is found that the scattering length a_s is negative, consistent with the fact that there is no singlet bound state. The experimental values of the scattering lengths and effective ranges are

$$a_t = 5.40\,\text{fm}, \quad r_{0t} = 1.73\,\text{fm}, \quad a_s = -23.7\,\text{fm}, \quad r_{0s} = 2.5\,\text{fm}.$$

It can be observed that a_s is large and negative, and that the neutron–proton system in the singlet state is very close to forming a bound state, showing the presence of a virtual state.

12.3 Inelastic scattering

12.3.1 The optical theorem

In general, in a collision particles can undergo not only elastic, but also inelastic scattering. For example, the scattering of a photon on an atom A in its ground state E_0 can leave the atom in an excited level A^* of energy E_1:

$$\gamma + A \rightarrow \gamma' + A^*,$$

the final photon having lost an energy $(E_1 - E_0)$ compared with the initial one (if the atomic recoil is neglected). It is also possible for the final particles to be different from the initial ones, as in

$$\pi^- + p \to K^0 + \Lambda$$

or

$$\pi^- + p \to \pi^- + \pi^+ + n.$$

We have seen that $|S_l(k)| = 1$ in the case of elastic scattering. We shall show that it is possible to generalize the expression for the scattering amplitude $f(\Omega)$ to the inelastic case if we allow $|S_l(k)| \leq 1$. This inequality follows from the condition that the modulus of the amplitude of the outgoing wave be smaller than that of the incoming wave, that is, the number of particles N_{out} leaving a large sphere of radius r enclosing the target must be smaller than the number N_{in} entering the sphere, because incident particles can only *disappear* in inelastic scattering. As we shall show below, this inequality holds for each partial wave, $N_{\text{out}}^l \leq N_{\text{in}}^l$, because the integration over the surface of the sphere eliminates interference between partial waves. If the scattering is purely elastic in the lth partial wave, $N_{\text{in}}^l = N_{\text{out}}^l$ and $|S_l(k)| = 1$. Let us evaluate N_{in}^l and N_{out}^l using the asymptotic form (12.22) of the wave function at $r \to \infty$. As in elastic scattering, only the outgoing wave term can be modified:

$$\frac{e^{ikr}}{r} \to S_l(k) \frac{e^{ikr}}{r},$$

from which we find the asymptotic behavior of $\psi(\vec{r})$:

$$\psi \simeq \frac{iA}{2kr} \sum_{l=0}^{\infty} (2l+1) P_l(\cos\theta) \left[(-1)^l e^{-ikr} - S_l e^{ikr} \right],$$

which gives for $f(\theta)$

$$f(\theta) = \frac{1}{2ik} \sum_{l=0}^{\infty} (2l+1) P_l(\cos\theta)(S_l - 1).$$

The total elastic cross section then is

$$\sigma_{\text{el}} = \int d\Omega |f(\theta)|^2$$

and the result of the integration over Ω generalizes (12.25):

$$\boxed{\sigma_{\text{el}} = \frac{\pi}{k^2} \sum_{l=0}^{\infty} (2l+1)|1 - S_l|^2} . \tag{12.50}$$

Let us calculate the number of incoming particles in the lth partial wave, N_{in}^l, by integrating the current entering through the surface of a sphere of radius $r \to \infty$ about the target.

Since the Legendre polynomials are orthogonal, there are no interference terms between different partial waves. We find

$$N_{\text{in}}^l = \left[\frac{(2l+1)^2|A|^2}{4k^2}\right]\left[\frac{2}{2l+1}\right]\left[\frac{\hbar k}{m}\right][2\pi] = \frac{\pi\hbar(2l+1)|A|^2}{mk}.$$

The first term comes from the normalization of $|\psi|^2$, the second from the orthogonality relation of the Legendre polynomials, the third from the expression for the current of the incoming wave, and the last from the integration over ϕ. A similar calculation gives N_{out}^l:

$$N_{\text{out}}^l = \frac{\pi\hbar(2l+1)|A|^2}{mk}|S_l|^2.$$

The condition $N_{\text{out}}^l \leq N_{\text{in}}^l$ implies that $|S_l| \leq 1$. The inelastic cross section in the lth partial wave is, up to the flux factor $\mathcal{F} = \hbar k|A|^2/m$, just the difference between the numbers of incoming and outgoing particles:

$$\sigma_{\text{inel}}^l = \frac{1}{\mathcal{F}}\left(N_{\text{in}}^l - N_{\text{out}}^l\right) = \frac{\pi\hbar(2l+1)|A|^2}{k^2}(1-|S_l|^2),$$

and the total inelastic cross section becomes

$$\sigma_{\text{inel}} = \frac{\pi}{k^2}\sum_{l=0}^{\infty}(2l+1)(1-|S_l|^2) \ . \tag{12.51}$$

If $N_{\text{in}}^l = N_{\text{out}}^l$, the number of outgoing particles is equal to the number of incoming ones, the scattering is elastic in the lth partial wave, and $|S_l(k)| = 1$, $S_l(k) = \exp[2i\delta_l(k)]$. The condition $|S_l| \leq 1$ implies $\sigma_{\text{inel}}^l \geq 0$, as it should. The sum of the elastic and inelastic cross sections is the total cross section:

$$\sigma_{\text{tot}} = \frac{2\pi}{k^2}\sum_{l=0}^{\infty}(2l+1)(1-\operatorname{Re} S_l) \ . \tag{12.52}$$

The presence of inelastic channels implies that $(1 - S_l) \neq 0$, and so in quantum physics it is not possible to have purely inelastic scattering, whereas in classical physics particles can be sent onto perfectly absorbing targets, without undergoing elastic scattering. If the absorption in the lth partial wave is total, which corresponds to $N_l^{\text{out}} = 0$ and therefore to $S_l = 0$, then

$$\sigma_{\text{el}} = \sigma_{\text{inel}}^l = \frac{\pi}{k^2}(2l+1). \tag{12.53}$$

By comparison, the maximum elastic cross section is

$$\sigma_{\text{el,max}}^l = \frac{4\pi}{k^2}(2l+1).$$

An important consequence of the intertwining of elastic and inelastic scattering is the optical theorem. Let us calculate the imaginary part of the forward scattering amplitude[10] Im $f(\theta = 0)$ using $P_l(1) = 1$:

$$\text{Im } f(\theta = 0) = \frac{1}{2k} \sum_{l=0}^{\infty} (2l+1)(1 - \text{Re } S_l).$$

Comparing this with (12.52) for σ_{tot}, we see that

$$\boxed{\sigma_{\text{tot}} = \frac{4\pi}{k} \text{Im} f(\theta = 0)}. \tag{12.54}$$

This relation is the *optical theorem*, which relates the total cross section to the imaginary part of the forward scattering. The proof of the theorem shows that it follows from probability conservation.

12.3.2 The optical potential

Inelastic scattering can be taken into account by introducing a complex potential in the Schrödinger equation. Actually, if we repeat the proof in Section 9.2.2 of the continuity equation for the current $\vec{\nabla} \cdot \vec{j} = 0$ in the case of a stationary wave $\psi_{\vec{k}}(\vec{r})$, we see that this equation is not satisfied if the potential is complex:

$$\vec{\nabla} \cdot \vec{j} = \frac{2}{\hbar} \text{Im } V(\vec{r})|\psi_{\vec{k}}(\vec{r})|^2. \tag{12.55}$$

Of course, we recover the result $\vec{\nabla} \cdot \vec{j} = 0$ in the case of the real potential used in Section 9.2.2. The number of particles absorbed per unit time is equal to the incident flux multiplied by the inelastic cross section. To calculate the number of absorbed particles, we imagine that the target is surrounded by a large sphere and calculate the flux of $\vec{j}$ through the surface $\mathcal{S}$ of the sphere:

$$-\int_{\mathcal{S}} \vec{j} \cdot d\vec{\mathcal{S}} = -\int_{\mathcal{V}} \vec{\nabla} \cdot \vec{j} \, d^3 r = -\frac{2}{\hbar} \int_{\mathcal{V}} \text{Im } V(\vec{r})|\psi_{\vec{k}}(\vec{r})|^2 \, d^3 r,$$

where $\mathcal{V}$ is the volume of the sphere and the minus sign corresponds to the fact that $d\vec{\mathcal{S}}$ points toward the outside. We then have

$$\sigma_{\text{in}} = -\frac{2m}{\hbar^2 k} \int \text{Im } V(\vec{r})|\psi_{\vec{k}}(\vec{r})|^2 \, d^3 r, \tag{12.56}$$

where we have integrated over all space because the potential is assumed to have finite range or to fall off sufficiently rapidly at infinity. From now on to the end of this chapter the potential $V(\vec{r})$ will be arbitrary, not necessarily invariant under rotation. Equation (12.56) implies that the imaginary part of $V(\vec{r})$ must be negative, Im $V(\vec{r}) \leq 0$.

---

[10] This quantity cannot be measured directly, because in the forward direction one finds mostly incident particles which have not undergone a collision. It is necessary to take the $\theta \to 0$ limit of $f(\theta)$. See also Footnote 3.

A complex potential with negative imaginary part $V(\vec{r})$ is called an *optical potential*. Such a potential is useful when we are interested not in the details of inelastic processes, but only in their effects on elastic processes. It is often used, in particular, in neutron–nucleus scattering. At low energies this complex potential can be represented as an effective potential of the type (12.41) with a complex scattering length $a = a_1 + ia_2$, $a_2 < 0$. Under these conditions $\operatorname{Im} f = -a_2$ and the total cross section is very large compared with the elastic cross section:

$$\sigma_{\text{tot}} \simeq \sigma_{\text{in}} \simeq \frac{4\pi}{k}|a_2| \gg \sigma_{\text{el}} = 4\pi a_1^2.$$

The proportionality of σ_{in} to $1/k$, or to $1/v$, where v is the speed of the incident neutrons, is an extremely important result: *the cross section for neutron absorption grows as $1/v$ when $v \to 0$.* This implies, for example, that neutrons must be slowed down in order to obtain sizable cross sections for uranium fission in a nuclear reactor. Another example is the use of cadmium to absorb neutrons: the scattering length is complex, with $a_1 = -3.8\,\text{fm}$ and $a_2 = -1.2\,\text{fm}$.

Let us rewrite the optical theorem using (12.56):

$$\operatorname{Im} f(\theta = 0) = \frac{k}{4\pi}\int |f(\Omega)|^2\,\mathrm{d}\Omega - \frac{m}{2\pi\hbar^2}\int \operatorname{Im} V(\vec{r})|\psi_{\vec{k}}(\vec{r})|^2\,\mathrm{d}^3r. \tag{12.57}$$

This equation can be generalized. We define the scattering amplitude $f(k\hat{r}, \vec{k})$ using the solution (12.9) of the Schrödinger equation:

$$\psi_{\vec{k}}^{(+)}(\vec{r}) = \mathrm{e}^{\mathrm{i}\vec{k}\cdot\vec{r}} + f(k\hat{r}, \vec{k})\frac{\mathrm{e}^{\mathrm{i}kr}}{r}.$$

Since the potential is not assumed to be invariant under rotation, the scattering amplitude depends on $\hat{r}$ and $\vec{k}$, and not only on k and the angle between $\hat{r}$ and $\hat{k}$. It is then possible to prove the *unitarity relation*:[11]

$$\frac{1}{2\mathrm{i}}\left[f(\vec{k}', \vec{k}) - f^*(\vec{k}, \vec{k}')\right] = \frac{k}{4\pi}\int f^*(k\hat{r}, \vec{k}')f(k\hat{r}, \vec{k})\,\mathrm{d}^2\hat{r}$$

$$-\frac{m}{2\pi\hbar^2}\int \operatorname{Im} V(\vec{r})\,[\psi_{\vec{k}'}^{(+)}(\vec{r})]^*\psi_{\vec{k}}^{(+)}(\vec{r})\mathrm{d}^3r. \tag{12.58}$$

Invariance under time reversal implies that $f(\vec{k}', \vec{k}) = f(-\vec{k}, -\vec{k}')$, and invariance under parity implies that $f(\vec{k}', \vec{k}) = f(-\vec{k}', -\vec{k})$. If these two invariances are valid, $f(\vec{k}', \vec{k}) = f(\vec{k}, \vec{k}')$ and

$$\frac{1}{2\mathrm{i}}\left[f(\vec{k}', \vec{k}) - f^*(\vec{k}, \vec{k}')\right] = \operatorname{Im} f(\vec{k}', \vec{k})$$

in (12.58). We then recover (12.57) by taking $\vec{k}' = \vec{k}$.

---

[11] See, for example, Landau and Lifschitz [1958], Section 124.

12.4 Formal aspects

12.4.1 The integral equation of scattering

In this section we shall take up several points that we have previously glossed over, in order to clarify certain arguments we have made above. First we shall prove an equation, the integral equation of scattering, which will allow us to justify the asymptotic expression (12.10) and will also prove useful for other aspects of scattering theory. The proof rests on the expression for the *Green's functions* $G(\vec{r})$ of the Schrödinger equation when $V = 0$, which satisfy

$$(\nabla^2 + k^2)G(\vec{r}) = \delta(\vec{r}). \tag{12.59}$$

In general, the Green's functions G of a wave equation $\mathcal{L}\psi = 0$ are defined from $\mathcal{L}G = \delta(\vec{r})$. The solution of an equation of this type is not unique and the precise form of function that must be used for a given problem is actually fixed by the boundary conditions. We shall need the Green's functions $G^{(\pm)}(\vec{r})$ corresponding to an outgoing spherical wave $[G^{(+)}(\vec{r})]$ and an incoming spherical wave $[G^{(-)}(\vec{r})]$. They are given by[12]

$$G^{(\pm)}(\vec{r}) = -\frac{1}{4\pi}\frac{e^{\pm ikr}}{r}. \tag{12.60}$$

We can immediately verify (12.59):

$$\nabla^2 \frac{e^{\pm ikr}}{r} = \nabla^2 \left[\frac{e^{\pm ikr} - 1}{r} \right] + \nabla^2 \frac{1}{r}$$

$$= \frac{1}{r}\frac{d^2}{dr^2}e^{\pm ikr} - 4\pi\delta(\vec{r})$$

$$= -k^2 \frac{e^{\pm ikr}}{r} - 4\pi\delta(\vec{r}),$$

where we have used (12.42) and the fact that the function $[\exp(ikr) - 1]/r$ is regular at $r = 0$.

Let us examine the behavior of the function $G^{(+)}(\vec{r} - \vec{r}')$ when $r \to \infty$ with r' remaining finite. In this limit

$$|\vec{r} - \vec{r}'| = r - \hat{r}\cdot\vec{r}' + O\left(\frac{r'^2}{r}\right)$$

and, defining $\vec{k}' = k\hat{r}$, we obtain

$$G^{(+)}(\vec{r} - \vec{r}') = -\frac{e^{ik|\vec{r} - \vec{r}'|}}{4\pi r} = -\frac{e^{ikr}\,e^{i\vec{k}'\cdot\vec{r}'}}{4\pi r} + O\left(k\frac{r'^2}{r^2}\right), \tag{12.61}$$

---

[12] Any combination $\lambda G^{(+)} + (1-\lambda)G^{(-)} + G_h$, where G_h is a solution of the homogeneous wave equation, also satisfies (12.59).

which shows that $G^{(+)}$ does behave as an outgoing spherical wave. The function $\psi_{\vec{k}}^{(+)}(\vec{r})$ defined implicitly as

$$\psi_{\vec{k}}^{(+)}(\vec{r}) = e^{i\vec{k}\cdot\vec{r}} + \frac{2m}{\hbar^2} \int G^{(+)}(\vec{r}-\vec{r}')V(\vec{r}')\psi_{\vec{k}}^{(+)}(\vec{r}')d^3r' \qquad (12.62)$$

obeys the Schrödinger equation. Actually, using (12.59) we have

$$(\nabla^2 + k^2)\psi_{\vec{k}}^{(+)}(\vec{r}) = \frac{2m}{\hbar^2} \int \delta(\vec{r}-\vec{r}')V(\vec{r}')\psi_{\vec{k}}^{(+)}(\vec{r}') = \frac{2m}{\hbar^2} V(\vec{r}) \psi_{\vec{k}}^{(+)}(\vec{r}).$$

Equation (12.62) is called the *integral equation of scattering*. The essential point is that $\psi_{\vec{k}}^{(+)}(\vec{r})$ does behave asymptotically as (12.9). Using (12.61) and (12.62) for $r \to \infty$, we find

$$\psi_{\vec{k}}^{(+)}(\vec{r}) \simeq e^{i\vec{k}\cdot\vec{r}} - \frac{m}{2\pi\hbar^2} \frac{e^{ikr}}{r} \int e^{-i\vec{k}'\cdot\vec{r}'} V(\vec{r}')\psi_{\vec{k}}^{(+)}(\vec{r}')d^3r'. \qquad (12.63)$$

We can immediately identify the scattering amplitude $f(\Omega)$ using (12.9):

$$f(\Omega) = f(\vec{k}',\vec{k}) = -\frac{m}{2\pi\hbar^2} \int e^{-i\vec{k}'\cdot\vec{r}'} V(\vec{r}')\psi_{\vec{k}}^{(+)}(\vec{r}')d^3r'. \qquad (12.64)$$

This equation is *exact*, but of course it is necessary to know $\psi_{\vec{k}}^{(+)}(\vec{r})$, and so we cannot avoid solving the Schrödinger equation! We can solve (12.63) approximately by iteration. The first iteration will be

$$\psi_{\vec{k}}^{(+)}(\vec{r}) = e^{i\vec{k}\cdot\vec{r}}.$$

Substituting this into (12.64), we obtain $f(\vec{k}',\vec{k})$ in the *Born approximation*:

$$f_{\mathrm{B}}(\vec{k}',\vec{k}) = -\frac{m}{2\pi\hbar^2} \int e^{-i\vec{q}\cdot\vec{r}} V(\vec{r})d^3r \quad . \qquad (12.65)$$

The vector $\vec{q} = \vec{k}' - \vec{k}$ is the wave vector transfer, $\hbar\vec{q}$ is the *momentum transfer*, and f_{B} is the Fourier transform of the potential with respect to $\vec{q}$. We note that

$$q = 2k\sin\frac{\theta}{2}$$

and that f_{B} depends only on the combination $k\sin(\theta/2)$ of k and θ if the potential is spherically symmetric. This feature is of course specific to the Born approximation. It is difficult to state the criteria for validity of the Born approximation precisely: generally speaking, the energy should be high or the potential should be weak. In the case of Coulomb scattering, the Born approximation gives the exact result for the *cross section* (but not the amplitude) at any energy, far outside its theoretical region of validity (Exercise 12.5.4).

12.4.2 Scattering of a wave packet

A second point that must be justified is the use of a stationary formalism, whereas particle scattering is fundamentally a time-dependent process. This forces us to study the scattering of a wave packet. We assume that we have a wave packet centered about a momentum $\hbar \vec{k}_0$ with a dispersion $\Delta k \ll k_0$, and we also assume that the dimension $\Delta r \sim 1/\Delta k$ of the wave packet is very small compared with the characteristic lengths in the experiment, for example the distance between the target and the detector. A free wave packet is described by an expression which is the three-dimensional generalization of (9.41):

$$\varphi(\vec{r}, t) = \int \frac{d^3 k}{(2\pi)^3} A(\vec{k}) \exp\left[i\vec{k} \cdot \vec{r} - i\omega_k t\right] \qquad (12.66)$$

with $\omega_k = \hbar k^2/2m$, the average frequency being $\omega_0 = \hbar k_0^2/2m$. In Section 9.1.4 we showed that if the condition $(\Delta k)^2 \hbar t/m \ll 1$ is satisfied (which is nearly always the case), we can neglect the spreading of the wave packet, and (12.66) in the form (9.48) generalized to three dimensions (with the change of notation $\bar{k} \to k_0$, $v_g \to v_0$) becomes

$$\varphi(\vec{r}, t) \simeq e^{i\omega_0 t} \varphi(\vec{r} - \vec{v}_0 t, t = 0), \qquad (12.67)$$

where the group velocity $\vec{v}_0 = \hbar \vec{k}_0/m$. This implies that $|\varphi(\vec{r}, t)|$ is negligible if $|\vec{r} - \vec{v}_0 t| \gg \Delta r$, that is, if $|\vec{r} - \vec{v}_0 t|$ is large compared with the extent Δr of the wave packet. The time-dependent wave function $\psi_{\vec{k}}^{(+)}(\vec{r}, t)$ in the presence of a potential $V(\vec{r})$ is obtained by replacing the plane wave $\exp(i\vec{k} \cdot \vec{r})$ in the expression for a wave packet (12.66) by $\psi_{\vec{k}}^{(+)}(\vec{r})$. The resulting expression is actually a solution of the time-dependent Schrödinger equation in the presence of the potential $V(\vec{r})$ with the behavior of an outgoing spherical wave. We decompose the wave function $\psi_{\vec{k}}^{(+)}(\vec{r}, t)$ into a free part and a scattered part:

$$\psi_{\vec{k}}^{(+)}(\vec{r}, t) = \varphi(\vec{r}, t) + \psi_{\text{scatt}}(\vec{r}, t).$$

When the wave packet is far from the target, $\psi_{\vec{k}}^{(+)}(\vec{r})$ can be replaced by its asymptotic form (12.63):

$$\psi_{\vec{k}}^{(+)}(\vec{r}) \to e^{i\vec{k} \cdot \vec{r}} + f(k\hat{r}, \vec{k}) \frac{e^{ikr}}{r},$$

and then

$$\psi_{\text{scatt}}(\vec{r}, t) = \int \frac{d^3 k}{(2\pi)^3} A(\vec{k}) f(k\hat{r}, \vec{k}) \frac{e^{ikr}}{r} e^{-i\omega_k t}.$$

We assume that $f(k\hat{r}, \vec{k})$ varies sufficiently slowly with $\vec{k}$.[13] Under these conditions

$$f(k\hat{r}, \vec{k}) \simeq f(k_0 \hat{r}, \vec{k}_0),$$

---

[13] This condition may not be satisfied in the presence of a resonance.

and the scattered part is

$$\psi_{\text{scatt}}(\vec{r}, t) \simeq \frac{f(k_0\hat{r}, \vec{k}_0)}{r} \int \frac{d^3k}{(2\pi)^3} \, A(\vec{k}) \exp[i(kr - \omega_k t)]. \qquad (12.68)$$

Next we note that

$$k = ([\vec{k}_0 + (\vec{k} - \vec{k}_0)]^2)^{1/2} = k_0 + \hat{k}_0 \cdot (\vec{k} - \vec{k}_0) + O\left[\frac{(\Delta k)^2}{k_0}\right] = \hat{k}_0 \cdot \vec{k} + O\left[\frac{(\Delta k)^2}{k_0}\right].$$

Since the characteristic time $t \sim r/v_0 = mr/\hbar k_0$, we have

$$\frac{(\Delta k)^2 \, r}{k_0} \simeq \frac{(\Delta k)^2 \hbar t}{m} \ll 1$$

and kr in (12.68) can be replaced by $r\hat{k}_0 \cdot \vec{k}$, which gives

$$\psi_{\text{scatt}}(\vec{r}, t) \simeq \frac{f(k_0\hat{r}, \vec{k}_0)}{r} \varphi(r\hat{k}_0, t) \simeq \frac{f(k_0\hat{r}, \vec{k}_0)}{r} \varphi[(r - v_0 t)\hat{k}_0, 0] e^{i\omega_0 t}.$$

When t is large and negative, $|(r - v_0 t)| \gg \Delta r$, and since $\varphi(r', 0)$ is negligible for $r' \gg \Delta r$, we have $\psi_{\text{scatt}} \to 0$ and the wave packet tends to a free wave packet: since the wave packet does not overlap with the potential, ψ_{scatt} is practically zero:

$$\lim_{t \to -\infty} \psi(\vec{r}, t) = \varphi(\vec{r}, t).$$

The wave packet interacts with the target for $t \sim 0$, and when $t \to +\infty$

$$\psi_{\text{scatt}}(\vec{r}, t) \simeq \frac{f(k_0\hat{r}, \vec{k}_0)}{r} \varphi[(r - v_0 t)\hat{k}_0, 0] \, e^{i\omega_0 t}.$$

We therefore recover the wave packet in a direction different from the initial one, modulated by the scattering amplitude $f(k_0\hat{r}, \vec{k}_0)$ and propagating radially with a speed v_0.

Now we can calculate the probability dp for triggering a detector of area $dS = r^2 d\Omega$ located in the direction $\vec{r}$. Since the current at time t is $v_0 |\psi_{\text{scatt}}|^2 \hat{r}$, the probability for triggering the detector is

$$dp = v_0 r^2 d\Omega \int_{-\infty}^{+\infty} |\psi_{\text{scatt}}(\vec{r}, t)|^2 dt$$

$$= v_0 \, d\Omega |f(k_0\hat{r}, \vec{k}_0)|^2 \int_{-\infty}^{+\infty} |\varphi[(r - v_0 t)\hat{k}_0, 0]|^2 dt.$$

On the other hand, the probability for the incident particle to cross a unit surface perpendicular to the incident beam is

$$\int_{-\infty}^{+\infty} |\varphi[(r - v_0 t)\hat{k}_0, 0]|^2 dt,$$

and from the definition (12.1) we find the cross section

$$\frac{d\sigma}{d\Omega} = |f(k_0\hat{r}, \vec{k}_0)|^2 = |f(\Omega)|^2, \qquad (12.69)$$

which completes the justification of (12.12).

12.5 Exercises

12.5.1 *The Gamow peak*

1. We wish to evaluate the cross section for the reaction

$$^2\text{H} + {}^3\text{H} \rightarrow {}^4\text{He} + \text{n} \tag{12.70}$$

occurring in the interior of a star at a temperature of the order of 10^7 K. We have chosen this particular reaction to be specific, but our discussion will apply to any nuclear reaction occurring in a star between light nuclei. Show that the kinetic energy of the incident ^{2}H and ^{3}H nuclei is of the order of keV. Why are the atoms completely ionized? The following relation is often useful in nuclear physics. In a system of units where $\hbar = c = 1$, the relation between the units fermi ($\equiv$ femtometer) and MeV can be written as

$$1 \, \text{fm}^{-1} \simeq 200 \, \text{MeV}.$$

Verify this relation. The potential $V(r)$ between the two incident nuclei is the repulsive Coulomb potential $V(r) = e^2/r$ for $r > R$ and an attractive nuclear potential for $r \le R$, with $R \simeq 1$ fm. Show that e^2/R is very large compared with the kinetic energy E of the incident nuclei.

2. Show that in classical physics the two nuclei cannot approach each other to distances less than $r_0 = e^2/E$, and the nuclear reaction (12.70) cannot occur. In quantum physics the reaction is possible owing to the tunnel effect. Using (9.106), show that the probability for tunneling is

$$\mathsf{p}_\text{T}(E) = \exp\left(-\frac{2}{\hbar} \int_R^{r_0} \left[2\mu\left(\frac{e^2}{r} - E\right)\right]^{1/2} dr\right),$$

where μ is the reduced mass: $E = \mu v^2/2$, v being the relative speed of the two nuclei. Show that $\mu \simeq (6/5)m_\text{p}$, where the proton mass $m_\text{p} \simeq 940 \, \text{MeV} \, c^{-2}$. To calculate $\mathsf{p}_\text{T}(E)$ we can make the change of variable

$$u^2 = \frac{e^2}{r} - E.$$

A useful integral is

$$\int \frac{u^2 du}{(u^2 + a^2)^2} = \frac{1}{2a} \tan^{-1} \frac{u}{a} - \frac{u}{2(u^2 + a^2)}.$$

Show that

$$\mathsf{p}_\text{T}(E) \simeq \exp\left(-\sqrt{\frac{E}{E_\text{B}}}\right), \quad E_\text{B} = 2\pi^2 \alpha^2 \mu c^2$$

with $\alpha = e^2/\hbar c \simeq 1/137$. Give the value of E_B in MeV.

3. Justify the approximate form of the cross section for the reaction (12.70):

$$\sigma(E) \sim \frac{4\pi}{k^2} \mathsf{p}_\text{T}(E)$$

assuming that the nuclear reaction occurs as soon as the nuclei come into contact with each other; k is the wave vector and $E = \hbar^2 k^2/2\mu$.

4. According to (12.1), the number of nuclear reactions (12.70) per unit time is $n_\text{i} n_\text{t} v\sigma(v)$, where n_i and n_t are the densities of the incident nuclei and the target nuclei. However, the speeds are

not fixed, and to obtain the reaction rate in a star it is necessary to average over the Maxwell velocity distribution:

$$p_M(v) = \left(\frac{\mu}{2\pi k_B T}\right)^{3/2} \exp\left(-\frac{\mu v^2}{2k_B T}\right).$$

The physically relevant quantity is the average $\langle v\sigma \rangle$. By integrating over angles, show that

$$\langle v\sigma \rangle = 4\pi \left(\frac{\mu}{2\pi k_B T}\right)^{3/2} \int_0^\infty dv\, v^3 \sigma(v) \exp\left(-\frac{\mu v^2}{2k_B T}\right).$$

Then, making the change of variable $v \to E$, deduce that

$$\langle v\sigma \rangle = \frac{16\pi^2 \hbar^2}{\mu^3} \left(\frac{\mu}{2\pi k_B T}\right)^{3/2} \int_0^\infty dE\, e^{-E/(k_B T)} e^{-\sqrt{E_B/E}}. \tag{12.71}$$

Show that the integrand in (12.72) has a sharp peak at an energy $E = E_0$ with

$$E_0 = \left(\frac{1}{2} k_B T \sqrt{E_B}\right)^{2/3},$$

and that the width of the peak ΔE is given by

$$\Delta E \propto E_B^{1/6} (k_B T)^{5/6}.$$

This peak is called the *Gamow peak*, and it determines the energy E_0 at which the reaction (12.70) has maximum probability: the reaction rate in the star is controlled by E_0. Obtain a numerical estimate of the position and width of the peak.

12.5.2 *Low-energy neutron scattering by a hydrogen molecule*

1. First let us consider the scattering of a particle by two different nuclei 1 and 2 of a diatomic molecule neglecting spin. The center of the molecule is located at the origin, and the detector is located at a distance r from the target. The nuclei 1 and 2 are located at the points $\vec{R}/2$ and $-\vec{R}/2$, with $R \ll r$. Show that the amplitude for scattering by the molecule is

$$f = a_1 \exp\left(-\frac{i}{2}\vec{q}\cdot\vec{R}\right) + a_2 \exp\left(\frac{i}{2}\vec{q}\cdot\vec{R}\right),$$

Denote by $\vec{k}$ the wave vector of the incident particles, $\vec{k}' = k\hat{r}$, $\hbar\vec{q} = \hbar(\vec{k}' - \vec{k})$ is the momentum transfer, and a_1 and a_2 are the scattering lengths for the nuclei 1 and 2. Sketch the cross section as a function of the angle θ between $\vec{k}'$ and $\vec{k}$ when $qR \sim 1$.

2. Now we consider the case of neutron scattering on a hydrogen molecule taking into account the neutron and proton spins. We assume that the energy is low enough that $qR \ll 1$. What must the energy be in eV for this condition to be satisfied? If the neutrons are produced in a reactor, to what temperature must they be cooled (cf. Section 1.4.2)? The total spin $\vec{S}$ of the molecule is defined as

$$\hbar\vec{S} = \frac{1}{2}\hbar(\vec{\sigma}_1 + \vec{\sigma}_2),$$

where $\vec{\sigma}_1$ and $\vec{\sigma}_2$ are the Pauli matrices describing the spins of the two protons. Show that the scattering amplitude is written in spin space as a function of the scattering lengths a_s and a_t as

$$\hat{f} = \frac{1}{2}(a_s + 3a_t)I + \frac{1}{2}(a_t - a_s)(\vec{\sigma}_n \cdot \vec{S}).$$

3. If the neutron–proton interaction is dealt with using an effective potential (12.41), the constant g will be fixed by the characteristics of the potential. Show that owing to a reduced-mass effect, it is necessary to use $4a/3$ for the scattering length on protons bound in a hydrogen molecule, where a is the scattering length for a neutron on a free proton. The cross section is therefore multiplied by a factor of 16/9; this is an effect of the chemical bond. This reduced-mass effect occurs as long as the neutron energy is so low that the vibrational levels of the molecule are not excited.

4. The hydrogen molecule can exist in two spin states: the parahydrogen state of spin zero and the orthohydrogen state of spin one. What is the neutron–parahydrogen total cross section? Is it sensitive to the sign of a_s?

5. Calculate the neutron–orthohydrogen total cross section assuming that the molecule is unpolarized. Hint: prove the identity

$$\mathrm{Tr}(A \otimes B)^2 = (\mathrm{Tr}\, A^2)(\mathrm{Tr}\, B^2).$$

12.5.3 Analytic properties of the neutron–proton scattering amplitude

The objective of this exercise is to relate the properties of bound states and resonances to the scattering amplitude. We shall limit ourselves to the s-wave. We neglect the neutron–proton mass difference and define $M \simeq m_p \simeq m_n$, so that the reduced mass is $M/2$. All spin effects are neglected.

1. Let $u(r)$ be the (real) radial wave function of a bound state, here the deuteron. It is characterized by its asymptotic behavior $\propto \exp(-\kappa r)$ and its asymptotic normalization N:

$$r \to \infty : u(r) \simeq Ne^{-\kappa r} \quad \text{with} \quad \int_0^\infty u^2(r)\,dr = 1.$$

Show that in the case of the spherical well of Fig. 12.4 of range R and depth V_0,

$$N^2 = \frac{2\kappa k'^2 e^{2\kappa r}}{(\kappa^2 + k'^2)(1 + \kappa R)}$$

with $k' = \sqrt{M(V_0 - B)}$ and $\kappa = \sqrt{MB}$, where B is the binding energy. Sketch $u(r)$ qualitatively.

2. Let $g(k, r)$ be a solution of the radial equation with the asymptotic behavior

$$r \to \infty : g(k, r) \propto e^{-ikr} \quad \text{with} \quad k = \frac{\sqrt{ME}}{\hbar}.$$

Show that the wave function $u(k, r)$ is given by

$$u(k, r) = g(-k, r)g(k) - g(k, -r)g(-k), \quad g(k) = g(k, r = 0),$$

and that the S-matrix element $S(k)$ is

$$S(k) = e^{2i\delta(k)} = \frac{g(k)}{g(-k)}.$$

3. We analytically continue $g(k, r)$ to complex values of k. Show that

$$g^*(k, r) = g(-k^*, r), \quad S^*(k^*) = \frac{1}{S(k)} = S(-k).$$

4. Calculate $g(k)$ and $S(k)$ for the spherical well and show that $g(k)$ is an entire function of k (that is, it is analytic for all k).

5. It can be proved that $g(k)$ is analytic in the half-plane $\text{Im}\, k < \mu/2$ for a potential which falls off more rapidly than $\exp(-\mu r)$ when $r \to \infty$. This result will be used in the rest of this exercise. Show that if $S(k)$ has a pole on the imaginary axis, $k = i\kappa$, $0 < \kappa < \mu/2$, this pole corresponds to a bound state of the potential. Show that if $S(k)$ has a pole at $k = h - ib$, $|b| < \mu/2$, then necessarily $b > 0$.

6. The case of the pole at $k = h - ib$, $b > 0$, is that of a *resonance*. Show that a choice for $S(k)$ satisfying the conditions of question 3 is

$$S(k) = \frac{(k - h - ib)(k + h - ib)}{(k - h + ib)(k + h + ib)} \simeq \frac{k - h - ib}{k - h + ib} \quad \text{for } k \sim h.$$

Assuming that $b \ll h$, find the behavior of the phase shift $\delta(k)$ as a function of k by showing that

$$\cot \delta = \frac{h - k}{b}.$$

Prove that δ passes through $\pi/2$ for $k = h$ and that the cross section can be written in the so-called *Breit–Wigner* form:

$$\sigma(E) = \frac{2\pi \hbar^2}{ME} \frac{\hbar^2 \Gamma^2/4}{(E - E_0)^2 + \hbar^2 \Gamma^2/4}. \tag{12.72}$$

Relate E_0 and Γ to b and h. Show that $h = 0$ corresponds to a virtual state.

7. Prove the relation

$$\left[u' \frac{\partial u}{\partial k} - u \frac{\partial u'}{\partial k} \right]_0^r = 2k \int_0^r u^2(r')\,dr', \quad u' = \frac{\partial u}{\partial r}.$$

By studying this expression for $r \to 0$ and $r \to \infty$, show that near a pole $k = i\kappa$

$$S(k) \simeq \frac{-iN^2}{k - i\kappa}.$$

8. Show that the function

$$k \cot \delta(k) = ik \frac{g(k) + g(-k)}{g(k) - g(-k)}$$

is analytic in k near $k = 0$, that it tends to a constant for $k \to 0$, and that it is an even function of k. Show that we can write

$$k \cot \delta(k) = -\frac{1}{a} + \frac{1}{2} r_0 k^2 + O(k^4).$$

Demonstrate the relations

$$r_0 = \frac{2}{\kappa}\left(1 - \frac{1}{\kappa a}\right), \quad N^2 = \frac{2\kappa}{1 - \kappa r_0}$$

between the deuteron parameters (κ, N) and the low-energy scattering characteristics (a, r_0). Calculate r_0 given that $B = 2.22$ MeV and $a = 5.40$ fm and compare this with the experimental result $r_0 = 1.73$ fm.

12.5.4 The Born approximation

1. Calculate the scattering amplitude $f_B(\vec{q})$, $\vec{q} = \vec{k}' - \vec{k}$, in the Born approximation when the potential has the so-called Yukawa form:

$$V(r) = V_0 \frac{e^{-\mu r}}{\mu r}.$$

Find $d\sigma/d\Omega$ and σ_{tot}.
2. Examine the limit $\mu \to 0$ with $V_0/\mu \to e^2 =$ const, where the Yukawa potential tends to the Coulomb potential $V(r) = e^2/r$. Show that

$$\frac{d\sigma}{d\Omega} = \frac{e^4}{16E^2 \sin^4 \theta/2}, \tag{12.73}$$

where $E = \hbar^2 k^2/2m$ is the incident energy. This result was obtained by Rutherford using arguments from classical mechanics (quantum mechanics did not yet exist!), and it is called the *Rutherford cross section*. This is also the result obtained by a rigorous treatment of the Coulomb potential in quantum mechanics. It is remarkable that the Born approximation, which is of more than doubtful validity in this case, gives the correct result for the cross section (but not for the amplitude $f(\theta)$).

12.5.5 Neutron optics

1. *Scattering by a thin plate.* We consider a low-energy neutron beam of vacuum wave vector k which passes through a very thin plate of thickness δ perpendicularly to the plate, and at first we neglect spin effects. The neutrons are detected after their passage through the plate at a point z on the axis Oz perpendicular to the plate, with the origin O chosen to lie at the center of the plate. If a neutron is scattered by a nucleus of the plate located a distance s from O, show that the probability amplitude for observing the scattered neutron at z is

$$\varphi_s = -\frac{a}{r} e^{ikr}, \quad r = \sqrt{s^2 + z^2},$$

where a is the scattering length. The probability amplitude for finding a neutron at z is the sum of the incident wave $\exp(ikz)$ and the wave scattered by the plate:

$$\varphi(z) = e^{ikz} - a \sum \frac{e^{ikr}}{r},$$

where the sum runs over all the nuclei of the plate. Show that

$$\varphi(z) = e^{ikz} - 2\pi a\rho\delta \left.\frac{e^{ikr}}{ik}\right|_z^\infty,$$

where ρ is the volume density of nuclei. The limit $r \to \infty$ gives zero if we average over oscillations, and we find

$$\varphi(z) = \left(1 - 2i\pi \frac{a\rho\delta}{k}\right) e^{ikz}.$$

2. *The index of refraction.* When the neutrons pass through the plate it behaves like a medium with index of refraction n, and so, as in optics, the wave vector is transformed as $k \to k' = nk$ or, equivalently, the wavelength $\lambda \to \lambda' = \lambda/n$. Comparing with the result of question 1 when $(n-1)k\delta \ll 1$, show that

$$n = 1 - \frac{2\pi a\rho}{k^2} = 1 - \frac{a\rho\lambda^2}{2\pi}.$$

When $n < 1$ a beam of neutrons arriving at grazing incidence on the flat surface of a crystal can undergo total reflection (the difference between the indices of refraction of the vacuum and air is negligible). If the angle of incidence is $(\pi/2 - \theta)$, $\theta \ll 1$, show that critical incidence is

$$\theta_c = \lambda \left(\frac{\rho a}{\pi}\right)^{1/2}.$$

Estimate θ_c numerically for the following typical values: $\lambda = 1\,\text{nm}$, $\rho = 10^{29}\,\text{m}^{-3}$, and $a = 10\,\text{fm}$. The property of total reflection is used to construct the neutron guides used in instruments for neutron optics.

3. *Spin effects: spin-1/2 nuclei.* In the following questions we study effects related to the neutron and nuclear spins. Taking the results of Exercise 3.3.9 and using (12.46), show that the amplitudes f_a, f_b, and f_c of this exercise are given as functions of the triplet and singlet scattering lengths a_t and a_s for spin-1/2 nuclei by

$$f_a = -\frac{1}{2}(a_t + a_s), \quad f_b = -\frac{1}{2}(a_t - a_s), \quad f_c = -a_t.$$

Show that the intensity scattered by the crystal is

$$\mathcal{I} = \frac{1}{16}(3a_t + a_s)^2 \sum_{i,j} e^{i\vec{q}\cdot(\vec{r}_i - \vec{r}_j)} + \frac{3\mathcal{N}}{16}(a_t - a_s)^2,$$

where $\mathcal{N}$ is the number of scattering nuclei. The first term of $\mathcal{I}$ corresponds to coherent scattering and the second to incoherent scattering (Exercise 1.6.8). By integrating $\mathcal{I}$ over angles we obtain the coherent and incoherent cross sections:

$$\sigma_{\text{coh}} = \frac{\pi}{4}(3a_t + a_s)^2, \quad \sigma_{\text{inc}} = \frac{3\pi}{4}(a_t - a_s)^2.$$

In the case of scattering by hydrogen, $a_t = 5.4$ fm and $a_s = -23.7$ fm. Evaluate σ_{coh} and σ_{inc} numerically and show that $\sigma_{\text{inc}} \gg \sigma_{\text{coh}}$. This property is peculiar to hydrogen, because in general the two cross sections are of the same order of magnitude. Show that the scattering length to be used in calculating the index of refraction is that defined by coherent scattering:

$$a_{\text{eff}} = \frac{3}{4} a_t + \frac{1}{4} a_s.$$

What is the physical interpretation of the weights 3/4 and 1/4? What is the sign of a_{eff} for hydrogen? Is it possible to obtain total reflection of neutrons on liquid hydrogen?

4. *Scattering by nuclei of spin j.* We assume that the nuclear scatterers have spin j. Let

$$\vec{I} = \vec{J} + \frac{\hbar}{2}\vec{\sigma}$$

be the total angular momentum of the nucleus + neutron system, where $\hbar\vec{\sigma}/2$ is the neutron spin operator. Show that the nucleus + neutron scattering amplitude is written in spin space as a function of the two lengths a and b as

$$-\hat{f} = a + \frac{b}{\hbar}\left(\vec{\sigma}\cdot\vec{J}\right).$$

Let $a_+ = a_{j+1/2}$ and $a_- = a_{j-1/2}$ be the two scattering lengths corresponding to scattering in the total angular momentum states $i_\pm = j \pm 1/2$. Show that

$$a_+ = a + bj, \quad a_- = a - b(j+1)$$

and, inversely,

$$a = \frac{1}{2j+1}\left[(j+1)a_+ + ja_-\right], \quad b = \frac{1}{2j+1}[a_+ - a_-].$$

5. *Coherent and incoherent scattering.* If the nuclei and neutrons are unpolarized, what are the probabilities that the scattering occurs in the states $i_+ = j+1/2$ and $i_- = j-1/2$? Using the results of Exercise 1.6.8, show that the coherent and incoherent cross sections are given by

$$\sigma_{\text{coh}} = \frac{4\pi}{(2j+1)^2}\left[(j+1)a_+ + ja_-\right]^2 = 4\pi a^2,$$

$$\sigma_{\text{inc}} = \frac{4\pi j(j+1)}{(2j+1)^2}[a_+ - a_-]^2 = 4\pi j(j+1)b^2.$$

Verify that the results of question 3 are recovered when $j = 1/2$.

12.5.6 The cross section for neutrino absorption

1. The goal of this exercise is to calculate the cross section for neutrino absorption by neutrons

$$\bar{\nu} + \text{p} \rightarrow \text{n} + e^+$$

in terms of the lifetime of the neutron, which decays via the reaction (1.2):

$$\text{n} \rightarrow \text{p} + e^- + \bar{\nu}.$$

The two processes are related because the same interaction, the weak interaction, is responsible for both phenomena. The transition matrix element for the calculation of the neutron lifetime can be written as

$$T_{fi} = G_F \mathcal{M}_{fi}\langle\varphi_f|\varphi_i\rangle,$$

where the initial- and final-state wave functions are plane waves normalized in a volume $\mathcal{V}$ and have the form

$$\frac{1}{\sqrt{\mathcal{V}}}e^{i\vec{p}\cdot\vec{r}/\hbar},$$

G_F is the *Fermi constant*, or the weak interaction coupling constant, and $\mathcal{M}_{fi}$ is a dimensionless spin-dependent matrix element.[14] The energy $E_0 = (m_n - m_p)c^2 \simeq 1.2$ MeV is the energy available in the decay (to an excellent approximation $m_\nu = 0$). Let $\vec{p}_n = 0$ (stationary neutron), $\vec{P} = \vec{p}_p$, $\vec{p} = \vec{p}_e$, and $\vec{q} = \vec{p}_\nu$ be the momenta in the initial and final states, and let $T = P^2/2m_p$ be the proton kinetic energy and E and cq be the total energies of the electron and the neutrino. Energy–momentum conservation can be written as

$$\vec{P} + \vec{p} + \vec{q} = 0, \quad T + E + cq = E_0.$$

Show that T can be neglected: $T \ll E, cq$. Let $d\Gamma/dE$ be the neutron decay rate per unit energy. It can be shown that there are no correlations between the electron and neutrino momenta. Show that under these conditions this rate is written as a function of the density of states $\mathcal{D}$ of the electron and the neutrino as

$$\frac{d\Gamma}{dE} = \frac{2\pi}{\hbar} G_F^2 \langle |\mathcal{M}_{fi}|^2 \rangle \mathcal{V}^{-2} \mathcal{D}_e(E) \mathcal{D}_\nu(E - E_0)$$

$$= \frac{2\pi}{\hbar} G_F^2 \langle |\mathcal{M}_{fi}|^2 \rangle \left[\frac{4\pi}{(2\pi\hbar)^3} \frac{pE}{c^2} \right] \left[\frac{4\pi}{(2\pi\hbar)^3} \frac{(E_0 - E)^2}{c^3} \right]$$

where $\langle |\mathcal{M}_{fi}|^2 \rangle$ represents the spin matrix element summed over the final spins and averaged over the initial spins. To obtain the lifetime $\tau = 1/\Gamma$ it is necessary to integrate over E. The integral

$$I(E_0) = \int_{m_e c^2}^{E_0} dE\, E(E_0 - E)^2 \sqrt{E^2 - m_e^2 c^2}$$

can be calculated exactly, but we shall just use an ultrarelativistic approximation neglecting the electron mass:

$$I(E_0) \simeq \int_0^{E_0} dE\, E^2(E_0 - E)^2 = \frac{E_0^5}{30}.$$

Find the expression for the lifetime:

$$\frac{1}{\tau} = \Gamma \sim \frac{G_F^2 E_0^5}{60\pi^3 \hbar (\hbar c)^6}.$$

What is the dimension of $G_F/(\hbar c)^3$? Estimate G_F from the lifetime $\tau \simeq 900$ s and compare with the exact value

$$\frac{G_F}{(\hbar c)^3} = 1.17 \times 10^{-5}\, \text{GeV}^{-2}.$$

2. Show that the differential cross section for neutrino absorption by neutrons is given by

$$\frac{d\sigma}{d\Omega} = \frac{2\pi}{\hbar c} G_F^2 \langle |\mathcal{M}_{fi}|^2 \rangle \frac{Ep}{(2\pi\hbar)^3 c^2}$$

where E is the energy of the positron e^+, and obtain

$$\sigma_{\text{tot}} \sim \frac{1}{\pi} \left[\frac{G_F}{(\hbar c)^3} \right]^2 (\hbar c)^2 E^2.$$

---

[14] $\mathcal{M}_{fi}$ also depends on two dimensionless constants of order unity, the vector coupling constant $g_V = 1$ and the axial coupling constant $g_A = 1.25$.

Verify that σ_{tot} does actually have the dimensions of area. Estimate σ_{tot} numerically for 8 MeV solar neutrinos, and show that the mean free path of solar neutrinos inside the Earth is measured in light-years.

3. The Fermi theory used in this exercise gives an isotropic cross section: the interaction occurs only in the s-wave, $l = 0$. Using (12.51), show that the result obtained for the absorption cross section cannot be valid at very high energy, and estimate the energy beyond which the Fermi theory must be modified. This modification is well known: it is the Glashow–Salam–Weinberg electroweak theory, a component of the Standard Model unifying the weak and electromagnetic interactions, with the Fermi constant related to the electron charge and the $W^{\pm}$- and Z^0-boson masses as $G_F \sim e^2/M_W^2$.

12.6 Further reading

A discussion of scattering theory more complete than that given here can be found in Merzbacher [1970], Chapters 11 and 19; Messiah [1999], Chapters X and XIX; and Landau and Lifschitz [1958], Chapters XVII and XVIII. Low-energy scattering theory is discussed by H. Bethe and Ph. Morrison, *Elementary Nuclear Theory*, New York: Wiley (1956), Chapters IX to XI, and in C. Pethick and H. Smith, *Bose–Einstein Condensation of Dilute Gases*, Cambridge: Cambridge University Press (2002), Chapter 5.

13

Identical particles

13.1 Bosons and fermions

13.1.1 Symmetry or antisymmetry of the state vector

Let us consider a state $|\Psi\rangle$ of two different particles, for example two different oxygen atoms ^{16}O and ^{18}O in their ground states, and let $|a_1\rangle$ and $|b_2\rangle$ be the respective states of these two atoms. The states $|a\rangle$ and $|b\rangle$ are, for example, eigenstates of the operators $\vec{P}$, $\vec{J}$, ... labeled by the momentum $\vec{p}$ of the atom, the atomic spin component j_z, and so on:[1]

$$|a\rangle = |\vec{p}, j_z, \ldots\rangle, \quad |b\rangle = |\vec{p}', j'_z, \ldots\rangle.$$

We use $|a_1 \otimes b_2\rangle$ to denote the two-particle state where particle 1 (^{16}O) is in the state $|a\rangle$ and particle 2 (^{18}O) is in the state $|b\rangle$; for example,[2] $|a_1 \otimes b_2\rangle = |\vec{p}_1 \otimes \vec{p}'_2\rangle$. For clarity, we can assume that the particles have interacted in the distant past and are in an entangled state $|\Psi\rangle$. The tests performed on particles 1 and 2 are clearly unrelated, as they take place in well-separated regions of space, like in the experiments discussed in Section 6.3.1. Two detectors D_1 and D_2 are used to determine $\vec{p}$, j_z, ... for each particle: D_1 detects an ^{16}O atom with momentum $\vec{p}$ and D_2 detects an ^{18}O atom with momentum $\vec{p}'$ (Fig. 13.1a), which makes it possible to perform an $|a_1 \otimes b_2\rangle$ test on the state $|\Psi\rangle$. The probability for the state $|\Psi\rangle$ to pass the $|a_1 \otimes b_2\rangle$ test is

$$p_{\Psi \to [a_1, b_2]} = |\langle a_1 \otimes b_2 | \Psi \rangle|^2. \tag{13.1}$$

One can also imagine the opposite configuration and measure the probability that the detector D_1 records an ^{18}O atom while D_2 records an ^{16}O atom (Fig. 13.1b). This is *different* from (13.1), as this probability corresponds to an $|a_2 \otimes b_1\rangle$ test, where the ^{18}O atom has momentum $\vec{p}$ and the ^{16}O atom has momentum $\vec{p}'$, so that except in special cases

$$p_{\Psi \to [a_2, b_1]} \neq p_{\Psi \to [a_1, b_2]}.$$

---

[1] The ^{16}O and ^{18}O atoms have spin 2 (the electronic state is $^3P_2$) and the ground state is five-fold degenerate. If necessary in a theoretical argument, this degeneracy can be lifted by the Zeeman effect in a magnetic field.

[2] This notation is not ideal. It suggests that particle 1 is in the momentum state $\vec{p}_1$, and not $\vec{p}$, and a better notation would be $|\vec{p}\rangle_1 \otimes |\vec{p}'\rangle_2$. However, there is no ambiguity in the case of two spins: $|+_1 \otimes -_2\rangle$, as in (13.14).

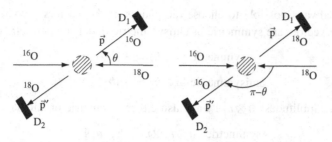

Fig. 13.1. ^{16}O–^{18}O scattering. (a) The scattering angle θ; (b) the scattering angle $(\pi - \theta)$.

Let us now assume that particles 1 and 2 are identical, for example that they are both ^{16}O atoms. If the energies involved in the interaction between these two particles are several eV, nothing will a priori distinguish this case from the preceding one, because ^{16}O–^{18}O and ^{16}O–^{16}O interactions are strictly identical. This is true up to energies of the order of MeV, where differences due to the nuclei begin to be important, and yet the two cases can differ radically, even at low energy. When the two particles are identical, it no longer makes sense to speak of an $|a_1 \otimes b_2\rangle$ test. It may be convenient to formally label the two particles and then speak of an $|a_1 \otimes b_2\rangle$ or $|a_2 \otimes b_1\rangle$ test, but such labeling has no physical significance. It is not physically acceptable to write a state in the form $|a_1 \otimes b_2\rangle$ (except if $a \equiv b$), because it cannot be stated that particle 1 is in state a and particle 2 in state b or vice versa, since the particles cannot be distinguished. The problem therefore is how to correctly define the state $|a \otimes b\rangle$. This state must be physically identical to $|b \otimes a\rangle$ and can only differ by a phase, which may depend on a and b:

$$|a \otimes b\rangle = e^{i\theta_{ab}} |b \otimes a\rangle,$$
$$|b \otimes a\rangle = e^{i\theta_{ba}} |a \otimes b\rangle. \tag{13.2}$$

These equations imply that

$$e^{i\theta_{ba}} e^{i\theta_{ab}} = 1. \tag{13.3}$$

We define the new vectors

$$|a \otimes b\rangle' = e^{i\theta_{ab}/2}|a \otimes b\rangle,$$
$$|b \otimes a\rangle' = e^{i\theta_{ba}/2}|b \otimes a\rangle. \tag{13.4}$$

Instead of (13.2) we have

$$|b \otimes a\rangle' = e^{-i\theta_{ba}/2}|b \otimes a\rangle = e^{i\theta_{ba}/2}|a \otimes b\rangle$$
$$= e^{i(\theta_{ab}+\theta_{ba})/2}|a \otimes b\rangle' = \pm|a \otimes b\rangle',$$

because according to (13.3)

$$e^{i(\theta_{ab}+\theta_{ba})/2} = \pm 1.$$

It is therefore always possible to choose the phases of the vectors $|a \otimes b\rangle$ and $|b \otimes a\rangle$ such that these vectors are symmetric or antisymmetric under the permutation $a \leftrightarrow b$:

$$\text{symmetric } |a \otimes b\rangle = +|b \otimes a\rangle, \tag{13.5}$$

$$\text{antisymmetric } |a \otimes b\rangle = -|b \otimes a\rangle. \tag{13.6}$$

As a result, the amplitudes $\langle a \otimes b|\Psi\rangle$ are also either symmetric or antisymmetric:

$$\text{symmetric } \langle a \otimes b|\Psi\rangle = \langle b \otimes a|\Psi\rangle, \tag{13.7}$$

$$\text{antisymmetric } \langle a \otimes b|\Psi\rangle = -\langle b \otimes a|\Psi\rangle. \tag{13.8}$$

This property of symmetry or antisymmetry is characteristic of the pair of identical particles under consideration. It *cannot* depend on the states $|\Psi\rangle$ or $|a \otimes b\rangle$. To show this, let us assume that for the same pair of particles we have a symmetric amplitude if $|\Psi\rangle = |\Phi_1\rangle$ and an antisymmetric one if $|\Psi\rangle = |\Phi_2\rangle$:

$$\langle a \otimes b|\Phi_1\rangle = \langle b \otimes a|\Phi_1\rangle,$$

$$\langle a \otimes b|\Phi_2\rangle = -\langle b \otimes a|\Phi_2\rangle.$$

The linearity of quantum mechanics also allows us to choose a state which is a linear combination of $|\Phi_1\rangle$ and $|\Phi_2\rangle$:

$$|\Psi\rangle = |\Phi_1\rangle\langle\Phi_1|\Psi\rangle + |\Phi_2\rangle\langle\Phi_2|\Psi\rangle,$$

where we assume for convenience that $\langle\Phi_1|\Phi_2\rangle = 0$. We then have

$$\langle a \otimes b|\Psi\rangle = \langle a \otimes b|\Phi_1\rangle\langle\Phi_1|\Psi\rangle + \langle a \otimes b|\Phi_2\rangle\langle\Phi_2|\Psi\rangle.$$

This probability amplitude is neither symmetric nor antisymmetric under the exchange $a \leftrightarrow b$, and it is physically unacceptable. It is necessary that $\langle\Phi_1|\Psi\rangle = 0$, or that $\langle\Phi_2|\Psi\rangle = 0$, for all states $|\Psi\rangle$. If $\langle\Phi_2|\Psi\rangle = 0$, transitions $\Psi \to \Phi_2$ are forbidden and $|\Phi_2\rangle$ does not belong to the space of two-particle states. As far as the behavior under the exchange of two states is concerned, there are two and only two classes of identical quantum particles, and they correspond to two types of amplitude:

• symmetric amplitudes (13.7), and the corresponding particles are called *bosons*;
• antisymmetric amplitudes (13.8), and the corresponding particles are called *fermions*.

The bosonic or fermionic nature of a particle space is called its *statistics*. As we shall see in an instant, electrons are an example of fermions, and it is also said that they obey *Fermi* (or *Fermi–Dirac*) *statistics*. Photons, which are bosons, obey *Bose* (or *Bose–Einstein*) *statistics*.

We have already noted that it is convenient to give artificial labels to particles: $1, 2, \ldots$ Equation (13.7) implies that the state vector of a system of two bosons will be symmetric under an exchange of labels $1 \leftrightarrow 2$:

$$|a \otimes b\rangle_{\text{B}} = \frac{1}{\sqrt{2}}\left(|a_1 \otimes b_2\rangle + |a_2 \otimes b_1\rangle\right), \tag{13.9}$$

and (13.8) implies that the state vector of two fermions must be antisymmetric:

$$|a \otimes b\rangle_F = \frac{1}{\sqrt{2}} \left(|a_1 \otimes b_2\rangle - |a_2 \otimes b_1\rangle \right). \tag{13.10}$$

If the particles have no internal degrees of freedom (spin, etc.), the particle state can be characterized by its wave function $\varphi_a(\vec{r}) = \langle \vec{r} | a \rangle$ and $\varphi_b(\vec{r}) = \langle \vec{r} | b \rangle$. The wave function of the system in the case of bosons is

$$\langle \vec{r}_1, \vec{r}_2 | a \otimes b \rangle_B = \frac{1}{\sqrt{2}} \left(\varphi_a(\vec{r}_1)\varphi_b(\vec{r}_2) + \varphi_a(\vec{r}_2)\varphi_b(\vec{r}_1) \right), \tag{13.11}$$

while in the case of fermions

$$\langle \vec{r}_1, \vec{r}_2 | a \otimes b \rangle_F = \frac{1}{\sqrt{2}} \left(\varphi_a(\vec{r}_1)\varphi_b(\vec{r}_2) - \varphi_a(\vec{r}_2)\varphi_b(\vec{r}_1) \right). \tag{13.12}$$

We have just written down the state vector, or wave function, of two *independent* identical particles without spin. When interactions are present, the wave function will be a linear combination of wave functions of the type (13.11) or (13.12), but even when interactions are absent the state vector, or wave function, will not be a simple tensor product.

The space of states for a pair of identical particles is therefore not the entire space $\mathcal{H}^{(1)} \otimes \mathcal{H}^{(2)}$, but only the subspace of vectors that are symmetric under exchange of labels in the case of two bosons, or antisymmetric under such exchange for two fermions. These two spaces are invariant under time evolution, because the Hamiltonian must be invariant under the exchange $1 \leftrightarrow 2$: $[H, P_{12}] = 0$, where P_{12} is the label permutation operator.

These results can be generalized immediately to the case of an arbitrary number N of identical bosons or fermions: the wave function of N bosons (fermions) must be symmetric (antisymmetric) under the exchange of any two labels of two particles. In the case of fermions, the wave function can therefore be written as a determinant. Let us write it out explicitly for three independent, identical fermions:

$$\langle \vec{r}_1, \vec{r}_2, \vec{r}_3 | a \otimes b \otimes c \rangle_F = \frac{1}{\sqrt{3!}} \begin{vmatrix} \varphi_a(\vec{r}_1) & \varphi_a(\vec{r}_2) & \varphi_a(\vec{r}_3) \\ \varphi_b(\vec{r}_1) & \varphi_b(\vec{r}_2) & \varphi_b(\vec{r}_3) \\ \varphi_c(\vec{r}_1) & \varphi_c(\vec{r}_2) & \varphi_c(\vec{r}_3) \end{vmatrix}. \tag{13.13}$$

If for example $\varphi_a = \varphi_b$ for fermions, the wave function vanishes. This is called the *Pauli principle*, although this "principle" actually follows from the antisymmetrization. It is often stated as follows: *it is impossible to put two or more fermions in the same state*. A spectacular effect of quantum statistics is described in Exercise 13.4.5.

13.1.2 Spin and statistics

In Equations (13.11) to (13.13) we have assumed that the particles do not have internal degrees of freedom, in particular, spin. When internal degrees of freedom are included, the exchange of labels must be done for *all* the quantum numbers characterizing the particle state. In particular, the spin degrees of freedom must be exchanged. It is remarkable that spin and statistics are intimately related by the *spin–statistics theorem*, which states that particles of integer spin $(0, \hbar, 2\hbar, \ldots)$ are bosons and those of half-integer spin $(\hbar/2, 3\hbar/2, \ldots)$ are fermions. Photons, which have spin 1, are bosons, and electrons, neutrinos, protons, and neutrons, which have spin 1/2, are fermions. The proof of the spin–statistics theorem uses relativistic quantum theory, or the relativistic theory of quantized fields, and requires an arsenal of sophisticated mathematics and the mastering of some difficult concepts. Therefore, it is unfortunately not possible to give even an intuitive idea of it here. It is frustrating to have to acknowledge that there is no elementary argument to justify this fundamental result which can be stated so simply.[3]

Having made this fundamental statement, we return to the state vectors (13.11) and (13.12). As we have just seen, spin-zero bosons can perfectly well exist (examples are π mesons, ^{4}He atoms, and so on) and there is no problem with using a state vector like (13.11) to represent the state of a system of two spin-zero bosons. On the other hand, the spin cannot be neglected for a system of two fermions and must be taken into account in writing down the state vector. The case of greatest practical importance is that of spin-1/2 fermions like electrons, protons, neutrons, and so on. According to the results of Section 10.6.1, using two spins 1/2 it is possible to construct angular momentum equal to unity with the three basis vectors $|jm\rangle$, collectively denoted χ_t:

$$|1,1\rangle = |+_1 \otimes +_2\rangle,$$
$$|1,0\rangle = \frac{1}{\sqrt{2}}\left(|+_1 \otimes -_2\rangle + |-_1 \otimes +_2\rangle\right), \qquad (13.14)$$
$$|1,-1\rangle = |-_1 \otimes -_2\rangle,$$

as well as angular momentum zero:

$$\chi_s = |0,0\rangle = \frac{1}{\sqrt{2}}\left(|+_1 \otimes -_2\rangle - |-_1 \otimes +_2\rangle\right). \qquad (13.15)$$

It is evident from (13.14) and (13.15) that the three states χ_t are symmetric under the exchange $1 \leftrightarrow 2$ while χ_s is antisymmetric. We recall that these states are respectively called triplet and singlet states, hence the notation χ_t and χ_s. The totally antisymmetric state vectors of a system of two fermions are therefore either antisymmetric in space and symmetric in spin,

$$\langle \vec{r}_1, \vec{r}_2 | a \otimes b\rangle_F = \frac{1}{\sqrt{2}}\left(\varphi_a(\vec{r}_1)\varphi_b(\vec{r}_2) - \varphi_a(\vec{r}_2)\varphi_b(\vec{r}_1)\right)\chi_t, \qquad (13.16)$$

or symmetric in space and antisymmetric in spin:

$$\langle \vec{r}_1, \vec{r}_2 | a \otimes b\rangle_F = \frac{1}{\sqrt{2}}\left(\varphi_a(\vec{r}_1)\varphi_b(\vec{r}_2) + \varphi_a(\vec{r}_2)\varphi_b(\vec{r}_1)\right)\chi_s. \qquad (13.17)$$

[3] For a proof, see R. Streater and A. Wightman, *PCT, Spin and Statistics and All That*, New York: Benjamin (1964). The situation is similar to that of the Fermat theorem, which can be stated very simply but, as shown by A. Wiles, is extremely complicated to prove. See, however, M. Berry and J. Robbins, Indistinguishability for quantum particles: spin, statistics and the geometric phase, *Proc. Roy. Soc. London A* **453**, 1771–1790 (1997).

As an application, let us assume that two spin-1/2 fermions are in a state of orbital angular momentum l in their center-of-mass frame. The angular part of the wave function of the relative particle is the spherical harmonic $Y_l^m(\hat{r})$, where $\vec{r} = \vec{r}_1 - \vec{r}_2$ is the vector joining the positions of the two fermions. Exchanging the labels is equivalent to $\vec{r} \to -\vec{r}$ or $\hat{r} \to -\hat{r}$. According to (10.71), the parity of the spherical harmonics is $(-1)^l$:

$$Y_l^m(-\hat{r}) = (-1)^l Y_l^m(\hat{r}). \tag{13.18}$$

In the center-of-mass frame, a system of two spin-1/2 fermions is in a state of even orbital angular momentum l if its spin state is a singlet, and in a state of odd orbital angular momentum l if its spin state is a triplet. It is usual to to denote the total spin as S, the total orbital angular momentum as L, the total angular momentum as J, and $^{2S+1}L_J$ the state of the two fermions. For example, a $^3P_2$ state corresponds to $S = 1$, $L = 1$, $J = 2$ and a $^1D_2$ state to $S = 0$, $L = 2$, $J = 2$. The case of two spin-zero bosons is even simpler: only states of even orbital angular momentum are allowed.

The symmetry properties of the state vector of two spins 1/2 can be generalized to the addition of any two spins $\vec{S}$ to form a total spin $\vec{F} = \vec{S}_1 + \vec{S}_2$, $0 \le F \le 2s$. The symmetry property of the Clebsch–Gordan coefficients[4]

$$C_{j_2 j_1; m_2 m_1}^{jm} = (-1)^{j_1 + j_2 - j} C_{j_1 j_2; m_1 m_2}^{jm}$$

shows that states of total spin $2F$, $2F - 2$, ... are symmetric under label exchange, while states $2F - 1$, $2F - 3$, ... are antisymmetric. As an application, let us show that these symmetry properties affect the rotational spectrum of a homonuclear diatomic molecule, that is, a molecule whose two nuclei are strictly identical, of the same isotope, for example the $^1\text{H}-^1\text{H} \equiv \text{H}_2$ molecule, in contrast to a heteronuclear molecule like $^1\text{H}-^2\text{H}$ or $\text{H}-\text{D}$, where a proton is replaced by a deuteron $\text{D} \equiv {}^2\text{H}$ (the deuterium is an isotope of hydrogen with nucleus formed of a proton and a neutron). The dynamics of the nuclei is that of a spherical rotator (cf. Section 10.3.1) whose wave functions are the spherical harmonics $Y_j^m(\hat{r})$, where $\vec{r}$ is the vector joining the two nuclei. The rotational levels, or rotational spectrum, are given as a function of j by (10.54):

$$E_j = \frac{j(j+1)}{2I},$$

where I is the moment of inertia.

If we choose the coordinate origin to lie at the center of the line joining the nuclei, the Hamiltonian H of the electrons is invariant under the parity operator Π taking $\vec{r} \to -\vec{r}$: $[\Pi, H] - 0$ (cf. Section 8.3.3). It is then possible to diagonalize Π and H simultaneously. Let $|\psi_{\text{el}}\rangle$ be an eigenvector of the electronic state common to H and Π. Since $\Pi^2 = I$, the eigenvalues of Π are ± 1, $\Pi|\psi_{\text{el}}\rangle = \pm|\psi_{\text{el}}\rangle$ (cf. (8.52)). In most cases, and in particular that of the hydrogen molecule, the electronic ground state corresponds to the $+$ sign, which is what we shall assume in the following discussion. The exchange of the labels of the

---

[4] See, for example, Cohen-Tannoudji *et al.* [1977], Complement B_X.

two nuclei corresponds to $\vec{r} \to -\vec{r}$, and in this operation the nuclear wave function is multiplied by the parity of the spherical harmonic $(-1)^j$. If the two nuclei have spin s, the total angular momentum F runs from zero to $2s$. The *complete* state vector of the molecule must be symmetric (antisymmetric) under the exchange of the labels of the two nuclei if the nuclei are bosons (fermions), and when they are bosons (integer s) there are two possible cases:

- F even and j even,
- F odd and j odd.

The result is the same when the two nuclei are fermions (half-integer s). The opposite situation could of course arise in rare cases where the parity of $|\psi_{\mathrm{el}}\rangle$ is negative. In the case of the hydrogen molecule, the proton spin is $s = 1/2$ and $F = 0$ (parahydrogen) or $F = 1$ (orthohydrogen). The value of F fixes the parity of j: $F = 1$ corresponds to odd j and $F = 0$ to even j. There are no restrictions on j in the case of the H–D molecule.

Another important consequence of the statistics is the appearance of *exchange forces*, which are responsible, in particular, for magnetism. Macroscopic magnetism corresponds to the alignment of a macroscopic number of electron spins in the same direction, and this alignment creates a macroscopic magnetic moment. If the alignment is produced by an external magnetic field and disappears in the absence of this field, the material is *paramagnetic*. If the alignment persists in the absence of the field, the material is *ferromagnetic* (examples are iron, cobalt, nickel, and so on). Ferromagnetism vanishes above a certain temperature, called the *Curie temperature* T_C. There is another type of magnetism, *antiferromagnetism*, where the spins are ordered but in alternating directions such that the magnetism is zero. This antiferromagnetic ordering also vanishes above a certain temperature, the *Néel temperature* T_N. For a material to be ferromagnetic or antiferromagnetic there must be an interaction between the spins which is strong enough to align them or arrange them in alternating order. In the absence of such an interaction the thermal motion tends to favor a state in which the spins are randomly oriented and the magnetism vanishes. This interaction does not originate in the coupling between the electron magnetic moments. A simple order-of-magnitude calculation shows that the Curie temperature, which is of order 10^3 K, would be no more than 1 K for this hypothesis. The interaction giving rise to magnetism is the Coulomb repulsion between the electrons in conjunction with the antisymmetrization of the state vector, which leads to a competition between the kinetic and (Coulomb) potential energy. Let us consider a pair of electrons. If they are in a triplet spin state, their spatial wave function is antisymmetric, which implies a weak Coulomb repulsion, because the wave function vanishes when the two electrons are close together. The kinetic energy is large, because the wave function must vary rapidly near the point where it vanishes. The reverse situation occurs when the spin state is a singlet. If it is preferable to minimize the potential energy, the two electrons will tend to align their spins, which implies a ferromagnetic type of interaction. If on the contrary the kinetic energy plays the leading role, we obtain an antiferromagnetic type of interaction with alternating ordering of the spins.

A consequence of the spin–statistics theorem is that spin-zero particles like ^{4}He, ^{16}O, and so on are bosons. However, these are composite particles, and it is interesting to check the consistency with the spin–statistics theorem starting from their elementary (or more elementary) constituents. Naturally, this only makes sense if the particle remains intact in the reactions it undergoes, for example because the energies involved are not high enough to dissociate the particle into its constituents. Instead of making completely general arguments, we shall content ourselves with studying a particular case, that of the deuteron. Let $|A\rangle$ be the deuteron state vector and $\langle a \otimes b|A\rangle = \langle ab|A\rangle$ be the amplitude for finding the proton in the state $|a\rangle$ and the neutron in the state $|b\rangle$ inside the deuteron, where we have suppressed the tensor product to simplify the notation. We introduce a second deuteron $|A_2\rangle$ assuming for now that there is a quantum number distinguishing the proton and neutron of this nucleus from those of the first nucleus. In the spirit of quantum chromodynamics, we imagine that we can assign a color to the protons and neutrons, green for the first nucleus and red for the second. We will then have a second amplitude $\langle a_2'b_2'|A_2'\rangle$, where the prime indicates that it involves red neutrons and protons, while the corresponding amplitude for the green neutrons and protons will be denoted $\langle a_1b_1|A_1\rangle$. Let us construct the two-deuteron state $|A_1A_2'\rangle$. The amplitude for finding the green proton and neutron in the states a_1 and b_1 and the red proton and neutron in the states a_2' and b_2' is, using the properties of the tensor product,

$$\langle a_1b_1a_2'b_2'|A_1A_2'\rangle = \langle a_1b_1|A_1\rangle\langle a_2'b_2'|A_2'\rangle.$$

However, we cannot really color protons and neutrons red and green, and so we must return to the real world, where the amplitude is given by $\langle a_1b_1a_2b_2|A_1A_2\rangle$. Since the proton and the neutron are fermions, this amplitude must be antisymmetric under the label exchanges $a_1 \leftrightarrow a_2$ and $b_1 \leftrightarrow b_2$:

$$\langle a_1b_1a_2b_2|A_1A_2\rangle = \langle a_1b_1|A_1\rangle\langle a_2b_2|A_2\rangle - \langle a_2b_1|A_1\rangle\langle a_1b_2|A_2\rangle$$
$$-\langle a_1b_2|A_1\rangle\langle a_2b_1|A_2\rangle + \langle a_2b_2|A_1\rangle\langle a_1b_1|A_2\rangle.$$

This amplitude is symmetric under the exchange $A_1 \leftrightarrow A_2$,

$$\langle a_1b_1a_2b_2|A_1A_2\rangle = \langle a_1b_1a_2b_2|A_2A_1\rangle, \tag{13.19}$$

and the deuteron is therefore a boson. In general, a particle composed of an even number of fermions is a boson, and one composed of an odd number is a fermion. The proton, made of three spin-1/2 quarks, is a fermion, while the π meson made of a quark and an antiquark is a boson. The ^{4}He atom, made of two protons, two neutrons, and two electrons, is a boson, whereas an isotope of it, namely the ^{3}He atom made of two protons, one neutron, and two electrons, is a fermion, which leads to completely different behaviors of these two isotopes at low temperatures. It should be noted that these results are compatible with the spin–statistics theorem, because given an odd number of particles of half-integer spin we can only make a particle of half-integer spin, a fermion, while given an even number of particles of half-integer spin we can only make a particle of integer spin, a boson.

13.2 The scattering of identical particles

Let us return to Fig. 13.1, which we can interpret as describing $^{16}\text{O}-^{18}\text{O}$ scattering in the center-of-mass frame. We assume that the ground-state degeneracy is lifted by a magnetic field, and the atoms are in the lowest Zeeman level (cf. Section 14.2.3). Let $f(\theta)$ be the amplitude for scattering at the angle θ in Fig. 13.1a; the two oxygen atoms are deflected by the angle θ. The scattering amplitude of Fig. 13.1b then is $f(\pi - \theta)$; the two oxygen atoms are deflected by the angle $\pi - \theta$. Let us assume the most plausible situation, namely that the detectors D_1 and D_2 do not distinguish between the two isotopes. The counting rate of detector D_1 (and also of D_2) will then be proportional to

$$\mathsf{p}(\theta) = |f(\theta)|^2 + |f(\pi - \theta)|^2. \tag{13.20}$$

This result also gives the differential cross section (12.12) $d\sigma/d\Omega$. In (13.20) we have added the *probabilities*, because the final states $[^{16}\text{O}$ in D_1, ^{18}O in $D_2]$ and $[^{16}\text{O}$ in D_2, ^{18}O in $D_1]$ are different final states, even if in practice the detectors are incapable of distinguishing between them. In calculating the total cross section we multiply (12.2) by $1/2$ in order to avoid double counting (or, equivalently, we restrict the integration over θ to the range $0 \le \theta \le \pi/2$):

$$\sigma_{\text{tot}} = \frac{1}{2} \int d\Omega \left(|f(\theta)|^2 + |f(\pi - \theta)|^2 \right). \tag{13.21}$$

Let us now turn to $^{16}\text{O}-^{18}\text{O}$ scattering. Although the atomic physics interactions between the two isotopes are strictly identical, the results in this case are totally different. The processes of Fig. 13.1a and 13.1b can no longer be distinguished, even in principle, and so the *amplitudes* must be added. The scattering amplitude $f(\theta)$ is defined by formally labeling the two particles, particles 1 and 2 being deflected by an angle θ. Exchange of the two atoms corresponds to $\theta \leftrightarrow \pi - \theta$. The total amplitude is obtained by adding $f(\theta)$ and $f(\pi - \theta)$, with the + sign being imposed by the symmetry under the exchange $\theta \leftrightarrow \pi - \theta$. Instead of (13.20), the probability for triggering D_1 is

$$\mathsf{p}(\theta) = |f(\theta) + f(\pi - \theta)|^2 \tag{13.22}$$

and the total cross section becomes

$$\sigma_{\text{tot}} = \frac{1}{2} \int d\Omega |f(\theta) + f(\pi - \theta)|^2 = \int_0^{\pi/2} \sin\theta \, d\theta \int_0^{2\pi} d\phi |f(\theta) + f(\pi - \theta)|^2. \tag{13.23}$$

The addition of the amplitudes suggests that the differential cross section can exhibit interference-like patterns, and this has actually been observed in numerous cases. We note that when the parity of the Legendre polynomials $P_l(-u) = (-1)^l P_l(u)$ is taken into account, only even values of l are involved in the partial-wave expansions of

$$f_{\text{tot}}(\theta) = f(\theta) + f(\pi - \theta), \quad f_{\text{tot}}(\theta) = f_{\text{tot}}(\pi - \theta).$$

In the above example we considered the scattering of two spin-zero bosons. The discussion becomes a bit more complicated when the particles have spin. Let us limit ourselves to the scattering of two identical spin-1/2 fermions, for example, two neutrons. In this case as in Section 12.2.4 we can define a scattering amplitude $\hat{f}(\theta)$ which is a 4×4 matrix in the tensor product space of the two spins. If $\mathcal{P}_t$ and $\mathcal{P}_s$ are the projectors on the triplet and singlet states, and if the scattering does not change the total spin, we can write

$$\hat{f}(\theta) = [f_s(\theta) + f_s(\pi - \theta)]\mathcal{P}_s + [f_t(\theta) - f_t(\pi - \theta)]\mathcal{P}_t, \tag{13.24}$$

which ensures the space + spin antisymmetry of the amplitude. If as in (12.16) we expand $[f_s(\theta) + f_s(\pi - \theta)]$ and $[f_t(\theta) - f_t(\pi - \theta)]$ in partial waves, the scattering will occur in the waves with $l = 0, 2, \ldots$ (or the $s, d, \ldots$ waves) for neutrons in the singlet state, and in the waves with $l = 1, 3, \ldots$ (or the $p, f, \ldots$ waves) for neutrons in the triplet state. The cross section is obtained as in Section 12.2.4. If the initial polarization of the set of two neutrons is denoted by α and the final polarization by β, the differential cross section will be

$$\frac{d\sigma_{\beta\alpha}}{d\Omega} = |\langle\beta|\hat{f}(\theta)|\alpha\rangle|^2. \tag{13.25}$$

If the polarization of the final neutrons is not measured we must sum over β, and if the initial state is an incoherent superposition of polarization states $|\alpha\rangle$ with probability p_α we have

$$\frac{d\sigma}{d\Omega} = \sum_\alpha p_\alpha \sum_\beta \langle\alpha|\hat{f}^\dagger|\beta\rangle\langle\beta|\hat{f}|\alpha\rangle$$

$$= \sum_\alpha p_\alpha \langle\alpha|\hat{f}^\dagger \hat{f}|\alpha\rangle = \mathrm{Tr}\left(\rho_{\mathrm{in}}\hat{f}^\dagger \hat{f}\right), \tag{13.26}$$

where ρ_{in} is the initial state operator of the spin states:

$$\rho_{\mathrm{in}} = \sum_\alpha p_\alpha |\alpha\rangle\langle\alpha|.$$

When the initial neutrons are not polarized, $\rho_{\mathrm{in}} = I/4$ and

$$\frac{d\sigma}{d\Omega}\bigg|_{\mathrm{unpol}} = \frac{1}{4}\mathrm{Tr}\left(\hat{f}^\dagger \hat{f}\right) = \frac{1}{4}\mathrm{Tr}\left[\left(f_s^{\mathrm{tot}*}\mathcal{P}_s + f_t^{\mathrm{tot}*}\mathcal{P}_t\right)\left(f_s^{\mathrm{tot}}\mathcal{P}_s + f_t^{\mathrm{tot}}\mathcal{P}_t\right)\right]$$

$$= \frac{1}{4}\mathrm{Tr}\left[|f_s^{\mathrm{tot}}|^2\mathcal{P}_s + |f_t^{\mathrm{tot}}|^2\mathcal{P}_t\right] = \frac{1}{4}|f_s^{\mathrm{tot}}|^2 + \frac{3}{4}|f_t^{\mathrm{tot}}|^2$$

$$= \frac{1}{4}|f_s(\theta) + f_s(\pi - \theta)|^2 + \frac{3}{4}|f_t(\theta) - f_t(\pi - \theta)|^2. \tag{13.27}$$

The weights 1/4 and 3/4 arise, of course, from the fact that there are one singlet state and three triplet states. The total cross section is obtained using (13.23). For spin-independent scattering $f_s = f_t = f$, which is the case in the Coulomb scattering of two charged particles, for example two electrons (Exercise 12.5.4):

$$\frac{d\sigma}{d\Omega}\bigg|_{\mathrm{unpol}} = |f(\theta)|^2 + |f(\pi - \theta)|^2 - \mathrm{Re}[f(\theta)f^*(\pi - \theta)],$$

and the interference term is reduced by a factor of two compared with that which would be obtained in the scattering of two spin-zero fermions (forbidden by the spin–statistics theorem!).

13.3 Collective states

The statistics has a decisive influence on the behavior of a system of N identical particles, $N \gg 1$, that is, on the collective behavior of such a system. Let us begin with fermions and examine the case of N fermions without interactions. We can, for example, assume that these N independent fermions are located in a potential well in which the energy levels ε_ℓ of an individual particle are labeled by an index ℓ. The index ℓ represents the *complete set* of quantum numbers needed to specify the ℓth state: the momentum, spin, and so on. It may perfectly well happen, and is the case in general, that several levels ε_ℓ correspond to the same energy. In other words, the energy levels of the Hamiltonian of a particle in the potential well are degenerate. Let us try to construct the ground-state level of the ensemble of N fermions. Since at most one fermion can be put in a state ε_ℓ, the state of lowest energy is obtained by filling the levels one by one starting from the lowest, until the N fermions have all been placed (Fig. 13.2). The state of highest energy $\varepsilon_{\ell,\max}$ that the last fermion is placed in is called the *Fermi level* and denoted as ε_F.[5] Let us take the potential well to be a cubic box of volume $\mathcal{V}$; a set of fermions in a box is called a *Fermi gas*. The quantum state of a fermion is then specified by its momentum $\vec{p}$ and spin component m_z: $\ell = \{\vec{p}, m_z\}$. In the absence of an external field the energy is purely kinetic, $\varepsilon = \vec{p}^2/2m$, and independent of m_z. Each value of $\vec{p}$ corresponds to $2s+1$ states of the same energy, and according to (9.152) the sum over ℓ becomes

$$\sum_\ell = \sum_{\vec{p}, m_z} = (2s+1) \sum_{\vec{p}} \rightarrow \frac{(2s+1)\mathcal{V}}{h^3} \int d^3 p. \tag{13.28}$$

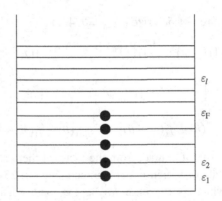

Fig. 13.2. Filling of the levels of a Fermi gas.

[5] From the viewpoint of thermodynamics, this system of fermions is a system at zero temperature $T = 0$. The Fermi level is also the chemical potential, because at zero temperature the chemical potential is the energy needed to add a particle. At nonzero temperature the occupation probability of the levels above the Fermi level is nonzero, and the chemical potential no longer coincides with the Fermi level.

To the Fermi energy ε_F corresponds the *Fermi momentum* p_F:

$$\varepsilon_F = \frac{p_F^2}{2m} \quad \text{or, in general,} \quad \varepsilon_F = \sqrt{p^2 c^2 + m^2 c^4} - mc^2. \tag{13.29}$$

Since the energy is an increasing function of p, all states $\{\vec{p}, m_z\}$ such that $p \leq p_F$ will have occupation number equal to unity. It is now straightforward to calculate the Fermi momentum:

$$N = \frac{(2s+1)\mathcal{V}}{h^3} \int_{p \leq p_F} d^3 p = \frac{(2s+1)\mathcal{V}}{h^3} \frac{4\pi}{3} p_F^3. \tag{13.30}$$

If $n = N/\mathcal{V}$ is the fermion density, then

$$p_F = \left[\frac{6\pi^2}{2s+1} \right]^{1/3} \hbar n^{1/3}. \tag{13.31}$$

This equation is valid at both nonrelativistic and relativistic energies. The sphere of radius p_F is called the *Fermi sphere* and its surface is the *Fermi surface*. These ideas can be generalized to solid-state physics, where the symmetry is no longer spherical symmetry, but a symmetry determined by the crystal lattice. The Fermi surface, which then has a shape more complicated than a sphere, is a fundamental object in the study of the electromagnetic properties of metals. From (13.31) we obtain the Fermi energy in the nonrelativistic case where $\varepsilon = p^2/2m$:

$$\varepsilon_F = \frac{p_F^2}{2m} = \left[\frac{6\pi^2}{2s+1} \right]^{2/3} \frac{\hbar^2}{2m} n^{2/3}. \tag{13.32}$$

The usual case is $s = 1/2$. The Fermi energy is the characteristic energy of a system of N fermions in a box of volume $\mathcal{V}$.

It is useful to perform an order-of-magnitude calculation in the most important particular case of a Fermi gas, that of the conduction electrons in a metal. Let us take the example of copper, with mass density 8.9 g cm^{-3} and atomic mass 63.5, which corresponds to a number density n of 8.4×10^{28} atoms per m^3. Since copper has one conduction electron per atom, this is also the electron number density. Substituting it into (13.32) with $s = 1/2$, for the Fermi level we find $\varepsilon_F \simeq 7.0$ eV. This is typical for the conduction electrons of a metal: the Fermi energy is several eV.

Let us now calculate the energy of the Fermi gas. According to (13.28) with $s = 1/2$, we have

$$E = \frac{\mathcal{V}}{\pi^2 \hbar^2} \int_0^{p_F} p^2 dp \left(\frac{p^2}{2m} \right) = \frac{3}{5} N \varepsilon_F, \tag{13.33}$$

where we have used (13.30) for p_F as a function of N in the case $s = 1/2$. Another interesting expression is that for the energy per particle E/N:

$$\frac{E}{N} = (3\pi^2)^{2/3} \frac{3\hbar^2}{10m} n^{2/3}. \tag{13.34}$$

The average kinetic energy of a particle grows as $n^{2/3}$. If we now take interactions into account in the case of an electron gas, the average potential energy is of order e^2/d, where $d \propto n^{-1/3}$ is the average distance between two electrons. The average potential energy per particle then is $\propto n^{1/3}$, and the denser the Fermi gas, the more the kinetic energy $\propto n^{2/3}$ wins over the potential energy. This result is the opposite to that for a classical gas: in contrast to the latter, *a Fermi gas approaches an ideal gas more closely the higher its density.*

An intuitive picture of a Fermi gas can be obtained by noting that the momentum dispersion Δp is of order p_F, whereas the order of magnitude of the position dispersion is $\mathcal{V}^{1/3}$. From (13.31) we then find

$$\Delta p \, \Delta x \sim \hbar N^{1/3}. \tag{13.35}$$

Owing to the Pauli principle, the $\hbar$ of the Heisenberg inequality is transformed into $\hbar N^{1/3}$.

The situation regarding bosons is more complicated than that of fermions. It is necessary to distinguish between the cases where the number of bosons is variable (photons, phonons, and so on) and where it is fixed (helium atoms). In the latter case, at strictly zero temperature the ground state is obtained by putting all the bosons in the lowest state ε_ℓ. The problem is to show that if the temperature is not zero, a *finite* fraction of the bosons remains in this ground state. This is called *Bose–Einstein condensation.* This condensation does not occur in all cases, for example it does not occur in a two-dimensional box, but it does occur in a three-dimensional one. The temperature at which Bose–Einstein condensation occurs can be estimated by noting that the two characteristic lengths of the problem, the thermal wavelength λ_T and the average distance between bosons $d \propto n^{-1/3}$, must be of the same order of magnitude: $\lambda_T \sim n^{-1/3}$. This estimate is confirmed by an exact calculation. Using

$$\lambda_T = \left(\frac{h^2}{2\pi mkT} \right)^{1/2}, \tag{13.36}$$

the condensation temperature is given by $\lambda_T = 2.61\, n^{-1/3}$.[6] Bose–Einstein condensation has recently been observed for gases of alkali atoms at very low temperature and for polarized hydrogen. We refer the interested reader to the References.

13.4 Exercises

13.4.1 The Ω^- particle and color

The Ω^- hyperon (of mass 1675 MeV c^{-2}) is a spin-3/2 particle composed of three strange quarks of spin 1/2. The quark model requires that the spatial wave function not vanish. Show that the three quarks cannot all be identical. In the early 1970s (in the early days of quantum chromodynamics) this observation provided one of the arguments in favor of the introduction of the concept of "color" making it possible to distinguish between quarks; the three quarks of the Ω^- have different colors.

---

[6] The wavelength λ_T is the de Broglie wavelength of a particle of energy $\sim k_B T$. The factor 2π is a convention.

13.4.2 Parity of the π meson

1. If low-energy π^- mesons are allowed to hit a deuterium target, the mesons can be captured and form bound states analogous to those of the hydrogen atom. Give the expression for the energy of these π-meson–deuteron bound states using the fact that the π-meson mass is of order 139 MeV c^{-2} and the deuteron mass is 1875 MeV c^{-2}. The π-meson is captured in a state of high principal quantum number n and terminates its radiative cascade in the $1s$ ground state[7] after emitting photons. Show that the energy of these photons must lie in the X-ray region.

2. Once it has arrived in the $1s$ state, the π meson undergoes a nuclear interaction which leads to the reaction

$$\pi^- + {}^2\mathrm{H} \rightarrow \mathrm{n} + \mathrm{n}$$

with two neutrons n in the final state. Using the fact that the spin of the deuteron is 1 and that of the π^- meson is zero, what is the initial angular momentum state of the reaction? Show that the two final neutrons can only be in a state of total orbital angular momentum $L = 1$ and total spin $S = 1$, that is, in the ${}^3P_1$ state. If, following convention, we assign positive parity to the nucleons (protons and neutrons) and use the fact that the deuteron orbital angular momentum is zero (the deuteron is a ${}^3S_1$ state),[8] show that the π meson has negative parity. Parity is conserved in the reaction.

13.4.3 Spin-1/2 fermions in an infinite well

We consider two identical spin-1/2 fermions in an infinite cubic well of side L. If these two fermions do not interact with each other, what are the possible eigenvalues of the total energy and the corresponding wave functions (space and spin)? We assume that the two fermions interact via a potential

$$V = V_0 \, \delta^{(3)}(\vec{r}_1 - \vec{r}_2),$$

where $\vec{r}_1$ and $\vec{r}_2$ are the positions of the two fermions. Show that triplet states are not affected by this potential.

13.4.4 Positronium decay

Positronium is an electron–positron $(\mathrm{e}^- - \mathrm{e}^+)$ bound state; the positron is a particle with the same mass m_e as the electron and opposite charge $-q_\mathrm{e}$.

1. In this question we neglect the spins of the two particles. Given that the energy levels of the hydrogen atom for an infinitely heavy proton have the form ($e^2 = q_\mathrm{e}^2/4\pi\varepsilon_0$)

$$E_n = \frac{E_0}{n^2} = -\frac{1}{2} \frac{m_\mathrm{e} e^4}{\hbar^2} \frac{1}{n^2}, \quad n = 1, 2, 3, \ldots,$$

what are the energy levels of positronium?

---

[7] The nuclear reaction also has a small probability of occurring in a state ns, $n \neq 1$, that is, for states where the probability density is nonzero at the origin. However, this does not change the argument.

[8] The deuteron also has a small d-wave component and therefore a ${}^3D_1$ component, but this does not affect the argument.

2. The electron and the positron have spin 1/2. The state of lowest energy, the ground state with $n = 1$, has orbital angular momentum $l = 0$ (s-wave). What are the possible values of the total angular momentum j of positronium in this $n = 1$ state?

3. Positronium in its ground state decays into two photons:[9]

$$e^- + e^+ \to 2\gamma.$$

In the positronium rest frame the two photons leave the decay point with opposite momenta. We choose the axis Oz to be the direction of the photon momentum. Using angular momentum conservation, show that the two photons necessarily have the same circular polarization, either right-handed or left-handed. Hint: sketch the decay.

4. By examining the effect of a rotation by π about the axis Oy and taking into account the fact that the two photons are identical, show that only one of the two states of angular momentum j of positronium can decay into two photons.[10]

5. Let Π be the parity operator acting on the state $|A\rangle$ of a particle A as $\Pi|A\rangle = \eta_A|A\rangle$, where η_A is the parity of A. It can be shown that $\eta_{e^-}\eta_{e^+} = -1$. Deduce that the parity of the ground state of positronium is -1. The two possible states of the two photons can be written as

$$(i) \ \ |\Phi_+\rangle = \frac{1}{\sqrt{2}}\left(|RR\rangle + |LL\rangle\right), \quad (ii) \ \ |\Phi_-\rangle = \frac{1}{\sqrt{2}}\left(|RR\rangle - |LL\rangle\right),$$

where $|R\rangle$ and $|L\rangle$ represent the right- and left-handed polarization states. Which of the states (i) or (ii) is obtained in positronium decay,[11] given that parity is conserved?

13.4.5 Quantum statistics and beam splitters

1. Let a and b be two identical modes of the electromagnetic field (e.g. identical wave packets), arriving at a beam splitter, one of them horizontally and the other one vertically. Using the results of Exercise 1.6.6, show that the beam splitter couples the two modes through an operator $U(\theta)$ as follows

$$U^\dagger(\theta)\, a\, U(\theta) = a\cos\theta + ib\sin\theta$$

$$U^\dagger(\theta)\, b\, U(\theta) = b\cos\theta + ia\sin\theta.$$

Here a and b are field operators which destroy photons in the modes a and b. Therefore, a transmitted photon has a nonzero probability amplitude of being exactly in the same mode as a reflected photon at the beam splitter output. A symmetric beam splitter (Exercise 2.4.12) has $\theta = \pi/4$.

2. Show that $U(\theta)$ can be written in the form of an evolution operator, $U(\theta) = \exp(-i\theta G)$, with

$$G = a^\dagger b + b^\dagger a.$$

G plays the role of an effective Hamiltonian for mode coupling. Hint: use (2.54) to compute

$$e^{i\theta G}\,[a/b]\,e^{-i\theta G}.$$

[9] The decay $e^- + e^+ \to \gamma$ is forbidden by energy–momentum conservation.
[10] The other state must decay into three photons.
[11] Correlations in the polarizations of the two photons have been measured by C. Wu and I. Shaknow, The angular correlations of scattered annihilation radiation, *Phys. Rev.* **77**, 136 (1950), who were able to verify that the parity of the ground state is indeed -1.

3. Assume that each of the modes contains exactly one photon, so that the initial state is

$$|\Psi_0\rangle = |1_a, 1_b\rangle.$$

Find the beam splitter output

$$|\Psi\rangle = U(\theta)|\Psi_0\rangle,$$

and show that for $\theta = \pi/4$

$$|\Psi\rangle = \frac{1}{\sqrt{2}}\left(|2_a, 0_b\rangle + |0_a, 2_b\rangle\right).$$

Out of the four possibilities at the beam splitter output, only two are physically realized, those in which the two photons stick together, while the situation where the two photons are in different beams does not occur (Fig. 13.3). This is an interference effect, a spectacular consequence of Bose–Einstein statistics,[12], which contradicts Dirac's statement: "a photon can only interfere with itself."

4. Suppose now that the incident particles are fermions, We neglect spin and interactions. Show that the antisymmetry of the state vector requires ($\{A, B\} = AB + BA$ is the anticommutator of A and B)

$$\{a, a^\dagger\} = \{b, b^\dagger\} = I,$$

all the other anticommutators being zero

$$\{a, a\} = \cdots = \{a, b^\dagger\} = 0.$$

Show that the preceding operator G is now replaced by

$$\overline{G} = ab^\dagger - a^\dagger b,$$

and compute the action of $U(\theta) = \exp(-i\theta\overline{G})$ on the operators a and b as in question 1. What happens if one starts with an initial state $|1_a, 1_b\rangle$ at the entrance of the beam splitter?

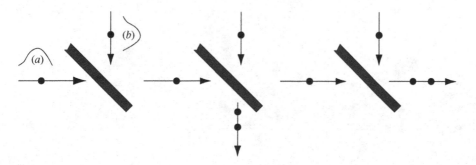

Fig. 13.3. If $\theta = \pi/4$, one cannot have one photon in mode a and the other in mode b at the exit of the beam splitter. Both photon must exit in the same mode.

[12] C. Santori, D. Fattal, J. Vukovic, G. Solomon and Y. Yamamoto, Indistinguishable photons from a single photon device, *Nature* **419**, 594 (2002). See also Ph. Grangier, Single photons stick together, *Nature* **419**, 577 (2002).

13.5 Further reading

An excellent discussion of identical particles accompanied by numerous examples is that of Lévy-Leblond and Balibar [1990], Chapter 7. See also Feynman *et al.* [1965], Vol. III, Chapter 4; Cohen-Tannoudji *et al.* [1977], Chapter XIV; and Basdevant and Dalibard [2002], Chapter 16. Collective states are studied by Le Bellac *et al.* [2004], Chapter 4, which contains an introduction and references to Bose–Einstein condensates of atomic gases. A very complete treatment of such condensates is given in C. Pethick and H. Smith, *Bose–Einstein Condensation of Dilute Gases*, Cambridge: Cambridge University Press (2002).

14

Atomic physics

This chapter is devoted to an introduction to atomic physics which will be mainly concerned with one-electron atoms. After a brief discussion of the perturbation and variational methods in Section 14.1, in Section 14.2 we study the fine and hyperfine structure of the energy levels as well as the effect of a magnetic field on these levels. In Section 14.3 we examine the coupling of an atom to an electromagnetic field and important applications of this coupling such as the photoelectric effect and the rate of spontaneous emission. In Section 14.4 we give a brief introduction to a subject which has been expanding enormously in the last twenty years, the laser manipulation of atoms, and we discuss Doppler cooling and magneto-optical traps. Finally, Section 14.5 is devoted to a short discussion of two-electron atoms.

14.1 Approximation methods

14.1.1 Generalities

In classical physics it is only in exceptional cases that it is possible to solve the Newton or Maxwell equations analytically given the initial conditions at time $t = t_0$ and, in the first case, the forces, or, in the second, the sources of electromagnetic field. In general, it is necessary to resort to an approximation method such as numerical integration of the equations, the perturbation method, or something else. The situation is no different in quantum physics: only in exceptional cases do we know how to "solve the Schrödinger equation" exactly, that is, how to obtain the time evolution of the state vector $|\varphi(t)\rangle$ as a function of its value $|\varphi(t_0)\rangle$ at initial time $t = t_0$. In the case where the Hamiltonian is time-independent, which is what we shall consider in this section, knowledge of this time evolution implies that we know how to diagonalize the Hamiltonian, that is, find its eigenvalues and eigenvectors. Except in some special cases (the square well, the harmonic oscillator, the hydrogen atom, and so on), we do not know how to diagonalize the Hamiltonian exactly, and approximation methods such as numerical integration or the perturbation method must be used.

In this section we shall present the method of time-independent perturbation theory. It consists of starting from a Hamiltonian H_0 which we know how to diagonalize exactly, and then perturbing it by adding a term W which gives the "exact" Hamiltonian $H = H_0 + W$ within some predefined domain of approximation (cf. Section 4.3). We write

$$H(\lambda) = H_0 + \lambda W, \tag{14.1}$$

where we have introduced a real parameter λ such that $H = H_0$ if $\lambda = 0$ and $H = H_0 + W$ if $\lambda = 1$. If $\lambda \to 0$, we can hope that the perturbation λW is in some sense "small" compared with H_0.[1] It may happen that it is possible to effectively vary λ. For example, if λW corresponds to the interaction of an atomic system with an external electromagnetic field, the value of this external field and therefore also λ can be varied at will, and $\lambda \to 0$ if the field is made to vanish. However, in general the perturbation is fixed by physical conditions that cannot be changed. In this case λ is a fictitious parameter that we vary artificially, and then at the end of the calculations we set it to its physical value $\lambda = 1$. We have already used this trick in the introduction to time-dependent perturbation theory in Section 9.6.3, where we wrote the perturbation as $\lambda W(t)$ and then took $\lambda = 1$ at the end of the calculation.

We therefore assume that the spectrum of H_0 is known. Let $E_0^{(n)}$ be its eigenvalues and $|n, r\rangle$ its eigenvectors, where r is the degeneracy index as in Section 2.3.1:

$$H_0|n, r\rangle = E_0^{(n)}|n, r\rangle. \tag{14.2}$$

We seek the eigenvalues and eigenvectors of $H(\lambda)$ in the form of series in powers of λ, called *perturbation series*. If $H(\lambda)|\varphi(\lambda)\rangle = E(\lambda)|\varphi(\lambda)\rangle$, we can write the series for the eigenvector $|\varphi(\lambda)\rangle$ and the energy $E(\lambda)$ as

$$|\varphi(\lambda)\rangle = |\varphi_0\rangle + \lambda|\varphi_1\rangle + \lambda^2|\varphi_2\rangle + \cdots, \tag{14.3}$$

$$E(\lambda) = E_0^{(n)} + \lambda E_1^{(n)} + \lambda^2 E_2^{(n)} + \cdots. \tag{14.4}$$

If $\lambda = 0$, $|\varphi(\lambda = 0)\rangle = |\varphi_0\rangle = |n, r\rangle$ and $E = E_0^{(n)}$. Our implicit hypothesis is that a series in λ with nonzero radius of convergence exists or, in other words, that the energy is an analytic function of λ at the point $\lambda = 0$. Two cases must be distinguished.

- The eigenvalue $E_0^{(n)}$ of H_0 is simple: then we have the case of *nondegenerate perturbation theory*.
- The eigenvalue $E_0^{(n)}$ of H_0 is degenerate with degeneracy N: then we have the case of *degenerate perturbation theory*.

We shall discuss these two cases one after the other, without entering into the details of the general method of calculating to all orders in λ. We limit ourselves to the lowest nontrivial order in λ and refer the reader to the classic texts for the general case.

---

[1] Rigorously proving that one operator is "small" compared with another is a most delicate mathematical problem.

14.1.2 Nondegenerate perturbation theory

We start from $H_0|n\rangle = E_0^{(n)}|n\rangle$ and set $|\varphi_0\rangle = |n\rangle$ with $\langle\varphi_0|\varphi_0\rangle = 1$, as well as $E_0 = E_0^{(n)}$ in order to simplify the notation. In practice, we shall be interested in the perturbative expansion (14.4) of the energy, treating the perturbative expansion (14.3) of the vector $|\varphi(\lambda)\rangle$ as auxiliary to the calculation permitting us to fix $|\varphi(\lambda)\rangle$ by a convenient condition: $\langle\varphi_0|\varphi(\lambda)\rangle = \langle\varphi_0|\varphi_0\rangle = 1$. With this condition $|\varphi(\lambda)\rangle$ is in general not a normalized vector, but it is always possible to make it one if we wish. Through order λ we have, on the one hand,

$$H(\lambda)|\varphi(\lambda)\rangle = H_0|\varphi_0\rangle + \lambda W|\varphi_0\rangle + \lambda H_0|\varphi_1\rangle,$$

while on the other

$$H(\lambda)|\varphi(\lambda)\rangle = (E_0 + \lambda E_1)|\varphi(\lambda)\rangle = E_0|\varphi_0\rangle + \lambda E_1|\varphi_0\rangle + \lambda E_0|\varphi_1\rangle,$$

from which, identifying the terms of order λ,

$$W|\varphi_0\rangle + H_0|\varphi_1\rangle = E_1|\varphi_0\rangle + E_0|\varphi_1\rangle.$$

Multiplying the two terms of this equation on the left by the bra $\langle\varphi_0|$ and using $\langle\varphi_0|H_0 = E_0\langle\varphi_0|$, we obtain[2]

$$E_1 = \langle\varphi_0|W|\varphi_0\rangle, \tag{14.5}$$

and so, denoting by ΔE_1 the energy difference between the cases $\lambda \neq 0$ and $\lambda = 0$ to first order in λ, we can write

$$\boxed{\Delta E_1 = \lambda\langle\varphi_0|W|\varphi_0\rangle}. \tag{14.6}$$

The order-λ^2 term is also found fairly easily (Exercise 14.6.1):

$$\boxed{\Delta E_2^{(n)} = \lambda^2 \sum_{k\neq n} \frac{|\langle k|W|n\rangle|^2}{E_0^{(n)} - E_0^{(k)}}}. \tag{14.7}$$

As an application, let us calculate the shift of the levels of the one-dimensional harmonic oscillator acted on by an anharmonic perturbation proportional to q^4:

$$\lambda W = \lambda\frac{m^2\omega^3}{\hbar} Q^4. \tag{14.8}$$

From the result of Exercise 11.5.1 and (14.6) we obtain the shift of the nth level through order λ:

$$\Delta E_1^{(n)} = \frac{3}{4}\lambda\hbar\omega\left(2n^2 + 2n + 1\right). \tag{14.9}$$

Even if λ is small, the result diverges for large n, because the larger n is the more important the wave function is at large values of q, and therefore the more important

---

[2] This expression (14.5) can be obtained directly using the Feynman–Hellmann theorem; see Exercise 4.4.3, Eq. (4.35).

the perturbation $\propto q^4$ is: the perturbation λQ^4 is never "small." We have begun with the hypothesis that there exists a series in powers of λ with nonzero radius of convergence. In practice, this hypothesis of analyticity at $\lambda = 0$ is not always satisfied, and the anharmonic oscillator we have just studied provides an example of this. Actually, it is easy to see in this case that $E^{(n)}$ cannot be analytic at $\lambda = 0$, because the nature of the Hamiltonian changes abruptly at this point. For $\lambda > 0$ it is bounded below and bound states are present, but for $\lambda < 0$ it is no longer bounded below and the problem becomes meaningless, unless one adds, for example, a $\lambda' q^6$, $\lambda' > 0$ term to avoid the difficulty. The perturbative series is therefore no longer meaningful for $\lambda < 0$, and it gives an example of an asymptotic series, which gives good results for $\lambda > 0$ if sufficiently few terms are kept, but which diverges if we try to keep too many. This type of series is well known in mathematics. A good example is the Stirling formula valid for $n \gg 1$:

$$\Gamma(n+1) = n! = \left(\frac{n}{e}\right)^n \sqrt{2\pi n} \left(1 + \frac{1}{12n} + \frac{1}{288n^2} + \cdots\right), \qquad (14.10)$$

which is a nonconvergent asymptotic series in powers of $1/n$. Sophisticated methods have been developed for summing such asymptotic series.[3]

14.1.3 Degenerate perturbation theory

Let us now consider the case of a degenerate level, using $\mathcal{H}^{(n)}$ to denote the subspace of dimension N of the eigenvalue $E_0^{(n)}$. The projector $\mathcal{P}^{(n)}$ on $\mathcal{H}^{(n)}$ is written as

$$\mathcal{P}^{(n)} = \sum_{r=1}^{N} |n, r\rangle\langle n, r|. \qquad (14.11)$$

In the subspace $\mathcal{H}^{(n)}$ the operator W is represented by an $N \times N$ matrix with elements $W_{sr}^{(n)} = \langle n, s|W|n, r\rangle$ which can be diagonalized. The eigenvectors $|\varphi_0^{(n,q)}\rangle$ of W in $\mathcal{H}^{(n)}$ are linear combinations of the $|n, r\rangle$:

$$|\varphi_0^{(n,q)}\rangle = \sum_{r=1}^{N} c_{qr} |n, r\rangle,$$

$$W|\varphi_0^{(n,q)}\rangle = E_1^{(n,q)} |\varphi_0^{(n,q)}\rangle.$$

The coefficients c_{qr} are of zeroth order in λ since W can be diagonalized without affecting H_0, which is a multiple of the identity in $\mathcal{H}^{(n)}$:

$$H_0|\varphi_0^{(n,q)}\rangle = E_0^{(n)} |\varphi_0^{(n,q)}\rangle.$$

The diagonalization of W in $\mathcal{H}^{(n)}$ gives the result for the energy through order λ. We recover the results of the nondegenerate case by taking the dimension of $\mathcal{H}^{(n)}$ equal to

---

[3] See, for example, J. Zinn-Justin, *Quantum Field Theory and Critical Phenomena*, Oxford: Oxford University Press (1989), Chapter 37.

unity. In summary, through order λ we can calculate the energy levels and eigenvectors as for a system with a finite number N of levels by diagonalizing the matrix representing $H_0 + \lambda W$ in $\mathcal{H}^{(n)}$. In fact, an approximation by a system with a finite number of levels is often obtained by neglecting the interactions between the subspaces $\mathcal{H}^{(n)}$. A final remark is that the quasi-degenerate case should also be treated by this method.

14.1.4 The variational method

We shall again limit ourselves to the study of a simple case, that of finding the ground-state energy, and leave the use of the variational method in other cases to the classic texts. Let E_0 be the ground-state energy of a Hamiltonian H and $|0\rangle$ be the corresponding eigenvector

$$H|0\rangle = E_0|0\rangle,$$

and let $|\varphi\rangle$ be an arbitrary unit vector in the Hilbert space of states. We write the expectation value of H in the state $|\varphi\rangle$ by decomposing $|\varphi\rangle$ on the basis of eigenstates $|n\rangle$ of H, $H|n\rangle = E_n|n\rangle$:

$$\langle\varphi|H|\varphi\rangle = \sum_{n,m} c_m^* c_n \langle m|H|n\rangle = \sum_n E_n |c_n|^2.$$

We find that

$$\langle\varphi|H|\varphi\rangle - \langle 0|H|0\rangle = \sum_n (E_n - E_0)|c_n|^2 \geq 0, \tag{14.12}$$

where we have used $\sum_n |c_n|^2 = 1$ and $E_n \geq E_0$. The variational method consists of specifying a trial vector $|\varphi(\alpha)\rangle$ depending on a parameter α, or several parameters α_i, which we try to choose to be as close as possible to the assumed form of $|0\rangle$. The result (14.12) shows that

$$\langle H \rangle(\alpha) = \langle\varphi(\alpha)|H|\varphi(\alpha)\rangle \geq E_0.$$

Within the framework of the chosen parametrization, the best result for E_0 will be obtained by seeking the minimum of $\langle H \rangle(\alpha)$:

$$\frac{\mathrm{d}}{\mathrm{d}\alpha}\langle H \rangle(\alpha)\Big|_{\alpha=\alpha_0} = 0, \tag{14.13}$$

and an upper bound on E_0 is

$$F_0 \leq \langle\varphi(\alpha_0)|H|\varphi(\alpha_0)\rangle \tag{14.14}$$

To compare two different choices $|\varphi(\alpha)$ and $|\tilde{\varphi}(\beta)\rangle$ we compare the two minima. The best choice will be the one that gives the smallest value of $\langle H \rangle$. The generalization to a vector depending on several parameters $\alpha_1, \ldots, \alpha_p$ is immediate: we seek the minimum of $\langle H \rangle$ using

$$\frac{\partial}{\partial\alpha_i}\langle H \rangle(\alpha_1, \ldots \alpha_p)\Big|_{\alpha_j = \alpha_{j0}} = 0.$$

As an example, let us study the variational calculation of the ground state of the harmonic oscillator, choosing the trial wave function to be a normalizable function of unit norm:

$$\langle x|\varphi(\alpha)\rangle = \varphi_\alpha(x) = \sqrt{\frac{2}{\pi}} \, \alpha^{3/2} \, \frac{1}{x^2 + \alpha^2}. \tag{14.15}$$

The integrals needed in the following calculations can be derived from the expression

$$I(\alpha) = \int_{-\infty}^{\infty} \frac{\mathrm{d}x}{x^2 + \alpha^2} = \frac{\pi}{\alpha} \tag{14.16}$$

by differentiating $I(\alpha)$ with respect to α^2. Starting from the form (11.9) of the Hamiltonian of the harmonic oscillator, we calculate $\langle H \rangle(\alpha)$:

$$\langle H \rangle(\alpha) = \frac{1}{2}\hbar\omega \int_{-\infty}^{\infty} \mathrm{d}x \left[\left(\frac{\mathrm{d}\varphi_\alpha}{\mathrm{d}x}\right)^2 + x^2 \varphi_\alpha^2(x) \right]$$

$$= \frac{1}{2}\hbar\omega \left[\frac{1}{2\alpha^2} + \alpha^2 \right].$$

The first term in the square brackets is the kinetic energy and the second is the potential energy. The value of $\langle H \rangle(\alpha)$ is a minimum for $\alpha^2 = \alpha_0^2 = 1/\sqrt{2}$ and

$$\langle H \rangle(\alpha_0) = \frac{\hbar\omega}{\sqrt{2}} > E_0 = \frac{\hbar\omega}{2}.$$

For $\alpha = \alpha_0$ the average kinetic energy and the average potential energy are equal:

$$\frac{1}{2m}\langle P^2 \rangle = \frac{1}{2}m\omega^2 \langle X^2 \rangle = \frac{\hbar\omega}{2\sqrt{2}}.$$

The choice (14.15) for the trial wave function is not very good (the error is $\sim 40\%$), because this wave function decreases much too slowly at infinity. If we use a Gaussian trial wave function, we of course find the exact result $\hbar\omega/2$. The power of the variational method will be illustrated in Section 14.5.1.

14.2 One-electron atoms

14.2.1 Energy levels in the absence of spin

In Chapter 10 we studied the spectrum of the hydrogen atom, which has a single electron. An immediate generalization can be made to the ions He^+, Li^{++}, etc. When there is more than one electron, it is no longer possible to analytically solve for the energy levels. It is necessary to resort to approximation methods, which are sometimes very accurate, as in the case of light atoms and, in particular, helium (Section 14.5). Alkaline atoms can also be treated using a simple approximation. Actually, to a first approximation an alkaline atom is an atom with a single outer electron subject to the effective potential produced by the nucleus and the other $(Z-1)$ electrons, called the inner-shell electrons. The spectrum is therefore similar to that of the hydrogen atom, with the difference that no degeneracy is observed between levels of different orbital angular momentum, because the effective potential does not behave as $1/r$. In the case of sodium, for example, the ground state is a $3s$ level, and the $3p$ level lies between the $3s$ and $4s$ levels (Fig. 10.7).

The spectra of Figs. 10.6 and 10.7 were obtained neglecting the spin of the outer electron as well as the nuclear spin. We are going to study the modifications introduced when these spins are taken into account, namely the *fine structure* due to the interaction between the electron orbital angular momentum and spin (Section 14.2.2), the *Zeeman effect* in the presence of an external magnetic field (Section 14.2.3), and the *hyperfine structure* due to the interaction between the nuclear spin and the electron spin and orbital angular momentum (Section 14.2.4).

14.2.2 The fine structure

The fine structure is an effect of relativistic origin whose correct study is based on a relativistic quantum wave equation which is valid for spin-1/2 particles, namely, the Dirac equation.[4] Within the framework of a classical description we are going to make an intuitive argument, which is not entirely correct, to justify the expression for the fine-structure Hamiltonian. In the reference frame where the nucleus is at rest, or the nucleus frame, the electromagnetic field is the gradient of the electrostatic potential $V(r)/q_e$ produced by the nucleus and the $(Z-1)$ inner-shell electrons, and the external electron moves with velocity $\vec{v}$ in this reference frame. In its rest frame, the electron sees the nucleus moving with velocity $-\vec{v}$ and an electromagnetic field which is the transform of the field in the nucleus frame. This transformed field consists of not only an electric field, but also a magnetic field given as a function of the electrostatic field $\vec{E}$ in the nucleus frame as[5]

$$\vec{B} \simeq -\frac{1}{c^2}\vec{v} \times \vec{E} \simeq \frac{1}{q_e c^2}\left[\frac{1}{r}\frac{dV(r)}{dr}\right]\left(\frac{\vec{p}}{m_e} \times \vec{r}\right). \tag{14.17}$$

This magnetic field interacts with the magnetic moment $\vec{\mu} = \gamma \vec{s}$ of the outer electron, leading to an interaction energy

$$W_{so} = -\vec{\mu} \cdot \vec{B} \simeq -\frac{q_e}{m_e}\vec{s} \cdot \vec{B}, \tag{14.18}$$

because the gyromagnetic ratio $\gamma \simeq q_e/m_e$. Combining these two equations and introducing the orbital angular momentum $\vec{l} = \vec{r} \times \vec{p}$, we derive the spin–orbit potential:

$$W_{so} = \frac{1}{m_e^2 c^2}\left[\frac{1}{r}\frac{dV(r)}{dr}\right]\vec{l} \cdot \vec{s}. \tag{14.19}$$

Our argument can be criticized because we have used the formulas for transformations between inertial reference frames, while the electron reference frame is accelerated with

---

[4] The Dirac equation is not the only relativistic quantum wave equation. Another important one is the Klein–Gordon equation, which describes particles of spin 0. However, neither of these equations is completely consistent, as the real unification of quantum mechanics and relativity requires quantized field theory.

[5] In (14.17) we have used the approximation $v \ll c$; the exact expression contains factors of $(1 - v^2/c^2)^{-1/2}$.

respect to the nuclear frame because the electron rotates about the nucleus. This rotational motion leads to the phenomenon of spin precession, called Thomas precession,[6] which reduces the result (14.19) by a factor of two. In the end, the correct quantum expression for the spin–orbit potential is obtained by correcting (14.19) by a factor of 1/2 and replacing the classical quantities $\vec{l}$ and $\vec{s}$ by the operators $\vec{L}$ and $\vec{S}$:

$$
W_{so} = \frac{1}{2m_e^2 c^2} \left[\frac{1}{r} \frac{dV(r)}{dr} \right] \vec{L} \cdot \vec{S} . \tag{14.20}
$$

Let us evaluate the order of magnitude of the correction to the energy levels for the hydrogen atom. Since $\vec{L}$ and $\vec{S}$ are of order $\hbar$ and $V(r) = -e^2/r$, in a state n we obtain

$$
\langle W_{so} \rangle \sim \frac{\hbar^2 e^2}{2m_e^2 c^2} \left\langle \frac{1}{r^3} \right\rangle \sim \frac{\hbar^2 e^2}{2m_e^2 c^2 n^3 a_0^3} = \left(\frac{e^2}{2a_0} \right) \left(\frac{e^2}{\hbar c} \right)^2 \frac{1}{n^3} = \frac{\alpha^2 R_\infty}{n^3},
$$

where we have introduced the Bohr radius a_0, the fine-structure constant α, and the Rydberg constant R_∞ (see (1.39)–(1.41)). The corrections to the energy levels are therefore of order α^2 in relative value, which is what we expect for relativistic corrections, because $(v/c)^2 \sim \alpha^2$.[7]

Let us examine the effect of the potential (14.20) on a level (nl) with principal quantum number n and orbital angular momentum l. Since the effect on the levels is small, $\sim \alpha^2$, we can use perturbation theory. Neither the orbital angular momentum $\vec{L}$ nor the spin $\vec{S}$ commutes with W_{so}. However, the scalar operator $\vec{L} \cdot \vec{S}$ commutes with the *total* angular momentum $\vec{J} = \vec{L} + \vec{S}$ and moreover, since $[\vec{L}^2, \vec{L}] = 0$ and $[\vec{L}^2, f(r)] = 0$, the potential (14.20) commutes with $\vec{L}^2$, which implies that levels with different l are not related. In summary, the spin–orbit potential is diagonal in the basis $|l\ 1/2\ jm_j\rangle$. In the absence of the spin–orbit potential, the degeneracy of a level (nl) is $2(2l+1)$ and it is necessary in principle to use degenerate perturbation theory. However, in the present case the situation is very simple, because we already know the basis $|l\ 1/2\ jm_j\rangle$ in which W_{so} is diagonal. The spin–orbit potential will partially lift the degeneracy. In fact, two values $j = l \pm 1/2$ of the total angular momentum are possible, and according to (10.138) and using $\vec{J}^2 = (\vec{L} + \vec{S})^2$,

$$
\vec{L} \cdot \vec{S} = \frac{\hbar^2}{2} \left[j(j+1) - l(l+1) - s(s+1) \right] \tag{14.21}
$$

or

$$
\begin{aligned}
\vec{L} \cdot \vec{S} &= -\frac{\hbar^2}{2} (l+1), \quad j = l - \frac{1}{2}, \\
\vec{L} \cdot \vec{S} &= +\frac{\hbar^2}{2} l, \qquad\quad j = l + \frac{1}{2}.
\end{aligned} \tag{14.22}
$$

---

[6] See, for example, E. Taylor and J. Wheeler, *Space-Time Physics*, New York: Freeman (1963), Section 103, or Jackson [1999], Section 11.8.

[7] For a nucleus of charge Z and a single electron, $(v/c)^2 \sim (Z\alpha)^2$.

The states of total angular momentum $j = l - 1/2$ and $j = l + 1/2$ therefore have different energies and the spin–orbit potential partially lifts the degeneracy. Naturally, each of the two corresponding energy levels still has a $(2j + 1)$-fold degeneracy. We note that the spin–orbit potential does not affect s-waves ($l = 0$).

As a special case, let us consider the $2p$ ($l = 1$) level of hydrogen. The two possible values of j are $j = 1/2$ and $j = 3/2$. The corresponding levels are denoted as $2p_{1/2}$ and $2p_{3/2}$. The $2p \rightarrow 1s$ transition is split, which is easily confirmed by spectroscopy. In the case of hydrogen, the $2s_{1/2}$ and $2p_{1/2}$ levels are degenerate in the approximation of the Dirac equation. They differ in energy from the $2p_{3/2}$ level by about 4.5×10^{-5} eV, which corresponds to a frequency difference of about 10 GHz. The order-of-magnitude calculation we have just done gives an energy difference $\sim \alpha^2 R_\infty / 8 \sim 10^{-4}$ eV, in qualitative agreement with experiment. Experiment shows that, in contrast to the prediction of the Dirac equation, the $2s_{1/2}$ and $2p_{1/2}$ levels are in fact nondegenerate: the $2p_{1/2}$ level is lower by about 5×10^{-5} eV, which corresponds to about 1 GHz. This difference, known as the Lamb shift, is explained by effects of quantum electrodynamics, the theory of the quantized electromagnetic and electron–positron fields.

The above notation $(nl)_j$ can be generalized to higher levels: for a d-wave ($l = 2$) the possible values of j are $3/2$ and $5/2$ and the levels are denoted as $nd_{3/2}$ and $nd_{5/2}$. For an f-wave ($l = 3$) we will have the $nf_{5/2}$ and $nf_{7/2}$ levels, and so on. A classic example in spectroscopy is the splitting of the yellow line of sodium, which corresponds to a $3p \rightarrow 3s$ transition; the two lines are called D_1 at 589.6 nm and D_2 at 589.0 nm. In general, the $j = l + 1/2$ level is higher than the $j = l - 1/2$ level because the expectation value $\langle dV/dr \rangle > 0$, but there are some exceptions. In the nuclear shell model, where the spin–orbit potential plays a crucial role, this order is systematically inverted.

14.2.3 The Zeeman effect

The $(2j + 1)$-fold degeneracy of the $(nl)_j$ level is lifted by placing the atom in a constant magnetic field $\vec{B}$. This is the *Zeeman effect*. It arises from the interaction of the magnetic field with the orbital magnetic moment due to the motion of the electron in its orbit, and also to the magnetic moment associated with the spin of this electron. The magnetic moment associated with $\vec{L}$ is given by the classical gyromagnetic ratio (3.30) $\gamma = q_e/2m_e$, and the gyromagnetic ratio due to the spin is roughly q_e/m_e. The interaction energy is derived from the coupling between the magnetic moment and the field:[8]

$$W = -\frac{q_e}{2m_e}(\vec{L} + 2\vec{S}) \cdot \vec{B} \,. \tag{14.23}$$

---

[8] However, this argument gives only the dominant term in the interaction; see Exercise 14.6.5 for a detailed justification of (14.23).

It is usual to choose $\vec{B}$ to be parallel to Oz:

$$W = -\frac{q_e B}{2m_e}(L_z + 2S_z). \tag{14.24}$$

When the Zeeman energy (14.23) is sufficiently small compared with the characteristic energy of the fine structure of the level under consideration, we can use degenerate perturbation theory for each level $(nl)_j$. If this is not the case, it is necessary to simultaneously diagonalize the Hamiltonian of the fine structure and that of the Zeeman effect; see Exercise 6.5.4. Let us consider the case of small Zeeman effect. The matrix elements of the perturbation (14.24) in the $(nl)_j$ level are

$$W^{nlj}_{mm'} = -\frac{q_e B}{2m_e}\langle nljm|L_z + 2S_z|nljm'\rangle. \tag{14.25}$$

The operators $\vec{L}$ and $\vec{S}$ are vector operators, and according to the Wigner–Eckart theorem (10.150) for these operators the matrix elements for, for example, L_z are given by

$$\langle nljm|L_z|nljm'\rangle = \frac{1}{\hbar^2 j(j+1)}\langle j||(\vec{J}\cdot\vec{L})||j\rangle\langle nljm|J_z|nljm'\rangle$$

$$= \frac{m}{\hbar j(j+1)}\langle(\vec{J}\cdot\vec{L})\rangle\,\delta_{mm'}.$$

Using

$$\vec{S}^2 = (\vec{J} - \vec{L})^2 \quad \text{and} \quad \vec{L}^2 = (\vec{J} - \vec{S})^2$$

to write out $\vec{J}\cdot\vec{S}$ and $\vec{J}\cdot\vec{L}$, we find

$$\langle\vec{J}\cdot(\vec{L} + 2\vec{S})\rangle = \frac{3}{2}\vec{J}^2 + \frac{1}{2}\vec{S}^2 - \frac{1}{2}\vec{L}^2$$

and then

$$\langle nljm|L_z + 2S_z|nljm'\rangle = \frac{m\hbar}{2j(j+1)}\left[3j(j+1) + \frac{3}{4} - l(l+1)\right]\delta_{mm'}$$

$$= m\hbar\left\{1 + \frac{1}{2j(j+1)}\left[j(j+1) + \frac{3}{4} - l(l+1)\right]\right\}\delta_{mm'}.$$

The final result can be written as

$$\boxed{W^{nlj}_{mm'} = -g\frac{q_e B}{2m_e}m\hbar\delta_{mm'}}\,. \tag{14.26}$$

Within our approximation the shifts of the Zeeman sublevels are linear in B. They are controlled by the *Landé g factor*:

$$\boxed{g = 1 + \frac{1}{2j(j+1)}\left[j(j+1) + \frac{3}{4} - l(l+1)\right]}\,. \tag{14.27}$$

The quantity $g q_e / 2 m_e$ can be interpreted physically as an effective gyromagnetic ratio. For a free electron in a magnetic field we have seen that the Landé g factor is 2. This is also the case for an s-wave, as can be verified by setting $l = 0$ in (14.27).

14.2.4 The hyperfine structure

An even smaller effect, of order 10^{-6} in relative value, arises from the interaction between the nuclear magnetic moment and the orbital and spin magnetic moments of the outer electron. The interaction between a nuclear magnetic dipole moment μ_n and an electron magnetic dipole is a priori weaker than that between two electric dipoles by a factor $\sim 10^{-3}$, as the nuclear Bohr magneton $\mu_N = q_p \hbar / 2 m_p$ is smaller than the Bohr magneton $\mu_B = |q_e| \hbar / 2 m_e$ by a factor of $m_p / m_e \sim 2000$. We recall the expressions for the electron and proton magnetic moment operators:

$$\vec{\mu}_e = \gamma_e \vec{S}_e \simeq -2 \mu_B \frac{\vec{S}_e}{\hbar}, \quad \vec{\mu}_p = \gamma_p \vec{S}_p \simeq 5.59 \mu_N \frac{\vec{S}_p}{\hbar}. \tag{14.28}$$

In classical electrodynamics it can be shown[9] that the magnetic field $\vec{B}(\vec{r})$ of a point dipole $\vec{\mu}_n$ at the origin is

$$\vec{B}(\vec{r}) = -\frac{\mu_0}{4\pi r^3} \left[\vec{\mu}_n - 3(\vec{\mu}_n \cdot \hat{r}) \hat{r} \right] + \frac{2\mu_0}{3} \vec{\mu}_n \delta(\vec{r}). \tag{14.29}$$

The energy of the orbital magnetic moment and the spin of the outer electron in this magnetic field can be written as in (14.23). We shall limit ourselves to the case of an s-wave electron, where only the spin magnetic moment needs to be taken into account, as in the s-wave there is no contribution from the orbital angular momentum to the atomic magnetic moment. Moreover, the term inside the square brackets in (14.29) gives a vanishing contribution. Actually, if we use perturbation theory to calculate the magnetic energy $\langle W' \rangle = -\langle \vec{\mu}_e \cdot \vec{B} \rangle$ corresponding to the interaction of the electron magnetic moment with the term inside the square brackets in (14.29) for an s-wave, where the wave function $\varphi(r)$ depends only on r, we find

$$\langle W' \rangle = \frac{\mu_0}{4\pi} \int d^3 r |\varphi(r)|^2 \frac{1}{r^3} \left[(\vec{\mu}_n \cdot \vec{\mu}_e) - 3(\vec{\mu}_n \cdot \hat{r})(\vec{\mu}_e \cdot \hat{r}) \right]$$

$$= \frac{\mu_0}{4\pi} \left\langle \frac{1}{r^3} \right\rangle \left[(\vec{\mu}_n \cdot \vec{\mu}_e) - 3 \sum_{i,j=1}^{3} \mu_{ni} \mu_{ej} I_{ij} \right].$$

To obtain the second line of the above equation we separated the radial part of the integral of the second term in the square brackets from the angular part by writing

$$\int d^3 r = 4\pi \int_0^\infty r^2 dr \int \frac{d\Omega}{4\pi}.$$

---

[9] See, for example, Jackson [1999], Section 5.6.

The radial integral gives

$$\int_0^\infty r^2 dr |\varphi(r)|^2 \frac{1}{r^3} = \left\langle \frac{1}{r^3} \right\rangle.$$

The angular integral I_{ij} is

$$I_{ij} = \int \frac{d\Omega}{4\pi} \hat{r}_i \hat{r}_j = \frac{1}{3} \delta_{ij}.$$

To prove this, we observe that the only rotationally invariant rank-2 tensor that can be constructed using the indices (i, j) is δ_{ij}:

$$I_{ij} = c \delta_{ij} \quad \text{and} \quad \sum_{ij} \delta_{ij} I_{ij} = 1,$$

which shows that $c = 1/3$. Thus, the term between square brackets in (14.29) gives a vanishing contribution and we are left with just the contact term:

$$W_{\text{cont}} = -\frac{2\mu_0}{3} \vec{\mu}_n \cdot \vec{\mu}_e \delta(\vec{r})$$

$$= -\frac{2\mu_0}{3} \gamma_n \gamma_e (\vec{S}_n \cdot \vec{S}_e) \delta(\vec{r}). \tag{14.30}$$

As an example, let us take the hyperfine structure of the ground state of the hydrogen atom: $\mu_n \to \mu_p$, $n = 1$, $l = 0$. The state vector is the tensor product of a spatial wave function derived from (10.94)

$$\varphi(\vec{r}) = \frac{1}{\sqrt{\pi a_0^3}} \exp\left(-\frac{r}{a_0}\right) \tag{14.31}$$

and a spin wave function, which itself is the tensor product of the state vectors in the electron and proton spin spaces. The spatial part and the spin part are completely decoupled. First we find the expectation value of the spatial part:

$$\langle W_{\text{cont}} \rangle_{\text{spat}} = -\frac{2\mu_0}{3} \gamma_n \gamma_e |\varphi(0)|^2 (\vec{S}_p \cdot \vec{S}_e)$$

$$= \frac{A}{\hbar^2} (\vec{S}_p \cdot \vec{S}_e). \tag{14.32}$$

The constant A is

$$A = \frac{2\mu_0}{3} (2\mu_B)(5.59\mu_N) \frac{1}{\pi a_0^3} \simeq 5.87 \times 10^{-6} \text{ eV}.$$

Then the effective Hamiltonian is $A(\vec{S}_p \cdot \vec{S}_e)/\hbar^2$, which acts in the four-dimensional Hilbert space that is the tensor product of the two spin spaces. In the absence of the hyperfine perturbation, the $1s_{1/2}$ ground state of the hydrogen atom is four-fold degenerate. It is necessary to diagonalize $A(\vec{S}_p \cdot \vec{S}_e)/\hbar^2$ in this subspace, which is straightforward if we introduce the total spin $\vec{S} = \vec{S}_p + \vec{S}_e$ and the identity

$$\vec{S}_p \cdot \vec{S}_e = \frac{1}{2} \left(\vec{S}^2 - \vec{S}_p^2 - \vec{S}_e^2 \right) = \frac{\hbar^2}{2} \left[s(s+1) - \frac{3}{2} \right]. \tag{14.33}$$

According to the results of Section 10.6.1, the two possible values of s are $s = 1$ (the triplet state) and $s = 0$ (the singlet state). The eigenvalues of the Hamiltonian are

$$s = 1, \quad \text{triplet state}: E_{\text{tr}} = E_0 + \frac{1}{4}A,$$

$$s = 0, \quad \text{singlet state}: E_{\text{sing}} = E_0 - \frac{3}{4}A,$$

where E_0 is the energy in the absence of the hyperfine effect, and the eigenvectors are given by (10.125) and (10.126). The two levels are separated by an amount $A \simeq 5.87 \times 10^{-6}$ eV, which corresponds to the emission of a photon of wavelength 21 cm when the atom makes a transition from the triplet to the singlet level. Although the lifetime of the triplet level is very long, 10^7 years, and a priori seems difficult to observe, it is of great importance in astrophysics. It has given fundamental information about the interstellar clouds of atomic hydrogen making up 10% to 50% of the mass of the galaxy, permitting measurements of mass and velocity distributions, magnetic fields, and so on.[10]

14.3 Atomic interactions with an electromagnetic field

14.3.1 The semiclassical theory

In this section we shall study the interaction between an electromagnetic field and an atom, modeled as before by an outer electron in a spherically symmetric potential. We shall begin with the *semi-classical approximation*, already introduced in Section 5.3.2, where the electromagnetic field is described classically while the atom is described in a quantum manner. In Section 5.3 we postulated a phenomenological interaction between an electromagnetic wave and an electric dipole responsible for transitions from one level to another. In this section we shall complete these results by justifying the dipole approximation and giving an explicit expression for the transition amplitude. At this point it is useful to summarize the various possible approximations which can be used to study interactions between an atom (or molecule) and the electromagnetic field (see Table 14.1). In principle, the atom and the field should both be treated in a quantum manner, but it may prove convenient to use a classical approximation for one or the other when it is clear that such an approximation is valid.

In the approach of Section 11.3.3, the classical electromagnetic wave is described in the Coulomb gauge $\vec{\nabla} \cdot \vec{A} = 0$ by a transverse vector potential $\vec{A}(\vec{r}, t)$. A plane wave of wave vector $\vec{k}$ and frequency ω can be written as

$$\vec{A}(\vec{r}, t) = \text{Re}\left[\vec{A}_0 e^{i(\vec{k} \cdot \vec{r} - \omega t)}\right], \quad \vec{k} \cdot \vec{A}_0 = 0. \tag{14.34}$$

Let us recall the action of the divergence and curl operators in the Fourier space:

$$\vec{\nabla} \cdot \rightarrow i\vec{k} \cdot, \quad \vec{\nabla} \times \rightarrow i\vec{k} \times, \tag{14.35}$$

---

[10] More details can be found in, for example, Basdevant and Dalibard [2002], Chapter 13.

Table 14.1 *Various approximation schemes*

Electromagnetic field	Atom	Examples
classical	classical	classical radiation Section 1.5.3
classical	quantum	absorption and stimulated emission Section 5.3.2, Sections 14.3.1 to 14.3.3
quantum	classical	coupling to a classical source Exercise 11.5.4
quantum	quantum	spontaneous emission Section 14.3.4, Section 14.4

which leads to the following electric and magnetic fields:

$$\vec{E}(\vec{r}, t) = -\frac{\partial \vec{A}}{\partial t} = \text{Re}\left[i\omega \vec{A}_0 e^{i(\vec{k}\cdot\vec{r}-\omega t)}\right], \tag{14.36}$$

$$\vec{B}(\vec{r}, t) = \vec{\nabla} \times \vec{A} = \text{Re}\left[i(\vec{k} \times \vec{A}_0) e^{i(\vec{k}\cdot\vec{r}-\omega t)}\right]. \tag{14.37}$$

The energy flux is given by the Poynting vector

$$\vec{S} = \varepsilon_0 c^2 \vec{E} \times \vec{B}, \tag{14.38}$$

and averaging over time using $\langle \cos^2(\omega t) \rangle = 1/2$ we find

$$\langle \vec{S} \rangle = \frac{1}{2}\varepsilon_0 c\omega^2 |\vec{A}_0|^2 \hat{k} = \mathcal{I}(\omega)\hat{k}. \tag{14.39}$$

The intensity $\mathcal{I}(\omega)$ is related to the photon flux $\mathcal{F}$ as

$$\mathcal{I}(\omega) = \hbar\omega\mathcal{F},$$

or, denoting the photon density by n,

$$\mathcal{F} = nc = \frac{1}{2\hbar}\varepsilon_0 c\omega |\vec{A}_0|^2. \tag{14.40}$$

According to (11.115), the Hamiltonian describing the interaction between the electron and the field is

$$H = \frac{1}{2m_e}\left[\vec{P} - q_e\vec{A}(\vec{R}, t)\right]^2 + V(\vec{R}), \tag{14.41}$$

where $V(\vec{R})$ represents the effective interaction of the outer electron with the nucleus and the $(Z-1)$ inner-shell electrons. The Hamiltonian (14.41) can be split into the unperturbed part H_0

$$H_0 = \frac{1}{2m_e}\vec{P}^2 + V(\vec{R}) \tag{14.42}$$

and a perturbation

$$W(\vec{R}, t) = -\frac{q_e}{2m_e}\left(\vec{P}\cdot\vec{A} + \vec{A}\cdot\vec{P}\right) + \frac{q_e^2}{2m_e}\vec{A}^2. \tag{14.43}$$

To first order in $q_e \vec{A}$ we can neglect the second term of (14.43), or the diamagnetic term $(q_e^2/2m_e)\vec{A}^2$ (Exercise 14.5.5). Moreover, the first term is simplified in the Coulomb gauge $\vec{\nabla} \cdot \vec{A} = 0$ because

$$\vec{P} \cdot [\vec{A} f(\vec{r})] = -i\hbar(\vec{\nabla} \cdot \vec{A}) f(\vec{r}) - i\hbar \vec{A} \cdot \vec{\nabla} f(\vec{r})$$
$$= -i\hbar \vec{A} \cdot \vec{\nabla} f(\vec{r}) = (\vec{A} \cdot \vec{P}) f(\vec{r}).$$

The perturbation $W(\vec{R}, t)$ is finally written as

$$W(\vec{R}, t) = -\frac{q_e}{m_e} \left[\vec{A}(\vec{R}, t) \cdot \vec{P} \right]. \tag{14.44}$$

We shall work in a representation in which $\vec{R}$ is diagonal, $\vec{R} \to \vec{r}$. Using (14.34), we have

$$W(\vec{r}, t) = -\frac{q_e}{2m_e} \left[e^{i(\vec{k} \cdot \vec{r} - \omega t)} \vec{A}_0 \cdot \vec{P} + e^{-i(\vec{k} \cdot \vec{r} - \omega t)} \vec{A}_0^* \cdot \vec{P} \right]. \tag{14.45}$$

Now we can use the results of Section 9.6.3: the term involving $\exp(-i\omega t)$ in (14.45) corresponds to energy absorption by the atom and the term involving $\exp(i\omega t)$ corresponds to energy emission. If there exist two energy levels E_i and E_f with $E_i < E_f$ corresponding to a resonance $E_f - E_i = \hbar\omega_0 \simeq \hbar\omega$, the atom will absorb energy $\hbar\omega_0$ in a transition $i \to f$, and emit energy $\hbar\omega_0$ in a transition $f \to i$. In a particle interpretation this would of course mean that the atom absorbs or emits a photon of energy $\hbar\omega_0$, but such an interpretation falls outside the framework of the semi-classical theory. According to (9.170), the probability per unit time of absorption $i \to f$ is given by

$$\Gamma_{fi} = \frac{2\pi}{\hbar} \left(\frac{q_e}{2m_e} \right)^2 \left| \langle f | \exp(i\vec{k} \cdot \vec{r}) \vec{A}_0 \cdot \vec{P} | i \rangle \right|^2 \delta(E_f - (E_i + \hbar\omega)). \tag{14.46}$$

14.3.2 The dipole approximation

Let us introduce a polarization unit vector $\vec{e}_s$, $\vec{e}_s^* \cdot \vec{e}_s = 1$, writing $\vec{A}_0 = |\vec{A}_0| \vec{e}_s$. The intensity $\mathcal{I}(\omega)$ per unit frequency is given by (14.39):

$$\mathcal{I}(\omega) = \frac{1}{2} \varepsilon_0 c \omega^2 |\vec{A}_0(\omega)|^2.$$

We rewrite (14.46) by integrating over ω and separating the squared modulus of the transition matrix element from the characteristics of the incident wave:

$$\Gamma_{fi} = \frac{2\pi}{\hbar} \left(\frac{q_e}{2m_e} \right)^2 \int d\omega |\vec{A}_0(\omega)|^2 \left| \vec{e}_s \cdot \langle f | \exp(i\vec{k} \cdot \vec{r}) \vec{P} | i \rangle \right|^2 \delta(E_f - (E_i + \hbar\omega))$$

$$= \frac{4\pi^2 \alpha}{\hbar \omega_0^2 m_e^2} \mathcal{I}(\omega_0) \left| \vec{e}_s \cdot \langle f | \exp(i\vec{k} \cdot \vec{r}) \vec{P} | i \rangle \right|^2. \tag{14.47}$$

The transition matrix element in (14.47) can be simplified by using the fact that the wavelength of the emitted or absorbed radiation, $0.1\,\mu\text{m} \lesssim \lambda \lesssim 1\,\mu\text{m}$, is very large compared with the atomic dimensions $a_0 \sim 0.1$ nm, which makes it possible to replace $\exp(i\vec{k} \cdot \vec{r})$ by unity because $\langle \vec{k} \cdot \vec{r} \rangle \sim k a_0 \sim a_0/\lambda \ll 1$:

$$\langle f | e^{i\vec{k}\cdot\vec{r}} \vec{P} | i \rangle = \int d^3 r \, \varphi_f^*(\vec{r}) e^{i\vec{k}\cdot\vec{r}} \left(-i\hbar\vec{\nabla} \right) \varphi_i(\vec{r})$$

$$\simeq \int d^3 r \, \varphi_f^*(\vec{r}) \left(-i\hbar\vec{\nabla} \right) \varphi_i(\vec{r}).$$

Moreover, $\vec{P}$ can be written as the commutator between $\vec{R}$ and H_0:

$$[\vec{R}, H_0] = \frac{i\hbar}{m_e} \vec{P},$$

which gives

$$\langle f | \vec{P} | i \rangle = \frac{m_e}{i\hbar} \langle f | \vec{R} H_0 - H_0 \vec{R} | i \rangle$$

$$= \frac{m_e}{i\hbar} (E_i - E_f) \langle f | \vec{R} | i \rangle = i m_e \omega_0 \langle f | \vec{R} | i \rangle. \tag{14.48}$$

In classical physics $\vec{r}$ is the vector joining the nucleus located at the origin to the outer electron and $q_e \vec{r}$ is the electric dipole moment $\vec{d}$ of the atom. The quantity $\langle f | q_e \vec{R} | i \rangle$ is therefore the *matrix element* $\vec{D}_{fi}$ *of the electric dipole moment operator* $\vec{D} = q_e \vec{R}$ between the states $|i\rangle$ and $|f\rangle$:

$$\vec{D}_{fi} = \langle f | \vec{D} | i \rangle = q_e \langle f | \vec{R} | i \rangle. \tag{14.49}$$

Substituting these results into (14.47), we obtain the transition probability per unit time for polarization $\vec{e}_s$:

$$\Gamma_{fi} = 4\pi^2 \left(\frac{|\vec{e}_s \cdot \vec{D}_{fi}|^2}{4\pi\varepsilon_0 \hbar^2 c} \right) \mathcal{I}(\omega_0) \tag{14.50}$$

$$= \frac{4\pi^2 \alpha}{\hbar} |\vec{e}_s \cdot \vec{R}_{fi}|^2 \mathcal{I}(\omega_0), \tag{14.51}$$

in agreement with (5.66). The dipole moment d introduced phenomenologically in Section 5.3.2 takes the following explicit form for a one-electron atom:

$$d^2 \to |\vec{e}_s \cdot \vec{D}_{fi}|^2 = q_e^2 |\vec{e}_s \cdot \vec{R}_{fi}|^2.$$

The expression (14.50) is more general than (14.51), and is valid for any atomic or molecular system when the selection rules for electric dipole transitions are satisfied: the transition probability is governed by the transition matrix element of the electric dipole moment of the system, which involves all the charged particles. By an identical calculation we find the rate of stimulated emission $\overline{\Gamma}_{if}$, which is also given by (14.50): $\overline{\Gamma}_{if} = \Gamma_{fi}$. Actually, to go from absorption to emission it is sufficient to replace D_{fi} by $D_{if} = D_{fi}^*$. Following the argument based on the Einstein relations of Section 5.4, from

Γ_{fi} we can deduce the probability for spontaneous emission of a photon by summing over the two possible polarization states $s = 1, 2$ and taking the average $\langle \bullet \rangle$ over angles and spins:

$$B' = \frac{4\omega_0^3}{c^2}\left(\frac{\frac{1}{2}\sum_{s=1}^2\langle|\vec{e}_s\cdot\vec{D}_{fi}|^2\rangle}{4\pi\varepsilon_0\hbar c}\right) = \frac{2\alpha\omega_0^3}{c^2}\left\langle\sum_{s=1}^2|\vec{e}_s\cdot\vec{R}_{fi}|^2\right\rangle. \tag{14.52}$$

The electric dipole moment operator $\vec{D}$, like the position operator $\vec{R}$, is a vector operator which is odd under a parity operation, that is, a polar vector. This property of $\vec{D}$ implies certain *selection rules* for electric dipole transitions. The Wigner–Eckart theorem for vector operators gives the matrix elements of the spherical components (10.145) D_q of $\vec{D}$: if j_i and j_f are the angular momenta of the initial state i and the final state f, and m_i and m_f are the magnetic quantum numbers, then from (10.149) we obtain

$$\langle j_f m_f|D_q|j_i m_i\rangle = C^{1 j_i}_{q m_i; j_f m_f}\langle j_f||D||j_i\rangle. \tag{14.53}$$

The Clebsch–Gordan coefficient can be nonzero only if $|j_i - 1| \leq j_f \leq j_i + 1$ and $m_f = q + m_i$. Moreover, the parities of the initial and final states must be opposite: $\Pi_i\Pi_f = -1$. Therefore, electric dipole transitions obey the following selection rules.

Selection rules for electric dipole transitions

$$|j_i - 1| \leq j_f \leq j_i + 1, \quad m_f = m_i + q, \; q = -1, 0, +1, \quad \Pi_i\Pi_f = -1.$$

These rules generalize the results obtained in Section 10.5.2 in the special case $j_i = 1$ and $j_f = 0$. The selection rules for the magnetic quantum number m are directly related to the conservation of the z component of the angular momentum, and some examples have already been given in Section 10.5.2 and Exercise 10.7.13.

14.3.3 The photoelectric effect

In the preceding subsection we studied a transition between two levels by generalizing the results of Section 5.3. Now let us consider a transition to the continuum. An electromagnetic wave of frequency $\omega > R_\infty/\hbar$ and polarization $\vec{e}_s$ arrives at a hydrogen atom in its ground state. In particle language the condition $\omega > R_\infty/\hbar$ implies that the photon energy is sufficient for ionizing the atom by ejecting its electron, which provides a very simple example of the photoelectric effect and is a case which can be completely solved analytically. According to the Fermi Golden Rule and the definition (12.1) of the cross section, to first order in perturbation theory in W the cross section for photoelectron production is

$$\frac{d\sigma}{d\Omega} = \frac{2\pi}{\hbar\mathcal{F}}|\langle f|W|i\rangle|^2\frac{\mathcal{V}m_e k_e}{(2\pi)^3\hbar^2} \tag{14.54}$$

where $\vec{k}_e$ is the wave vector of the final electron and the last factor is the electron density of states (9.151) in a volume $\mathcal{V}$. When $\hbar\omega \gg R_\infty$ (but $\hbar\omega \ll m_e c^2$ in order to preserve the nonrelativistic kinematics and prevent electron–positron pair production[11]), we can neglect the interaction of the final electron with the proton and take a plane wave for the final state, thus obtaining the Born approximation:

$$\langle \vec{r} | f \rangle = \varphi_f(\vec{r}) = \frac{1}{\sqrt{\mathcal{V}}} e^{i\vec{k}_e \cdot \vec{r}}.$$

We note that the dipole approximation is not valid under the kinematic conditions defined above. The initial state is described by the wave function (14.31) of the ground state of the hydrogen atom. The matrix element $\langle f | W | i \rangle$ is given by (14.46):

$$\langle f | W | i \rangle = \left(-\frac{q_e}{2m_e} \right) |\vec{A}_0| \vec{e}_s \cdot \int d^3 r \frac{1}{\sqrt{\mathcal{V}}} e^{i(\vec{k} - \vec{k}_e) \cdot \vec{r}} \left(-i\hbar \vec{\nabla} \varphi_i(\vec{r}) \right)$$

or, integrating by parts and using the fact that $\vec{e}_s \cdot \vec{k} = 0$,

$$\langle f | W | i \rangle = \left(-\frac{q_e}{2m_e} \right) |\vec{A}_0| \frac{\hbar(\vec{e}_s \cdot \vec{k}_e)}{\sqrt{\mathcal{V} \pi a_0^3}} \int d^3 r \, e^{i(\vec{k} - \vec{k}_e) \cdot \vec{r}} \, e^{-r/a_0}$$

$$= \left(-\frac{q_e}{2m_e} \right) |\vec{A}_0| \frac{\hbar(\vec{e}_s \cdot \vec{k}_e)}{\sqrt{\mathcal{V} \pi a_0^3}} \frac{8\pi/a_0}{(q^2 + 1/a_0^2)^2}, \tag{14.55}$$

where we have defined $\vec{q} = \vec{k} - \vec{k}_e$, so that $\hbar q$ is the momentum transfer between the initial photon and the final electron. To calculate the integral in (14.55) we have used the formula

$$\int d^3 r \, e^{i\vec{q} \cdot \vec{r}} \, e^{-\lambda r} = 2\pi \int_0^\infty r^2 \, dr \, e^{-\lambda r} \int_{-1}^{+1} e^{iqr\cos\theta} d\cos\theta = \frac{2\pi}{q} \int_0^\infty r \, dr \, \sin qr \, e^{-\lambda r}$$

$$= \frac{2\pi}{q} \operatorname{Im} \int_0^\infty r \, dr \, e^{iqr} e^{-\lambda r} = \frac{4\pi}{q} \operatorname{Im} \frac{1}{(\lambda - iq)^2} = \frac{8\pi\lambda}{(\lambda^2 + q^2)^2}.$$

Assembling all the factors in (14.54), we obtain

$$\boxed{\frac{d\sigma}{d\Omega} = \frac{32\alpha\hbar}{m_e \, \omega \, a_0^5} \frac{|\vec{e}_s \cdot \hat{k}_e|^2 \, k_e^3}{\left[(\vec{k} - \vec{k}_e)^2 + 1/a_0^2 \right]^4}}. \tag{14.56}$$

Let us make (14.56) explicit by choosing $\vec{k}$ to be parallel to Oz and taking linear polarization $\vec{e}_s$ parallel to Ox. Let $(\Omega = \theta, \phi)$ be the polar angles defining $\hat{k}_e$:

$$(\vec{e}_x \cdot \hat{k}_e)^2 = \sin^2 \theta \cos^2 \phi.$$

This quantity is a maximum when $\vec{e}_x$ and $\vec{k}_e$ are parallel, or $\theta = \pi/2$ and $\phi = 0$ or π. The denominator in (14.56) varies slowly with θ, because, with the kinematical conditions defined above, from energy conservation we have

$$\frac{k}{k_e} \simeq \frac{\hbar k_e}{2m_e c} = \frac{v_e}{2c} \ll 1,$$

---

[11] This approximation is valid in the case of X-rays, where the energy varies from 1 to 100 keV.

where v_e is the speed of the photoelectrons and

$$(\vec{k} - \vec{k}_e)^2 \simeq k_e^2 \left(1 - \frac{v_e}{c} \cos \theta\right).$$

Therefore, the electrons are preferentially ejected in a plane perpendicular to the incident wave vector and parallel to the electric field of the wave. If the incident wave is not polarized, the contributions of the polarizations in the x and y directions must be added incoherently and averaged over:

$$\frac{1}{2} \left[(\vec{e}_x \cdot \hat{k}_e)^2 + (\vec{e}_y \cdot \hat{k}_e)^2\right] = \frac{1}{2} \sin^2 \theta .$$

This still gives preferential emission in the plane perpendicular to $\vec{k}$:

$$\frac{d\sigma}{d\Omega}\bigg|_{\text{unpol}} = \frac{16\alpha\hbar}{m_e \omega a_0^5} \frac{k_e^3 \sin^2 \theta}{\left[(\vec{k} - \vec{k}_e)^2 + 1/a_0^2\right]^4} \simeq \frac{16\alpha\hbar}{m_e \, \omega \, a_0^5 \, k_e^5} \frac{\sin^2 \theta}{\left[1 - \frac{v_e}{c} \cos \theta\right]^8}. \tag{14.57}$$

Under the chosen kinematical conditions we can neglect $1/a_0^2$ compared to q^2, because

$$\frac{\hbar^2 k_e^2}{2m_e} \gg R_\infty = \frac{e^2}{2a_0} \Rightarrow (k_e a_0)^2 \gg \frac{m_e e^2}{\hbar^2} a_0 = 1.$$

It is important to note that we have managed to treat the photoelectric effect in a semiclassical approach without introducing the photon, contradicting the widespread belief that the photon concept is necessary to explain the threshold effect (Section 1.3.2). In the semiclassical approach the threshold effect arises from the resonance condition: the photoelectric effect is appreciable only if the light wave is in resonance with the ground state E_0 and a level E_C of the continuum: $E_C - E_0 = \hbar\omega$. The photoelectric effect can be explained without the photon, but not without $\hbar$!

14.3.4 The quantized electromagnetic field: spontaneous emission

We have often had recourse to the concept of the photon in order to interpret intuitively the results of the semiclassical theory, whereas strictly speaking this concept is foreign to this theory. Unless we use an indirect argument[12] like that of Section 5.4, it is not possible to calculate the probability of spontaneous emission by an atom in an excited state, because there is no pre-existing classical electromagnetic field and the interaction term $\propto \vec{A} \cdot \vec{P}$ is zero. It is necessary to resort to the concept of quantized electromagnetic field developed in Section 11.3.3, because the annihilation and creation operators $a_{\vec{k}s}$ and $a_{\vec{k}s}^\dagger$ are capable of changing the number of photons. More precisely, if $n_{\vec{k}s}$ is the number of photons in the mode of wave vector $\vec{k}$ and polarization s, we are interested in transitions with the emission of a photon $n_{\vec{k}s} \to n_{\vec{k}s} + 1$ or the absorption of a photon $n_{\vec{k}s} \to n_{\vec{k}s} - 1$,

---

[12] The argument uses the Planck distribution, which implicitly involves the concept of photon: the occupation probability of a mode of the electromagnetic field is given by the quantum theory of the harmonic oscillator. It is therefore not surprising that it is possible to calculate spontaneous emission.

with spontaneous emission in the mode $(\vec{k}, s)$ corresponding to $n_{\vec{k}s} = 0$. Let us recall the expansion of the quantized electromagnetic field (11.79) at $t = 0$ in a volume $L^3 = \mathcal{V}$:

$$\vec{A}(\vec{r}) = \sqrt{\frac{\hbar}{2\varepsilon_0 \mathcal{V}}} \sum_{\vec{k}} \sum_{s=1}^{2} \frac{1}{\sqrt{\omega_k}} \left(a_{\vec{k}s} \vec{e}_s(\hat{k}) \, e^{i\vec{k}\cdot\vec{r}} + a_{\vec{k}s}^\dagger \vec{e}_s^*(\hat{k}) \, e^{-i\vec{k}\cdot\vec{r}} \right).$$

The coupling between the electromagnetic field[13] and the atom is, to first order in $\vec{A}$,

$$W = -\frac{q_e}{m_e} \vec{A} \cdot \vec{P}. \tag{14.58}$$

This *time-independent* coupling $\vec{A} \cdot \vec{P}$ brings in the terms

$$a_{\vec{k}s} \, e^{i\vec{k}\cdot\vec{r}} \left(\vec{e}_s(\hat{k}) \cdot \vec{P} \right), \tag{14.59}$$

and

$$a_{\vec{k}s}^\dagger \, e^{-i\vec{k}\cdot\vec{r}} \left(\vec{e}_s^*(\hat{k}) \cdot \vec{P} \right). \tag{14.60}$$

The term (14.59) destroys a photon and the term (14.60) creates a photon in the mode $(\vec{k}, s)$. Let $|i, n_{\vec{k}s}\rangle$ be the initial state with i labeling the state of the atom and let $|f, n_{\vec{k}s} \pm 1\rangle$ be the final state. The nonzero matrix elements of $a_{\vec{k}s}$ and $a_{\vec{k}s}^\dagger$ are given by (11.16) and (11.17):

$$\langle n_{\vec{k}s} + 1 | a_{\vec{k}s}^\dagger | n_{\vec{k}s} \rangle = \sqrt{n_{\vec{k}s} + 1},$$

$$\langle n_{\vec{k}s} - 1 | a_{\vec{k}s} | n_{\vec{k}s} \rangle = \sqrt{n_{\vec{k}s}}. \tag{14.61}$$

We shall examine spontaneous emission corresponding to the case $n_{\vec{k}s} = 0$ and return briefly to absorption and stimulated emission at the end of this subsection. The interesting physical quantity is the probability per unit time for an atom to emit a photon of wave vector approximately equal to $\vec{k}$ and polarization s at a solid angle $d\Omega$, $\Omega = (\theta, \phi)$, about $\vec{k}$.[14] To obtain this probability we need the photon density of states:

$$\frac{\mathcal{V}}{(2\pi)^3} \, d^3k = \frac{\mathcal{V}}{(2\pi)^3 c^3} \, \omega^2 \, d\omega \, d\Omega \tag{14.62}$$

---

[13] It is necessary to take the electromagnetic field (11.79) at $t = 0$, that is, in the Schrödinger picture $\vec{A}_S = \vec{A}_H(t = 0) = \vec{A}$, because we are using the Schrödinger picture in the perturbative calculations and the *operators* $\vec{A}$ and $\vec{P}$ must also be in this picture. In Subsections 14.1.1 to 14.1.3 the time dependence of the classical field is fixed by an external source, that which produces the incident electromagnetic wave, whereas the quantized field is independent of any external source.

[14] To be rigorous, we should note that we are working in the reference frame where the initial atom is at rest. Energy conservation implies that

$$E_i - E_f = \hbar\omega + \frac{\hbar^2 k^2}{2M_{at}}$$

in this reference frame. The second term is the recoil energy, which will be discussed in (14.106). In general, this recoil energy is negligible; everything happens as though the atom were infinitely heavy, $M_{at} \to \infty$.

with $\omega_k \to \omega$. The transition probability per unit time is given by the Fermi Golden Rule (9.170) with a final photon $(\vec{k}, s)$ of energy $\hbar\omega$:

$$d\Gamma_{fi}^s(\vec{k}, s) = \frac{2\pi}{\hbar} |\langle f, n_{\vec{k}s} = 1|W|i, n_{\vec{k}s} = 0\rangle|^2 \delta[\hbar\omega - (E_i - E_f)] \frac{V}{(2\pi)^3 c^3} \omega^2 \, d\omega \, d\Omega,$$

(14.63)

with the matrix element $\langle f, n_{\vec{k}s} = 1|W|i, n_{\vec{k}s} = 0\rangle$ given by (in contrast to Section 14.3.2, we now have $E_i > E_f$ and $\hbar\omega_0 = E_i - E_f$)

$$\langle f, n_{\vec{k}s} = 1|W|i, n_{\vec{k}s} = 0\rangle = -\frac{q}{m_e} \langle f, n_{\vec{k}s} = 1|\vec{A} \cdot \vec{P}|i, n_{\vec{k}s} = 0\rangle$$

$$= -\frac{q}{m_e} \sqrt{\frac{\hbar}{2\varepsilon_0 V\omega}} \langle f, n_{\vec{k}s} = 1|a_{\vec{k}s}^\dagger e^{i\vec{k}\cdot\vec{r}} (\vec{e}_s^*(\hat{k}) \cdot \vec{P})|i, n_{\vec{k}s} = 0\rangle$$

$$\simeq iq_e\omega_0 \sqrt{\frac{\hbar}{2\varepsilon_0 \omega V}} \langle f|(\vec{e}_s^*(\hat{k}) \cdot \vec{R})|i\rangle.$$

(14.64)

We have used the dipole approximation $\exp(i\vec{k} \cdot \vec{r}) \simeq 1$ and expressed the matrix element of $\vec{P}$ using (14.48).

To obtain the probability for photon emission in a solid angle $d\Omega$ we must integrate (14.63) over ω. The δ function fixes the photon energy to

$$\hbar\omega = E_i - E_f = \hbar\omega_0,$$

which, using (14.64) and defining $\vec{R}_{fi} = \langle f|\vec{R}|i\rangle$, gives

$$\frac{d\Gamma_{fi}^s}{d\Omega} = \frac{\alpha\omega_0^3}{2\pi c^2} |\vec{e}_s^*(\hat{k}) \cdot \vec{R}_{fi}|^2.$$

(14.65)

An equivalent expression involves the dipole moment $\vec{D} = q_e\vec{R}$:

$$\frac{d\Gamma_{fi}^s}{d\Omega} = \left(\frac{1}{4\pi\varepsilon_0}\right)\left(\frac{\omega_0^3}{2\pi\hbar c^3}\right) |\vec{e}_s^*(\hat{k}) \cdot \vec{D}_{fi}|^2.$$

(14.66)

To obtain the total transition probability Γ, which is the inverse of the lifetime τ of the excited state $\tau = 1/\Gamma$, it is necessary to integrate over Ω and sum over the two polarization states:

$$\Gamma = \frac{1}{\tau} = \sum_{s=1}^{2} \int \frac{d\Gamma_{fi}^s}{d\Omega} \, d\Omega.$$

(14.67)

In order to calculate the matrix element in, for example, the form (14.65) we work in a representation where $\vec{R}$ is diagonal:[15]

$$\vec{R}_{fi} = \int d^3r \, \varphi_f^*(\vec{r}) \, \vec{r} \, \varphi_i(\vec{r}),$$

(14.68)

---

[15] To simplify the formulas we neglect spin, which can easily be shown to play no role at all.

and we separate the r-dependent radial part and the $\hat{r}$-dependent angular part in the integral (14.68) by writing $\vec{r} = r\hat{r}$. To deal with a specific case, we take the example of the $2p \rightarrow 1s$ transition of the hydrogen atom.[16] The initial wave function is written as the product of its radial part (10.96) and its angular part, which is the spherical harmonic $Y_1^m(\hat{r})$:

$$\varphi_i^m(\vec{r}) = \frac{1}{\sqrt{4!a_0^5}} \, r \exp\left(-\frac{r}{2a_0}\right) Y_1^m(\hat{r}), \qquad (14.69)$$

and the final wave function is given by (14.31). It is convenient to introduce the spherical components (10.64) of the vectors $\vec{e}_s(\hat{k})$ and $\vec{r}$ noting that the scalar product $\vec{e}_s^* \cdot \hat{r}$ is[17]

$$\vec{e}_s^* \cdot \hat{r} = (\vec{e}_s \cdot \hat{r})^* = \left(\sum_{q=\pm 1,0} e_{sq}^* \hat{r}_q\right)^* = \sum_{q=\pm 1,0} e_{sq} \hat{r}_q^*.$$

On the other hand, the projector (11.80) orthogonal to $\vec{k}$ is written in spherical coordinates as

$$\sum_{s=1}^{2} e_{sq}(\hat{k}) e_{sq'}^*(\hat{k}) = \delta_{qq'} - \hat{k}_q \hat{k}_{q'}^* \,,$$

which gives for the angular part

$$\left|\sum_s \vec{e}_s^*(\hat{k}) \cdot \langle f|\hat{r}|i\rangle\right|^2 = \sum_{qq'} (\delta_{qq'} - \hat{k}_q \hat{k}_{q'}^*)\langle f|\hat{r}_q^*|i\rangle\langle i|\hat{r}_{q'}|f\rangle.$$

The matrix element $\langle f|\hat{r}_q^*|i\rangle$ is easily calculated by noting that according to (10.64) $\hat{r}_q$ is proportional to the spherical harmonic $Y_1^q(\hat{r})$. If the magnetic quantum number of the initial state is m, then

$$\langle f|\hat{r}_q^*|i, m\rangle = \sqrt{\frac{4\pi}{3}} \int d^2\hat{r} \, [Y_1^q(\hat{r})]^* \, Y_1^m(\hat{r}) = \sqrt{\frac{4\pi}{3}} \, \delta_{qm},$$

where we have used the orthogonality relations (10.55) of the spherical harmonics. This then gives

$$\left|\sum_s \vec{e}_s^*(\hat{k}) \cdot \langle f|\hat{r}|i\rangle\right|^2 = \frac{4\pi}{3}(1 - |\hat{k}_m|^2).$$

The factor $(1 - |\hat{k}_m|^2)$ becomes $(1 - \frac{1}{2}\sin^2\theta)$ for $m = \pm 1$ and $(1 - \cos^2\theta)$ for $m = 0$, which gives the angular distribution of the emitted photon if the initial state has a

---

[16] In the general case of an initial state i of angular momentum (j_i, m_i) and a final state f of angular momentum (j_f, m_f), we can use the Wigner–Eckart theorem to express the matrix element of the spherical components D_q of $\vec{D}$ in the form (14.53).

[17] The scalar product of two vectors $\vec{a}$ and $\vec{b}$ is given as a function of their spherical coordinates by

$$\vec{a} \cdot \vec{b} = \sum_{q=\pm 1,0} a_q^* b_q = \sum_{q=\pm 1,0} (-1)^q a_{-q} b_q.$$

well-defined value of m. If the initial state is unpolarized, the angular distribution is of course isotropic because there is no preferred direction:

$$\frac{1}{3}\left[2\left(1-\frac{1}{2}\sin^2\theta\right)+(1-\cos^2\theta)\right]=\frac{2}{3}.$$

To obtain the total transition probability (14.67) we integrate over Ω, the result being the same for the three cases $m=\pm 1,0$:

$$\int d\Omega\left(1-\frac{1}{2}\sin^2\theta\right)=\int d\Omega\,(1-\cos^2\theta)=\frac{8\pi}{3}.$$

The angular part gives an overall factor of $32\pi^2/9$. According to (14.31) and (14.69), the radial part of the matrix element is

$$\langle f|r|i\rangle=\frac{1}{a_0^4\sqrt{4!\pi}}\int_0^\infty r^4 dr\exp\left(-\frac{3r}{2a_0}\right)=\sqrt{\frac{4!}{\pi}}\left(\frac{2}{3}\right)^5 a_0.$$

The combination of all these results gives the transition probability $\Gamma(2p\to 1s)$:

$$\Gamma(2p\to 1s)=\left[\frac{\alpha\omega_0^3}{2\pi c^2}\right]\left[\frac{4!}{\pi}\left(\frac{4}{9}\right)^5 a_0^2\right]\left[\frac{32\pi^2}{9}\right], \qquad (14.70)$$

and using

$$\omega_0=\frac{3}{4}\frac{R_\infty}{\hbar}=\frac{3}{8}\frac{\alpha^2 m_e c^2}{\hbar}\quad\text{and}\quad a_0=\frac{\hbar^2}{m_e e^2}=\frac{\hbar}{\alpha m c},$$

we can write the result in the final form, recalling (cf. Section 1.5.3) that $\hbar/m_e c^2 = 1.29\times 10^{-21}$ s:

$$\Gamma(2p\to 1s)=\alpha^5\left(\frac{m_e c^2}{\hbar}\right)\left(\frac{4}{9}\right)^4\simeq 6.2\times 10^8\text{ s}^{-1}, \quad \tau=\frac{1}{\Gamma}\simeq 1.6\times 10^{-9}\text{ s}. \quad (14.71)$$

Let us return to the qualitative aspects of these results. Starting from (14.52) or (14.65) with $|\vec{e}_s^*\cdot R_{fi}|\sim a$, where a is the typical atomic scale ($a\simeq 10^{-10}$ m), we obtain the estimate

$$\Gamma\sim\frac{\alpha\omega_0^3}{c^2}a^2=\alpha\left(\frac{a\omega_0}{c}\right)^2\omega_0\sim\alpha^3\omega_0\sim\alpha^5\left(\frac{m_e c^2}{\hbar}\right).$$

In fact, the speed v of the electron in its orbit is $v\sim\alpha c$.[18] The characteristic frequency ω_0 is given by $\hbar\omega_0\sim 1$ eV or $\omega_0\sim 1.5\times 10^{15}$ rad s^{-1}, and the lifetime τ of the excited state is estimated to be

$$\tau=\frac{1}{\Gamma}\sim 2\times 10^{-9}\text{ s}.$$

The lifetimes of excited states that de-excite by an electric dipole transition essentially lie between $\sim 10^{-7}$ s and $\sim 10^{-9}$ s. It is instructive also to study the case of an excited

---

[18] The factors that are assumed to be "near unity" in this type of estimate are not always so; the above estimate differs from the exact value (14.71) by a factor of $(8/3)(4/9)^4\simeq 1/10$.

level of a nucleus which decays by emitting a photon γ. The typical energy of such a photon is ~ 1 MeV, which corresponds to a wavelength $\lambda \simeq 10^{-12}$ m. Since the nuclear dimensions are of the order of a fermi (or femtometer), $R \simeq 10^{-15}$ m, the use of $R/\lambda \ll 1$ and the electric dipole approximation is a priori justified. To estimate the lifetime, the result from atomic physics must be multiplied by a factor of 10^{-18} to take into account the change of energy scale 1 eV$\rightarrow$ 1 MeV, and by a factor of 10^{10} to take into account the change of dimension, $a \rightarrow R$, making a factor of 10^{-8} altogether. The estimated lifetime of a nuclear excited state is then

$$\tau_{\text{nucl}} \sim 10^{-8}\,\tau_{\text{atom}} \sim 10^{-15} \text{ s}.$$

An example is the decay of an excited state of an isotope of nitrogen, ^{13}N:

$$^{13}\text{N}^* \rightarrow {}^{13}\text{N} + \gamma \ (2.38 \text{ MeV}),$$

where the lifetime is 10^{-15} s, in qualitative agreement with our estimate.

Let us conclude this discussion by returning briefly to emission and stimulated absorption. If we take into account the factors (14.61)–(14.62) for absorption and stimulated emission, the absorption probability (14.50) is not modified. On the other hand, if the atom is located in a cavity of volume $\mathcal{V}$ containing $\mathcal{N}_{\vec{k}s}$ photons in the mode $(\vec{k}, s)$, the semiclassical emission probability is proportional to the photon density $n_{\vec{k}s} = \mathcal{N}_{\vec{k}s}/\mathcal{V}$, while the use of the quantized field gives a factor of $(\mathcal{N}_{\vec{k}s} + 1)/\mathcal{V}$. The correction is in general negligible, except in the case of superconducting microwave cavities where the number of photons is small (Appendix B).

14.4 Laser cooling and trapping of atoms

14.4.1 The optical Bloch equations

It has long been known that light exerts forces on matter, the best-known example being radiation pressure. However, when the light comes from conventional sources these forces are very weak. It is only in the last twenty years that the use of lasers has made it possible to exert sizable forces on atoms, forces which can be up to 10^5 times their weight. A particularly interesting application is *laser cooling*, and we shall give an elementary example of it in Section 14.4.3. We shall use the model of the two-level atom: two atomic levels E_a and E_b ($E_b > E_a$) are separated by $E_b - E_a = \hbar\omega_0$. We assume E_a is the ground state of the atom, or at least a metastable state of lifetime long enough not to be involved in the discussion. The atom is placed in an electromagnetic wave produced by a laser whose wave vector $\vec{k}$ is parallel to Oz and whose frequency ω is close to resonance: $\omega \simeq \omega_0$. As in Section 5.2.2, we call the difference $\delta = \omega - \omega_0$ the *detuning*. The electric field at the position of the atom is of the form

$$\vec{E} = \vec{e}_s E_0 \cos \omega t. \tag{14.72}$$

For the time being we ignore the translational degrees of freedom of the atom, assuming it to be infinitely heavy.[19] Under these conditions, the Hamiltonian H is given by (5.52) written as

$$H = \begin{pmatrix} -\dfrac{\hbar}{2}\omega_0 & -dE_0\cos\omega t \\ -dE_0\cos\omega t & \dfrac{\hbar}{2}\omega_0 \end{pmatrix}. \tag{14.73}$$

The rows and columns are ordered as (a, b), the zero of the energy in the absence of the field has been chosen to lie midway between E_a and E_b, and d is the matrix element $(\vec{D}\cdot\vec{e}_s)_{ab}$ of the $\vec{e}_s$ component of the electric dipole moment operator between the two levels. As in Section 5.3.2, we introduce the Rabi frequency ω_1:

$$\omega_1 = -\frac{dE_0}{\hbar}. \tag{14.74}$$

The minus sign takes into account the negative charge of the electron, so that $\omega_1 > 0$. With this definition we can rewrite H as a function of the Pauli matrices σ_1 and σ_3:

$$\frac{1}{\hbar}H = \begin{pmatrix} -\dfrac{1}{2}\omega_0 & \omega_1\cos\omega t \\ \omega_1\cos\omega t & \dfrac{1}{2}\omega_0 \end{pmatrix} = -\frac{1}{2}\omega_0\sigma_3 + (\omega_1\cos\omega t)\sigma_1. \tag{14.75}$$

In general, the quantum state of the atom will be described by a state operator ρ. In fact, the atom is in continuous interaction with the quantized electromagnetic field, and even if the field + atom ensemble were in a pure state, the state of the atom would not be pure, because the atom is not a closed quantum system. As we have seen in Section 6.2.3, its state is described by taking the partial trace over the field variables, and the result is not a vector of the atom space of states, but a state operator, the reduced state operator of the atom represented by a 2×2 matrix acting in the two-dimensional space of the two-level atom. We recall that the state matrix must be Hermitian $\rho = \rho^\dagger$, it must have unit trace $\text{Tr}\,\rho = 1$, and it must be positive. The results of Section 6.2.2 allow us to write the most general state matrix as a function of a real vector $\vec{b}$, the Bloch vector (6.24), such that $\vec{b}^2 \leq 1$:

$$\rho = \frac{1}{2}\left(I + \sum_{i=1}^{3}\sigma_i b_i\right) = \frac{1}{2}\left(I + \vec{\sigma}\cdot\vec{b}\right). \tag{14.76}$$

Conforming to the usual notation, we use $(u, v, -w)$ to denote the components of the Bloch vector: $u = b_1$, $v = b_2$, and $w = -b_3$. We can also write ρ in the explicit matrix form

$$\rho = \begin{pmatrix} \frac{1}{2} - \frac{1}{2}(\rho_{bb} - \rho_{aa}) & \rho_{ab} \\ \rho_{ba} & \frac{1}{2} + \frac{1}{2}(\rho_{bb} - \rho_{aa}) \end{pmatrix} = \frac{1}{2}\begin{pmatrix} 1 - w & u - iv \\ u + iv & 1 + w \end{pmatrix}. \tag{14.77}$$

---

[19] More precisely, the translational degrees of freedom are treated classically, assuming that $\hbar\Gamma \gg E_R$, where Γ is the linewidth and E_R is the recoil energy (14.106). We also assume that the medium is dilute enough that collisions between atoms can be neglected.

In this expression for ρ we have taken into account the condition $\rho_{aa} + \rho_{bb} = 1$. The quantity $w = \rho_{bb} - \rho_{aa}$ measures the population difference between the levels E_b and E_a: if we have a collection of $\mathcal{N}$ atoms, on average $\mathcal{N}\rho_{aa}$ will be in the state E_a and $\mathcal{N}\rho_{bb}$ in the state E_b. The off-diagonal matrix elements $\rho_{ab} = \rho_{ba}^*$ are the coherences. The presence of nonzero coherences, that is, phase-dependent effects, is a sure signal of quantum effects.

If we first ignore the quantized electromagnetic field, the evolution equation for ρ is given by (6.37):

$$i\dot{\rho} = \left[\frac{1}{\hbar} H, \rho \right].\qquad(14.78)$$

The commutator can be calculated directly by multiplying the matrices, but it is more elegant to use the Bloch form and the commutation relations (3.52) of the Pauli matrices. We find

$$\dot{u} = \omega_0 v,$$
$$\dot{v} = -\omega_0 u + 2\omega_1 w \cos \omega t,\qquad(14.79)$$
$$\dot{w} = -2\omega_1 v \cos \omega t.$$

To complete these equations and justify an approximation which will be made below, it is convenient to rewrite them as a function of the coherence $r = \rho_{ab} = (u - iv)/2$:

$$\dot{w} = -2i\omega_1 (r - r^*) \cos \omega t,\qquad(14.80)$$
$$\dot{r} = i\omega_0 r - i\omega_1 w \cos \omega t.\qquad(14.81)$$

These equations for the evolution of the state matrix are Hamiltonian, that is, they are governed by a law of the type (14.78) depending on a Hamiltonian. This evolution is unitary, because (14.78) is equivalent to

$$\rho(t) = U(t, 0)\rho(t = 0)U^\dagger(t, 0),$$

where $U(t)$ is the unitary evolution operator (4.14). Actually, though, these equations are incomplete. The interaction of the atom with its environment leads to equations that are not of the form (14.78), and so to an evolution which is non-Hamiltonian. It is the ensemble atom + environment that obeys a unitary evolution, and if we are interested only in the atomic degrees of freedom, the evolution is no longer Hamiltonian. This phenomenon is familiar in statistical mechanics, where we consider the interaction of a system with a heat bath, and nonunitary evolution is closely related to dissipation.[20] We are going to consider the case of an atomic environment limited to a quantized electromagnetic field, which is an excellent approximation for atoms trapped by lasers, which form a dilute medium. However, there could also be other sources of non-Hamiltonian evolution, such

---

[20] See, for example, Le Bellac *et al.* [2004], Chapter 2.

as collisions with other atoms in a dense medium.[21] The calculation based on (14.78) takes into account the interaction with the laser field and therefore absorption and stimulated emission, but it does not include the interaction with the quantized field and so spontaneous emission is neglected. Owing to spontaneous emission, an atom in the level E_b tends to return to the level E_a by emitting a photon with probability per unit time Γ (cf. (14.67)). The differential equation giving $\dot{\rho}_{bb}$ must therefore include a term $-\Gamma\rho_{bb}$ on the right-hand side, which in the absence of a laser field leads to exponentially decreasing population of the level E_b, $\exp(-\Gamma t)$. One then deduces that the right-hand side of the differential equation for w contains a term $-\Gamma(w+1)$. The coherences must also decrease because, in the absence of a laser field, the atom returns to its ground state E_a for $t \gg \tau = 1/\Gamma$, and the only nonzero element of the density matrix is $\rho_{aa} = 1$. It will be shown in Section 15.2.4 that in our approximation of a diluted medium the decay rate for coherences is $\Gamma/2$. Therefore, equations (14.80) and (14.81) become

$$\dot{w} = -2\mathrm{i}\omega_1(r - r^*)\cos\omega t - \Gamma(w+1), \tag{14.82}$$

$$\dot{r} = \mathrm{i}\omega_0 r - \mathrm{i}\omega_1 w \cos\omega t - \frac{\Gamma}{2} r. \tag{14.83}$$

Let us transform these equations using the rotating wave approximation of Section 5.3.2. We note that if $\omega_1 \ll \omega_0$, (14.83) implies that $r \sim \exp(\mathrm{i}\omega_0 t)$ whereas w varies slowly. Writing $\cos\omega t$ as a sum of complex exponentials and neglecting the rapidly varying terms $\propto \exp(\pm\mathrm{i}(\omega+\omega_0)t)$ in the rotating wave approximation, Equations (14.82)–(14.83) become

$$\dot{w} = -\mathrm{i}\omega_1(\mathrm{e}^{-\mathrm{i}\omega t} r - \mathrm{e}^{\mathrm{i}\omega t} r^*) - \Gamma(w+1), \tag{14.84}$$

$$\dot{r} = \mathrm{i}\omega_0 r - \frac{\mathrm{i}}{2}\omega_1 w\left(\mathrm{e}^{\mathrm{i}\omega t} + \mathrm{e}^{-\mathrm{i}\omega t}\right) - \frac{\Gamma}{2} r. \tag{14.85}$$

All the terms on the right-hand side of (14.84) vary slowly. To display the time evolution of the terms on the right-hand side of (14.85) we set

$$\mathrm{e}^{-\mathrm{i}\omega t} r = r', \qquad \mathrm{e}^{-\mathrm{i}\omega t}\dot{r} = \mathrm{i}\omega r' + \dot{r}',$$

which, multiplying (14.85) by $\exp(-\mathrm{i}\omega t)$, gives

$$\dot{r}' = \mathrm{i}(\omega_0 - \omega)r' - \frac{\mathrm{i}}{2}\omega_1 w\left(1 + \mathrm{e}^{-2\mathrm{i}\omega t}\right) - \frac{\Gamma}{2} r'.$$

The rotating wave approximation consists of neglecting the rapidly varying term $\propto \exp(-2\mathrm{i}\omega t)$ in this expression. We then end up with the system of differential equations

$$\dot{w} = -\mathrm{i}\omega_1(r' - r'^*) - \Gamma(w+1), \tag{14.86}$$

$$\dot{r}' = \mathrm{i}(\omega_0 - \omega)r' - \frac{\mathrm{i}}{2}\omega_1 w - \frac{\Gamma}{2} r'. \tag{14.87}$$

---

[21] An example is the active medium for a laser, which is described by optical Bloch equations analogous to (14.82)–(14.83), but with two unrelated relaxation rates for populations and coherences; see, for example, Mandel and Wolf [1995], Chapter 18.

14.4.2 Dissipative forces and reactive forces

When the atom interacts with the laser field during a time interval $t \gg \tau$, a stationary regime $\dot{w} = \dot{r} = 0$ is reached where it is easy to write down the solution of the system of differential equations (14.86)–(14.87). Passing through the intermediate stage

$$r'_{\text{st}} = \frac{i\omega_1 w_{\text{st}}/2}{i(\omega_0 - \omega) - \Gamma/2},$$

we obtain for the stationary value w_{st} of w

$$w_{\text{st}} = -\frac{(\omega - \omega_0)^2 + \Gamma^2/4}{(\omega - \omega_0)^2 + \Gamma^2/4 + \omega_1^2/2}. \tag{14.88}$$

We then find $\rho_{bb} = (1 + w_{\text{st}})/2 < 1/2$: there cannot be a population inversion, that is, a situation where the excited level is more populated than the ground state. The stationary result for r' is

$$r'_{\text{st}} = \frac{i\omega_1}{2} \frac{\Gamma/2 - i(\omega - \omega_0)}{(\omega - \omega_0)^2 + \Gamma^2/4 + \omega_1^2/2}. \tag{14.89}$$

It is convenient to introduce the *saturation parameter* s proportional to the intensity $\mathcal{I}$ of the laser (we recall that the detuning $\delta = \omega - \omega_0$):

$$s = \frac{\omega_1^2/2}{\delta^2 + \Gamma^2/4} \propto \mathcal{I}_{\text{laser}}, \tag{14.90}$$

so that we can write

$$\rho_{bb,\text{st}} = \frac{1}{2}(1 + w_{\text{st}}) = \frac{s}{2(1+s)}, \quad r'_{\text{st}} = \frac{i}{\omega_1}\left(\frac{s}{1+s}\right)\left(\frac{\Gamma}{2} - i\delta\right). \tag{14.91}$$

These results allow us to obtain the forces exerted by the laser light on an atom in the stationary regime. The equivalent of the radiation pressure on the atom can be found by a simple argument. Since in the stationary regime the probability of finding an atom in the excited state E_b is $\rho_{bb,\text{st}}$, the average number of photons spontaneously emitted per unit time is

$$\left\langle \frac{dN}{dt} \right\rangle = \Gamma \rho_{bb,\text{st}} = \frac{\Gamma}{2} \frac{s}{1+s}. \tag{14.92}$$

These photons are emitted isotropically and contribute to the disordered motion of the atom, which we shall study in the following subsection. However, once the atom has returned to its ground state it absorbs a photon of the laser field, and the momenta of these photons $\hbar \vec{k}$ are all in the same direction. The number of photons absorbed is the same as the number of photons spontaneously emitted, and the atom is subject to a force due to photon absorption which is equal to the change of momentum per unit time:

$$\boxed{\vec{F}_{\text{diss}} = \hbar \vec{k} \left\langle \frac{dN}{dt} \right\rangle = \hbar \vec{k} \frac{\Gamma}{2}\left(\frac{s}{1+s}\right) = \hbar \vec{k} \frac{\Gamma}{2} \frac{\omega_1^2/2}{\delta^2 + \Gamma^2/4 + \omega_1^2/2}.} \tag{14.93}$$

When the saturation parameter $s \gg 1$, the acceleration $\vec{a}$ approaches its maximum value

$$\vec{a}_{\max} = \frac{\hbar \vec{k}}{M} \frac{\Gamma}{2},\tag{14.94}$$

where M is the mass of the atom. In the case of the D_2 line of sodium, $\Gamma^{-1} = 1.6 \times 10^{-8}$ s and $a_{\max} \sim 10^6$ m s^{-2}, which is about 10^5 times the gravitational acceleration.

Now let us rederive the result (14.93) for the dissipative force by examining the force exerted by the electromagnetic field (14.72) on an atomic dipole. The form of the dipole operator in the two-dimensional space of the two-level atom is $D = d\sigma_1$, and according to (6.21) its expectation value is

$$\begin{aligned}
\langle D \rangle &= d \operatorname{Tr}(\rho \sigma_1) = d(\rho_{ab} + \rho_{ab}^*) \\
&= d(r + r^*) = d\left(r' e^{i\omega t} + r'^* e^{-i\omega t}\right) \\
&= \frac{2ds}{\omega_1(1+s)}\left[-\frac{\Gamma}{2}\sin \omega t + \delta \cos \omega t\right],
\end{aligned}\tag{14.95}$$

where we have used the expression (14.91) for r' in the stationary regime. This expectation value of the dipole operator contains a term $\propto \cos \omega t$ in phase with the field (14.72) and a term out of phase by $\pi/2 \propto \sin \omega t$. The work dW/dt done on the dipole per unit time by the field (14.72), that is, the power supplied to the atom,[22] is

$$\frac{dW}{dt} = E_0 \cos \omega t \frac{d\langle D \rangle}{dt}.$$

Using (14.95) immediately gives $d\langle D \rangle/dt$ and we find

$$\frac{dW}{dt} = -\frac{2ds\omega E_0}{\omega_1(1+s)}\left[\frac{\Gamma}{2}\cos^2 \omega t + \delta \sin \omega t \cos \omega t\right].\tag{14.96}$$

Taking the time average, we obtain

$$\left\langle \frac{dW}{dt} \right\rangle = -\frac{2ds\omega E_0}{\omega_1(1+s)}\frac{\Gamma}{4} = \frac{\hbar \omega s}{1+s}\frac{\Gamma}{2}.\tag{14.97}$$

The number of photons absorbed per second is

$$\left\langle \frac{dN}{dt} \right\rangle = \frac{1}{\hbar \omega}\left\langle \frac{dW}{dt} \right\rangle = \frac{\Gamma}{2}\left(\frac{s}{1+s}\right),$$

in agreement with (14.92). Elementary study of the forced harmonic oscillator shows that it is the component involving the displacement out of phase by $\pi/2$ with the external force which is responsible for the frictional dissipation, which gives rise to the expression "dissipative force" for the radiation pressure. The part in phase with the field is called

---

[22] It is useful to recall the elementary forced harmonic oscillator in one dimension:

$$\ddot{x} + \gamma \dot{x} + \omega_0^2 x = f \cos \omega t.$$

The power supplied to the oscillator is $f \cos \omega t \, dx/dt$. The correspondence with the present problem is given by $f \to E_0$ and $x \to \langle D \rangle$.

the "reactive" part. The model we have studied does not contain any spatial dependence, and so the average value of the term of $\langle D \rangle$ in phase with the field does not produce any work. In order to obtain a nonzero result, a spatial dependence must be introduced. It can then be shown (Exercise 14.6.7) that the reactive component of the force depends on the gradient of the Rabi frequency:

$$\vec{F}_{\text{react}} = -\frac{\hbar\delta}{2} \frac{\vec{\nabla}\omega_1^2(\vec{r})/2}{\delta^2 + \Gamma^2/4 + \omega_1^2/2}. \tag{14.98}$$

The reactive force is zero in a plane wave, where the Rabi frequency ω_1 is independent of $\vec{r}$. It does not transmit any energy to the atoms. If, for example, the spatial variation of the Rabi frequency is due to the use of several laser waves, the effect of the reactive force is to redistribute the energy among the various waves. In contrast to the dissipative force, the reactive force is not saturated when $s \to \infty$.

The reactive force is derived from a potential

$$\vec{F}_{\text{react}} = -\vec{\nabla}U(\vec{r}), \quad U(\vec{r}) = \frac{\hbar\delta}{2}\ln\left(1 + \frac{\omega_1^2(\vec{r})/2}{\delta^2 + \Gamma^2/4}\right).$$

For $\delta < 0$, a region in which $\omega_1^2(\vec{r})$ is a maximum appears as an attractive potential well for the atom. In a nonuniform laser field the atom is attracted toward the regions of stronger intensity. This has allowed the development of numerous practical applications where microscopic objects are manipulated. An example is the creation of "optical tweezers" for manipulating segments of DNA.

14.4.3 Doppler cooling

An important application of the dissipative force (14.93) is the *Doppler cooling* of atoms. The atoms are modeled as above by a system of two levels separated by $\hbar\omega_0$. They are localized in laser beams coming from opposite directions and having identical frequencies ω close to the resonance frequency ω_0, but with $\omega < \omega_0$, that is, with a detuning $\delta = \omega - \omega_0 < 0$. In order to simplify the discussion we limit ourselves to cooling along an axis which we shall choose to be the z axis, and use two laser beams with wave vectors $\vec{k} \parallel \hat{z}$ and $-\vec{k} \parallel -\hat{z}$ (Fig. 14.1). Cooling in three spatial dimensions requires the use of six lasers, two on each axis, with opposite wave vectors. We shall take the case of saturation parameter $s \ll 1$, which will permit us to neglect the term ω_1^2 in the denominator of (14.93).

laser beam (+)   atoms   laser beam (−)

Fig. 14.1. The principle of Doppler cooling.

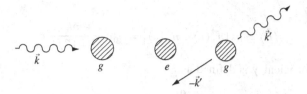

Fig. 14.2. The fluorescence cycle.

An atom in the field of the lasers undergoes *fluorescence cycles*. A fluorescence cycle consists of the absorption of a photon from one of the two lasers by an atom in its ground state so that it makes a transition to its excited state. This is followed by the spontaneous emission of a photon, returning the atom to its ground state (Fig. 14.2). Let $n_+(v)$ be the number of fluorescence cycles per second that an atom of speed v (in the z direction since our discussion is limited to one dimension) undergoes with absorption of photons of wave vector $+\vec{k}$, and let $n_-(v)$ be the number of fluorescence cycles with absorption of a photon of wave vector $-\vec{k}$. If an atom is moving toward the left ($v < 0$), owing to the Doppler effect it will see photons of frequency $\omega - kv$ coming from the $+\vec{k}$ laser and photons of frequency $\omega + kv$ coming from the $-\vec{k}$ laser. Because of the negative detuning ($\omega < \omega_0$), the photons of wave vector $+\vec{k}$ are closer to resonance and are absorbed in greater numbers than the photons of wave vector $-\vec{k}$ which are farther from resonance. This will give a force pointing toward the right for these atoms. Conversely, for atoms moving toward the right ($v > 0$) the force will be toward the left. In summary, atoms moving toward the left will preferentially absorb photons of wave vector $+\vec{k}$ and atoms moving toward the right will preferentially absorb photons of wave vector $-\vec{k}$. In both cases the atoms will be slowed down and a viscosity-like force will appear. This is the reason for the term "optical molasses." The average force on an atom of speed v is

$$\langle \vec{F} \rangle = \hbar \vec{k}[n_+(v) - n_-(v)] \tag{14.99}$$

with

$$n_\pm(v) = \frac{\Gamma}{4} \frac{\omega_1^2}{(\delta \mp kv)^2 + \Gamma^2/4}. \tag{14.100}$$

Let us expand (14.100) in powers of the velocity through order v:

$$n_\pm(v) \simeq \frac{\Gamma \omega_1^2/4}{\delta^2 + \Gamma^2/4} \left(1 \pm \frac{2\delta kv}{\delta^2 + \Gamma^2/4} \right). \tag{14.101}$$

This equation gives the average number of fluorescence cycles per second $2n_0$:

$$n_0 = \frac{1}{2}[n_+(v) + n_-(v)] = \frac{\Gamma \omega_1^2/4}{\delta^2 + \Gamma^2/4} = \frac{\Gamma}{2} s \tag{14.102}$$

and the force proportional to

$$n_+(v) - n_-(v) = n_0 \frac{4\delta kv}{\delta^2 + \Gamma^2/4},$$

which becomes

$$\langle \vec{F} \rangle = \hbar \vec{k} [n_+(v) - n_-(v)] = n_0 v \frac{4\hbar \delta k^2}{\delta^2 + \Gamma^2/4} \hat{k}. \tag{14.103}$$

The viscosity coefficient γ is defined as

$$\frac{dv}{dt} = -\gamma v \tag{14.104}$$

and its value is obtained from (14.103):

$$\boxed{\gamma = -\frac{\langle F \rangle}{Mv} = -n_0 \frac{4\hbar k^2}{M} \frac{\delta}{\delta^2 + \Gamma^2/4}}, \tag{14.105}$$

which is positive because $\delta < 0$. Taking n_0 to be constant, the viscosity coefficient is a maximum for $\delta = -\Gamma/2$:

$$\gamma_{\max} = \frac{4\hbar k^2}{M\Gamma} n_0 = \frac{8n_0}{\hbar \Gamma} \frac{\hbar^2 k^2}{2M} = \frac{8n_0}{\hbar \Gamma} E_R. \tag{14.106}$$

The energy $E_R = Mv_R^2/2$ is called the *recoil energy*: it is the recoil kinetic energy when the atom emits a photon of momentum $\hbar k$, and it is also the energy acquired by an atom at rest that absorbs a photon of momentum $\hbar k$. The speed v_R is the recoil velocity. Let us give some numerical values for the D_2 line of rubidium. The transition wavelength is $\lambda = 0.78 \, \mu\text{m}$, the linewidth is $\Gamma = 3.7 \times 10^7 \, \text{s}^{-1}$, and the atomic mass is $M = 1.41 \times 10^{-25}$ kg. These values correspond to energy $\hbar \Gamma = 2.4 \times 10^{-8}$ eV, recoil velocity $v_R = \hbar k/m = 5.8 \times 10^{-3} \, \text{m s}^{-1}$, and recoil energy $E_R = 1.5 \times 10^{-11}$ eV, and therefore to *recoil temperature* $T_R = E_R/k_B = 1.7 \times 10^{-7}$ K.

Using these typical numerical values, we find

$$\gamma \simeq 5 \times 10^{-3} n_0 = 2.5 \times 10^{-3} \Gamma s.$$

We can take the saturation parameter $s \ll 1$ such that

$$\Gamma^{-1} \ll n_0^{-1} \ll \gamma^{-1}.$$

Under these conditions, there are three distinct time scales in the problem (Fig. 14.3). The relation $\Gamma^{-1} \ll n_0^{-1}$ shows that the fluorescence cycles do not overlap and are independent. Let us consider a time interval δt, with $\Gamma^{-1} \ll \delta t \ll \gamma^{-1}$. Let $N_{\pm}$ be the number of fluorescence cycles $\pm k$ in this interval δt. The condition $\delta t \ll \gamma^{-1}$ implies that the speed v of the atom does not have the time to vary appreciably under the action of the viscosity force during the interval δt and so we can average over this interval, with $\langle N_{\pm} \rangle = n_{\pm}(v) \delta t$. Let $\mathsf{p}(N_+, N_-; \delta t)$ be the probability of observing N_+ $(+k)$ cycles and N_- $(-k)$ cycles during the interval δt. Since the fluorescence cycles are independent, this probability obeys a Poisson law:

$$\mathsf{p}(N_+, N_-; \delta t) = \frac{\langle N_+ \rangle^{N_+} \langle N_- \rangle^{N_-} \exp[-(\langle N_+ \rangle + \langle N_- \rangle)]}{N_+! \, N_-!}.$$

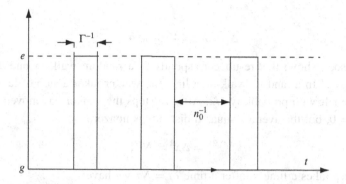

Fig. 14.3. A sequence of fluorescence cycles.

We use $\hbar q_1, \ldots, \hbar q_{(N_+ + N_-)}$ to designate the $N_+ + N_-$ momenta of photons emitted spontaneously by the atom during the interval δt and $\hbar Y$ to denote their sum:

$$\hbar Y = \hbar q_1 + \cdots + \hbar q_{(N_+ + N_-)}.$$

The emitted photons are not correlated with each other and $\langle Y \rangle = 0$. The average variation of the momentum during the time δt is due only to the absorbed photons:

$$\langle p(\delta t) \rangle = (n_+(v) - n_-(v)) \hbar k \delta t. \tag{14.107}$$

Let us now evaluate the variance of $p(\delta t)$,

$$\Delta p^2(\delta t) = \left\langle p^2(\delta t) - \langle p(\delta t) \rangle^2 \right\rangle.$$

Since the spontaneous photons are not correlated with the absorbed photons, $\langle Yk \rangle = 0$ and the two contributions can be treated separately. The contribution to the variance from the absorbed photons is

$$\Delta p^2(\delta t)\big|_{\text{abs}} = \hbar^2 k^2 \left\langle (N_+ - N_-)^2 - (\langle N_+ \rangle - \langle N_- \rangle)^2 \right\rangle = 2\hbar^2 k^2 n_0 \delta t,$$

where we have used the classical property of the Poisson distribution $\Delta N_\pm^2 = \langle N_\pm \rangle$ as well as the fact that the $+$ and $-$ cycles are independent: $\langle N_+ N_- \rangle = \langle N_+ \rangle \langle N_- \rangle$. The contribution from the emitted photons is

$$\Delta p^2(\delta t)\big|_{\text{em}} = \hbar^2 \langle Y^2 \rangle = \hbar^2 \sum_{i=1}^{N_+ + N_-} q_i^2 = \hbar^2 k^2 \langle N \rangle = 2 n_0 \hbar^2 k^2 \delta t.$$

Since we have reduced the kinematics to one dimension, we have assumed that the emitted photons have momentum $\pm \hbar k$ with probabilities of $1/2$.[23] Adding these two contributions,

---

[23] For three-dimensional kinematics and isotropic photon emission we would have $\langle \hbar^2 Y^2 \rangle = \hbar^2 k^2 / 3$.

we find[24]

$$\Delta p^2(\delta t) = 4n_0 \hbar^2 k^2 \delta t. \tag{14.108}$$

As we shall soon show, this result corresponds to a random walk in one-dimensional momentum space. In a random walk on a line, the walker takes a step of length l to the right or to the left with probability 1/2. After N steps the walker has moved an average distance $\langle x \rangle = 0$, but the average squared distance is nonzero:

$$\langle x^2 \rangle = \Delta x^2 = N l^2,$$

and if each step takes a time τ, after a time $\delta t = N\tau$ we have

$$\Delta x^2 = \frac{l^2}{\tau} \delta t = 2D \, \delta t. \tag{14.109}$$

This equation defines the *diffusion coefficient* D. The proportionality of Δp^2 to δt in (14.108) justifies the expression "random walk in momentum space" with diffusion coefficient $D = 2n_0 \hbar^2 k^2$.

In this random walk the kinetic energy E of the atom increases by $\Delta p^2(\delta t)/2M$. The diffusion therefore tends to increase the kinetic energy. By analogy with statistical mechanics, we define a fictitious temperature T as

$$E = \frac{1}{2} k_B T, \tag{14.110}$$

where k_B is the Boltzmann constant. If E increases, T increases, and it can be said that the atoms are heated by the spontaneous emission, which creates a disordered motion analogous to thermal motion. However, the temperature is actually fictitious, because there is no thermodynamical equilibrium: the temperature (14.110) is perfectly well defined for an isolated atom. The viscosity tends to slow the atoms down, and thus to "cool" them. When the two effects are in equilibrium, we obtain an "equilibrium temperature" which is the fictitious temperature of the atoms in the stationary regime. This temperature in fact provides an intuitive way of measuring their average speed. According to (14.104), the viscosity gives the following contribution to the time variation of the energy:

$$\frac{dE}{dt}\bigg|_{\text{visc}} = \frac{1}{2} M \frac{d}{dt} v^2 = -M\gamma v^2 = -\frac{\gamma p^2}{M}, \tag{14.111}$$

and adding the effect of spontaneous emission, we find

$$\frac{dE}{dt} = \frac{2n_0 \hbar^2 k^2}{M} - \frac{\gamma p^2}{M}.$$

---

[24] In three-dimensional kinematics

$$\Delta p^2(\delta t) = \frac{8}{3} n_0 \hbar^2 k^2 \delta t.$$

The condition for the regime to be stationary $dE/dt = 0$ gives the equilibrium value p_{eq}^2 of p^2, and choosing $\gamma = \gamma_{max}$ in (14.106), we have

$$p_{eq}^2 = \frac{2n_0\hbar^2 k^2}{\gamma_{max}} = \frac{1}{2}\hbar\Gamma M,$$

which gives for the temperature $T = T_D$

$$\boxed{k_B T_D = \frac{p_{eq}^2}{M} = \frac{1}{2}\hbar\Gamma}.$$ (14.112)

This temperature, which is of the order of $100\,\mu K$ for the D_2 line of rubidium, is called the *Doppler temperature*. The equilibrium condition $dE/dt = 0$ can also be written as a function of the momentum diffusion coefficient:

$$D = \gamma p_{eq}^2 = M\gamma k_B T.$$ (14.113)

This equation relating the diffusion coefficient D and the viscosity coefficient γ to temperature is very general[25] and is well known as the *Einstein relation*. In the case of Brownian motion, viscosity forces and diffusion have a common origin, namely, collisions of the Brownian particle with the fluid molecules, and it is not surprising that the diffusion and viscosity coefficients are not independent. Diffusion and viscosity are both dissipative processes. In our case the origin of the dissipative process is spontaneous emission, which we have seen corresponds to nonunitary evolution.

14.4.4 A magneto-optical trap

Doppler cooling is the maximum cooling that can be obtained if we limit ourselves to the model of the two-level atom. To go farther, and in particular to consider cooling mechanisms which are even more effective, allowing temperatures of microkelvins and lower to be obtained, it is necessary to bring into play the level substructure, both fine and hyperfine. Let us consider an elementary example, taking a ground state $j = 0$ and an excited state $j = 1$ which we split into three sublevels using the Zeeman effect. This will permit us to trap atoms not only in velocity, as in Doppler cooling, but also in space. Since a magnetic field must be used to obtain the Zeeman effect, such a trap is called a *magneto-optical trap* (MOT). We use a nonuniform, z-dependent magnetic field pointing in the z direction, $B(z) = -bz$, $b > 0$. According to (14.26), the Zeeman levels of the excited state (e) with magnetic quantum number[26] m_e are given by

$$W_{m_e} = -\mu B m_e = -g\frac{q_e \hbar B}{2m}m_e \quad \text{with} \quad \mu = g\frac{q_e\hbar}{2m} < 0.$$

The Zeeman levels of the excited state then have energies $-\mu bz$ ($m_e = -1$), 0 ($m_e = 0$), and $+\mu bz$ ($m_e = 1$), with Oz taken as the angular momentum quantization axis.

[25] See, for example, Le Bellac *et al.* [2004], Chapter 5.
[26] m_e should not be confused with the electron mass m_e.

We again take the configuration of laser beams used above for Doppler cooling, but now assuming that these beams are left-hand circularly polarized. Angular momentum conservation along Oz (cf. (10.106)–(10.107)) implies that $m_e = -1$ if the atom absorbs a photon of wave vector $+\vec{k}$ and $m_e = +1$ if it absorbs one of wave vector $-\vec{k}$; see Fig. 14.4. We assume that $\delta < 0$. For $z > 0$ the sign of B implies that the level $m_e = +1$ is lower than the level $m_e = -1$ and therefore closer to resonance (Fig. 14.4). This implies that the atom will preferentially absorb photons of wave vector $-\vec{k}$ and be pushed toward the left. The opposite occurs if the atom is in the region $z < 0$ where the level $m_e = -1$ is lower than the level $m_e = +1$: the atom preferentially absorbs photons of wave vector $+\vec{k}$ and is pushed to the right. The action of the two beams is equivalent to the existence of two forces, a viscosity force $-\gamma M v$ and a restoring force $-\kappa z$:

$$F = -\gamma M v - \kappa z, \tag{14.114}$$

to which we must add the diffusion in momentum space. The atoms are not only slowed down, but they are also confined by the recoil force in the region $z \simeq 0$; this is the principle of the magneto-optical trap. In practice, we want to confine atoms in three-dimensional space, and so it is necessary to use six polarized laser beams (Fig. 14.5).

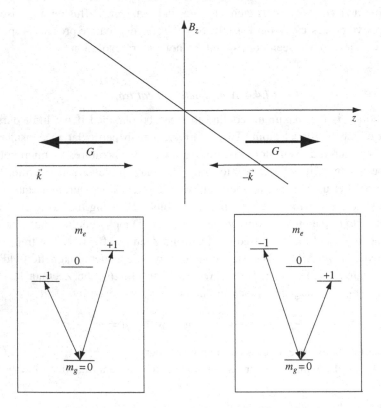

Fig. 14.4. Zeeman levels for $z < 0$ and $z > 0$.

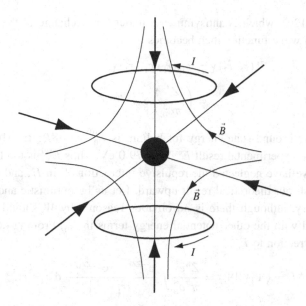

Fig. 14.5. Laser configuration for a magneto-optical trap.

14.5 The two-electron atom

14.5.1 The ground state of the helium atom

The helium atom is a two-electron atom with a nucleus of charge $2q_e$, which we write as Zq_e, $Z = 2$, so that our theory also applies, for example, to the Li$^+$ ion with $Z = 3$. Assuming the nucleus to be infinitely heavy (an approximation better than 0.1%), in a representation where the position operator is diagonal the Hamiltonian H reads as

$$H = -\frac{\hbar^2}{2m_e}\nabla_1^2 - \frac{\hbar^2}{2m_e}\nabla_2^2 - \frac{Ze^2}{r_1} - \frac{Ze^2}{r_2} + \frac{e^2}{|\vec{r}_1 - \vec{r}_2|}. \tag{14.115}$$

The vectors $\vec{r}_1$ and $\vec{r}_2$ are the positions of electrons 1 and 2. We write $H = H_0 + W$, where H_0 is the free Hamiltonian describing the electrons interacting with the nucleus,

$$H_0 = -\frac{\hbar^2}{2m_e}\nabla_1^2 - \frac{\hbar^2}{2m_e}\nabla_2^2 - \frac{Ze^2}{r_1} - \frac{Ze^2}{r_2}, \tag{14.116}$$

and W is a perturbation, whose physical origin is the electrostatic repulsion between the two electrons:

$$W = \frac{e^2}{|\vec{r}_1 - \vec{r}_2|}. \tag{14.117}$$

Let us seek the lowest energy level by first neglecting W. This level is clearly a $1s^2$ level, where the two electrons are in a $1s$ state; the superscript counts the number of electrons in a given state. However, electrons are fermions, and the two electrons cannot be in the same state. Fortunately, spin saves the situation, since the electrons can be put in a singlet spin state χ_s (10.126), which is antisymmetric under the exchange of the two electrons. Our space + spin wave function then becomes

$$\Psi(\vec{r}_1, \vec{r}_2)\chi = \varphi_{1s}(\vec{r}_1)\varphi_{1s}(\vec{r}_2)\chi_s$$
$$= \left(\frac{Z^3}{\pi a_0^3}\right) e^{-Zr_1/a_0} e^{-Zr_2/a_0} \chi_s. \tag{14.118}$$

The corresponding ground-state energy for helium is $E_0^{(0)} = -8R_\infty \simeq -108.8$ eV, to be compared with the experimental result $E^{\text{exp}} = -79.0$ eV. Thus $E_0^{(0)}$ is too low by roughly 30%. However, we have neglected the repulsive interaction W in H, and we expect that this term will push our theoretical result upward. Let us be optimistic and blindly apply perturbation theory, although there is no obvious reason why W should be considered "small" compared with the other potential energy terms in H_0. From (14.6) we compute the first-order correction to E_0:

$$\Delta E = \langle \Psi | W | \Psi \rangle = \frac{Z^6 e^2}{\pi^2 a_0^6} \int \frac{e^{-2Zr_1/a_0} e^{-2Zr_2/a_0}}{|\vec{r}_1 - \vec{r}_2|} d^3 r_1 d^3 r_2. \tag{14.119}$$

To compute this six-dimensional integral, we use the following representation of $1/r$:[27]

$$\frac{1}{r} = \frac{4\pi}{(2\pi)^3} \int \frac{d^3 k}{k^2} e^{i\vec{k}\cdot\vec{r}},$$

and we find

$$\Delta E = \frac{Z^6 e^2}{2\pi^4 a_0^6} \int \frac{dk}{k^2} \left[\int e^{-2Zr/a_0} e^{i\vec{k}\cdot\vec{r}} d^3 r\right]^2. \tag{14.120}$$

The integral in the square brackets has already been encountered in (14.55):

$$\int e^{-2Zr/a_0} e^{i\vec{k}\cdot\vec{r}} d^3 r = \frac{16\pi Z/a_0}{[k^2 + (2Z/a_0)^2]^2}, \tag{14.121}$$

and plugging this result into (14.120) gives

$$\Delta E = \frac{4Ze^2}{\pi a_0} \int_0^\infty \frac{dx}{(1+x^2)^4} = \frac{4Ze^2}{\pi a_0} \times \frac{5\pi}{32} = \frac{5}{4} Z R_\infty. \tag{14.122}$$

As expected, ΔE is positive and

$$E_0^{(0)} + \Delta E \simeq -74.8 \text{ eV}, \tag{14.123}$$

which is much closer to the experimental value than we had a right to expect.

The variational method will give an even better result. As our trial function for one electron we take

$$\varphi(\vec{r}) = \left(\frac{z^3}{\pi a_0^3}\right)^{1/2} e^{-zr/a_0}, \tag{14.124}$$

---

[27] To check this formula, compute the Fourier transform of $(k^2 + \alpha^2)^{-1}$ and take the limit $\alpha \to 0$.

where z is the variational parameter. In order to compute the expectation values, we write

$$\int d^3 r \, \varphi^*(\vec{r}) \left(-\frac{1}{2m_e} \nabla^2 - \frac{ze^2}{r} \right) \varphi(\vec{r}) = -z^2 R_\infty \,, \tag{14.125}$$

since (14.124) is the ground-state solution of the Schrödinger equation for a one-electron atom in a Coulomb potential $-ze^2/r$. Since the potential energy is twice the total energy, we also have

$$\int d^3 r \left(-\frac{ze^2}{r} \right) |\varphi(\vec{r})|^2 = -2ze^2 R_\infty. \tag{14.126}$$

Equations (14.125) and (14.126) allow us to compute the expectation value of H_0:

$$\langle H_0 \rangle = -2(2zZ - z^2) R_\infty.$$

The expectation value of W has just been computed in the perturbative approach:

$$\langle W \rangle = \frac{5}{4} z R_\infty.$$

Collecting all the contributions we find

$$E_0(z) = 2 \left(z^2 - 2Zz + \frac{5}{8} z \right) R_\infty. \tag{14.127}$$

The optimal value of z is obtained from $dE(z)/dz = 0$, so that $z = Z - 5/16$ and

$$E_0^{\text{var}} = -2 \left(Z - \frac{5}{16} \right)^2 R_\infty. \tag{14.128}$$

In the case of helium, we find $E_0^{\text{var}} \simeq -77.5$ eV, which is closer to the experimental result than the perturbative estimate. We can also check that $E_0^{\text{var}} > E_0^{\text{exp}}$, as must be the case. For the same volume of calculations, we see that the variational method with a good guess for the trial wave function gives much better results than the perturbative approach!

14.5.2 The excited states of the helium atom

As we have just seen, the ground state of the helium atom has zero orbital angular momentum and zero spin. Using the notation $^{2S+1}L_J$, where S is the total spin, L the total orbital angular momentum, and J the total angular momentum, the ground state of the helium atom is therefore a $^1S_0$ state. The next lowest energy levels are the $1s^1 2s^1$ and $1s^2 2p^2$ states. These levels are degenerate if H_0 (14.116) is used as the Hamiltonian. However, it is a better strategy to try to take into account, at least approximately, the effect of the repulsion W by using not the Coulomb potential $-Ze^2/r$, but an *effective* one-electron potential $V_{\text{eff}}(r)$ which can be determined from self-consistency arguments. Therefore, instead of H_0 we use a Hamiltonian H_0':

$$H_0' = -\frac{\hbar^2}{2m_e}\nabla_1^2 - \frac{\hbar^2}{2m_e}\nabla_2^2 + V_{\text{eff}}(r_1) + V_{\text{eff}}(r_2), \qquad (14.129)$$

and instead of W a perturbation W':

$$W' = \frac{e^2}{|\vec{r}_1 - \vec{r}_2|} + \left[-\frac{Ze^2}{r_1} - V_{\text{eff}}(r_1)\right] + \left[-\frac{Ze^2}{r_2} - V_{\text{eff}}(r_2)\right]. \qquad (14.130)$$

With H_0' as the Hamiltonian, the $2s$ and $2p$ levels are no longer degenerate (see Fig. 10.7), and the $2p$ level lies above the $2s$ level. An important remark is that W' is invariant under spatial rotations, so that it commutes with the total orbital angular momentum $\vec{L}$: $[\vec{L}, W'] = 0$, although, for example, $[\vec{L}_1, W'] \neq 0$. W' therefore has vanishing matrix elements between the $1s^1 2s^1$ and $1s^1 2p^2$ states, which have total orbital angular momentum $L = 0$ and $L = 1$, respectively. Thus, although these levels are not far from being degenerate, we can use nondegenerate perturbation theory within each of the levels.

Let us begin with the $1s^1 2s^1$ state, which is the first excited level. We can build symmetric and antisymmetric wave functions:

$$\Psi_{\pm}(\vec{r}_1, \vec{r}_2) = \frac{1}{\sqrt{2}}\left[\varphi_{1s}(\vec{r}_1)\varphi_{2s}(\vec{r}_2) \pm \varphi_{1s}(\vec{r}_2)\varphi_{2s}(\vec{r}_1)\right]. \qquad (14.131)$$

The one-electron terms of W' are independent of the symmetry of Ψ, but the W contribution is symmetry-dependent:

$$\langle\Psi_{\pm}|W|\Psi_{\pm}\rangle = e^2 \int d^3r_1\, d^3r_2 \frac{|\varphi_{1s}(\vec{r}_1)|^2|\varphi_{2s}(\vec{r}_2)|^2}{|\vec{r}_1 - \vec{r}_2|}$$

$$\pm e^2 \int d^3r_1\, d^3r_2\, \varphi_{1s}(\vec{r}_1)\varphi_{2s}(\vec{r}_2)\frac{1}{|\vec{r}_1 - \vec{r}_2|}\varphi_{1s}(\vec{r}_2)\varphi_{2s}(\vec{r}_1) \qquad (14.132)$$

$$= K \pm J.$$

The integral K is clearly positive, and it can be shown that J, called the *exchange integral*, is also positive, so that the energy of the antisymmetric wave function is lower than that of the symmetric one. This is easy to understand: since the antisymmetric wave function vanishes at $\vec{r}_1 = \vec{r}_2$, the expectation value of $|\vec{r}_1 - \vec{r}_2|$, which is a maximum (and in fact infinite) at $\vec{r}_1 - \vec{r}_2$, is lower in the antisymmetric case. These considerations are completely independent of the fermionic nature of the electrons, and would also hold if we had two kinds of electron in the helium atom, a red one and a green one. What the Pauli principle implies is that *the symmetry of the spatial wave function is related to that of the spin state*. Then the lowest energy state is a $^3S_1$ state, and the highest is a $^1S_0$ state (Fig. 14.6a). If the electrons were red and green, the total spin would not be related to the symmetry of the wave function.

In the $1s^1 2p^1$ state the total angular momentum is $L = 1$ and the possible states are $^1P_1$ in the singlet spin state and $^3P_0$, $^3P_1$, and $^3P_2$ in the triplet spin state. The exchange

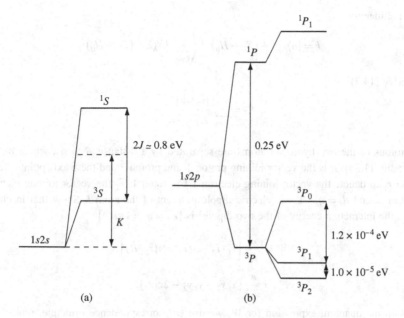

Fig. 14.6. The first two excited states of the helium atom. After Cohen-Tannoudji *et al.* [1977], Complement B_{XIV}.

integral is again positive, so that the triplet states lie lower than the singlet state. The level scheme is sketched in Fig. 14.6b.

14.6 Exercises

14.6.1 Second-order perturbation theory and van der Waals forces

The van der Waals forces between two neutral atoms arise from the interactions between the induced dipole moments. We wish to evaluate them in the case of two hydrogen atoms in their ground states $|\varphi_0\rangle$. To do this we shall need to use second-order perturbation theory.

1. *Second-order perturbation theory.* First we determine $|\varphi_1\rangle$ assuming that $|\varphi_0\rangle$ is nondegenerate; the notation is the same as in Section 14.1.2. Show that

$$(E_0 - H_0)|\varphi_1\rangle = (W - E_1)|\varphi_0\rangle.$$

Keeping the term of second order in λ in the series (14.3) and (14.4), show that

$$E_2 = \langle\varphi_0|W|\varphi_1\rangle.$$

We recall that $|\varphi_0\rangle \equiv |n\rangle$ and

$$H_0|n\rangle = E_0^{(n)}|n\rangle, \quad H_0|k\rangle = E_0^{(k)}|k\rangle.$$

Prove the identity

$$I = |n\rangle\langle n| + (E_0 - H_0)^{-1} \left(\sum_{k \neq n} |k\rangle\langle k| \right) (E_0 - H_0)$$

and derive (14.7):

$$E_2 = \sum_{k \neq n} \frac{|\langle n|W|k\rangle|^2}{E_0^{(n)} - E_0^{(k)}}.$$

2. The protons of the two hydrogen atoms are separated by a distance $R \gg a_0$, where a_0 is the Bohr radius (1.34); $\vec{R}$ is the vector joining proton 1 and proton 2 and the z axis points along $\vec{R}$. We use $\vec{r}_1$ to denote the vector joining electron 1 to proton 1, $\vec{r}_2$ the vector joining electron 2 to proton 2, and $\vec{d}_i = q_e \vec{r}_i$ is the electric dipole moment of the atom i. Show that in classical physics the interaction energy of the two dipoles is $[e^2 = q_e^2/(4\pi\varepsilon_0)]$

$$W = \frac{e^2}{R^3} \left[\vec{r}_1 \cdot \vec{r}_2 - 3(\vec{r}_1 \cdot \hat{R})(\vec{r}_2 \cdot \hat{R}) \right]$$

$$= \frac{e^2}{R^3} \left[x_1 x_2 + y_1 y_2 - 2 z_1 z_2 \right].$$

3. To obtain the quantum expression for W, we use the correspondence principle, replacing the numbers $x_1, \dots, z_2$ by the operators $X_1, \dots, Z_2$:

$$W = \frac{e^2}{R^3} \left[X_1 X_2 + Y_1 Y_2 - 2 Z_1 Z_2 \right].$$

Show that the expectation value of W vanishes in first-order perturbation theory:

$$E_1 = \langle \varphi_{01} \varphi_{02} | W | \varphi_{01} \varphi_{02} \rangle = 0.$$

4. In second order, if $|\varphi_\alpha\rangle$ designates an excited state or a continuum state of energy E_α, then

$$E_2 = \sum_{\alpha_1, \alpha_2} \frac{|\langle \varphi_{\alpha_1} \varphi_{\alpha_2} | W | \varphi_{01} \varphi_{02} \rangle|^2}{-2R_\infty - E_{\alpha_1} - E_{\alpha_2}},$$

where R_∞ is the Rydberg constant (1.35). To obtain the order of magnitude of E_2 we neglect E_{α_1} and E_{α_2} in the denominator. Show that

$$E_2 \sim -6 \frac{e^2}{R} \left(\frac{a_0}{R} \right)^5.$$

The interaction energy varies as R^{-5} and the force as R^{-6}. Show that the preceding estimate is no longer valid if $R \gtrsim \hbar c/R_\infty$. Show that the force law is R^{-7} for distances $R \gg \hbar c/R_\infty$.

14.6.2 *Order-α^2 corrections to the energy levels*

Hint. In both this problem and the following one, it is recommended that for numerical work the energies be written in dimensionless form by using the factor $R_\infty = 13.61$ eV.

In addition to the fine structure, there exist two other $O(v/c)^2$ corrections to the energy levels of the hydrogen atom (or, more generally, one-electron atoms).

1. *The kinematical correction.* The relativistic form of the electron kinetic energy is

$$K = \sqrt{p^2 c^2 + m_e^2 c^4} = m_e c^2 + \frac{p^2}{2m_e} - \frac{1}{8} \frac{p^4}{m_e^3 c^2} + O\left(\frac{p^6}{m_e^5 c^4}\right).$$

Verify this series in powers of $p/m_e c$ valid for $p/m_e c \ll 1$. The first term is the mass energy, a simple additive constant, and the second is the nonrelativistic form of the kinetic energy used in solving the Schrödinger equation. The objective is to evaluate the corrections due to the third term $O(p^4)$. Show that this term gives a correction $\Delta E_K \propto \alpha^2 (v/c)^2 = O(\alpha^4)$ to the energy levels. In order to evaluate this correction precisely, we use perturbation theory. Show that in first order

$$\Delta E_K = -\frac{1}{8m_e^3 c^2} \int d^3 p \, p^4 \, |\tilde{\varphi}(\vec{p})|^2,$$

where $\tilde{\varphi}(\vec{p})$ is the Fourier transform of the wave function $\varphi(\vec{r})$:

$$\tilde{\varphi}(\vec{p}) = \frac{1}{(2\pi\hbar)^{3/2}} \int d^3 r \, e^{i\vec{p}\cdot\vec{r}/\hbar} \varphi(\vec{r}).$$

Calculate ΔE_K for the $1s$ level of the hydrogen atom. The necessary integrals can be derived from

$$I(x) = \int_0^\infty \frac{dq}{q^2 + x} = \frac{\pi}{2} x^{-1/2}$$

by differentiating with respect to x ($x > 0$).

2. *The Darwin term.* The second correction arises from the fact that in the nonrelativistic approximation of the Dirac equation, the electron cannot be localized to better than within $\hbar/m_e c$, the electron Compton wavelength. To take this spatial extent into account, the potential energy is written as

$$E_{\text{pot}} = \int d^3 u \, f(|\vec{u}|) V(\vec{r} + \vec{u}),$$

where V is the usual potential energy and $f(u)$, which is spherically symmetric, has extent $\sim \hbar/m_e c$ and is normalized by

$$\int d^3 u \, f(u) = 1.$$

Expanding $V(\vec{r} + \vec{u})$ about $u = 0$, show that

$$E_{\text{pot}} = V(\vec{r}) + O\left[\left(\frac{\hbar}{m_e c}\right)^2\right] \nabla^2 V + O\left(\frac{\hbar}{m_e c}\right)^4.$$

The Dirac equation gives the exact coeffficient:

$$E_{\text{pot}} = V(\vec{r}) + \frac{\hbar^2}{8m_e^2 c^2} \nabla^2 V + O\left(\frac{\hbar}{m_e c}\right)^4.$$

The second term in E_{pot} is called the *Darwin term*. Show that this term affects only s-waves and gives

$$\Delta E_D = \frac{\pi e^2 \hbar^2}{2m_e^2 c^2} |\varphi(\vec{r} = 0)|^2.$$

Evaluate ΔE_D numerically for the $1s$ level of hydrogen.

14.6.3 Muonic atoms

The muon (μ) is a lepton completely identical to the electron except that its mass is $m_\mu \simeq$ 105.7 MeV $c^{-2} \simeq 206.8 m_e$ (cf. Section 1.1.3). An atom can capture a negative muon μ^- into an orbit about the nucleus just like an electron, to form a "muonic atom."

1. Calculate the Bohr radius a_μ^Z of the muon, as a function of the atomic number Z, the ratio m_μ/m_e, and $a_0 = \hbar^2/m_e e^2$, for an atom of atomic number Z by writing

$$a_\mu^Z = \frac{1}{Z\alpha(A)} \, a_0.$$

   The reduced mass is used in the calculation of $\alpha(A)$. Compare a_μ^Z to the nuclear radius R for aluminum ($Z = 13$, $A = 27$) and lead ($Z = 82$, $A = 208$). We recall that R is given by $R \simeq 1.2 \times A^{1/3}$ fm, where A is the number of nucleons.
2. Let $\Delta E_e^{Z=1} = \Delta E_e = E_{2p} - E_{1s}$ be the energy difference between the $2p$ and $1s$ levels of the hydrogen atom. Calculate the corresponding quantity ΔE_μ^Z for an atom of atomic number Z as a function of ΔE_e and m_μ/m_e. Compare to the experimental values:

$$\text{Aluminum}: \Delta E_\mu^{13} = 0.3443 \, \text{MeV}, \quad \text{Lead}: \Delta E_\mu^{82} = 5.96 \, \text{MeV}.$$

   What type of photon is emitted in these transitions?
3. Show that the screening of the inner-shell electrons is negligible. In contrast, an important correction comes from the finite size of the nucleus. Show that the potential seen by the muon is not $-Ze^2/r$ but

$$V(r) = \frac{Ze^2}{2R}\left[\left(\frac{r}{R}\right)^2 - 3\right], \quad r < R,$$

$$V(r) = -\frac{Ze^2}{r}, \qquad\qquad r > R.$$

   We wish to calculate the level shift using first-order perturbation theory starting from the solution for the exact Coulomb potential. What perturbation $W(r)$ should be used? Show qualitatively that the finite size of the nucleus is negligible except for s states, and that in this case for small Z and an orbit of principal quantum number n with radius large compared to R the shift will be

$$\Delta E_n = \frac{2\pi Z e^2}{5} \, R^2 |\varphi_n(\vec{r} = 0)|^2,$$

   where $\varphi_n(\vec{r})$ is the unperturbed wave function. Show that for the $1s$ state

$$\Delta E = \frac{4}{5} R_\infty \left(\frac{Z^2 m_\mu'}{m_e}\right)\left(\frac{R}{a_\mu^Z}\right)^2,$$

   where m_μ' is the reduced mass. Find the numerical value of this shift for aluminum.[28] Is the correction in the right direction? Is it reasonable to apply the method to the case of lead?

---

[28] Aside from the correction due to the finite size of the nucleus, the most important correction comes from the vacuum polarization due to virtual electron–positron pairs. The correction for the $1s$ state of aluminum is -2.25 keV. The sign of this correction is negative; in fact, at short distances α is larger than $1/137$ and the muon, which sees a larger charge, is more tightly bound than if α were constant. This behavior of α was mentioned in Footnote 36 of Chapter 1: α grows with energy and, according to the Heisenberg inequality, short distance implies large momentum and therefore high energy.

4. Show that the ratio of the typical fine-structure energies to the typical level energies is the same for the electron and the muon. Show that this ratio, however, is larger by a factor m_μ/m_e for the hyperfine structure.

14.6.4 Rydberg atoms

The results of Exercise 10.7.9 allow us to write down the radial wave functions $u_{nl}(r)$ of the hydrogen atom in the form

$$u_{nl}(r) = \sum_{q=0}^{n-l-1} c_q \left(\frac{r}{a_0}\right)^{q+l+1} \exp\left(-\frac{r}{na_0}\right).$$

To write down the formula for the coefficients c_q, it is convenient to define $k = n - l$:

$$c_q = \left(-\frac{2}{n}\right)^q \frac{(k-1)!(2l+1)!}{q!(q+2l+1)!(k-q-1)!},$$

where c_0 is fixed by the normalization condition of the wave function. We are interested in values $n \gg 1$, typically $n \sim 50$.

1. Show that if l takes its maximum value $l = l_{max} = n - 1$, the radial wave function displays a narrow peak near the point $r = a_0 n^2$. What is the width Δr of this peak? Hint: study the function

$$f_n(x) = x^n e^{-x/n}$$

and show that for $x \simeq x_0 = n^2$,

$$f_n(x) \simeq f_n(x_0) \exp\left[-\frac{1}{2n^3}(x-x_0)^2\right].$$

Show qualitatively that if $l < n - 1$, the dispersion Δr is larger than for $l = n - 1$.

2. We are now interested in the angular part. According to (10.53),

$$Y_l^m(\theta, \phi) = e^{im\varphi} f_l^m(\theta).$$

Using $L_+ Y_l^l = 0$ and the expression (10.48) for L_+, show that

$$Y_l^l(\theta, \phi) \propto e^{il\phi} \sin^l \theta.$$

Show that if $l \gg 1$, $|Y_l^l(\theta, \phi)|^2$ is nonzero only near the xOy plane (that is, for $\theta = \pi/2$) and calculate the dispersion $\Delta\theta$. What happens if $|m| \neq l$?

3. Using the first two questions, show that for $n \gg 1$ the states $l = n - 1$ and $|m| = l$ are localized in a horizontal torus of radius $n^2 a_0$ whose cross section is a circle of radius $a_0 n^{3/2}$. Compare with the orbits (1.33) obtained using the Bohr prescriptions of Section 1.5.2.

14.6.5 The diamagnetic term

When we derived the form of the Hamiltonian (14.23) of the Zeeman effect, we neglected a term $\propto \vec{A}^2$ called the *diamagnetic term*. To justify this approximation, let us consider the case of a uniform, constant magnetic field $\vec{B}$, a possible expression for $\vec{A}$ being (cf. Section 11.4.2)

$$\vec{A} = \frac{1}{2}\vec{B} \times \vec{r}.$$

1. Show that the quantum Hamiltonian of an electron of charge q in this magnetic field can be written as

$$H = \frac{1}{2m_e}(\vec{P} - q\vec{A})^2$$

$$= \frac{\vec{P}^2}{2m_e} - \frac{q}{2m_e}\vec{B} \cdot \vec{L} + \frac{q^2}{8m_e}\left[\vec{R}^2\vec{B}^2 - (\vec{R} \cdot \vec{B})^2\right]$$

$$= H_0 + H_Z + H_D,$$

where $\vec{L} = \vec{R} \times \vec{P}$ is the orbital angular momentum. Carefully justify the operator commutations.
2. Identify H_Z as the part of the Zeeman Hamiltonian (14.23) of orbital origin and give the order of magnitude of this term for a magnetic field of 1 T when the electron is bound in an atom. The diamagnetic term H_D can be written as

$$H_D = \frac{q^2 B^2}{8m_e}\vec{R}_{\perp}^2,$$

where $\vec{R}_{\perp}$ is the component of $\vec{R}$ perpendicular to $\vec{B}$. What can we take for the order of magnitude of $\langle \vec{R}_{\perp}^2 \rangle$? Show that $|\langle H_D \rangle| \ll |\langle H_Z \rangle|$ for an electron bound in an atom, and that the diamagnetic term can be neglected in calculating the Zeeman effect. However, this term cannot be neglected in calculating the Landau levels, because the radius of the electron orbits is macroscopic in that case.

14.6.6 Vacuum Rabi oscillations

Let us assume that the eigenfrequency ω of a cavity is close to the frequency $\omega_0 = (E_e - E_g)/\hbar$ of a transition between two levels e and g of an atom, and use $\delta = \omega - \omega_0$ to denote the detuning. If the atom interacts with the quantized electromagnetic field inside the cavity, we can to an excellent approximation limit the expansion (11.136) of the quantized field to a single frequency mode ω, because this mode is the only one that interacts with the atom in a resonant fashion. We work in one dimension, keeping only the dependence on z and the polarization in the x direction, so that the field can be treated as a scalar.

1. Using (11.136), show that for the quantized field E we can write

$$E_H(z, t) = i\sqrt{\frac{\hbar\omega}{\varepsilon_0 V}}\left(ae^{-i\omega t} - a^{\dagger}e^{i\omega t}\right)\sin kz.$$

We assume that the atom always moves along the line of constant phase $\sin kz = 1$.

2. The atom + field Hamiltonian is

$$H = H_{\text{atom}} + H_{\text{field}} + W,$$

where W represents the interaction between the atom and the field. We take $|g\rangle$ to be the zero-energy state with no photons. Derive the form of H

$$H = \hbar\omega_0 |e\rangle\langle e| + \hbar\omega N + W,$$

where N is the number operator for photons in the mode of frequency ω. Give the spectrum of H first neglecting W and assuming that $|\delta| \ll \omega_0$, but $\delta \neq 0$. Let $\mathcal{H}^{(n)}$ be the subspace of the Hilbert space formed from the following basis states, where n is the number of photons in the cavity:

$$\varphi_n^e = |e \otimes (n-1)\rangle, \quad \varphi_n^g = |g \otimes n\rangle.$$

Show that these states are nearly degenerate if W is neglected.

3. We define the operators

$$b = |g\rangle\langle e|, \quad b^\dagger = |e\rangle\langle g|$$

and the dipole moment of the atom (cf. Section 5.2.2)

$$D = d(b + b^\dagger).$$

Write down the interaction term W explicitly in the dipole approximation. Show that if W is constrained to the subspaces $\mathcal{H}^{(n)}$, then

$$W = -i\frac{\hbar\Omega_R}{2}(ab^\dagger - a^\dagger b)$$

with

$$\hbar\Omega_R = 2d\sqrt{\frac{\hbar\omega}{\varepsilon_0 V}}.$$

The frequency Ω_R is called the *vacuum Rabi frequency*. What terms have been neglected in the approximate expression for W and how can this approximation be justified? The atom + field Hamiltonian involving the approximate expression for W is called the *Jaynes–Cummings Hamiltonian*.

4. What are the values of E_n and the corresponding eigenstates when W is taken into account? We shall take (cf. Section 2.3.2)

$$\tan 2\theta_n = \frac{\Omega_R\sqrt{n}}{\delta} = \frac{\Omega_n}{\delta}.$$

Qualitatively sketch the spectrum of the first few levels of H as a function of δ.

5. The atom in the excited state $|e\rangle$ is sent to the empty cavity along a trajectory such that $\sin kz = 1$. We take the resonant case $\delta = 0$. Show that the probability $\mathsf{p}_e(t)$ of finding the atom in the state $|e\rangle$ after a time t spent in the cavity is a periodic function of t. We obtain Rabi oscillations, and since these oscillations arise from the interaction of the atom with the vacuum fluctuations, they

are called vacuum Rabi oscillations. The experimental observation of these oscillations provides direct proof of the quantization of the electromagnetic field. The numerical values are[29]

$$d = 1.1 \times 10^{-26} \, \text{C m}, \quad \frac{\omega}{2\pi} = 5.0 \times 10^{10} \, \text{Hz}, \quad \mathcal{V} = 1.87 \times 10^{-6} \, \text{m}^3.$$

Compare to the experimental value $\Omega_R/2\pi = 47$ kHz.

6. Calculate $\mathsf{p}_e(t)$ away from resonance, and show that the oscillation frequency is now (always in the case where there are no photons in the cavity)

$$\Omega = \sqrt{\delta^2 + \Omega_R^2}.$$

Show that for the detuning $\Omega_R \ll |\delta| \ll \omega_0$ the atom nearly always remains in its excited state: spontaneous emission is inhibited by the presence of the cavity.

7. How should the results of the two preceding questions be modified if the cavity contains exactly n photons? If $\delta = 0$, what happens when the cavity contains a coherent state of the field?

14.6.7 Reactive forces

We take the Jaynes–Cummings Hamiltonian of the preceding Exercise 14.5.6 for an atom with two levels $|g\rangle$ and $|e\rangle$ immersed in the quantized electric field of a cavity:

$$\mathsf{E} = i\mathcal{E}(a - a^\dagger)\sin kz, \quad \mathcal{E} = \sqrt{\frac{\hbar\omega}{\varepsilon_0 \mathcal{V}}},$$

with the notation of the preceding exercise. The Hamiltonian is given by

$$H = \hbar\omega_0|e\rangle\langle e| + \hbar\omega N + W$$

with[30]

$$W = \frac{1}{2}\hbar\Omega_1(ab^\dagger + a^\dagger b),$$

where $b = |g\rangle\langle e|$ and $b^\dagger = |e\rangle\langle g|$. The frequency Ω_1 defined as

$$\Omega_1(z) = 2\frac{d\mathcal{E}}{\hbar}\sin kz$$

is a function of z.

1. In the two-dimensional subspace $\mathcal{H}^{(n)}$ in which the states $|g \otimes n\rangle$ and $|e \otimes (n-1)\rangle$ form an orthonormal basis, show that up to an additive constant the Hamiltonian takes the form

$$H = \frac{1}{2}\hbar\begin{pmatrix} \delta & \Omega_1\sqrt{n} \\ \Omega_1\sqrt{n} & -\delta \end{pmatrix},$$

where $\delta = \omega - \omega_0$ is the detuning. We set

$$\Omega_{1n}(z) = \sqrt{\delta^2 + n\Omega_1^2(z)} = \sqrt{\delta^2 + \Omega_n^2(z)}$$

---

[29] M. Brune, *et al.* Quantum Rabi oscillations: a direct test of field quantization in a cavity, *Phys. Rev. Lett.* **76**, 1800 (1996).
[30] A suitable choice of phase for the vectors $|e\rangle$ and $|g\rangle$ has allowed us to eliminate the factors of i of the preceding exercise.

and define the angle $\theta_n(z)$ as

$$\cos 2\theta_n(z) = \frac{\delta}{\Omega_{1n}(z)}, \quad \sin 2\theta_n(z) = \frac{\Omega_n(z)}{\Omega_{1n}(z)}.$$

Show that the eigenvectors of H restricted to $\mathcal{H}^{(n)}$ are

$$|\chi_{1n}(z)\rangle = -\sin\theta_n(z)|g \otimes n\rangle + \cos\theta_n(z)|e \otimes (n-1)\rangle,$$

$$|\chi_{2n}(z)\rangle = \cos\theta_n(z)|g \otimes n\rangle + \sin\theta_n(z)|e \otimes (n-1)\rangle.$$

What are the eigenvalues of H? Calculate the force on an atom at rest at z when this atom is in the state $|\chi_{1n}\rangle$ or the state $|\chi_{2n}\rangle$.

2. In what follows we assume that the field inside the cavity is that of a laser in a coherent state with an average number of photons $\langle n \rangle \gg 1$ such that $\Delta n \ll \langle n \rangle$. We can then write down a classical expression for this field:

$$E_L(t, z) = \mathcal{E}_0 \cos \omega t \sin kz.$$

Using (11.93), show that

$$\hbar \Omega_1(z) \sqrt{\langle n \rangle} = \hbar \omega_1(z), \quad \omega_1(z) = \frac{d\mathcal{E}_0}{\hbar} \sin kz,$$

where $\omega_1(z)$ is the usual Rabi frequency (cf., for example, (14.74)). In the preceding discussion we have neglected spontaneous emission, which has the effect of depopulating the laser mode in favor of the vacuum mode. The rate of transitions between the states with n and $n-1$ photons is given by

$$\Gamma_{ij}(z) = \Gamma |\langle \chi_{i,n-1}(z)|b + b^\dagger |\chi_{jn}(z)\rangle|^2$$

with $(i, j) = 1, 2$. Calculate $\Gamma_{ij}(z)$ as a function of the angles $\theta_n(z)$ and $\theta_{n-1}(z)$. In what follows we assume that the laser is intense, $n \gg 1$ and $\Omega_{1n} \simeq \Omega_{1\langle n \rangle}(z) = \sqrt{\delta^2 + \omega_1^2(z)}$.

3. The populations $\mathsf{p}_i(z)$ are defined as

$$\mathsf{p}_i(z) = \sum_n \langle \chi_{in}(z)|\rho|\chi_{in}(z)\rangle,$$

where ρ is the state operator of the atom dressed by the field. Show that if $\Omega_{1\langle n \rangle} \gg \Gamma$ the populations obey the master equation

$$\dot{\mathsf{p}}_1(z) = -\Gamma_{21}(z)\mathsf{p}_1(z) + \Gamma_{12}(z)\mathsf{p}_2(z),$$

$$\dot{\mathsf{p}}_2(z) = \Gamma_{21}(z)\mathsf{p}_1(z) - \Gamma_{12}(z)\mathsf{p}_2(z).$$

What are the stationary values of the populations p_i^{st} as a function of z? Show that an atom at rest feels a force

$$F(z) = \frac{1}{4}\hbar \frac{\partial \omega_1^2(z)}{\partial z} \frac{1}{\sqrt{\delta^2 + \omega_1^2(z)}} \left(\mathsf{p}_1^{st}(z) - \mathsf{p}_2^{st}(z)\right).$$

Substitute the values of p_1^{st} and p_2^{st} into this result and compare with (14.98).

14.6.8 Radiative capture of neutrons by hydrogen

NB *It is useful to reread Sections 12.2.3 and 12.2.4.* In a boiling-water or pressurized-water nuclear reactor a fraction of the neutrons is absorbed by the hydrogen of the water in the reaction

$$n + p \rightarrow D + \gamma,$$

where n is a neutron, p is a proton, D is a deuteron, and γ is a photon. This reaction, called radiative capture, has the drawback of decreasing the number of neutrons available for fission. The deuteron is a neutron–proton bound state of total angular momentum $J = 1$ and binding energy $B = 2.23$ MeV. It is a mixture of the $^3S_1$ and $^3D_1$ states, but to simplify the discussion we shall take into account only the $^3S_1$ state. The goal is to calculate the radiative capture cross section. In the numerical calculations it will be convenient to use a system of units in which $\hbar = c = 1$. In this system the mass, momentum, and energy have the dimensions of inverse length, and the conversion factor is

$$1 \, (\text{fm})^{-1} \simeq 200 \text{ MeV}.$$

1. The reactor neutrons have very low energy ($\ll 1$ MeV), and so the n–p potential in the S-wave can to a good approximation be represented by a delta function $\delta(\vec{r})$ (see (12.44)). The bound-state wave function is given by (12.45), with $a \rightarrow a_t \simeq 5.40$ fm. Calculate the normalization constant C and κ^{-1} in fm. We note that κ^{-1} fixes the length scale of the problem.
   The scattering states of interest to us will be the $^1S_0$ states, where the scattering length is a_s, $a_s \simeq -23.7$ fm. It is convenient to fix the normalization by writing

$$\psi(r) = \frac{\sin[pr + \delta(p)]}{pr}.$$

Show that for $p \rightarrow 0$

$$\psi(r) \simeq -\frac{a}{r}\left(1 - \frac{r}{a}\right), \quad a = a_t \text{ or } a = a_s.$$

2. The neutron of the capture reaction is very slow, and, owing to the centrifugal barrier, the reaction occurs in the S-wave, which a priori presents two possibilities:

$$(n - p : {}^3S_1) \rightarrow D({}^3S_1) + \gamma, \quad (n - p : {}^1S_0) \rightarrow D({}^3S_1) + \gamma.$$

Electric dipole transitions are negligible because they would correspond to initial state in a P-wave (why?). The reaction comes from the coupling $\vec{\mu} \cdot \vec{B}$ between the deuteron magnetic moment $\vec{\mu}$ and the quantized magnetic field $\vec{B}$, with

$$\begin{aligned}
\vec{\mu} &= \frac{1}{2}\mu_N \left(g_p \vec{\sigma}_p + g_n \vec{\sigma}_n\right) \\
&= \frac{1}{4}\mu_N \left[(g_p + g_n)(\vec{\sigma}_p + \vec{\sigma}_n) + (g_p - g_n)(\vec{\sigma}_p - \vec{\sigma}_n)\right],
\end{aligned} \quad (14.133)$$

where $\mu_N = q_p\hbar/2M$. The quantities $g_p \simeq 5.59$ and $g_n \simeq -3.83$ are related to the proton and neutron gyromagnetic ratios and $\vec{\sigma}$ are the Pauli matrices. Show that the coupling to the quantized electromagnetic field responsible for the reaction is

$$W' = -\frac{i}{c}\sqrt{\frac{\hbar\omega}{2\varepsilon_0 V}}\,\vec{\mu}\cdot[\hat{k}\times\vec{e}_\lambda^*(\hat{k})]a_{k\lambda}^\dagger\, e^{-i\vec{k}\cdot\vec{r}},$$

where the photon has wave vector $\vec{k}$ and frequency $\omega = ck$, $\hat{k} = \vec{k}/k$, $\vec{e}_\lambda$ is a polarization unit vector which we can take to be real, and V is the normalization volume. Neglecting the deuteron recoil and noting that the incident neutron energy $\varepsilon \ll B$, calculate k in fm^{-1} and show that it is possible to make the approximation $\exp(-i\vec{k}\cdot\vec{r}) \simeq 1$.

3. Justify the various factors in the following expression for the cross section, where Ω is the emission direction of the photon with wave vector $\vec{k}$ and $\mathcal{F}$ is the incident neutron flux:

$$\frac{d\sigma}{d\Omega} = \frac{2\pi}{\hbar\mathcal{F}}|\langle f|W|i\rangle|^2\delta(\hbar\omega - (E_i - E_f))\frac{V\omega^2 d\omega}{(2\pi)^3 c^3}$$

with

$$W = -\frac{i}{c}\sqrt{\frac{\hbar\omega}{2\varepsilon_0 V}}\,\vec{\mu}\cdot\vec{e}_\lambda'(\hat{k}), \quad \vec{e}_\lambda'(\hat{k}) = \hat{k}\times\vec{e}_\lambda.$$

Here $|i\rangle$ is the initial n–p state and $|f\rangle$ is the deuteron state.

4. The matrix element $\langle f|W|i\rangle$ breaks up into a spin part and a spatial part, because the total state vector $\Psi_{i,f}$ is a product of the spin vector $\chi_{i,f}$ and the spatial wave function $\psi_{i,f}(\vec{r})$:

$$\Psi_{i,f} = \psi_{i,f}(\vec{r})\,\chi_{i,f}.$$

(a) If χ_t^m, $m = \pm 1, 0$, and χ_s denote the triplet and singlet spin states, the spin part of $\langle f|W|i\rangle$ will be

$$W_{\text{spin}} = \frac{1}{4}\mu_N\langle\chi_f|(g_p + g_n)(\vec{\sigma}_p + \vec{\sigma}_n)\cdot\vec{e}_\lambda' + (g_p - g_n)(\vec{\sigma}_p - \vec{\sigma}_n)\cdot\vec{e}_\lambda'|\chi_i\rangle,$$

where $\chi_f = \chi_t^m$ and $\chi_i = \chi_t^m$ or χ_s. Show that

$$(\vec{\sigma}_p + \vec{\sigma}_n)|\chi_s\rangle = 0, \quad \langle\chi_s|\vec{\sigma}_p|\chi_s\rangle = 0.$$

(b) The spatial part will involve the integral

$$I_{fi} = \int d^3r\,\psi_f^*(\vec{r})\psi_i(\vec{r}) = \int d^3r\,\psi_D^*(\vec{r})\psi_i(\vec{r}).$$

Show without calculation that $I_{fi} = 0$ if ψ_i and ψ_f are the $L = 0$ wave functions of the triplet state. Calculate I_{fi} explicitly if ψ_i is a singlet wave function using the approximations of question 1.

5. The above results can be summarized as

$$W_{\text{spin}} = \frac{1}{4}\mu_N(g_p - g_n)\langle\chi_t^m|(\vec{\sigma}_p - \vec{\sigma}_n)\cdot\vec{e}_\lambda'|\chi_s\rangle$$

$$\rightarrow \frac{1}{2}\mu_N(g_p - g_n)\langle\chi_t^m|\vec{\sigma}_p\cdot\vec{e}_\lambda'|\chi_s\rangle = \frac{1}{2}\mu_N(g_p - g_n)W_{\text{spin}}'.$$

Atomic physics

It is necessary to square this, sum over the final photon polarizations ($\sum_\lambda$), sum over the final deuteron spins ($\sum_m$), and average over the initial spins (the factor of 1/4). Show that

$$\langle |W'_{\text{spin}}|^2 \rangle = \frac{1}{4} \sum_m \sum_\lambda |\langle \chi_t^m | \vec{\sigma}_p \cdot \vec{e}_\lambda{}' | \chi_s \rangle|^2$$

$$= \frac{1}{4} \sum_{i,j} (\delta_{ij} - \hat{k}_i \hat{k}_j) \langle \chi_s | \sigma_{pi} \sigma_{pj} | \chi_s \rangle.$$

Hint: show that $\sum_m |\chi_t^m\rangle \langle \chi_t^m|$ can be replaced by the identity operator in spin space. Obtain the result $\langle |W'_{\text{spin}}|^2 \rangle = 1/2$.

6. Assemble all these factors to show that

$$\frac{1}{4} \sum_{\text{spins}} \sum_\lambda |\langle f | W | i \rangle|^2 = \frac{\hbar}{16\varepsilon_0 V} \frac{\omega}{c^2} \mu_N^2 (g_p - g_n)^2 I_{fi}^2.$$

Taking into account the normalization of the spatial wave functions, it can be shown that the flux factor is $\mathcal{F} = \sqrt{2\varepsilon/M}$. Derive the total cross section for the capture reaction ($\alpha = q_p^2/4\pi\varepsilon_0\hbar c$):

$$\sigma_{\text{tot}} = \int d\Omega \frac{d\sigma}{d\Omega} = \frac{\alpha\pi\hbar^2}{2c^4} \frac{B^{3/2}}{\sqrt{2\varepsilon}} \frac{1}{M^3} (g_p - g_n)^2 (1 - \kappa a_s)^2.$$

Compare to the experimental result for thermal neutrons at 300 K:

$$\sigma_{\text{tot}} = 0.329 \pm 0.006 \times 10^{-28} \text{ m}^2 = 32.9 \pm 0.6 \text{ fm}^2.$$

14.7 Further reading

Perturbation theory and the variational method are described in all the classic texts. A source for further details about the energy level structure is Cohen-Tannoudji *et al.* [1977]: fine structure, Chapter XII; Zeeman effect, Complement D_{VII}; hyperfine structure, Chapter XII. See also B. Bransden and C. Joachain, *Physics of Atoms and Molecules*, Harlow: Longman Scientific and Technical (1983). Cohen-Tannoudji's course, 'Atomic motion in laser light', in *Optical Coherence and Quantum Optics*, Les Houches School, Amsterdam North-Holland (1992), contains a very complete discussion of the laser manipulation of atoms. See also D. Suter, *The Physics of Laser–Atom Interactions*, Cambridge: Cambridge University Press (1997). The helium atom is treated in great detail by Cohen-Tannoudji *et al.* [1977], Complement B_{XIV}.

15

Open quantum systems

Most textbooks on quantum mechanics deal exclusively, or almost exclusively, with the time evolution of closed systems, and up to now this book has been no exception, apart from a glimpse of nonunitary evolution in Section 14.4.1. The time evolution of closed systems is governed by the Schrödinger equation (4.11) or its integral form (4.14). However, a closed system is an idealization, and in practice all quantum systems (except maybe the Universe as a whole) are in contact with some kind of environment. The Hilbert space of states is then a tensor product $\mathcal{H}_A \otimes \mathcal{H}_E$, where $\mathcal{H}_A$ ($\mathcal{H}_E$) is the Hilbert space of states of the system $\mathcal{A}$ (environment $\mathcal{E}$). In Chapter 6 we learned that the state operator ρ_A of $\mathcal{A}$ is obtained by taking the trace over the degrees of freedom of $\mathcal{E}$ (see (6.30)), and the time evolution of ρ_A is not unitary: it is not governed by (6.37) with a Hermitian Hamiltonian. The von Neumann entropy $\text{Tr}[\rho_A \ln \rho_A]$, which is constant for unitary evolution, is time-dependent when the system is not closed. In general, it increases because information is leaking into the environment, and irreversible behavior is observed because we are not able to control the degrees of freedom of the environment. As just mentioned, in Section 14.4.1 we gave a first example of nonunitary evolution, the system being a two-level atom and the environment the quantized electromagnetic field. In the present chapter we wish to give a general approach to the theory of quantum systems which are not closed, or *open quantum systems*.

Let us introduce the subject by looking at a specific (but very important) case, the time evolution of an open two-level system. In order that consistent notation be used throughout this chapter, we borrow the notation of quantum information (Section 6.4.2) and call $|0\rangle$ and $|1\rangle$ the basis vectors of the two-level system, with a "free" Hamiltonian H_0

$$H_0 = -\frac{1}{2} \hbar \omega_0 \sigma_z, \tag{15.1}$$

so that the eigenstates of H_0 are $|0\rangle$ and $|1\rangle$:

$$H_0|0\rangle = -\frac{1}{2} \hbar \omega_0 |0\rangle, \quad H_0|1\rangle = \frac{1}{2} \hbar \omega_0 |1\rangle. \tag{15.2}$$

Then $\hbar \omega_0$ is the energy difference between the ground and excited states. The matrix elements ρ_{00} and $\rho_{11} = 1 - \rho_{00}$ of the state operator describe the populations of levels

507

$|0\rangle$ and $|1\rangle$, while $\rho_{01} = \rho_{10}^*$ describes the coherences. At thermal equilibrium with temperature T, the populations ρ_{00}^{eq} and ρ_{11}^{eq} are fixed by Boltzmann's law

$$\frac{\rho_{11}^{\mathrm{eq}}}{\rho_{00}^{\mathrm{eq}}} = \exp\!\left(-\frac{\hbar\omega_0}{k_{\mathrm{B}}T}\right). \tag{15.3}$$

For the sake of definiteness, we specialize to the NMR case (Section 5.2). If the proton spins were isolated, their time evolution would be governed by (6.37), where the Hamiltonian depends on the constant magnetic field $\vec{B}_0$ and the radiofrequency field $\vec{B}_1(t)$. As in Section 14.4.1, it is convenient to use the Bloch vector $\vec{b} = (u, v, -w)$ (6.24); $w = (\rho_{11} - \rho_{00})$ describes the population difference and (u, v) the coherences, $\rho_{01} = r = (u - iv)/2$. If the spins were isolated from any kind of environment, the evolution equation (6.37) for ρ with the Hamiltonian (5.23) in terms of populations and coherences would read as

$$\dot{w} = i\omega_1\left(r^* e^{i\omega t} - r\,e^{-i\omega t}\right),$$
$$\dot{r} = i\omega_0 r + \frac{i\omega_1}{2}\,w\,e^{i\omega t}, \tag{15.4}$$

where ω_1 is the Rabi frequency. The slight differences from (14.80)–(14.81) drop out in the rotating-wave approximation. In order to take into account the interaction with the environment in a phenomenological way, we follow Section 14.4.1 and supplement these equations by two relaxation terms

$$\dot{w} = i\omega_1\left(r^* e^{i\omega t} - r\,e^{-i\omega t}\right) - \Gamma_1(w - w_{\mathrm{eq}}),$$
$$\dot{r} = i\omega_0 r + \frac{i\omega_1}{2}\,w\,e^{i\omega t} - \Gamma_2\,r. \tag{15.5}$$

These equations are the *Bloch equations* of NMR. The form of the relaxation term is not the most general one, but the approximations leading to (15.5) are usually justified: see the comments following (15.113). In order to give a physical interpretation of the new terms, let us assume that the radiofrequency field has been switched off at $t = 0$, so that $\omega_1 = 0$ for $t > 0$. Then the solution of (15.5) is

$$w(t) - w_{\mathrm{eq}} = [w(t = 0) - w_{\mathrm{eq}}]\,e^{-\Gamma_1 t},$$
$$r(t) = r(t = 0)\,e^{i\omega_0 t}\,e^{-\Gamma_2 t}. \tag{15.6}$$

The populations return to equilibrium with a relaxation time $T_1 = 1/\Gamma_1$, the longitudinal relaxation time, and the coherences with a relaxation time $T_2 = 1/\Gamma_2$, the transverse relaxation time introduced in Section 5.2. The main difference from (14.84)–(14.85) is that we now have two independent relaxation times,[1] while in (14.82)–(14.83) we had $\Gamma_2 = \Gamma_1/2 = \Gamma/2$. In the NMR case, T_1 and T_2 are of the order of a few seconds, and $T_2 \lesssim T_1$ (with $T_2 \ll T_1$ in most cases, for example $T_2 \sim 1\,\mathrm{ms}$ and $T_1 \sim 1\,\mathrm{s}$; see Levitt [2001]).

---

[1] Bloch equations with two independent relaxation times are also encountered in laser physics; see, e.g., Mandel and Wolf [1995], Chapter 18.

The chapter is organized as follows. In Section 15.1 we give some additional results on entanglement to supplement the more elementary approach of Chapter 6 by introducing the Schmidt decomposition of entangled states and the concept of positive operator-valued measure (POVM). Section 15.2 is devoted to establishing the general expression for the reduced state operator at time t as a function of its value at time $t = 0$, which we shall write in the Kraus form. Section 15.3 will address the particular but very important case where one is able to write the time evolution of the state operator in the form of a first-order differential equation in time, called a master equation. Finally, Section 15.4 will be devoted to the study of two models where the system of interest interacts with a thermal bath of harmonic oscillators. The first example will be that of a two-level atom and the second that of a Brownian particle. We shall derive master equations in both cases and examine their physical implications. The case of Brownian motion will be particularly important, as there we shall be able to understand the decoherence of the initially coherent superposition of two wave packets in the case of heavy particles, an example of a Schrödinger's cat.

15.1 Generalized measurements

15.1.1 Schmidt's decomposition

In this subsection, we give some further mathematical results on entangled states living in a Hilbert space of states[2] $\mathcal{H}_A \otimes \mathcal{H}_B$, in order to supplement the discussion of Chapter 6. Here $\mathcal{H}_A$ and $\mathcal{H}_B$, of dimensions N_A and N_B, are the Hilbert spaces of states of $\mathcal{A}$ and $\mathcal{B}$. The full state operator acting in $\mathcal{H}_A \otimes \mathcal{H}_B$ is denoted ρ_{AB}. We use Latin indices for $\mathcal{H}_A$ and Greek indices for $\mathcal{H}_B$, so that the matrix elements of ρ_{AB} are $\rho_{m\mu;n\nu}^{AB}$.[3] We have seen in (6.30) that the reduced state operator ρ_A of A is obtained by taking the trace over the B variables:

$$\rho_A = \mathrm{Tr}_\mathcal{B}\rho_{AB}, \qquad \rho_{mn}^A = \sum_\mu \rho_{m\mu;n\mu}^{AB}. \tag{15.7}$$

Let $|\varphi_{AB}\rangle \in \mathcal{H}_A \otimes \mathcal{H}_B$ be a pure state of the coupled $\mathcal{AB}$ system, and let $\{|m_a\rangle\}$ and $\{|\mu_B\rangle\}$ be two othonormal bases of $\mathcal{H}_A$ and $\mathcal{H}_B$. The most general decomposition of $|\varphi_{AB}\rangle$ on the basis $\{|m_A \otimes \mu_B\rangle\}$ of $\mathcal{H}_A \otimes \mathcal{H}_B$ reads

$$|\varphi_{AB}\rangle = \sum_{m,\mu} c_{m\mu}|m_A \otimes \mu_B\rangle. \tag{15.8}$$

Defining the vectors $|\tilde{m}_B\rangle \in \mathcal{H}_B$ as

$$|\tilde{m}_B\rangle = \sum_\mu c_{m\mu}|\mu_B\rangle,$$

---

[2] For the time being we do not think of system $\mathcal{B}$ as necessarily being an environment $\mathcal{E}$ for system $\mathcal{A}$.
[3] For clarity of notation, we use superscript AB when writing matrix elements.

we can rewrite (15.8) as

$$|\varphi_{AB}\rangle = \sum_m |m_A \otimes \tilde{m}_B\rangle. \tag{15.9}$$

Note that the set $\{|\tilde{m}_B\rangle\}$ need not form an orthonormal basis of $\mathcal{H}_B$. Now let us choose as a basis of $\mathcal{H}_A$ a set $\{|m_A\rangle\}$ which diagonalizes the reduced state operator ρ_A:

$$\rho_A = \mathrm{Tr}_{\mathcal{B}}|\varphi_{AB}\rangle\langle\varphi_{AB}| = \sum_{m=1}^{N_S} \mathsf{p}_m|m_A\rangle\langle m_A|. \tag{15.10}$$

If the number N_S of nonzero coefficients p_m is smaller than the dimension N_A of $\mathcal{H}_A$, we complete the set $\{|m_A\rangle\}$ by a set of $(N_A - N_S)$ orthonormal vectors, chosen to be orthogonal to the space spanned by the vectors $|m_A\rangle$ in (15.10). We use (6.34) to compute ρ_A from (15.9):

$$\rho_A = \sum_{m,n}\langle\tilde{n}_B|\tilde{m}_B\rangle|m_A\rangle\langle n_A|. \tag{15.11}$$

On comparing (15.10) and (15.11) we see that

$$\langle\tilde{n}_B|\tilde{m}_B\rangle = \mathsf{p}_m\delta_{mn},$$

and with our choice of basis $\{|m_A\rangle\}$ it turns out that the vectors $\{|\tilde{m}_B\rangle\}$ are, after all, orthogonal. To obtain an orthonormal basis, we only need to rescale the vectors $|\tilde{n}_B\rangle$

$$|n_B\rangle = \mathsf{p}_n^{-1/2}|\tilde{n}_B\rangle,$$

where we may assume that $\mathsf{p}_n > 0$ because, as explained above, it is always possible to complete the basis of $\mathcal{H}_B$ by a set of $(N_B - N_S)$ orthonormal vectors. We finally obtain Schmidt's decomposition of $|\varphi_{AB}\rangle$ on an orthonormal basis of $\mathcal{H}_A \otimes \mathcal{H}_B$:

$$\boxed{|\varphi_{AB}\rangle = \sum_n \mathsf{p}_n^{1/2}|n_A \otimes n_B\rangle.} \tag{15.12}$$

Any pure state $|\varphi_{AB}\rangle$ may be written in the form (15.12), but the bases $\{|n_A\rangle\}$ and $\{|n_B\rangle\}$ will of course depend on the state under consideration. If some of the p_n are equal, then the decomposition (15.12) is not unique, as is the case for the spectral decomposition of a Hermitian operator with degenerate eigenvalues. The reduced state operator ρ_B is readily computed from (6.35) using the orthogonality condition $\langle m_A|n_A\rangle = \delta_{mn}$:

$$\rho_B = \mathrm{Tr}_A|\varphi_{AB}\rangle\langle\varphi_{AB}| = \sum_n \mathsf{p}_n|n_B\rangle\langle n_B|. \tag{15.13}$$

Comparing (15.10) and (15.13), we see that ρ_A and ρ_B have the same eigenvalues. The *Schmidt number* N_S is the number of nonzero eigenvalues of ρ_A (or ρ_B). A state $|\varphi_{AB}\rangle$ is a tensor product if and only if its Schmidt number is exactly equal to one. It is entangled whenever $N_S \geq 2$. If $N_A = N_B = N$, a *maximally entangled state* corresponds to $N_S = N$, $\mathsf{p}_n = 1/N$:

$$|\varphi_{AB}^{\max}\rangle = \frac{1}{\sqrt{N}}\sum_n e^{i\alpha(n)}|n_A \otimes n_B\rangle, \tag{15.14}$$

where $\exp[i\alpha(n)]$ is a phase factor. The Bell states

$$|\Phi_\pm\rangle = \frac{1}{\sqrt{2}}\left(|0_A \otimes 0_B\rangle \pm |1_A \otimes 1_B\rangle\right),$$

$$|\Psi_\pm\rangle = \frac{1}{\sqrt{2}}\left(|0_A \otimes 1_B\rangle \pm |1_A \otimes 0_B\rangle\right) \tag{15.15}$$

provide an example of maximally entangled states for $N_A = N_B = 2$. It can be verified directly that maximally entangled states have the property that the individual reduced state operators are proportional to the identity operators I_A and I_B. An important result is that a *local evolution* described by a unitary operator of the form $U_A \otimes U_B$ does not change the Schmidt number, because

$$(U_A \otimes U_B)|\varphi_{AB}\rangle = \sum_n \mathsf{p}_n^{1/2}|n'_A \otimes n'_B\rangle$$

with

$$|n'_A\rangle = U_A|n_A\rangle, \quad |n'_B\rangle = U_B|n'_B\rangle.$$

As a consequence, *a product state* (tensor product) *cannot be transformed into an entangled state through a local evolution* in which systems $\mathcal{A}$ and $\mathcal{B}$ evolve independently.[4] One needs nonlocal evolution, involving an interaction between the two systems, in order to entangle a state which is initially a product state. Conversely, one needs nonlocal evolution to disentangle an entangled state into a product state.

15.1.2 Positive operator-valued measures

In Chapter 4 we defined a maximal test of a quantum system whose state vector lives in a Hilbert space of dimension N as being a test with exactly N mutually exclusive outcomes, whose probabilities add up to one. Mathematically, a maximal test corresponds to defining N one-dimensional orthogonal projectors $\mathcal{P}_a$ adding up to the identity operator:

$$\mathcal{P}_a\mathcal{P}_b = \delta_{ab}, \quad \sum_{a=1}^{N}\mathcal{P}_a = I. \tag{15.16}$$

Because its eigenvalues are zero and one, $\mathcal{P}_a$ is a positive operator. If a physical property M_A of system A with nondegenerate eigenvalues λ_a is built up as

$$M_A = \sum_{a=1}^{N}\lambda_a\mathcal{P}_a,$$

then measuring M_A is equivalent to performing a maximal test. A set of projectors (15.16) is called a *von Neumann*, or *orthogonal*, *measurement*. Let ρ be the initial state operator

---

[4] In this chapter, "local" and "nonlocal" have the following meanings. Acting locally on a particle means that there is no interaction with the other particles, for example because the particle is far away from the others. Acting nonlocally means that there must be an interaction between this particle and other particles of the ensemble.

of a quantum system and let us perform a von Neumann measurement of M_A with result λ_a (or simply a). Recall from Section 6.2.5 that the probability $\mathsf{p}(a)$ of obtaining result a is

$$\mathsf{p}(a) = \mathrm{Tr}\,(\rho\,\mathcal{P}_a). \tag{15.17}$$

Then from the WFC postulate in the form given in Section 6.2.5 the state operator is transformed into

$$\rho \to \rho'' = \frac{\mathcal{P}_a\,\rho\,\mathcal{P}_a}{\mathrm{Tr}\,(\mathcal{P}_a\,\rho\,\mathcal{P}_a)}. \tag{15.18}$$

The denominator in (15.18) ensures that $\mathrm{Tr}\,\rho'' = 1$. If the measurement is not read (if λ_a is not observed), then the measurement destroys the coherences (see Appendix B):

$$\rho \to \rho' = \sum_{a=1}^{N} \mathcal{P}_a\,\rho\,\mathcal{P}_a. \tag{15.19}$$

The most efficient way of obtaining information on a quantum system is not always a von Neumann measurement (or a maximal test). We shall introduce generalized measurements by incorporating system $\mathcal{A}$ into a larger system $\mathcal{AB}$ and performing a joint measurement of a physical property M_{AB} acting in $\mathcal{H}_A \otimes \mathcal{H}_B$, assuming that the quantum state of $\mathcal{AB}$ is prepared as a tensor product[5] $\rho_A \otimes \rho_B$. Let us write a complete set of orthogonal projectors $\mathcal{P}_a$ (15.16) acting in $\mathcal{H}_A \otimes \mathcal{H}_B$; the probability of outcome a is

$$\mathsf{p}(a) = \mathrm{Tr}_A\mathrm{Tr}_B[\mathcal{P}_a(\rho_A \otimes \rho_B)] = \mathrm{Tr}_A[\mathcal{Q}_a\rho_A], \tag{15.20}$$

with

$$\mathcal{Q}_a = \mathrm{Tr}_B[\mathcal{P}_a\rho_B], \tag{15.21}$$

or, in terms of matrix elements (see Footnote 3),

$$\mathcal{Q}^a_{mn} = \sum_{\mu,\nu} \mathcal{P}^a_{m\mu;n\nu}\,\rho^B_{\nu\mu}. \tag{15.22}$$

The operators $\mathcal{Q}_a$ act in $\mathcal{H}_A$, and it is easy to check the following properties.

1. Hermiticity: $\mathcal{Q}_a = \mathcal{Q}_a^{\dagger}$. Indeed,

$$(\mathcal{Q}^a_{nm})^* = \sum_{\mu,\nu}(\mathcal{P}^a_{n\mu;m\nu})^*\,(\rho^B_{\nu\mu})^* = \sum_{\mu,\nu}\mathcal{P}^a_{m,\nu;n\mu}\,\rho^B_{\mu\nu} = \mathcal{Q}^a_{mn},$$

   where we have used the Hermiticity of $\mathcal{P}_a$ and ρ_B.
2. Positivity: $\mathcal{Q}_a \geq 0$. In a basis which diagonalizes ρ_B

$$\rho_B = \sum_{\mu} \mathsf{p}_\mu|\mu_B\rangle\langle\mu_B|,$$

---

[5] However, the space of states of $\mathcal{AB}$ need not be a tensor product. From a mathematical point of view, the space of states may be a direct sum $\mathcal{H}_A \oplus \mathcal{H}_B$ (see Exercise 15.5.1), although, in practice, it seems difficult to implement the POVM in that case. We shall therefore limit our discussion to the case of tensor products.

we have

$$\langle \psi_A | \mathcal{Q}_a | \psi_A \rangle = \sum_\mu \mathsf{p}_\mu \langle \psi_A \otimes \mu_B | \mathcal{P}_a | \psi_A \otimes \mu_B \rangle \geq 0$$

because $\mathcal{P}_a$ is a positive operator.

3. Completeness:

$$\sum_a \mathcal{Q}_a = \mathrm{Tr}_{\mathcal{B}} \left[\sum_a (\mathcal{P}_a) \rho_B \right] = I_A,$$

because $\sum_a \mathcal{P}_a = I_{AB}$ and $\mathrm{Tr}_{\mathcal{B}} \rho_B = 1$.

In contrast to the projectors (15.16), the $\mathcal{Q}_a$ need not be orthogonal: $\mathcal{Q}_a \mathcal{Q}_b \neq \delta_{ab}$. In general, one defines a *positive operator-valued measure* (POVM) as a set of operators $\mathcal{Q}_a$ acting in $\mathcal{H}_a$ which obey

$$\boxed{ \mathcal{Q}_a = \mathcal{Q}_a^\dagger, \quad \mathcal{Q}_a \geq 0, \quad \sum_a \mathcal{Q}_a = I. } \tag{15.23}$$

We can now generalize (15.17)–(15.19) to the POVM case. From (15.20) the probability of result a is

$$\mathsf{p}(a) = \mathrm{Tr}(\mathcal{Q}_a \rho). \tag{15.24}$$

If the measurement is performed but the result is not read, the state operator transforms as

$$\rho \to \rho' = \sum_a \mathcal{Q}_a \rho \mathcal{Q}_a, \tag{15.25}$$

while if the result of the measurement is read

$$\rho \to \rho'' = \frac{\mathcal{Q}_a \rho \mathcal{Q}_a}{\mathrm{Tr}(\mathcal{Q}_a \rho \mathcal{Q}_a)}. \tag{15.26}$$

We have introduced the POVM starting from an orthogonal measurement in $\mathcal{H}_A \otimes \mathcal{H}_B$. This is indeed the most general case, at least if the POVM involves rank-one operators: it follows from Neumark's theorem[6] that any POVM defined by (15.23) in $\mathcal{H}_A$ can be realized as a von Neumann measurement in some Hilbert space $\mathcal{H}_A \otimes \mathcal{H}_B$.

15.1.3 Example: a POVM with spins 1/2

Let us give as an example a POVM with spins 1/2. Let $\{\hat{n}_\alpha\}$ be a set of unit vectors in $\mathbb{R}^3$ and $\{c_\alpha\}$ a set of real coefficients such that

$$\sum_\alpha c_\alpha \hat{n}_\alpha = 0, \quad 0 \leq c_\alpha \leq 1, \quad \sum_\alpha c_\alpha = 1, \tag{15.27}$$

---

[6] See, e.g., Peres [1993], Chapter 9 for a proof.

and let us define the following operators for a spin 1/2:

$$\mathcal{Q}_\alpha = c_\alpha(I + \vec{\sigma} \cdot \hat{n}_\alpha) = 2c_\alpha \mathcal{P}(\hat{n}_\alpha), \tag{15.28}$$

where $\mathcal{P}(\hat{n}_\alpha)$ is the projector on the spin state $|\hat{n}_\alpha\rangle$, which is an eigenvector of $(\vec{\sigma} \cdot \hat{n}_\alpha)$ with eigenvalue $+1$:

$$(\vec{\sigma} \cdot \hat{n}_\alpha)|\hat{n}_\alpha\rangle = |\hat{n}_\alpha\rangle.$$

The state $|\hat{z}\rangle$ is identified with $|0\rangle$, and the state $|\hat{n}_\alpha\rangle$ is obtained from $|0\rangle$ by the rotation of angle θ_α around the y axis which brings the vertical unit vector $\hat{z}$ onto $\hat{n}_\alpha$:

$$|\hat{n}_\alpha\rangle = \exp\left(-\frac{i}{2}\theta_\alpha\sigma_y\right)|0\rangle. \tag{15.29}$$

From (15.27) one sees that the $\mathcal{Q}_\alpha$ are positive operators which obey the completeness relation $\sum_\alpha \mathcal{Q}_\alpha = I$, but are not in general orthogonal. They are therefore an example of a POVM. The simplest illustration of a POVM that is not a von Neumann measurement is obtained by choosing three vectors $\{\hat{n}_\alpha\} = (\hat{n}_a, \hat{n}_b, \hat{n}_c)$ with, for example,

$$c_a = c_b = c_c = \frac{1}{3}, \quad \hat{n}_a + \hat{n}_b + \hat{n}_c = 0.$$

Then the $\mathcal{Q}_\alpha$ are

$$\mathcal{Q}_\alpha = \frac{1}{3}(I + \vec{\sigma} \cdot \hat{n}_\alpha) = \frac{2}{3}\mathcal{P}(\hat{n}_\alpha). \tag{15.30}$$

If we choose unit vectors $\hat{n}_\alpha$ in the xOz plane, a possible symmetric choice is as follows: $\hat{n}_a$ lying along the z axis and $\hat{n}_b$ and $\hat{n}_c$ making angles of $4\pi/3$ and $8\pi/3$ with the z axis, so that (15.29) leads to

$$|a\rangle := |\hat{n}_a\rangle = |0\rangle,$$

$$|b\rangle := |\hat{n}_b\rangle = -\frac{1}{2}|0\rangle + \frac{\sqrt{3}}{2}|1\rangle, \tag{15.31}$$

$$|c\rangle := |\hat{n}_c\rangle = -\frac{1}{2}|0\rangle - \frac{\sqrt{3}}{2}|1\rangle.$$

Our first goal is to give an explicit verification of Neumark's theorem by constructing the POVM (15.28) from orthogonal projectors in a larger space, a space $\mathcal{H}_A \otimes \mathcal{H}_B$ of two spins 1/2. The auxiliary spin, $\mathcal{B}$, is called an *ancilla*. We build the following orthonormal basis of entangled states in $\mathcal{H}_A \otimes \mathcal{H}_B$, $\alpha = (a, b, c)$:

$$|\alpha_{AB}\rangle = \sqrt{\frac{2}{3}}|\alpha_A \otimes 0_B\rangle + \sqrt{\frac{1}{3}}|0_A \otimes 1_B\rangle, \tag{15.32}$$

$$|\beta_{AB}\rangle = |1_A \otimes 1_B\rangle.$$

The orthogonality of the basis is easily checked by using the scalar products:

$$\langle a|b\rangle = \langle a|c\rangle = \langle b|c\rangle = -\frac{1}{2}.$$

Let us call $\mathcal{P}_\alpha$, $\alpha = a, b, c$, the set of orthogonal projectors on the basis vectors (15.32):

$$\mathcal{P}_\alpha = |\alpha_{AB}\rangle\langle\alpha_{AB}|, \quad \mathcal{P}_\beta = |\beta_{AB}\rangle\langle\beta_{AB}|,$$

and let us choose spin $\mathcal{B}$ in the state $|0_B\rangle$, $\rho_B = |0_B\rangle\langle 0_B|$. Then we find for the POVM $\mathcal{Q}_\alpha$ and $\mathcal{Q}_\beta$

$$\mathcal{Q}_\alpha = \mathrm{Tr}_{\mathcal{B}}(\rho_B\mathcal{P}_\alpha) = \frac{2}{3}|\alpha_A\rangle\langle\alpha_A|,$$

$$\mathcal{Q}_\beta = 0. \tag{15.33}$$

These equations give an explicit verification of Neumark's theorem in this particular case: we have been able to construct the set $\{\mathcal{Q}_\alpha\}$ from a set of orthogonal projectors in $\mathcal{H}_A \otimes \mathcal{H}_B$.

Let us now describe a possible strategy to implement the POVM. Define the unit vector $\hat{u}$ in the xOz plane making an angle θ with the z axis such that

$$\cos\frac{\theta}{2} = -\sqrt{\frac{1}{3}}, \quad \sin\frac{\theta}{2} = \sqrt{\frac{2}{3}}$$

and the spin states $|\hat{u}\rangle$ and $|-\hat{u}\rangle$

$$|\hat{u}\rangle = \exp\left(-\frac{i}{2}\theta\sigma_y\right)|0\rangle, \quad |-\hat{u}\rangle = \exp\left(-\frac{i}{2}\theta\sigma_y\right)|1\rangle.$$

The vectors $|\alpha_{AB}\rangle$, $\alpha = a, b, c$, may be written in terms of $|\pm\hat{u}_B\rangle$:

$$|a_{AB}\rangle = |0_A \otimes -\hat{u}_B\rangle,$$

$$|(b/c)_{AB}\rangle = \frac{1}{\sqrt{2}}|0_A \otimes \hat{u}_B\rangle \pm \frac{1}{\sqrt{2}}|1_A \otimes 0_B\rangle, \tag{15.34}$$

where the $+$ $(-)$ sign corresponds to b (c). To disentangle the states in the second line of (15.34), we use a basic component of quantum information, the control-U or cU gate, which has the following action in our particular case:[7]

$$cU|0_A \otimes 0_B\rangle = |0_A \otimes 0_B\rangle, \quad cU|0_A \otimes 1_B\rangle = |0_A \otimes 1_B\rangle,$$

$$cU|1_A \otimes 0_B\rangle = |1_A \otimes \hat{u}_B\rangle, \quad cU|1_A \otimes 1_B\rangle = |1_A \otimes -\hat{u}_B\rangle. \tag{15.35}$$

In other words, cU leaves spin $\mathcal{B}$ unchanged if spin $\mathcal{A}$ is in state $|0_A\rangle$, and it rotates spin $\mathcal{B}$ by an angle θ if spin $\mathcal{A}$ is in state $|1\rangle$. The unitary operator cU *is a nonlocal interaction: it is not a tensor product* $U_A \otimes U_B$. Let us apply cU to $|\alpha_{AB}\rangle$:

$$cU|a_{AB}\rangle = |0_A \otimes -\hat{u}_B\rangle = \frac{1}{\sqrt{2}}(|\hat{x}_A \otimes -\hat{u}_B\rangle + |-\hat{x}_A \otimes -\hat{u}_B\rangle),$$

$$cU|(b/c)_{AB}\rangle = \frac{1}{\sqrt{2}}(|0_A\rangle \pm |1_A\rangle) \otimes |\hat{u}_B\rangle = |\pm\hat{x}_A \otimes \hat{u}_B\rangle, \tag{15.36}$$

---

[7] A cU gate leaves spin $\mathcal{B}$ unchanged if spin $\mathcal{A}$ is in state $|0_A\rangle$, and it performs a unitary transformation $|\varphi_B\rangle \to U_B|\varphi_B\rangle$ on spin $\mathcal{B}$ if spin $\mathcal{A}$ is in state $|1_A\rangle$. The cU gate generalizes the cNOT gate defined in (6.73), which corresponds to the choice $U_B = \sigma_{Bx}$.

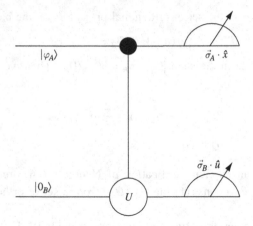

Fig. 15.1. Graphical representation of (15.36).

where the states $|\pm\hat{x}\rangle = (|0\rangle \pm |1\rangle)/\sqrt{2}$ are eigenvectors of $\sigma_x = (\vec{\sigma}\cdot\hat{x})$ with eigenvalues ± 1. If we measure $(\sigma_A \cdot \hat{x})$ and $(\vec{\sigma}_B \cdot \hat{u})$ after applying the cU gate (Fig. 15.1), the results of the measurements of the pair $(\sigma_A \cdot \hat{x}; \vec{\sigma}_B \cdot \hat{u})$ lead to the following correspondence (0 and 1 refer to the values of the qubits measured along $\hat{x}$ or $\hat{u}$):

$$(0; 1) \quad \text{and} \quad (1; 1) \rightarrow a,$$

$$(0; 0) \rightarrow b,$$

$$(1; 0) \rightarrow c.$$

Let us show that a POVM can in some cases give better results than a von Neumann measurement, in the sense that the former allows a better discrimination between different states of system $\mathcal{A}$ when these states are not orthogonal. Assume that Alice sends Bob a sequence of particles of spin 1/2 which are randomly distributed with equal probabilities in the states $|a_\perp\rangle$ and $|b_\perp\rangle$:[8]

$$|a_\perp\rangle = |1\rangle, \quad |b_\perp\rangle = |-\hat{n}_b\rangle = \frac{\sqrt{3}}{2}|0\rangle + \frac{1}{2}|1\rangle.$$

What is the best strategy that Bob can follow to tell *with certainty* whether a given spin was sent by Alice in state $|a_\perp\rangle$ or $|b_\perp\rangle$? Bob can perform a von Neumann measurement, by taking a Stern–Gerlach filter oriented along $\hat{z}$. If the spin is deflected upward, he can tell with certainty that spin $\mathcal{A}$ was in the state $|b_\perp\rangle$, and this occurs with probability 3/8. Thus, in 37.5% of the cases, Bob is able to decide with certainty between the states $|a_\perp\rangle$ and $|b_\perp\rangle$. He can do better by performing a POVM measurement, as we are going to

---

[8] We choose $|a_\perp\rangle$ and $|b_\perp\rangle$ rather than $|a\rangle$ and $|b\rangle$ in order to use the cU gate (15.35).

demonstrate. Bob entangles spin $\mathcal{A}$ with an ancilla spin $\mathcal{B}$ in the state $|0_B\rangle$ using the cU gate (15.35). An easy calculation gives

$$\text{cU}|1_A \otimes 0_B\rangle = \frac{1}{\sqrt{2}}|\hat{x}_A \otimes \hat{u}_B\rangle - \frac{1}{\sqrt{2}}|-\hat{x}_A \otimes \hat{u}_B\rangle,$$

$$\text{cU}|-\hat{n}_{bA} \otimes 0_B\rangle = -\frac{1}{\sqrt{2}}|-\hat{x}_A \otimes \hat{u}_B\rangle + \frac{1}{2}|\hat{x}_A \otimes \hat{u}_B\rangle + \frac{1}{2}|-\hat{x}_A \otimes -\hat{u}_B\rangle, \qquad (15.37)$$

$$\text{cU}|-\hat{n}_{cA} \otimes 0_B\rangle = -\frac{1}{\sqrt{2}}|\hat{x}_A \otimes \hat{u}_B\rangle + \frac{1}{2}|\hat{x}_A \otimes -\hat{u}_B\rangle + \frac{1}{2}|-\hat{x}_A \otimes -\hat{u}_B\rangle.$$

If spin $\mathcal{A}$ is in the initial state $|a_\perp\rangle = |1\rangle$, then we find the following probabilities when measuring the pair $(\sigma_A \cdot \hat{x}; \vec{\sigma}_B \cdot \hat{u})$:

$$\mathsf{p}(0;0) = \mathsf{p}(1;0) = \frac{1}{2}, \quad \mathsf{p}(0;1) = \mathsf{p}(1;1) = 0,$$

while if it is in the state $|b_\perp\rangle = |-\hat{n}_b\rangle$ we have

$$\mathsf{p}(0;0) = 0, \quad \mathsf{p}(1;0) = \frac{1}{2}, \quad \mathsf{p}(0;1) = \mathsf{p}(1;1) = \frac{1}{4}.$$

Then, if Bob's measurement gives $(0;0)$, he knows with certainty that spin $\mathcal{A}$ was initially in the state $|a_\perp\rangle$, while if he measures $(0;1)$ or $(1;1)$ he can be sure that it was in the state $|b_\perp\rangle$. If he measures $(1;0)$, he cannot decide. This occurs in 50% of the cases, so that he is able to distinguish between the two states with a 50% probability, instead of the 37.5% in the case of a von Neumann measurement. The same results are obtained by using the POVM $(\mathcal{Q}_a, \mathcal{Q}_b, \mathcal{Q}_c)$ (see Exercise 15.5.2). It can be shown that this is the best result Bob can achieve: a general theorem states that optimal POVMs consist of rank-one operators.[9]

15.2 Superoperators

15.2.1 Kraus decomposition

We have seen in the preceding section how an orthogonal measurement on a bipartite system whose state vector lies in a Hilbert space $\mathcal{H}_A \otimes \mathcal{H}_B$ is translated into a POVM on $\mathcal{A}$ alone. In the present section, which is, as we shall see later on, closely related to the preceding one, we shall attempt to answer the following question: if a state of $\mathcal{H}_A \otimes \mathcal{H}_B$ undergoes a unitary evolution U_{AB} from $t=0$ to t, what is the general expression for the (generally nonunitary) evolution of the state operator for $\mathcal{A}$? The answer is provided by the Kraus representation, which we are going to derive. We assume that the state operator at $t=0$ is a tensor product, with $\rho_B = |0_B\rangle\langle 0_B|$ a pure state, a kind of "reference state":

$$\rho_{AB}(t=0) = \rho_A \otimes \rho_B = \rho_A \otimes |0_B\rangle\langle 0_B|. \qquad (15.38)$$

---

[9] E. Davies, *IEEE Trans. Inform. Theory* **IT-24**, 596 (1978).

We shall comment on this apparently very restrictive assumption later on. The bipartite system $\mathcal{AB}$ evolves during a time interval t according to

$$\rho_{AB}(t=0) = \rho_{AB} \rightarrow \rho_{AB}(t) = \rho'_{AB} = U_{AB}\rho_{AB}(t=0)\,U^{\dagger}_{AB}, \qquad (15.39)$$

where U_{AB} is obtained by solving (4.17) in $\mathcal{H}_A \otimes \mathcal{H}_B$. In order to find the state operator $\rho_A(t) = \rho'_A$ of system A, we perform a partial trace (see Footnote 3):

$$\rho'_{mn}{}^{A} = \sum_{\mu} U^{AB}_{m\mu;k0}\,\rho^{A}_{kl}\,(U^{AB})^{\dagger}_{l0;n\mu}, \qquad (15.40)$$

where we have made explicit use of the peculiar form (15.38) of the initial state operator ρ_{AB}. The matrix elements of U_{AB} are

$$U^{AB}_{m\mu;n\nu} = \langle m_A \otimes \mu_B | U_{AB} | n_A \otimes \nu_B \rangle.$$

Equation (15.40) can be written in operator form by introducing the *superoperator M_μ* acting in $\mathcal{H}_A$ through

$$M_\mu = \langle \mu_B | U_{AB} | 0_B \rangle. \qquad (15.41)$$

Writing $\rho'_A = \mathcal{K}(\rho_A)$, (15.40) becomes (see Fig. 15.2)

$$\mathcal{K}(\rho_A) = \sum_{\mu} M_\mu\,\rho_A\,M^{\dagger}_{\mu}. \qquad (15.42)$$

The unitarity of U_{AB} implies that the set of superoperators M_μ obeys the completeness relation (note the order of the operators; $\sum_{\mu} M_\mu M^{\dagger}_{\mu}$ has no simple expression in the general case):

$$\sum_{\mu} M^{\dagger}_{\mu} M_\mu = \sum_{\mu} \langle 0_B | U^{\dagger}_{AB} | \mu_B \rangle \langle \mu_B | U_{AB} | 0_B \rangle = I_A. \qquad (15.43)$$

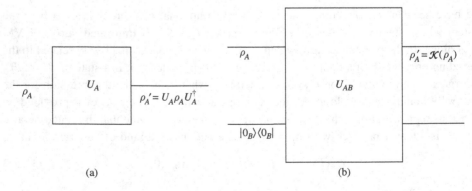

Fig. 15.2. Graphical representation of unitary evolution (a) and the evolution (15.42) (b).

Equation (15.42) is the *Kraus representation* of ρ'_A. This Kraus representation, together with the completeness relation (15.43), defines a linear map $\rho_A \to \rho'_A = \mathcal{K}(\rho_A)$. The operator ρ'_A obeys the three necessary conditions for being a bona fide state operator:

(i) ρ'_A is obviously Hermitian;
(ii) $\mathrm{Tr}\,\rho'_A = 1$ owing to (15.43);
(iii) ρ'_A is positive: indeed, with $|\psi^\mu_A\rangle = M^\dagger_\mu|\varphi_A\rangle$,

$$\langle\varphi_A|\rho'_A|\varphi_A\rangle = \sum_\mu((\langle\varphi_A|M_\mu)\rho_A(M^\dagger_\mu|\varphi_A\rangle)) = \sum_\mu\langle\psi^\mu_A|\rho_A|\psi^\mu_A\rangle \geq 0.$$

Conversely, any Kraus representation (15.42) can always be derived from a unitary representation in some Hilbert space $\mathcal{H}_A \otimes \mathcal{H}_B$, as we now show. Let us choose as $\mathcal{H}_B$ a Hilbert space whose dimension is at least the number of terms in (15.42), and let $\{|\mu_B\rangle\}$ be an orthonormal basis in $\mathcal{H}_B$ and $|0_B\rangle$ one particular vector of this basis. *Define* the action of U_{AB} on the vector $|\varphi_A \otimes 0_B\rangle$, where $|\varphi_A\rangle$ is an arbitrary vector of $\mathcal{H}_A$, as

$$U_{AB}|\varphi_A \otimes 0_B\rangle = \sum_\mu (M_\mu \otimes I_B)|\varphi_A \otimes \mu_B\rangle. \tag{15.44}$$

Equation (15.44) describes a *quantum jump*: in the time interval $[0, t]$, the AB system "jumps" from $|\varphi_A \otimes 0_B\rangle$ to a superposition of states $M_\mu|\varphi_A\rangle \otimes |\mu_B\rangle$. The operator U_{AB} preserves the scalar product

$$\left(\sum_\mu\langle\psi_A \otimes \mu_B|[M^\dagger_\mu \otimes I_B]\right)\left(\sum_\nu [M_\nu \otimes I_B]|\varphi_A \otimes \nu_B\rangle\right) = \langle\psi_A|\left(\sum_\mu M^\dagger_\mu M_\mu\right)|\varphi_A\rangle = \langle\psi_A|\varphi_A\rangle,$$

and therefore U_{AB}, which is a priori defined only on a subset of $\mathcal{H}_A \otimes \mathcal{H}_B$, can be extended as a unitary operator on the full $\mathcal{H}_A \otimes \mathcal{H}_B$. Taking a partial trace, we find

$$\mathrm{Tr}_\mathcal{B}\left(U_{AB}|\varphi_A \otimes 0_B\rangle\langle\varphi_A \otimes 0_B|U^\dagger_{AB}\right) = \sum_\mu M_\mu|\varphi_A\rangle\langle\varphi_A|M^\dagger_\mu,$$

so that any state operator ρ_A, which can be written as $\sum_i \mathsf{p}_i|\varphi^i_A\rangle\langle\varphi^i_A|$, transforms according to (15.42). We shall see later on that any "reasonable" evolution law for ρ_A is of the form (15.42). Then the fact that one can always find a unitary representation (15.44) is somewhat surprising at first sight: in principle, at $t = 0$, systems A and B are entangled, so that assuming an initial state of the form (15.38) looks like a very restrictive assumption. But it seems that, *for the purpose of describing the evolution of ρ_A*, it is always possible to find a model environment such that there is no initial entanglement between the system and its (fictitious) environment.

Let us conclude this subsection by stating some general results on the Kraus representation. As some of the proofs are rather technical, we shall omit them and refer the reader to the bibliography. One interesting question is the following: under what "reasonable" conditions can the Kraus representation (15.42) be proved? A priori, it would seem that one should require that

(i) $\mathcal{K}$ is a linear operation:[10]

$$\mathcal{K}(\lambda \rho_A + \mu \rho_B) = \lambda \mathcal{K}(\rho_A) + \mu \mathcal{K}(\rho_B).$$

(ii) $\mathcal{K}(\rho_A)$ is Hermitian:

$$\mathcal{K}(\rho_A) = [\mathcal{K}(\rho_A)]^\dagger.$$

(iii) $\mathcal{K}$ is trace-preserving: $\text{Tr}[\mathcal{K}(\rho_A)] = 1$.
(iv) $\mathcal{K}(\rho_A)$ is a positive operator: $\mathcal{K}(\rho_A) \geq 0$.

But condition (iv) is actually too weak. Suppose $\mathcal{A}$ is coupled to a system $\mathcal{B}$ and there is a third system $\mathcal{C}$, totally uncoupled to $\mathcal{A}$, of which we are unaware. If $\mathcal{A}$ evolves and $\mathcal{C}$ does not, then $\mathcal{K}(\rho_A) \otimes I_C$ must be a positive operator. Thus $\mathcal{K}(\rho_A)$ should obey not (iv) but the stronger condition (iv′):

(iv′) $\mathcal{K}(\rho_A)$ is *completely positive*: $\mathcal{K}(\rho_A) \otimes I_C \geq 0$, for any system $\mathcal{C}$.

An example of an operator that obeys (i) to (iii), but not (iv′), is the transposition (Exercise 15.5.4)

$$\mathcal{K}(\rho_A) = \rho_A^T.$$

We can now state without proof the *Kraus representation theorem*: any operator $\rho \rightarrow \mathcal{K}(\rho)$ in a space of dimension N which obeys the conditions (i) to (iii) and (iv′) can be written in the form

$$\boxed{\mathcal{K}(\rho) = \sum_{\mu=1}^{K} M_\mu \rho M_\mu^\dagger, \quad \sum_{\mu=1}^{K} M_\mu^\dagger M_\mu = I,} \tag{15.45}$$

where the number of terms in the sum is bounded by $K \leq N_A^2$, with N_A the dimension of $\mathcal{H}_A$; K is the *Kraus number*. There always exists an expression for $\mathcal{K}(\rho)$ with a number of terms $\leq N_A^2$, independently of the dimension of the Hilbert space $\mathcal{H}_B$ of the environment, even if this dimension is infinite.

The Kraus representation is not unique, but any two representations may be related through a unitary transformation: if

$$\mathcal{K}(\rho) = \sum_{\mu=1}^{K} M_\mu \rho M_\mu^\dagger = \sum_{\mu=1}^{L} N_\mu \rho N_\mu^\dagger,$$

then N_ν is related to M_μ by a unitary transformation:

$$N_\nu = \sum_\mu U_{\nu\mu} M_\mu.$$

---

[10] However, see, e.g., J. Preskill, *Quantum Computation*, http://www.theory.caltech.edu/˜ preskill/(1999), Section 3.2 for a discussion; the arguments that nonlinear evolution should be excluded are not entirely compelling.

As some of the matrix elements $U_{\nu\mu}$ may vanish, the number of nonzero terms need not be the same in both decompositions: it may happen that $K \neq L$.

Let us finally make the link with POVM by showing that a unitary transformation that entangles $\mathcal{A}$ with $\mathcal{B}$ followed by an orthogonal measurement on $\mathcal{B}$ can be described as a POVM. If a state $|\varphi_A \otimes 0_B\rangle$ evolves according to (15.44), then an orthogonal measurement on $\mathcal{B}$ which projects onto the $\{|\mu_B\rangle\}$ basis has a probability $\mathsf{p}(\mu)$ of finding result μ

$$\mathsf{p}(\mu) = \sum_{\nu,\tau} \mathrm{Tr}\Big[(I_A \otimes |\mu\rangle\langle\mu|) M_\nu |\varphi_A \otimes \nu\rangle\langle\varphi_A \otimes \tau| M_\tau^\dagger \Big]$$

$$= \langle\varphi_A| M_\mu^\dagger M_\mu |\varphi_A\rangle.$$

Writing the state operator of $\mathcal{A}$ as $\rho_A = \sum_i \mathsf{p}_i |i\rangle\langle i|$, we find in the general case

$$\mathsf{p}(\mu) = \mathrm{Tr}(\mathcal{Q}_\mu \rho_A), \quad \mathcal{Q}_\mu = M_\mu^\dagger M_\mu = \mathcal{Q}_\mu^\dagger \geq 0. \tag{15.46}$$

Furthermore, $\sum_\mu \mathcal{Q}_\mu = \sum_\mu M_\mu^\dagger M_\mu = I$, so that the $\mathcal{Q}_\mu$ form a POVM. Conversely, let $\mathcal{R}_\mu$ be a set of Hermitian and positive operators which obey $\sum_\mu \mathcal{R}_\mu = I$ and $\mathsf{p}(\mu) = \mathrm{Tr}(\mathcal{R}_\mu \rho)$. A POVM which modifies the state operator according to

$$\rho \rightarrow \rho' = \sum_\mu \sqrt{\mathcal{R}_\mu}\, \rho \sqrt{\mathcal{R}_\mu} \tag{15.47}$$

gives $\sqrt{\mathcal{R}_\mu}$ as a special case of superoperator. Then, from the unitary representation (15.44), one can find a unitary operator U_{AB} such that

$$U_{AB}|\varphi_A \otimes 0_B\rangle = \sum_\mu \sqrt{\mathcal{R}_\mu}\, |\varphi_A \otimes \mu_B\rangle. \tag{15.48}$$

By performing an orthogonal measurement on $\mathcal{B}$ which projects onto the basis $\{|\mu_B\rangle\}$, we obtain an implementation of the POVM. However, this is not in general the most economic way to proceed, because the dimension of $\mathcal{H}_B$ is at least K, the number of different POVMs. For example, in Section 15.1.3, we had $K = 3$, so that $N_B = 3$, while we were able to use an ancilla living in a two-dimensional Hilbert space, $N_B = 2$. Thus we have (at least) two ways of implementing a POVM: (i) associate with $\mathcal{A}$ an ancilla $\mathcal{B}$ and perform a *nonlocal* measurement on $\mathcal{AB}$; (ii) entangle $\mathcal{A}$ with $\mathcal{B}$ and perform a *local* measurement on $\mathcal{B}$.

Let us apply the notion of superoperators to three important examples of physical mechanisms leading to nonunitary evolution of a two-level system. In all three examples, conventionally called "channels," a two-level system is coupled to an environment $\mathcal{E}$, so that the unitary evolution takes place in a Hilbert space $\mathcal{H}_A \otimes \mathcal{H}_E$. In what follows we wish to think of $\mathcal{B}$ as an environment $\mathcal{E}$: $\mathcal{B} \rightarrow \mathcal{E}$. In the three examples we shall start from a unitary evolution in $\mathcal{H}_A \otimes \mathcal{H}_E$ in the form of quantum jumps, from which we shall derive the explicit form of the Kraus operators. Our three examples will be (i) the depolarizing channel; (ii) the phase-damping channel; (iii) the amplitude-damping channel.

The terminology will be justified in each of the corresponding subsections.

15.2.2 The depolarizing channel

In this example, $\mathcal{H}_E$ has dimension $N_E = 4$ and an orthonormal basis is formed by a reference state $|0_E\rangle$ and three states $|i_E\rangle$, $i = 1, 2, 3$. The quantum jump (15.44) is assumed to take the form

$$U_{AE}|\varphi_A \otimes 0_E\rangle = \sqrt{1-p}\,|\varphi_A \otimes 0_E\rangle + \sqrt{\frac{p}{3}}\left[\sum_{i=1}^{3}(\sigma_{iA} \otimes I_E)|\varphi_A \otimes i_E\rangle\right]. \tag{15.49}$$

Therefore, the initial state $|0_E\rangle$ is unchanged with probability $(1-p)$, and a change $|\varphi_A\rangle \to \sigma_i|\varphi_A\rangle$, which occurs with probability $p/3$, is accompanied by a change $|0_E\rangle \to |i_E\rangle$. On comparing (15.49) with (15.44) we find

$$M_0 = \sqrt{1-p}\,I, \quad M_i = \sqrt{\frac{p}{3}}\,\sigma_i. \tag{15.50}$$

These four superoperators obey the completeness relation (15.43)

$$\sum_{\mu=0}^{3} M_\mu^\dagger M_\mu = \left[(1-p)I + 3\frac{p}{3}I\right] = I,$$

where we have used $\sigma_i^2 = I$. The state operator of the system evolves according to (15.45):

$$\rho \to \mathcal{K}(\rho) = (1-p)\rho + \frac{p}{3}\sum_{i=1}^{3}(\sigma_i\rho\sigma_i). \tag{15.51}$$

Let us write the Bloch form (6.24) of the state operator with a Bloch vector $\vec{b}$ (recall that $\vec{b}$ is the polarization in the case of a spin 1/2)

$$\rho = \frac{1}{2}\left(I + \vec{\sigma}\cdot\vec{b}\right) = \frac{1}{2}\left(I + \sum_{j=1}^{3}\sigma_j b_j\right). \tag{15.52}$$

The identities (3.49) for the Pauli matrices lead to the relation

$$\sigma_i\sigma_j\sigma_i = 2\sigma_j\delta_{ij} - \sigma_j,$$

so that

$$\rho' = \frac{1}{2}\left(I + \vec{\sigma}\cdot\vec{b}'\right), \quad \vec{b}' = \left(1 - \frac{4p}{3}\right)\vec{b}. \tag{15.53}$$

This transformation corresponds to a simple rescaling of the polarization by a factor $(1-4p/3)$: if the initial state is a pure state with polarization $|\vec{b}| = 1$, we see that the polarization is reduced from $|\vec{b}| = 1$ to $|1-4p/3|$, hence the terminology *depolarizing channel*. Note that in all cases the norm of $\vec{b}$ is scaled down by a factor $|1-4p/3| \le 1$, and that the orientation of $\vec{b}$ changes if $3/4 < p \le 1$.

15.2.3 The phase-damping channel

In this case $\mathcal{H}_E$ is of dimension 3, and the unitary evolution (quantum jump) is assumed to be of the form

$$U_{AE}|0_A \otimes 0_E\rangle = \sqrt{1-\mathsf{p}}\,|0_A \otimes 0_E\rangle + \sqrt{\mathsf{p}}\,|0_A \otimes 1_E\rangle,$$
$$U_{AE}|1_A \otimes 0_E\rangle = \sqrt{1-\mathsf{p}}\,|1_A \otimes 0_E\rangle + \sqrt{\mathsf{p}}\,|1_A \otimes 2_E\rangle. \tag{15.54}$$

Unlike the preceding case, system $\mathcal{A}$ does not make any transition. The Kraus decomposition is readily written from (15.44):

$$M_0 = \sqrt{1-\mathsf{p}}\,I, \quad M_1 = \sqrt{\mathsf{p}}\begin{pmatrix} 1 & 0 \\ 0 & 0 \end{pmatrix}, \quad M_2 = \sqrt{\mathsf{p}}\begin{pmatrix} 0 & 0 \\ 0 & 1 \end{pmatrix}, \tag{15.55}$$

and the transformed state matrix is

$$\mathcal{K}(\rho) = (1-\mathsf{p})\rho + \mathsf{p}\begin{pmatrix} \rho_{00} & 0 \\ 0 & \rho_{11} \end{pmatrix} = \begin{pmatrix} \rho_{00} & (1-\mathsf{p})\rho_{01} \\ (1-\mathsf{p})\rho_{10} & \rho_{11} \end{pmatrix}. \tag{15.56}$$

We note that the operations affect only the coherences (the off-diagonal matrix elements of ρ), hence the terminology *phase-damping channel*. Furthermore, if we apply $\mathcal{K}$ twice we get

$$\mathcal{K}^2(\rho) = \mathcal{K}[\mathcal{K}(\rho)] = \begin{pmatrix} \rho_{00} & (1-\mathsf{p})^2\rho_{01} \\ (1-\mathsf{p})^2\rho_{10} & \rho_{11} \end{pmatrix}.$$

Assume now that the quantum jump takes place in a short time interval Δt, with a probability proportional to Δt: $\mathsf{p} = \Gamma\Delta t \ll 1$.[11] Let us write $t = n\Delta t$, $n \gg 1$, and make n iterations of $\mathcal{K}$:

$$\mathcal{K}^n(\rho) = \begin{pmatrix} \rho_{00} & (1-\mathsf{p})^n\rho_{01} \\ (1-\mathsf{p})^n\rho_{10} & \rho_{11} \end{pmatrix} \to \begin{pmatrix} \rho_{00} & \rho_{01}\,e^{-\Gamma t} \\ \rho_{10}\,e^{-\Gamma t} & \rho_{11} \end{pmatrix}. \tag{15.57}$$

The relaxation time of the coherences (the transverse relaxation time T_2 of NMR) is $T_2 = 1/\Gamma$. If the two-level system is prepared at $t = 0$ in a pure state which is a *coherent* superposition of $|0\rangle$ and $|1\rangle$

$$|\varphi\rangle = a|0\rangle + b|1\rangle, \quad \rho_{00} = 1 - \rho_{11} = |a|^2, \quad \rho_{01} = \rho_{10}^* = ab^*,$$

then, after a time $t \gg 1/\Gamma$, the quantum state is transformed from (15.57) into an *incoherent* superposition of $|0\rangle$ and $|1\rangle$:

$$t \gg 1/\Gamma: \rho(t) \to |a|^2|0\rangle\langle 0| + |b|^2|1\rangle\langle 1|.$$

As an application, let us give a heuristic discussion of the decoherence of a quantum superposition involving macroscopic systems. Let us identify $|0_A\rangle$ and $|1_A\rangle$ with the position eigenstates $|x\rangle$ and $|-x\rangle$ (or, more realistically, with narrow nonoverlapping

---

[11] Note that this is a rather bold assumption. In general, we expect *amplitudes* to be proportional to Δt if transitions take place toward a single state; see (5.62). One needs transitions to a continuous set of states, as in the Fermi Golden Rule (9.170), in order to obtain *probabilities* proportional to Δt.

wave packets centered at x and $-x$) of a "dust particle,"[12] that elastically scatters photons initially in state $|0_E\rangle$. Scattering by the dust particle in state $|x\rangle$ ($|-x\rangle$) sends the photons into states $|1_E\rangle$ ($|2_E\rangle$), while the dust particle remains in its initial state. If the distance $2|x|$ between the centers of the wave packets is large compared with the photon wavelength, the states $|1_E\rangle$ and $|2_E\rangle$ will be approximately orthogonal, $\langle 1_E|2_E\rangle \simeq 0$, because photon scattering is localized in space. We are therefore in the situation described by (15.54). If the dust particle is initially in a coherent superposition of the two wave packets

$$|\varphi\rangle = \frac{1}{\sqrt{2}}(|x\rangle + |-x\rangle),$$

the coherence between the wave packets will be destroyed after a time $\sim 1/\Gamma$. The relaxation rate Γ is proportional to the elastic cross section σ. A rough estimate for σ is $\sigma \propto R^2 \propto M^{2/3}$, where R is the radius of the dust particle and M is its mass. The decoherence time τ_{dec} is proportional to $M^{-2/3}$

$$\tau_{\text{dec}} \simeq \frac{1}{\Gamma} \propto \frac{1}{M^{2/3}}.$$

The decoherence time is much shorter than the damping time τ characteristic of the motion of the dust particle, the time taken by the dust particle to change its momentum under photon scattering: $\tau_{\text{dec}} \ll \tau$; as in Section 6.4.1, τ_{dec} is controlled by the scattering of one photon and τ by the scattering of a large number of photons. This result has important consequences for the Young's slit experiment: if the time taken by the particle to travel from the slits to the screen is larger than τ_{dec}, no interference is possible, because the coherence of the two wave packets leaving slits is destroyed before the particle arrives at the screen. "Which path" information is encoded in the environment.

As we have seen in Section 6.4.1, a quantum superposition of two macroscopic states is called a Schrödinger's cat. We have just seen that this Schrödinger's cat is destroyed over a time $\sim \tau_{\text{dec}}$, and we are left with an incoherent mixture. The mechanism responsible for decoherence selects a preferred basis: photon scattering selects a basis of position states, because the photons scattered by different position eigenstates are sent into orthogonal states. Actually, it is not necessary that the final photon states be orthogonal. Assume, for example, that the states $|1_E\rangle$ and $|2_E\rangle$ satisfy $\langle 1_E|2_E\rangle = 1 - \varepsilon$; then the probability p would be replaced by $\mathsf{p} \to \varepsilon\mathsf{p}$ and the decoherence time by $\tau_{\text{dec}} \to \varepsilon^{-1}\tau_{\text{dec}}$.

15.2.4 The amplitude-damping channel

This is a schematic model for describing the spontaneous decay of a two-level atom with the emission of one photon. By detecting the emitted photon, we perform a POVM which gives us information about the state of the atom. The system is the two-level atom, and the environment is the quantized electromagnetic field. If the atom and the field are in their respective ground states $|0_A\rangle$ and $|0_E\rangle$, nothing can happen. If the atom is in its

---

[12] A "dust particle" is large by microscopic standards and small by macroscopic standards.

excited state $|1_A\rangle$ and the field is in $|0_E\rangle$, there is a probability p that the atom emits a photon and is left in $|0_A\rangle$. The unitary representation of the quantum jump then is

$$U_{AE}|0_A \otimes 0_E\rangle = |0_A \otimes 0_E\rangle,$$

$$U_{AE}|1_A \otimes 0_E\rangle = \sqrt{1-\mathsf{p}}\,|1_A \otimes 0_E\rangle + \sqrt{\mathsf{p}}\,|0_A \otimes 1_E\rangle. \tag{15.58}$$

where $|1_E\rangle$ is a one photon state.

The Kraus operators from (15.44) are

$$M_0 = \begin{pmatrix} 1 & 0 \\ 0 & \sqrt{1-\mathsf{p}} \end{pmatrix}, \quad M_1 = \begin{pmatrix} 0 & \sqrt{\mathsf{p}} \\ 0 & 0 \end{pmatrix}, \tag{15.59}$$

and $\rho' = \mathcal{K}(\rho)$ is given by

$$\mathcal{K}(\rho) = \begin{pmatrix} 1 + (1-\mathsf{p})\rho_{11} & \sqrt{1-\mathsf{p}}\,\rho_{01} \\ \sqrt{1-\mathsf{p}}\,\rho_{10} & (1-\mathsf{p})\rho_{11} \end{pmatrix}. \tag{15.60}$$

As in the preceding example, we take $\mathsf{p} = \Gamma \Delta t \ll 1$, $\Delta t = t/n$, and make n iterations of $\mathcal{K}$:

$$\mathcal{K}^n(\rho) = \begin{pmatrix} 1 + (1-\mathsf{p})^n \rho_{11} & (1-\mathsf{p})^{n/2} \rho_{01} \\ (1-\mathsf{p})^{n/2} \rho_{10} & (1-\mathsf{p})^n \rho_{11} \end{pmatrix} \rightarrow \begin{pmatrix} 1 - e^{-\Gamma t} \rho_{11} & e^{-\Gamma t/2} \rho_{01} \\ e^{-\Gamma t/2} \rho_{10} & e^{-\Gamma t} \rho_{11} \end{pmatrix}. \tag{15.61}$$

In this model, $T_2 = 2T_1 = 2/\Gamma$, which explains why we chose Γ and $\Gamma/2$ as the relaxation rates of populations and coherences, respectively, in the optical Bloch equations (14.82)–(14.83).

In contrast to the preceding example, where we could not envisage detecting the photons scattered by the dust particles, it may be possible in the present case to detect the emitted photon. A coherent superposition of the two atomic states evolves as

$$\left(a|0_A\rangle + b|1_A\rangle \right) \otimes |0_E\rangle \rightarrow \left(a|0_A\rangle + b\sqrt{1-\mathsf{p}}\,|1_A\rangle \right) \otimes |0_E\rangle + b\sqrt{\mathsf{p}}\,|0_A \otimes 1_E\rangle.$$

If we detect the photon, we know with certainty that the initial state of the atom was $|1_A\rangle$. If we detect no photon, then we have prepared the (unnormalized) atomic state

$$a|0_A\rangle + b\sqrt{1-\mathsf{p}}\,|1_A\rangle.$$

The atomic state has evolved owing to our failure to detect a photon! As we have seen, a unitary transformation which entangles $\mathcal{A}$ and $\mathcal{E}$, followed by an orthogonal measurement on $\mathcal{E}$, can be described as a POVM on $\mathcal{A}$. From (15.46), the POVM are $\mathcal{R}_\mu = M_\mu^\dagger M_\mu$, so that

$$\mathcal{R}_0 = M_0^\dagger M_0 = \begin{pmatrix} 1 & 0 \\ 0 & 1-\mathsf{p} \end{pmatrix}, \quad \mathcal{R}_1 = M_1^\dagger M_1 = \begin{pmatrix} 0 & 0 \\ 0 & \mathsf{p} \end{pmatrix}, \tag{15.62}$$

and $\mathsf{p}(\mu) = \mathrm{Tr}(\mathcal{R}_\mu \rho_A)$ if the atom is initially in the mixed state ρ_A.

15.3 Master equations: the Lindblad form

15.3.1 The Markovian approximation

The system of Bloch equations (15.5), which we wrote down from heuristic arguments, is typical of what is called a master equation: the time evolution of the state operator is given by a first-order differential equation in time, or, in other words, the evolution is local in time. For example, the optical Bloch equations (14.86)–(14.87) are equivalent to (15.61) if we ignore the unitary part $(-i/\hbar)[H, \rho]$ of the evolution, since (15.61) may be written as

$$\frac{d\rho}{dt} = -\frac{\Gamma}{2}\begin{pmatrix} -2\rho_{11}(t) & \rho_{01}(t) \\ \rho_{10}(t) & 2\rho_{11}(t) \end{pmatrix}. \tag{15.63}$$

It is readily checked that (15.63) can be cast in the form

$$\frac{d\rho}{dt} = \frac{\Gamma}{2}\left[2\sigma_+\rho\,\sigma_- - \{\sigma_-\sigma_+, \rho\}\right], \tag{15.64}$$

where $\sigma_\pm = (\sigma_x \pm i\sigma_y)/2$,[13] and $\{A, B\}$ denotes the *anticommutator* of two operators A and B:

$$\{A, B\} = AB + BA. \tag{15.65}$$

Actually, one can readily supplement (15.63) by a unitary evolution as in (15.5):

$$\frac{d\rho}{dt} = -\frac{i}{\hbar}[H, \rho] + \frac{\Gamma}{2}\left[2\sigma_+\rho\,\sigma_- - \{\sigma_-\sigma_+, \rho\}\right]. \tag{15.66}$$

For later purposes, it will be useful to introduce the interaction picture:

$$\begin{aligned} \tilde{\rho}(t) &= e^{iH_0 t/\hbar}\,\rho(t)\,e^{-iH_0 t/\hbar}, \\ \tilde{\sigma}_\pm(t) &= e^{iH_0 t/\hbar}\,\sigma_\pm\,e^{-iH_0 t/\hbar} = e^{\mp i\omega_0 t}\sigma_\pm, \end{aligned} \tag{15.67}$$

where H_0 is the free Hamiltonian (15.1). Equation (15.66) is an example of a master equation in the Lindblad form, of which we now give a general derivation.

In the preceding example, we have been able to go from the Kraus form (15.45) to the master equation in the Lindblad form (15.66). However, our derivation depends on the crucial (and strong) assumption that the probability p is proportional to Δt. In the general case, it is far from obvious that it is possible to obtain a differential equation for the nonunitary evolution of $\tilde{\rho}(t)$, because we expect *memory effects* to be present (a priori, a local time evolution can be valid only for $\tilde{\rho}$, not ρ). Information flows from the system to the environment, but, conversely, information also flows from the environment to the system. Schematically, an equation with memory effects has the form of an integro-differential equation which is *nonlocal in time*:

$$\frac{d\tilde{\rho}}{dt} = -\int_{-\infty}^{t} \gamma(t - t')\tilde{\rho}(t')dt', \tag{15.68}$$

---

[13] Beware of the fact that Nielsen and Chuang [2000] use the opposite convention for $\sigma_\pm$. We have chosen a convention consistent with the definition (10.4) of the angular momentum operators $J_\pm$, and which is moreover consistent with that of field theory, since σ_+ (σ_-) is a positive (negative) frequency operator like a ($a^\dagger$); see (11.67).

where $\gamma(t - t')$ is the memory function, or memory kernel.[14] If the characteristic relaxation time τ_* of $\gamma(t - t')$ is much smaller that the typical evolution time τ of $\tilde{\rho}$, $\tau_* \ll \tau$, we may write (15.68) in the approximate form

$$\frac{d\tilde{\rho}}{dt} \simeq -\rho(t) \int_{-\infty}^{t} \gamma(t') \, dt' = -\Gamma \tilde{\rho}(t), \tag{15.69}$$

and we obtain a master equation. The short-memory approximation in (15.69) is also called the *Markovian approximation*: $d\tilde{\rho}/dt$ depends only on $\tilde{\rho}$ at time t, and not on its value at earlier times $t' < t$. The Markovian approximation will hold if there are two widely separated time scales: τ, the typical evolution time of $\tilde{\rho}$, and τ_*, the typical relaxation time of the memory function, with $\tau_* \ll \tau$. The assumption of two widely separated time scales is a very common one in nonequilibrium statistical mechanics.

Let us examine the conditions under which we may hope to derive a master equation from the Kraus representation. The first step is to use a coarse-graining approximation with a typical time Δt which obeys

$$\tau_* \ll \Delta t \ll \tau. \tag{15.70}$$

Assuming this condition to be valid, we write the Kraus representation for the evolution between t and $t + \Delta t$ as

$$\frac{d\rho_A}{dt} \simeq \frac{\Delta \rho_A}{\Delta t} = \frac{1}{\Delta t}[\rho_A(t + \Delta t) - \rho_A(t)] = \frac{1}{\Delta t}[\mathcal{K}_{t,t+\Delta t}(\rho_A(t)) - \rho_A(t)].$$

In order to derive a master equation, we need to satisfy two conditions.

(i) The state operator of the bipartite $\mathcal{AE}$ system must factorize: $\rho_{AE}(t) \simeq \rho_A(t) \otimes \rho_E(t)$. This condition is needed to write the Kraus representation at time $t + \Delta t$.

(ii) The superoperator $\mathcal{K}_{t,t+\Delta t}$ must depend only on Δt, and not on t.

Further comments on these conditions will be made in the next section, in the context of a specific model for $\mathcal{E}$ and its interaction with $\mathcal{A}$. A general statement is that both (i) and (ii) are valid provided $|\mathcal{V}|\tau_*/\hbar \ll 1$, where $|\mathcal{V}|$ is a typical matrix element of the $\mathcal{AE}$ interaction.

15.3.2 The Lindblad equation

Let us *assume* that conditions (i) and (ii) hold. Then we can write, from (15.45),

$$\mathcal{K}_{\Delta t}[\rho_A(t)] = \sum_{\mu} M_{\mu}(\Delta t)\rho_A(t)M_{\mu}^{\dagger}(\Delta t), \tag{15.71}$$

and $[\mathcal{K}_{\Delta t}[\rho_A(t)] - \rho_A(t)]$ is first-order in Δt:

$$\mathcal{K}_{\Delta t}(\rho_A(t)) = \rho_A(t) + \mathcal{O}(\Delta t).$$

---

[14] In general, (15.68) takes a matrix form, and some additional terms are present; see the references in "Further reading."

It follows that one of the M_μ, which we call by convention M_0, must have an expansion of the form

$$M_0(t) = I_A + \left(-\frac{i}{\hbar}H - K\right)\Delta t + O(\Delta t)^2, \tag{15.72}$$

where H and K are Hermitian operators. Then the first term in (15.71) reads

$$M_0(\Delta t)\rho_A(t)M_0^\dagger(\Delta t) = \rho_A(t) - \frac{i\Delta t}{\hbar}[H, \rho_A] - \Delta t\{K, \rho_A\} + O(\Delta t)^2, \tag{15.73}$$

(see (15.65)). The other terms in (15.71) must be of order $\sqrt{\Delta t}$

$$M_\mu(\Delta t) = L_\mu\sqrt{\Delta t}, \tag{15.74}$$

and the completeness relation in (15.45) leads to

$$I_A = M_0 M_0^\dagger + \sum_{\mu>0} M_\mu^\dagger M_\mu = I_A - 2K\Delta t + \left(\sum_{\mu>0} L_\mu^\dagger L_\mu\right)\Delta t,$$

which implies

$$K = \frac{1}{2}\sum_{\mu>0} L_\mu^\dagger L_\mu. \tag{15.75}$$

Combining (15.71), (15.73), and (15.75) and from now on suppressing the subscript A, we find the *Lindblad equation* for the state operator ρ of $\mathcal{A}$:

$$\boxed{\frac{d\rho}{dt} = -\frac{i}{\hbar}[H, \rho] + \sum_{\mu>0}\left(L_\mu\rho L_\mu^\dagger - \frac{1}{2}\{K, \rho\}\right).} \tag{15.76}$$

The operators L_μ are the *quantum jump operators*. They describe how the state of $\mathcal{A}$ is modified by an orthogonal measurement on the environment. Provided the $L\mu S$ are bounded operators, the Lindblad equation is the most general (Markovian) master equation which preserves the positivity of the state operator.

It is instructive to rederive the Bloch equations (15.64) from the Lindblad form (15.76). Using the expressions (15.59) of M_0 and M_1 for the amplitude-damping channel, we can write M_0 and M_1 in the form

$$M_0 = \begin{pmatrix} 1 & 0 \\ 0 & 1 - \frac{\Gamma}{2}\Delta t \end{pmatrix}, \quad M_1 = \begin{pmatrix} 0 & \sqrt{\Gamma\Delta t} \\ 0 & 0 \end{pmatrix},$$

or, in terms of Pauli matrices,

$$M_0 = I - \frac{\Gamma\Delta t}{2}\sigma_-\sigma_+, \quad M_1 = \sqrt{\Gamma\Delta t}\,\sigma_+.$$

The operators K and L_1 then are

$$K = \frac{\Gamma}{2}\sigma_-\sigma_+, \quad L_1 = \sqrt{\Gamma}\,\sigma_+, \tag{15.77}$$

and we recover (15.64).

15.3.3 Example: the damped harmonic oscillator

Consider a harmonic oscillator coupled to the quantized electromagnetic field, assuming for simplicity that the system is at zero temperature; the case of nonzero temperature will be dealt with in Section 15.4.3. If the oscillator is initially in an excited state, it can only cascade down due to spontaneous photon emission; it cannot absorb photons, as no photons are available at zero temperature. Hence there is only one quantum jump operator L_1, which must be proportional to a (recall the analogy between the annihilation operator a (11.6) and σ_+; see Footnote 13):

$$L_1 = \sqrt{\Gamma}\, a. \tag{15.78}$$

Then by inspection we can write down the Lindblad equation, by comparison with (15.66):

$$\frac{d\rho}{dt} = -\frac{i}{\hbar}[H_0, \rho] + \frac{1}{2}\Gamma\left[2a\rho a^\dagger - \{a^\dagger a, \rho\}\right], \tag{15.79}$$

where $H_0 = \hbar\omega_0 a^\dagger a$ is the free Hamiltonian. In this derivation, we missed the radiative renormalization of the energy levels of the harmonic oscillator due to the interaction between the oscillator and the quantized electromagnetic field, which is an example of a Lamb shift, computed explicitly in Section 15.4.3. Moreover, a full derivation of (15.79) shows that it is only valid under the condition $\Gamma \ll \omega_0$. This condition allows us to ignore the coupling between matrix elements of the state operator that evolve with different eigenfrequencies, for example the coupling between populations and coherences. We get rid of the commutator by going to the interaction picture (compare with (15.67)):

$$\begin{aligned}\tilde{a}(t) &= e^{iH_0 t/\hbar}\, a\, e^{-iH_0 t/\hbar} = a e^{-i\omega_0 t},\\ \tilde{a}^\dagger(t) &= e^{iH_0 t/\hbar}\, a^\dagger\, e^{-iH_0 t/\hbar} = a^\dagger e^{i\omega_0 t},\end{aligned} \tag{15.80}$$

whence

$$\frac{d\tilde{\rho}}{dt} = \Gamma\left[\tilde{a}\tilde{\rho}\tilde{a}^\dagger - \frac{1}{2}\{\tilde{a}^\dagger\tilde{a}, \tilde{\rho}\}\right] = \Gamma\left[a\tilde{\rho}a^\dagger - \frac{1}{2}\{a^\dagger a, \tilde{\rho}\}\right]. \tag{15.81}$$

Here we have used (15.80) to obtain the second expression. In the absence of damping ($\Gamma = 0$), the average value of the operator

$$\bar{a} = e^{-iH_0 t/\hbar}\, a\, e^{iH_0 t/\hbar}$$

is time-independent. If $\Gamma \neq 0$, from (15.81) we derive the evolution equation for its average value:

$$\frac{d}{dt}\langle\bar{a}\rangle = \frac{d}{dt}\mathrm{Tr}\,(\bar{a}\rho) = \frac{d}{dt}\mathrm{Tr}\left(a\, e^{iH_0 t}\, \rho\, e^{-iH_0 t}\right) = \frac{d}{dt}\mathrm{Tr}\,(a\tilde{\rho}) = \mathrm{Tr}\left(a\frac{d\tilde{\rho}}{dt}\right),$$

while from (15.81)

$$\begin{aligned}\mathrm{Tr}\left(a\frac{d\tilde{\rho}}{dt}\right) &= \frac{\Gamma}{2}\mathrm{Tr}\left[2a^2\tilde{\rho}a^\dagger - aa^\dagger a\tilde{\rho} - a\tilde{\rho}a^\dagger a\right]\\ &= \frac{\Gamma}{2}\mathrm{Tr}\left[[a^\dagger, a]a\tilde{\rho}\right] = -\frac{\Gamma}{2}\langle\bar{a}\rangle,\end{aligned}$$

so that we find the decay law

$$\langle \overline{a}(t) \rangle = e^{-\Gamma t/2} \langle \overline{a}(t=0) \rangle. \tag{15.82}$$

An analogous computation shows that the average occupation number $n(t) = \langle a^\dagger a \rangle$ decays with a relaxation time $1/\Gamma$:

$$n(t) = e^{-\Gamma t} n(t=0). \tag{15.83}$$

As shown in Exercise 15.5.7, if the initial state of the oscillator is a coherent state (Section 11.2) $|z\rangle$ at $t = 0$, time evolution leads to

$$|z\rangle \rightarrow |z e^{-i\omega_0 t} e^{-\Gamma t/2}\rangle,$$

so that the coherent state does not become entangled with its environment, although it decays slowly ($\Gamma \ll \omega_0$) toward the vacuum state. However, if one starts from a coherent superposition of coherent states $|z_1\rangle$ and $|z_2\rangle$

$$|\Psi\rangle = \frac{1}{\sqrt{2}} \left(|z_1\rangle + |z_2\rangle \right),$$

as shown in Exercise 15.5.7, the off-diagonal terms of the state matrix decay as

$$\exp\left[-\frac{1}{2} \Gamma |z_1 - z_2|^2 t \right].$$

The decoherence rate Γ_{dec} is much larger than the damping rate Γ if $|z_1 - z_2|^2 \gg 1$:

$$\Gamma_{\text{dec}} = \frac{1}{2} \Gamma |z_1 - z_2|^2. \tag{15.84}$$

It is proportional to the square of the "distance" $|z_1 - z_2|$ between the two coherent states.

15.4 Coupling to a thermal bath of oscillators

15.4.1 Exact evolution equations

In order to derive more detailed properties of the master equation, in this section we choose specific models for the system and the reservoir: in Section 15.4.3 system $\mathcal{A}$ will be a two-level system, in Section 15.4.4 it will be a Brownian particle, and in both cases the environment will be modeled by a large number of uncoupled harmonic oscillators in thermal equilibrium at temperature T. In deference to the standard terminology of thermodynamics, the environment will be called the "reservoir": $\mathcal{E} \rightarrow \mathcal{R}$. Our reservoir is thus a thermal bath of harmonic oscillators, whose Hamiltonian H_R is

$$H_R = \sum_\lambda \hbar \omega_\lambda a_\lambda^\dagger a_\lambda. \tag{15.85}$$

It is important that the frequencies ω_λ form a quasi-continuuum in a large frequency interval $\sim 1/\tau_*$. The state operator of the uncoupled reservoir is given by the Boltzmann law (1.12):

$$\rho_R(t=0) = \frac{e^{-H_R/k_B T}}{\mathrm{Tr}\,(e^{-H_R/k_B T})}. \tag{15.86}$$

We shall need the following equilibrium average values, which are immediately derived from (15.86):

$$\langle a_\lambda \rangle = \langle a_\lambda^\dagger \rangle = 0, \quad \langle a_\lambda^\dagger a_\mu \rangle = n_\lambda \delta_{\lambda\mu}, \quad \langle a_\lambda a_\mu^\dagger \rangle = (n_\lambda + 1)\delta_{\lambda\mu}, \tag{15.87}$$

where the average occupation number n_λ of oscillator λ is (see (1.20))

$$n_\lambda = \frac{1}{e^{\hbar\omega_\lambda/k_B T} - 1}. \tag{15.88}$$

The system–reservoir coupling V is assumed to be of the form

$$V = AR, \quad R = R^\dagger = \sum_\lambda (g_\lambda a_\lambda + g_\lambda^* a_\lambda^\dagger), \tag{15.89}$$

where $A = A^\dagger$ is an operator acting in $\mathcal{H}_A$ and the total Hamiltonian H_{AR} is

$$H_{AR} = H_A + H_R + V = H_T + V, \quad H_T = H_A + H_R. \tag{15.90}$$

The evolution equation for the state operator, first written in the Schrödinger picture

$$\frac{d\rho_{AR}}{dt} = -\frac{i}{\hbar}[H_{AR}, \rho_{AR}],$$

is transformed into the interaction picture, defined as previously by

$$\tilde{\rho}_{AR}(t) = e^{iH_T t/\hbar}\, \rho_{AR}\, e^{-iH_T t/\hbar}.$$

In this picture the evolution equation reads

$$\frac{d\tilde{\rho}_{AR}}{dt} = -\frac{i}{\hbar}[V(t), \tilde{\rho}_{AR}(t)] = -\frac{i}{\hbar}[A(t)R(t), \tilde{\rho}_{AR}(t)], \tag{15.91}$$

where $A(t)$ and $R(t)$ are given by[15]

$$A(t) = e^{iH_T t/\hbar}\, A\, e^{-iH_T t/\hbar} = e^{iH_A t/\hbar}\, A\, e^{-iH_A t/\hbar},$$

$$R(t) = e^{iH_T t/\hbar}\, R\, e^{-iH_T t/\hbar} = e^{iH_R t/\hbar}\, R\, e^{-iH_R t/\hbar} = \sum_\lambda \left(g_\lambda a_\lambda\, e^{-i\omega_\lambda t} + g_\lambda^* a_\lambda^\dagger\, e^{i\omega_\lambda t}\right). \tag{15.92}$$

The last expression in both lines of (15.92) is valid because H_R (H_A) does not act on the degrees of freedom of $\mathcal{A}$ ($\mathcal{R}$). The quantity that will play a central role in what follows is the *equilibrium autocorrelation function* $g(t')$ of $R(t)$:

$$g(t') = \langle R(t)R(t-t') \rangle = \langle R(t')R(0) \rangle, \tag{15.93}$$

---

[15] We have suppressed the tilde to simplify the notation.

where the average $\langle \bullet \rangle$ is taken with respect to the *equilibrium* state operator (15.86) of the reservoir. From time-translation invariance at equilibrium, g depends only on t' and not on t and t' separately (hence the second expression in (15.93)), while from the Hermiticity of R we have $g(t') = g^*(-t')$. The autocorrelation function $g(t')$ plays a fundamental role in *linear response theory*,[16] where it is customary to write its real and imaginary parts $C(t')$ and $-\chi(t')/2$ separately:

$$C(t') = \frac{1}{2}\langle\{R(t'), R(0)\}\rangle, \tag{15.94}$$

$$\chi(t') = \frac{i}{\hbar}\langle[R(t'), R(0)]\rangle\theta(t'), \tag{15.95}$$

where the second line contains a step function $\theta(t')$, because we are interested only in the case $t' \geq 0$. The function $\chi(t')$ is called the *dynamical susceptibility* of the reservoir. In linear response theory, one shows that if the reservoir is submitted to a perturbation $-f(t)R$ (in the Schrödinger picture), where $f(t)$ is a *classical* function, then, to first order in f, the *nonequilibrium average* $\overline{R}(t)$ is

$$\overline{R}(t) = \int dt'\, \chi(t') f(t-t'). \tag{15.96}$$

As a consequence, if $f(t') = f\theta(-t')$, that is, we have a constant perturbation $-fR$ for $t' < 0$, the return to equilibrium $[f(t') = 0]$ is governed by the *equilibrium* time fluctuations, a result known as the *Onsager principle*.

Using (15.87) and (15.89), it is easy to derive explicit expressions for $g(t')$, $C(t')$, and $\chi(t')$:

$$g(t') = \sum_\lambda |g_\lambda|^2 [n_\lambda e^{i\omega_\lambda t'} + (n_\lambda+1)e^{-i\omega_\lambda t'}], \tag{15.97}$$

$$C(t') = \sum_\lambda |g_\lambda|^2 (2n_\lambda+1)\cos\omega_\lambda t', \tag{15.98}$$

$$\chi(t') = \frac{2\theta(t')}{\hbar}\sum_\lambda |g_\lambda|^2 \sin\omega_\lambda t'. \tag{15.99}$$

We observe that the dynamical susceptibility does not depend on the state of the reservoir: it is independent of n_λ. Because the reservoir is large and because the frequencies ω_λ are closely spaced in a frequency interval $\sim 1/\tau_*$, we expect the correlation function to decay with a characteristic time τ_*:

$$|g(t')| \sim e^{-|t'|/\tau_*}. \tag{15.100}$$

Indeed, $g(t')$ is a superposition of a large number of complex exponentials oscillating at different frequencies, and these exponentials interfere destructively once $|t'| \gtrsim \tau_*$.

---

[16] See "Further reading" for references on linear response theory. In these references, the "interaction picture" is called "Heisenberg picture," because coupling to another quantum system is not of interest.

Having examined the properties of the autocorrelation function, we may now revert to the evolution equation (15.91), which can be written in integral form as

$$\tilde{\rho}_{AR}(t) = \rho_{AR}(0) - \frac{i}{\hbar} \int_0^t dt' \left[V(t'), \tilde{\rho}_{AR}(t') \right].$$

We iterate this expression once

$$\tilde{\rho}_{AR}(t) = \rho_{AR}(0) - \frac{i}{\hbar} \int_0^t dt' \left[V(t'), \rho_{AR}(0) \right]$$

$$- \frac{1}{\hbar^2} \int_0^t dt' \int_0^{t'} dt'' \left[V(t'), \left[V(t''), \tilde{\rho}_{AR}(t'') \right] \right],$$

and differentiate with respect to t to obtain

$$\frac{d\tilde{\rho}_{AR}}{dt} = -\frac{i}{\hbar} \left[V(t), \rho_{AR}(0) \right] - \frac{1}{\hbar^2} \int_0^t dt' \left[V(t), \left[V(t'), \tilde{\rho}_{AR}(t') \right] \right]. \tag{15.101}$$

As usual, we assume a factorized form for $\rho_{AR}(t = 0)$

$$\rho_{AR}(t = 0) = \rho(t = 0) \otimes \rho_R(t = 0), \tag{15.102}$$

and take the partial trace over the reservoir degrees of freedom. Then the first term in (15.101) gives (Exercise 15.5.6)

$$\mathrm{Tr}_{\mathscr{R}} \left[V(t), \rho_{AR}(0) \right] = \left[A(t), \rho_A(0) \right] \mathrm{Tr}_{\mathscr{R}} \left(R(t) \rho_R \right) = 0,$$

where we have made use of (15.87). Under the factorization assumption (15.102), we finally obtain an *exact* equation for the state operator $\tilde{\rho}_A(t) = \tilde{\rho}(t)$ of system $\mathscr{A}$:

$$\frac{d\tilde{\rho}}{dt} = -\frac{1}{\hbar^2} \int_0^t dt' \, \mathrm{Tr}_{\mathscr{R}} \left(\left[V(t), \left[V(t'), \tilde{\rho}_{AR}(t') \right] \right] \right). \tag{15.103}$$

15.4.2 The Markovian approximation

The derivation of a master equation from the exact equation (15.103) relies on the following crucial assumption: for all times t' that are relevant for the integral in (15.103) (and not only for $t = 0$ as in (15.102)!), we can use for $\tilde{\rho}_{AR}(t)$ a factorized form similar to that in the initial state (15.102):

$$\tilde{\rho}_{AR}(t) \simeq \tilde{\rho}(t) \otimes \rho_R(t = 0). \tag{15.104}$$

There are two different points to be emphasized in explaining the physical origin of (15.104).

(i) All the system–reservoir correlations which arise from third- and higher-order terms in V are neglected.
(ii) The modifications to the state of the reservoir induced by its coupling to the system are neglected.

Both items (i) and (ii) are physically reasonable if the reservoir is much "larger" than the system $\mathcal{A}$: the back action of the system on the reservoir and higher-order terms in the perturbative expansion may be neglected. One can indeed show that the true small parameter in an expansion in powers of V is $|\mathcal{V}|\tau_*/\hbar$, where $\mathcal{V}$ is a typical matrix element of V. The condition for the validity of (i) and (ii) is then $|\mathcal{V}|\tau_*/\hbar \ll 1$.[17] In particular, it can be shown that

$$|\tilde{\rho}_{AR}(t) - \tilde{\rho}_A(t) \otimes \tilde{\rho}_R(t)| = O\left(\frac{|\mathcal{V}|\tau_*}{\hbar}\right)^2.$$

Plugging (15.104) into (15.102), we obtain an equation of motion for $\tilde{\rho}$ which depends only on A and g (Exercise 15.5.6):

$$\frac{d\tilde{\rho}}{dt} = \frac{1}{\hbar^2}\int_0^t dt'\, g(t')\Big[A(t-t')\tilde{\rho}(t-t')A(t) - A(t)A(t-t')\tilde{\rho}(t-t')\Big] + \text{H.c.}, \quad (15.105)$$

where H.c. = Hermitian conjugate and we have made the change of variable $t' \to t - t'$.

Equation (15.105) is still an integro-differential equation containing memory effects, and not a master equation. To obtain a master equation, we note from (15.100) that the times t' which contribute significantly to the integral are bounded by τ_*, $t' \lesssim \tau_*$. Hence the difference $[\tilde{\rho}(t-t') - \tilde{\rho}(t)]$ is bounded by

$$|\tilde{\rho}(t-t') - \tilde{\rho}(t)| \lesssim O\left(\frac{|\mathcal{V}|\tau_*}{\hbar}\right),$$

and we can replace $\tilde{\rho}(t'-t)$ by $\tilde{\rho}(t)$ in a manner which is consistent with the preceding approximation: the error is of higher order in the small parameter. In this way, we have justified a Markovian approximation, and $\tilde{\rho}$ is given by a first-order differential equation. Taking $t \gg \tau_*$, we can send the upper limit in the integral to infinity and write

$$\frac{d\tilde{\rho}}{dt} = \frac{1}{\hbar^2}\int_0^\infty dt'\, g(t')\Big[A(t-t')\tilde{\rho}(t)A(t) - A(t)A(t-t')\tilde{\rho}(t)\Big] + \text{H.c.}$$

Actually, this equation can be derived from perturbation theory limited to second order. It is convenient (but by no means necessary) to revert to the Schrödinger picture and to write the master equation in its final form:

$$\boxed{\frac{d\rho}{dt} = -\frac{i}{\hbar}[H_A, \rho] + \frac{1}{\hbar^2}\left(W\rho A + A\rho W^\dagger - AW\rho - \rho W^\dagger A\right),} \quad (15.106)$$

where the operator W is given by

$$\boxed{W = \int_0^\infty g(t')A(-t')dt'.} \quad (15.107)$$

---

[17] See C. Cohen-Tannoudji, J. Dupont-Roc, and G. Grynberg, *Atom–Photon Interactions*, New York: Wiley (1992), Chapter IV for a detailed discussion.

We see that the characteristic evolution time of $\tilde{\rho}$ is

$$\tau \sim \frac{\hbar^2}{|\mathcal{V}|^2 \tau_*} = \left(\frac{\hbar}{|\mathcal{V}| \tau_*}\right)^2 \tau_* \gg \tau_*.$$

The characteristic time τ is even much larger than the "natural time" $\hbar/|\mathcal{V}|$, which one would expect if $\mathcal{A}$ were coupled to a *single* mode of the reservoir:

$$\tau_* \ll \frac{\hbar}{|\mathcal{V}|} \ll \tau \sim \frac{\hbar^2}{|\mathcal{V}|^2 \tau_*}. \tag{15.108}$$

The effective coupling is reduced owing to the fact that $\mathcal{A}$ is coupled to a large number of independent modes, a phenomenon called *motion narrowing*.

15.4.3 Relaxation of a two-level system

Let us now apply the preceding results to the case where system $\mathcal{A}$ is a two-level system coupled to a thermal bath of independent harmonic oscillators: photons, phonons... The free Hamiltonian H_A of the two-level system is now H_0 (15.2)

$$H_A \equiv H_0 = -\frac{\hbar \omega_0}{2} \sigma_z, \tag{15.109}$$

and our goal is to understand its relaxation properties. The system–reservoir interaction V must be able to induce transitions between the two levels, and a possible choice is

$$V = \sigma_x R = \sigma_x \sum_\lambda \left(g_\lambda a_\lambda + g_\lambda^* a_\lambda^\dagger\right). \tag{15.110}$$

Thus $A = \sigma_x = (\sigma_+ + \sigma_-)$. The operator W (15.107) acting on the two-level system is

$$W = \int_0^\infty g(t') A(-t') dt' = G_+(\omega_0) \sigma_+ + G_-(\omega_0) \sigma_-, \tag{15.111}$$

with

$$G_\pm(\omega_0) = G_\mp(-\omega_0) = \int_0^\infty g(t') e^{\pm i\omega_0 t'} dt', \tag{15.112}$$

where we have used (15.67). Plugging (15.112) into (15.106) and using $\sigma_+^2 = \sigma_-^2 = 0$, we obtain

$$\frac{d\rho}{dt} = \frac{i}{2} \omega_0 [\sigma_z, \rho]$$

$$+ (G_+ + G_+^*) \sigma_+ \rho \, \sigma_- - G_+ \sigma_- \sigma_+ \rho - G_+^* \rho \, \sigma_- \sigma_+$$

$$+ (G_- + G_-^*) \sigma_- \rho \, \sigma_+ - G_- \sigma_+ \sigma_- \rho - G_-^* \rho \, \sigma_+ \sigma_- \tag{15.113}$$

$$+ (G_+ + G_-^*) \sigma_+ \rho \, \sigma_+ + (G_- + G_+^*) \sigma_- \rho \, \sigma_-.$$

Using the invariance of the trace under cyclic permutations, we can check that the equation has been written in such a way that each of the first three lines in the right-hand side has zero trace. The fourth line does not contribute to the evolution of the populations (Exercise 15.5.8), only to that of the coherences. But even this contribution to the coherences may be neglected in the rotating-wave approximation, using the same argument as in Section 14.4.1: in the interaction picture, $\tilde{\sigma}_\pm(t) \sim \exp(\mp i\omega_0 t)$, and the last term in (15.113) varies as $\exp(\mp 2i\omega_0 t)$. It is rapidly oscillating and assumed to average to zero. This is a general result: if the relaxation terms are written as

$$\frac{\mathrm{d}\tilde{\rho}_{ij}}{\mathrm{d}t} = \sum_{k,l=0}^{1} \gamma_{ijkl}\tilde{\rho}_{kl},$$

it can be shown that the coefficients γ_{ijkl} can be neglected if $|\omega_{ij} - \omega_{kl}| \gg \Gamma$, $\hbar\omega_{ij} = E_i - E_j$.[18] This is called the *secular approximation*, and it allows us to justify the form of the Bloch equations (15.6).

Let us compute $G_\pm(\omega_0)$ explicitly:

$$G_+(\omega_0) = \sum_\lambda |g_\lambda|^2 \left((n_\lambda + 1) \frac{i}{\omega_0 - \omega_\lambda + i\eta} + n_\lambda \frac{i}{\omega_0 + \omega_\lambda + i\eta} \right),$$

$$G_-(\omega_0) = \sum_\lambda |g_\lambda|^2 \left(n_\lambda \frac{-i}{\omega_0 - \omega_\lambda - i\eta} + (n_\lambda + 1) \frac{-i}{\omega_0 + \omega_\lambda - i\eta} \right),$$

$$(15.114)$$

where $\eta \to 0^+$. Using the standard formula

$$\frac{i}{x \pm i\eta} = i\frac{\mathbb{P}}{x} \pm \pi\delta(x), \tag{15.115}$$

where $\mathbb{P}$ denotes a Cauchy principal value, for $G_+(\omega_0)$ we find

$$G_+(\omega_0) = \frac{1}{2}\Gamma_+ - i\Delta_+,$$

$$\Gamma_+ = 2\pi \sum_\lambda |g_\lambda|^2 (n_\lambda + 1)\delta(\omega_0 - \omega_\lambda),$$

$$\Delta_+ = -\sum_\lambda |g_\lambda|^2 \left((n_\lambda + 1)\frac{\mathbb{P}}{\omega_0 - \omega_\lambda} + n_\lambda \frac{\mathbb{P}}{\omega_0 + \omega_\lambda} \right),$$

$$(15.116)$$

while $G_-(\omega_0)$ is given by

$$G_-(\omega_0) = \frac{1}{2}\Gamma_- - i\Delta_-,$$

$$\Gamma_- = 2\pi \sum_\lambda |g_\lambda|^2 n_\lambda \delta(\omega_0 - \omega_\lambda),$$

$$\Delta_- = -\sum_\lambda |g_\lambda|^2 \left(n_\lambda \frac{\mathbb{P}}{\omega_0 - \omega_\lambda} + (n_\lambda + 1)\frac{\mathbb{P}}{\omega_0 + \omega_\lambda} \right).$$

$$(15.117)$$

[18] See C. Cohen-Tannoudji *et al.*, *Atom–Photon Interactions*, New York: Wiley (1992), Chapter IV for a detailed discussion.

Substituting the two preceding equations into (15.113) and making use of

$$\sigma_+\sigma_- = \frac{1}{2}(1+\sigma_z), \quad \sigma_-\sigma_+ = \frac{1}{2}(1-\sigma_z),$$

we obtain the final expression for the master equation in the Lindblad form:

$$
\begin{aligned}
\frac{d\rho}{dt} = &\frac{i}{2}(\omega_0+\Delta)[\sigma_z,\rho] \\
&+\frac{1}{2}\Gamma_+(2\sigma_+\rho\sigma_- - \{\sigma_-\sigma_+,\rho\}) \\
&+\frac{1}{2}\Gamma_-(2\sigma_-\rho\sigma_+ - \{\sigma_+\sigma_-,\rho\}).
\end{aligned}
\tag{15.118}
$$

The two quantum jump operators of the Lindblad equation are

$$L_+ = \sqrt{\frac{\Gamma_+}{2}}\,\sigma_+, \quad L_- = \sqrt{\frac{\Gamma_-}{2}}\,\sigma_-. \tag{15.119}$$

The energy shift Δ (or Lamb shift)

$$\Delta = \Delta_- - \Delta_+ = \sum_\lambda |g_\lambda|^2(2n_\lambda+1)\left(\frac{\mathbb{P}}{\omega_0-\omega_\lambda} + \frac{\mathbb{P}}{\omega_0+\omega_\lambda}\right) \tag{15.120}$$

represents the radiative correction to the energy-level difference ω_0 due to the interaction of the two-level system with the thermal bath of oscillators. Equation (15.118) generalizes (15.64) obtained at $T=0$, where only spontaneous emission was taken into account and the Lamb shift could not be computed. At nonzero temperature photon absorption must also be taken into account: the relaxation rate Γ_+ describes the transitions $|1\rangle \to |0\rangle$ and Γ_- the transitions $|0\rangle \to |1\rangle$ (Fig. 15.3). It is easy to check (Exercise 15.5.8) that $\Gamma = \Gamma_+ + \Gamma_-$ is the relaxation rate for the populations, while that for the coherences is $\Gamma/2$: the relation $T_2 = 2T_1$ also holds at nonzero temperature. In the same exercise, it is

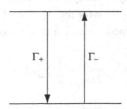

Fig. 15.3. Transition rates Γ_+ and Γ_-.

shown that in the long-time limit the populations of the levels $|0\rangle$ and $|1\rangle$ are given by Boltzmann's law (15.3), with temperature T equal to that of the thermal bath. This is a quite satisfactory result, as it shows that the system is in equilibrium with the bath in the long-time limit. The total width Γ is given explicitly by

$$\Gamma = \Gamma_+ + \Gamma_- = \frac{2\pi}{\hbar} \sum_\lambda |g_\lambda|^2 (2n_\lambda + 1)\delta(\hbar\omega_0 - \hbar\omega_\lambda). \qquad (15.121)$$

This provides a nice check of the calculation, as (15.121) can be written in the form of the Fermi Golden Rule (9.170):

$$\Gamma = \frac{2\pi}{\hbar} |g(\omega_0)|^2 (2n_0 + 1)\mathcal{D}(\hbar\omega_0),$$

where $\mathcal{D}(\hbar\omega_0)$ is the density of states of the reservoir. The ratio Γ_+/Γ_- is given by a Boltzmann law

$$\frac{\Gamma_+}{\Gamma_-} = e^{\hbar\omega/k_B T}.$$

The master equation (15.118) allows us to write by inspection (recall the correspondence $a \to \sigma_+$, $a^\dagger \to \sigma_-$; see Footnote 12) the $T \neq 0$ generalization of (15.79), which gives the master equation for a harmonic oscillator coupled to the quantized electromagnetic field at nonzero temperature:

$$\frac{d\rho}{dt} = -\frac{i}{\hbar}[H_0, \rho] + \frac{1}{2}\Gamma_+\big[2a\rho a^\dagger - \{a^\dagger a, \rho\}\big] + \frac{1}{2}\Gamma_-\big[2a^\dagger \rho a - \{aa^\dagger, \rho\}\big]. \qquad (15.122)$$

Detailed derivations of the preceding equation can be found in textbooks on quantum optics (see "Further reading").

15.4.4 Quantum Brownian motion

Our last example will be that of a heavy free particle with mass M coupled to a thermal bath of harmonic oscillators with masses m_λ and frequencies ω_λ. This is a typical situation for Brownian particle motion. A heavy particle interacts with a thermal bath of light particles (molecules), and one may identify two widely separated time scales: the time scale τ_* for the bath and the time scale τ for the motion of the heavy particle, with $\tau_* \ll \tau$. The full Hamiltonian H_{AR} is assumed to have a translation-invariant form

$$H_{AR} = \frac{P^2}{2M} + \sum_\lambda \frac{P_\lambda^2}{2m_\lambda} + \frac{1}{2}\sum_\lambda m_\lambda \omega_\lambda^2 (X - X_\lambda)^2, \qquad (15.123)$$

where (P, P_λ) and (X, X_λ) are momentum and position operators for the particle and the oscillators. For the sake of simplicity, we have limited ourselves to one-dimensional motion, without losing any essential physics. The decomposition (15.90) of H_{AR} reads

$$H_A = \frac{P^2}{2M}, \qquad (15.124)$$

$$H_R = \sum_\lambda \left(\frac{P_\lambda^2}{2m_\lambda} + \frac{1}{2} \omega_\lambda^2 X_\lambda^2 \right) = \sum_\lambda \hbar \omega_\lambda a_\lambda^\dagger a_\lambda, \qquad (15.125)$$

$$V = \frac{1}{2} \kappa X^2 + XR = H_{CT} + X \left[-\sum_\lambda g_\lambda \left(a_\lambda + a_\lambda^\dagger \right) \right], \qquad (15.126)$$

with $g_\lambda = \sqrt{\hbar m_\lambda \omega_\lambda^3/2}$, $\kappa = \sum_\lambda m_\lambda \omega_\lambda^2$, and CT standing for "counter-term" for reasons to be explained below. The operator A is therefore to be identified with the position operator of the Brownian particle, and we have neglected the zero-point energy of the oscillators. It may appear that translation invariance has been broken in (15.126), but this is of course an artefact of the decomposition: as we shall see later on, the contribution of the translation-noninvariant counter-term

$$H_{CT} = \frac{1}{2} \kappa X^2 \qquad (15.127)$$

is canceled by another contribution from the interaction. It will be convenient but by no means necessary (see the comments following (15.142)) to work in the high-temperature limit where (15.88) becomes

$$n_\lambda \simeq n_\lambda + 1 \simeq \frac{k_B T}{\hbar \omega_\lambda} \gg 1. \qquad (15.128)$$

We recall that the frequencies ω_λ are assumed to be closely spaced in an interval $\sim 1/\tau_*$, so that the sums over λ can be replaced by integrals over ω. It is convenient to define the spectral function $J(\omega)$:

$$J(\omega) = \frac{\pi}{\hbar} \sum_\lambda |g_\lambda|^2 \delta(\omega - \omega_\lambda) = \frac{\pi}{2} \sum_\lambda m_\lambda \omega_\lambda^3 \delta(\omega - \omega_\lambda). \qquad (15.129)$$

From (15.98), (15.99), and (15.126) we find the expressions for the real and imaginary parts of the autocorrelation function $g(t')$:

$$C(t') = \frac{2k_B T}{\pi} \int_0^\infty \frac{d\omega}{\omega} J(\omega) \cos \omega t',$$

$$\chi(t') = \frac{2\theta(t')}{\pi} \int_0^\infty d\omega \, J(\omega) \sin \omega t' = -\frac{\theta(t')}{k_B T} \frac{dC}{dt'}. \qquad (15.130)$$

In order to proceed further, we must now choose a specific form for the spectral function $J(\omega)$. The typical frequency scale for $J(\omega)$ being $\omega_* = 1/\tau_*$, we choose $J(\omega)$ to vanish for $\omega \gg \omega_*$: ω_* plays the role of a frequency cutoff. The most convenient model for analytic calculations is that of Caldeira and Leggett,[19] where $J(\omega)$ is linear for $\omega \leq \omega_*$ and vanishes for $\omega > \omega_*$:

$$J(\omega) = M\gamma\omega, \quad 0 \leq \omega \leq \omega_*,$$
$$J(\omega) = 0, \qquad \omega > \omega_*. \tag{15.131}$$

The coefficient γ has the dimension of a frequency, and will be interpreted physically as a friction coefficient, as in the equation of motion (14.104) $\dot{v} = -\gamma v$. We expect that the results do not depend qualitatively on the exact shape of $J(\omega)$, the only important feature being the existence of a high-frequency cutoff ω_*. In Exercise 15.5.10 it is shown that equivalent results are obtained using

$$J(\omega) = M\gamma\,\omega\left(\frac{\omega_*^2}{\omega^2 + \omega_*^2}\right).$$

With the choice (15.131), $C(t')$ has a simple analytic form:

$$C(t') = 2k_B TM\gamma\,\frac{\sin \omega_* t'}{\pi t'}. \tag{15.132}$$

The function $\sin \omega_* t'/\pi t'$ has a peak of height ω_*/M and width $\sim 1/\omega_* = \tau_*$ at $t' = 0$, and it becomes a delta function in the limit $\omega_* \to \infty$. We shall call $\delta_*(t')$ a smeared delta function of width τ_*. With this notation, the autocorrelation function reads

$$g(t') = 2M\gamma k_B T\delta_*(t') + i\hbar M\gamma\delta_*'(t') = 2D\delta_*(t') + i\hbar M\gamma\delta_*'(t'), \tag{15.133}$$

where we have used Einstein's relation (14.113) linking the momentum diffusion coefficient D to γ and T, $D = M\gamma k_B T$.

After these preliminaries, we are now ready to give an explicit form for the general master equation (15.106), which in the present case becomes

$$\frac{d\rho}{dt} = -\frac{i}{\hbar}\left[\frac{P^2}{2M}, \rho\right] - \frac{i}{\hbar}\left[\frac{1}{2}\kappa X^2, \rho\right] - \frac{1}{\hbar^2}\left(W\rho X + X\rho W - XW\rho - \rho W^\dagger X\right), \tag{15.134}$$

with

$$W = \int_0^\infty g(t')\tilde{X}(-t')dt'. \tag{15.135}$$

The operator $\tilde{X}$ in the interaction picture is given by

$$\tilde{X}(t') = \exp\left[\frac{iP^2 t'}{2M\hbar}\right] X \exp\left[-\frac{iP^2 t'}{2M\hbar}\right] = X + \frac{Pt}{M}, \tag{15.136}$$

[19] A. Caldeira and A. Leggett, Path integral approach to quantum Brownian motion, *Physica* **121**A, 587 (1983).

a result which is immediately derived from (8.67). The term proportional to D on the right-hand side of the master equation involves the integral

$$\frac{2D}{\hbar^2} \int_0^\infty \delta_*(t') \left[\left(X - \frac{Pt'}{M} \right) \rho(t) X + X\rho(t) \left(X - \frac{Pt'}{M} \right) \right.$$

$$\left. - X \left(X - \frac{Pt'}{M} \right) \rho(t) - \rho(t) \left(X - \frac{Pt'}{M} \right) X \right] dt'.$$

Owing to the narrow width of $\delta_*(t')$, the terms proportional to Pt'/M are negligible and we are left with the double commutator

$$-\frac{D}{\hbar^2}[X, [X, \rho(t)]]. \tag{15.137}$$

The term proportional to $M\gamma$ is

$$\frac{iM\gamma}{\hbar} \int_0^\infty \delta'_*(t') \left[\left(X - \frac{Pt'}{M} \right) \rho(t) X + X\rho(t) \left(X - \frac{Pt'}{M} \right) \right.$$

$$\left. - X \left(X - \frac{Pt'}{M} \right) \rho(t) - \rho(t) \left(X - \frac{Pt'}{M} \right) X \right] dt'. \tag{15.138}$$

The two integrals that we need are

$$\text{(i)} \int_0^\infty \delta'_*(t')t'\, dt' = -\frac{1}{2}, \tag{15.139}$$

$$\text{(ii)} \int_0^\infty \delta'_*(t')dt' = \int_0^\infty dt' \frac{d}{dt'} \left(\frac{\sin \omega_* t'}{\pi t'} \right) = -\frac{\omega_*}{\pi}.$$

Equation (15.138) can be written as a sum of two terms. The first one, which depends on (i), is

$$\frac{\gamma}{2i\hbar}[X, \{P, \rho(t)\}], \tag{15.140}$$

and the second one depending on (ii) is

$$\frac{iM\gamma\omega_*}{\pi\hbar}[X^2, \rho(t)] = \frac{i}{\hbar}\left[\frac{1}{2}\kappa X^2, \rho(t)\right], \tag{15.141}$$

because in the Caldeira–Leggett model κ is given by

$$\kappa = \sum_\lambda m_\lambda \omega_\lambda^2 = \frac{2}{\pi} \int_0^\infty \frac{d\omega}{\omega} J(\omega) = \frac{2M\gamma\omega_*}{\pi}.$$

Then the term in (15.141) exactly cancels the contribution of H_{CT} to the evolution of the state operator. Collecting all the contributions to $d\rho/dt$, we finally obtain the master equation describing the quantum evolution of the Brownian particle:

$$\boxed{\frac{d\rho}{dt} = -\frac{i}{\hbar}\left[\frac{P^2}{2M}, \rho(t)\right] - \frac{i\gamma}{2\hbar}[X, \{P, \rho(t)\}] - \frac{D}{\hbar^2}[X, [X, \rho(t)]].} \tag{15.142}$$

Equation (15.142) is one of the basic results in the theory of open quantum systems. It should be observed that this equation is not of the Lindblad form, although it preserves the positivity of the state operator. The first term gives the unitary evolution of the wave packet, the second one describes friction, and the last one governs decoherence, as we shall see in detail in the next subsection. A Fokker–Planck equation for the probability distribution of p can be derived from (15.142); see Exercise 15.5.11.

Since the model defined in Eq. (15.123) is linear (that is, its classical equations of motion are linear), it can be solved exactly without taking the high-temperature limit. This is done in practice using path-integral methods. One can even put the Brownian particle in a harmonic potential well with frequency Ω. The exact solution at time t is

$$\frac{d\rho}{dt} = -\frac{i}{\hbar}\left[\frac{P^2}{2M} + \frac{1}{2}M\Omega^2(t)X^2, \rho(t)\right] - \frac{i\gamma(t)}{2\hbar}[X, \{P, \rho(t)\}]$$

$$-\frac{D(t)}{\hbar^2}[X, [X, \rho(t)]] - \frac{f(t)}{\hbar}[X, [P, \rho]].$$

We note the presence of a fourth term, called anomalous diffusion, which is negligible in the long-time limit $t \to \infty$. The functions $\Omega(t)$, $\gamma(t)$, $D(t)$, and $f(t)$ are given by integrals which must, in general, be computed numerically. In the long-time limit which has been taken in (15.142), analytical evaluation of the integrals is sometimes possible.

15.4.5 Decoherence and Schrödinger's cats

The preceding results are of the utmost importance, because they exhibit precise mechanisms for decoherence. A Brownian particle is a large object by macroscopic standards, and by constructing a quantum state of the particle which is a coherent superposition of two nonoverlapping wave packets we exhibit an example of a Schrödinger's cat. To be specific, let us assume that at $t = 0$ we have a coherent superposition of two Gaussian wave packets centered at $x = \pm a$ and having width $\sigma \ll a$, so that the overlap of the two wave packets is negligible. The initial wave function of the Brownian particle then is

$$\varphi(x) \simeq \frac{1}{\sqrt{2}}\left(\frac{1}{\pi\sigma^2}\right)^{1/4}\left(\exp\left[-\frac{(x-a)^2}{2\sigma^2}\right] + \exp\left[-\frac{(x+a)^2}{2\sigma^2}\right]\right). \tag{15.143}$$

The Fourier transform $\tilde{\varphi}(p)$ of (15.143) is readily computed and the momentum probability distribution $|\tilde{\varphi}(p)|^2$ is found to be

$$|\tilde{\varphi}(p)|^2 = \frac{2\sigma}{\hbar\sqrt{\pi}}\exp\left(-\frac{\sigma^2 p^2}{\hbar^2}\right)\cos^2\frac{pa}{\hbar}. \tag{15.144}$$

$|\tilde{\varphi}(p)|^2$ is a Gaussian of width $\sim \hbar/\sigma$ modulated by fast oscillations of period $\pi\hbar/a \ll \hbar/\sigma$. These oscillations originate in the coherence of the two wave packets in (15.143). Before exploiting (15.142), let us give a qualitative physical explanation for decoherence. The Brownian particle undergoes a large number of collisions with the light particles (molecules) in the thermal bath. Because of these collisions the particle follows a random walk in momentum space[20] with a diffusion coefficient D (14.113), and the momentum dispersion Δp is

$$\Delta p^2 = 2Dt. \qquad (15.145)$$

Each of the peaks in $|\tilde{\varphi}(p)|^2$ is broadened under the influence of collisions, and the peaks will be completely blurred out after a decoherence time τ_{dec} found from (15.145) as

$$\Delta p^2 \sim \left(\frac{\pi \hbar}{a}\right)^2 = 2D\tau_{\text{dec}},$$

or

$$\tau_{\text{dec}} \sim \frac{\hbar^2}{Da^2}. \qquad (15.146)$$

Let us derive this result from the master equation (15.142). We limit ourselves to short times, so that the motion of the Brownian particle can be neglected.[21] This is equivalent to taking the limit $M \to \infty$ in the master equation, and in this limit only the last term on the right-hand side survives (see Exercise 15.5.11 for a study of the general case). The off-diagonal matrix elements of the state operator obey the differential equation

$$\frac{\partial}{\partial t}\langle x|\rho(t)|x'\rangle = -\frac{D}{\hbar^2}(x-x')^2\langle x|\rho(t)|x'\rangle. \qquad (15.147)$$

The off-diagonal matrix elements of ρ decay with a relaxation time τ_{dec}:

$$\boxed{\tau_{\text{dec}} \simeq \frac{\hbar^2}{4Da^2},} \qquad (15.148)$$

because $|x - x'| \simeq 2a$, in agreement with the preceding heuristic estimate.

Let us give a very rough estimate for a typical decoherence time. Consider a Brownian particle of radius $R \simeq 1\,\mu\text{m}$ in air with viscosity $\eta \sim 10^{-5}$. The friction coefficient γ is given by the Stokes law $\gamma = 6\pi\eta R/M$. For $a = 10\,\mu\text{m}$ we find $\tau_{\text{dec}} \sim 10^{-27}$ s. "Large" Schrödinger's cats are really quite short-lived! In Appendix B we describe experiments in which one is able to build Schrödinger's cats small enough that decoherence can be observed and τ_{dec} measured, thus allowing an experimental verification of the decoherence mechanism.

There are other ways of writing the result (15.148). Using $D = M\gamma k_B T$ and introducing the thermal wavelength

$$\lambda_T = \frac{h}{\sqrt{2\pi M k_B T}},$$

that is, the de Broglie wavelength at temperature T, (15.148) becomes

$$\tau_{\text{dec}} \sim \frac{1}{\gamma}\left(\frac{\lambda_T}{a}\right)^2. \qquad (15.149)$$

---

[20] Not to be confused with diffusion in position space!

[21] This is a general result. In the short-time limit, Brownian motion is dominated by diffusion; see "Further reading."

The results of the present chapter allow us to give a general picture of decoherence. The first general feature is that one finds privileged states in the Hilbert space of states: coherent states in the case of Section 15.3.3 and position states in that of Section 15.4.4. These states are called *pointer states*, and they define the preferred basis of Section 6.4.1. A generic state of the Hilbert space is not stable when the system is put in contact with an environment but decays into an incoherent superposition of pointer states, which do not become entangled with their environment and are therefore the stable states. The stability of the pointer states can be traced back to the form the system interaction with the environment. For example, the pointer states of the Brownian particle are position states, because the interaction with its environment is proportional to the position operator X. As already mentioned in Section 15.3.3, a mode of the quantized electromagnetic field in a coherent state remains in a coherent state, because of the form of its interaction with a $T = 0$ environment. Coherent states are therefore also pointer states. The second general feature is that the decoherence time is inversely proportional to the square of the "distance" between pointer states: this distance is the ordinary one in the case of the position states of Section 15.4.4, and $|z_1 - z_2|$ in the case of the coherent states $|z_1\rangle$ and $|z_2\rangle$ of Section 15.3.3. The decoherence time is nothing other than the lifetime of Schrödinger's cats, and this lifetime is extremely short for macroscopic, and even mesoscopic, objects. As explained in Appendix B, decoherence is very likely an essential ingredient in the theory of quantum measurements. It explains why the measurement apparatus cannot be found in a quantum superposition, but can only exist in one of its pointer states.

15.5 Exercises

15.5.1 POVM as projective measurement in a direct sum

Let us consider the POVM defined by (15.30) and (15.31). Define the unnormalized vectors $|\tilde{\alpha}\rangle = \sqrt{2/3}\,|\alpha\rangle$, $\alpha = a, b, c$, and use these three vectors belonging to $\mathcal{H}^{(2)}$ to construct two vectors belonging to a three-dimensional space $\mathcal{H}^{(3)}$, written as the first two *rows* of a 3×3 matrix M. Complete M by a third vector orthogonal to the two preceding ones to obtain

$$M = \begin{pmatrix} \sqrt{2/3} & -\sqrt{1/6} & -\sqrt{1/6} \\ 0 & \sqrt{1/2} & -\sqrt{1/2} \\ \sqrt{1/3} & \sqrt{1/3} & \sqrt{1/3} \end{pmatrix}.$$

Why is M an orthogonal matrix, $M^T M = I$? Consider a projective measurement in $\mathcal{H}^{(3)}$ built from the three *columns* $|u_\alpha\rangle$ of M. Show that an observer unaware of the third component of $|u_\alpha\rangle$ will conclude that she has performed a POVM in $\mathcal{H}^{(2)}$.

15.5.2 Using a POVM to distinguish between states

Assume that Alice sends Bob qubits that can be either in state $|a_\perp\rangle$ or in state $|b_\perp\rangle$, each with 50% probability (the notation is that of Section 15.1.3). Bob performs a POVM with elements $\mathcal{Q}_a$, $\mathcal{Q}_b$, and $\mathcal{Q}_c$ (15.30). Show that if he finds the result b (a), he can be sure that the qubit was in state $|a_\perp\rangle$ ($|b_\perp\rangle$), but if he finds result c, he cannot decide. Show that in 50% of the cases Bob will be able to decide with certainty between the two states.

15.5.3 A POVM on two arbitrary qubit states

1. Consider the following two qubit states:

$$|a\rangle = \cos\alpha\,|0\rangle + \sin\alpha\,|1\rangle, \quad |b\rangle = \sin\alpha\,|0\rangle + \cos\alpha\,|1\rangle.$$

What are the projectors $\mathcal{P}_{a_\perp}$ and $\mathcal{P}_{b_\perp}$ onto the states $|a_\perp\rangle$ and $b_\perp\rangle$ respectively orthogonal to $|a\rangle$ and $|b\rangle$? Let $|c\rangle$,

$$|c\rangle = \lambda|0\rangle + \mu|1\rangle,$$

be a third qubit state vector. Build a POVM with $\mathcal{P}_{a_\perp}$, $\mathcal{P}_{b_\perp}$, and $\mathcal{P}_c$ by writing

$$A(\mathcal{P}_{a_\perp} + \mathcal{P}_{b_\perp}) + B\mathcal{P}_c = I.$$

Show that A and B can be expressed in terms of the scalar product $S = \langle a|b\rangle = \sin 2\alpha$.
2. Assume that Alice sends Bob a random sequence of states $|a\rangle$ and $|b\rangle$, each occurring with 50% probability. Bob performs a POVM

$$\{\mathcal{Q}_{a_\perp}, \mathcal{Q}_{b_\perp}, \mathcal{Q}_c\}.$$

What is the probability that he can be sure that Alice sent $|a\rangle$ or $b\rangle$? Application: in the quantum cryptography setup explained in Section 3.1.3, $\alpha = \pi/8$. Show that Eve can fool Bob in 79% of the cases.

15.5.4 Transposition is not completely positive

Let $\mathcal{H}_A$ and $\mathcal{H}_B$ be two Hilbert spaces of dimension N. Consider the maximally entangled state

$$|\varphi_{AB}\rangle = \frac{1}{\sqrt{N}} \sum_{m=1}^{N} |m_A \otimes m'_B\rangle,$$

and the corresponding state operator $\rho_{AB} = |\varphi_{AB}\rangle\langle\varphi_{AB}|$. The transposition operator $\mathcal{T}_A$ in $\mathcal{H}_A$ has the following action on ρ_{AB}:

$$(\mathcal{T}_A \otimes I_B)\rho_{AB} = \frac{1}{N} \sum_{m,n} (|n\rangle\langle m|)_A \otimes (|m'\rangle\langle n'|)_B.$$

Define $\mathcal{N} = N\rho_{AB}$ and show that applying $\mathcal{N}$ to a vector $|\varphi_A \otimes \psi_B\rangle$ has the result

$$\mathcal{N}|\varphi_A \otimes \psi_B\rangle = |\psi_A \otimes \varphi_B\rangle.$$

Show that $\mathcal{N}^2 = 1$. Write the explicit form of $\mathcal{N}$ in the case $N = 2$: this is the so-called SWAP matrix. Show, first in the case $N = 2$ and then in general, that $\mathcal{N}$ must have negative eigenvalues.

15.5.5 Phase and amplitude damping

1. Let us examine the following model with simultaneous phase and amplitude damping for a two-level system. Three Kraus operators are given by

$$
M_0 = \begin{pmatrix} 1 & 0 \\ 0 & \sqrt{1-\lambda-\gamma} \end{pmatrix}, \quad M_1 = \begin{pmatrix} 0 & \sqrt{\gamma} \\ 0 & 0 \end{pmatrix}, \quad M_2 = \begin{pmatrix} 0 & 0 \\ 0 & \sqrt{\lambda} \end{pmatrix}.
$$

Check that

$$
\sum_{\mu=0}^{2} M_\mu^\dagger M_\mu = I.
$$

What are the restrictions on λ and γ?

2. Show that the transformed state matrix $\mathcal{K}[\rho]$ is

$$
\begin{pmatrix} \rho_{00} + \gamma\rho_{11} & \rho_{01}\sqrt{1-\lambda-\gamma} \\ \rho_{10}\sqrt{1-\lambda-\gamma} & \rho_{11}(1-\gamma) \end{pmatrix}.
$$

3. What is the result after n iterations of the Kraus operator? Setting

$$
\gamma = \frac{\Gamma t}{n}, \quad \lambda = \frac{\Lambda t}{n}, \quad n \gg 1,
$$

show that

$$
\rho(t) = \begin{pmatrix} 1 - \rho_{11}e^{-\Gamma t} & \rho_{01}e^{-(\Lambda+\Gamma)t/2} \\ \rho_{10}e^{-(\Lambda+\Gamma)t/2} & \rho_{11}e^{-\Gamma t} \end{pmatrix}.
$$

What are the relaxation times T_1 and T_2? Check that $T_2 \le 2T_1$.

15.5.6 Details of the proof of the master equation

1. Show that if $\rho_{AR}(0) = \rho_A(0) \otimes \rho_R(0)$, then

$$
\mathrm{Tr}_{\mathcal{R}}\left[V(t), \rho_{AR}(0)\right] = [A(t), \rho_A(0)]\mathrm{Tr}\left(R(t)\rho_R(0)\right) = 0.
$$

2. Fill in the details of the calculations leading from (15.101) to (15.105).

15.5.7 Superposition of coherent states

We wish to study the decoherence of a superposition of two coherent states in the damped oscillator model of Section 15.3.3. The time evolution of the state operator is given by (15.79):

$$
\frac{d\rho}{dt} = -\frac{i}{\hbar}[H_0, \rho] + \frac{1}{2}\Gamma\left[2a\rho a^\dagger - \{a^\dagger a, \rho\}\right].
$$

In this problem it is instructive to keep the H_0 part of the evolution.

1. Let us consider eigenstates $|n\rangle$ of the free Hamiltonian $H_0 = \omega_0 a^\dagger a$, and let ρ_{nm} be the matrix element $\langle n|\rho|m\rangle$ of ρ. Show that the diagonal matrix element ρ_{nn} obeys

$$\frac{d\rho_{nn}}{dt} = -n\Gamma\rho_{nn} + (n+1)\Gamma\rho_{n+1,n+1}.$$

Can you give a physical interpretation for the two terms of this equation? Argue that Γ is the rate for spontaneous emission of a photon (or a phonon). What is the evolution equation for the coherence $\rho_{n+1,n}$?

2. Let us introduce the function $C(\lambda, \lambda^*; t)$ by

$$C(\lambda, \lambda^*; t) = \text{Tr}\left(\rho e^{\lambda a^\dagger} e^{-\lambda^* a}\right).$$

Show that partial derivatives with respect to λ have the following effect in the trace:

$$\frac{\partial}{\partial\lambda} \to \rho a^\dagger, \qquad \left(\frac{\partial}{\partial\lambda} - \lambda^*\right) \to a^\dagger \rho.$$

Hint: use the identity (2.54) to commute $a^\dagger$ and $\exp(-\lambda^* a)$. What are the corresponding identities for $\partial/\partial\lambda^*$?

3. Show that $C(\lambda, \lambda^*; t)$ obeys the partial differential equation

$$\left[\frac{\partial}{\partial t} + \left(\frac{\Gamma}{2} - i\omega_0\right)\frac{\partial}{\partial\ln\lambda} + \left(\frac{\Gamma}{2} + i\omega_0\right)\frac{\partial}{\partial\ln\lambda^*}\right]C(\lambda, \lambda^*; t) = 0.$$

This equation is solved by the method of characteristics. The solution is (derive it or check it!)

$$C(\lambda, \lambda^*; t) = C_0\left(\lambda\exp[-(\Gamma/2 - i\omega_0)t], \lambda^*\exp[-(\Gamma/2 + i\omega_0)t]\right)$$

with

$$C(\lambda, \lambda^*; t = 0) = C_0(\lambda, \lambda^*).$$

4. Assume that the initial state $t = 0$ is a coherent state $|z\rangle$:

$$|z\rangle = e^{-|z|^2/2} e^{za^\dagger} |0\rangle.$$

Show that in this case

$$C_0 = \exp(\lambda z^* - \lambda^* z),$$

and that the state at time t is the coherent state $|z(t)\rangle$ with

$$z(t) = z e^{-i\omega_0 t} e^{-\Gamma t/2}.$$

Therefore, a coherent state remains a coherent state when $\Gamma \neq 0$ (compare with (11.38)), but $|z(t)| \to 0$ for $t \gg 1/\Gamma$. In the complex plane, $z(t)$ spirals to the origin. As $\Gamma \ll \omega_0$, one observes many turns around the origin.

5. Let us now consider a superposition of two coherent states at $t = 0$:

$$|\Phi\rangle = c_1|z_1\rangle + c_2|z_2\rangle.$$

Show that at $t = 0$

$$C_{12}(t = 0) = \text{Tr}\left(|z_1\rangle\langle z_2|e^{\lambda a^\dagger} e^{-\lambda^* a}\right) = \langle z_2|z_1\rangle e^{\lambda z_2^*} e^{-\lambda^* z_1}.$$

What is the interpretation of $C_{12}(t)$? Let us define

$$\eta(t) = \frac{\langle z_2 | z_1 \rangle}{\langle z_2(t) | z_1(t) \rangle}$$

and write $C_{12}(t)$ in the form

$$C_{12}(t) = \eta(t) \langle z_2(t) | z_1(t) \rangle \, e^{\lambda z_2^*(t)} \, e^{-\lambda^* z_1(t)}.$$

Show that

$$|\eta(t)| = \exp\left[-\frac{1}{2} |z_1 - z_2|^2 \left(1 - e^{-\Gamma t}\right) \right] \simeq \exp\left[-\frac{\Gamma}{2} |z_1 - z_2|^2 \right],$$

where the last expression holds for $\Gamma t \ll 1$. The decoherence time is therefore

$$\tau_{\text{dec}} = \frac{2}{\Gamma |z_1 - z_2|^2}.$$

6. Let us choose $z_1 = 0$ (ground state of the oscillator) and $z_2 = z$. From question 1, the average time for the emission of *one* photon is $\sim (\Gamma |z_2|^2)^{-1}$. Argue that taking the trace over the environment (here the radiation field) shows that the coherence between the components $z_1 = 0$ and z of $|\Psi\rangle$ will be lost after the spontaneous emission of a single photon.

15.5.8 Dissipation in a two-level system

1. Starting from (15.113), derive the evolution equation for the matrix elements of the state operator ρ:

$$\frac{d\rho_{00}}{dt} = \left(G_+ + G_+^*\right)\rho_{11} - \left(G_- + G_-^*\right)\rho_{00},$$

$$\frac{d\rho_{01}}{dt} = i\omega_0 \rho_{01} - \left(G_+^* + G_-\right)\rho_{01} + \left(G_+ + G_-^*\right)\rho_{10}.$$

The last line in (15.113) therefore does not contribute to the evolution of populations.

2. Argue that in the rotating-wave approximation one can neglect the term $\left(G_+ + G_-^*\right)\rho_{10}$ in the evolution of ρ_{01}. Within this approximation, rewrite the evolution equations in terms of $\Gamma_\pm$ and $\Delta_\pm$ (15.116)–(15.117). Check that the relaxation rate is $\Gamma = \Gamma_+ + \Gamma_-$ for the populations and $\Gamma/2$ for the coherences.

3. From the expressions for Γ_+ and Γ_-, show that at equilibrium the relative populations of the levels $|0\rangle$ and $|1\rangle$ are

$$p_0 = \frac{\Gamma_-}{\Gamma}, \quad p_1 = \frac{\Gamma_+}{\Gamma},$$

and that their ratio is given by Boltzmann's law

$$\frac{p_1}{p_0} = \exp\left(-\frac{\hbar\omega_0}{k_B T}\right).$$

15.5.9 Simple models of relaxation

1. In the first model a two-level atom A is prepared in a superposition of ground ($|0_A\rangle$) and excited ($|1_A\rangle$) states at $t = 0$. The electromagnetic field is assumed to be in its ground (vacuum) state $|0_B\rangle$, so that the initial state vector is

$$|\Psi(t=0)\rangle = (\lambda|0_A\rangle + \mu|1_A\rangle) \otimes |0_B\rangle, \quad |\lambda|^2 + |\mu|^2 = 1.$$

Guided by the Wigner–Weisskopf method (Appendix C), we write the state vector at time t as

$$|\Psi(t)\rangle = \lambda|0_A \otimes 0_B\rangle + \alpha\mu|0_A \otimes 1_B\rangle + \mu\,e^{-(i\omega_0 + \Gamma/2)t}|1_A \otimes 0_B\rangle,$$

where $|1_A\rangle$ is a normalized one-photon state. Use the conservation of the norm $\||\Psi(t)\rangle\|^2 = 1$ to compute α and deduce from your computation the matrix elements of the state operator at time t. Compare with the damping models of Section 15.2 and find the Kraus operators. Show that $T_2 = 2T_1$.

2. In the second model the state $|1\rangle$ is assumed to be stable, but the resonance frequency is time-dependent. This will be the case, for example, in NMR where a spin 1/2 is submitted to a fluctuating magnetic field $\vec{B}_0(t)$. The state vector of the spin system at time t is

$$|\Psi(t)\rangle = \lambda(t)|0\rangle + \mu(t)|1\rangle,$$

with $\lambda(t)$ and $\mu(t)$ given by

$$i\dot{\lambda}(t) = -\frac{1}{2}\omega_0(t)\lambda(t), \quad i\dot{\mu}(t) = \frac{1}{2}\omega_0(t)\mu(t), \quad \lambda(0) = \lambda_0, \quad \mu(0) = \mu_0.$$

The solution is

$$\lambda(t) = \lambda_0 \exp\left[\frac{i}{2}\int_0^t \omega_0(t')dt'\right], \quad \mu(t) = \mu_0 \exp\left[-\frac{i}{2}\int_0^t \omega_0(t')dt'\right].$$

Assume that $\omega_0(t)$ is a Gaussian stationary random function with connected autocorrelation function

$$C(t') = \langle\omega_0(t+t')\omega_0(t)\rangle - \langle\omega_0\rangle^2,$$

where $\langle\bullet\rangle$ is an ensemble average over all realizations of the random function. Also assume that

$$C(t') \simeq C\exp\left(-\frac{|t'|}{\tau}\right).$$

Show that the populations ρ_{00} and ρ_{11} are time-independent, but that the coherences are given by

$$\rho_{01}(t) = \rho_{01}(t=0)e^{i\langle\omega_0\rangle t}\,e^{-C\tau t}, \quad t \gg \tau.$$

Which of the models in Section 15.2 corresponds to this situation?

15.5.10 Another choice for the spectral function $J(\omega)$

Instead of (15.131), we use another choice for the spectral function $J(\omega)$, namely

$$J(\omega) = M\gamma\omega \frac{\omega_*^2}{\omega^2 + \omega_*^2}.$$

Show that the real part $C(t)$ of the autocorrelation function is

$$C(t) = k_B T M \gamma \omega_* e^{-\omega_* |t|}.$$

Show that all the steps leading to (15.142) remain valid with this new spectral function.

15.5.11 *The Fokker–Planck–Kramers equation for a Brownian particle*

1. Let $\rho(t)$ be the state operator of the Brownian particle of Section 15.4.4. Let us define the *Wigner function* $w(x, p; t)$ by

$$w(x, p; t) = \frac{1}{2\pi\hbar} \int_{-\infty}^{+\infty} e^{-ipy/\hbar} \langle x + \frac{y}{2} | \rho(t) | x - \frac{y}{2} \rangle dy.$$

Show that another expression for $w(x, p; t)$ is

$$w(x, p; t) = \frac{1}{2\pi\hbar} \int_{-\infty}^{+\infty} e^{-ixz/\hbar} \langle p + \frac{z}{2} | \rho(t) | p - \frac{z}{2} \rangle dz.$$

Show that integrating the Wigner function over $x[p]$ gives the probability density $w_x(x; t)$ $[w_p(p; t)]$.

2. Unlike $w_x(x; t)$ and $w_p(p; t)$, the Wigner function, although real, is not necessarily positive and cannot be interpreted in a straightforward way as a probability distribution in phase space. First, compute the Wigner function for a Gaussian wave packet and check that it is positive in this particular case. Then compute the Wigner function of the superposition (15.143) of two wave packets and check that it is not positive everywhere.

3. Derive from (15.142) the following partial differential equation for $w(x, p; t)$:

$$\frac{\partial w}{\partial t} + \frac{p}{M} \frac{\partial w}{\partial x} = \gamma \frac{\partial}{\partial p} [pw] + D \frac{\partial^2 w}{\partial p^2}.$$

4. Integrate over x to obtain a Fokker–Planck equation for the probability density $w_p(p; t)$:

$$\frac{\partial w_p}{\partial t} = \gamma \frac{\partial}{\partial p} [pw_p] + D \frac{\partial^2 w_p}{\partial p^2}.$$

Show that the long-time limit of w_p is a Maxwell distribution and recover the Einstein relation between γ and $k_B T$.

15.6 Further reading

The present chapter has drawn on the following sources: Peres [1993], Chapter 9; J. Preskill, *Quantum Computation*, http://www.theory.caltech.edu/~preskill/ (1999), Chapter 3; Nielsen and Chuang [2000], Chapters 2 and 8; J. Dalibard, *Cohérence quantique et dissipation*, graduate lecture notes, Ecole Normale Supérieure, Paris (2003); S. Haroche, *Superpositions mésoscopiques d'états*, Collège de France lectures, 2003/2004. These last two references (in French) are available from the website http://www.lkb.ens.fr. Levitt [2001], Chapter 16, provides a study of the relaxation mechanisms in NMR. The concept of open quantum systems is widely used in quantum optics: see for example

H. Carmichael, *An Open System Approach to Quantum Optics*, Berlin: Springer-Verlag (1993) or M. Scully and M. Zubairy, *Quantum Optics*, Cambridge: Cambridge University Press (1997). Memory effects, linear response, and Brownian motion are studied in detail in, e.g., D. Foerster, *Hydrodynamics Fluctuations, Broken Symmetry and Correlation Functions*, New York: Benjamin (1975), Chapters 1 to 6, or Le Bellac *et al.* [2004], Chapter 9. The model in Section 15.4 was first introduced by A. Caldeira and A. Leggett, *Physica* **121** A, 587 (1983); see also C. Cohen-Tannoudji, J. Dupont-Roc, and G. Grynberg, *Atom–Photon Interactions*, New York: Wiley (1992), Chapter IV. Recent references to the Caldeira–Leggett model can be traced from the review article by Zurek [2003].

Appendix A The Wigner theorem and time reversal

In this Appendix we shall demonstrate the Wigner theorem stated in Section 8.1.2 and study invariance under time reversal, which is special because the operator that realizes this symmetry in the Hilbert space $\mathcal{H}$ is antiunitary rather than unitary, in contrast to all the other cases we have encountered so far. Let us recall the definition (see Section 8.1.1) of a ray in Hilbert space: a ray is a vector up to a phase factor. Two unit vectors φ and $\overline{\varphi}$ differing by a phase factor $\varphi = \exp(i\alpha)\,\overline{\varphi}$ belong to the same equivalence class, which is precisely a ray $\tilde{\varphi}$ of $\mathcal{H}$. Since the modulus of the scalar product is independent of the representation in the equivalence class

$$|(\overline{\varphi}, \overline{\chi})| = |(\varphi, \chi)|,$$

the *modulus* of the scalar product of two rays $\tilde{\varphi}$ and $\tilde{\chi}$ is well defined by choosing two arbitrary representatives in each equivalence class:

$$|(\tilde{\varphi}, \tilde{\chi})| = |(\varphi, \chi)|, \tag{A.1}$$

but it is quite clear that it makes no sense to speak of the scalar product of two rays. We shall use the notation $(\bullet, \bullet)$ for the scalar product in order to avoid the ambiguities of the Dirac notation, which would be particularly cumbersome in this appendix.

Let there be in $\mathcal{H}$ a correspondence between rays

$$\tilde{\varphi} \rightarrow T\tilde{\varphi} \tag{A.2}$$

such that the modulus of the scalar product is invariant:

$$|(\tilde{\varphi}, \tilde{\chi})| = |(T\tilde{\varphi}, T\tilde{\chi})|. \tag{A.3}$$

The Wigner theorem states that it is always possible to choose the phases of vectors such that the correspondence between *rays* becomes a correspondence between *vectors*:

$$\varphi \rightarrow U\varphi, \quad |(U\varphi, U\chi)| = |(\varphi, \chi)|, \tag{A.4}$$

where the transformation U is either linear and unitary

$$(U\varphi, U\chi) = (\varphi, \chi) \tag{A.5}$$

or antilinear and unitary (= antiunitary):

$$(U\varphi, U\chi) = (\chi, \varphi) = (\varphi, \chi)^*. \tag{A.6}$$

A.1 Proof of the theorem

Let $\{\chi_i\}$, $i = 1, \ldots, N$, be an orthonormal basis of $\mathcal{H}$ assumed to have dimension N, $(\chi_i, \chi_k) = \delta_{ik}$. We shall assign a special role to the first basis vector: by convention, the indices i and k will vary between 1 and N and the indices j and l between 2 and N. We choose a representative $\chi_1'' \equiv \chi_1'$ in the class of $T\tilde{\chi}_1$ and a representative χ_j'' in the class of $T\tilde{\chi}_j$, $j = 2, \ldots, N$. According to (A.3), the set $\{\chi_1'', \chi_j''\}$ also forms a basis of $\mathcal{H}$ because

$$|(\chi_i'', \chi_k'')| = |(\chi_i, \chi_k)| = \delta_{ik}.$$

Let us consider the set of vectors φ_j

$$\varphi_j = \chi_1 + \chi_j, \quad j = 2, \ldots, N, \tag{A.7}$$

and let $T\tilde{\varphi}_j$ be the transform of the ray $\tilde{\varphi}_j$. If φ_j'' is a representative of $T\tilde{\varphi}_j$, we will have

$$|(\chi_1', \varphi_j'')| = |(\chi_1, \varphi_j)| = 1,$$
$$|(\chi_j'', \varphi_l'')| = |(\chi_j, \varphi_l)| = \delta_{jl}.$$

A representative φ_j'' of $T\tilde{\varphi}_j$ will then have components only along χ_1' and χ_j'':

$$\varphi_j'' = c_j\chi_1' + d_j\chi_j'',$$

and these components will have unit modulus: $|c_j| = |d_j| = 1$. We can now choose representatives φ_j' and χ_j'

$$\varphi_j' = \frac{1}{c_j}\varphi_j'', \quad \chi_j' = \frac{d_j}{c_j}\chi_j'' \tag{A.8}$$

such that

$$\varphi_j' = \frac{1}{c_j}(c_j\chi_1' + d_j\chi_j'') = \chi_1' + \chi_j'. \tag{A.9}$$

We have thus defined an operation on *vectors* of $\mathcal{H}$

$$\chi_1 + \chi_j \to (\chi_1 + \chi_j)' = \chi_1' + \chi_j'$$

such that $\chi_1' \in T\tilde{\chi}_1$, $\chi_j' \in T\tilde{\chi}_j$, and $\chi_1' + \chi_j' \in T(\widetilde{\chi_1 + \chi_j})$. Let us now try to determine if it is possible that an arbitrary vector ψ transforms as

$$\psi = \sum_{k=1}^{N} c_k\chi_k \to \psi' = \sum_{k=1}^{N} c_k'\chi_k'.$$

If such a transformation law is valid, we must have, on the one hand,

$$|c'_k| = |(\chi'_k, \psi')| = |(\chi_k, \psi)| = |c_k|,$$

and on the other,

$$(\chi_1 + \chi_j, \psi) = c_1 + c_j, \quad (\chi'_1 + \chi'_j, \psi') = c'_1 + c'_j,$$

which according to (A.3) implies that

$$|c_1 + c_j| = |c'_1 + c'_j|. \tag{A.10}$$

The two pairs of complex numbers (c_1, c_j) and (c'_1, c'_j) must be such that $|c_1| = |c'_1|$ and $|c_j| = |c'_j|$ and they must also satisfy (A.10). We set

$$c_1 = |c_1| e^{i\theta_1}, \quad c_j = |c_j| e^{i\theta_j},$$

$$c'_1 = |c'_1| e^{i\theta'_1}, \quad c'_j = |c'_j| e^{i\theta'_j}.$$

The angles (θ_1, θ_j) and (θ'_1, θ'_j) are related by the equation

$$\cos(\theta_1 - \theta_j) = \cos(\theta'_1 - \theta'_j), \tag{A.11}$$

which has two solutions

$$\theta_1 - \theta_j = \theta'_1 - \theta'_j, \tag{A.12}$$

$$\theta_1 - \theta_j = -(\theta'_1 - \theta'_j). \tag{A.13}$$

Let us examine the first case. We can redefine the phase of ψ' such that $c'_1 = c_1$ and then $\theta'_1 = \theta_1$. In this case $\theta'_j = \theta_j$ and $c'_j = c_j$, and so

$$\psi' = \sum_k c_k \chi'_k.$$

If we consider another vector $\eta = \sum_k d_k \chi_k$ again with $d'_1 = d_1$, we will have

$$(\lambda\psi + \mu\eta)' = \sum_k (\lambda c_k + \mu d_k) \chi'_k = \lambda\psi' + \mu\eta'.$$

By a suitable choice of phase the transformation T can be chosen to be linear, and since it conserves the modulus of the scalar product, it is also unitary: $T \to U$ with $U^\dagger U = UU^\dagger = I$.

In the second case we redefine the phase of ψ' such that $c'_1 = c_1^*$. We then have $c'_j = c_j^*$ and

$$\psi' = \sum_k c_k^* \chi'_k.$$

The transform of $\lambda\psi + \mu\eta$ then is

$$(\lambda\psi + \mu\eta)' = \left[\sum_k (\lambda c_k + \mu d_k) \chi_k\right]' = \lambda^*\psi' + \mu^*\eta', \tag{A.14}$$

and the transformation law of the scalar product is

$$(\psi', \eta') = (\psi, \eta)^* = (\eta, \psi).$$ (A.15)

The transformation takes $T \to V$, where V is termed *antiunitary*. It is antilinear and conserves the norm.

The preceding proof is actually incomplete. In fact, it should be shown that it is not possible to have (A.12) for c_j and (A.13) for c_l, $l \neq j$. The proof that this cannot happen is cumbersome and we leave it to the reader;[1] it requires examining the behavior of the transform of a vector $\psi = \chi_1 + \chi_j + \chi_l$.

A.2 Time reversal

In classical mechanics, Newton's equation

$$m\frac{d^2\vec{r}(t)}{dt^2} = \vec{F}(\vec{r}(t))$$

is invariant under time reversal $t \to -t$. If we take $\vec{r}'(t) = \vec{r}(-t)$, then

$$m\frac{d^2\vec{r}'(t)}{dt^2} = m\frac{d^2\vec{r}(-t)}{dt^2} = \vec{F}(\vec{r}(-t)) = \vec{F}(\vec{r}'(t)).$$

One observes that $\vec{r}'(t)$ also obeys Newton's equation. The reason is obviously that this equation depends only on the second derivative of $\vec{r}$ with respect to time and not on the first derivative.[2] An intuitive image of time reversal is the following: we imagine that we follow the trajectory of a particle from $t = -\infty$ to $t = 0$ and that at $t = 0$ we abruptly reverse the direction of the momentum (or velocity): $\vec{p}(0) \to -\vec{p}(0)$. Under these conditions the particle "retraces" its trajectory, passing at time t the position it had at time $-t$ with momentum in the opposite direction (Fig. A.1):

$$\vec{r}'(t) = \vec{r}(-t), \quad \vec{p}'(-t) = -\vec{p}(t).$$ (A.16)

The position vector $\vec{r}$ is even under time reversal while $\vec{p}$ is odd. Invariance under time reversal is called *microreversibility*. If we film the motion of some particles and then run the film backward, then microreversibility implies that this projection appears to be physically possible.[3] We know that this is not the case in everyday life, which is fundamentally irreversible, and it is not yet completely clear,[4] even to this day, how a

---

[1] See Weinberg [1995], Chapter 2, where all the subtleties of the proof are explained in detail.

[2] An equation like that of the damped harmonic oscillator

$$m\ddot{x} + \gamma\dot{x} + m\omega^2 x = 0$$

is not invariant under time reversal, but the viscosity force $-\gamma\dot{x}$ is an effective force, phenomenologically representing the effect of collisions with the fluid molecules on the particle of mass m.

[3] The analogy with parity conservation is obvious; the image of an experiment in a mirror appears to be physically possible if parity is conserved.

[4] As already shown by the heated discussions between Boltzmann and his adversaries. See, for example, Balian [1991], Chapter 15, or Le Bellac *et al.* [2004], Chapter 2.

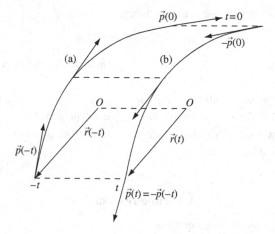

Fig. A.1. Time reversal on a classical trajectory.

dynamics which is reversible at the microscopic scale can lead to phenomena which are irreversible at the macroscopic scale.

Let us now return to quantum mechanics, using Θ to denote the operator that performs time reversal in $\mathcal{H}$. This operator must transform $\vec{R}$, $\vec{P}$, and $\vec{J}$ as

$$\Theta \vec{R} \Theta^{-1} = \vec{R},$$

$$\Theta \vec{P} \Theta^{-1} = -\vec{P}, \tag{A.17}$$

$$\Theta \vec{J} \Theta^{-1} = -\vec{J}.$$

Actually, $\vec{J}$ must transform as $\vec{R} \times \vec{P}$, which is odd under time reversal: the angular momentum defines a sense of rotation which is reversed by time reversal. Examination of the Θ transformation of the canonical commutation relations shows that Θ must be antiunitary. Let us calculate a matrix element of the commutator $[X_i, P_j] = i\hbar \delta_{ij} I$ in two different ways:

$$(\Theta\varphi, \Theta[X_i, P_j]\psi) = (\Theta\varphi, \Theta i\hbar\delta_{ij}I\psi) = \delta_{ij}(\varphi, i\hbar\psi)^* = -i\hbar\delta_{ij}(\varphi, \psi)^*$$

$$= (\Theta\varphi, \Theta[X_i, P_j]\Theta^{-1}\Theta\psi) = (\Theta\varphi, -i\hbar\delta_{ij}I\Theta\psi) = -i\hbar\delta_{ij}(\varphi, \psi)^*,$$

where in the second line we have used the transformation laws (A.17) for X_i and P_j:

$$\Theta[X_i, P_j]\Theta^{-1} = -[X_i, P_j].$$

The two lines of the preceding equation are compatible, which would not be the case if the Θ transformation were unitary.

There is another very instructive argument proving the antiunitarity of Θ. Let $\varphi(t)$ be the state vector of a quantum system at time t, and let $\varphi = \varphi(t = 0)$ be its state at time $t = 0$:

$$\varphi(t) = \exp\left(-\frac{i}{\hbar} Ht\right) \varphi.$$

Invariance under time reversal implies that the state transformed from $\varphi(-t)$ by time reversal, $\Theta\varphi(-t)$, coincides with the state obtained by the time evolution of $\Theta\varphi(0)$:

$$\Theta\varphi(-t) = \exp\left(-\frac{i}{\hbar}Ht\right)\Theta\varphi,$$

and since these equations are valid for all φ,

$$\Theta\exp\left(\frac{i}{\hbar}Ht\right) = \exp\left(-\frac{i}{\hbar}Ht\right)\Theta. \tag{A.18}$$

If Θ were unitary, this would imply that

$$\Theta H = -H\Theta,$$

and to any eigenvector φ_E of H with energy E there would correspond an eigenvector $\Theta\varphi_E$ of energy $-E$. Under these conditions the energy would not be bounded below and a fundamental instability would exist. If on the contrary Θ is antiunitary, since

$$\Theta iH = -i\Theta H,$$

(A.18) implies that

$$\Theta H = H\Theta \quad \text{or} \quad \Theta H\Theta^{-1} = H. \tag{A.19}$$

This equation expresses the invariance of H under time reversal. However, in contrast to the parity operator Π, Θ does not lead to a conserved quantity, because (8.17) implies that the operator A is Hermitian, which is not the case with Θ. It is known that all the fundamental interactions of physics are invariant under time reversal except for an extremely weak interaction whose effects are seen only in the K^0–$\overline{K}^0$-meson system (Exercise 4.4.8), and also very recently in the system of B mesons, which are formed of an ordinary quark and a bottom (b) antiquark or vice versa.

A double time reversal obviously does not have any effect, and the state $\Theta^2\varphi$ is equivalent to φ, $\Theta^2 = cI$, where c is a phase factor. The chain of equalities

$$(\Theta\varphi_a, \varphi_b) = (\Theta\varphi_b, \Theta^2\varphi_a) = c(\Theta\varphi_b, \varphi_a) = c(\Theta\varphi_a, \Theta^2\varphi_b) = c^2(\Theta\varphi_a, \varphi_b)$$

shows that $c^2 = 1$, so that $c = \pm 1$. In the case where $c = -1$, the choice $\varphi_a = \varphi_b$ in the preceding equation implies that

$$(\Theta\varphi_a, \varphi_a) = 0. \tag{A.20}$$

If $c = -1$ and H is invariant under time reversal, the eigenstates of H can be arranged as pairs of states which are degenerate under time reversal. Let φ be an eigenvector of H, $H\varphi = E\varphi$. Then

$$H(\Theta\varphi) = \Theta(H\varphi) = E(\Theta\varphi)$$

and $\Theta\varphi$ is an eigenvector of H with eigenvalue E: if $(\Theta\varphi, \varphi) = 0$, there exist (at least) two eigenstates of H with eigenvalue E. This property is called *Kramers degeneracy*.

Taking into account the transformation properties of $\vec{J}$ (A.17), we must have

$$\Theta|jm\rangle = e^{i\alpha}(-1)^{j-m}|j,-m\rangle, \tag{A.21}$$

where by applying J_+ and J_- it can be shown that α can depend on j, but not on m. The antilinearity of Θ can be used to show that

$$\Theta^2|jm\rangle = (-1)^{2j}|jm\rangle \tag{A.22}$$

and so $\Theta^2 = I$ if j is an integer or $\Theta^2 = -I$ if j is a half-integer. The Kramers degeneracy then implies that a system with an odd number of electrons possesses energy levels that are doubly degenerate in the absence of a magnetic field. The presence of a magnetic field breaks the invariance under time reversal, because in order to respect this invariance it would be necessary to reverse the direction of the currents producing this field. The reason the Zeeman effect completely lifts the level degeneracy is that the magnetic field breaks the invariance under time reversal.

Invariance under time reversal implies that for a transition amplitude $\mathcal{T}_{a \to b}$

$$\mathcal{T}_{a \to b} = \mathcal{T}_{\Theta b \to \Theta a}, \tag{A.23}$$

where Θa (Θb) is the state obtained from a (b) by time reversal, by reversing all the momenta and angular momenta. We can derive, for example, the relation for the scattering amplitude used in Section 12.3.2:

$$f(\vec{k}', \vec{k}) = f(-\vec{k}, -\vec{k}'),$$

and, more generally, for a reaction in which the incident particles have momenta $(\vec{p}_1, \vec{p}_2)$ and spin projections (m_1, m_2) and the final particles have $(\vec{p}_3, \vec{p}_4)$ and (m_3, m_4),

$$f_{m_1, m_2; m_3, m_4}(\vec{p}_1 + \vec{p}_2 \to \vec{p}_3 + \vec{p}_4) = f_{-m_3, -m_4; -m_1, -m_2}(-\vec{p}_3 - \vec{p}_4 \to -\vec{p}_1 - \vec{p}_2).$$

For a particle without spin, the operation of time reversal is simply complex conjugation. If $\psi(\vec{r}, t)$ satisfies the Schrödinger equation

$$i\hbar \frac{\partial \psi(\vec{r}, t)}{\partial t} = -\frac{\hbar^2}{2m} \nabla^2 \psi(\vec{r}, t) + V(\vec{r})\psi(\vec{r}, t),$$

the function $\Theta \psi(\vec{r}, t) = \psi^*(\vec{r}, -t)$ satisfies

$$i\hbar \frac{\partial \psi^*(\vec{r}, -t)}{\partial t} = -\frac{\hbar^2}{2m} \nabla^2 \psi^*(\vec{r}, -t) + V(\vec{r})\psi^*(\vec{r}, -t)$$

provided the potential $V(\vec{r})$ is real. This property has been used in Sections 9.4.1 and 9.4.3 to restrict the form of the transmission matrix M and of the S matrix.

As a final example, let us examine the impact of invariance under time reversal on the neutron electric dipole moment. Since the dipole moment operator $\vec{D}$ is odd under parity,

$$\Pi \vec{D} \Pi^{-1} = -\vec{D},$$

the dipole moment[5] of a particle is zero if this particle has definite parity, which will be the case if its interactions conserve parity. This is why atoms in their ground state do not have a permanent dipole moment. However, parity is not conserved in the weak interactions, and this can a priori restore the possibility of a dipole moment. In fact, it is also necessary that invariance under time reversal be violated. The only vector at our disposal is the neutron spin $\hbar \vec{\sigma}/2$, and we must have $\vec{D} = \lambda \vec{\sigma}$, where λ is a constant. We note that $\lambda \neq 0$ implies parity violation because $\vec{D}$ is a vector and $\vec{\sigma}$ is a pseudovector. The coupling $\vec{D} \cdot \vec{E}$ of a dipole to an electric field is odd under time reversal and must vanish if there is invariance under time reversal, because according to (A.17) $\vec{\sigma}$ is odd and $\vec{E}$ is even; under time reversal, charges are not changed (but currents are reversed, as we have seen above). If we send a neutron possessing an electric dipole moment in nonuniform, constant electric and magnetic fields and at $t = 0$ reverse the neutron velocity and the currents creating the magnetic field, then, in contrast to the case in Fig. A.1, the neutron will not "retrace" its trajectory.

Let us try to estimate the neutron dipole moment by a dimensional argument. This dipole moment must involve weak interactions and therefore the Fermi constant G_F (Exercise 12.5.6), or, more precisely, the combination $G_F/(\hbar c)^3$, and a dimensionless parameter ε measuring the importance of the violation of time-reversal invariance. Its order of magnitude can be estimated to be about 10^{-3} based on study of neutral K mesons. We also have a mass at our disposal, the neutron mass $m_n \simeq 1 \, \text{GeV} c^{-2}$. By dimensional analysis the only possible solution is

$$d \sim q_e \frac{G_F}{(\hbar c)^3} \varepsilon \, m_n (\hbar c^3).$$

It is convenient to use a system of units in which $\hbar = c = 1$, $200 \, \text{MeV} \simeq 1 \, \text{fm}^{-1}$ (Exercise 12.5.1), or $1 \, \text{fm} \simeq 5 \, \text{GeV}^{-1}$:

$$d \sim q_e \times 10^{-5} \times 10^{-3} \times 1 = q_e \times 10^{-8} \, \text{GeV}^{-1} \sim q_e \times 10^{-9} \, \text{fm} = q_e \times 10^{-24} \, \text{m}.$$

The most precise measurements of the neutron dipole moment have been made at the research reactor of the Laue–Langevin Institute in Grenoble and give the upper bound

$$d \lesssim q_e \times 10^{-27} \, \text{m},$$

which strongly disagrees with our naive estimate! In fact, owing to a technical feature of the Standard Model,[6] the neutron dipole moment must be proportional to G_F^2:

$$d \sim G_F^2 \varepsilon \, m_n^3 (\hbar c^7) \simeq q_e \times 10^{-29} \, \text{m}.$$

---

[5] For a particle to have nonzero electric dipole moment, it is imperative that its angular momentum be nonzero. If that is not the case, rotational invariance is incompatible with the existence of a dipole moment.

[6] See, for example, J. Donoghue, E. Golowich, and B. Holstein, *Dynamics of the Standard Model*, Cambridge: Cambridge University Press (1992), Chapter IX.

The theoretical estimates of the neutron dipole moment are not very accurate and generally lie somewhere near $q_e \times 10^{-32}$ m; our estimate is reduced by a factor of $\sim 10^{-3}$ because perturbative calculations using the Standard Model lead to a multiplicative numerical factor of $\pi^{-4} \simeq 10^{-2}$ and suggest that a typical mass of order 0.3 GeV be used instead of m_n.

Appendix B Measurement and decoherence

In this appendix we shall describe in more detail how the experiment of Brune *et al.* mentioned in Section 6.4.1 provided evidence of the phenomenon of decoherence in an entirely controlled manner. In addition to its intrinsic interest, this experiment is a prime example of actual experiments which allow the fundamentals of quantum mechanics to be tested with a precision undreamed of by its founders, and the study of this experiment constitutes a beautiful exercise in quantum physics. It will also allow us to give a small sample of the current ideas on the notion of measurement in quantum mechanics.[1] We shall first return to the interference experiment of Section 1.4.4, this time discussing it within the framework of an elementary theory of measurement. Then we shall examine the realization of Ramsey fringes using Rydberg atoms, and show how the interaction of these atoms with an electromagnetic field progressively blurs these fringes when we try to answer the "which of the two trajectories?" question. Finally, we shall show how the use of a pair of atoms allows decoherence to be tested.

B.1 An elementary model of measurement

Let us return to the discussion of the Young's slit experiment with the trajectories labeled as in Fig. 1.13, enlarging on it with the introduction of a mathematical formulation. Let $c_1(x)$ [$c_2(x)$] be the complex probability amplitude for an atom to be localized at a point x on the screen after having passed through slit 1 [2]. The (arbitrary) normalization is fixed by $|c_1(x)|^2 = |c_2(x)|^2 = 1$. In the absence of any device for observing the trajectories, the probability of arriving at a point x on the screen is

$$p(x) = \frac{1}{2}\left(|c_1(x)|^2 + |c_2(x)|^2 + 2\,\mathrm{Re}\,[c_1(x)c_2^*(x)]\right). \tag{B.1}$$

The last term in (B.1) is of course the interference term. The probability amplitude $c_1(x)$ is the product of the amplitude[2] $\langle\varphi_1|S\rangle$ for an atom emitted by the source S to be localized

---

[1] A very complete discussion of measurement theory can be found in the 1989–1990 course at the Collège de France by C. Cohen-Tannoudji (in French, available from the website www.lkb.ens.fr). The current ideas on measurement owe a great deal to the work of W. Zurek, a pedagogical discussion of which can be found in W. Zurek, *Physics Today*, October 1991, p. 36.

[2] With, for example, $\langle\varphi_1|S\rangle \propto \exp(ikr_{1S})/r_{1S}$, where k is the modulus of the wave vector of the atom and r_{1S} is the modulus of the vector joining the source to slit 1; cf. Feynman *et al.* [1965], Vol. III, Chapter 3.

at slit 1 and the amplitude $\langle x|\varphi_1\rangle$ for an atom emitted by slit 1 to be localized at x on the screen. There is an analogous expression for $c_2(x)$, and so

$$c_1(x) = \langle x|\varphi_1\rangle\langle\varphi_1|S\rangle, \quad c_2(x) = \langle x|\varphi_2\rangle\langle\varphi_2|S\rangle.$$

It is convenient to include the amplitudes $\langle\varphi_1|S\rangle$ and $\langle\varphi_2|S\rangle$ in the definition of the atomic states $|\varphi_1\rangle$ and $|\varphi_2\rangle$ and write simply

$$c_1(x) = \langle x|\varphi_1\rangle, \quad c_2(x) = \langle x|\varphi_2\rangle. \tag{B.2}$$

The states $|\varphi_1\rangle$ and $|\varphi_2\rangle$ are assumed to be normalized and orthogonal, because they are localized at different slits and their wave functions do not overlap. Let us now place the cavities C_1 and C_2 of Fig. 1.13 in front of the slits and let $|\chi_{10}\rangle$ be the state where C_1 contains one photon and C_2 zero photons, and $|\chi_{01}\rangle$ be the state describing the opposite situation. The atom + photon state is then an entangled state $|\Psi\rangle$:

$$|\Psi\rangle = \frac{1}{\sqrt{2}}\left(|\varphi_1\rangle \otimes |\chi_{10}\rangle + |\varphi_2\rangle \otimes |\chi_{01}\rangle\right), \tag{B.3}$$

and the corresponding state operator is

$$\rho_{\text{tot}} = |\Psi\rangle\langle\Psi| = \frac{1}{2}\Big[(|\varphi_1\rangle\langle\varphi_1|) \otimes (|\chi_{10}\rangle\langle\chi_{10}|) + (|\varphi_2\rangle\langle\varphi_2|) \otimes (|\chi_{01}\rangle\langle\chi_{01}|)$$
$$+ (|\varphi_1\rangle\langle\varphi_2|) \otimes (|\chi_{10}\rangle\langle\chi_{01}|) + (|\varphi_2\rangle\langle\varphi_1|) \otimes (|\chi_{01}\rangle\langle\chi_{10}|)\Big]. \tag{B.4}$$

Let us now seek the reduced state operator of the atom alone using (6.34):

$$\rho_{\text{at}} = \text{Tr}_{\text{phot}}\, \rho_{\text{tot}} = \frac{1}{2}\Big[|\varphi_1\rangle\langle\varphi_1| + |\varphi_2\rangle\langle\varphi_2| + (\langle\chi_{01}|\chi_{10}\rangle|\varphi_1\rangle\langle\varphi_2| + \text{H.c.})\Big], \tag{B.5}$$

where H.c. denotes the Hermitian-conjugate expression. In the basis $\{|\varphi_1\rangle, |\varphi_2\rangle\}$ the matrix form of this result is

$$\rho_{\text{at}} = \frac{1}{2}\begin{pmatrix} 1 & \langle\chi_{01}|\chi_{10}\rangle \\ \langle\chi_{10}|\chi_{01}\rangle & 1 \end{pmatrix}. \tag{B.6}$$

We recall that the off-diagonal elements of ρ_{at} are called coherences. In the scheme of Fig. 1.13, the states $|\chi_{10}\rangle$ and $|\chi_{01}\rangle$ are orthogonal: $\langle\chi_{10}|\chi_{01}\rangle = 0$, which reflects the localization of the photons in two different cavities such that their wave functions do not overlap. Under these conditions the state matrix (B.6) is diagonal. It is instructive to consider a more general situation, where the photon associated with passage of an atom through slit 1 is not completely localized in the cavity C_1, but has a certain probability of leaking toward C_2, and vice versa for the photon associated with passage of an atom through slit 2. Under these conditions the observation of a photon in C_1 or C_2 does not allow a definite labeling of the atomic trajectory. We easily obtain the probability $p(x)$ of arriving at a point x on the screen:

$$p(x) = \text{Tr}\Big(|x\rangle\langle x|\rho_{\text{at}}\Big) = \frac{1}{2}\Big(|c_1(x)|^2 + |c_2(x)|^2 + 2\,\text{Re}\,[c_1(x)c_2^*(x)\langle\chi_{01}|\chi_{10}\rangle]\Big). \qquad (\text{B.7})$$

The photon emitted in C_1 or C_2 performs a *measurement = labeling of the trajectory*, or, more precisely, a *premeasurement*, and this premeasurement corresponds to the formation of an entangled state (B.3), that is, to the establishment of quantum correlations between the atom (the system) and the photon (the measuring device). The possible interferences are contained in the coherences of the reduced state matrix (B.6), and these interferences vanish if the coherences are zero, when $|\chi_{10}\rangle$ and $|\chi_{01}\rangle$ are orthogonal. In this case the measurement of the trajectory is unambiguous. It is not necessary for the photon to be observed, or, in other words, for the measurement result to be recorded, in order to obtain (B.7). It is the entanglement of the photon with the atom and the orthogonality of the states $|\chi_{10}\rangle$ and $|\chi_{01}\rangle$ that destroy the coherences. On the contrary, if the states $|\chi_{10}\rangle$ and $|\chi_{01}\rangle$ are not orthogonal, the measurement of the trajectory is not unambiguous, and the interferences are only partially blurred, the blurring being more important the closer $\langle\chi_{10}|\chi_{01}\rangle$ is to zero. In the limit where $|\langle\chi_{10}|\chi_{01}\rangle| = 1$, the device gives no information on the trajectories, the interferences are completely re-established, and we recover (B.1).

The preceding discussion can be generalized to an elementary measurement model. Let us suppose that we wish to make a measurement on a quantum system S which can be found in one of N states $|\varphi_n\rangle$ belonging to the space of states $\mathcal{H}_S^{(N)}$ of dimension N. The first phase of the measurement, which we shall call the *premeasurement phase*, is performed using an interaction between S and another quantum system M, the "measuring device." In the above example, the atom is the system S and the photon is the measuring device M. If S is initially in the state $|\varphi_n\rangle$ and M is in the state $|\chi\rangle$, we assume that the interaction between S and M has the following effect:

$$|\varphi_n\rangle \otimes |\chi\rangle \Longrightarrow |\varphi_n\rangle \otimes |\chi_n\rangle,$$

where $|\chi\rangle$ and $|\chi_n\rangle$ belong to a Hilbert space $\mathcal{H}_M$. An explicit mechanism giving this type of result is described in Exercise 9.7.14. It is crucial to note that the evolution during the premeasurement phase where S and M interact is unitary and governed by an evolution equation of the type (4.11) with a Hamiltonian H_{S+M}. The reading of the final state of M makes it possible to recover the initial state $|\varphi_n\rangle$ of S: M is a "needle" whose "position" $|\chi_n\rangle$ gives the state of S. The linearity of quantum mechanics implies that if the initial state of S is the linear superposition $|\varphi\rangle = \sum_{n=1}^{N} c_n|\varphi_n\rangle$, the result of the premeasurement is given by

$$|\varphi\rangle \otimes |\chi\rangle = \left(\sum_{n=1}^{N} c_n|\varphi_n\rangle\right) \otimes |\chi\rangle \Longrightarrow \sum_{n=1}^{N} c_n|\varphi_n\rangle \otimes |\chi_n\rangle.$$

The result is an entangled state of $S + M$. We easily calculate the reduced density operator of S using (6.34):

$$\rho_S = \text{Tr}_M \, \rho_{S+M} = \sum_{n,m=1}^{N} c_n c_m^* |\varphi_n\rangle\langle\varphi_m| \langle\chi_m|\chi_n\rangle. \tag{B.8}$$

If the states $|\chi_n\rangle$ are orthogonal, $\langle\chi_n|\chi_m\rangle = \delta_{nm}$, the result of the measurement is unambiguous, because the observation of M determines the state of S uniquely, and the coherences of ρ_S vanish:

$$\rho_S = \sum_{n=1}^{N} |c_n|^2 |\varphi_n\rangle\langle\varphi_n|. \tag{B.9}$$

The reduced state operator ρ_S is completely different from the initial state operator ρ_S^{in} of S:

$$\rho_S^{\text{in}} = \sum_{n,m=1}^{N} c_n c_m^* |\varphi_n\rangle\langle\varphi_m|. \tag{B.10}$$

The coherences have vanished in going from (B.10) to (B.9). Only the information on the *probabilities* $|c_n|^2$ of finding S in the state $|\varphi_n\rangle$ is conserved. However, the situation is still reversible: as long as the system $S + M$ remains closed, only a premeasurement has been made, not a true measurement, and the information on the phases has not been lost in the full $S + M$ system. Moreover, it is possible to use a basis of $\mathcal{H}_M$ other than the basis $\{|\chi_n\rangle\}$; this new basis is coupled to a basis of $\mathcal{H}_S$ which is different from the basis $\{|\varphi_n\rangle\}$, and physical properties different from those measured in the former case are associated with it. There are therefore ambiguities in the physical properties of S which are measured by M. However, the interactions of M with its environment, which have not been taken into account up to now, will select a preferred basis of pointer states, thus lifting the ambiguities.

B.2 Ramsey fringes

Let us now discuss the experiment of Brune *et al.* The experimental setup is shown in Fig. 6.10. Rubidium atoms in a *circular Rydberg state* (Exercise 14.6.4) are prepared at O. A Rydberg state of rubidium (which is an alkali atom) is an atomic state in which the outer electron of the atom is located in an orbit of very high principal quantum number n, and so the size of the atom is very large compared with the Bohr radius a_0. Moreover, the orbital angular momentum is made to take its maximum value $l = n - 1$, as is the magnetic quantum number $|m| = l$. Under these conditions a circular Rydberg state is obtained, that is, a state in which the orbit is circular and the electron is confined in a very thin torus about the average radius of the orbit $\simeq n^2 a_0 \simeq 125$ nm. In the experiment the two Rydberg states that are used correspond to $n = 50$ (denoted $|g\rangle$) and $n = 51$ (denoted $|e\rangle$). These states are separated in energy by 0.21 meV, which corresponds to a frequency $\omega_0 = 3.21 \times 10^{11}$ rad s^{-1} ($\nu = 51.1$ GHz). Owing to the choice of circular orbits, these states have a very long lifetime, of order 30 ms, on the atomic scale, and the probability of spontaneous decay during their flight between O and the detectors D is negligible. The atoms are detected by *selective* ionization detectors D_e and D_g, because the states e and g are ionized by different fields. The efficiency of the detectors, that is, the probability that D_e is triggered by e and D_g by g, is of order 40%, while the probability of triggering by the "wrong" state is a few percent.

At first the cavity C is empty and the atoms are subjected to a radiofrequency field

$$\mathcal{E}(t) = E_0 \cos(\omega t - \phi) \tag{B.11}$$

in the cavities R_1 and R_2, where the value of ϕ depends on the cavity. The frequency ω is close to the resonance frequency ω_0 and the detuning is $\delta = \omega - \omega_0$. To an excellent approximation the atom+field system is described by a two-level system $|e\rangle$ and $|g\rangle$ interacting with a classical field (B.11). This system has been studied in detail in Chapter 5, and we can immediately use Equations (5.32) with only trivial modifications to take into account the phase ϕ in (B.11). It is convenient to revert to the notation of Chapter 5 and to define

$$|g\rangle \to |+\rangle, \quad |e\rangle \to |-\rangle, \quad E_e - E_g \to E_- - E_+ = \hbar\omega_0 > 0.$$

The solution of the evolution equations with the initial conditions $\gamma_+(0) = 1$, $\gamma_-(0) = 0$ is, when $\phi \neq 0$,

$$\gamma_+(t) = \cos\frac{\omega_1 t}{2},$$

$$\gamma_-(t) = -i\,e^{i\phi} \sin\frac{\omega_1 t}{2}, \tag{B.12}$$

where the functions $\gamma_\pm(t)$ are defined in (5.26). The solution of the evolution equations with the initial conditions $\gamma_+(0) = 0$, $\gamma_-(0) = 1$ is obtained without calculation by noting that it is sufficient to make the substitutions $+ \leftrightarrow -$ and $\phi \to -\phi$ in (B.12):

$$\gamma_+(t) = -i\,e^{-i\phi} \sin\frac{\omega_1 t}{2},$$

$$\gamma_-(t) = \cos\frac{\omega_1 t}{2}. \tag{B.13}$$

If the time to cross the cavity R is adjusted such that $\omega_1 t/2 = \pi/4$, that is a $\pi/2$ pulse, an atom entering the cavity in the state $|+\rangle$ leaves according to (B.12) in the state $|\varphi_+\rangle$,

$$|\varphi_+\rangle = \frac{1}{\sqrt{2}}\left(|+\rangle - i e^{i\phi}|-\rangle\right), \tag{B.14}$$

and one entering in the state $|-\rangle$ leaves according to (B.13) in the state $|\varphi_-\rangle$,[3]

$$|\varphi_-\rangle = \frac{1}{\sqrt{2}}\left(-i e^{-i\phi}|+\rangle + |-\rangle\right). \tag{B.15}$$

---

[3] We can get rid of the factors of i by redefining the phase of the states $|\pm\rangle$ and returning to the phase conventions of Section 6.4.1.

The two cavities are fed symmetrically by the same source S and are therefore exactly in phase. It is always possible to choose $\phi = 0$ for R_1, but for R_2 we must take into account the time T to travel between R_1 and R_2, with $\phi = -\omega T$. Although we are at resonance $\omega = \omega_0$, we shall formally retain the two frequencies ω and ω_0 for later use. Taking into account (B.14) and the different free time evolution of the states $|+\rangle$ and $|-\rangle$ during time T, if an atom enters the cavity R_1 in the state $|+\rangle$, it will arrive in the cavity R_2 in the state $|\varphi'\rangle$:

$$|\varphi'\rangle = \frac{1}{\sqrt{2}} \left(|+\rangle - \mathrm{i} \mathrm{e}^{-\mathrm{i}\omega_0 T} |-\rangle \right). \tag{B.16}$$

Now using (B.14) and (B.15) and the value $\phi = -\omega T$, we can state that the atom leaves R_2 in a state $|\psi\rangle$:

$$|\psi\rangle = \frac{1}{2} \left[\left(1 - \mathrm{e}^{\mathrm{i}\delta T}\right) |+\rangle - \mathrm{i} \mathrm{e}^{-\mathrm{i}\omega_0 T} \left(1 + \mathrm{e}^{-\mathrm{i}\delta T}\right) |-\rangle \right], \tag{B.17}$$

since as already mentioned we have formally retained δ even though $\delta = 0$ at resonance. Actually, the two frequencies ω and ω_0 play different roles: ω_0 controls the free time evolution and ω controls the phase ϕ. It is therefore possible to identify their respective roles. If we take $\delta = 0$ in (B.17), the global effect is $|\psi\rangle \propto |-\rangle$, which was to be expected because we have effectively applied a π-pulse to the atom, thus transforming a state $|+\rangle$ into a state $|-\rangle$.

The evolution equations in the nonresonant case have been solved in Section 5.2.2, and the result for the initial conditions $\gamma_+(0) = 1$, $\gamma_-(0) = 0$ is

$$\gamma_+(t) = \frac{\mathrm{e}^{\mathrm{i}\delta t/2}}{\Omega} \left[\Omega \cos \frac{\Omega t}{2} - \mathrm{i}\delta \sin \frac{\Omega t}{2} \right],$$
$$\gamma_-(t) = -\frac{\mathrm{i}\omega_1}{\Omega} \mathrm{e}^{\mathrm{i}(\phi - \delta t/2)} \sin \frac{\Omega t}{2}. \tag{B.18}$$

The result for the initial conditions $\gamma_+(0) = 0$, $\gamma_-(0) = 1$ is again obtained without calculation by making the substitutions $+ \leftrightarrow -$, $\delta \to -\delta$, and $\phi \to -\phi$ in (B.18).

We choose the detuning δ to be nonzero, but sufficiently small that $|\delta|/\Omega \ll 1$. Then δ can be neglected, except in terms involving $\exp(\mathrm{i}\delta T)$, because there is no reason for δT to be small compared with unity. We then recover the results of the nonresonant case, but δ in (B.17) is explicitly nonzero. If the atom has entered the cavity R_1 in the state $|+\rangle$, the probability p_{++} of finding it in the state $|+\rangle$ at the exit from R_2 is given from (B.17):

$$\mathsf{p}_{++} = \frac{1}{4} \left| 1 - \mathrm{e}^{\mathrm{i}\delta T} \right|^2 = \frac{1}{2} \sin^2 \frac{\delta T}{2}. \tag{B.19}$$

We therefore predict that p_{++} varies with δ with period $T = 2\pi/\delta$, a phenomenon called *Ramsey fringes*. Experiment confirms the existence of these fringes with a good contrast of about 55% (Fig. B.1a).

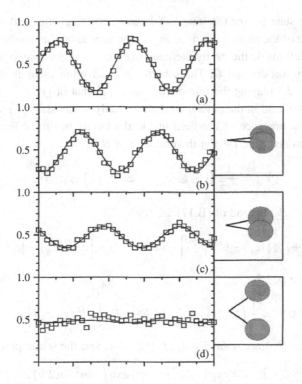

Fig. B.1. Ramsey fringes. (a) Empty cavity; (b) to (d) average number of photons $\langle n \rangle = 9.5$. The column on the right gives the overlap of the coherent states $|z_{\pm}\rangle$; see Fig. 6.11. From M. Brune *et al.*, *Phys. Rev. Lett.* **77**, 4887 (1996).

B.3 Interaction with a field inside the cavity

The superconducting cavity C now contains an electromagnetic field in a coherent state (11.31) of an eigenmode of frequency $\omega_C \simeq \omega_0$ of the cavity, with a detuning $\delta_C = \omega_C - \omega_0 \neq 0$:

$$|z\rangle = \exp\left(-\frac{|z|^2}{2}\right)\exp(a^{\dagger}z)|0\rangle, \tag{B.20}$$

where $|0\rangle$ is the vacuum of the electromagnetic field in the mode under consideration (the zero-photon state). The complex number $z_{\pm}$ is defined as

$$z_{\pm} = \alpha e^{\pm i\Phi},$$

where α is a real positive number; α^2 is the average number of photons $\langle n \rangle$ in the cavity, $\alpha^2 = \langle n \rangle$. Since the atom and the field are not in resonance, there is no photon emission or absorption during the passage of the atom through C. Away from resonance the atom acts like a medium of index of refraction $\neq 1$ and the passage of the atom through C has the effect of changing the phase of the field by an angle $\pm\Phi$ depending on whether the atom is in the state $|+\rangle$ or the state $|-\rangle$:[4] the electromagnetic field inside the cavity measures the state of the atom, with Φ acting as the needle on the measuring device. The state of the field left inside the cavity theoretically[5] makes it possible to identify the state of the atom which has crossed C. The scheme is exactly the same as that described in Section B.1, with $|\pm\rangle$ playing the role of $|\varphi_n\rangle$ and $|z_\pm\rangle$ that of $|\chi_n\rangle$.

When there is no field in the cavity, an atom initially in the state $|+\rangle$ will arrive at R_2 in $|\varphi'\rangle$ (B.16). The presence of the field inside the cavity has the effect of creating an atom + field entangled state $|\Psi'\rangle$ at the entrance of R_2:

$$|\Psi'\rangle = \frac{1}{\sqrt{2}}\left(|+\rangle \otimes |z_+\rangle - ie^{-i\omega_0 T}|-\rangle \otimes |z_-\rangle\right),$$

while at the exit of R_2, instead of (B.17) we find

$$|\Psi\rangle = \frac{1}{2}\left(|+\rangle \otimes \left[|z_+\rangle - e^{i\delta T}|z_-\rangle\right] - ie^{-i\omega_0 T}|-\rangle \otimes \left[e^{-i\delta T}|z_+\rangle + |z_-\rangle\right]\right). \qquad (B.21)$$

We can then obtain p_{++}:[6]

$$p_{++} = \frac{1}{4}\left(\langle z_+|z_+\rangle + \langle z_-|z_-\rangle - 2\,\mathrm{Re}\left[e^{i\delta T}\langle z_+|z_-\rangle\right]\right).$$

The coherent states $|z_\pm\rangle$ are normalized, $\langle z_\pm|z_\pm\rangle = 1$, and the scalar product $\langle z_+|z_-\rangle$ is

$$\langle z_+|z_-\rangle = \exp\left(-2\alpha^2 \sin^2 \Phi\right)\exp\left(-i\alpha^2 \sin 2\Phi\right). \qquad (B.22)$$

Substituting these values into p_{++}, we obtain the final result

$$p_{++} = \frac{1}{4}\left[1 - \exp(-2\alpha^2 \sin^2 \Phi)\cos(\delta T - \alpha^2 \sin 2\Phi)\right]. \qquad (B.23)$$

This expression shows that the Ramsey fringes are blurred and that the factor responsible for this blurring is $\exp(-2\alpha^2 \sin^2 \Phi)$ coming from $\langle z_+|z_-\rangle$: the fringes are more blurred the larger the average number of photons α^2 and the larger the phase shift Φ (Fig. B.1). The quantity $2\alpha \sin \Phi$ has an interesting geometrical interpretation: it is the distance in the complex z plane between the centers of the circles representing the quantum fluctuations of the electric fields (Fig. 6.11). These results can be interpreted in terms of paths in a Hilbert space. The probability amplitude a_{++} of observing an atom in the state $|+\rangle$ at the exit of R_2 when this atom entered R_1 in the state $|+\rangle$ is the sum of two terms

---

[4] Φ can be calculated explicitly:

$$\Phi = \sqrt{\frac{\pi}{2}}\,\frac{\Omega_R^2}{\delta_C}\,T_C,$$

where Ω_R is the vacuum Rabi frequency in the cavity (Exercise 14.6.6) and T_C is the duration of the passage through C. Using the experimental numbers, $\Omega_R \simeq 47$ kHz, $\delta_C/2\pi \simeq 100$ kHz, $T_C \simeq 20\,\mu$s, and $\Phi \simeq 0.7$ rad. Φ can be varied by varying the detuning δ_C.

[5] But not in practice using current technology! However, it is not necessary that the measurement actually be made; it is sufficient that we can imagine making it.

[6] The careful reader can verify this result by calculating the reduced state matrix of the atom.

corresponding to the possible intermediate states $|+\rangle$ and $|-\rangle$ given by (B.17) when C is empty:

$$a_{++} = a_{+++} + a_{+-+} = \frac{1}{2} - \frac{1}{2}\exp(i\delta T).$$

These two paths (in Hilbert space!) are indistinguishable when there is no field in the cavity C. When there is a field in the cavity, the passage of the atom through the cavity leaves a trace by changing the phase of the field by $\pm\Phi$, and this trace is different depending on the state $|\pm\rangle$ of the atom. We can therefore distinguish between the two paths and the interference pattern is blurred. The degree of blurring is controlled by the overlap of the states $|z_+\rangle$ and $|z_-\rangle$. In the limit where these states are orthogonal, the paths are completely distinct and the fringes are destroyed. In the opposite limit, when the angle $|\Phi| \ll 1$, the state of the field does not allow the paths to be distinguished and the fringes remain. This experiment is a concrete realization of one proposed by Feynman for distinguishing between the two trajectories in a Young slit experiment (cf. the discussion of Section 1.4.4). However, our discussion makes it evident that the destruction of the interference pattern does not arise from any perturbation of the atomic trajectories, but from the possibility of *labeling* the two paths.

B.4 Decoherence

Let us return to the connection with the general discussion of measurement in Section B.1. The system S is the atom which crosses the experimental apparatus, the measuring device M is the field, and the position of the needle is the phase shift $\pm\Phi$ of the field after passage of the atom in the state $|\pm\rangle$. According to (B.21), after the atom has passed through the cavity the measuring device is left in the state

$$|Z\rangle = \frac{1}{\sqrt{2}}\left[|z_+\rangle \mp e^{i\delta T}|z_-\rangle\right], \tag{B.24}$$

depending on the result of the detection. For an operation to truly be a measurement, it is necessary that there be a one-to-one correspondence between the state of the measuring device (the field) and the system (the atom):

$$|+\rangle \longleftrightarrow |z_+\rangle, \quad |-\rangle \longleftrightarrow |z_-\rangle.$$

However, this is not always the case: after passage through the cavity the state $|+\rangle$ corresponds to a linear superposition (B.24) of the states $|z_+\rangle$ and $|z_-\rangle$. This is a symptom of the ambiguity mentioned at the end of Section B.1.

Moreover, the state (B.24) is a Schrödinger's cat (Section 6.4.1),[7] that is, a linear superposition of two positions of the needle on the meter of the measuring device. If we compare the states $|z_\pm\rangle$ to classical states, which will be correct when there are a large number of photons in the cavity, and if Φ is the position of the needle, the state (B.24)

---

[7] Since the measuring device is at best mesoscopic, it is a kitten rather than a cat.

is a linear superposition of two positions of this needle. To explain why such states are never observed, at least in the limit where the measuring device is macroscopic, it is necessary to consider the coupling of M to the environment E. In fact, it can be shown in a general way that an interaction $S + M$ is not sufficient for making a measurement of S: it is also necessary to introduce a coupling of M to the environment in order to make a real measurement. As long as no information is leaked to the environment the situation remains reversible, that is, we remain in the premeasurement stage, and the entanglement $S + M$ can be manipulated as above. It is the leakage of information to the environment that makes the measurement irreversible. The coupling of the measuring device to the environment leads to the phenomenon of *decoherence*: the quantum coherences of M with S are destroyed in a very short time such that only the states of a preferred basis in the Hilbert space of M are physically observable, and linear superpositions of such states, the Schrödinger's cats, are eliminated. The states of the preferred basis are the classical states of M, which are fixed by the form of the interaction of M with the environment. Consequently, the physical properties of S that are measured by M are also well determined: the quantum correlations between M and S are transformed into classical correlations and the ambiguity in the physical properties measured by M is removed.

In Section 15.4.5 we proved the following results for a linear superposition of two wave packets describing a Brownian particle:

- the decoherence time is inversely proportional to the diffusion coefficient;
- this time is the shorter the larger the "distance" a between the two linearly superimposed states in (B.24).

For a sufficiently large particle the decoherence time is infinitesimally short compared with the characteristic time of the quantum evolution. Moreover, the environment selects the basis of the position states as the privileged basis, because it very rapidly destroys the coherences between different position states.

In the experiment of Brune *et al.*, the decoherence time T_D is estimated as follows. The lifetime of a photon inside the cavity is $T_r = Q/\omega \simeq 160\,\mu s$, where Q is the quality factor. The leakage of a single photon is sufficient to destroy the coherence of the superposition (B.24) and this occurs after a time $T_D \sim T_r/\langle n \rangle$. More precisely, the "distance" between the two superimposed states is $a = 2\langle n \rangle^{1/2} \sin \Phi$ (Fig. B.2), and according to (15.148) we expect the decoherence time to be inversely proportional to a^2:

$$T_D \simeq \frac{T_r}{4\langle n \rangle \sin^2 \Phi}. \tag{B.25}$$

The principle of measuring T_D is the following. A second atom is sent into the cavity C with a variable delay τ after the first in order to probe the field in the state in which it has been left by the passage of the first atom. Let $\mathsf{p}_{[\varepsilon_1 \varepsilon_2]}$ be the joint probability of detecting the first atom in the state $\varepsilon_1 = \pm$ and the second atom in the state $\varepsilon_2 = \pm$ at the exit of R_2. The passage of the first atom in the state $|+\rangle_1$ shifts the phase of the field by $+\Phi$, and that of the second atom in the state $|-\rangle_2$ shifts it by $-\Phi$, so that the total phase shift is zero. It is clear that a phase shift of zero is also obtained when the order is reversed, $|-\rangle_1$ followed by $|+\rangle_2$, as the trajectories (in the Hilbert space) $[1+, 2-]$ and $[1-, 2+]$ are indistinguishable, and this property is what leads to the interferences in the joint probabilities $p_{[\varepsilon_1 \varepsilon_2]}$. It can be shown that the quantity

$$\eta = \frac{p_{[1-,2-]}}{p_{[1-,2-]} + p_{[1-,2+]}} - \frac{p_{[1+,2-]}}{p_{[1+,2-]} + p_{[1+,2+]}} \tag{B.26}$$

is $1/2$ if the two states of the field are coherent and zero if they form a statistical mixture. Measurement of η, which is controlled by the coherences of the state matrix, permits recovery to the degree of partial coherence preserved after a time τ (Fig. B.2). The experimental results confirm the expected properties in every point.

Returning to the general analysis of Section B.1, once the measurement has been completed, the state operator of S is given by an incoherent superposition (B.9). In this sense the WFC postulate is a consequence of the measurement operation, and this postulate is convenient but not independent of the other postulates. In the case of two consecutive measurements, if the interaction of the measuring devices with the environment is taken into account it is possible to calculate the probabilities of results of the second measurement without resorting to the wave-function collapse postulate, and the results will be the same as those obtained when this postulate is used. On the other hand, postulate **II** remains completely outside the scope of decoherence:[8] this postulate tells us that the *probability* of the result $|\varphi_n\rangle$ is $|c_n|^2$, but it is a *unique* result which is obtained

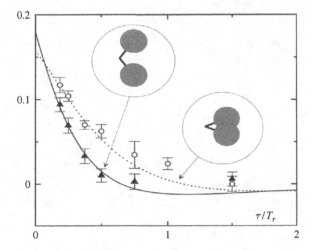

Fig. B.2. Time falloff of the coherence. The solid line corresponds to a large angle 2Φ between the two coherent states, and the dotted line corresponds to two strongly overlapping states. From M. Brune *et al.*, *Phys. Rev. Lett.* **77**, 4887 (1996).

---

[8] We recall that the prescription of the partial trace, which we have used intensively in this appendix, is a *consequence* of postulate **II**.

in a particular measurement, with a certain probability. All the possible results appear in (B.9), but there is nothing that can explain why one particular result will emerge from a particular experiment. No unitary evolution of the type (4.11) can explain this uniqueness of the result, and so far there is no known justification – assuming that this term even makes sense in this context – of postulate **II**.

Appendix C The Wigner–Weisskopf method

The derivation of the Fermi Golden Rule in Section 9.6.3 is limited to sufficiently short times $t \ll \tau_2$, and the exponential decay law (9.171) cannot be justified using only the arguments of that section. A method due to Wigner and Weisskopf permits this law to be justified for long times with the help of another approximation scheme.[1] Let us consider the following situation. A state of an isolated system a of energy E_a decays to a continuum of states b of energy E_b. Examples of such a situation are the de-excitation of an excited state of an atom, a molecule, a nucleus, and so on with the emission of a photon, or the decay of an elementary particle. The states of energy E_a and E_b are the eigenstates of a Hamiltonian $H^{(0)}$:

$$H^{(0)}|a\rangle = E_a|a\rangle, \quad H^{(0)}|b\rangle = E_b|b\rangle, \qquad (C.1)$$

and a time-independent perturbation W is responsible for the transition $a \to b$; in the case of spontaneous photon emission, W is given by (14.58). The states a and b are *not* stationary states of the *total* time-independent Hamiltonian $H = H^{(0)} + W$. We can assume that the diagonal matrix elements of W are zero:[2] $W_{aa} = W_{bb} = 0$ and we use $|\psi(t)\rangle$ to denote the state vector of the system, the initial state being $|\psi(t = 0)\rangle = |a\rangle$. Let us decompose the state $|\psi(t)\rangle$ on the states $|a\rangle$ and $|b\rangle$ using the density of states $\mathcal{D}(E_b)$:

$$|\psi(t)\rangle = \gamma_a(t)e^{-iE_a t/\hbar}|a\rangle + \int dE_b\, \mathcal{D}(E_b)\, \gamma_b(t)\, e^{-iE_b t/\hbar}. \qquad (C.2)$$

The Schrödinger equation applied to the decomposition (C.2)

$$i\hbar \frac{d|\psi(t)\rangle}{dt} = \left(H^{(0)} + W\right)|\psi(t)\rangle$$

---

[1] The method of Wigner and Weisskopf is described by Cohen-Tannoudji *et al.* [1977], Complement D_{XIII}, and by Basdevant and Dalibard [2002], Chapter 17; a detailed and rigorous treatment is given by Messiah [1999], Chapter XXI.

[2] If this were not the case, we could redefine $H^{(0)}$:

$$H^{(0)} \to H^{(0)'} = H^{(0)} + |a\rangle\, W_{aa}\, \langle a| + \int dE_b\, \mathcal{D}(E_b)\, |b\rangle\, W_{bb}\, \langle b|.$$

leads to the system of differential equations (cf. (9.163))

$$i\hbar\dot{\gamma}_a(t) = \int e^{i\omega_{ab}t} W_{ab}\, \gamma_b(t)\mathcal{D}(E_b)dE_b, \tag{C.3}$$

$$i\hbar\dot{\gamma}_b(t) = e^{-i\omega_{ab}t} W_{ab}^*\, \gamma_a(t), \tag{C.4}$$

with $\omega_{ab} = (E_a - E_b)/\hbar$. We know empirically that $|\gamma_a(t)|^2$ is given by an exponential law:

$$|\gamma_a(t)|^2 = e^{-\Gamma t}, \tag{C.5}$$

which suggests that we try the function

$$\gamma_a(t) = \exp\left(-\frac{i\delta}{2}t\right), \qquad \delta = \delta_1 - i\Gamma, \tag{C.6}$$

where δ_1 is real. Substitution of (C.6) into (C.4) with the initial conditions $\gamma_b(t=0) = 0$ gives after integration over t

$$\gamma_b(t) = \frac{W_{ab}^*}{\hbar[\omega_{ab} + \delta/2]}\left[\exp\left(-i(\omega_{ab} + \delta/2)t\right) - 1\right]. \tag{C.7}$$

For long times $t \gg \Gamma^{-1}$, the exponential in (C.7) tends rapidly to zero and

$$\lim_{t\to\infty} |\gamma_b(t)|^2 = \frac{|W_{ab}|^2}{\hbar^2\left[(\omega_{ab} + \delta_1/2)^2 + \Gamma^2/4\right]}. \tag{C.8}$$

To verify that our initial hypothesis is consistent with the evolution equations (C.3) and (C.4), we substitute (C.6) into (C.3), which gives

$$\frac{\hbar\delta}{2} = \int dE_b\, \mathcal{D}(E_b)\,|W_{ab}|^2 \frac{1 - \exp(i[\omega_{ab} + \delta/2]t)}{\hbar(\omega_{ab} + \delta/2)}. \tag{C.9}$$

The constant δ must be a solution of the integral equation (C.9). To be specific, let us study a transition from an excited state i of energy E_i to the ground state f of energy E_f of an atom, with the emission of a photon of energy $\hbar\omega$. To an excellent approximation we can neglect the recoil kinetic energy of the final atom, which simply has energy E_f in the reference frame chosen to be that where the atom in its initial state is at rest (cf. the discussion of Section 14.3.4). The density of final states to be used in (C.9) is that of the photon (14.62). In summary, we have $|a\rangle = |i\rangle$ and $|b\rangle$ is the atom in the state $f+$ photon $|f \otimes \vec{k}s\rangle$, as well as energy conservation

$$\hbar\omega_{ab} = E_a - E_b = E_i - (E_f + \hbar\omega) = \hbar(\omega_0 - \omega) \tag{C.10}$$

with $\hbar\omega_0 = E_i - E_f$. Choosing ω as the integration variable instead of E_f, with $dE_b = \hbar d\omega$, Equation (C.9) becomes

$$\frac{\hbar\delta}{2} = \int_0^\infty d\omega\, \mathcal{D}(\omega)\,|W_{ab}(\omega)|^2 \frac{1 - \exp(i[\omega_0 - \omega + \delta/2]t)}{\omega_0 - \omega + \delta/2}. \tag{C.11}$$

We are interested in the behavior of this equation at long times, and we shall need the behavior for $t \to \infty$ of the function $f(t, x)$ considered as a distribution:

$$f(t, x) = \frac{1 - e^{itx}}{x}.$$

When x is real, its Fourier transform is

$$\tilde{f}(t, u) = i[\theta(u) - \theta(t + u)] \tag{C.12}$$

because

$$-i \int_{-t}^{0} dx \, e^{-iux} = \frac{1 - e^{itx}}{x}$$

and the $t \to \infty$ limit of $\tilde{f}(t, u)$ is simply $-i\theta(-u)$, which gives the following for $f(t, x)$:

$$\lim_{t \to \infty} f(t, x) = \lim_{\eta \to 0^+} \frac{1}{x + i\eta} = \mathbb{P} \frac{1}{x} - i\pi\delta(x). \tag{C.13}$$

This result is also valid when x has a small imaginary part, $x = \mathrm{Re}\,x \pm i\eta$, $\eta \to 0^+$. To see this it is sufficient to integrate over x in the complex plane, completing the integration contour by a semicircle whose radius tends to infinity. If δ_1 and Γ are small compared with the typical ranges of variation of the functions $\mathcal{D}(\omega)$ and $|W_{ab}(\omega)|^2$, we can substitute (C.13) into (C.11) and find the value of δ:

$$\delta = \frac{2}{\hbar} \mathbb{P} \int_{-\infty}^{\infty} d\omega \, \frac{\mathcal{D}(\omega)|W_{ab}(\omega)|^2}{\omega_0 - \omega} - \frac{2i\pi}{\hbar} \mathcal{D}(\omega_0)|W_{ab}(\omega_0)|^2. \tag{C.14}$$

The second term on the right-hand side of (C.14) confirms that $\Gamma = -i\,\mathrm{Im}\,\delta$ is really given by the Fermi Golden Rule:

$$\Gamma = \frac{2\pi}{\hbar} \mathcal{D}(\omega_0)|W_{ab}(\omega_0)|^2, \tag{C.15}$$

while the first term corresponds to the shift of the energy level:

$$\mathrm{Re}\,\delta = \delta_1 = \frac{2}{\hbar} \mathbb{P} \int_{-\infty}^{\infty} d\omega \, \frac{\mathcal{D}(\omega)|W_{ab}(\omega)|^2}{\omega_0 - \omega}. \tag{C.16}$$

This shift could have been obtained by a calculation using second-order time-independent perturbation theory; it is zero in first-order according to our hypothesis $W_{aa} = W_{bb} = 0$. The phase δ_1 can be absorbed in a redefinition of ω_0: $\omega_0 \to \omega_0 + \delta_1$, and then according to (C.8) the probability of observing a photon of frequency ω is

$$p(\omega) \, d\omega \simeq \frac{\mathcal{D}(\omega_0)|W_{ab}(\omega_0)|^2}{\hbar[(\omega - \omega_0)^2 + \Gamma^2/4]} \, d\omega. \tag{C.17}$$

This probability is correctly normalized to unity

$$\int_{-\infty}^{+\infty} \mathsf{p}(\omega)\,d\omega = 1 \quad \text{because} \quad \int_{-\infty}^{+\infty} \frac{dx}{x^2 + a^2} = \frac{\pi}{a}$$

taking into account the value (C.15) of Γ. The curve representing $\mathsf{p}(\omega)$ is a Lorentzian (also known as a Breit–Wigner curve):

$$\mathsf{p}(\omega) = \frac{\mathcal{D}(\omega_0)|W_{ab}(\omega_0)|^2}{\hbar[(\omega - \omega_0)^2 + \Gamma^2/4]}. \tag{C.18}$$

The frequency of the final photon is not sharply defined, but has a spread $\Delta\omega = \Gamma$,[3] which is the width at half-max of the curve $\mathsf{p}(\omega)$:

$$\mathsf{p}\left(\omega_0 \pm \frac{1}{2}\Delta\omega\right) = \frac{1}{2}\mathsf{p}(\omega = \omega_0).$$

In other words, the frequency spectrum of the emitted photon is not monochromatic. The quantity Γ is called the *linewidth* or sometimes the natural linewidth, as there are also other causes of this broadening such as the Doppler effect or collisions. Owing to (C.5), the lifetime of the excited state is the inverse of the linewidth, $\tau = 1/\Gamma$. The energy spread of the final photon shows, from energy conservation, that the energy of the excited state

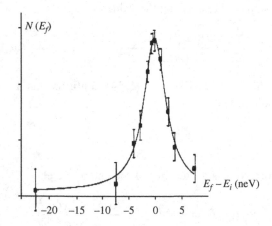

Fig. C.1. Photon spectrum $N(E_f)$ from the decay $^{57}\mathrm{Fe}^* \rightarrow {}^{57}\mathrm{Fe} + \mathrm{photon}$, as a function of the difference $E_f - E_i$ between the initial and final energies. After Basdevant and Dalibard [2002].

---

[3] $\Delta\omega$ is not a dispersion because the integral

$$\int_0^\infty d\omega(\omega - \omega_0)^2 \mathsf{p}(\omega)$$

is divergent, and so strictly speaking a dispersion cannot be defined; see Exercise 4.4.5.

has a spread $\Delta E = \hbar\Gamma$ about a central value E_i, and from it we derive the relation (4.30) between the lifetime and the energy spread:

$$\tau\,\Delta E = \hbar. \tag{C.19}$$

However, there is in principle no limit to the precision with which this central value can be measured. Figure C.1 shows the experimental curve of $p(\omega)$ for the decay of an excited level of ^{57}Fe*:

$$^{57}\text{Fe*} \rightarrow\ ^{57}\text{Fe} + \text{ photon } (14\,\text{keV}),$$

where the lifetime is $\tau \simeq 1.4 \times 10^{-7}$ s.

References

Balian, R. (1991) *From Microphysics to Macrophysics*, Berlin: Springer.

Ballentine, L. (1998) *Quantum Mechanics*, Singapore: World Scientific.

Basdevant, J. L. and Dalibard, J. (2002) *Quantum Mechanics*, Berlin/Heidelberg: Springer.

Cohen-Tannoudji, C., Diu, B. and Laloë, F. (1977) *Quantum Mechanics*, NewYork: John Wiley.

Feynman, R., Leighton, R. and Sands, M. (1965) *The Feynman Lectures on Physics*, Reading: Addison-Wesley.

Grynberg, G., Aspect, A. and Fabre, C. (2006) *Introduction to Lasers and Quantum Optics*, Cambridge: Cambridge University Press.

Isham, C. (1995) *Lectures on Quantum Theory*, London: Imperial College Press.

Jackson, J. D. (1999) *Classical Electrodynamics*, 3rd edn. NewYork: John Wiley.

Jauch, J. (1968) *Foundations of Quantum Mechanics*, Reading: Addison Wesley.

Kittel, C. (1996) *Introduction to Solid State Physics*, NewYork: John Wiley.

Landau, L. and Lifschitz, E. (1958) *Quantum Mechanics*, London: Pergamon Press.

Laloë, F. (2001) Do we really understand quantum mechanics? Strange correlations, paradoxes and theorems, *Am. Journ. Phys.* **69**, 655.

Le Bellac, M. (1991) *Quantum and Statistical Field Theory*, Oxford: Clarendon Press.

Le Bellac, M., Mortessagne, F. and Batrouni, G. (2004) *Equilibrium and Non-Equilibrium Statistical Thermodynamics*, Cambridge: Cambridge University Press.

Levitt, M. H. (2001) *Spin Dynamics, Basics of Nuclear Magnetic Resonance*, NewYork: John Wiley.

Lévy-Leblond, J. M. and Balibar, F. (1990) *Quantics: Rudiments of Quantum Physics*, NewYork: North Holland.

Mandel, L. and Wolf, E. (1995) *Optical Coherence and Quantum Optics*, Cambridge: Cambridge University Press.

Merzbacher, E. (1970) *Quantum Mechanics*, NewYork: John Wiley.

Messiah, A. (1999) *Quantum Mechanics*, Minneola: Dover Publications.

Nielsen, M. and Chuang, I. (2000) *Quantum Computation and Quantum Information*, Cambridge: Cambridge University Press.

Omnès, R. (1999) *Understanding Quantum Mechanics*, NewYork: Princeton University Press.

Peres, A. (1993) *Quantum Theory, Concepts and Methods*, Boston: Kluwer.

Weinberg, S. (1995) *The Quantum Theory of Fields*, Cambridge: Cambridge University Press.

Wichman, E. H. (1967) *Quantum Physics*, Berkeley Physics Course Vol. 4, NewYork: McGraw Hill.

Zurek, W. H. (2003) Decoherence, einselection and the quantum origin of the classical, *Rev. Mod. Phys.* **75**, 715.

Index

absorption 145, 150
addition theorem (of spherical harmonics) 321
alpha-radioactivity 279
ammonia molecule 139
amplitude damping channel 524
ancilla 514
angular momentum 228
  addition theorem 342
  conservation of 222
  quantization axis 308
  standard basis of 310
annihilation operator 359, 371
anticommutator 453, 526
antiferromagnetism 444
asymptotic series 458
atom 4, 28
  dressed 503
atomic nucleus 4
atomic number 4
autocorrelation function 531

beam (of particles) 404
beam splitter 59, 452
Bell inequality 174, 175, 203
Bell measurement 197
Bell state 511
benzene molecule 128
beta-radioactivity 5
binding energy 4
birefringent plate 63
black body 13
Bloch equations (of NMR) 508
Bloch theorem 284
Bloch vector 165, 479
Bohr atom 29
Bohr frequency 29
Bohr magneton 465
  nuclear 465
Bohr radius 30, 260, 327
Boltzmann constant 11

Boltzmann law 10, 138, 508
Boltzmann weight 11
Born approximation 426, 433
Born–Oppenheimer approximation 154
Born rule 97
Bose–Einstein condensation 450
boson 440
bound state 4, 27, 265
bra 47
Bragg angle 36
Breit–Wigner curve 432, 576
Brownian motion 538
butadiene molecule 152

Caldeira–Leggett model 541
canonical commutation relations 114, 235
  representation of 236
canonical transformation 383
Casimir effect 399
Cauchy series 210
cell 37
center-of-mass 248
central extension (of a Lie algebra) 234
centrifugal barrier 325
chemical bonding 125
chemical potential 448
chemical reaction 4
chemical shift 138
classical source (or force) 392
Clebsch–Gordan (C–G) coefficients 343
closed quantum system 106
coherences (of a state matrix) 166, 191, 480
coherent state 189, 365, 381, 390, 567
coherent superposition 101, 166
commutation relations 231
  of angular momentum 232, 307
commutator 51
complementary bases 71, 130, 255
complete set of compatible (or commuting)
  operators 52, 103

complete vector space   210
completeness relation   48, 218, 250, 289
Compton wavelength   31, 114, 497
computational basis   193
conjugate momentum   373, 378
connected   234
   simply   234
conservation law   228
contextuality   104, 185
continuity equation   262, 290
continuum limit   372
control bit   194
control-not (c-NOT) gate   194
control-U (c-U) gate   515
Copenhagen interpretation   185
convergence
   strong (or in the norm)   211
   weak   211
correspondence principle   114, 243
cosmic microwave background   15
Coulomb gauge   376, 395, 467
Coulomb law   7
counterfactual   178
coupling constant   8
covariant derivative   385, 398
creation operator   359, 371
cross section
   coherent   435
   differential   405
   elastic   421
   incoherent   435
   inelastic   422
   total   405, 422
crystal lattice   3
Curie temperature   444
current   262, 407
current density   262

de Broglie wavelength   18
Debye frequency   395
Debye model   395
decoherence   187, 190, 570
decoherence time   543, 548
delayed choice experiment   190
delocalization energy   128
density matrix: *see* state matrix
density of states   292, 436, 573
density operator: *see* state operator
depolarizing channel   522
detuning   134, 144, 478
deuterium   4
deuteron   5, 350, 420, 431, 443, 504
Deutsch algorithm   207
diamagnetic term   499
diffraction   18

diffusion coeffcient   488
dimension
   of a Lie group   227
   of a vector space   42, 210
dipole approximation   469, 475
Dirac equation   461
Dirac notation   47
Dirac picture: *see* interaction picture
dispersion   24, 104
dispersion law   369
dissipative force   483
dissociation energy   5
domain (of an operator)   213
Doppler cooling   484
Doppler temperature   489
dynamical susceptibility   532

effective mass   288
effective potential   324, 417
effective range   416
Ehrenfest theorem   111, 229
eigenstate   80
eigenvalue   48
   degenerate   48
   subspace of an   49
eigenvector   48, 217
Einstein–Podolsky–Rosen (EPR) argument   171
Einstein relation   489
electric dipole moment   141, 470
   neutron   559
electric dipole transition   149, 336, 352
electromagnetic current   384
electromagnetic interactions   7
electron   4
electroweak interactions   7, 398, 437
elementary excitation   371
element of reality   173
energy band   284
energy level   27, 30, 271
entangled state   160
environment   187, 521, 570
ethylene molecule   125
evolution equation   106
evolution operator   108
exchange force   444
exchange integral   494
expectation value (of an operator)   81, 99, 100
exponential decay law   112, 120, 297, 573
extension (of an operator)   214

factorization rule   69
Fermi constant   436, 559
Fermi gas   448
Fermi Golden Rule   297, 471
Fermi level   448

Fermi momentum 449
fermion 440
Fermi sphere 449
Fermi surface 449
ferromagnetism 444
Feynman diagram (or graph) 113
Feynman–Hellmann theorem 117
field operator 373
fine structure 461
fine structure constant 32
first Brillouin zone 287, 369
fluorescence cycle 485
flux 405, 468
Fock space 371
Fokker–Planck equation 550
forbidden band 287
formaldehyde molecule 152
Fourier transform 255, 290
    discrete, or lattice 368
fullerene 21, 33
functional space 213

Galilean transformation 240, 305
Gamow peak 430
gauge boson 399
gauge field 398
gauge group 397
gauge symmetry 397
gauge transformation 376
    global 384
    local 242, 254, 385
Gaussian integral 57
Gaussian wave packet 299
gluon 7, 399
gravitational constant 9
gravitational interaction 9
graviton 7
Greenberger–Horne–Zeilinger (GHZ) state 182
Green function 425
ground state 30, 127, 259
group property 108
group velocity 259, 427
gyromagnetic ratio 76, 465

Hadamard gate (or matrix) 60, 194
Hamiltonian 87, 106, 228
Hamiltonian (or unitary) evolution 106, 169
hard sphere scattering 406
harmonic oscillator 13, 358
    damped 529
    forced 379
Heisenberg inequality 24, 35, 105, 257, 259, 299, 367, 382
    temporal 88, 111
Heisenberg picture 114, 122
Heisenberg uncertainty principle: *see* Heisenberg inequality

helicity 333
helicity amplitude 339
helium 491
Hermite polynomial 363
Hermitian conjugate 44, 215
hidden variables 68, 177
Higgs boson 9
Hilbert space 42, 209
Hilbert space of states 70
homomorphism 234
hydrogen atom 29, 327
hydrogen molecular ion 154
hyperfine structure 466

impact parameter 406, 412
incoherent superposition 166
incoming wave 281, 407
incompatible bases 71, 81
independent particle approximation 128
indistiguishable paths 24, 569
inertial reference frame 223, 240
infinitesimal generator 219, 228
    of Galilean transformations 240
    of rotations 232
    of time-translations 110
    of translations 251
infinitesimal rotation 231
input register 194
integral equation of scattering 426
intensity of a light wave 62
interaction picture 392, 529
interference 18
internal symmetry 397
ionization energy 5, 30
irreducible tensor operator 347, 357
isometry 45

Jaynes–Cummings Hamiltonian 501
Jones vector 67

ket 47
kinetic energy 242, 257
Klein–Gordon equation 461
Kramers degeneracy 557
Kraus representation 519, 520
Kraus number 520

(lambda-mu) polarizer 66
Lamb–Dicke parameter 401
Lamb shift 380, 463
Landau level 388
Landé g-factor 464
Larmor frequency 77, 87, 133, 387
Larmor precession 77, 133, 170
laser 146

laser cooling   478
laser trapping: *see* magneto-optical trap
Legendre polynomial   320, 409
Lennard–Jones potential   300
lepton   6
level density   292
level spectrum   30
Lie algebra   230, 247
Lie group   227, 246
lifetime of an excited state   33, 112, 149
Lindblad equation   528
linear response theory   532
linewidth   479, 576
local evolution   511
local realism   174, 184
longitudinal relaxation time   138, 508
long-range force (or law)   8
long-range order   3
Lorentz force (or law)   10
Lorentz gauge   379

Mach–Zehnder interferometer   37, 206
magnetic dipole transtion   336
magnetic moment   76
magnetic quantum number   308
magnetic resonance imaging (MRI)   138
magneto-optic trap (MOT)   489
magnon   201
Malus law   63
Markovian approximation   527, 534
maser   146
mass number   4
master equation   526, 537, 538, 541
matrix   44
   normal   57
   positive   58
   strictly positive   58
maximally entangled states   510
maximal test: *see* test
Maxwell equations   10, 375
measurement   186, 561
   ideal   100
   von Neumann (or orthogonal)   511
memory effect   526
memory kernel   527
microreversibility   555
minimal coupling   386
mixture   162
molecular orbital   126
molecule   4
   diatomic   33, 301, 318, 443
momentum   10
   conservation of   222
momentum operator   228, 250
momentum transfer   426, 472

muon   498
muonic atom   498

Néel temperature   444
Neumark theorem   513
neutrino   7, 435
neutrino oscillations   121
neutron   4
   cold   18
   thermal   18
neutron diffaction   18, 35
neutron interferometer   38, 93
neutron optics   433
no-cloning theorem   191
node   271, 326, 363
non-Abelian gauge theory   386, 397
nonlocal evolution   511
nonseparability (of the state vector)   179
norm
   of a vector   43, 209
   of an operator   213
normalized vector   47, 97
normal mode   367, 369
   magnetic resonance imaging (MRI)   138
nuclear magnetic resonance (NMR)   132, 201, 508
nuclear reaction   4
nucleon   4
number of levels   115
number operator   360
nutation frequency   113

observable: *see* physical property
occupation number   371
offset frequency   134
open quantum system   507
operator
   antilinear   553
   antiunitary   225, 553
   bounded   213
   compatible   72
   Hermitian (self-adjoint)   45, 98, 215
   incompatible   72
   linear   44
   scalar   232, 345
   unbounded   213
   unitary   45, 52, 219
   vector   233, 345
optical Bloch equations   478, 525
optical molasses   485
optical potential   424
optical theorem   423, 424
optical tweezers   484
orbital angular momentum   317
orthohydrogen   431, 444
orthonormal basis   43, 210

outgoing wave 281, 407
output register 195

parahydrogen 431, 444
paramagnetism 444
partial wave 324
parity 237, 322, 443
  negative 240
  positive 240
partial wave expansion 409, 411
Pauli matrices 85
Pauli principle 128, 441
periodic boundary conditions 291
periodic potential 283, 302
perturbation theory 492
  degenerate 456, 458
  nondegenerate 456, 457
  second-order 495
  time-dependent 294
  time-independent 455
perturbation series 456
phase damping channel 523
phase factor 98, 163
phase shift 410
phase space 291
phenomenological law 11
phonon 33, 371
photo-electric effect 16, 471
photon 7, 16, 378
physical property 70, 98, 163
  compatible 72, 103
  incompatible 104, 308
pi electron 125
pi-meson 113, 445
pi-pulse 135, 566
pi/2-pulse 135, 565
Planck's constant 15
Planck–Einstein relation 17
pointer state 544
Poisson distribution (or law) 365, 394, 487
Polarization
  circular 64
  elliptic 67
  left-handed circular 65, 333
  linear 62, 70
  of light 61, 166
  of a photon 68, 166, 322
  right-handed circular 65, 333
polarized 81
  completely 165
  partially 165
population inversion 138, 145, 482
position operator 228, 250, 254
positive operator valued measure (POVM) 513
positron 7, 199, 451

positronium 199, 451
potential 27, 261
potential barrier 27, 277
potential scattering 404
potential well 27, 265
Poynting vector 34, 148
preferred basis 187, 544
premeasurement 186, 563
preparation 71, 99, 164
principal quantum number 310, 327
probability amplitude 23, 68, 97
probability current 263, 290
probability density 127, 254, 290
projection operator (or projector) 46, 252
proton 4
public key 73
pure state (or pure case) 98, 162, 164, 166

quantization in a box 291, 377
quantization of energy levels 29, 271
quantized electromagnetic field 377
quantized field 371, 373
quantum bit (qubit) 193
quantum chromodynamics (QCD) 8, 398
quantum computing 192
quantum cryptography 73
quantum electrodynamics (QED) 116, 463
quantum field theory 8
quantum fluctuations (of the electromagnetic field) 379,
  380
quantum information 191
quantum jump 519
quantum jump operator 528, 537
quantum key distribution (QKD) 73
quantum logic gate 193
quark 6
quasi-momentum 284
quasi-particle: *see* elementary excitation
quasi-resonant approximation: *see* rotating wave
  approximation

Rabi frequency 133, 144, 479
Rabi oscillation 135
radial equation 324, 351, 409
radial quantum number 327
radial wave function 324
radiation gauge: *see* Coulomb gauge
radiative capture 504
radiative decay (or transition) 331
Ramsey fringes 566
ray 98, 223, 552
reactive force 484, 502
reciprocal lattice 36
recoil energy 474, 486, 574
recoil temperature 486

reduced mass   31, 34, 248
reduced matrix element   345, 347
reflection coefficient   269
relaxation time   527
renormalization   8, 32
representation (of a group)   247
  irreducible   314
  projective   226, 249
  spinor   227
  vector   227
reservoir   530
resolvent   54, 217
resonance   132, 432
resonance curve   136
resonance frequency   143
rotating (reference) frame   134, 155
rotating wave (or quasi-resonant) approximation   145, 156,
     401, 481
rotation matrix   86, 312, 344
rotational levels   319
RSA encryption   73
Rutherford cross section   433
Rydberg atom (or state)   499, 564
Rydberg constant   30, 327

saturation parameter   482
scalar field   317, 371
scalar product   42, 209
scanning tunneling microscopy (STM)   280
scattering
  coherent   40, 434
  elastic   412
  incoherent   40, 434
  inelastic   420
scattering amplitude   407, 426
scattering angle   404
scattering experiment   404
scattering length   95, 413
scattering of identical particles   446
scattering state   27, 265, 273
Schmidt decomposition   510
Schmidt number   510
Schrödinger's cat   187, 190, 524, 542, 569
Schrödinger equation
  time-dependent   261
  time-independent   261, 264, 290
Schrödinger picture   114, 122
Schwarz inequality   43, 211, 212
secret key   73
secular approximation   536
selection rules (for electric dipole transitions)   471
semi-classical aproximation   32, 149, 467
separability (of a Hilbert space)   210
short-range force   8
sigma electron   125

singlet state   340, 419, 442
S-matrix (or scattering matrix)   280
S-matrix element   411
$SO(3)$ group   227
source (of the electromagnetic field)   10, 375
source (of particles)   264
space of states   42, 70, 97
space of polarization states   64, 70
spectral decomposition   50, 218
spectral function   539, 549
spectrum (of an operator)   217
  continuous   217
  discrete   217
spherical Bessel function   410
spherical component (of a vector)   321, 345, 476
spherical harmonic   318
spherical rotator   318, 443
spherical well   350, 413
spin   76
spin echo   201
spin 1/2   77
spin orbit coupling   343
spin orbit potential   462
spin statistics theorem   442
spontaneous emission   149, 473
square well
  finite   271
  infinite   270
squeezed state   383, 394
standard model (of particle physics)   9
state matrix   164
state operator   162, 163
  reduced   167, 509
state vector   70, 97
stationary state   88, 110
stationary phase approximation   258
statistics   440
  Bose–Einstein   440
  Fermi–Dirac   440
Stern–Gerlach experiment   77, 172, 303
Stern–Gerlach filter   79
stimulated emission   145, 150
Stone theorem   219
strong interactions   8
$SU(2)$ group   86, 233
superoperator   518
superposition principle   42, 62, 97
superselection rule   98
survival probability   112, 119
symmetry   222
system with a finite number of levels   125

target   404
target bit   194
teleportation   195

tensor product
  of two vector spaces   158
  of two operators   160
test   71
  ideal   100
  maximal   103, 166, 511
thermal wavelength   450
Thomas precession   462
time reversal   239, 274, 282, 555, 556
T-matrix (or transition matrix)   335
trace (of an operator)   48, 55
  partial   167
transformation
  active   223
  passive   223
transition probability   148, 296
transmission coefficient   269, 278
transmission matrix   276
transverse relaxation time   138, 508
trapped ions   195, 400
triplet state   340, 419, 442
tunnel effect   140, 278, 301, 429
turning point   279
two-level atom   114, 149

unitarity relation   424
unit vector   8
unpolarized   165, 420

vacuum energy   379
vacuum Rabi frequency   501
vacuum Rabi oscillations   502
vacuum state   380
van der Waals force   495

vanishing boundary conditions   291
variational method   117, 459, 492
vector
  axial (or pseudo)   237
  polar   237
virtual (or antibound) state   415
von Neumann (or statistical) entropy   168, 507
von Neumann (or orthogonal) measurement   511
von Neumann measurement theory   304
von Neumann theorem   237

wave equation   261, 372, 377
wave function   127, 250, 254, 290
  in the p-representation   255
wave function collapse (WFC)   102, 571
wave mechanics   250, 362
wave packet   256, 407
wave-packet scattering   427
wave-packet spreading   259, 298
wave vector   13, 18, 257
W-boson   7, 399
weak interactions   8, 238, 435
width of a state   112
Wigner–Eckart theorem   347
Wigner function   550
Wigner matrix: *see* rotation matrix
Wigner theorem   225, 552
Wigner–Weisskopf method   573

Z-boson   7, 112, 399
Zeeman effect   463, 489
Zeeman level   87, 133, 464, 489
zero-point (or vacuum) energy   361, 379, 400